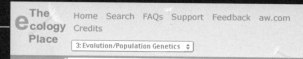

Free Student Aid.

Log on.

Explore.

Succeed.

To help you succeed in the study of Ecology, your professor has arranged for you to enjoy access to a great media resource, the Ecology Place. You'll find that the Ecology Place that accompanies your textbook will enhance your course materials.

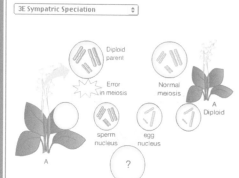

The Ecology Place

Home   Search   FAQs   Support   Feedback   aw.com   Credits

3: Evolution/Population Genetics

Objectives
Web Links
Activities
Quiz
Glossary
Syllabus Manager
Calculator

3E Sympatric Speciation

previous | replay | next
3 of 8

A chromosomal mutation during meiosis can result in a diploid (2N) rather than a haploid (1N) gamete.

What would happen if a diploid sperm nucleus fertilized an egg nucleus from a normal member of the species?

Select your answers to the following questions:

—How many total chromosomes would be present in the zygote?

—How many sets of

Diploid parent
Error in meiosis
Normal meiosis
sperm nucleus
egg nucleus
Diploid
A
A
?

Zygote with 3 , 6 , 9 , or 12 chromosomes?
1 , 2 , 3 , or 4 sets of chromosomes?

Got technical questions?
For technical support, please visit www.aw.com/techsupport, send an email to online.support@pearsoned.com (for web site questions), or send an email to media.support@pearsoned.com (for CD-ROM questions) with a detailed description of your computer system and the technical problem. You can also call our tech support hotline at 1-800-677-6337 Monday-Friday, 8 a.m. to 5 p.m. CST.

What your system needs to use these media resources:

WINDOWS
166 MHz Intel Pentium processor or greater
Windows 95, 98, NT4, 2000
32 MB RAM installed
800 X 600 screen resolution
Browser: Internet Explorer 5.0 or higher or Netscape Communicator 4.7
NOTE: THIS SITE DOES NOT SUPPORT NETSCAPE NAV. 6.0
Plug-Ins: Shockwave Player 8, Flash Player 4, QuickTime 4

MACINTOSH
120 MHz PowerPC
OS 8.6 or higher
24 MB RAM available
800 x 600 screen resolution, thousands of colors
NOTE: THIS SITE DOES NOT SUPPORT NETSCAPE NAV. 6.0
Browser: Internet Explorer 5.0 or higher or Netscape Communicator 4.7
Plug-Ins: Shockwave Player 8, Flash Player 4, QuickTime 4

Here's your personal ticket to success:

✂

How to log on to www.ecologyplace.com:

1. Go to www.ecologyplace.com
2. Click "Enter"
3. Click "Register Here."
4. Enter your pre-assigned access code exactly as it appears below.
5. Complete online registration form to create your own personal Login

Name and Password.
6. Once your personal Login Name and Password are confirmed by email, go back to www.ecologyplace, type in your new Login Name and Password, and click "Enter."

Your Access Code is:

USECO-TRUSS-REBEL-DYFED-CONTO-PRIES

Record your new User ID and Password on the back of this card.

Cut out this card and keep it handy. It's your ticket to valuable information.

Important: Please read the License Agreement, located on the launch screen before using the Ecology Place. By using the website or CD-ROM, you indicate that you have read, understood and accepted the terms of this agreement.

0-321-06878-5

The Cross-Platform Prep Course

McGraw-Hill Education's multi-platform course gives you a variety of tools to help you raise your test scores. Whether you're studying at home, in the library, or on-the-go, you can find practice content in the format you need—print, online, or mobile.

Print Book

This print book gives you the tools you need to ace the test. In its pages you'll find smart test-taking strategies, in-depth reviews of key topics, and ample practice questions and tests. See the Welcome section of your book for a step-by-step guide to its features.

Online Platform

The Cross-Platform Prep Course gives you additional study and practice content that you can access *anytime, anywhere*. You can create a personalized study plan based on your test date that sets daily goals to keep you on track. Integrated lessons provide important review of key topics. Practice questions, exams, and flashcards give you the practice you need to build test-taking confidence. The game center is filled with challenging games that allow you to practice your new skills in a fun and engaging way. And, you can even interact with other test-takers in the discussion section and gain valuable peer support.

Getting Started

To get started, open your account on the online platform:

Go to www.xplatform.mhprofessional.com

↓

Enter your access code, which you can find on the inside back cover of your book

↓

Provide your name and e-mail address to open your account and create a password

↓

Click "Start Studying" to enter the platform

It's as simple as that. You're ready to start studying online.

Your Personalized Study Plan

First, select your test date on the calendar, and you'll be on your way to creating your personalized study plan. Your study plan will help you stay organized and on track and will guide you through the course in the most efficient way. It is tailored to *your* schedule and features daily tasks that are broken down into manageable goals. You can adjust your test date at any time and your daily tasks will be reorganized into an updated plan.

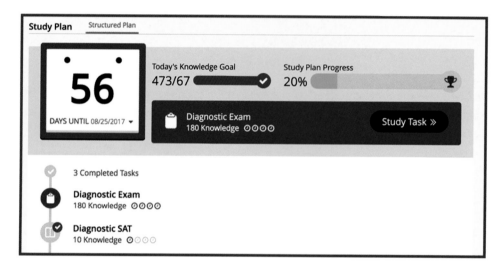

You can track your progress in real time on the Study Plan Dashboard. The "Today's Knowledge Goal" progress bar gives you up-to-the minute feedback on your daily goal. Fulfilling this every time you log on is the most efficient way to work through the entire course. You always get an instant view of where you stand in the entire course with the Study Plan Progress bar.

If you need to exit the program before completing a task, you can return to the Study Plan Dashboard at any time. Just click the Study Task icon and you can automatically pick up where you left off.

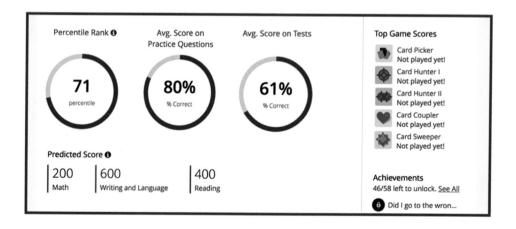

Practice Tests

One of the first tasks in your personalized study plan is to take the Diagnostic Test. At the end of the test, a detailed evaluation of your strengths and weaknesses shows the areas where you need the most focus. You can review your practice test results either by the question category to see broad trends or question-by-question for a more in-depth look.

The full-length tests are designed to simulate the real thing. Try to simulate actual testing conditions and be sure you set aside enough time to complete the full-length test. You'll learn to pace yourself so that you can work toward the best possible score on test day.

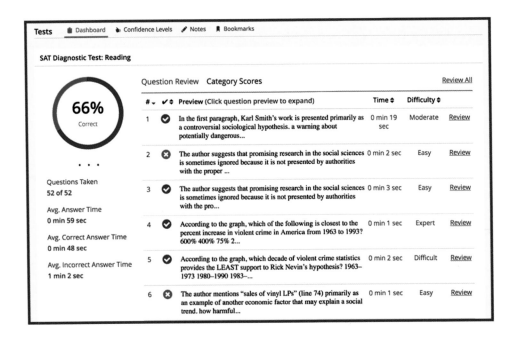

Lessons

The lessons in the online platform are divided into manageable pieces that let you build knowledge and confidence in a progressive way. They cover the full range of topics that you're likely to see on your test.

After you complete a lesson, mark your confidence level. (You must indicate a confidence level in order to count your progress and move on to the next task.) You can also filter the lessons by confidence levels to see the areas you have mastered and those that you might need to revisit.

Use the bookmark feature to easily refer back to a concept or leave a note to remember your thoughts or questions about a particular topic.

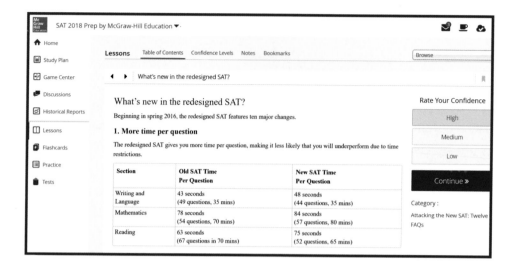

Practice Questions

All of the practice questions are reflective of actual exams and simulate the test-taking experience. The "Review Answer" button gives you immediate feedback on your answer. Each question includes a rationale that explains why the correct answer is right and the others are wrong. To explore any topic further, you can find detailed explanations by clicking the "Help me learn about this topic" link.

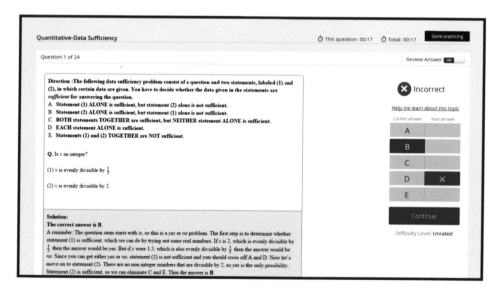

You can go to the Practice Dashboard to find an overview of your performance in the different categories and sub-categories.

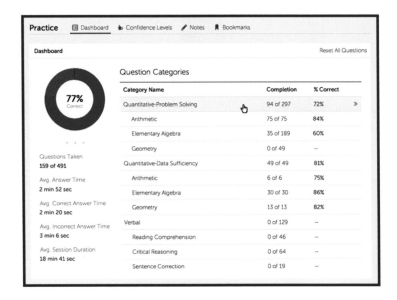

Dashboard

The dashboard is constantly updating to reflect your progress and performance. The Percentile Rank icon shows your position relative to all the other students enrolled in the course. You can also find information on your average scores in practice questions and exams.

A detailed overview of your strengths and weaknesses shows your proficiency in a category based on your answers and difficulty of the questions. By viewing your strengths and weaknesses, you can focus your study on areas where you need the most help.

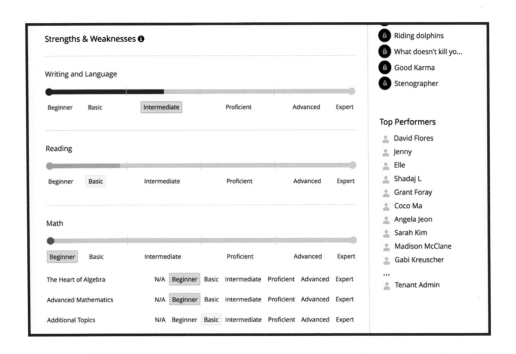

Flashcards

The hundreds of flashcards are perfect for learning key terms quickly, and the interactive format gives you immediate feedback. You can filter the cards by category and confidence level for a more organized approach. Or, you can shuffle them up for a more general challenge.

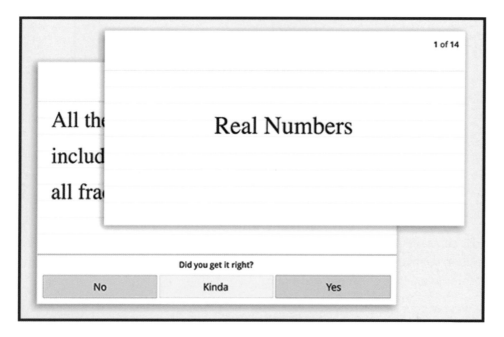

Another way to customize the flashcards is to create your own sets. You can either keep these private or share or them with the public. Subscribe to Community Sets to access sets from other students preparing for the same exam.

Game Center

Play a game in the Game Center to test your knowledge of key concepts in a challenging but fun environment. Increase the difficulty level and complete the games quickly to build your highest score. Be sure to check the leaderboard to see who's on top!

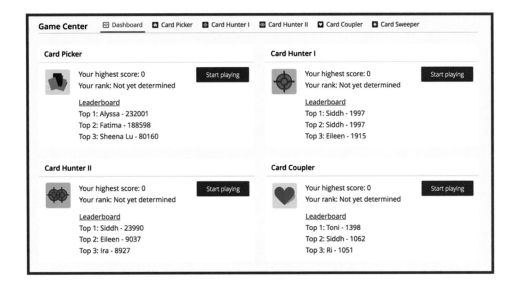

Social Community

Interact with other students who are preparing for the same test. Start a discussion, reply to a post, or even upload files to share. You can search the archives for common topics or start your own private discussion with friends.

Mobile App

The companion mobile app lets you toggle between the online platform and your mobile device without missing a beat. Whether you access the course online or from your smartphone or tablet, you'll pick up exactly where you left off.

Go to the iTunes or Google Play stores and search "McGraw-Hill Education Cross-Platform App" to download the companion iOS or Android app. Enter your e-mail address and the same password you created for the online platform to open your account.

Now, let's get started!

FIFTH EDITION

Elements of Ecology

Robert Leo Smith
West Virginia University, Emeritus

Thomas M. Smith
University of Virginia

Benjamin Cummings

San Francisco   Boston   New York
Cape Town   Hong Kong   London   Madrid   Mexico City
Montreal   Munich   Paris   Singapore   Sydney   Tokyo   Toronto

For Our Family

Acquisitions Editor: Elizabeth Fogarty
Assistant Editor: Jeanne Zalesky
Developmental Editors: Elizabeth Zayatz, Meredith Nightingale
Text Designer: Carolyn Deacy
Cover Designer: Yvo Riezebos
Artists: Robert L. Smith, Jr., Ned Smith
Photo Researchers: Maureen Spuhler, Kathleen Olson, Travis Amos
Manufacturing Coordinator: Vivian McDougal
Project Coordination: Gina Linko, Isabel Fiore, Elm Street Publishing
Production Editor: Steven Anderson
Copy editor: Sybil Sosin
Proofreader: Amy Schneider
Illustrator: Scientific Illustrators
Composition: TSI Graphics
Marketing Manager: Josh Frost
Cover Printer: Phoenix Color
Printer and Binder: R.R. Donnelley

Cover Photo: ©Getty/Art Wolfe

ISBN 0-8053-4473-X

Library of Congress Cataloging-in-Publication Data
Smith, Robert Leo.
    Elements of ecology / by Robert Leo Smith and Thomas M. Smith.— 5th ed.
      p. cm.
    Includes bibliographical references (p. 627).
    ISBN 0-8053-4473-X
    1. Ecology. I. Smith, T. M. (Thomas Michael), 1955- II. Title.
QH541 .S624 2002
577—dc21   2002004865

Benjamin Cummings

1 2 3 4 5 6 7—DOR—06 05 04 03 02

www.aw.com/bc

Brief Contents

Contents

5 Soil 79

PART III

The Organism and Its Environment 99

6 Plant Adaptations to the Environment 101

7 Decomposers and Decomposition 123

8 Animal Adaptations to the Environment 137

16 Parasitism and Mutualism 309

17 Processes Shaping Communities 329

18 Human Interactions Within Communities 349

19 Landscape Ecology 367

PART VI

Ecosystems 393

Preface

The first edition of *Elements of Ecology* appeared in 1976 as a short version of *Ecology and Field Biology*. Since that time, *Elements of Ecology* has evolved into a textbook for a much broader audience than the originally intended readership of students majoring in the life sciences. That evolution, driven by our belief that ecology should be part of a liberal education, continues with this fifth edition. We believe that students who major in such diverse fields as economics, sociology, engineering, political science, law, history, English, languages, and the like should have some basic understanding of ecology for the simple reason that it impinges on their lives. An informed opinion about such highly politicized environmental issues as air and water quality, habitat preservation, declining biological diversity, the management of public lands, sustainability, and global warming requires a grounding in ecological concepts.

The fourth edition marked a major departure from the organization and style of previous editions and, for that matter, from other ecology textbooks. Our goal was to make ecology more accessible to students from different academic fields who may not have had a course in biology and who want to take ecology as a science elective. In this fifth edition, we retain the modified modular approach by discussing a key concept in each section and introducing it with a conceptual statement. These sentence headings become the focal point of the section. The chapter summaries are organized into a hierarchy of related topics identified by brief headings. Such groupings enable students to see how the concepts presented in the chapter fit together. Because it is a mixed-majors text, we deliberately hold to a minimum discussions requiring a background in mathematics, chemistry, and physics. Where they are essential to the topic, we try to explain them clearly. In a number of places, especially in chapters dealing with the demography of populations, we have placed this material in special boxes entitled Quantifying Ecology.

An Integrated Approach

Further changes serve to shift this new edition from the traditional function of cataloging and illustrating concepts to a more integrative approach. The theme of adaptation and natural selection provides a unifying concept that explicitly links the various topics presented throughout the text. To do so, we have restructured the presentation. Much of the material that was previously in chapters on adaptation and population genetics now takes the form of a historical narrative that develops the theme of adaptation and natural selection. To provide a background on the primary features of the physical environment that function as agents of natural selection, we extracted and rewrote the material on temperature, light, moisture, and other aspects of the physical environment and created a new chapter on the abiotic environment. In the fourth edition, the discussion of the interaction of organisms with the physical environment was organized by major environmental factors: light, temperature, water, nutrients. To present a more integrated perspective, we now examine plant and animal adaptations to variations in the physical environment separately. This change eliminated a fault common to ecology texts in which students and instructors have to jump from plant to animal adaptation within the context of a specific environmental feature, such as temperature. Our new approach allows for a better understanding of the interdependent nature of features of the environment and species adaptations. This is especially true for plants, where light, temperature, moisture, and nutrients play an integrated role in photosynthesis and carbon allocation. By focusing on the tradeoffs involved in the adaptation of organisms to differing environmental conditions, we set the stage for exploring the consequences of adaptations at the level of the organism to population, community, and ecosystem dynamics in later chapters.

We have incorporated the presentation of species interactions into the discussion of community ecology. This treatment allows a greater emphasis on the role of interspecific relations on community structure. The discussion of spatial and temporal changes in community structure now integrates the role of species adaptations to the physical environment and the modifying effects of species interactions with changing environmental conditions.

The major change in the discussion of ecosystem processes is an expanded view of nutrient cycling. As with the discussion of communities, we strive to explore the link between adaptations at the level of the individual and the collective processes of productivity and nutrient cycling within ecosystems, and how those patterns vary with changes in the physical environment.

The Guide to Ecosystems section of the fourth edition now forms the core material for a new part that explores the topics of biogeography and biodiversity. The material on ecosystem functions previously found in those chapters has been integrated into earlier chapters in this edition, and the descriptive material has been shortened and rearranged to give greater emphasis to global biodiversity.

Hallmark Features

The special features found in the previous edition that make learning easy have been retained.

- **Unique modular format.** With a complete concept statement at the beginning of each module, the text helps students focus on the core concepts by dividing chapter material into manageable amounts of information.
- **Dynamic four-color art** engages and maintains student interest.
- **The text's clear descriptive approach** helps students appreciate and understand ecology without overwhelming them with excessive amounts of quantitative information.
- **Pedagogy,** including chapter-opening objectives, chapter summaries, study questions, boldface key terms, and an extensive glossary, increases student retention and understanding of key concepts.
- **Focus on Ecology boxes** contain real-world examples of ecological principles.
- **Quantifying Ecology boxes** clarify mathematical or quantitative aspects of ecology.
- **Ecological Application Essays** that appear throughout the text demonstrate the real-world

relevance of ecological concepts. For example, the application essay "Cheating Nature" at the end of Part IV explores the history of human population growth.

- **References** for each chapter at the end of the text encourage further exploration. The references include the source material for the chapter as well as selected books and journal articles. Annotation helps students choose among them.
- **Cross-references** throughout the text tie both concepts and chapters together, emphasizing the interrelated ecological principles.

Organization

We have divided the fifth edition into seven related parts with numerous cross-references and an ever-broadening focus. Part I sets the stage. Chapter 1 explains what ecology is, how it relates to other sciences, and how ecologists use scientific methods. Chapter 2 introduces the basic concepts of natural selection, heritability, adaptation, evolution, and speciation.

Part II concerns the physical or abiotic environment within which life exists. Chapter 3 introduces climate, a major determinant of the physical environment in which all organisms live. Chapter 4 examines the major features of the physical environment to which organisms must adapt. These include the light, thermal, moisture, and chemical environments. Chapter 5 explores soil, the foundation of all terrestrial environments.

Part III explores how life has adapted to the diverse physical environmental conditions that are found on Earth. Chapter 6 looks at the role of light, temperature, and moisture in photosynthesis and how the constraints imposed on plants by the physical environment influence plant adaptations. Chapter 7 examines the process of decomposition, the organisms involved, and how decomposition relates to both plant and animal life. Chapter 8 looks at animal adaptations: homeostasis, energy exchange as influenced by ectothermy and endothermy, and responses to the light environment.

Part IV focuses on populations of organisms. Chapter 9 introduces the population and its major properties: density, distribution, and age structure. Chapter 10 explores population growth as a function of the demographic processes of mortality, natality, and survivorship. Regulation of population growth involves intraspecific competition, covered in Chapter 11. Introduced in this chapter are the concepts of density and growth in plants, dispersal, and social behavior. Chapter 12 looks at the relationships among individuals reflected in various life history patterns, including mating and reproductive strategies. Much of this chapter falls into the category of behavioral ecology.

Part V deals with communities. The community is structured by interspecific interrelations. Chapter 13 introduces community structure in its broadest sense, including the concepts of dominance and diversity, vertical and horizontal structure, and the spatial and temporal dynamics of the community structure. Influencing community structure are interspecific competition, predation, parasitism, and mutualism, the topics of Chapters 14 through 16. Chapter 17 integrates these interactions as the processes shaping communities. Having a major impact on communities and community structure are human interactions, especially with natural populations. Chapter 18 considers exploitation, restoration, conservation, and pest control. It introduces the concepts of sustained yield and integrated pest management. Concluding Part V is a new chapter on landscape ecology that considers processes that create a variety of patches on the landscape, and the influences of landscape patterns on population and community dynamics.

Part VI explores ecosystem dynamics. Chapter 20 presents the concept of the ecosystem and primary and secondary production and how energy flows through the ecosystem. Chapter 21 details the nutrient cycles from a functional point of view, considering inputs, outputs, and internal cycling. It contrasts nutrient cycling in terrestrial, aquatic, and marine ecosystems. Chapter 22 describes the major biogeochemical cycles. Chapter 23 examines how humans have intruded upon them.

Part VII considers biogeography and biodiversity. Chapter 24 introduces some of the basic ideas of biogeography as it applies to a region-to-global scale, while an overview of Earth's major ecosystems is presented in Chapters 25 through 29. Chapter 25 covers wooded ecosystems: forests, woodlands, and savannas. Chapter 26 examines grassland, shrubland, desert, and tundra ecosystems. Chapter 27 examines the physical and biological characteristics of lakes, ponds, and estuarine ecosystems. Chapter 28 explores marine ecosystems from the intertidal zone to the open sea. Chapter 29 describes ecosystems that are halfway between aquatic and terrestrial environments, the wetlands, both freshwater and saltwater. Part VII ends with an overview of issues relating to global warming. This chapter explores the influence of forest clearing and the use of fossil fuels on the global carbon cycle, and the potential impact of the changing atmospheric chemistry on global climate dynamics. It examines how global climate change might affect natural ecosystems, agricultural production, and human health, drawing upon many of the concepts discussed throughout the text to help students understand the issue of global environmental change.

Illustration Program

This fifth edition features full-color illustrations. Retained in black and white are the outstanding original pen-and-ink drawings by the late Ned Smith that date back to the first edition, as well as a number of pen-and-ink illustrations by Robert Leo Smith, Jr., that are most effective in their original format. All the remaining art has been redesigned and redrawn by Robert Leo Smith, Jr. New color photographs have been carefully selected to supplement the text.

Supplements

Print Supplements

A set of free supplementary materials supports the instructor:

- Instructor's Art CD-ROM: All of the art from the text available in PowerPoint.
  0-321-06887-4

- A set of 100 four-color Transparency Acetates of the key figures in the text.
  0-8053-4846-8

- A combined Instructor's Manual and Test Bank with new multiple-choice questions and updated commentary.
  0-8053-4845-X

- Computerized Test Bank for Macintosh and Windows.
  0-8053-4847-6

Media Supplements

- The Ecology Place Web site
  www.ecologyplace.com
  Access to the Web site is free with every copy of the text. The Ecology Place includes 27 interactive field activities and tutorials, multiple-choice quiz questions, Web links, and glossary.

- Biology Labs On-Line: Ecology Version
  0-8053-7052-8
  www.biologylab.awlonline.com
  The Ecology Version of this on-line resource includes five labs and a print manual, including Evolution Lab, Population Genetics Lab, Leaf Lab, Demography Lab, and Population Ecology Lab.

Acknowledgments

No textbook is a product of the authors alone. The material this book covers represents the work of hundreds of ecological researchers who have spent lifetimes in the field and the laboratory. Their published experimental results, observations, and conceptual thinking provide the raw material out of which the textbook is fashioned.

Revision of a textbook depends heavily on the input of users who point out mistakes and opportunities. We took these suggestions seriously and incorporated many of them. We are deeply grateful to the following reviewers for their helpful comments and suggestions on how to improve this edition for the broader audience for which it is intended: Earl Aagard, Pacific Union College; John Baccus, Southwest Texas State University; Claude Baker, Indiana University; Edmund Bedecarrax, City College of San Francisco; Emma Benenati, Northern Arizona University; Jim Bever, Indiana University; Judy Bluemer, Morton College; Steven Blumenshine, Fresno State University; Paul Bologna, Fairleigh Dickinson University; Mark Brenner, Michelle Briggs, Lycoming College; Evert Brown, Casper College; Joseph Bruseo, Amherst College; Warren Burggren, University of North Texas; Willodean Burton, Austin Peay State; Guy Cameron, University of Cincinnati; George Cline, Jacksonville State University; Todd Crowl, Utah State University; John Cruzan, Geneva College; Elissa Miller Derrickson, Loyola College; Howard Epstein, University of Virginia; George Estabrook, University of Michigan; Frank Gilliam, Marshall University; Joseph Hendricks, State University of West Georgia; Cheryl Hogue, California State University-Northridge; Donald Keith, Tarleton State University; Tim Knight, Ouachita Baptist University; George Kraemer, SUNY Purchase College; Shannon Kuchel, Colorado Christian University; Ralph Larson, San Francisco State University; Todd Livdahl, Clark University; Richard E. MacMillan, University of California-Irvine; Andy McCollum, Cornell College; Matthew Moran, Hendrix College; Larry Mueller, University of California-Irvine; John Mullins, Richard Niesenbaum, Muhlenburg College; John O'Brien, University of North Carolina Greensboro; Fatima Pale, Thiel College; David Pindel, Corning Community College; Rick Relyea, University of Pittsburgh; Robin Richardson, Winona State University; Philip Robertson, Southern Illinois University Carbondale; Irene Rossell, University of North Carolina at Asheville; Brian Schultz, Hampshire College; Stuart Skeate, Lees-McRae College; George Sideris, Long Island University; Pamela Silver, Pennsylvania State University Erie; Jeffrey Smallwood, California State University-Northridge; Alan Stam, Capital University; Jack Stout, University of Central Florida; Merrill Sweet, Texas A&M University; Robert Tamarin, University of Massachusetts-Lowell; Robert Tinnin, Portland State University; Conrad Toepfer, Millikin University; Robert Twilley, University of Louisiana; Cynthia Walter, Saint Vincent College; Fred Wasserman, Boston University; David Webster, University of North Carolina-Wilmington; Mary Wicksten, Texas A&M University; Robert Winget, Brigham Young University-Hawaii; Bruce Wunder, Colorado State University.

Several of the Ecological Application essays that appeared in the fourth edition have survived into this edition (III, V, VI, and VII). The three new essays (I, II, and IV) represent a collaborative effort between the authors and Jeanne Zalesky, Elizabeth Zayatz, and Kay Ueno, whose constructive editing guided our often-diffuse ideas back to the central message.

The publication of a modern textbook requires the work of many editors to handle the specialized tasks of development, photography, graphic design, illustration, copy editing, and production, to name only a few. Overseeing this team of specialists is one individual whose job it is to make it all come together. That individual was Jeanne Zalesky, Assistant Editor. It was her effort, organization, and calming effect during the harried pace of production that ultimately made this project not only possible, but also enjoyable. Words cannot adequately express our appreciation and respect.

Through it all our families, especially our spouses Alice and Nancy, had to endure the throes of book production. Their love, understanding, and support provide the balanced environment that makes our work possible.

Robert Leo Smith
Thomas M. Smith

PART 1

Introduction

Researcher sampling fish in the Water Conservation Area of Everglades National Park.

The Nature of Ecology

OBJECTIVES

On completion of this chapter, you should be able to:

- Define ecology.
- Define ecosystem, community, and population.
- Relate ecology to the other biological and physical sciences.
- Describe how an ecologist conducts research.
- Define hypothesis and discuss the role of hypothesis testing in science.
- Discuss the role of uncertainty in science.

*E*cology—for years the term was familiar only to specialists in an obscure field. Overshadowed by molecular biology, the subject was scarcely recognized by the academic world. Then came the environmental movement of the late 1960s and early 1970s. Suddenly ecology was thrust into the limelight, and the term became a household word appearing in newspapers, magazines, and books—although it was often misused. Even now people confuse it with the environment and with environmentalism. Ecology is neither one; yet a thorough understanding of ecological concepts is essential for making sound environmental decisions.

1.1 Ecology Is a Science

So what is ecology? According to the usual definition, **ecology** is the scientific study of the relationship between organisms and their environment. That definition is satisfactory so long as you consider relationship and environment in their fullest meaning. *Environment* includes not only the physical conditions, but also the biological or living components that make up an organism's surroundings. *Relationship* includes interactions with the physical world, as well as with members of the same and other species.

The term *ecology* comes from the Greek words *oikos,* meaning "the family household," and *logy,* meaning "the study of." Literally, ecology is the study of the household. It has the same root word as economics, or "management of the household." We could consider ecology to be the study of the economics of nature. In fact, some economic concepts, such as resource allocation and cost-benefit ratios, have crept into ecology. The German zoologist Ernst Haeckel originally coined the term in 1866. He called it *Oecologie* and defined its scope as the study of the relationship of animals to their environment.

1.2 The Major Unit of Ecology Is the Ecosystem

Organisms interact with their environment within the context of the **ecosystem.** The *eco* part of the word relates to the environment. The *system* part implies that the ecosystem is a system. A system is a collection of related parts that function as a unit. The automobile engine is a system; subparts of the engine, such as the ignition, are also systems. In a

FIGURE 1.1 A tropical rain forest ecosystem in northeast Australia. Note the layers of vegetation and the diversity of plants.

similar fashion, the ecosystem has interacting parts that support a whole. Broadly, the ecosystem consists of two basic interacting components, the living or **biotic,** and the physical or **abiotic.**

Consider a natural ecosystem, a forest (Figure 1.1). The physical (abiotic) component of the forest consists of the atmosphere, climate, soil, and water. The biotic component includes the many different plants and animals that inhabit the forest. Relationships are complex; each organism not only responds to the physical environment but also modifies it and, in doing so, becomes part of the environment itself. The trees in the canopy of a forest intercept the sunlight and use this energy to fuel the process of photosynthesis. In doing so, they modify the environment of plants below them, reducing the sunlight and lowering air temperature. Birds foraging on insects in the litter layer of fallen leaves reduce insect numbers and modify the environment for other organisms that depend on this shared food resource. We will explore these complex interactions between the living and the physical environment in greater detail in succeeding chapters.

1.3 Ecosystem Components Form a Hierarchy

The various kinds of organisms that inhabit our forest make up populations. The term *population* has many uses and meanings in other fields of study. In ecology, a **population** is a group of individuals of the same species occupying a given area. Populations of plants and animals in the ecosystem do not function independently of each other. Some populations compete with other populations for limited resources, such as food, water, or space. In other cases, one population is the food resource for another. Two populations may mutually benefit each other, each doing better in the presence of the other. All populations of different species living and interacting within an ecosystem are referred to collectively as a **community.**

We can now see that the ecosystem, consisting of the biotic community and the physical environment, has many levels. On one level, individual organisms, including humans, both respond to and influence the physical environment. At the next level, individuals of the same species form populations, such as a population of white oak trees or gray squirrels within the forest, that can be described in terms of number, growth rate, and age distribution. Further, individuals of these populations interact among themselves and with individuals of other species to form a community. Herbivores consume plants; predators eat prey; and individuals compete for limiting resources. When individuals die, their remains decompose, releasing nutrients consumed and incorporated into their tissues back into the soil to be recycled. Ecology is the study of all these relationships—the complex web of interactions between organisms and their environment.

Combined, the ecosystems of Earth form the planetary ecosystem or **biosphere.** Organisms within the biosphere not only adapt to the environment, but also interact to modify and control chemical and physical conditions of the biosphere. This view of a self-sustaining biosphere, in which every organism is linked to the other, is known as the Gaia hypothesis, developed by James Lovelock. Although not all ecologists agree with this hypothesis, it does serve as a model to illustrate the interconnected nature of the physical and biological components of our planet.

1.4 Ecology Has Strong Ties to Other Disciplines

The complex interactions taking place within the ecosystem involve all varieties of physical and biological processes. To study these interactions, ecolo-

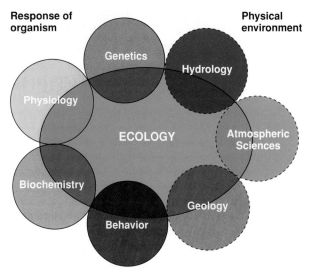

FIGURE 1.2 Ecology is an interdisciplinary science. It overlaps with many elements of physical and biological sciences.

gists have to draw on other sciences. This dependency makes ecology an interdisciplinary science (Figure 1.2).

In the following chapters, we will discuss aspects of biochemistry, physiology, and genetics. We do so only in the context of understanding the interplay of organisms with their environment. The study of how plants take up carbon dioxide and lose water (Chapter 6), for example, belongs to plant physiology. Ecology looks at how these processes respond to variations in rainfall and temperature. This information is crucial to understanding the distribution and abundance of plant populations, and the structure and function of ecosystems on land. Likewise, we must dip into many of the physical sciences, such as geology, hydrology, and meteorology. They will help us chart other ways organisms and environment interact. For instance, as plants take up water, they influence soil moisture and the patterns of surface water flow. As they lose water to the atmosphere, they increase atmospheric water content and influence regional patterns of precipitation. The geology of an area influences the availability of nutrients and water for plant growth. In each example, other scientific disciplines are critical to understanding how individuals both respond to and shape their environment. (See Focus on Ecology 1.1: A Brief History of Ecology.)

As we have made the transition from the 20th to the 21st century, ecology has entered a new frontier, one that requires expanding our view of ecology to include the dominant role of humans in nature. Among the many environmental problems facing humanity, four broad and interrelated areas can be identified as critical: human population growth,

A BRIEF HISTORY OF ECOLOGY

The genealogy of most sciences is direct. Tracing the roots of mathematics, chemistry, and physics is relatively easy. The science of ecology is different. Its roots are complex, intertwined with a wide array of scientific advances in other disciplines within the biological and physical sciences.

You can argue that ecology goes back to the ancient Greek scholar Theophrastus, a friend of Aristotle, who wrote about the relations between organisms and the environment. On the other hand, ecology as we know it today has its early roots in plant geography and natural history.

In the 1800s, botanists began exploring and mapping the world's vegetation. Early plant geographers such as Carl Ludwig Willdenow (1765–1812) and Friedrich Heinrich Alexander von Humboldt (1769–1859) pointed out that regions of the world with similar climates supported vegetation that was similar in form, even though the species were different. The recognition that the form and function of plants within a region reflected the constraints imposed by the physical environment led the way for a new generation of scientists who explored the relationship between plant biology and plant geography.

Among this new generation of scientists was Johannes Warming (1841–1924) at the University of Copenhagen, who studied the tropical vegetation of Brazil. He wrote the first text on plant ecology, *Plantesamfund*. In this book Warming integrated plant morphology, physiology, taxonomy, and biogeography into a coherent whole. The book had a tremendous influence on the early development of ecology.

Meanwhile, activities in other areas of natural history were assuming an important role. One was the voyage of Charles Darwin (1809–1882) on the *Beagle*. Working for years on notes and collections from this trip, Darwin compared similarities and dissimilarities among the current species inhabiting the regions he explored and the fossils he had collected. To quote his opening paragraph of *The Origin of Species* (1859): "When on board H.M.S. Beagle, as naturalist, I was much struck with certain facts in the distribution of the organic beings inhabiting South America, and in the geographical relations of the present to the past inhabitants of that continent. These facts . . . seem to throw some light on the origin of species—that mystery of mysteries, as it has been called by one of our greatest philosophers."

Developing his theory of evolution and the origin of species, Darwin came across the writings of Thomas Malthus (1766–1834). An economist, Malthus advanced the principle that populations grow in a geometric fashion, doubling at regular intervals until they outstrip the food supply. Ultimately, the population will be restrained by a "strong, constantly operating force such as sickness and premature death." From this concept, Darwin developed the idea of "the survival of the fittest" as a mechanism of natural selection and evolution (see Chapter 2).

Meanwhile, unknown to Darwin, an Austrian monk, Gregor Mendel (1822–1884), was studying the transmission of characteristics from one generation of pea plants to another in his garden. Mendel's work on inheritance and Darwin's work on natural selection provided the foundation for the study of evolution and adaptation, the field of **population genetics.** Darwin's theory of natural selection, combined with the new understanding of genetics, the means by which characteristics are transmitted from one generation to the next, provided the mechanism for understanding the link between organisms and their environment—the focus of ecology.

Early ecologists, particularly plant ecologists, were concerned with observing the patterns of organisms in nature, attempting to understand how patterns were formed and maintained by interactions with the physical environment. Some, notably Frederic E. Clements, sought some system of organizing nature. He proposed that the plant community behaves as a complex organism or superorganism that grows and develops through stages to a mature or climax stage (see Chapter 13). His idea was accepted and advanced by other ecologists. A few ecologists, however, notably Arthur G. Tansley (1871–1955), did not share this view. In its place he advanced a holistic and integrated ecological concept that combined living organisms and their physical environment into a system, which he called an ecosystem (see Chapter 20).

While the early plant ecologists were concerned mostly with terrestrial vegetation, another group of European biologists was interested in the relationship between aquatic plants and animals and their environment. These scientists advanced the ideas of organic nutrient cycling and feeding levels, using the terms *producers* and *consumers*. Their work influenced a young limnologist at the University of Minnesota, Raymond A. Lindeman. He traced "energy-available" relationships within a lake community. Together with the writings of Tansley, Lindeman's 1942 paper, *The Trophic-Dynamic Aspects of Ecology,* marked the beginning of **ecosystem ecology,** the study of whole living systems.

Animal ecology was initially largely independent of the early developments in plant ecology. The beginnings of animal ecology can be traced to two Europeans, R. Hesse of Germany and Charles Elton of England. Elton's *Animal Ecology* (1927) and Hesse's *Tiergeographie auf logischer grundlage* (1924), translated into English as *Ecological Animal Geography,* strongly influenced the development of animal ecology in the United States. Charles Adams and Victor Shelford were two pioneering U.S. animal ecologists. Adams published the first textbook on animal ecology, *A Guide to the Study of Animal Ecology* (1913). Shelford wrote *Animal Communities in Temperate America* (1913).

Shelford gave a new direction to ecology by stressing the interrelationship of plants and animals. Ecology became a science of communities. Some earlier European ecologists, particularly the marine biologist Karl Mobius, had developed the general concept of the community. In his essay *An Oyster Bank Is a Biocenose* (1877), Mobius explained that the oyster bank, although dominated by one animal, was really a complex community of many interdependent organisms. He proposed the word *biocenose* for such a community. The word comes from the Greek, meaning "life having something in common."

The appearance in 1949 of the encyclopedic *Principles of Animal Ecology* by five second-generation ecologists from the University of Chicago (W. C. Allee, A. E. Emerson, Thomas Park, Orlando Park, and K. P. Schmidt) pointed the direction modern ecology was to take. It emphasized feeding relationships and energy budgets, population dynamics, and natural selection and evolution.

The writings of Thomas Malthus that were so influential in the development of Darwin's ideas about the origin of species also stimulated the study of natural populations. The study of populations in the early 20th century branched into two fields. One, **population ecology,** is concerned with population growth (including birthrates and death rates), fluctuation, spread, and interactions. The other, **evolutionary ecology,** is concerned with the natural selection and evolution of populations. Closely associated with population ecology and evolutionary ecology—and often difficult to separate clearly and completely—is **community ecology.**

Physiological ecology arose at the same time. It is concerned with the responses of individual organisms to temperature, moisture, light, and other environmental conditions. Natural history observations also spawned **behavioral ecology.** Nineteenth-century behavioral studies included those on ants by William Wheeler and on South American monkeys by Charles Carpenter. Later, Konrad Lorenz and Niko Tinbergen gave a strong impetus to the field with their pioneering studies on the role of imprinting and instinct in the social life of animals, particularly birds and fish.

Other observations led to investigations of chemical substances in the natural world. Scientists began to explore the use and nature of chemicals in animal recognition, trail-making, and courtship, and in plant and animal defense. Such studies make up the specialized field of **chemical ecology.**

With advances in biology, physics, and chemistry throughout the latter part of the 20th century, new areas of study emerged in ecology. The development of aerial photography, and later the launching of satellites by the space program, provided scientists with a new perspective of Earth's surface through the use of remotely sensed data. Ecologists began to explore spatial processes that linked adjacent communities and ecosystems in the newly emerging field of **landscape ecology.** A new appreciation for the impacts of changing land use on natural ecosystems led to the development of **conservation ecology,** which applies principles of many different fields, from ecology to economics and sociology, to the maintenance of biological diversity. The application of principles of ecosystem development and function to the restoration and management of disturbed lands has given rise to **restoration ecology.**

As we have made the transition from the 20th to the 21st century, ecology has entered a new frontier. Understanding Earth as a system is the newest area of ecological study—**global ecology.**

---

biological diversity, sustainability, and global climate change. As the human population increased from approximately 500 million to over 6 billion in the past two centuries, dramatic changes in land use altered Earth's surface. Clearing forests for agriculture has destroyed many natural habitats, resulting in a rate of species extinction that is unprecedented in Earth's history. In addition, the expanding human population is exploiting natural resources at unsustainable levels. Due to growing demand for energy from fossil fuels needed to sustain economic growth, the chemistry of the atmosphere is changing in ways that may alter Earth's climate. These environmental problems are ecological in nature, and the science of ecology is essential to understanding their causes and identifying the means to mitigate their impacts. Addressing these issues, however, requires a broader interdisciplinary framework to better understand their historical, social, legal, political, and ethical dimensions. That broader framework is known as environmental science. Environmental science examines the impact of humans on the natural environment, and as such covers a wide range of topics including agronomy, soils, demography, agriculture, energy, and hydrology, to name but a few.

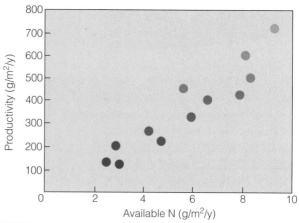

FIGURE 1.3 The response of grassland production to nitrogen availability. Nitrogen, the independent variable, goes on the *x* axis; grassland productivity, the dependent variable, goes on the *y* axis.

1.5 Ecologists Use Scientific Methods

To investigate the relation of organisms to their environment, ecologists have to undertake experimental studies in the laboratory and in the field. All these studies have one thing in common; they involve the collection of data to test hypotheses. A **hypothesis** is an "educated guess" that a scientist poses to explain an observed phenomenon; it should be a statement of cause and effect that can be tested. A hypothesis may be based on an observation in the field or laboratory, or on previous investigations.

For example, an ecologist might hypothesize that the availability of the nutrient nitrogen is the major factor limiting growth and productivity of plants in the prairie grasslands of North America. To test this hypothesis, the ecologist can gather data in a number of ways. The first approach might be a field study. The ecologist would examine the correlation between available nitrogen and grassland productivity across a number of locations. Both factors vary across the landscape. If nitrogen is controlling grassland productivity, productivity should increase with nitrogen. The ecologist would measure nitrogen availability and grassland productivity at a number of sites in the region. Then the relationship between these two variables, nitrogen and productivity, could be expressed graphically.

The graph in Figure 1.3 shows nitrogen availability on the horizontal or *x* axis and plant productivity on the vertical or *y* axis. The reason for this arrangement is important. The scientist is assuming that nitrogen is the cause and that plant productivity is the effect. We call *x* the independent variable and *y* the dependent variable.

Although the data suggest that nitrogen does control grassland production, they do not prove it. It might well be that some other factor that varies with nitrogen availability, such as moisture or acidity, is actually responsible for the observed relationship. To test the hypothesis a second way, the scientist may choose to undertake an experiment. In designing the experiment, the scientist will try to isolate the presumed causal agent—in this case, nitrogen availability.

The scientist may choose to do a field experiment, adding nitrogen to some natural sites and not to others. The investigator controls the independent variable, levels of nitrogen, in a predetermined way and monitors the response of the dependent variable, plant growth. By observing the differences in productivity between grasslands that were fertilized with nitrogen and those that were not, the scientist tries to test whether nitrogen is the causal agent. However, in choosing the experimental sites, the scientist must try to locate areas where other factors that may influence productivity, such as moisture and acidity, are similar. Otherwise the scientist cannot be sure which factor is responsible for the observed differences in productivity among the sites.

Finally, the scientist might try a third approach, a series of laboratory experiments. The advantage of laboratory experiments is that the scientist has much more control over the environmental conditions. For example, the scientist can grow the native grasses in the greenhouse under conditions of controlled temperature, soil acidity, and water availability. If the plants exhibit increased growth under higher nitrogen fertilization, the scientist has evidence in support of the hypothesis. Nevertheless, the scientist faces a limitation common to all laboratory experiments; the results are not directly applicable in the field. The response of grass plants under controlled laboratory conditions may not be the same as their response under natural conditions in the field. In the field, the plants are part of the ecosystem, interacting with other plants, animals, and the physical environment. Despite this limitation, the scientist now knows the basic growth response of the plants to nitrogen availability and goes on to design both laboratory and field experiments to explore new questions about the cause-and-effect relationship.

1.6 Experiments Can Lead to Predictions

Scientists use the understanding derived from observation and experiments to make models. Data are limited to the special case of what happened when the measurements were made. Like photographs, data represent a given place and time. Models use the understanding gained from the data to predict what will happen in some other place and time.

Models are abstract, simplified representations of real systems. They allow us to predict some behavior or response using a set of explicit assumptions. Models can be mathematical, like computer simulations; or they can be verbally descriptive, like Darwin's theory of evolution by natural selection. Hypotheses are such word-based models. Our hypothesis about nitrogen availability is a model. It predicts that plant productivity will increase with increasing nitrogen availability. However, this prediction is qualitative—it does not predict how much. In contrast, mathematical models offer quantitative predictions. For example, from the data in Figure 1.3, we can develop a regression equation, a form of statistical model that predicts the amount of plant productivity per unit of nitrogen in the soil (Figure 1.4).

All of the approaches discussed above—observation, experimentation, hypothesis testing, and models—appear in the following chapters to illustrate basic concepts and relationships. They are the tools of science.

1.7 Uncertainty and Debate Are Key Features of Science

Collecting observations, developing and testing hypotheses, and constructing predictive models all form the backbone of the scientific method. It is a continuous process of testing and correcting concepts in order to arrive at explanations for the variation we

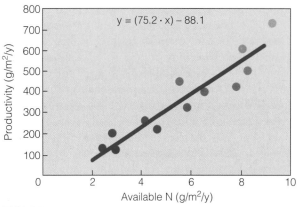

FIGURE 1.4 A simple linear regression model to predict plant productivity (*y* axis) from nitrogen availability (*x* axis).

observe in the world around us, providing unity in observations that upon first inspection seem unconnected. The difference between science and art is that, although both involve the creation of concepts, in science their exploration is limited to the facts. In science there is no test of concepts other than their empirical truth.

However, there is no permanence to scientific concepts because they are only our interpretations of natural phenomena. We are limited to inspecting only a part of nature because we have to simplify in order to understand. As discussed in Section 1.5, when we design experiments we control the pertinent factors and attempt to eliminate others that may confuse the results. Our intent is to focus on a subset of nature from which we can establish cause and effect. The tradeoff is that whatever cause and effect we succeed in identifying represents only a partial connection to the nature we hope to understand. For that reason, when experiments and observations support our hypotheses, and when the predictions of the models are verified, our job is still not complete. We work to loosen the constraints imposed by the necessity to simplify in order to understand. We expand our hypothesis to cover a broader range of conditions, and once again begin testing its ability to explain our new observations.

It may sound odd at first, but the truth is that science is a search for evidence that proves our concepts wrong. There is never a single hypothesis that provides an explanation for our observations. As a result, determining that observations are consistent with a hypothesis is not sufficient to prove that the hypothesis is true. The real goal of hypothesis testing is the elimination of incorrect ideas. Thus we must follow a process of elimination, searching for evidence that proves a hypothesis wrong. Science is essentially a self-correcting activity, dependent on the continuous process of debate. Dissent is the activity of science, fueled by free inquiry and independence of thought. To the outside observer, this essential process of debate may appear to be a shortcoming. After all, we are dependent on science for our technology, our ability to solve problems. In the case of current environmental issues, the solutions may well involve difficult ethical, social, and economic decisions. For that reason, the uncertainty that is inherent to science is discomforting. However, we must not mistake uncertainty for confusion, nor should we allow disagreement among scientists to become an excuse for inaction. Instead, we need to understand the uncertainty so that we may balance it against the costs of inaction.

1.8 We Begin with the Individual

As we noted in the previous discussion, ecology encompasses a broad area of investigation, from individuals to ecosystems. There are many points from which we can depart to begin our study. We have chosen to begin with the individual organism, to examine the processes it uses and constraints it meets in maintaining life under varying environmental conditions. It is here that we can begin to understand the mechanisms that give rise to the diversity of life and ecosystems on Earth.

CHAPTER REVIEW

Summary

An Overview of Ecology (1.1–1.3) Ecology is the scientific study of the relationship between organisms and their environment (**1.1**). Organisms interact with their environment within the context of the ecosystem. Broadly, the ecosystem consists of two basic interacting components, the living or biotic, and the physical or abiotic (**1.2**). The components of an ecosystem form a hierarchy. Organisms of the same kind that inhabit a given physical environment make up a population. Populations of different kinds of organisms interact with members of their own species as well as with individuals of other species. These interactions range from competition for shared resources to predation to mutual benefit. Interacting populations make up a biotic community. Community plus the physical environment make up an ecosystem (**1.3**).

An Interdisciplinary Science (1.4–1.7) Ecology is an interdisciplinary science because the interactions of organisms with their environment and with each other involve physiological, behavioral, and physical responses. The study of these responses draws upon such fields as physiology, biochemistry, genetics, geology, hydrology, and meteorology (**1.4**). The study of patterns and processes within ecosystems involves field study or experiments. Experimentation begins with formulating a hypothesis. A hypothesis is a statement about a cause and effect that we can test experi-

mentally (**1.5**). From research data, ecologists develop models. Models are abstractions and simplifications of natural phenomena. Such simplification is necessary to understand natural processes (**1.6**). Uncertainty is an inherent feature of scientific study, arising from the limitation that we can focus on only a small subset of nature, resulting in an incomplete perspective (**1.7**).

Study Questions

1. Define: ecology, ecosystem, population, community.

2. Why is ecology an interdisciplinary science?

3. What role does ecology play in addressing current environmental issues?

4. Why must ecosystems be considered systems?

5. What are the differences between a field study and a field experiment? What are the differences between a field experiment and a laboratory experiment? How are they related?

6. What is a hypothesis? How does it relate to a scientific study?

7. What is a model?

8. Why are uncertainty and debate essential features of science?

A male large ground finch *(Geospiza magnarostris)*—one of the Darwin finches—on Santa Cruz Island, Galápagos Islands.

Adaptation and Evolution

OBJECTIVES

On completion of this chapter, you should be able to:

* Explain the relationships among adaptation, natural selection, and evolution.
* Distinguish among directional, disruptive, and stabilizing selection.
* Discuss the sources of genetic variation within a population.
* Contrast the concepts of morphological and biological species.
* Describe the variety of mechanisms that can maintain reproductive isolation.
* Contrast the allopatric and sympatric models of speciation.

Do you remember your first childhood visit to a zoo? You were probably amazed by the diversity of strange and wonderful animals: the giraffe with its long neck, the polar bear's snow-white coat, and the orangutan's exceedingly long arms. These animals seemed as if they were from another world, so unlike the animals that inhabit the environment we know. Among the widely dispersed, umbrella-shaped trees of the savannas of Africa, the ice flows of the Arctic, and the canopy of the tropical rain forest in Borneo, however, these animals look as natural as the birds at our backyard feeders, or the deer that emerge from the forest's edge at dusk. What appear to be peculiarities in the context of one environment appear as advantages—characteristics that enable the organisms to thrive in another environment. The long neck of the giraffe allows it to feed in areas of the tree inaccessible to other browsing animals in the savanna. The white coat of the polar bear makes it virtually invisible to potential prey on the snowy landscape of the Arctic. The long arms of the orangutan are essential for life in the canopy, where balance requires more than a sure step. These characteristics that enable an organism to thrive in a given environment are called **adaptations.**

Prior to the mid-19th century, examples such as these served to illustrate the "wise laws that brought about the perfect adaptation of all organisms one to another and to their environment." Adaptation, after all, implied design, and design a designer. Natural history was the task of cataloging the creations of the divine architect. By the mid-1800s, however, a revolutionary idea emerged that would forever change our view of nature.

> In considering the origin of species, it is quite conceivable that a naturalist . . . might come to the conclusion that species had not been independently created, but had descended, like varieties, from other species. Nevertheless, such a conclusion, even if well founded, would be unsatisfactory, until it could be shown how the innumerable species, inhabiting this world, have been modified, so as to acquire that perfection of structure and coadaptation which justly excites our admiration.

The pages that followed in Charles Darwin's *The Origin of Species*, first published on November 24, 1859, altered the history of science and brought into question a view of the world that had been held for millennia (Figure 2.1). It affected not only the long-held view on the origin of the diversity of life on Earth, but also the view on the very origin of the human species. What Charles Darwin put forward in those pages was the *theory of natural selection*. Its beauty lay in its simplicity; the mechanism of natural

FIGURE 2.1 Charles Darwin (1809–1882).

selection is the simple elimination of "inferior" individuals. The philosopher Herbert Spencer, a contemporary of Darwin, described the theory with the now familiar term *survival of the fittest.*

2.1 Natural Selection Requires Two Conditions

Stated more precisely, **natural selection** is the differential success (survival and reproduction) of individuals within the population resulting from their interaction with their environment. As outlined by Darwin, natural selection is a product of two conditions: (1) variation among individuals within a population in some characteristic, (2) which results in differences among individuals in their survival and reproduction. Natural selection is a numbers game. Darwin wrote:

> Among those individuals that do reproduce, some will leave more offspring than others. These individuals are considered more fit

than the others because they contribute the most to the next generation. Organisms that leave few or no offspring contribute little or nothing to the succeeding generations and so are considered less fit.

The **fitness** of an individual is measured by the proportionate contribution it makes to future generations. Under a given set of environmental conditions, individuals that possess certain characteristics that enable them to survive and reproduce, eventually passing those characteristics on to the next generation, are selected for. Individuals that do not are selected against.

The work of Peter and Rosemary Grant provides an excellent documented example of natural selection. The Grants spent more than two decades studying the birds of the Galápagos Islands, the same islands whose diverse array of animals so influenced the young Charles Darwin when he was the naturalist aboard the expeditionary ship *H.M.S. Beagle.* Among other events, their research documented a dramatic shift in a physical characteristic of finches inhabiting some of these islands during a period of extreme climate change.

Figure 2.2 shows variation in beak size in Darwin's ground finch *(Geospiza fortis)* that inhabits the 40-hectare islet of Daphne Major, one of the Galápagos Islands off the coast of Ecuador. Beak size is a trait that influences the feeding behavior of these seed-eating birds. Individuals with large beaks can feed on a wide range of seeds, from small to large, while those individuals with smaller beaks are limited to feeding on smaller seeds.

During the early 1970s, the island received regular rainfall (127–137 mm), supporting an abundance of seeds and a large finch population (1500 birds). In 1977, however, only 24 mm of rain fell. In the drought, seed production declined drastically. Small seeds declined in abundance faster than large seeds, increasing the average size and hardness of seeds available. The finches, which normally fed on small seeds, had to turn to larger ones. Small birds had difficulty finding food. Large birds, especially males with large beaks, survived best because they were able to crack large, hard seeds. Females suffered heavy mortality. Overall, the population declined 85 percent from mortality and possibly from emigration (Figure 2.3a). The increased survival rate of larger individuals resulted in a dramatic shift in the distribution of beak size in the population (Figure 2.3b). This type of natural selection, where the mean value of the trait is shifted toward one extreme over another (Figure 2.4a) is called **directional selection.** In other cases, natural selection may favor individuals near the population mean at the expense of the two

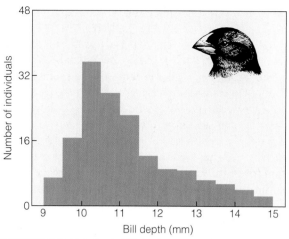

FIGURE 2.2 Variation in bill size (as measured by depth) in the population of Galápagos ground finch *(Geospiza fortis)* on the island of Daphne Major. Bill depth has a direct influence on the size of seeds that can be eaten by individual birds (Grant 1999).

extremes, referred to as **stabilizing selection** (Figure 2.4b). When natural selection favors both extremes simultaneously, although not necessarily to the same degree, it can result in a bimodal distribution of the characteristic(s) in the population (Figure 2.4c). Such selection, known as **disruptive selection,** occurs when members of a population are subject to different selection pressures. A classic example of disruptive selection is the difference in coloration between the sexes of many bird species (termed sexual dimorphism) (Figure 2.5). Males of the species are characteristically more colorful, a feature related to courtship and the attraction of potential mates (a detailed discussion of this phenomenon is presented in Chapter 12).

2.2 Heritability Is an Essential Feature of Natural Selection

Two assumptions are basic to the theory that natural selection brings about the change of one species into another. The first is that an individual's characteristics are passed on from one generation to the next: from parent to offspring. This point was not controversial. For millennia, it had been observed that offspring resemble parents. How this inheritance occurred, however, remained a mystery. Not having an understanding of genetics when developing his theory of natural selection, Darwin accepted the hypothesis of inheritance that was in favor at his time—blending inheritance. It was thought that the characteristics of the parents were blended in their

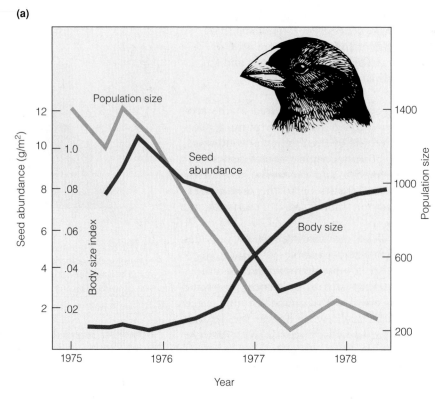

(a)

(b)

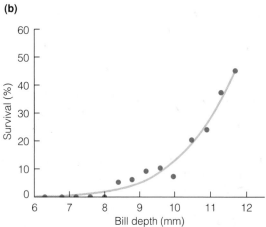

FIGURE 2.3 Evidence of natural selection in Darwin's medium ground finch, *Geospiza fortis*. (a) The yellow line represents the population estimate on the island of Daphne Major based on the censuses of marked populations, and the green line estimates seed abundance, excluding two species of seeds never eaten by any Galápagos finches. Populations declined in the face of seed scarcity during a prolonged drought. The brown line plots changes in body size. Note how the body size of surviving birds increased during the drought period, suggesting that small-bodied birds were being selected against and large-bodied birds were being favored. The selection for larger body size is also reflected in the relationship between bill size and survival over this same period. (b) Results suggest that the most intense selection in a species occurs under unfavorable environmental conditions. (Reprinted with permission from P. T. Boag and P. R. Grant, "Intense Natural Selection in a Natural Population of Darwin's Finches," *Science* 214 (1981): 83. Copyright © 1981 American Association for the Advancement of Science.)

offspring, producing characteristics that were intermediate in form to the parents.

The second assumption, that the characteristics of a species could change through time, is where the controversy arose. The prevailing view was that species were independently created and invariable; the character of a species did not change from generation to generation—or at all for that matter.

When Charles Darwin proposed the theory of natural selection, the concept that species were mutable,

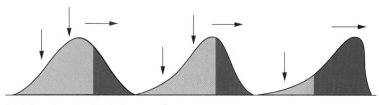

(a) Directional selection

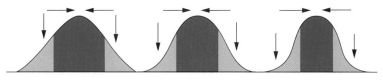

(b) Stabilizing selection

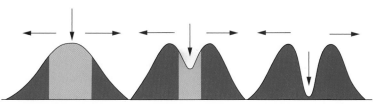

(c) Disruptive selection

FIGURE 2.4 Three types of selection. (a) Directional selection moves the mean of the population toward one extreme. (b) Stabilizing selection favors organisms with values close to the population mean. Little or no change takes place. (c) Disruptive selection increases the frequencies of both extremes. Downward arrows represent selection pressures; horizontal arrows represent the direction of evolutionary change.

FIGURE 2.5 Many species of passerine birds, such as the male and female northern cardinal *(Cardinalis cardinalis)* shown, exhibit sexual dimorphism in plumage color, with the male of the species typically being more colorful.

subject to change, was an idea that, although not widely accepted, had existed for some time. Charles Darwin's own grandfather, Erasmus Darwin (1731–1802) published *The Laws of Organic Life (Zoonomia)* in 1794. Like his predecessor, the French philosopher Diderot, the elder Darwin believed that if an animal experienced a need, this need would pro-

vide for the formation of an "organ that satisfied the need." He believed that the modification of species was brought about by the "satisfaction of wants due to lust, hunger and danger" and "that many of these acquired forms or propensities are transmitted to their posterity" (offspring).

French soldier and natural historian Jean-Baptiste Lamarck (1744–1829) arrived at an independent and similar hypothesis to that of Erasmus Darwin. Lamarck supposed that as the environment imposed new "needs" upon an animal, its "inner feeling" set into motion (unknown) processes that produced new organs to satisfy those needs—in other words, adapting the animal to its environment. These organs were then transmitted by heredity to its offspring.

Such was the state of thinking when Charles Darwin wrote *The Origin of Species*. His theory of natural selection was a brilliant example of deductive reasoning. He had arrived at an explanation of how the characteristics of a species could change from generation to generation through the differential reproduction and survival of individuals within the population. There was no need to invoke the "inner feeling" or "living force" of his predecessors. Applying this mechanism to the origin of species, however, still required a huge leap of faith. Natural selection is bounded by the variation that is contained within the population. In the example of natural selection in Darwin's ground finch presented in Section 2.1, the shift in the distribution of beak size during the period of drought was still limited to the range of "potential beak sizes" present within the population. How does the range of characteristics get extended? How do new characteristics arise? For Darwin, mutation was the primary mechanism for maintaining variation within the population: "the sudden and considerable deviation of structure" sometimes seen in offspring. Darwin also noted that, under domestication, mutations of an extreme type—"monstrosities"—sometimes arise, such as a two-headed calf or a "pig born with a sort of proboscis." The acceptance of blending inheritance required that Darwin invoke an unreasonably high mutation rate in order to maintain the patterns of variation that he observed in natural populations. Darwin was painfully aware of this limitation on his theory. What Darwin did not realize was that the answer to these questions regarding variation, as well as the means by which characteristics were transmitted from generation to generation, had already begun to unfold in the work of a contemporary.

Unknown to Darwin, an Austrian monk, Gregor Mendel (1822–1884), was studying the transmission of characteristics from one generation of pea plants to another in his garden. Mendel lived and worked in the abbey in Brunn, Austria (now the city of Brno in the Czech Republic). The pea plants that Mendel chose for his experiments had two characteristics that made them ideal for his work. First, they formed many varieties that were easily distinguishable based on characteristics such as flower color, the size and shape of seeds and seedpods, and stem length. Secondly, Mendel was able to maintain strict control over which plant mated with which. He could either let a plant self-fertilize (pollen fertilizes the egg of the same flower) or cross-fertilize it with another selected individual.

Mendel worked with his plants until he was certain that he had varieties for which self-fertilization produced offspring that were identical in characteristics to the parent. For example, he identified a purple-flowered variety that, when self-fertilized, produced plants with purple flowers only. In the terminology of plant breeders, these plants would "breed true." Mendel was then ready to examine what would happen when he crossed varieties with different characteristics. For example, what would happen if he crossed a plant with purple flowers with a plant having white flowers (Figure 2.6). Mendel discovered that the plants produced by this crossing (mating), termed the F_1 generation (F referring to *filial* for the Latin word for son), all had purple flowers. The question Mendel faced was what had happened to the characteristic of white flowers?

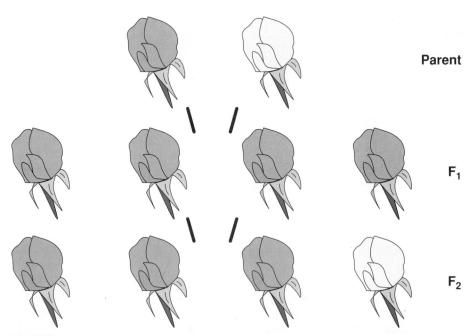

Parent

F_1

F_2

FIGURE 2.6 Mendel tracked the heritable character of flower color for three generations. Purple and white parent plants that "bred true" were crossed to produce the F_1 generation, all of which had purple flowers. When individuals from the F_1 generation were crossed, the offspring in the resulting F_2 generation were a 3:1 ratio of purple- to white-flowered individuals.

When Mendel then crossed individuals from the F_1 generation, he had his answer. Out of the 929 plants produced by these crossings (termed the F_2 generation), 705 (approximately 75 percent) had purple flowers, and 224 (approximately 25 percent) had white flowers. From these experiments, Mendel concluded that the heritable factor for white flowers had not been lost in the F_1 generation, but rather that the heritable factor for purple flowers was controlling flower color. He also deduced that, since all of the F_1 plants had purple flowers yet were able to pass on the trait of white flowers to the next generation, the plants must carry two factors for the characteristic of flower color. From his experimental results, Mendel drew the following conclusions:

1. There are alternate forms of the units that control heritable traits (such as flower color).

2. For each inherited characteristic, an organism has two units, one from each parent (one each from the egg and sperm). These units may be the same (purple and purple, or white and white) or different (purple and white).

3. When the two units are different, one is fully expressed while the other has no noticeable effect on the organism's outward appearance. The unit that is expressed is called the **dominant** (purple), while the other is called the **recessive** (white).

By now you may have noticed that we have avoided using the term *gene* (or *allele*), choosing the word *unit* instead. This is intentional. Although Mendel's work had established a set of rules for the inheritance of characteristics, the nature of these units of transmission remained unknown. But as is often the case in science, although the landscape that lay ahead was unknown, the course of exploration had been set by the results of Mendel's experiments.

2.3 Genes Are the Units of Inheritance

Mendel's experiments had provided the rules, but what was the unit of transmission? The search for the answer to this question left a trail of accumulated understanding that went on to form the basis of modern genetics. This trail finally ended in April 1953 when James Watson and Francis Crick published a two-page manuscript in the journal *Nature* outlining their model for the structure of **DNA** (deoxyribonucleic acid).

At the root of all similarities and differences among organisms is the information contained within the molecules of DNA. All cells have DNA, and the manner in which information is encoded is the same for all organisms. The structure of DNA is a double helix (Figure 2.7) composed of four kinds of chemical compounds called nucleotides. All DNA is composed of the same four nucleotides, with the blueprint of each species differing only in their sequence. DNA is contained in threadlike structures called **chromosomes.**

Chromosomes come in matched pairs, called homologous chromosomes. Each chromosome carries DNA that is organized into discrete subunits, the **genes,** which form the informational units of the DNA molecule (see Figure 2.7). Genes existing in alternative forms are called **alleles,** the units of heredity that controlled the expression of flower color in

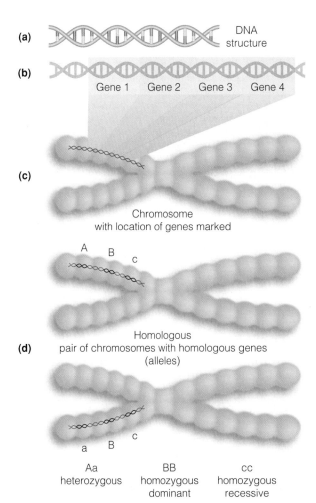

FIGURE 2.7 (a) The double helix structure of DNA. (b) Genes are specific stretches of the DNA molecule that program the production and function of proteins. (c) Genes are contained in threadlike structures called chromosomes. (d) Chromosomes come in matched pairs, called homologous chromosomes. Because chromosomes are paired within the cells, alleles are also paired. Members of the pair of alleles occupy the same locus on homologous chromosomes.

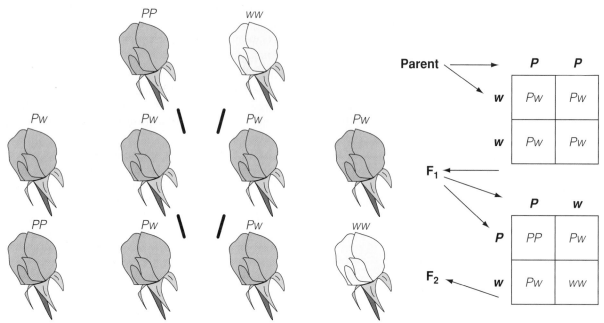

FIGURE 2.8 A more detailed depiction of Mendel's experiment outlined in Figure 2.6. The genotype of each plant is identified in the three generations (parent, F_1, and F_2). The purple flower allele is dominant, and the white flower allele is recessive. Each plant has two alleles controlling flower color. When two plants are crossed, each parent contributes one allele to the offspring. A homozygous parent can therefore only contribute one form of allele (P or w), while a heterozygous parent has an equal probability of contributing either a dominant or recessive allele. Note that although the proportion of different genotypes varies from generation to generation in this experiment, the gene frequency of each generation (the proportion of P and w alleles) is the same.

Mendel's experiments. Because chromosomes are paired within the cells, alleles are also paired. The position an allele occupies on a chromosome is its **locus.** Members of the pair of alleles occupy the same locus on homologous chromosomes. If the alleles occupying the same loci on homologous chromosomes affect a given trait in the same manner, the individuals possessing them are called **homozygous.** If the alleles affect a trait differently, the individual is called **heterozygous.** In this case, one allele is fully expressed and the other has no noticeable effect. The allele fully expressed is the dominant allele, as was the allele for purple flowers in Mendel's experiment. The hidden, unexpressed allele is the recessive one (the allele for white flowers).

We can now revisit the results of Mendel's experiments outlined in Figure 2.6, using the terminology of modern genetics (Figure 2.8). Since the initial parent plants "breed true," we know that they are both homozygous. Defining the dominant allele, the gene for purple flowers, as P, and the recessive allele for white flowers as w, the white flower plants are ww and the purple flower plants are PP. This situation ensures that the offspring, which receive one allele from each parent, will all be heterozygous—Pw. The

gene for purple flower being dominant, all of the resulting plants in the F_1 generation will have the physical expression of purple flowers.

When individuals of the F_1 generation are crossed, the possible outcome is different. Since both parents are heterozygous *(Pw)*, each has an equal probability of contributing a gene for either purple *(P)* or white *(w)*. The result of this crossing (F_2 generation), expressed in terms of proportion of total offspring, will be 0.5 heterozygous *Pw*, 0.25 homozygous dominant *PP*, and 0.25 homozygous *ww*. In terms of physical expression, this will translate into the proportions of 0.75 purple-flowered individuals and 0.25 white-flowered individuals, as was observed by Mendel (see Section 2.2, Figure 2.6).

The sum of hereditary information (genes) carried by the individual is the **genotype.** The total collection of genes across all individuals in the population at any one time is referred to as the **gene pool.** The genotype directs development and produces the individual's morphological, physiological, and behavioral makeup. The external, observable expression of the genotype is the **phenotype,** such as the flower color in Mendel's pea plants. The ability of a genotype to give rise to a range of phenotypic expressions under

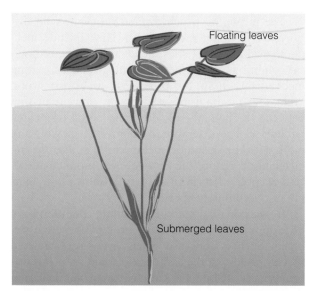

Floating leaves

Submerged leaves

FIGURE 2.9 Plasticity in leaf shape in response to environmental conditions. Pondweed *(Potamogeton)* has lance-shaped underwater leaves, responsive to underwater movements, and broad, heart-shaped leaves that float on the surface.

different environmental conditions is **phenotypic plasticity.** Some genotypes have a narrow range of reaction to environmental conditions and therefore give rise to fairly constant phenotypic expressions. Some of the best examples of phenotypic plasticity occur among plants. The size of the plant, the ratio of reproductive tissue to vegetative tissue, and even the shape of the leaf may vary widely at different levels of nutrition, light, and moisture (Figure 2.9).

2.4 Genetic Variation Is the Essential Ingredient for Natural Selection

One of Charles Darwin's many brilliant insights was his focus on variation among individuals within a species. Darwin's contemporaries viewed variation among individuals as an exception rather than the rule. Although the variation among individuals was used by breeders to create the many varieties of domestic plants and animals, these varieties were seen as more or less unstable—the product of careful selection and cross-breeding of individuals (see Ecological Application: Taking the Uncertainty out of Sex). Left to themselves, these varieties would return to the normal form of the parent species. In fact, the unstable nature of domestic varieties was used as an argument in support of the original and permanent distinctiveness of species. Darwin saw variation among individuals differently; it was not the exception, but rather the norm. Variation was the essential ingredient of natural selection.

The major sources of genetic variation are mutation and genetic recombination in sexual reproduction. **Mutations** are inheritable changes in a gene or a chromosome. Point mutations are chemical changes that occur in just one nucleotide in a single gene (see Figure 2.7). If a point mutation occurs in a gamete (egg or sperm), it can be transmitted to offspring and to a succession of future generations. Most mutations of a single gene have little or no apparent effect. Mutations of single genes that produce larger effects are usually harmful. Gene mutations are important because they add variation to the gene pool. An example of a single gene mutation is albinism in mice. Albinism is the complete absence of pigment in the fur and the iris of eyes, causing a white coat and pink eyes. This gene is recessive and breeds true when albinos mate.

Chromosomal mutations may result from a change in the structure of a chromosome or a change in the number of chromosomes. Structural changes involve duplication, transposition (change in order or position), or deletion of part of a chromosome. Such changes result in abnormal phenotypic conditions. A change in chromosomal number can arise in two ways: (1) the complete or partial duplication of chromosomes, or (2) the deletion of one or more chromosomes. Polyploidy, the duplication of entire sets of chromosomes, is discussed in Section 2.6.

By far the greatest amount of genetic variation among individuals within a population occurs among species that reproduce sexually. In sexual reproduction, two individuals produce haploid (one-half the normal number of chromosomes) gametes—egg and sperm—that combine to form a diploid cell, or zygote, which has a full complement of chromosomes. Because the possible number of gene recombinations is enormous, recombination is an immediate and major source of genetic variability among offspring. In the case of humans, the number of possible unique combinations of chromosomes that could be produced in either a single egg or a single sperm is approximately 8 million. However, not all organisms reproduce sexually.

Asexual reproduction is the production of offspring by a single parent, without the participation of egg and sperm. Asexual reproduction takes many forms, but in all cases creates offspring that are genetically identical to the parent (see Chapter 12 for further discussion and examples). In the absence of mutation, as the age-old saying goes, like really does beget like.

Although Darwin understood that heritable variation is what makes natural selection possible, an exact explanation for the mechanism that gave rise to the observable variations among individuals within a species eluded him. Although the work of Gregor Mendel was published only 6 years after the publication of *The Origin of Species,* his discovery went largely unknown until 1900, some 18 years after Darwin's death.

2.5 The Concept of Species Is Based on Genetic Isolation

Field guide in hand, we can distinguish a robin from a wood thrush or a white oak from a red oak. Each has physical characteristics that set it apart. Each is an entity, a discrete unit, to which a name has been given. That is the way Carl von Linne (1707–1778), who gave us our system of binomial classification (Focus on Ecology 2.1: Classifying Organisms), saw plants and animals. He, like others of his day, regarded the many different organisms as fixed and unchanging units, the so-called products of special creation. They differed in color, pattern, structure, proportion, and other characteristics. By these criteria, naturalists such as Charles Darwin described, separated, and arranged species into groups. Each species was discrete. Some variation was permissible, but those variants were considered accidental. This classic **morphological species** concept is still alive, useful, and necessary for classifying the vast number of plants and animals. It is the basis of the descriptions of organisms found in field guides.

The understanding of the genetic basis for heredity altered the concept of species. The evolutionary biologist Ernst Mayr advanced the idea of the **biological species,** a group of populations whose individuals have the potential to interbreed and produce fertile offspring. This definition of species implied reproductive isolation; and since reproduction is the means of transmitting genetic information (DNA), it also implied genetic isolation. But what maintains reproductive isolation among populations, giving rise to distinct species?

Each spring, there is a rush of courtship and mating activity in woods and fields. Fish move to spawning areas; amphibians migrate to breeding pools; birds sing. During this frenzy of activity, each species remains distinct. Song sparrows mate with song sparrows, brook trout with brook trout, wood frogs with wood frogs, and few mistakes occur, even between species similar in appearance.

The means by which diverse species remain distinct are isolating mechanisms. They include morphological characteristics, behavioral traits, ecological conditions, and genetic incompatibility. Isolating mechanisms may be premating or postmating. Premating mechanisms—those that prevent mating between individuals of different species—include habitat selection, temporal isolation, behavior, and mechanical or structural incompatibility. Postmating mechanisms reduce the survival or reproductive success of those offspring that may arise from the mating of two different species.

If two potential mates in breeding condition have little opportunity to meet, they are not likely to interbreed. Such is the case when two closely related species coexist in the same geographic area, but utilize different habitats (local environments). Isolation through differences in habitat is common among frogs and toads. Different calling and mating sites among concurrently breeding frogs and toads tend to keep the species separated. The upland chorus frog *(Pseudoacris triseriata feriarum)* and the closely related southern chorus frog *(Pseudoacris nigrata)* breed in the same pools, but calling males tend to segregate themselves in different locations in the pond (Figure 2.10). The southern chorus frog calls from a concealed position at the base of grass clumps or among vegetation debris, whereas the upland chorus frog calls from a more open location.

Temporal isolation (differences in the timing of breeding and flowering seasons) can function to isolate species that occur in the same geographic area of habitat. The American toad *(Bufo americanus)* breeds early in the spring, whereas Fowler's toad *(Bufo woodhousei fowleri)* breeds a few weeks later.

Behavioral barriers (differences in courtship and mating behavior) are the most important isolating mechanisms. The males of animals have specific courtship displays, to which, in most instances, only females of the same species respond. These displays involve visual, auditory, and chemical stimuli. Some insects, such as certain species of butterflies and fruit flies, and some mammals possess species-specific scents. Birds, frogs and toads, some fish, and such singing insects as crickets, grasshoppers, and cicadas have specific calls that attract the "correct" species. Visual signals are highly developed in birds and some fish (see Figure 2.5). Among insects, the flight patterns and flash patterns of fireflies on a summer night are most unusual visual stimuli. The light signals emitted by various species differ in timing, brightness, and color, which may be white, blue, green, yellow, orange, or red.

Mechanical isolating mechanisms make copulation or pollination between closely related species impossible. Although evidence for such mechanical

CLASSIFYING ORGANISMS

Although humans have been classifying plants and animals for centuries, there was no uniform scheme. Then came a young Swedish naturalist and medical doctor, Carl von Linne, who Latinized his name to Carolus Linnaeus (1707–1778). Linnaeus had a passion for classifying. He developed a binomial or two-name system for naming plants and animals. This system finally brought order out of chaos. Linnaeus named all living organisms in Latin, then the universal language of science. Because it was a dead language, it would never change in grammar, syntax, or vocabulary. That gave it a great degree of stability.

Linnaeus explored Lapland and northern Europe; he had collectors send him plants from other parts of the world. Out of this research came a book in 1735, *Species Plantarum*. In it he listed and named all plants then known to science, giving each plant two Latin names—a genus name and a species name. In 1758, he published a tenth edition, *Systema Naturae*, in which he consistently applied the same idea to animals.

Before Linnaeus could assign names to each organism, he had to sort plants and animals into broad general groups. First, all living things, he reasoned, belonged to two great groups or kingdoms, plant and animal. The animal kingdom he subdivided into six parts—mammals, birds, reptiles, fishes, insects, and a catchall group, vermes (for worms).

He then divided each group again. Among the mammals, for example, he noted that some organisms were similar. It was not too difficult to identify a cat, although there are many different kinds. All have retractile claws, rounded heads, and well-developed, but relatively small, rounded ears. Some have long tails; others have short tails, ear tufts, and a ruff on the cheeks and throat. All the long-tailed cats, then, were placed in one genus, to which Linnaeus give the Latin name *Felis*, for "cat." To the short-tailed cats he gave the name *Lynx*. Doglike mammals, too, are readily distinguishable, and he broke them down into groups based on similar appearance: the sharp-nosed little foxes, and the doglike wolves and their descendants, the true dogs. To wolves and dogs, Linnaeus gave the name *Canis*, Latin for "dog." Each genus then contained a number of different animals, with some broad structural characteristics in common, yet with differences in finer details—color, size, thickness of hair, and the like. Thus, the genus could be broken down into still finer groups, the individual species. Linnaeus gave a second part to the name of species. The wolf became *Canis lupus*, the mountain lion *Felis concolor*, and the African lion *Felis leo*.

Later, all catlike animals were grouped together into a larger classification, the family, which takes its name from one of the genera it embraces. In the case of cats, the genus *Felis* was used. By adding the standardized suffix *idae* onto the root word *Felis*, the cat family became Felidae; and by a similar process the dog family became Canidae. Dogs and cats, skunks and weasels (Mustelidae), raccoons (Procyonidae), bears (Ursidae), seals (Phocidae), and so on were classified into families. All of them have several general features in common—five toes with claws, shearing teeth, well-developed canine teeth or fangs, and other skeletal structures. Therefore, these animals were grouped together and placed in an even higher category, the order, in this case Carnivora. All animals giving birth to living young, possessing body hair, and nourishing young with milk were placed in an even higher category, the class Mammalia. All animals having a backbone were grouped into a phylum (in this instance Chordata) belonging to the animal kingdom.

The final result was a hierarchy of classification in which similar species were grouped in a genus, similar genera were grouped in a family, and similar families were grouped into orders. The principal classification of the mountain lion *(Felis concolor)* looks like this:

Kingdom: Animalia
  Phylum: Chordata
    Class: Mammalia
      Order: Carnivora
        Family: Felidae
          Genus: *Felis*
            Specific epithet: *concolor*

As time went on, the number of known animals swelled, and the need arose to redefine this hierarchy. Phyla were divided into subphyla, classes into subclasses and infraclasses, orders into suborders, families into subfamilies (and even enlarged into superfamilies), and species into subspecies.

The principal categories used in plant classification are somewhat different. For example, here are the categories for the American beech *(Fagus grandifolia)*:

Kingdom: Plantae
  Division: Tracheophyta
    Subdivision: Pteropsida
      Class: Angiospermae
        Subclass: Dicotyledonae
          Order: Fagales
            Family: Fagaceae
              Genus: *Fagus*
                Specific epithet: *grandifolia*

(a)

(b)

FIGURE 2.10 (a) The upland chorus frog *(Pseudoacris triseriata feriarum)* and the closely related (b) southern chorus frog *(Pseudoacris nigrata)* breed in the same pools, but calling males tend to segregate themselves in different locations in the pond.

isolation among animals is scarce, differences in floral structure and intricate mechanisms for cross-pollination are common among plants.

Even when two individuals of differing species are able to mate, genetic differences generally do not allow fertilization to occur. Should fertilization occur, however, other genetic barriers may function as postmating mechanisms maintaining isolation between the two gene pools. The offspring that results from the mating of two different species are called **hybrids.** In most cases, hybrid individuals are inviable and do not survive. In other situations, offspring survive, but are sterile—unable to produce offspring of their own. Such is the case with the mule, which is a cross between a female horse and a male donkey (Figure 2.11). Since the mule cannot reproduce, it is not able to transmit its genetic information to future generations.

One additional factor is a key element of reproductive isolation and the maintenance of distinct species—geographic isolation. Species may be sympatric or allopatric. **Sympatric** species occupy the same area at the same time, so they have the opportunity to interbreed. **Allopatric** species occupy areas separated by time or space. Because individuals of allopatric species do not come into physical contact with each other, there is no indication whether they are capable of interbreeding. Only if the barriers are broken, allowing them to come together, can reproductive isolation be tested. Often, geographic isolation is the only mechanism of reproductive isolation for two similar, but allopatric species, and without barriers the two species function as one. Such is the case, for example, with the red-shafted

flicker *(Colaptes cafer)* and the yellow-shafted flicker *(Colaptes auratus)* (Figure 2.12). If you were to refer to early guides to birds, such as the Audubon or Peterson series (about 1940 through the 1960s), you would find pictures and descriptions of both species. If you were to refer to the current editions, however, you would see that the two species have now been lumped into a single species referred to as the northern flicker *(Colaptes auratus)*. Historically, the two species had different geographic ranges, with the red-shafted flicker's range being to the west of the yellow-shafted flicker. With time, the distributions of the two species expanded and overlapped (became sympatric), and the two species interbred and produced viable offspring (hybrids). As a result, they are now classified as a single species. As well as being a

FIGURE 2.11 A mule (left) is a hybrid cross between a female horse (right) and a male donkey (center).

(a)

(b)

FIGURE 2.12 Flickers. (a) The eastern form of the "yellow-shafted" flicker has yellowish wing linings and undertail surfaces, and a brown face with a black mustache. (b) The western form of the "red-shafted" flicker has pinkish-red wing linings and undertail surfaces, and a gray face with a red mustache. Once considered separate species, they are now regarded as one species because they intergrade and interbreed.

mechanism for maintaining reproductive isolation among closely related species, geographic isolation is also an important factor in the process through which new species arise.

2.6 The Process of Speciation Involves the Development of Reproductive Isolation

As Darwin had hypothesized in *The Origin of Species,* speciation involves the divergence of existing species through natural selection. The critical step in the origin of a species is the point at which the gene pool of a population becomes isolated from other populations of the parent species and the exchange of genes no longer occurs. Once the gene pool of the subpopulation is isolated, it will follow its own independent course as genetic changes occur due to natural selection and mutation. The process of speciation can be classified into two models based on the geo-

graphic relationship of the subpopulation to its parent population. If the subpopulation is geographically isolated from the original population, it is termed **allopatric speciation.** If the subpopulation becomes reproductively isolated in the presence of the parent population, it is termed **sympatric speciation.**

The first step in allopatric speciation, or geographic speciation as it is sometimes called, is the splitting of a single interbreeding population into two spatially isolated populations, each taking its own evolutionary route (Figure 2.13a). Imagine a piece of land, warm and dry, occupied by species A. At some point in geological time, mountains uplift and land sinks and floods with water. That action splits the piece of land and separates a segment of species A from the rest of the population. The newly isolated population will become subpopulation A′. It now occupies an area with a cool, moist climate.

Because it represents only a subset of the gene pool for species A, population A′ will be slightly different genetically. The climatic conditions and the selective forces under which this population now lives are different. Natural selection will favor individuals best adapted to a cool, moist climate. Similar selection for a warm, dry climate will continue in

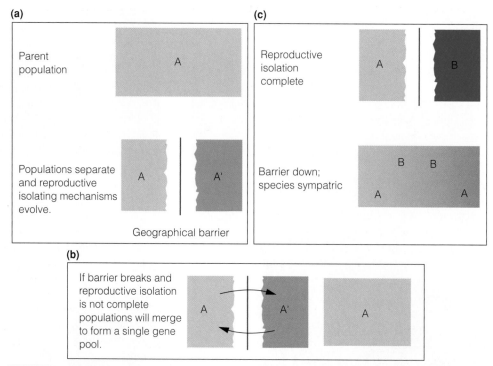

FIGURE 2.13 Types of speciation. (a) Allopatric or geographic speciation takes place when two populations are isolated from each other over a long period of time. The first step in speciation comes about when species A becomes divided into two populations by a geographical barrier: A and A'. Over time under different selection pressures, A' diverges even further. (b) At this point, if sufficient differences evolve before the barrier breaks down but isolating mechanisms are incomplete, hybrids form. (c) When reproductive isolating mechanisms are complete, A' has evolved into a new species B. If the barrier should break down, species A and B can exist sympatrically.

population A on the original landmass. With different selection forces acting on them, the two populations will diverge. Accompanying genetic divergence will be possible changes in physiology, morphology, color, and behavior resulting in ever-increasing external differences (see Focus on Ecology 2.2: Geographic Variation within Species). If the geological barrier breaks down and the two populations rejoin before natural selection has resulted in differences that cause reproductive isolation, the two populations may simply intermingle and merge once again into a single gene pool (Figure 2.13b). But if sufficient differences occur to inhibit interbreeding, the two populations will function as separate (sympatric) species even if they come together again (Figure 2.13c).

Sympatric speciation occurs without geographic isolation. Many species of plants appear to have arisen from a form of sympatric speciation in which a new species arises spontaneously. The most common method of abrupt speciation is a form of chromosomal mutation called **polyploidy,** the doubling of the number of chromosomes. An organism that has a double number of chromosomes in its gametes (egg or sperm) cannot produce fertile offspring with a member of its parent population, but it can do so by

mating with another polyploid. Thus, a polyploid has already achieved reproductive isolation.

Many of our common cultivated plants—potatoes, wheat, alfalfa, coffee, and grasses, to mention a few—are polyploids. Polyploidy is widespread among native plants, in which it produces a complex of species, such as blackberries *(Rubus)*, willows *(Betula)*, and birches *(Salix)*. The common blue flag *(Iris versicolor)* of northern North America is a polyploid that probably originated from two other species, *Iris virginica* and *I. setosa,* when the two, once widely distributed, met during the retreat of the Wisconsin ice sheet. The sequoia *(Sequoiadendron giganta)* is a relict polyploid, its parent species having become extinct.

2.7 Evolution Is a Change in Gene Frequency

Darwin did not use the word *evolution* in the original edition of *The Origin of Species.* Rather, he spoke of "descent with modification." *Evolution* (taken from the Latin word *evolvere* meaning "to unroll") had become a common English word in Darwin's time that implied the appearance of a series of events in an orderly succession. The term was tied to the concept of progress—directional change that represented the development from simple to complex. It implied the same concept of "directed purpose" that had been invoked by his predecessors, the very concept Darwin was intent on dispelling with his theory of natural selection (see Section 2.2).

Although the common application of the term *evolution* still implies directional change, the biological definition specifically applies to genetic change and does not imply direction or purpose. In the broadest sense, **evolution** is a change in gene frequencies within a population (or species) over time. Genes are transmitted from individual to individual, from parent to offspring, but individuals do not evolve—populations evolve. The focus of evolution is the gene pool, the collective. To illustrate this point, we can return to Mendel's experiments in Figure 2.8.

The proportion of individuals in each of the three possible genotypes (as represented by the trait of flower color) changes from generation to generation (Parent, F_1, F_2). The gene frequencies (the proportion of P and w genes in the population), however, do not change. Both P and w remain in the same proportions from generation to generation: 0.5 P and 0.5 w. No matter how many generations forward we calculate, the frequency of each gene (P and w) will remain constant unless acted on by other agents. This rule is known as the Hardy-Weinberg principle, named for the two scientists who derived it in 1908. More specifically, the principle states that the gene frequencies will remain the same in successive generations of a sexually reproducing population if the following conditions hold: (1) mating is random; (2) mutations do not occur; (3) the population is large, so that changes by chance in gene frequencies are insignificant; (4) natural selection does not occur; and (5) migrations (movement of individuals and thus genes into or out of the population) do not occur.

In the population of plants represented by Mendel's experiments, the frequencies of different genotypes changed through time (generations), as did the phenotypes, giving rise to a succession of colors—from purple and white to purple only and back again to purple and white. But as Mendel deduced, the change in the color of his garden did not imply a change in gene frequency (or units of inheritance in Mendel's terminology).

In natural populations, however, the conditions required by the Hardy-Weinberg principle are never fully met. Mutations do occur. Mating is not random. Individuals move between populations, and natural selection does take place. All of these circumstances change gene frequencies from one generation to another, the result being evolution. Natural selection is in effect the selection (differential survival and reproduction) of one genotype over another, resulting in a change in gene frequencies. Speciation by natural selection is a form of evolution. The process of speciation requires changes in the gene frequencies of a population that result in reproductive isolation: the evolution of characteristics that sever the exchange of genes with the parent population.

The beauty of evolution by natural selection is that it does not require a direction or final cause. It is guided only by random changes in the sequence of the four nucleotides of the DNA that is the blueprint of each individual organism, constrained only by the necessity of survival and reproduction—the transmission of that DNA to succeeding generations. To quote the evolutionary biologist Ernst Mayr: "By adopting natural selection, Darwin settled the several-thousand-year-old argument among philosophers over chance or necessity. Change on the earth is a result of both; the first step is dominated by randomness, the second by necessity."

2.8 Adaptations Reflect Tradeoffs and Constraints

The heritable characteristics an organism possesses, it owes to past generations. Its ancestors, in effect, experienced the process of natural selection that produced the heritable characteristics owned by present individuals. The possession of these characteristics enables an organism to match the features of its environment. As long as environmental conditions under which individuals of the current generation exist are similar to those experienced by past generations, the organism is adapted to the environment. If environmental conditions change significantly, then the fitness and even the survival of individuals will be in jeopardy (see Section 2.1, Figure 2.3). Adaptation, then, is any heritable behavioral, morphological, or physiological trait that maintains or increases the fitness of an organism under a given set of environmental conditions.

If Earth were one large homogeneous environment, perhaps a single genotype, a single set of characteristics, might bestow upon all living organisms the ability to survive, grow, and reproduce. But this is not the case; environmental conditions that directly influence life vary in both space and time; likewise, the objective of selection changes with environmental circumstances—in both space and time. The response of an organism to its environment is not a repertoire

GEOGRAPHIC VARIATION WITHIN SPECIES

Species having a wide geographic distribution often encounter a broader range of environmental conditions than do species whose distribution is more restricted. The variation in environmental conditions often gives rise to a corresponding variation in many morphological, physiological, and behavioral characteristics. Significant differences often exist among populations of a single species inhabiting different regions. The greater the distance between populations, the more pronounced the differences may become, with each population adapting to the locality it inhabits. Geographic variation within a species can result in clines, ecotypes, and geographic isolates.

A **cline** is a measurable, gradual change over a geographic region in the average of some phenotypic character, such as size and coloration, or a gradient in genotypic frequency. Clines are usually associated with an environmental gradient such as temperature, moisture, light, or altitude. Continuous variation results from gene flow from one population to another along the gradient. Because environmental constraints influencing natural selection vary along the gradient, any one population along the gradient will differ genetically to some degree from another, the difference increasing with the distance between the populations. For this reason, populations at the two extremes along the gradient may functionally behave as different species.

Clinal differences exist in size, body proportions, coloration, and physiological adaptations among animals. For example, white-tailed deer in North America exhibit a clinal variation in body weight. Deer in Canada and the northern United States are heaviest, weighing on the average more than 136 kg. Deer weigh 93 kg in Kansas, 60 kg in Louisiana, and 46 kg in Panama. The smallest, the Key deer in Florida, weighs less than 23 kg.

Clinal variations can show marked discontinuities. Such abrupt changes, or step clines, often reflect abrupt changes in local environments. Such variants are called **ecotypes.** An ecotype is a population adapted to its unique local environmental conditions. For example, a population inhabiting a mountaintop may differ from a population of the same species in the valley below. Often,

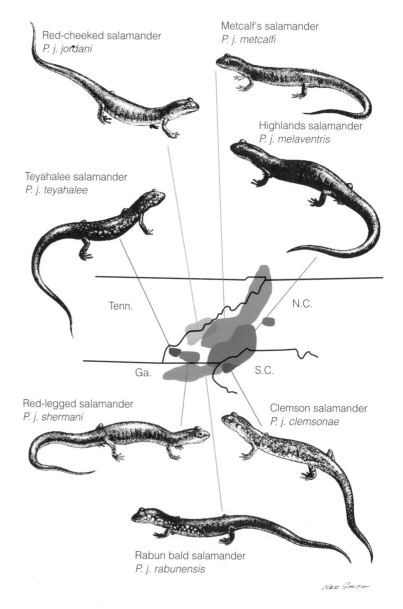

Red-cheeked salamander
P. j. jordani

Metcalf's salamander
P. j. metcalfi

Highlands salamander
P. j. melaventris

Teyahalee salamander
P. j. teyahalee

Tenn.

N.C.

Ga.

S.C.

Red-legged salamander
P. j. shermani

Clemson salamander
P. j. clemsonae

Rabun bald salamander
P. j. rabunensis

FIGURE A Geographical isolates in *Plethodon jordani* of the Appalachian highlands. These salamanders originated when the population of the salamander *P. yonahlossee* became separated by the French Broad River valley. The eastern population developed into Metcalf's salamander, which spread northward, the only direction in which any group member could find suitable ecological conditions. To the south, southwest, and northwest, the mountains end abruptly, limiting the remaining *jordani*. Metcalf's salamander is the most specialized and ecologically divergent and the least competitive. Next, the red-cheeked salamander became isolated from the red-legged and the rest of the group by the deepening of the Little Tennessee River. Remaining members are still somewhat connected.

ecotypes will be scattered like a mosaic across the landscape. That frequently is the situation when several habitats to which the species is adapted recur throughout the range of the species. Some ecotypes may have evolved independently from different local populations.

Yarrow, *Achillea millefolium,* blankets the temperate and subarctic Northern Hemisphere with an exceptional number of ecotypes. It exhibits considerable variation, an adaptive response to different climates at various latitudes. Populations at lower altitudes are tall and have high seed production. Montane populations have a distinctive small size and low seed production.

In a classic study, J. Clausen, D. D. Keck, and W. M. Hiesey planted seeds of yarrows collected from different elevations in a transect across the Sierra Nevada mountains in transplant gardens at several elevations from lowland to mountain. Yarrow from the high elevations retained their short stature, no matter where planted. The other ecotypes showed some phenotypic plasticity in growth, but were dwarfed at high elevations.

The southern Appalachian Mountains are noted for their diversity of salamanders, fostered in part by a rugged terrain, an array of environmental conditions, and the limited ability of salamanders to disperse (Figure A). Populations become isolated from one another, preventing a free flow of genes. One species of salamander, *Plethodon jordani,* breaks into a number of semi-isolated populations, each characteristic of a particular part of the mountains. Groups of such populations make up **geographic isolates**, prevented by some extrinsic barrier—in the case of the salamanders, rivers and mountain ridges—from effecting a free flow of genes with other subpopulations. The degree of isolation depends upon the efficiency of the extrinsic barrier, but rarely is the isolation complete. These geographic isolates make up subspecies. Strong isolates with little gene flow between them may be in the first stages of speciation. Unlike the situation with clines, we can draw a geographic line that will separate them all into subspecies.

of infinite possibilities. It is bounded, constrained within a range of environmental tolerances (Figure 2.14). These tolerances are a function of tradeoffs imposed by constraints that can ultimately be traced to the laws of physics and chemistry. Stated simply: *the characteristics that enable an individual to do well under one set of conditions limit its performance under a different set of conditions.* This general, but important concept is obvious to sports fans. Figure 2.15 is a photograph of two great sports figures. Wilt Chamberlain was probably the greatest center in basketball. Willie Shoemaker was probably the best

FIGURE 2.14 The response of an organism to an environmental gradient such as temperature. The end points of the curve represent the upper and lower limits for survival. Within this gradient are more restricted ranges within which the organism can grow and reproduce.

FIGURE 2.15 (a) Willie Shoemaker and (b) Wilt Chamberlain each excelled in a sport for which the other was physically unsuited.

THE NICHE

The word *niche* in everyday terms means a recess in a wall where you place some object, or a place or activity for which a person or thing is best fitted. Joseph Grinnell, a California ornithologist, was the first to propose its use in ecology in 1917. He defined the niche as the ultimate distributional unit within which a species is restrained by the limitations of its physical structure and its physiology. What Grinnell had defined was a species's *habitat*—the place where individuals of the species live. In 1927 an English animal ecologist, Charles Elton, considered the niche the basic role of an organism in the community—what it does, its relationship to its food and enemies. In other words, he defined the niche as the species's *occupation*.

In 1958 G. E. Hutchinson, a limnologist, expanded the idea for the niche to its current form. Now the **niche** includes all the physical and biological variables that affect an organism's well-being. Hutchinson's approach to the niche is similar to the graph in Figure 2.14. However, the niche does not fall along a single axis or environmental factor. All environmental factors to which an organism responds are part of it. Rather than a two-dimensional graph, a surface, Hutchinson's niche is a multidimensional response called a **hypervolume.**

We can begin to visualize a multidimensional niche by creating a three-dimensional one. Consider three niche-related variables for a hypothetical organism: temperature, humidity, and food size (Figure A). If the organism can live

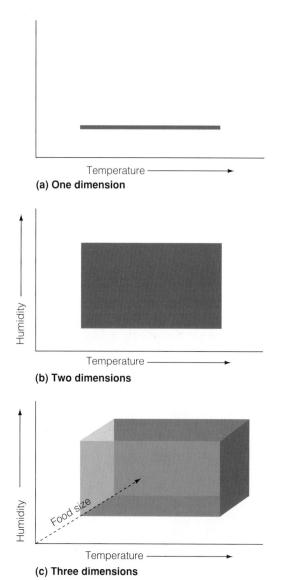

(a) One dimension

(b) Two dimensions

(c) Three dimensions

FIGURE A An illustration of niche dimension. Assume three elements comprising the hypothetical organism's niche: temperature, humidity, and food size. (a) A one-dimensional niche involving only temperature. (b) A second dimension, humidity, has been added. Enclosing that space, we have a two-dimensional niche. (c) Adding a third axis, food size, and enclosing all those points gives a three-dimensional niche space, or volume, for the organism. A fourth element would create a hypervolume.

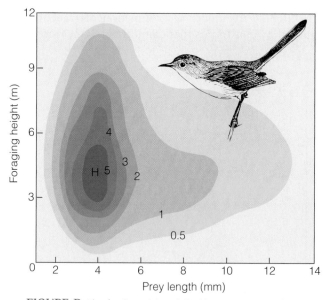

FIGURE B The feeding niche of the blue-gray gnatcatcher (*Polioptila caerulea*) is based on two variables, size of prey and foraging height. The contour lines map the feeding frequencies for adult gnatcatchers during the incubation period from July to August in California oak woodlands. The maximum response level is at H. Contour lines spreading out from this optimum represent decreasing response levels. The outermost line represents the boundary of the niche for these two variables (Root 1967).

only within a certain range of temperature, we plot that range on one axis. Temperature is one dimension of its niche. Next, suppose that our organism can survive and reproduce only within a certain range of humidity. We plot humidity on a second axis. Enclosing that space, we have defined a two-dimensional niche. Now suppose our organism can eat only a certain range of food size. Food size is plotted on a third axis. Enclosing the new space, we come up with a volume, a three-dimensional niche. An example of a two-dimensional niche is the feeding niche of the blue-gray gnatcatcher *(Polioptila caerulea)*, a bird of open woods and brushy edges in the eastern United States (Figure B). Two variables, foraging height and size of insect prey, define the species's niche in this example. The niche is a reflection of the species's adaptations, and is therefore the product of natural selection.

jockey to ride a racehorse. At the height of 1.49 meters, Willie Shoemaker could never have played center for the Los Angeles Lakers, and at 2.15 meters, Wilt Chamberlain could never have ridden to victory in the Kentucky Derby. The set of physical characteristics that enables you to excel at one of these sports precludes your ability to do well at the other. So, too, the characteristics of organisms constrain them (see Chapter 8, Figure 8.1, and Focus on Ecology 2.3: The Niche).

Throughout the following chapters, particularly in Part III (Chapters 6–8), we will examine this basic principle as it applies to the adaptation of species, and explore how the nature of adaptations changes with changing environmental conditions—both physical and biotic. Later chapters will emphasize the consequences of those adaptations as environmental conditions change in space and time, giving rise to the patterns and processes observed in communities and ecosystems. First, however, we will examine the stage upon which this drama will play—the physical environment.

CHAPTER REVIEW

Summary

Natural Selection (2.1–2.2) Natural selection is the differential success (survival and reproduction) of individuals within the population resulting from their interaction with their environment. Natural selection is a product of two conditions: (1) variation among individuals within a population in some characteristic that (2) results in differences among individuals in their survival and/or reproduction. The fitness of an individual is measured by the proportionate contribution it makes to future generations.

When natural selection acts to shift the mean value of the trait toward one extreme over another, it is called directional selection. When natural selection favors individuals near the population mean at the expense of the two extremes, it is referred to as stabilizing selection. When natural selection favors both extremes simultaneously, it can result in a bimodal distribution of the characteristic(s) in the population. Such selection, known as disruptive selection, occurs when members of a population are subject to different selection pressures (**2.1**).

Natural selection requires the characteristic to be heritable, able to be passed on from parent to offspring (**2.2**).

Genes (2.3–2.4) The units of heredity are genes carried on chromosomes. The alternative forms of a gene are alleles. Individuals that possess like pairs of alleles are homozygous. If the alleles are unlike, the individuals are heterozygous. In the case of heterozygosity, the allele that is expressed is the dominant, and that which is not expressed is the recessive.

The sum of heritable information carried by the individual is the genotype; its physical expression, on which natural selection acts, is the phenotype. The range of phenotypic expression under different environmental conditions is phenotypic plasticity (**2.3**).

Most inherited variation arises from recombination of genes in sexual reproduction. Some genetic material is altered by mutation. Gene mutations alter sequences of nucleotides. Chromosomal mutations change the structure or number of chromosomes. The duplication of entire sets of chromosomes is polyploidy, which is most common in plants. Single gene mutations rarely are visible. Most gene mutations are neutral and maintain genetic variability in a population (2.4).

Evolution and Speciation (2.5–2.7) Taxonomists describe plants and animals as morphological species, discrete entities that exhibit little variation. Variation over space and time has given rise to the concept of the biological species, a group of interbreeding individuals living together in a similar environment in a given region. Species may be sympatric or allopatric. Sympatric species inhabit the same region at the same time and have the opportunity to interbreed. Allopatric species, separated by time and space, have no opportunity to interbreed.

Species maintain their identities by means of isolating mechanisms. Premating isolating mechanisms include distinctive behavioral patterns, habitat preferences, the timing of breeding, and physical differences that inhibit mating between species. Postmating isolating mechanisms reduce mating success, as when hybrids are sterile or at selective disadvantage (2.5).

Species arise by the interaction of heritable variations, by natural selection, and by barriers to gene flow between populations. The most widely accepted mechanism of speciation is allopatric or geographic speciation. A single interbreeding population splits into spatially isolated populations, which diverge into distinct species. When reproductive isolation precedes differentiation and the process takes place within a population, we have sympatric speciation. The most common form of sympatric speciation is polyploidy among plants (2.6).

The outcome of natural selection is evolution, a change in gene frequency through time. Speciation by natural selection is a form of evolution. The process of speciation requires changes in the gene frequencies of a population that result in reproductive isolation: the evolution of characteristics that sever the exchange of genes with the parent population (2.7).

Tradeoffs and Constraints (2.8) Environmental conditions that directly influence life vary in both space and time; likewise, the objective of selection changes with environmental circumstances in both space and time. The characteristics that enable a species to survive, grow, and reproduce under one set of conditions limit its ability to do equally well under different environmental conditions.

Study Questions

1. What is natural selection?
2. What conditions are necessary for natural selection to occur?
3. Who are Charles Darwin and Gregor Mendel?
4. Contrast directional, stabilizing, and disruptive selection.
5. What is the relationship between DNA, genes, and chromosomes?
6. Define allele, locus, homozygous, heterozygous.
7. Distinguish between a genotype and a phenotype. What is phenotypic plasticity?
8. What is the gene pool?
9. What are the major sources of variation in the gene pool?
10. What is mutation? Distinguish between a gene mutation and a chromosomal mutation.
11. What is the difference between the morphological and biological species concepts?
12. What are the major mechanisms maintaining reproductive isolation between populations?
13. Distinguish between an allopatric species and a sympatric species.
14. What is speciation? Distinguish between allopatric and sympatric speciation.
15. What is the definition of evolution, and how does evolution relate to natural selection?

Taking the Uncertainty out of Sex

The genetic variation produced by sexual reproduction brought about the abounding diversity of life on Earth. It is required for the evolution of species diversity through the process of natural selection. Indeed, one might say, sex adds variety to life.

Humans have used the genetic variation made possible by sexual reproduction as raw material from which to produce a variety of domestic plants and animals. For example, all breeds of domestic dog, from the Chihuahua to the Saint Bernard, are descendents of the same species of wild ancestral dog. The different breeds of domestic dog were produced through **selective breeding.** By selecting individuals with a desired trait and mating them with individuals exhibiting the same trait (or traits), dog breeders produce populations of dogs with specific physical and behavioral characteristics. The process of selective breeding is analogous to natural selection—the differential fitness of individuals within the population resulting from the differences in some heritable characteristic(s). Unlike natural selection, however, humans rather than the environment function as agents of selection. Charles Darwin referred to selective breeding as "artificial selection."

The process of selective breeding requires one other important constraint imposed by humans—reproductive isolation. To maintain the unique characteristics of the various breeds or varieties, mating among individuals of different breeds must not be allowed. If it is, they will once again merge into a single gene pool. A visit to any animal shelter provides sufficient proof that, given the opportunity, the many varieties of domestic dog interbreed readily. Selective breeding represents a human version of creating new species or, in other words, our own home-grown method of evolution in which changes in gene frequency and reproductive isolation are directed by humans rather than natural selection.

Selective breeding is the earliest example of "biotechnology"—the use of biological processes to solve problems or make useful products. As with any process of mass production, success depends on a level of consistency in the product, otherwise known as quality assurance. While sexual reproduction produces the genetic variation from which we create the many domestic breeds of plants and animals, it also reduces the certainty of maintaining the desired characteristics for which the breeds were developed into future generations, therefore causing a fundamental problem. A major objective of the science and technology of selective breeding is to produce an organism with a consistent set of characteristics—that is, to reduce the uncertainty resulting from sex. The breeding of domestic cattle provides an excellent example of the use of technology to reduce this uncertainty. Bulls are selected as breeding stock based on the characteristics of their offspring, namely high weight gain in beef cattle and high milk production in dairy cows. The technology of artificial insemination makes it possible for an individual bull with the desired characteristics to father literally tens of thousands of offspring. The genes that control the desired trait come to dominate successive generations, reducing the genetic variation within the population, in effect providing a consistent product.

Although various technologies have improved the process, the basic approach of selective breeding changed little over the thousands of years that humans have bred and domesticated animals, until Dolly arrived. In 1997, a group of research geneticists at the Roslin Institute, led by British scientist Ian Wilmut, made news around the world when a sheep named Dolly was born at their research facility in Scotland. The birth of Dolly marked the first successful cloning of a mammal and heralded the promise of a new era in the science of selective breeding.

The creation of Dolly had its roots in an experimental technique scientists had developed in the 1950s for cloning amphibians such as frogs and salamanders. The procedure involves destroying the nucleus of an egg cell from the species to be cloned and injecting in its place a nucleus removed from an adult body cell of an individual (the donor) of the same species. Since the nucleus from the donor cell has a full complement of chromosomes (is diploid), the egg develops into an animal that is genetically identical to the donor—a clone. Applying the technique to mammals proved exceedingly difficult, and almost half a century passed before the birth of Dolly. Since then, scientists from around the world have used a similar technique to successfully produce clones of mice, cattle, and other animals.

Cloning provides the ultimate solution to quality control, in contrast to traditional approaches of selective breeding that only reduce the genetic variability that results from sexual reproduction. Traditional approaches shift the odds, but do not guarantee the desired results in the offspring. The problem is further compounded when the female is the carrier of the desired trait(s), since she can birth only a limited number of offspring in her lifetime. By way of cloning, reproduction is no longer limited to "ideal" females. Since the nucleus is removed from the female gametes, the eggs can be harvested from many individuals of the species. The male gamete (sperm) becomes unnecessary, since the full complement of genes is supplied by the nucleus of the donor individual. In theory, a single male and female with the desired characteristics can produce identical offspring—via surrogate females—for an indefinite number of generations into the future. In effect, the technology of cloning has allowed us to change the mode of reproduction in organisms that naturally reproduce sexually to a form of asexual reproduction. We can utilize the genetic diversity created by sexual reproduction to produce different varieties of plants and animals, and then use the technology of cloning to switch the mode of reproduction to asexual—eliminating further genetic variation among individuals in future generations.

Since the birth of Dolly, the science of cloning has progressed, but not without some setbacks. A recent study reported by Wilmut and his colleagues showed that Dolly's cells appeared prematurely old. That finding raised fears that cloned animals would have shorter lifespans. At $5\frac{1}{2}$ years of age, less than half the normal 12-year lifespan of a sheep, Dolly has a crippling form of arthritis and no longer appears normal and healthy. In addition, Wilmut and his colleagues produced only one successful Dolly out of 277 attempts. The other 276 embryos that were produced as part of the procedure died at various stages of development. Even for the scientists who created Dolly, cloning success is the exception, not the rule. A vast majority of efforts fail, even in species that have at one time or another been cloned. The day when scientists can easily and routinely produce cloned animals seems far in the future.

Now that cloning mammals is a possibility, however, researchers are starting to develop different ways to use this technology. One of the more imaginative applications of cloning, which makes *Jurassic Park* seem more like science than fiction, is the preservation of endangered species. Late last year in Iowa, an ordinary cow named Bessie gave birth to the world's first cloned endangered species, a baby bull named Noah. Noah is a gaur, a rare and endangered species of large oxlike animals indigenous to India, Indochina, and Southeast Asia. Although Noah died within 48 hours of birth as a result of common dysentery, this remarkable animal provided proof that one animal can carry and give birth to the clone of an animal of a different species. Scientists are currently attempting to clone a variety of other endangered species, including the African bongo antelope, the Sumatran tiger, and the giant panda. Scientists also hope to reincarnate some species that are already extinct, such as the bucardo mountain goat of Spain. The last individual died recently in captivity, but Spanish scientists have preserved some of its cells.

"Selective breeding represents a human version of creating new species or, in other words, our own home-grown method of evolution in which changes in gene frequency and reproductive isolation are directed by humans rather than natural selection."

Since the birth of Dolly, it has been doubtful that science could keep the process of cloning down on the farm. The medical applications of cloning animals are too obvious and too promising not to be pursued by other scientists. Researchers at the Roslin Institute have incorporated into sheep the gene for human factor IX, a blood-clotting protein used to treat hemophilia. The current production of this clotting protein requires the use of large quantities of human blood, restricting its production and making it extremely expensive to produce. Another hopeful area for development is the rapid production of large animals carrying genetic defects that mimic human illnesses, such as cystic fibrosis, enabling researchers to evaluate forms of treatment for use in humans. Although these applications of cloning raise serious ethical questions regarding the treatment and use of animals, a new era of biotechnology—combining cloning technologies with genetic engineering—is creating even more of an ethical dilemma.

Genetic engineering is the technology based on the artificial manipulation and transfer of genetic material. This technology can move genes and the traits they determine from one individual into another. Cells modified by these techniques pass the new genes and traits on to their offspring. The amazing feature of this technique is that the individuals involved in the transfer need not be of the same species. In fact, genes have been successfully transferred not only from one type of animal or plant to another, but also from plants to animals and from animals to plants. Prior to the development of this technology, a breeder who wanted to develop a red sheep could do so only if the necessary red gene existed somewhere in a sheep. A genetic engineer has no such restrictions. If red genes exist anywhere in nature, such as in a fish, a bird, or even a rose, those genes could potentially be used to produce a red sheep. This ability to transfer genes allows genetic engineers to create gene combinations that exist nowhere in nature, in effect creating "designer organisms" that can then be duplicated through cloning technologies. Novel organisms, however, have the potential to create novel risks as well as the intended benefits. Given the extraordinary ability to create organisms that would never be found in nature, it is easy to see why groups that oppose the application of these new technologies to agriculture use terms such as "Frankenfood" to describe genetically engineered crops and livestock.

Humans have come a long way from the initial domestication of plants and animals through selective breeding. We now have the technology to create not only new species, but also the potential to create new life-forms—something previously done only in the exclusive realm of nature and chance. Despite the incredible potential that our new technologies hold, perhaps it is wise to advance in a thoughtful and cautious manner. Otherwise, the difficulty we face may not be taking the uncertainty out of sex, but rather living with the uncertainty of unintended consequences.

PART 11

The Physical Environment

A fierce winter wind blows across the taiga (Waterton Lakes National Park, Alberta, Canada).

Climate

OBJECTIVES

On completion of this chapter, you should be able to:

- Describe the fate of solar energy reaching Earth.
- Describe how the atmosphere heats and circulates.
- Tell how solar radiation influences seasonal temperatures.
- Explain the Coriolis effect on atmospheric circulation and ocean currents.
- Discuss the measures of atmospheric moisture.
- Describe microclimates and their ecological effects.

From space, planet Earth appears as a bright ball of blue and white. Its unique color comes from its cover of water—ice, snow, oceans, and clouds—reflecting the sunlight in which it is immersed. Earth's atmosphere intercepts sunlight on its outer edge. The resulting molecular interactions create heat and cause thermal patterns which, coupled with Earth's rotation and movement, generate the prevailing winds and ocean currents. These movements of air and water in turn influence Earth's weather patterns, including the distribution of rainfall. In this chapter, we explore in more detail how this thin envelope of air, water, and surface land are linked by the essential flow of energy originating in the Sun.

What determines whether a particular geographical region will be a tropical forest, a grassy plain, or a barren landscape of sand dunes? The aspect of the physical environment that most influences a particular ecosystem by placing the greatest constraint on organisms is climate. Climate is one of those terms we tend to use loosely. In fact, people sometimes confuse climate with weather. **Weather** is the combination of temperature, humidity, precipitation, wind, cloudiness, and other atmospheric conditions occurring at a specific place and time. **Climate** is the long-term average pattern of weather and may be local, regional, or global. In the following chapter, we learn how climate determines the availability of heat and water on Earth's surface and influences the amount of solar energy that plants can capture. Thus it controls the distribution and abundance of plants and animals.

3.1   Earth Intercepts Solar Radiation

Solar radiation, the electromagnetic energy emanating from the Sun, travels more or less unimpeded through the vacuum of space until it reaches Earth's atmosphere. Scientists conceptualize solar radiation as a stream of photons, packets of energy, that—in

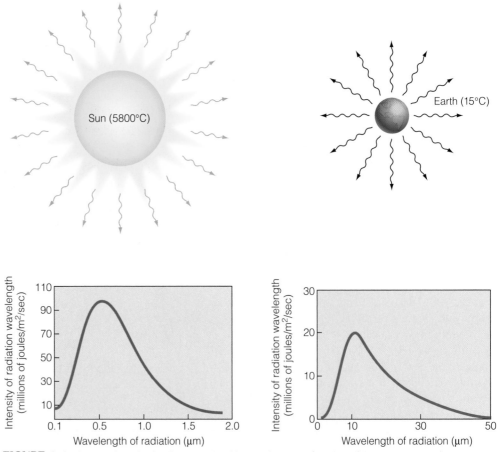

FIGURE 3.1 The wavelength of radiation emitted by an object is a function of its temperature. The Sun, with an average temperature of 5800°C, emits relatively shortwave radiation as compared to Earth, with an average surface temperature of 15°C, which emits relatively long-wave radiation.

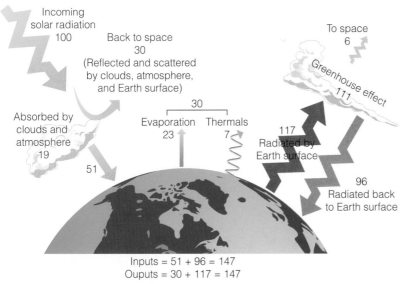

Incoming
solar radiation
100

Back to space
30
(Reflected and scattered
by clouds, atmosphere,
and Earth surface)

To space
6

Greenhouse effect
111

Absorbed by
clouds and
atmosphere
19

30
Evaporation   Thermals
23   7

117
Radiated by
Earth surface

51

96
Radiated back
to Earth surface

Inputs = 51 + 96 = 147
Ouputs = 30 + 117 = 147

FIGURE 3.2 Disposition of solar energy reaching Earth's atmosphere.

one of the great paradoxes of science—behave either as waves or particles, depending on the manner in which they are observed. Scientists characterize waves of energy in terms of their wavelength (λ), the physical distances between successive crests, and frequency (ν), the number of crests that pass a given point per second (Figure 3.1). An object typically emits energy across a wide range of wavelengths, but the exact nature of the energy emitted depends on the object's temperature. The hotter the object, the more energetic the emitted photons and the shorter the wavelength. A very hot surface such as that of the Sun (>5000°C) gives off primarily shortwave radiation. In contrast, cooler objects such as Earth's surface (average temperature of 15°C) emit radiation of longer wavelengths, or long-wave radiation.

However bright the Sun may seem, only 51 percent of the solar energy traveling to Earth makes it through the atmosphere to Earth's surface. What happens to all the incoming energy? If you take the amount of solar radiation that reaches the atmosphere as 100 units, on average, clouds and the atmosphere reflect and scatter 26 units, while the Earth's surface reflects an additional 4 units, giving a total of 30 units being reflected back to space (Figure 3.2). Together the atmosphere and clouds absorb another 19 units, leaving 51 units of direct and indirect solar radiation to be absorbed by Earth's surface.

Of the 51 units that reach the surface, 23 units are used to evaporate water and another 7 units are lost by way of direct heating of air in contact with the warm surface (thermals), leaving 21 units to heat the planet's land masses and oceans. These 21 units

are eventually emitted back to the atmosphere as long-wave radiation. The Earth's surface, however, actually radiates upward some 117 units. How is this possible? It does so because, although it receives solar (shortwave) radiation only during the day, it constantly emits long-wave radiation during both day and night. Additionally, the atmosphere above only allows a small fraction of this energy (6 units) to pass through into space. The majority (111 units) is absorbed by the water vapor and CO_2 in the atmosphere, and by clouds. Much of this energy (96 units) is radiated back to Earth, producing the greenhouse effect (see Chapter 30), which is critical to maintaining the surface warmth of the Earth. As a result, the Earth's surface receives nearly twice as much long-wave radiation from the atmosphere as it does shortwave radiation from the Sun. In all of these exchanges, the energy lost at the Earth's surface (117 + 30 = 147 units) is exactly balanced by the energy gained (51 + 96 = 147 units) (see Figure 3.2). The radiation budget of the Earth is in balance.

3.2 Intercepted Solar Radiation Varies over Earth's Surface

The amount of solar energy intercepted at any point on Earth's surface varies markedly with latitude (Figure 3.3). Two factors influence this variation. First, at higher latitudes, radiation hits the surface

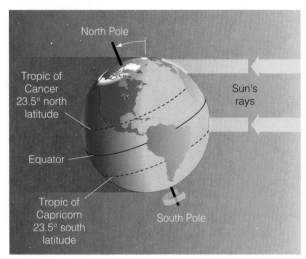

FIGURE 3.3 Solar radiation striking Earth at high latitudes arrives at an oblique angle and spreads over a wide area. Therefore, it is less intense than energy arriving vertically at the equator.

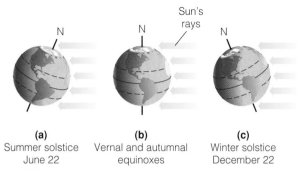

(a) Summer solstice June 22 **(b)** Vernal and autumnal equinoxes **(c)** Winter solstice December 22

FIGURE 3.4 Angle of the Sun and circle of illumination at the equinoxes and solstices.

at a steeper angle, spreading sunlight over a larger area. Second, radiation that penetrates the atmosphere at a steep angle must travel through a deeper layer of air. In the process, it encounters more particles in the atmosphere, which reflect more of it back into space.

What gives rise to the seasons on Earth? Why do the hot days of summer give way to the changing colors of fall, or the freezing temperatures and snow-covered landscape of winter to the blanket of green with the onset of spring? Although the variation in solar radiation reaching Earth's surface with latitude can explain the gradient of decreasing temperature from the equator to the poles, it does not explain the systematic variation that occurs over the course of a year. The answer is quite simple. It is because Earth does not stand up straight. It tilts on its side.

Earth, like all planets, is subject to two distinct motions. While it orbits around the Sun, Earth rotates about an axis that passes through the North and South Poles, giving rise to the brightness of day followed by the darkness of night (the diurnal cycle). Earth travels about the Sun in a plane called the ecliptic (a plane traveled by all other planets with the exception of Pluto). As chance would have it, Earth's axis of spin is not perpendicular to the ecliptic, but is tilted at an angle of 23.5° (Figure 3.4). It is this tilt (inclination) that is responsible for the seasonal variations in temperature and daylength.

Only at the equator are there exactly 12 hours of daylight and darkness every day of the year. At the spring or vernal equinox (approximately March 21)

and fall or autumnal equinox (approximately September 22), solar radiation falls directly on the equator (Figure 3.4a). At this time, the equatorial region is heated most intensely, and every place on Earth receives the same twelve hours each of daylight and night.

At the summer solstice (June 22) in the Northern Hemisphere, solar rays fall directly on the Tropic of Cancer (23.5° north latitude) (Figure 3.4b). This time is when days are longest in the Northern Hemisphere and the Sun heats the surface most intensely. In contrast, the Southern Hemisphere experiences winter at this time. At winter solstice (December 22) in the Northern Hemisphere, solar rays fall directly on the Tropic of Capricorn (23.5° south latitude) (Figure 3.4c). This period is summer in the Southern Hemisphere, while the Northern Hemisphere is enduring shorter days and colder temperatures. Thus, the summer solstice in the Northern Hemisphere coincides with the winter solstice in the Southern Hemisphere.

The seasonality of solar radiation, temperature, and daylength increases with latitude. At the Arctic and Antarctic Circles (66.5° north and south latitude), daylength varies from 0 to 24 hours over the course of the year. The days shorten until the winter solstice, a day of continuous darkness. The days lengthen with spring, and on the day of the summer solstice the Sun never sets.

Figure 3.5 shows how annual, seasonal, and daily solar radiation vary over Earth. Although in theory every location on Earth receives the same amount of daylight over the course of a year, the tilt of the earth exposes equatorial regions to the most solar radiation. In high latitudes where the Sun is never positioned directly overhead, the annual input of solar radiation is the lowest. This pattern of varying exposures to solar radiation controls mean annual temperature around the globe (Figure 3.6). Like annual

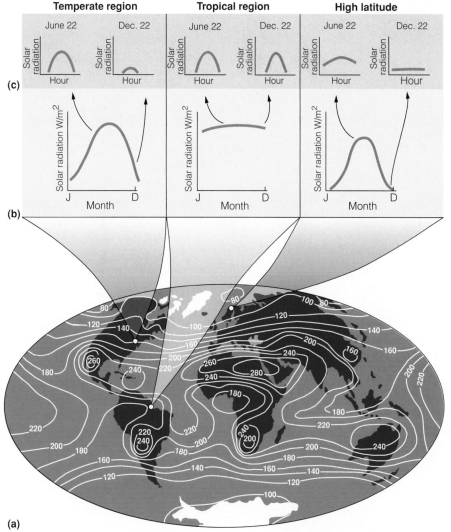

FIGURE 3.5 Annual variation in solar radiation on Earth. (a) Global mean solar radiation. (b) Variations in solar radiation from summer solstice to winter solstice at three locations: a temperate region, a tropical region, and a high latitude region. (c) Diurnal variations in solar radiation on two days in the year: the summer solstice and the winter solstice.

solar radiation, mean annual temperatures are highest in tropical regions and decline as one moves toward the poles.

3.3 Air Temperature Decreases with Altitude

Whereas varying degree and length of exposure to solar radiation may explain changes in latitudinal, seasonal, and daily temperatures, they do not explain why air gets cooler with increasing altitude. Mount Kilimanjaro, for example, rises from the hot plain of tropical East Africa, but its peak is capped with ice and snow (Figure 3.7). The answer to this apparent oddity of snow in the tropics lies in the physical properties of air.

The weight of all the air molecules surrounding Earth is a staggering 5600 trillion tons. The air's weight acts as a force upon Earth's surface, and the amount of force exerted over a given area of surface is called **atmospheric pressure** or air pressure. Envision a vertical column of air. The pressure at any point in the column can be measured in terms of the total mass of air above that point. As we climb in elevation, the mass of air above us decreases, and therefore pressure declines. Although atmospheric pressure decreases continuously, the rate of decline slows with increasing altitude (Figure 3.8).

Because of the greater air pressure at Earth's surface, the density of air (the number of molecules

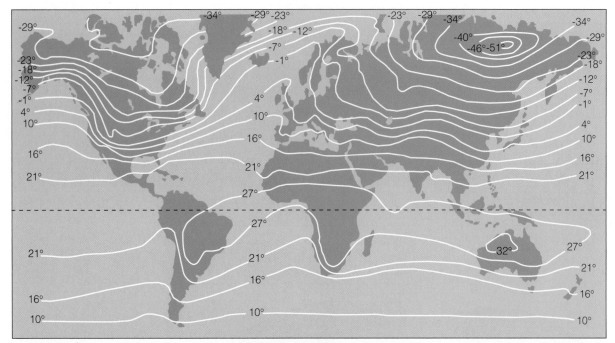

(a) January isotherms (lines of equal temperature) around the earth

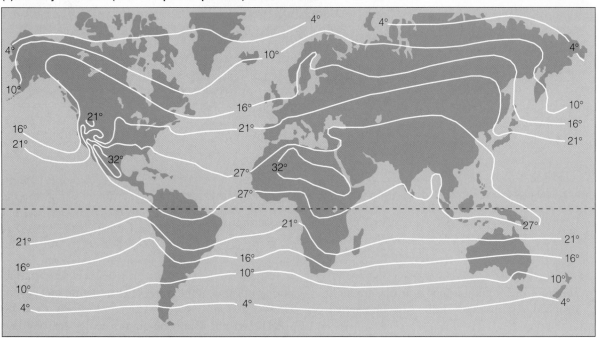

(b) July isotherms (lines of equal temperature) around the earth

FIGURE 3.6 Mean annual global temperatures change with latitude and season. (a) Mean sea-level temperatures (°C) in January. (b) Mean sea-level temperatures (°C) in July.

per unit volume) is high, decreasing in parallel with air pressure as we climb in altitude. As altitude above sea level increases, both air pressure and density decrease. Although by an altitude of 50 km, air pressure is only 0.1 percent of that measured at sea level, the atmosphere continues to extend upward for many hundreds of kilometers, gradually becoming thinner and thinner until it merges into outer space.

Although both air pressure and density decrease systematically with height above sea level, air temperature has a more complicated vertical profile. Air temperature normally decreases from Earth's surface up to an altitude of approximately 11 km (nearly

FIGURE 3.7 Although near the equator, Mount Kilimanjaro in Africa is snowcapped and supports tundralike vegetation near its summit. Global warming is causing a rapid melting of this snowcap.

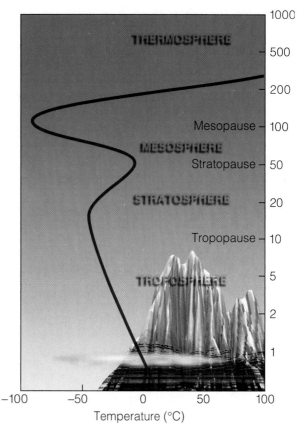

FIGURE 3.9 Changes in atmospheric temperature (global average) with altitude above sea level. The regions of the atmosphere are labeled, and Mount Everest (the highest mountain peak on Earth) is drawn for perspective. From T. Graedel and P. Crutzen, *Atmosphere, Climate & Change*, Scientific American Library.

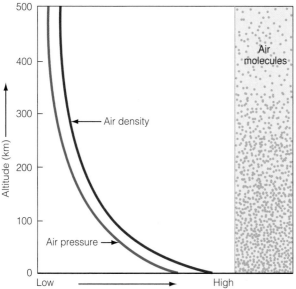

FIGURE 3.8 Both air pressure and air density decrease with increasing altitude above sea level.

36,000 feet). The rate at which temperature decreases with altitude is called the **lapse rate.**

The decrease in air temperature as one moves farther from Earth's surface is caused by two factors. Because of the higher density of air at the surface, air molecules collide, generating heat. The decrease in air density with altitude results in fewer collisions, thus generating less heat. The primary reason for the decrease in air temperature with increasing altitude, however, is the corresponding decline in the "warming effect" of Earth's surface. The absorption of solar

radiation functions to warm Earth's surface. Energy is emitted upward from the surface, heating the air above it. This process of transfer continues upward as heat flows spontaneously from warmer to cooler areas, but at a continuously declining rate as the energy emitted from the surface is dissipated.

Unlike air pressure and density, air temperature does not decline continuously with increasing height above Earth's surface. In fact, at certain heights in the atmosphere, a change in altitude can result in an abrupt change in temperature. Atmospheric scientists use these altitudes at which temperatures change abruptly to distinguish different regions in the atmosphere (Figure 3.9). Beginning at Earth's surface, the regions are called the troposphere, the stratosphere, the mesosphere, and the thermosphere. The boundary zones between these four regions of the atmosphere are the tropopause, stratopause, and mesopause respectively. The two most important regions in terms of climate and, therefore, life on Earth are the troposphere and stratosphere.

So far, the discussion of the change in air temperature with increasing altitude has assumed no vertical movement of air from the surface to the top of the atmosphere. When a volume of air at the surface warms, however, it becomes buoyant and rises (just as does a hot air balloon). As the volume of air (referred to as a parcel of air) rises, the decreasing pressure causes it to expand and cool. The decrease in air temperature through expansion, rather than through heat loss to the surrounding atmosphere, is called **adiabatic cooling.** The same process works in an air conditioner, where a coolant is compressed. As the coolant moves from the compressor to the coils, the drop in pressure causes it to expand and cool.

The rate of adiabatic cooling depends on how much moisture is in the air. The adiabatic cooling of dry air is approximately 10°C per 1000 meters elevation. Moist air cools more slowly. The rate of temperature change with elevation is called the **adiabatic lapse rate.**

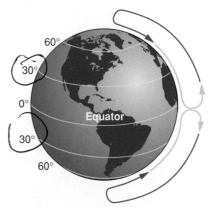

FIGURE 3.10 Circulation of air cells and prevailing winds on an imaginary, nonrotating Earth. Air heated at the equator rises and moves north and south. Cooling at the poles, it descends and moves back toward the equator.

3.4 Air Masses Circulate Globally

The blanket of air that surrounds the planet, the atmosphere, is not static. It is in a constant state of movement, driven by the rising and sinking of air masses and the rotation of Earth on its axis. The equatorial region receives the largest annual input of solar radiation. Warm air rises because it is less dense than the cooler air above it. Air heated at the equatorial region rises to the top of the atmosphere, establishing a zone of low pressure down at the surface (Figure 3.10). More air rising beneath it forces the air mass to spread north and south toward the poles. As air masses move poleward, they cool, become heavier, and sink. The sinking air raises surface air pressure (high pressure zone). The cooled, heavier air then flows toward the equator, replacing the warm air rising over the tropics.

If Earth were stationary and without irregular land masses, the atmosphere would circulate as shown in Figure 3.10. Earth, however, spins on its axis from west to east. Although each point on Earth's surface makes a complete rotation every 24 hours, the speed of rotation varies with latitude (and circumference). At a point on the equator (its widest circumference— 40,176 km), the speed of rotation is 1674 km per hour. In contrast, at 60° north or south, Earth's circumference is approximately half that at the equator (20,130 km), and the speed of rotation is 839 km per hour. According to the law of angular motion, the momentum of an object moving from a greater circumference to a lesser circumference will deflect in the direction of the spin, and an object moving from a lesser circumference to a greater circumference will deflect in the

direction opposite that of the spin. As a result, air masses and all moving objects in the Northern Hemisphere are deflected to the right, and in the Southern Hemisphere to the left. This deflection in the pattern of air flow is the **Coriolis effect,** named after a 19th-century French mathematician, G. C. Coriolis, who first analyzed the phenomenon (Figure 3.11).

The Coriolis effect, then, prevents a direct, simple flow of air from the equator to the poles. Instead, it creates a series of belts of prevailing winds, named for the direction from which they come. These belts break the simple flow of surface air toward the equator and the flow aloft to the poles into a series of six cells, three in each hemisphere. They produce areas of high and low pressure as air masses descend toward and ascend from the surface respectively (Figure 3.12). To trace the flow of air as it circulates between the equator and poles, we begin at Earth's equatorial region, which receives the largest annual input of solar radiation.

Air heated in the equatorial zone rises upward, creating a low pressure zone near the surface, the equatorial low. Sailors call this region of calm the **doldrums.** For centuries mariners have feared this equatorial region because the calm conditions caused by the warm rising air would stall ships on their journey. The region is also prone to violent weather shifts, with calm winds giving way to gale conditions within a matter of minutes.

The upward flow of air in the region of the doldrums is balanced by a flow of air from the north and south toward the equator. As the warm air mass rises, it begins to spread, diverging northward and southward toward the North and South Poles, cooling as it goes. In the Northern Hemisphere, the Coriolis effect forces air in a westerly direction, slowing its progress north. At about 30° north latitude,

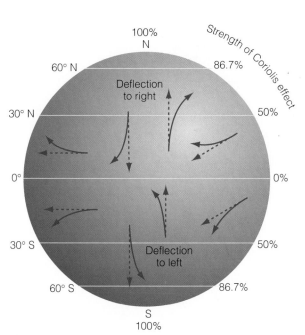

FIGURE 3.11 Effect of the Coriolis force on wind direction. The effect is absent at the equator, where the linear velocity is the greatest, 465 meters per second (1040 miles per hour). Any object on the equator is moving at the same rate. The Coriolis effect increases regularly toward the poles. If an object, including an air mass, moves northward from the equator at a constant speed, it speeds up because Earth moves more slowly (403 meters per second at 30° latitude, 233 meters per second at 60° latitude, and 0 meters per second at the poles). As a result, the path of the object appears to deflect to the right or east in the Northern Hemisphere, and to the left or west in the Southern Hemisphere.

winds and generally fair weather is known as the **horse latitudes.** The name originates from the merchant sailors who would sometimes throw overboard horses that were carried as cargo in a desperate attempt to lighten the ship's load and use the available gentle breezes to reach their destination.

Having descended, the cool air warms and splits into two currents flowing over the surface. One moves northward toward the pole, diverted to the right by the Coriolis effect to become the prevailing **westerlies.** Meanwhile the other current moves southward toward the Equator. Also deflected to the right, this southward flowing stream becomes the strong reliable winds that were called **trade winds** by the 17th-century merchant sailors, who used them to reach the Americas from Europe. In the Northern Hemisphere these winds are known as the northeast trades. In the Southern Hemisphere where similar flows take place, these winds are known as the southeast trades.

As the mild air of the westerlies moves poleward, it encounters cold air moving down from the pole (approximately 60° N). These two air masses of contrasting temperature do not readily mix. They are separated by a boundary called the polar front, a zone of low pressure (the subpolar low) where surface air converges and rises. Some of the rising air moves southward toward the horse latitudes, where it sinks back to the surface closing the second of the three cells, the Ferrel cell, named after the American meteorologist William Ferrel.

As the northward moving air reaches the pole, it slowly sinks to the surface and flows back (southward) toward the polar front, completing the last of the three cells, the polar cell. This southward moving air is deflected to the right by the Coriolis effect, giving rise to the **polar easterlies.** Similar flows take place in the Southern Hemisphere (see Figure 3.12).

the now cool air sinks, closing the first of the three cells, the Hadley cells, named for the Englishman George Hadley who first described this pattern of circulation in 1735. The descending air forms a semi-permanent high pressure belt encircling Earth, the subtropical high. Over the oceans, this region of light

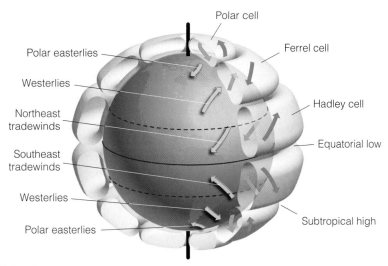

FIGURE 3.12 Belts and cells of air circulation about a rotating Earth. This circulation gives rise to the trade, westerly, and easterly winds.

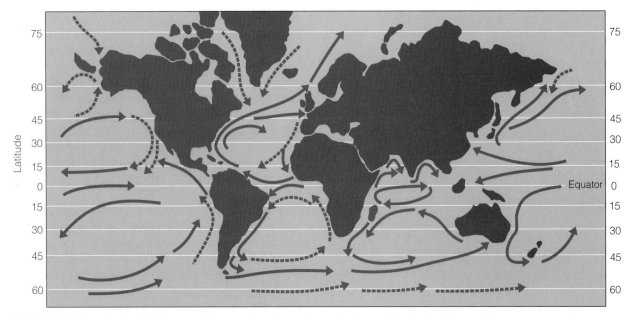

FIGURE 3.13 Ocean currents of the world. Notice how the circulation is influenced by the Coriolis force and continental land masses, and how oceans are connected by currents. Dashed arrows represent cool water, and solid arrows warm water.

3.5 Solar Energy, Wind, and Earth's Rotation Create Ocean Currents

The global pattern of the prevailing winds plays a critical role in determining major patterns of surface water flow in Earth's oceans. These systematic patterns of water movement are called **currents.** In fact, the oceans' major currents generally mimic the movement of the wind currents above until they encounter one of the continents.

Each ocean is dominated by two great circular water motions, or **gyres.** Within each gyre, the ocean current moves clockwise in the Northern Hemisphere and counterclockwise in the Southern Hemisphere (Figure 3.13). Along the equator, trade winds push warm surface waters westward. When these waters encounter the eastern margins of the continents, they split into north- and south-flowing currents along the coasts, forming north and south gyres. As the currents move farther from the equator, the water cools. Eventually they encounter the westerly winds at higher latitudes (30–60° north and south), which produce eastward-moving currents. When these eastward-moving currents encounter the western margins of the continents, they form cool currents flowing along the coastline toward the equator. Just north of the Antarctic continent, ocean waters circulate unimpeded around the globe.

3.6 Temperature Influences the Amount of Moisture Air Can Hold

Air temperature plays a critical role in the exchange of water between the atmosphere and Earth's surface. The amount of water that a given volume of air can hold is a function of its temperature. Water vapor, which is water in a gaseous state, gets into the air by evaporation. The process of transforming water from a liquid to a gaseous state requires energy, referred to as the **latent heat of evaporation.** In the air, water vapor acts as an independent gas that, like air, has weight and exerts pressure. The amount of pressure water vapor exerts independent of the pressure of dry air (see Section 3.4) is called **vapor pressure.** Vapor pressure is typically defined in units of megapascals (MPa). Because the quantity of water in the air is measured in terms of pressure, the maximum amount of water that can be held in a given volume of air is called the **saturation vapor pressure.** Warm air can hold more water vapor than cold air; thus, the saturation vapor pressure increases as air temperature increases (Figure 3.14). The difference between saturation vapor pressure and actual (ambient) vapor pressure at any given temperature is called the **vapor pressure deficit.**

The amount of water in a given volume of air is its absolute humidity. The most familiar measure is

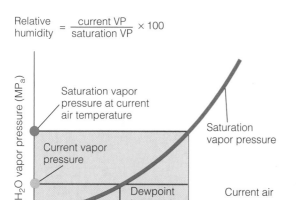

$$\text{Relative humidity} = \frac{\text{current VP}}{\text{saturation VP}} \times 100$$

FIGURE 3.14 Saturation vapor pressure as a function of air temperature. For a given air temperature, the relative humidity is the ratio of actual vapor pressure to saturation vapor pressure. For a given vapor pressure, the temperature at which saturation vapor pressure occurs is called the dew point.

relative humidity, or the amount of water vapor in the air expressed as a percentage of the saturation vapor pressure. At saturation vapor pressure, the relative humidity is 100 percent. If air cools while the actual amount of moisture it holds (water vapor pressure) remains constant, then relative humidity increases because cold air cannot hold as much water as warm air can (lower saturation vapor pressure). If the air cools beyond the saturation vapor pressure, moisture in the air will condense and form clouds. As soon as particles of water or ice in the air become too heavy to remain suspended, precipitation falls.

For a given water content of a parcel of air (vapor pressure), the temperature at which saturation vapor pressure is achieved is called the **dew point temperature.** Think about finding dew or frost on a cool fall morning. As nightfall approaches, temperatures drop and relative humidity rises. If cool night air temperatures reach the dew point, water condenses and dew forms, lowering the amount of water in the air. As the sun rises, air temperature warms and the amount of moisture that the air can hold increases. The dew evaporates, increasing vapor pressure in the air.

3.7 Precipitation Has a Global Pattern

By bringing together patterns of temperature, winds, and ocean currents, we are ready to understand the global pattern of precipitation (Figure 3.15). As

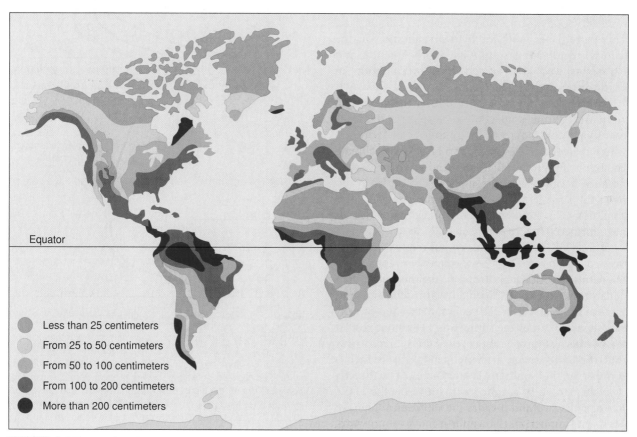

Equator

- Less than 25 centimeters
- From 25 to 50 centimeters
- From 50 to 100 centimeters
- From 100 to 200 centimeters
- More than 200 centimeters

FIGURE 3.15 Annual world precipitation. Relate the wettest and driest areas to mountain ranges, ocean currents, and winds.

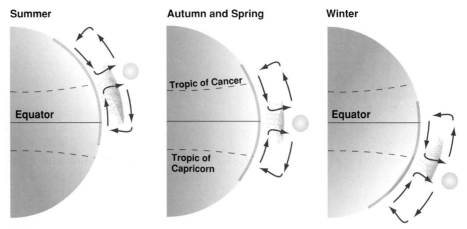

Summer

Equator

Autumn and Spring

Tropic of Cancer

Tropic of Capricorn

Winter

Equator

FIGURE 3.16 Shifts of the intertropical convergence, producing rainy seasons and dry seasons. Note that as the distance from the equator increases, the dry season is longer and the rainfall is less. These oscillations result from changes in the altitude of the Sun between the equinoxes and the solstices as diagrammed in Figure 3.4. Patterns of air circulation are shown in Figure 3.12.

the warm trade winds (see Figures 3.12 and 3.13) move across the tropical oceans, they gather moisture. Near the equator, the northeasterly trade winds meet the southeasterly trade winds. This narrow region where the trade winds meet is the **intertropical convergence zone (ITCZ)**, characterized by high amounts of precipitation (Figure 3.16). Where the two air masses meet, air piles up, and the warm humid air rises and cools. When the dew point is reached, clouds form, and precipitation falls as rain. This pattern accounts for high precipitation in the tropical regions of eastern Asia, South America, and Africa, as well as relatively high precipitation in southeastern North America.

Having lost much of its moisture, the ascending air mass continues to cool as it splits and moves northward and southward. In the horse latitudes, where the cool air descends, two belts of dry climate encircle the globe. The descending air warms. Because it can hold more moisture, it draws water from the surface through evaporation, causing arid conditions. In these belts, the world's major deserts have formed (Chapter 26).

The ITCZ is not stationary, but tends to migrate toward regions of the globe with the warmest surface temperature. Although tropical regions about the equator are always exposed to warm temperatures, the Sun is directly over the geographical equator only two times a year, at the spring and fall equinoxes. At the northern summer solstice, the Sun is directly over the Tropic of Cancer; at the winter solstice (which is summer in the Southern Hemisphere), it is directly over the Tropic of Capricorn. As a result the ITCZ moves poleward and invades the subtropical highs in northern summer; in the winter it moves southward, leaving clear dry weather behind. As it migrates southward, it brings rain to the southern summer.

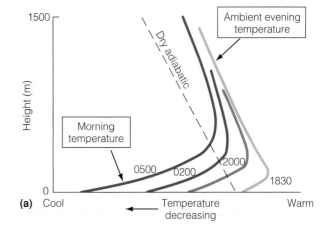

(a) Cool ← Temperature decreasing → Warm

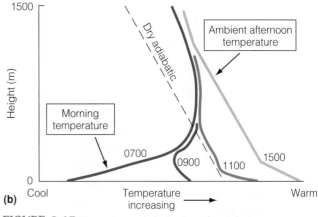

(b) Cool ← Temperature increasing → Warm

FIGURE 3.17 Formation and elimination of a nighttime surface inversion typical on clear cool nights. (a) After sunset (1830 hours, or 6:30 pm) the ground cools rapidly and in turn cools the surface air above it. A shallow surface inversion forms. As cooling continues during the night, the inversion deepens from the surface upward, reaching its maximum depth just before dawn (0500). (b) After sunrise (0700) solar radiation begins to warm the surface and the nighttime inversion gradually disappears during the forenoon of a clear day as the surface and air above heat to a midafternoon (1500) maximum.

Thus as the ITCZ shifts north and south, it brings on the wet and dry seasons in the tropics (see Figure 3.16). Because air and water heat slowly, a time lag of about one month develops between the change in the vertical orientation of the Sun and the shift in the intertropical convergence. For this reason, the intertropical convergence does not move as far north and south as does the Sun.

3.8 Inversions Occur When Surface Air Loses More Energy Than It Receives

Have you experienced the great banks of fog that fill valleys and lower elevations when the land above is immersed in early morning sunlight? What you are witnessing is an **inversion,** an atmospheric condition in which air temperature rises rather than falls as elevation increases.

Solar energy heats Earth's surface and the air above it by day. As night approaches, the surface air above the ground cools as it loses more heat energy than it receives. This surface layer of cooler air deepens (Figure 3.17a) as the night progresses and forms a **nighttime surface inversion** in which the temperature increases, rather than decreases with elevation (see Section 3.3). By morning, solar energy once again heats the surface and gradually eliminates this nighttime inversion (Figure 3.17b).

Such inversions are particularly pronounced in hilly and mountainous country during summer when the air mass is stable and the weather is calm and clear (Figure 3.18). At night in the valley, the air next to the ground cools, forming a weak surface inversion. At the same time, cold, dense air flows down slopes from the hill or mountaintop. Together they cause the inversion to become deeper and stronger, and the cold, dense air is trapped beneath a layer of warm air. In mountainous areas, the top of the night inversion is usually below the main ridge. If air is sufficiently cool and moist, fog may form in the valley. Smoke from industry and other heated pollutants released during such an inversion will rise only until their temperature equals that of the surrounding air, a point called the thermal belt or warm slope zone. At that point, smoke flattens out and spreads horizontally just below the thermal belt. These inversions

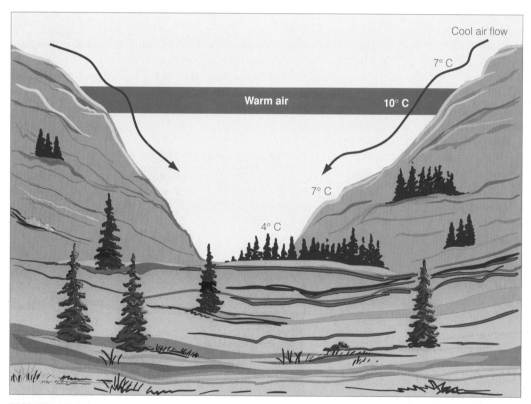

FIGURE 3.18 Topography can produce daily climatic extremes in valleys and depressions in the ground. At night, air cools next to the ground, forming a weak surface inversion in which the temperature increases, rather than decreases, with height. At the same time, cool air moves downslope, deepening the inversion. When air is sufficiently cool and moist, fog forms in the valley. Smoke or air pollution released in such a situation will rise only until its temperature equals that of the surrounding air. Then it will flatten out just below the layer of warm air.

Surface air must flow out as subsidence progresses

Cool air from aloft begins to settle

Warm, very dry air approaches the surface

Figure 3.19 Descent of a subsidence inversion.

break up when surface air warms during the day to create vertical convections and turbulence, or when a new air mass moves in.

A second, more widespread type of inversion occurs when a surface high pressure air mass stalls over a region. Air in the high pressure mass spreads outward and flows clockwise in an ever-widening spiral. The air flowing outward from the center must be replaced, and the only source for replacement is less dense air from above. The sinking movement of cool air from aloft is called subsidence.

Similar but more widespread inversions occur when a high pressure area stagnates over a region. In a high pressure area, the airflow is clockwise and spreads outward. The air flowing away from the high must be replaced, and the only source for replacement air is from above. Thus, surface high pressure areas are regions of sinking air movements from aloft.

When winds traveling at high altitudes in the atmosphere slow down, the cold air tends to sink. As the parcel of air descends, it is compressed, heats, and becomes drier. The result is a layer of warm dry air that has developed at a high level in the atmosphere with no chance to descend. It hovers several hundred to several thousand feet above Earth's surface, forming a **subsidence inversion** (Figure 3.19). Subsidence inversions may persist for days and function to increase the intensity of air pollution. Pollutants

trapped below the high pressure cap cannot dissipate vertically and increase in concentration.

Along the west coast of the United States, and occasionally along the east coast, the warm seasons often produce a coastal or **marine inversion** (Figure 3.20). In

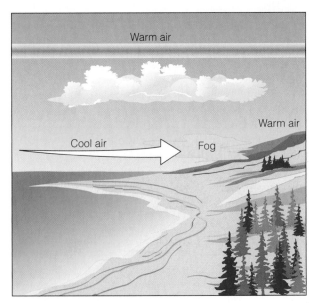

FIGURE 3.20 During a marine inversion, cool air from the ocean moves in beneath the heated layer above (After Schroeder and Buck 1970).

URBAN MICROCLIMATES

The density of urban structures and the activity of their occupants create urban microclimates. In an urban complex, materials such as brick, stone, asphalt, and concrete, which are used in buildings and in pavement, share a high capacity for absorbing and reradiating heat. They replace natural vegetation, which is characterized by low conductivity. Rain falling on these impervious surfaces rapidly drains away before evaporation can cool the air. In addition, metabolic heat from masses of humans combines with waste heat from buildings and from combustion engines, raising the temperature of the surrounding air. Industrial activities, power production, and vehicles pour great quantities of water vapor, gases, and particulate matter into the atmosphere.

These processes create a **heat dome,** a region of the atmosphere above urban areas in which reradiated heat is stored, around cities large and small. Within this dome, temperatures may be 6°C to 8°C higher than in the surrounding countryside. Heat domes have high temperature gradients. The highest temperatures occur in areas of highest population density and activity, whereas temperatures decline markedly toward the periphery of the city. Although they are detectable throughout the year, heat domes are most pronounced during the summer and early winter. They are most noticeable at night, when heat stored by pavements and buildings reradiates into the air.

During summer months, buildings and pavements of the inner city absorb and store considerably more heat than does vegetation of the countryside. Walls of tall buildings and narrow streets radiate heat toward each other instead of skyward. At night these structures slowly give off heat stored during the day.

In winter, solar radiation is considerably less because of the low angle of the Sun. Nevertheless, heat still accumulates from human and animal metabolism, home heating, transportation, industry, and power generation. In fact, heat from these sources is 2.5 times that from solar radiation. Warming the atmosphere directly or indirectly, urban heat makes winter milder in the city than in the countryside.

Throughout the year, urban areas are blanketed with particulate matter, carbon dioxide, and water vapor. This haze reduces the amount of solar radiation reaching the city, which may receive 10 to 20 percent less than the surrounding countryside. At the same time, the haze absorbs part of the heat radiating upward and reflects it back, warming both the air and the ground. The higher the concentration of pollutants, the more intense the heat dome.

Particulate matter has other microclimatic effects. Because of the city's low evaporation rate and lack of vegetation, relative humidity in urban areas is lower than in surrounding rural areas. However, the particulate matter also acts as condensation nuclei for water vapor in the air, producing fog and haze. Fog is much more frequent in urban areas than in the country, especially in winter.

---

this case cool, moist air from the ocean spreads over low-lying coastal land. The resulting layer of air typically thickens and advances inland during the night and early morning hours, before retreating to the sea or "burning off" around midday. Varying in depth from a hundred to several thousand meters, it is topped by warmer, drier air that traps pollutants in the lower layer. A good example is the mix of smoke and fog—referred to as smog—for which Los Angeles is famous.

3.9 Most Organisms Live in Microclimates

Most organisms live in local conditions that do not match the general climate profile of the larger region surrounding them. For example, today's weather report may state that the temperature is 28°C and the sky is clear. However, your weather forecaster is only painting a general picture. Actual conditions of specific environments will be quite different depending on whether they are underground versus on the surface, beneath vegetation or on exposed soil, on mountain slopes or at the seashore. Light, heat, moisture, and air movement all vary greatly from one part of the landscape to another, creating a wide range of localized climates. These **microclimates** define the conditions in which organisms live. (See Focus on Ecology 3.1: Urban Microclimates.)

On a sunny but chilly day in early spring, flies may be attracted to sap oozing from the stump of a maple tree. The flies are active on the stump in spite of the near-freezing air temperature because, during the day, the surface of the stump absorbs solar radiation, heating a thin layer of air above the surface. On a still day, the air heated by the tree stump remains close to the surface, and temperatures decrease sharply above and below this layer. A similar phenomenon occurs when the frozen surface of the ground absorbs solar radiation and thaws. On a sunny late winter day, you walk on muddy ground, even though the air about you is cold.

By altering soil temperatures, moisture, wind movement, and evaporation, vegetation moderates microclimates, especially areas near the ground. For example, areas shaded by plants have lower temperatures at ground level than do places exposed to the sun. On fair summer days in locations 25 mm (1 inch) above the ground, dense forest cover can reduce the daily range of temperatures by 7°C to 12°C below soil temperature in bare fields. Under the shelter of heavy grass and low plant cover, the air at ground level is completely calm. This calm is an outstanding feature of microclimates within dense vegetation at Earth's surface. It influences both temperature and humidity, creating a favorable environment for insects and other ground-dwelling animals.

Microclimates are a matter of scale, and two very different locations, such as a south-facing and a north-facing slope in the Northern Hemisphere, often exhibit microclimatic extremes. South-facing slopes receive the most solar energy, whereas north-facing slopes receive the least. At other slope positions, energy varies between these extremes, depending upon their compass direction.

Different exposure to solar radiation at south- and north-facing sites has a marked effect on the amount of moisture and heat present. Microclimate conditions range from warm, dry, variable conditions on the south-facing slope to cool, moist, more uniform conditions on the north-facing slope. Because

high temperatures and associated high rates of evaporation draw moisture from soil and plants, the evaporation rate at south-facing slopes is often 50 percent higher; the average temperature is higher; and soil moisture is lower. Conditions are driest on the top of south-facing slopes where air movement is greatest, and dampest at the bottom of north-facing slopes.

The same microclimatic conditions occur on a smaller scale on north-facing and south-facing slopes of large anthills, mounds of soil, dunes, and small ground ridges in otherwise flat terrain, as well as on the north-facing and south-facing sides of buildings, trees, and logs. The south-facing sides of buildings are always warmer and drier than the north-facing sides, a consideration for landscape planners, horticulturists, and gardeners. North sides of tree trunks are cooler and moister than south sides, a fact reflected in more vigorous growth of moss. In winter, temperature of the north-facing side of a tree may be below freezing while the south side, heated by the sun, is warm. This temperature difference may cause frost cracks in the bark as sap, thawed by day, freezes at night. Bark beetles and other wood-dwelling insects that seek cool, moist areas in which to lay their eggs prefer north-facing locations. Flowers on the south side of tree crowns often bloom sooner than those on the north side.

Microclimatic extremes also occur in depressions in the ground and on the concave surfaces of valleys where the air is protected from the wind. Heated by sunlight during the day and cooled by terrestrial vegetation at night, this air often becomes stagnant. As a result, these sheltered sites experience lower night-time temperatures (especially in winter), higher day-

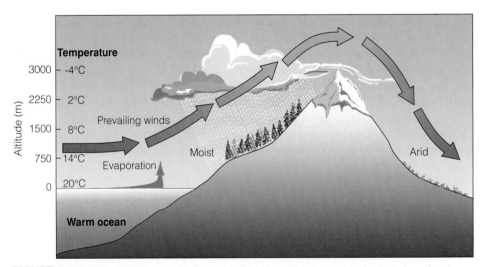

FIGURE 3.21 Formation of a rain shadow. Air is forced to go over a mountain. As it rises, the air mass cools and loses its moisture as precipitation on the windward side. The descending air, already dry, picks up moisture from the leeward side.

(a)

(b)

FIGURE 3.22 Rain shadow on the mountains of Maui, Hawaiian Islands. (a) The windward, east-facing slopes, intercepting the trade winds, are cloaked with wet forest. (b) Low–growing shrubby vegetation of the dry western side.

time temperatures (especially in summer), and higher relative humidity. If the temperature drops low enough, frost pockets form in these depressions. The microclimates of the frost pockets often display the same phenomenon, supporting different kinds of plant life than found on surrounding higher ground.

Mountainous topography influences local and regional microclimates by changing patterns of precipitation. Mountains intercept air flow. As an air mass reaches a mountain, it ascends, cools, becomes saturated with water vapor (because cold air holds less moisture than warm air), and releases much of its moisture at upper altitudes of the windward side. This phenomenon is called a **rain shadow** (Figure 3.21). As the now cool, dry air descends the leeward side, it warms again and gradually picks up moisture. As a result, the windward side of a mountain supports denser, more vigorous vegetation and different species of plants and associated animals than does the leeward side, where in some areas dry, desertlike

conditions exist. Thus in North America, the westerly winds that blow over the Sierra Nevada and Rocky Mountains, dropping their moisture on west-facing slopes, support vigorous forest growth. By contrast, the eastern slopes exhibit semidesert or desert conditions.

Some of the most pronounced effects of this same phenomenon occur in the Hawaiian islands. There plant cover ranges from scrubby vegetation on the leeward side of an island to moist, forested slopes on the windward side (Figure 3.22). In the central Appalachian Mountains, within several square kilometers, western, windward slopes support mesic forest vegetation dominated by yellow-poplar (*Liriodendron tulipifera*), red oak (*Quercus rubra*), white oak (*Q. alba*), and black cherry (*Prunus serotina*); whereas dry eastern, leeward slopes are dominated by scarlet oak (*Q. coccinea*), black oak (*Q. velutina*), and chestnut oak (*Q. prinus*).

CHAPTER REVIEW

Summary

Interception of Solar Radiation (3.1–3.2) Earth intercepts solar energy in the form of shortwave radiation, which relatively easily passes through the atmosphere, and emits much of it back as long-wave radiation. However, energy of longer wavelengths cannot readily pass through the atmosphere and so is returned to Earth, producing the greenhouse effect (**3.1**). The amount of solar radiation intercepted varies markedly with latitude. Tropical regions near the equator receive the greatest amount of solar radiation, and high latitudes the least. Because Earth tilts on its axis, parts of Earth receive seasonal differences in solar radiation. These differences give rise to seasonal variations in temperature and rainfall. There is a global gradient in mean annual temperature; it is warmest in the tropics and declines toward the poles (**3.2**).

Dynamics of the Atmosphere (3.3–3.5) Heating and cooling, influenced by energy emitted from Earth's surface and by atmospheric pressure, cause air masses to rise and sink. This movement of air masses involves an adiabatic process in which heat is neither gained from nor lost to the outside (3.3). Vertical movements of air masses give rise to global patterns of atmospheric circulation. The spin of Earth on its axis deflects air and water currents to the right in the Northern Hemisphere and to the left in the Southern Hemisphere. Three cells of global air flow occur in each hemisphere (3.4). The global pattern of winds and the Coriolis effect cause major patterns of ocean currents. Each ocean is dominated by great circular water motions, or gyres. These gyres move clockwise in the Northern Hemisphere and counterclockwise in the Southern Hemisphere (3.5).

Atmospheric Moisture and Precipitation (3.6–3.7) Atmospheric moisture is expressed in terms of relative humidity. The maximum amount of moisture the air can hold at any given temperature is called the saturation vapor pressure, which increases with temperature. Relative humidity is the amount of water in the air expressed as a percentage of the maximum amount the air could hold at a given temperature (3.6). Wind, temperature, and ocean currents produce global patterns of precipitation. They account for regions of high precipitation in the tropics and belts of dry climate in the horse latitudes (~30°N and S) (3.7).

Inversions (3.8) Under certain conditions, the temperature of air masses increases with height rather than decreasing. Such an air mass is very stable, creating a temperature inversion that can trap atmospheric pollutants and hold them close to the ground. Inversions break up when air close to the ground gains heat, causing it to circulate and rise through the inversion, or when a new air mass moves into the area.

Microclimates (3.9) The actual climatic conditions under which organisms live vary considerably within one climate. These local variations or microclimates reflect topography, vegetative cover, exposure, and other factors on every scale. In mountainous areas the windward sides receive more precipitation than the leeward sides. Angles of solar radiation cause marked differences between north-facing and south-facing slopes, whether on mountains, sand dunes, or ant mounds.

Study Questions

1. What is the fate of the Sun's energy when it reaches Earth's atmosphere?

2. Why does the equator receive more solar energy than the polar regions?

3. How does the tilt of Earth's axis influence the seasons?

4. How does Earth maintain a heat balance?

5. What is the Coriolis effect? How does it affect atmospheric circulation?

6. What causes ocean currents? What are gyres?

7. What are the relationships among saturation vapor pressure, relative humidity, and dew point?

8. What is a microclimate, and how is it created?

9. Why and how does a north-facing slope differ from a south-facing slope?

CHAPTER 4

A rainstorm over the ocean—a part of the water cycle.

The Abiotic Environment

OBJECTIVES

On completion of this chapter, you should be able to:

- Describe the nature of light as it reaches Earth.
- Explain the fate of visible light in the plant canopy and in water.
- Discuss the relationship between daily and seasonal variation of light on rhythms of life.
- Explain how the structure of water affects its properties.
- Trace the water cycle.
- Explain how the movement of water affects the nature of aquatic environments.
- Discuss how temperature and oxygen vary with water depth.
- Discuss ways in which heat transfers between organisms and their environment.
- Distinguish between macronutrients and micronutrients.
- Discuss how acidity and salinity affect the environment of organisms.

The interaction between organisms and their environment includes both the **biotic** (living) and **abiotic** (nonliving), often referred to as the biological and physical environment respectively. The physical environments of organisms are characterized by light, temperature, moisture, and nutrients. Solar radiation directly and indirectly determines the nature of much of the physical environments. Solar radiation, as outlined in Chapter 3, involves both light energy and thermal energy. The visible portion of solar radiation, with its daily and seasonal periodicity, is the energy source of photosynthesis, the basic process supporting life on Earth. The infrared portion of solar radiation, the primary source of Earth's heat budget, influences the thermal environment in which organisms live. Solar energy drives the movement of water between Earth and atmosphere and its presence, distribution, behavior, and availability in the environment.

4.1 Solar Radiation Includes Visible Light

Of the total range of solar radiation reaching Earth's atmosphere (Chapter 3), the wavelengths of approximately 400 to 700 nm (a nanometer is one-billionth of a meter) make up visible light (Figure 4.1). Collectively, these wavelengths are also known as **photosynthetically active radiation (PAR)** because they include the wavelengths plants use in photosynthesis (Chapter 6). Wavelengths shorter than the visible range are ultraviolet or UV light. There are two types of ultraviolet light: UV-A with wavelengths from 315 nm to 380 nm, and UV-B with wavelengths from 280 to 315 nm. Radiation with wavelengths longer than the visible range is infrared. Near infrared has wavelengths of approximately 740 to 4000 nm, and far infrared or thermal radiation from 4000 to 100,000 nm.

Light that reaches the earth's surface is not quite the same light that arrives at the top of the earth's atmosphere (Figure 4.2). The ozone layer in the upper atmosphere (stratosphere) absorbs nearly all wavelengths, but especially the violets and blues of visible light. Molecules of atmospheric gases scatter these shorter wavelengths, giving a bluish color to the sky and causing Earth to shine in space. Water vapor scatters all wavelengths, giving clouds their white appearance. Dust scatters long wavelengths to produce reds and yellows in the sky. Because of the scattering, part of solar radiation reaches Earth as diffuse light from the sky, known as skylight. The relative contributions of direct and diffuse light change not only over the course of a single day but globally as a function of latitude. Diffuse light represents a larger proportion of total intercepted light reaching Earth's surface at higher latitudes as a result of low sun angles (see Chapter 3, Figure 3.3).

Light intercepted by Earth is reflected, absorbed, or transmitted through objects at the surface. Of greatest ecological interest is light reaching vegetation. Leaves reflect about 6 to 12 percent of photosynthetically active radiation, whereas they reflect about 70 percent of infrared light and only about 3 percent of ultraviolet light striking them directly. The degree of reflection varies with the nature of the leaf surface. Because leaves preferentially absorb violet, blue, and red light and strongly reflect green light, they appear green. The light not reflected or absorbed is transmitted through the leaf. How much light is transmitted depends upon the thickness and structure of the leaf. A leaf may transmit up to 40 percent of the light it receives, but 10 to 20 percent is more usual. Transmitted light is primarily green and far red. At ground level in a dense forest, even green light may be extinguished.

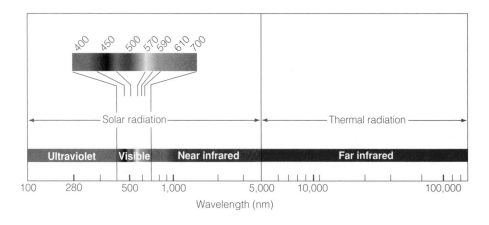

FIGURE 4.1 A portion of the electromagnetic spectrum, separated into solar and thermal radiation. Wavelengths on a spectrum are bunched irregularly. Ultraviolet, visible, and infrared light waves represent only a small portion of the spectrum. To the left of ultraviolet radiation are X rays and gamma rays (not shown). (After Halverson and Smith 1979.)

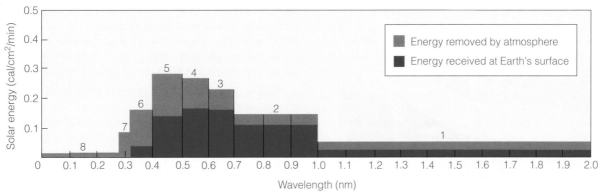

FIGURE 4.2 Energy in the solar spectrum before and after depletion by the atmosphere, given a solar altitude of 30°. Figures above the bars indicate (1) near infrared wavelengths over 1 micron; (2) near infrared, 0.7–1.0 micron; (3–5) visible light: (3) red; (4) green, yellow, and orange; (5) violet and blue; (6–8) ultraviolet. Note the strong reduction in ultraviolet. Nearly all the ultraviolet wavelengths are absorbed by the ozone. The region of peak energy shifts toward the red end of the spectrum. Visible light in the blue wavelength is scattered rather than absorbed, producing the blue light of the sky. (From Reifsnyder and Lull 1965.)

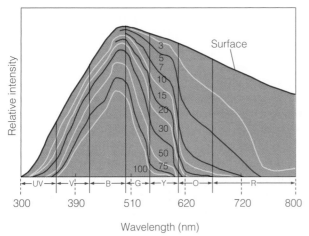

FIGURE 4.3 The spectral distribution of solar energy at Earth's surface and after it has been modified by passage through varying depths of pure water, measured in meters. Note how rapidly red wavelengths are attenuated. At approximately 10 meters, red light is depleted; but at 100 meters, blue wavelengths still retain nearly one-half their intensity. (Clark 1939.)

4.2 Plant Cover Intercepts Light

When you walk beneath a stand of trees in summer, you notice a decrease in light (Figure 4.4a). If you could examine the lowest layer in a grassland or an old field, you would observe much the same effect (Figure 4.4b). The amount of light that does penetrate a stand of vegetation to reach the ground varies with both the quantity and position of the leaves.

The amount of light at any depth in the canopy is a function of the number of leaves above. As you move down through the canopy, the number of leaves above you increases, so the amount of light decreases. However, because leaves vary in size and shape, the number of leaves is not the best measure of quantity.

The quantity of leaves, or foliage density, is generally expressed in terms of leaf area. Because most leaves are flat, the leaf area is the surface area of one or both sides of the leaf. When the leaves are not flat, the entire surface area is sometimes measured. To quantify the changes in light environment with increasing area of leaves, we need to define the area of leaves per unit ground area (m^2 leaf area/m^2 ground area). This measure is the **leaf area index,** or LAI for short. A leaf area index of 3 (LAI = 3) means that there are 3 square meters of leaf area over each 1 square meter of ground area. The greater the leaf area index above any surface, the lower the quantity of light reaching that surface. Figure 4.5 shows this relationship. As you move from the top of the canopy to the ground in a forest stand, the cumulative leaf area and LAI increase. Correspondingly, light

Light that enters water is absorbed rapidly (Figure 4.3). Only about 40 percent reaches a depth of 1 meter into clear lake water. Moreover, water absorbs some wavelengths more than others. First to be absorbed are visible red light and infrared radiation in wavelengths greater than 750 nm. This absorption reduces solar energy by one-half. In clear water, yellow disappears next, followed by green and violet, leaving only blue wavelengths to penetrate deeper water. A fraction of blue light is lost with increasing depth. In the clearest of seawater, only about 10 percent of blue light reaches to more than 100 meters in depth.

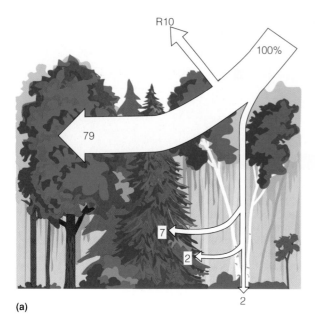

(a)

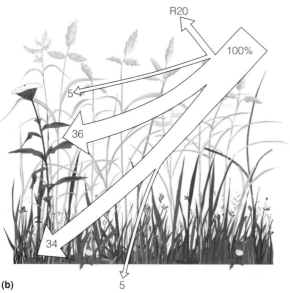

(b)

FIGURE 4.4 Attenuation of light by plant canopies. Values represent the percentage of incoming solar radiation at the top of the canopy (100 percent). (a) A boreal mixed forest reflects (R) about 10 percent of the incident photosynthetically active radiation (PAR) from the upper crown, and it absorbs most of the remainder within the crown. (b) A meadow reflects 20 percent of the photosynthetically active radiation from the upper surface. The middle and lower regions, where the leaves are most dense, absorb most of the rest. Only 2 to 5 percent of PAR reaches the ground. (Adapted from Larcher 1980.)

decreases. The general relationship between available light and leaf area index is described by Beer's law (see Quantifying Ecology 4.1: Beer's Law and the Attenuation of Light).

The arrangement of leaves on plants influences the attenuation of light with increasing leaf area.

Densely grown plants with vertically angled leaves and a high LAI intercept more total light than those with horizontal leaves (Figure 4.6) because horizontal leaves shade each other. Thus, leaf angle influences the vertical distribution of light through the canopy as well as the total amount of light absorbed and reflected.

Although light decreases downward through the forest canopy, some direct sunlight does penetrate openings in the crown and reaches the forest floor as sunflecks, changing small patterns of sunlight on the ground. Sunflecks can account for 70 to 80 percent of solar energy reaching the forest floor.

Only about 1 to 5 percent of the light that strikes the canopy of a typical temperate deciduous forest (LAI 3–5) in the summer reaches the forest floor. More light travels through a stand of pine trees (LAI 2–4)—about 10 to 15 percent. In a tropical rain forest (LAI 6–10), only 0.25 to 2 percent gets through. Relatively open woodlands, with trees such as birch and oaks, allow light to filter through. There, light attenuates gradually throughout the canopy. Likewise, in grassland, where the top of the canopy is relatively open, the middle and lower layers intercept the most light (see Figure 4.4b).

In many environments, seasonal changes strongly influence leaf area. For example, in the temperate regions of the world many forest tree species are deciduous, shedding their leaves during the winter months. In these cases, the amount of light that penetrates a stand of vegetation varies with the season (Figure 4.7). In early spring in temperate regions, when leaves are just expanding, 20 to 50 percent of the incoming light may reach the forest floor.

Growth of phytoplankton and rooted aquatic plants decrease light in aquatic environments. Heavy growth of phytoplankton and cyanobacteria can shut out light to deeper waters just as effectively, if not more so, than the dense canopy of a deciduous or tropical rain forest. Other environmental conditions, however, are also involved. Freshwater lakes, streams, and ponds and coastal waters have a certain degree of turbidity, which greatly affects light penetration. Because of the high level of yellow substances such as clay colloids and fine dead organic material washed into water from terrestrial ecosystems, blue wavelengths are the most strongly attenuated and are removed at very shallow levels. Green penetrates most deeply, and where the concentration of yellow material is high, red wavelengths may penetrate as far as the green. In very yellow water, red is the last wavelength to be extinguished. Waters with organic stains, such as bog lakes and ponds, appear dark; and nutrient-poor lakes and ponds, low in phytoplankton, are clear and bluish. Waters supporting a dense growth of phytoplankton take on a decidedly greenish appearance.

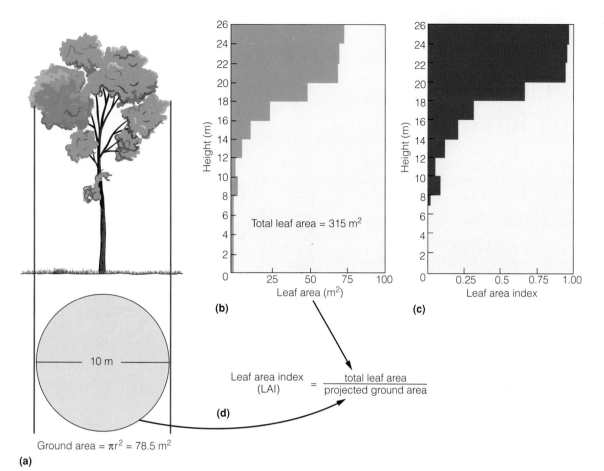

(b)

Total leaf area = 315 m²

(c)

(d)

$$\text{Leaf area index (LAI)} = \frac{\text{total leaf area}}{\text{projected ground area}}$$

10 m

Ground area = πr² = 78.5 m²

(a)

FIGURE 4.5 The concept of the leaf area index. (a) A tree with a 10-meter-wide crown projects the same size circle on the ground. (b) The foliage density of the crown at various heights above the ground. (c) The contributions of layers in the crown to the leaf area index. (d) Calculation of LAI. The total leaf area is 315 m². The projected ground area is 78.5 m². The LAI is 4.

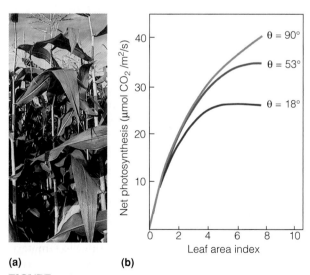

(a)   **(b)**

FIGURE 4.6 (a) The sharply angled leaves of corn plants with their high LAI allow dense planting. (b) The relationship of photosynthetic capacity of a barley plant to leaf angle (θ) from the ground surface and to leaf area index. (From Leopold and Kriedemann 1975.)

4.3 Daily and Seasonal Patterns of Light Govern Life's Activities

Birdsong signals the arrival of dawn. Butterflies, dragonflies, and bees warm their wings; hawks begin to circle; and chipmunks and tree squirrels become active. Even the pattern of human activity quickens with daylight. At dusk, daytime animals retire, waterlilies fold, moonflowers open, and animals of the night appear. Foxes, raccoons, flying squirrels, owls, and luna moths take over niches others occupy during the day. Human activity also changes as evening begins.

As seasons progress, daylength changes, and activities shift. Spring brings migrant birds and initiates the reproductive cycles of many plants and animals.

Beer's Law and the Attenuation of Light

Equations can make a complicated picture easy to grasp. For instance, to describe the reduction, or attenuation, of light through a stand of plants we can use Beer's law:

$$AL_i = e^{-LAI_i * k}$$

The subscript i refers to the vertical height of the canopy. For example, a value of $i = 5$ refers to a height of 5 meters above the ground. The value AL_i is the light reaching any vertical position i in the stand, expressed as a proportion of the light reaching the top of the plants (a value from 0 to 1.0); e is the natural logarithm; LAI_i is the leaf area index above height i; k is the light extinction coefficient. The light extinction coefficient is a measure of the degree to which leaves absorb and reflect light.

For the yellow-poplar stand in Figure 4.7, we can construct a curve describing the available light at any height in the canopy. In Figure A, the light extinction coefficient has a value of $k = 0.6$. We label vertical positions from the top of the canopy to ground level on the curve. Knowing the amount of leaves (LAI) above any position in the canopy, we can use the equation to calculate the amount of light there.

The light levels and rates of light-limited photosynthesis for each of the vertical canopy positions are shown in the curve in Figure B. Light levels are expressed as a proportion of values for fully exposed leaves at the top of the canopy. As you move from the top of the canopy downward, the amount of light reaching the leaves and the corresponding rate of photosynthesis decline.

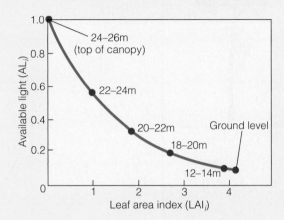

Figure A

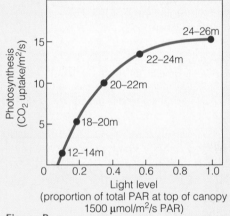

Figure B

In fall the deciduous trees of temperate regions become dormant; insects and herbaceous plants disappear; summer birds return south; and winter birds arrive.

These rhythms are driven by the daily rotation of Earth on its axis and its 365-day revolution about the Sun. Through time, life has become attuned to the daily and seasonal changes in the environment.

Day-to-day activities conform to a 24-hour cycle. The correlation of the onset of activity with light and dark cycles suggests that light has a direct or indirect regulatory effect. This innate rhythm of activity and inactivity covering approximately 24 hours is characteristic of all living organisms except bacteria.

Because these rhythms approximate, but seldom match, the periods of Earth's rotation, they are called **circadian rhythms** (from the Latin *circa*, "about," and *dies*, "day"). The period of the circadian rhythm, the number of hours from the beginning of activity one day to the beginning of activity on the next, is its free-running cycle. In other words, the rhythm of activity exhibits a self-sustained oscillation even under constant conditions of either dark or light.

Thus two daily periodicities—the external rhythm of 24 hours and the internal circadian rhythm of approximately 24 hours—influence the activities of plants and animals. If the two rhythms are to be in phase, some external cue or time-setter must adjust

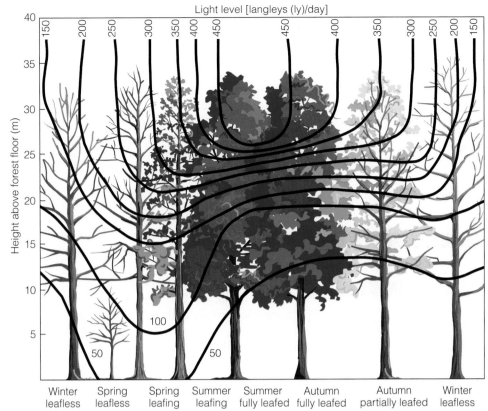

FIGURE 4.7 Light levels within and above a yellow-poplar (*Liriodendron tulipifera*) stand over a year. The greatest intensity of solar radiation occurs in summer, but the canopy attenuates most of the light, so little reaches the forest floor. Most illumination reaches the forest floor in spring, when trees are still leafless. The forest receives the least radiation in winter with its lower solar elevations and shorter daylengths. As a result, the amount of solar radiation reaching the forest floor is little more than that of midsummer. (Hutchinson and Matt 1977.)

the internal rhythm to the environmental rhythm. The most obvious time-setters are temperature, light, and moisture. Of the three, the master time-setter in the temperate zone is light. It brings the circadian rhythm of organisms into phase with the 24-hour photoperiod of their external environment, an effect called **entrainment.**

In the middle and upper latitudes of the Northern and Southern Hemispheres, the daily periods of light and dark lengthen and shorten with the seasons. The activities of plants and animals are geared to the changing seasonal rhythms of night and day.

The signal for these responses is critical daylength (Figure 4.8). When the duration of light (or dark) reaches a certain portion of the 24-hour day, it inhibits or promotes a photoperiodic response. Critical daylength varies among organisms, but it usually falls somewhere between 10 and 14 hours. Through the year, plants and animals compare critical daylength with the actual length of day or night and respond appropriately. Some organisms can be classified as **day-neutral**; they are not controlled by

daylength, but by some other influence such as rainfall or temperature. Others are short-day or long-day organisms. **Short-day organisms** are those whose reproductive or other seasonal activity is stimulated by daylengths shorter than their critical daylength. **Long-day organisms** are those whose seasonal responses, such as flowering and reproduction, are stimulated by daylengths longer than the critical daylength.

4.4 The Structure of Water Is Based on Hydrogen Bonds

Water is the essential substance of life. Covering some 75 percent of Earth's surface, it is the dominant component of all living organisms. Because of the physical arrangement of its hydrogen atoms and hydrogen bonds, water is a unique substance. A molecule of

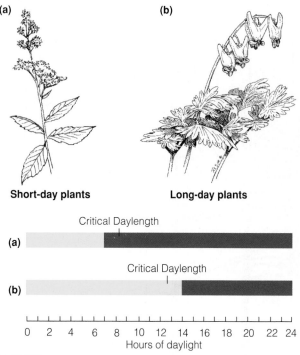

(a) Short-day plants   **(b)** Long-day plants

Critical Daylength

(a)

Critical Daylength

(b)

0 2 4 6 8 10 12 14 16 18 20 22 24
Hours of daylight

FIGURE 4.8 The influence of photoperiod (daylength) on flowering in short-day and long-day plants. Each species has a critical daylength, the duration of light (or dark) during a 24-hour day that promotes a response. (a) Flowering in short-day plants is stimulated by daylengths shorter than their critical daylength. (b) Long-day plants flower when stimulated by daylengths greater than their critical daylength.

water consists of one atom of oxygen and two smaller atoms of hydrogen bonded together covalently (Figure 4.9a). Covalence is the sharing of electrons. Each hydrogen proton shares its single electron with oxygen, leaving unbonded two pairs of oxygen electrons (Figure 4.9b). The repulsive force between these two pairs and between them and electrons in the O-H bonds pushes the bonds toward each other. Instead of a linear arrangement of the atoms of H-O-H, the arrangement is V-shaped (Figure 4.9a) with an angle of 105°. The shared hydrogen atoms are closer to the oxygen atom than they are to each other. If the molecules were linear, then the two bonds' polarities—the pull they exert—would cancel each other out. Because of the V-shaped arrangement, the side of the molecule on which the H atoms are located has a positive charge and the opposite side has a negative charge, polarizing the water molecule (Figure 4.9c).

Because of its polarity, each water molecule becomes weakly bonded with its neighboring molecules (Figure 4.9d). The H or positive end of one molecule attracts the negative or opposite end of the other. The 105° angle of association between the

hydrogen atoms encourages an open tetrahedral-like arrangement of water molecules. This situation, in which hydrogen atoms act as connecting links between water molecules, is hydrogen bonding. The simultaneous bonding of an H atom to two different water molecules accounts for the lattice arrangement of water.

In ice the lattice is complete. Each oxygen atom is hydrogen-bonded to two hydrogen atoms. Each hydrogen atom is covalently bonded to one oxygen atom and is hydrogen-bonded to another oxygen atom (Figure 4.9d). Ultimately, each oxygen atom connects to four other oxygen atoms by means of hydrogen atoms. One such unit built upon the other gives rise to a lattice with large open spaces (Figure 4.9e). Water molecules so structured occupy more space than they would in liquid form. As a result, water expands upon freezing, and ice floats.

As ice melts, the hydrogen bonds break and the latticework partially collapses. The volume occupied by the water molecules decreases and the density increases, until water achieves its greatest density at 3.98°C. At this temperature, contraction of the molecules, brought about by the partial collapse of the lattice structures, balances normal thermal expansion of warming molecules. As water heats, more hydrogen bonds break, converting the water to a liquid state, which is a mixture of individual and aggregated molecules (Figure 4.9f). Further heating results in a complete breakdown of the hydrogen bonds, separating aggregates of water molecules into individuals. At this point water enters the gaseous state, as water vapor.

Seawater behaves somewhat differently. Seawater is a solution of salts dominated by sodium chloride. Although seawater is defined as containing 34.5 grams of salt per 1000 grams of water (34.5 **practical salinity units** or **psu**), it can vary in salinity. For this reason, saltwater has no definitive freezing point. Seawater begins to freeze at about 2°C. As pure water freezes out, the remaining unfrozen water becomes higher in salinity; the density of seawater increases; and its freezing point lowers. Finally, a solid block of ice crystals and salt forms.

4.5 Water Has Important Physical Properties

Water has a number of unique properties related to its hydrogen bonds. One property is high **specific heat**—that is, the number of calories necessary to raise 1 gram of water 1°C. The specific heat of water is defined as a value of 1, and other substances are given a value relative to water. Water can store

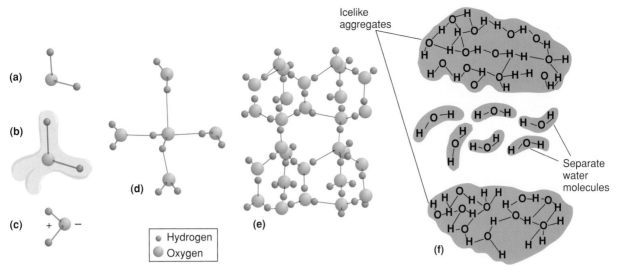

FIGURE 4.9 The structure of water. (a) An isolated water molecule, showing the angular arrangement of the hydrogen atoms. (b) Lone pairs of electrons in a water molecule. (c) Polarity of water. (d) Hydrogen bonds to one molecule of water in ice. (e) The open lattice structure of ice. (f) The structure of liquid water.

Legend:
- Hydrogen
- Oxygen

Labels: Icelike aggregates; Separate water molecules

tremendous quantities of heat energy with a small rise in temperature. It is exceeded in this capacity only by ammonium, liquid hydrogen, and lithium.

Great quantities of heat must be absorbed before the temperature of natural waters, such as ponds, lakes, and seas, rises just 1°C. These bodies of water warm up slowly in spring and cool off just as slowly in the fall. This behavior prevents the wide seasonal fluctuations in the temperature of aquatic habitats that are characteristic of air temperatures, and it moderates the temperatures of local and worldwide environments.

Because of the high specific heat of water, large quantities of heat energy must be removed before water can change from a liquid to a solid, and must be absorbed before ice can convert to a liquid. Collectively, the energy released or absorbed in the transformation of water from one state to another is called **latent heat**. It takes approximately 80 calories of heat to convert 1 gram of ice at 1°C to a liquid state. The same amount of heat would raise 80 grams of water 1°C.

Evaporation, the process by which a liquid changes into a gas, occurs at the interface between air and water at all ranges of temperature. Here again, considerable amounts of heat are involved: it takes 536 calories to overcome the attraction between molecules and convert 1 gram of water at 100°C into vapor. That is as much heat as is needed to raise 536 grams of water 1°C. When evaporation occurs, the source of thermal energy may be the Sun, the water itself, or objects in and around the water.

Viscosity is the resistance of a liquid to flow. Because of the energy in hydrogen bonds, the viscosity of water is high. Imagine liquid flowing through a glass tube. The liquid behaves as if it consists of a series of parallel concentric layers flowing over one another. The rate of flow is greatest at the center; because of internal friction between layers, the flow decreases toward the sides of the tube.

Viscosity is the source of frictional resistance to objects moving through water. The frictional resistance of water is 100 times greater than that of air. A mucous coating and streamlined body shape help fish reduce this frictional resistance. Replacement of water in the space left behind by the moving animal adds additional drag on the body. An animal streamlined in reverse, with a short, rounded front and a rapidly tapering body, meets the least resistance in the water. The acme of such streamlining is the sperm whale (*Physeter catodon*).

Within all substances, similar molecules are attracted to one another. Water is no exception. Water molecules surround each other below the surface. The forces of attraction are the same on all sides. At the water's surface, there is a different set of conditions. Below the surface, molecules of water are strongly attracted to one another. Above is the much weaker attraction between water molecules and air. Therefore, molecules on the surface are drawn downward, resulting in a surface that is taut like an inflated balloon. This condition, called **surface tension,** is important in the lives of aquatic organisms.

The surface of water is able to support small objects and animals, such as the water strider (Gerridae) and water spiders (*Dolomedes* spp.) that run across the pond's surface. To other small organisms, surface tension is a barrier when they wish to penetrate the water below or escape into the air above. For some the surface tension is too great to break; for others it is a trap to avoid while skimming the surface to feed or to lay eggs. If caught in the surface tension, a small insect may flounder on the surface. The nymphs of mayflies (Ephemeroptera) and caddisflies (Trichoptera) that live in the water and transform into winged adults find surface tension a handicap in their efforts to emerge from the water. Slowed down at the surface, these insects become easy prey for fish.

Surface tension is associated with capillary action, or capillarity—the rise and fall of liquids within narrow tubes. Capillarity affects the movement of water in soil and the transport of water to all parts of a plant (Chapter 7).

4.6 Transfer of Water Between the Environment and Organisms Involves Osmosis

For water to move from the external environment into an organism, it must pass through the cell membranes and cell walls. Some membranes are permeable; they do not impede the movement of substances. Other membranes are selectively permeable; they allow some substances to pass through them, but not others. They are fairly permeable to water and are more or less permeable to various other substances.

The general tendency of molecules is to move from a region of high concentration to one of low concentration. This movement or **diffusion** accounts for the spread of a solute (dissolved substance) throughout a solvent (the medium that dissolves the solute). To understand the movement of water and solutes into the living cell, consider this example. Suppose we enclose a solute such as salt (sodium chloride) in high concentration (and water in low concentration) in a funnel sealed with a semipermeable membrane and lower it into a beaker of distilled water. The membrane is permeable to water but not to salt. The volume of the fluid within the funnel will increase and move up the tube as water moves across the membrane into the solution by a process called **osmosis** (Figure 4.10). Water continues to move across the membrane until the **osmotic pressure** of the solute—decreasing as the solute becomes more diluted by the pure water—is balanced by the physical pressure the fluid exerts on the tube.

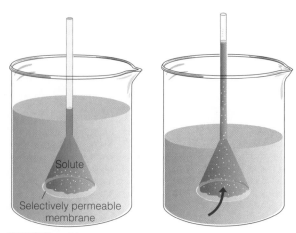

FIGURE 4.10 Demonstration of osmotic potential. Fluid within the funnel increases in volume and moves up the tube as water passes through the membrane into the solution by osmosis. Water continues to cross the membrane until the osmotic pressure of the solute, decreasing with dilution, is balanced by the gravitational pressure exerted by the fluid in the tube.

Osmotic pressure accounts for the internal pressure (called turgor) plant cells achieve when the water supply is adequate. As plants lose water, the concentration of water molecules in the cells decreases, so water, when available, moves from the soil solution into the plant. The tendency of solutes in a solution to cause water molecules to move from areas of high to low concentration is called **osmotic potential.** The osmotic potential of a solution depends upon its concentration. The higher the concentration of solutes, the lower its osmotic potential and the greater its tendency to gain water.

Osmosis and osmotic potential are involved in maintaining fluid balance not only in plants but also in all living things from single-cell protozoans to vertebrates. They play a particularly important role in freshwater and marine life with different internal solute concentrations from the water around them.

4.7 Water Cycles Between Earth and the Atmosphere

Water is the medium by which elements and other materials make their never-ending odyssey through the ecosystem. Without the cycling of water, ecosystems could not function, and life could not persist.

Solar energy, heating Earth's atmosphere (see Chapter 3) and evaporating water, is the driving force behind the water cycle (Figure 4.11). Water vapor circulating in the atmosphere, eventually falls in some form of **precipitation.** Some of the water falls

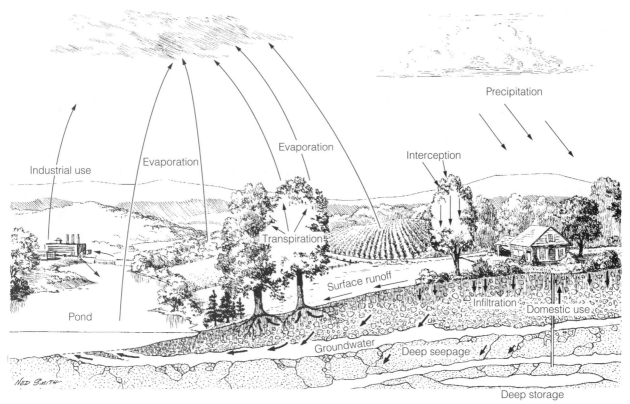

Precipitation

Evaporation

Evaporation

Interception

Industrial use

Transpiration

Surface runoff

Infiltration

Domestic use

Pond

Groundwater

Deep seepage

NED SMITH

Deep storage

FIGURE 4.11 The water cycle on a local scale, showing the major pathways of water movement.

directly on the soil and bodies of water. Some is intercepted by vegetation, dead organic matter on the ground, and urban structures and streets.

Because of **interception,** which can be considerable, various amounts of water never infiltrate the ground but evaporate directly back to the atmosphere. In urban areas a great portion of rain falls on roofs, sidewalks, roads, and other paved areas, which are impervious to water. Water runs down gutters and drainage ditches, to be directed off to rivers.

Precipitation that reaches the soil moves into the ground by **infiltration.** The rate of infiltration depends on the type of soil (Chapter 5), slope, vegetation, and characteristics of the precipitation. During heavy rains when the soil is saturated, excess water flows across the surface of the ground as surface runoff or overland flow. It concentrates into depressions and gullies, and the flow changes from sheet to channelized flow, a process you can observe on city streets as water moves across the pavement into gutters. Because of low infiltration, runoff from urban areas might be as much as 85 percent of the precipitation.

Some water entering the soil seeps down to an impervious layer of clay or rock to collect as **groundwater** (see Figure 4.11). From here water finds its way into springs, streams, and eventually rivers and seas. Humans draw a great portion of groundwater and surface water for domestic and industrial pur-

poses and then discharge it into streams and rivers or the atmosphere to reenter the water cycle.

Water remaining on the surface of the ground, in the upper layers of the soil, and on the surface of vegetation, as well as water in the surface layers of streams, lakes, and oceans, returns to the atmosphere by evaporation. The rate of evaporation is governed by how much moisture the air contains (see Section 3.6). Additional water losses from the soil take place through plants. Plants take in water from the soil through their roots and lose it through their leaves and other organs in a process called **transpiration.** Transpiration is the evaporation of water from internal surfaces of leaves, stems, and other living parts and its diffusion from the plant (Chapter 6). The total flux of evaporating water—from the surfaces of the ground and vegetation—is called **evapotranspiration.**

4.8 Water Movements Shape Freshwater and Marine Environments

The movement of water—currents in streams and waves in open bodies of water and breaking on their shores—determines the nature of many aquatic

(a)

(b)

FIGURE 4.12 (a) A fast mountain stream. The gradient is steep and the bottom is largely bedrock. (b) A slow stream meanders through a growth of willows.

environments. The velocity of a current molds the character and structure of a stream. The shape and steepness of the stream channel, its width, and depth, the roughness of the bottom, and the intensity of rainfall and rapidity of snowmelt affect velocity. Fast streams are those whose velocity of flow is 50 cm per second or higher. At this velocity, the current will remove all particles less than 5 mm in diameter and will leave behind a stony bottom. High water increases the velocity; it moves bottom stones and rubble, scours the streambed, and cuts new banks and channels. As the gradient decreases and the width, depth, and volume of water increase, silt and decaying organic matter accumulate on the bottom. Thus the character of a stream changes from fast water to slow (Figure 4.12).

Wind generates waves on large lakes and open seas. The frictional drag of the wind on the surface of smooth water ripples it. As the wind continues to blow, it applies more pressure to the steep side of the ripple, and wave size begins to grow. As the wind becomes stronger, short, choppy waves of all sizes appear; and as they absorb more energy, they continue to grow. When the waves reach a point at which the energy supplied by the wind is equal to the energy lost by the breaking waves, they become whitecaps. Up to a certain point, the stronger the wind, the higher the waves.

The waves that break on a beach are not composed of water driven in from distant seas. Each particle of water remains largely in the same place and follows an elliptical orbit with the passage of the wave form. As a wave moves forward, it loses energy to the waves behind and disappears, its place taken by another.

As the waves approach land, they advance into increasingly shallow water. The height of each wave rises until the wave front grows too steep and topples over. As the waves break on shore, they dissipate their energy, pounding rocky shores or tearing away sandy beaches at one point and building up new beaches elsewhere (Figure 4.13).

In coastal regions, winds blowing parallel to the coast cause surface waters to be blown offshore. Water moving upward from the deep, a process known as **upwelling,** replaces this surface water (see Chapter 21).

FIGURE 4.13 Waves breaking on a rocky shore.

4.9 Tides Dominate the Marine Environment

Tides profoundly influence the rhythm of life on ocean shores. Tides come about from the gravitational pulls of the Sun and the Moon. Each causes two bulges in the waters of the oceans. The two caused by the Moon occur at the same time on opposite sides of Earth on an imaginary line extending from the Moon through the center of Earth. The tidal bulge on the Moon side is due to gravitational attraction; the bulge on the opposite side occurs because the gravitational force there is less than at the center of Earth. As Earth rotates eastward on its axis, the tides advance westward. Thus, Earth will in the course of one daily rotation pass through two of the lunar tidal bulges, or high tides, and two of the lows, or low tides, at right angles to the high tides. Since the Moon revolves in a $29\frac{1}{2}$-day orbit around Earth, the average period between successive high tides is approximately 12 hours 25 minutes.

The Sun also causes two tides on opposite sides of Earth, and these tides have a relation to the Sun like that of the lunar tides to the Moon. Because the gravitational pull of the Sun is less than that of the Moon, solar tides are partially masked by lunar tides except for two times during the month—when the Moon is full and when it is new. At these times, Earth, Moon, and Sun are nearly in line, and the gravitational pulls of the Sun and the Moon are additive. This combination causes the high tides of those periods to be exceptionally large, with maximum rise and fall. These are the fortnightly spring tides, a name derived from the Saxon *sprungen,* which refers to the brimming fullness and active movement of the water. When the Moon is at either quarter, its pull is at right angles to the pull of the Sun, and the two forces interfere with each other. At this time, the differences between high and low tide are exceptionally small. These are the neap tides, from an old Scandinavian word meaning "barely enough."

Tides are not entirely regular, nor are they the same all over Earth. They vary from day to day in the same place, following the waxing and waning of the Moon. They may act differently in several localities within the same general area. In the Atlantic, semidaily tides are the rule. In the Gulf of Mexico and the Aleutians of Alaska, the alternate highs and lows more or less efface each other, and flood and ebb follow one another at about 24-hour intervals to produce one daily tide. Mixed tides in which successive or low tides are of significantly different heights through the cycle are common in the Pacific and Indian oceans. These tides are combinations of daily and semidaily tides in which one partially cancels out the other.

Local tides around the world are inconsistent for many reasons. These reasons include variations in the gravitational pull of the Moon and the Sun due to the elliptical orbit of Earth, the angle of the Moon in relation to the axis of Earth, onshore and offshore winds, the depth of water, the contour of the shore, and wave action.

4.10 All Organisms Live in a Thermal Environment

All organisms live in a thermal environment characterized by heat and temperature. There is a difference between the two. **Temperature** is a measure of the average speed or kinetic energy of the atoms and molecules in a substance. Higher temperatures correspond to faster average speeds. It is the degree of hotness or coldness of a substance as measured by a thermometer. In contrast, **heat** is energy in the process of being transferred from one object to another because of the temperature difference between the objects. Heat is transferred from warmer objects (high temperature) to cooler objects (low temperature). After it is transferred from one object to another, heat is once again stored as kinetic energy, raising the temperature of the object.

The environmental temperatures (air, land, and water) experienced by most organisms result, directly or indirectly, from solar radiation. The amount of solar radiation reaching any point on Earth at any time varies with the time of year, time of day, topographic position (slope and aspect), cloud cover, and other factors. Seasonal fluctuations can be extreme from one location to another and are related to the season changes in the amount of intercepted solar radiation, as discussed in Chapter 3 (Section 3.2). In Fargo, North Dakota (latitude 47°N), for example, where the annual mean temperature is 5°C, air temperatures fluctuate from a low of –43°C in winter to 45°C in summer. In Manaus, Brazil, located just south of the equator (latitude 3°S), mean annual temperature is 27°C. Due to the lack of any strong seasonality in solar radiation, mean daily air temperatures throughout the year only fluctuate between 24°C and 38°C.

Environmental temperatures in any area differ between sunlight and shade and between daylight and dark. In temperate regions, surface temperatures of soil may be 30°C higher in the sunlight than in the shade. Daytime temperatures are often 17°C higher than nighttime temperatures; on deserts this spread may be as high as 40°C. Temperatures on tidal flats may rise to 38°C when exposed to direct sunlight and drop to 10°C within a few hours when the flats are covered by water.

4.11 Aquatic Environments Experience Seasonal Shifts in Temperature

Life in lakes and larger ponds have a more stable temperature environment, but they do experience seasonal shifts. The heating and cooling of surface waters change temperature levels throughout the basin. In late spring and early summer, increasingly direct solar radiation and warming air temperatures heat the surface water faster than deeper water. Because water reaches its maximum density at 4°C, the surface water becomes lighter as its temperature increases. Soon a layer of lighter, warm water, called the **epilimnion,** rests on top of a heavier mixed layer of cooler water (Figure 4.14). This layer, known as the **metalimnion,** becomes cooler with depth. For approximately every 1 meter downward, the temperature declines 1°C. This transition zone where temperature drops quickly is called the **thermocline.** (If you dive into deep water, you become suddenly aware of the thermocline.) When the temperature of the water reaches 4°C and its greatest density, it lies as a layer of cold water on the bottom that is called the **hypolimnion.** The thermocline acts as a barrier between the epilimnion and hypolimnion. The lake basin is much like a sandwich, with the epilimnion and hypolimnion forming the top and bottom of the roll and the metalimnion as the filling in between. The filling is thick enough to prevent any contact between the top and bottom water, and little circulation takes place.

By fall, conditions begin to change, and a turnabout takes place. Air temperatures and sunlight decrease, and the surface water starts to cool. As it does, the water becomes denser and sinks, displacing the warmer water below to the surface, where it cools in turn. This cooling continues until the temperature is uniform throughout the basin (see Figure 4.14). Now pond and lake water circulates throughout the basin. This circulation is called the fall **overturn.** Stirred by wind, the overturn may last until ice forms.

Then comes winter, and the surface water cools to below 4°C, becomes lighter again, and remains on the surface. (Remember, water becomes lighter above and below 4°C (see Section 4.4 for discussion). If the winter is cold enough, surface water freezes; otherwise it remains close to 0°C. Now the warmest place in the pond or lake is on the bottom. A slight tem-

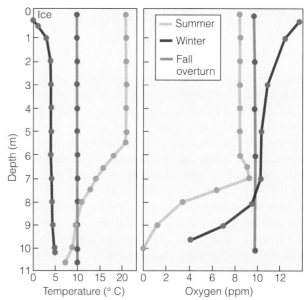

FIGURE 4.14 Oxygen stratification in Mirror Lake, New Hampshire, in winter, summer, and late fall. The late fall overturn results in both a uniform temperature and uniform distribution of oxygen throughout the lake basin. In summer, a pronounced stratification of both temperature and oxygen exists. Oxygen declines sharply in the thermocline and is nonexistent on the bottom because of decomposition taking place in the sediments. In winter, oxygen is also stratified, but it is present at a low concentration in deep water. (Likens 1985.)

perature inversion develops in which the water becomes warmer, up to 4°C, with depth. Water beneath the ice may be warmed by solar radiation through the ice. Because that increases its density, this water flows to the bottom, where it mixes with water warmed by heat conducted from the bottom mud. The result is a higher temperature on the bottom, although the overall stability of the water is undisturbed. With the spring breakup of ice and the heating of surface water up to 4°C, another overturn occurs. As the season wears on, the lake water again becomes stratified into the three familiar layers.

Not all lakes experience such seasonal changes in stratification, and you should not consider this phenomenon to be characteristic of all deep bodies of water. In shallow lakes and ponds, temporary stratification of short duration may occur; in others stratification may exist, but no thermocline develops. In some very deep lakes, the thermocline may simply descend during periods of overturn and not disappear at all. In such lakes, the bottom water never becomes mixed with the top layer. However, some form of thermal stratification occurs in all very deep lakes, including those of the tropics.

The temperature of a stream, on the other hand, is variable (Figure 4.15). Small, shallow streams tend to follow, but lag behind, air temperatures. They warm and cool with the seasons but rarely fall below

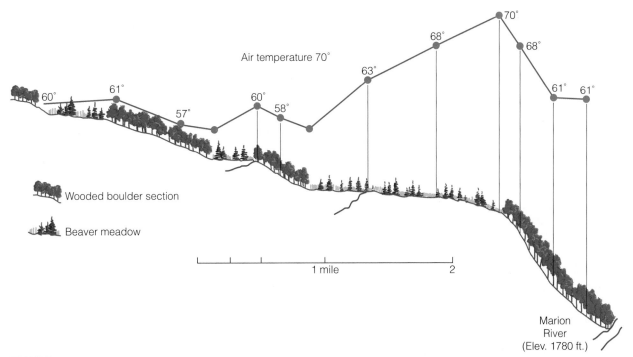

FIGURE 4.15 Profile of Bear Brook (Adirondack Mountains, New York) and a graph of its water temperatures, showing the warming effect of open beaver meadows and the cooling effects of a wooded boulder stream. (New York State Conservation Dept. 1934.)

freezing in winter. Streams with large areas exposed to sunlight are warmer than those shaded by trees, shrubs, and high banks. That fact is ecologically important because temperature affects the stream community, influencing the presence or absence of cool-water and warm-water organisms.

4.12 Temperatures of the Marine Environment Can Be Variable

The range of temperature in the marine environment is far less than that on land, although it is considerable—from 2°C in Antarctic waters to warmer than 27°C in parts of the western Pacific. In general, seawater is never more than 2° to 3° below the freezing point of freshwater nor warmer than 27°C. The mean depth of the oceans is 3800 meters (or 12,464 feet), ranging from the shallow near-shore waters to deep-sea trenches that exceed 10,000 meters in depth. At any given place, the temperature of deep water (below 1000 meters) is almost constant and cold. Seawater has no definite freezing point, although there is a temperature for seawater of any given salinity at which ice crystals form (Section 4.4). Thus, pure water freezes out, leaving even more saline water behind. Eventually, it becomes a frozen

block of mixed ice and salt crystals. With rising temperatures, the process is reversed.

Unlike freshwater, seawater (defined by a salinity of 24.7 practical salinity units or higher) becomes heavier as it cools and does not reach its greatest density at 4°C; therefore the limitation of 4°C as the temperature of bottom water does not apply to the sea. The temperature of the sea bottom generally averages around 2°C even in the tropics if the water is deep enough.

Water from all streams and rivers eventually drains into the sea. The place where this freshwater joins and mixes with the salt is called an **estuary** (Chapter 29). Temperatures in estuaries fluctuate considerably, both daily and seasonally. Sunlight and inflowing and tidal currents heat the water. High tide on the mudflats may heat or cool the water, depending on the season. The upper layer of estuarine water may be cooler in winter and warmer in summer than the bottom, a condition that, as in a lake, will cause spring and autumn overturns (Section 4.11).

In the tidal zone, the area lying between the water lines of high and low tide (Chapter 29), temperatures vary widely with the ebb and flow of the tide. As the tide recedes, the uppermost layers of life are exposed to air, wide temperature fluctuations, intense solar radiation, and desiccation for a considerable period, while the lowest fringes on the intertidal shore may be exposed only briefly before the high tide submerges them again. Organisms living within the sand

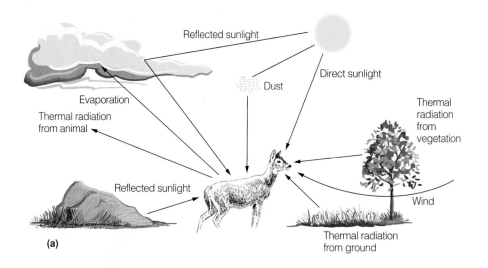

(a)

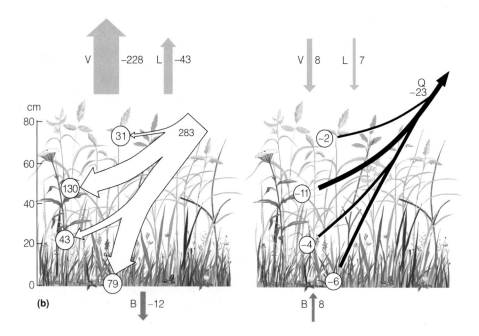

(b)

FIGURE 4.16 (a) Sources of energy exchange between a deer and its environment. (b) Energy exchange—absorption (positive values) and emission (negative values)—in a meadow on a sunny day (left) and at night (right). All values are in calories/cm². During the day, the upper layer of the canopy absorbs approximately 45 percent (31 + 130 + 43) of the net radiation (Q = 283). The lower layer absorbs 28 percent (79). By day, 80 percent of the radiant energy absorbed is emitted as latent heat (the evaporation of water: V = −228) and 15 percent through convective heat loss (L = −43); 5 percent is transferred to the soil (B = −12), raising soil temperatures. At night, net radiation and heat exchange is reversed. (Cernusa 1976.)

and mud do not experience the same violent fluctuations in temperature as those on rocky shores. Although the surface temperature of the sand at midday may be 10°C or more higher than the returning seawater, the temperature a few centimeters below the surface of the sand remains almost constant throughout the year.

4.13 Living Organisms Exchange Energy with the Environment

All organisms exchange energy with the external environment, a process that directly influences their body temperature. They absorb solar radiation, which may be direct, diffused from the sky, or reflected from the ground, as well as thermal radiation from rocks, soil, vegetation, and the atmosphere (Figure 4.16). In addition, organisms produce heat from metabolic processes such as respiration (Chapter 6) and lose heat as infrared radiation. To regulate body temperature, organisms must both lose heat to and gain heat from the environment.

The transfer of heat between organisms takes place in several ways. An important transfer is the loss of moisture through **evaporation**, the conversion of a liquid to a gas (water vapor). Evaporation requires 600 calories of energy to convert a gram of water to vapor at room temperature. This use of heat to evaporate water results in cooling. The rate of evaporation depends upon the difference in vapor pressure between the air and the object as well as the resistance of the surface or boundary layer to the loss of moisture. If the humidity of the air is high, little water evaporates and little heat is lost (see Section 3.8). As air becomes drier, the rate of evaporation and thus the rate of heat dissipation increase. You experience evaporative cooling when you step out of warm water into cooler air, or when you sweat. To dissipate heat by evaporation, an organism needs to take in water.

Heat also transfers between two solid objects, moving from the warmer object to the cooler one. This method of heat transfer is **conduction.** How fast heat moves between the two objects depends upon the degree of contact, their temperature difference, and the resistance of the objects to the transfer of heat. The greater the temperature difference, the thinner the conducting layer; and the greater the conductivity (the ability to transfer heat), the faster heat moves. When you sit on a cold stone or walk across the hot sand on a beach, you experience heat loss or gain by conduction.

Convection takes place when fluids (air or water) move over an object. Heat transfers more rapidly by convection than by conduction for a given temperature difference. When you stand in front of a rotating fan, you cool yourself by convection.

A fourth important method of heat transfer is the emission of long-wave **radiation,** or **thermal radiation.** Radiant energy striking a surface is absorbed, increasing the kinetic energy of the atoms and molecules in the object, and its temperature. You experience that type of heat transfer when you stand in front of a fire.

When the surrounding or ambient temperature is lower than the temperature of the organism, the organism loses heat to the environment. When the surrounding temperature is higher, heat moves from the environment to the organism. The problem, then, is for the organism to balance heat gains with heat losses to regulate internal temperature.

4.14 Essential Nutrients Are Either Macronutrients or Micronutrients

Living organisms require at least 30 to 40 chemical elements for growth, development, and metabolism (Table 4.1). Some of these elements are needed in large amounts. Known as **macronutrients,** they include carbon, oxygen, hydrogen, nitrogen, phosphorus, calcium, potassium, magnesium, sulfur, sodium, and chlorine, among others. Other elements are needed in lesser, often minute quantities. These elements are called **micronutrients** or trace elements. They include iron, copper, zinc, iodine, boron, cobalt, molybdenum, manganese, and selenium.

The prefixes *micro-* and *macro-* refer only to the quantity in which the nutrients are needed, not their importance to the organism. If micronutrients are lacking, plants and animals fail as completely as if they lacked nitrogen, calcium, or any other macronutrient. Some micronutrients are essential to all organisms; others appear to be essential to only a few. All of the micronutrients, especially the heavy metals, can be toxic in quantities greater than needed. The macronutrients and micronutrients that plants and animals require come from either the atmosphere or geological sources—rocks and minerals. Because the chemical composition of the atmosphere is relatively homogeneous, varying little from location to location, most spatial variation in nutrients occurs due to differences in the underlying geology (composition of rocks and minerals) and climate. Climate has a direct influence on the release of nutrients from rocks and minerals through the process of weathering, a topic discussed in detail in the following chapter (Section 5.2).

Biological as well as physical processes influence the chemical environment. For this reason, a more complete discussion of nutrients appears in Chapters 21 and 22, after we have examined the key biological processes involved in the uptake, transformation, and release of nutrients by plants and animals.

4.15 Oxygen Can Be Limiting in Aquatic Environments

Rarely is oxygen limiting in terrestrial environments except in soils; in aquatic environments the supply of oxygen, even at saturation levels, is meager and problematic. Oxygen enters the water by diffusion from the atmosphere and as a product of photosynthesis (Chapter 6). The amount of oxygen and other gases that water can hold depends on temperature, pressure, and salinity. Cold water holds more oxygen than warm water because the solubility (ability to stay in solution) of a gas in water decreases as the temperature rises. However, solubility increases as atmospheric pressure increases, and it decreases as salinity increases, which is of little consequence in freshwater.

Oxygen absorbed by surface water is mixed with deeper water by turbulence and internal currents. In shallow, rapidly flowing water and in wind-driven sprays, oxygen may reach and maintain saturation

TABLE 4.1

Some Essential Elements

Macronutrients	
Element	**Role**
Carbon (C) Hydrogen (H) Oxygen (O)	Basic constituents of all organic matter.
Nitrogen (N)	Used only in a fixed form: nitrates, nitrites, ammonium. Component of chlorophyll and enzymes; building block of protein.
Calcium (Ca)	In animals, needed for acid-base relationships, clotting of blood, contraction and relaxation of heart muscles. Controls movement of fluid through cells; gives rigidity to skeletons of vertebrates; forms shells of mollusks, arthropods, and one-celled Foraminifera. In plants, combines with pectin to give rigidity to cell walls; essential to root growth.
Phosphorus (P)	Necessary for energy transfer in living organisms; major component of nuclear material of cells. Animals require a proper ratio of Ca:P, usually 2:1 in the presence of vitamin D. Wrong ratio in vertebrates causes rickets. Deficiency in plants arrests growth, stunts roots, and delays maturity.
Magnesium (Mg)	In all living organisms, essential for maximum rates of enzymatic reactions in cells. Integral part of chlorophyll; involved in protein synthesis in plants. In animals, activates more than 100 enzymes. Deficiency in ruminants causes a serious disease, grass tetany.
Sulfur (S)	Basic constituent of protein. Plants use as much sulfur as they do phosphorus. Excessive sulfur is toxic to plants.
Sodium (Na)	Needed for maintenance of acid-base balance, osmotic homeostasis, formation and flow of gastric and intestinal secretions, nerve transmission, lactation, growth, and maintenance of body weight. Toxic to plants along roadsides when used to treat icy highways.
Potassium (K)	In plants, involved in osmosis and ionic balance; activates many enzymes. In animals, involved in synthesis of protein, growth, and carbohydrate metabolism.
Chlorine (Cl)	Enhances electron transfer from water to chlorophyll in plants. Role in animals similar to that of sodium, with which it is associated in salt (NaCl).

Micronutrients	
Element	**Role**
Iron (Fe)	In plants, involved in the production of chlorophyll; is part of the complex protein compounds that activate and carry oxygen and transport electrons in mitochondria and chloroplasts. In animals, iron-rich respiratory pigment hemoglobin in blood of vertebrates and hemolymph of insects transports oxygen to every organ and tissue. Synthesized into hemoglobin and hemolymph throughout life. Deficiency results in anemia.
Manganese (Mn)	In plants, enhances electron transfer from water to chlorophyll and activates enzymes in fatty-acid synthesis. In animals, necessary for reproduction and growth.
Boron (B)	Fifteen functions are ascribed to boron in plants, including cell division, pollen germination, carbohydrate metabolism, water metabolism, maintenance of conductive tissue, translation of sugar. Deficiency causes stunted growth in leaves and roots and yellowing of leaves.
Cobalt (Co)	Required by ruminants for the synthesis of vitamin B_{12} by bacteria in the rumen.
Copper (Cu)	In plants, concentrates in chloroplasts, influences photosynthetic rates, activates enzymes. Excess interferes with phosphorus uptake, depresses iron concentration in leaves, reduces growth. Deficiency in vertebrates causes poor utilization of iron, resulting in anemia and calcium loss in bones.
Molybdenum (Mo)	In free-living nitrogen-fixing bacteria and cyanobacteria, a catalyst for the conversion of gaseous nitrogen to usable form. High concentration in ruminants causes a disease characterized by diarrhea, debilitation, and permanent fading of hair color.
Zinc (Zn)	In plants, helps form growth substances (auxins); associated with water relationships; component of several enzyme systems. In animals, functions in several enzyme systems, especially the respiratory enzyme carbonic anhydrase in red blood cells. Deficiency in animals causes dermatitis, parakeratosis.
Iodine (I)	Involved in thyroid metabolism. Deficiency results in goiter, hairlessness, and poor reproduction.
Selenium (Se)	Closely related to vitamin E in function. Prevents white-muscle disease in ruminants. Borderline between requirement level and toxicity is narrow. Excess results in loss of hair, sloughing of hooves, liver injury, and death.

and even supersaturated levels because of the increase of absorptive surfaces at the air-water interface. Oxygen is lost from the water as temperatures rise and decrease solubility, and through the uptake of oxygen by aquatic life.

During the summer, oxygen, like temperature (Section 4.11), may become stratified in lakes and ponds. The amount of oxygen is usually greatest near the surface, where an interchange between water and atmosphere, further stimulated by the stirring action of the wind, takes place (see Figure 4.14). The quantity of oxygen decreases with depth because of the oxygen demand by decomposer organisms living in the bottom sediments. In some lakes, oxygen varies little from top to bottom; every layer is at the saturated level defined by its temperature and pressure. Water in some lakes is so clear that light penetrates below the depth of the thermocline and encourages the growth of phytoplankton. In these situations, the combined effects of oxygen release during photosynthesis and the greater solubility of colder water may lead to a greater oxygen content in deep water than in surface waters.

During spring and fall overturn, when water recirculates through the lake, oxygen becomes replenished in deep water. In winter the reduction of oxygen in unfrozen water is slight because the demand for oxygen by organisms is reduced by the cold and the greater solubility of oxygen at low temperatures. Under ice, however, oxygen depletion may be serious because of the lack of diffusion from the atmosphere to the surface waters.

The constant churning and swirling of stream water over riffles and falls give greater contact with the atmosphere; therefore, the oxygen content of the water is high, often near saturation for the prevailing temperature. Only in deep holes or in polluted waters does dissolved oxygen show any significant decline.

4.16 Acidity Has a Pervasive Influence on Terrestrial and Aquatic Environments

Acidity influences the availability and uptake of nutrients and restricts the distribution of organisms sensitive to acid conditions. The measurement of acidity is pH, calculated as the negative logarithms (base 10) of the concentrations of hydrogen in solution. The scale is based on the pH of pure water. Water is a weak electrolyte, a small fraction of which dissociates into ions: $H_2O \longrightarrow H^+ + OH^-$. In pure water, the ratio of H^+ ions to OH^- ions is 1. Because both occur in a concentration of 10^{-7} moles per liter, a neutral solution has a pH of 7 [$-\log(10^{-7}) = 7$]. A solution departs from neutral when one ion increases and the other decreases. Customarily, we use the negative logarithm of the hydrogen ion to describe a solution as an acid or a base. Thus, a gain of hydrogen ions to 10^{-6} moles per liter means a decrease of OH^- ions to 10^{-8} moles per liter, and the pH of the solution is 6. The negative logarithmic pH scale goes from 0 to 14. A pH greater than 7 denotes an alkaline solution (greater OH^- concentration), a pH of less than 7 an acidic solution (greater H^+ concentration). Beacause the scale is based on $\log_{10}$, a solution with a pH of 5 has 10 times the hydrogen concentration of one of pH 6; a solution with a pH of 4 has 10 times as many hydrogen ions as one of pH 5, and 100 times the hydrogen concentration of a solution of pH 6.

The buffering action of calcium and other ions in soil solution and exchange sites on clay particles (see Chapter 5, Section 5.8) influence acidity and alkalinity in the soil. In aquatic environments, there is a close relationship between the degree of acidity or alkalinity and concentration of carbon dioxide.

Like oxygen, carbon dioxide diffuses into the water surface from the atmosphere. Upon entering the water, carbon dioxide reacts chemically to produce carbonic acid:

$$CO_2 + H_2O \longleftrightarrow H_2CO_3$$

Carbonic acid further dissociates into a hydrogen ion and a bicarbonate ion:

$$H_2CO_3 \longleftrightarrow HCO_3^- + H^+$$

Bicarbonate may further dissociate into another hydrogen ion and a carbonate ion:

$$HCO_3^- \longleftrightarrow H^+ + CO_3^{2-}$$

The carbon dioxide–carbonic acid–bicarbonate system is a complex chemical system that tends to stay in equilibrium. (Note that the arrows on the equations go in both directions.) Therefore, if CO_2 is removed from the water, the equilibrium is disturbed and the equations will shift to the left, with carbonic acid and bicarbonate producing more CO_2 until a new equilibrium is produced.

These chemical reactions result in the production and absorption of free hydrogen ions (H^+). Since the abundance of hydrogen ions in solution is the measure of acidity, the dynamics of the carbon dioxide–carbonic acid–bicarbonate system has a direct influence on the pH of aquatic ecosystems. In general, the carbon dioxide–carbonic acid–bicarbonate system functions as a buffer to keep the pH of water within a narrow range. It does this by absorbing hydrogen ions

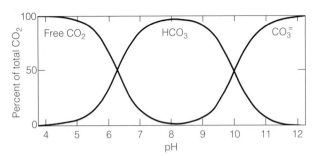

FIGURE 4.17 Theoretical percentages of CO_2 in each of the three forms of water in relation to pH.

TABLE 4.2

Composition of Seawater of 35 practical salinity units (psu)

Elements	g/kg	Milli-moles/kg	Milli-equiva-lents/kg
Cations			
Sodium	10.752	467.56	467.56
Potassium	0.395	10.10	10.10
Magnesium	1.295	53.25	106.50
Calcium	0.416	10.38	20.76
Strontium	0.008	0.09	0.18
			605.10
Anions			
Chlorine	19.345	545.59	545.59
Bromine	0.066	0.83	0.83
Fluorine	0.0013	0.07	0.07
Sulphate	2.701	28.12	56.23
Bicarbonate	0.145	2.38	—
Boric acid	0.027	0.44	—
			602.72

in the water when they are in excess (producing carbonic acid and bicarbonates) and producing them when they are in short supply (producing carbonate and bicarbonate ions).

At neutrality (pH 7), most of the CO_2 is present as HCO_3^- (Figure 4.17). At a high pH, more CO_2 is present as CO_3^{2-} than at a low pH, where more CO_2 occurs in the free condition. Addition or removal of CO_2 affects pH, and a change in pH affects CO_2.

The pH of natural waters ranges between the extremes of 2 and 12. Waters draining from watersheds dominated geologically by limestone will have a much higher pH and be well buffered compared with waters from watersheds dominated by acid sandstone and granite (Chapter 27).

4.17 Salinity Dominates Marine and Arid Environments

Life in oceans, estuaries, and arid lands and associated bodies of water exists in a salty environment. The salinity of the open sea is fairly constant, averaging about 35 practical salinity units, psu (35/00). In contrast, the salinity of freshwater ranges from 0.065 to 0.30 psu. Two elements, sodium and chlorine, make up some 86 percent of sea salt. These, along with other major elements such as sulfur, magnesium, potassium, and calcium whose relative proportions vary little, comprise 99 percent of sea salts (Table 4.2). Determination of the most abundant element, chlorine, is used as an index of salinity of a given volume of seawater. Salinity is expressed in 0/00 (parts per thousand) as the grams of chlorine in a kilogram of seawater.

The salinity of parts of the ocean varies because of physical processes. Salinity is affected by evaporation and precipitation, which are most pronounced at the interface of sea and air; by the movement of water masses; by the mixing of water masses of dif-ferent salinities, especially near coastal areas; by the formation of insoluble precipitates that sink to the ocean floor; and by the diffusion of one water mass into another.

In the estuarine environment, where freshwater meets the sea, the interaction of inflowing freshwater and tidal saltwater influences the salinity. Salinity varies vertically and horizontally, often within one tidal cycle (Figure 4.18). Salinity may be the same from top to bottom, or it may be completely stratified, with a layer of freshwater on top and a layer of dense, salty water on the bottom. Salinity is homogeneous when currents are strong enough to mix the water from top to bottom. The salinity in some estuaries is homogeneous at low tide, but at flood tide a surface wedge of seawater moves upstream more rapidly than the bottom water. Salinity is then unstable, and density is inverted. The seawater on the surface tends to sink as lighter freshwater rises, and mixing takes place from the surface to the bottom. This phenomenon is known as **tidal overmixing.** Strong winds, too, tend to mix saltwater with freshwater in some estuaries; but when the winds are still, the river water flows seaward on a shallow surface over an upstream movement of seawater, more gradually mixing with the salt.

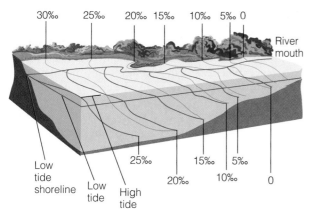

30‰ 25‰ 20‰ 15‰ 10‰ 5‰ 0

River mouth

Low tide shoreline
Low tide
High tide
25‰ 20‰ 15‰ 10‰ 5‰ 0

FIGURE 4.18 Vertical and horizontal stratification of salinity (0/00) in an estuary at both high and low tide. At high tide the incoming seawater increases the salinity toward the river mouth. At low tide, salinity is reduced. Note how salinity increases with depth, because the lighter freshwater flows over the denser salt water.

Horizontally, the least saline waters are at the river mouth and the most saline at the sea (see Figure 4.18). Incoming and outgoing currents deflect this configuration. In all estuaries of the Northern Hemisphere, outward-flowing freshwater and inward-flowing seawater are deflected to the right because of Earth's rotation. As a result, salinity is higher on the left side; the concentration of metallic ions carried by rivers varies from drainage to drainage; and the salinity and chemistry of estuaries differ. The portion of dissolved salts in the estuarine waters remains about the same as in seawater, but the concentration varies in a gradient from freshwater to sea.

Exceptions to these conditions exist in regions where evaporation from the estuary exceeds the inflow of freshwater from river discharge and rainfall (a negative estuary). This condition causes the salinity to increase in the upper end of the estuary, and horizontal stratification is reversed.

Terrestrial saline environments are characteristic of arid and semiarid regions where soils contain excessive amounts of soluble salts. Drainage of water from higher lands raises ground-level water near the soil surface on surrounding lower land. With little rainfall available to leach and transport materials away and with high evaporation rates, salt accumulates at or near the surface. Soil becomes enriched in salts faster than they can be leached. These saline soils support only salt-tolerant vegetation.

CHAPTER REVIEW

Summary

Light (4.1–4.3) Light has a major effect on almost all living organisms. Visible light, that part of the electromagnetic spectrum between the wavelengths of 400 to 700 nm, is known as photosynthetically active radiation (PAR). Short wavelengths between 280 and 380 nm are ultraviolet light; wavelengths longer than 740 nm are infrared. In addition to its spectral qualities, light also possesses intensity, duration, and directionality, all of which vary daily and seasonally. Light impinging on an object may be reflected, absorbed, or transmitted through it. Plants reflect green light most strongly and absorb violet, blue, and red wavelengths used in photosynthesis. Plants transmit the wavelengths they do not reflect or absorb. Light passing through a canopy of vegetation or through water becomes attenuated. Certain wavelengths drop out before others. On a clear day in a forest, green and far-red wavelengths pass through relatively unaltered. In pure water, red and infrared light are absorbed first, followed by yellow, green, and violet; blue penetrates the deepest (4.1).

The density and orientation of leaves in a plant canopy influence the amount of light reaching the ground. Foliage density is expressed as leaf area index (LAI), the area of leaves per unit of ground area. The amount of light reaching the ground in terrestrial vegetation varies with the season. In forests only about 1 to 5 percent of light striking the canopy reaches the ground. Sunflecks on the forest floor enable plants to endure shaded conditions. The amount of light reaching a plant influences its photosynthetic rate (4.2).

Light influences the daily and seasonal activities of living organisms. Living organisms, except bacteria, have an innate rhythm of activity and inactivity. This rhythm is free-running under constant conditions, with an oscillation that deviates slightly from 24 hours. For that reason it is called a circadian rhythm. Under natural conditions, external time cues, notably light and dark (day and night), set or entrain the circadian rhythm to the 24-hour day. This setting synchronizes the activity of plants and animals with the environment. The onset and cessation of activity depend upon whether the organisms are light-active or dark-active (4.3)

Structure and Properties of Water (4.4–4.5) Water has a unique molecular structure. One large oxygen atom and two smaller hydrogen atoms are bonded covalently. The two hydrogen atoms are separated by an 105° angle, which results in a V-shaped molecule. Because of this arrangement, the hydrogen atom side has a positive charge, and the oxygen side has a negative charge, polarizing the water molecule. Because of their polarity, water molecules become coupled with neighboring water molecules to produce a latticelike structure with unique properties (4.4).

Depending upon its temperature, water may be in the form of a liquid, solid, or gas. It absorbs or releases considerable quantities of heat with a small rise or fall in temperature. Water has a high viscosity that affects its flow. It exhibits high surface tension, caused by a stronger attraction of water molecules for each other than for the air above the surface. These properties are important ecologically and biologically (4.5).

Osmosis and Osmotic Pressure (4.6) Water moves from the environment into organisms through permeable cell membranes or conductive tissues. Molecules move from areas of high concentration to areas of low concentration until both sides are in equilibrium. This movement of molecules across membranes, called osmosis, generates pressure that slows the movement. The amount of pressure needed to counteract the movement is osmotic pressure. The tendency of a solution to attract water molecules from areas of high concentration to areas of low concentration is osmotic potential.

The Water Cycle (4.7) Water, on which life depends, follows a local cycle. It moves through cloud formation in the atmosphere, precipitation, interception, and infiltration into the ground. It eventually reaches groundwater, springs, streams, and lakes from which evaporation takes place, bringing water back to the atmosphere in the form of clouds.

Water Movement (4.8–4.9) Currents in streams and rivers and waves in open sea and breaking on ocean shores determine the nature of many aquatic and marine environments. Velocity of currents shapes the environment of flowing water. Waves pound rocky shores and tear away and build up sandy beaches (4.8). Rising and falling tides shape the environment and influence the rhythm of life of the coastal intertidal zones (4.9).

Thermal Environment (4.10) All organisms live in a thermal environment characterized by heat and temperature. Heat is a form of energy that results from random motion of molecules within a substance. Temperature is a measure of a substance's tendency to give up heat. Environmental temperatures result, directly or indirectly, from solar radiation and vary seasonally and daily. All organisms require a certain range of temperatures to carry on their metabolic processes.

Temperature in Aquatic Environments (4.11–4.12) Lakes and ponds experience seasonal shifts in temperature. In summer, a lake has a surface layer of warm, circulating water, the epilimnion; the middle zone, the metalimnion, that contains the thermocline in which the temperature drops rapidly; and a bottom layer, the hypolimnion, of dense water at approximately 4°C. When the surface waters cool in the fall, the temperature becomes uniform throughout the basin, and water circulates throughout the lake. A similar mixing takes place in the spring when the water warms. These seasonal overturns recirculate nutrients and mix the bottom water with the top. Temperature of flowing water is variable, warming and cooling with the season, Within the stream or river, temperatures vary with

depth, amount of shading, and exposure to sun (4.11). In the marine environment, seawater ranges from −2°C to −3°C to 27°C, depending on global location. The temperature of the sea bottom averages around 2°C. Temperatures in estuaries fluctuate daily and seasonally. In tidal zones, temperatures vary widely with the ebb and flow of tides (4.12).

Heat Balance (4.13) All organisms must balance heat inputs and outputs between themselves and their environment. Heat gains come from direct and reflected sunlight, diffuse radiation, long-wave infrared radiation, convection, conduction, and metabolism. Organisms lose heat to the environment by infrared radiation, conduction, convection, and evaporation. Organisms can have a net loss of heat to the environment only when the environmental temperature is less than body-core temperature.

Macronutrients and Micronutrients (4.14) All living organisms require certain nutritive elements essential to survival, growth, and reproduction. Some they need in large quantities, including carbon, oxygen, nitrogen, calcium, phosphorus, potassium, and sodium. This group makes up the macronutrients. Organisms require other nutrients in lesser, often minute quantities. They are micronutrients or trace elements. They include copper, zinc, and iron. Without trace elements, organisms die or become impaired.

Oxygen (4.15) Oxygen, rarely limiting in terrestrial environments, can be limiting in aquatic environments. The amount of oxygen that water can hold depends on temperature, pressure, and salinity. In lakes, oxygen absorbed by surface water mixes with deeper water by turbulence. During the summer, oxygen may become stratified, decreasing with depth because of decomposition in bottom sediments. During spring and fall overturn, oxygen becomes replenished in deep water. Constant swirling of stream water gives it greater contact with the atmosphere, and it thus maintains a high oxygen content.

Acidity (4.16) Acidity influences the availability of nutrients and restricts the environment of organisms sensitive to acid situations. The measurement of acidity is pH, the negative logarithm of the concentration of hydrogen ions in solution. Acidity or alkalinity in the soil relates to the buffering action of calcium and other ions in soil solution. In aquatic environments, a close relationship exists between carbon dioxide and the degree of acidity and alkalinity.

Salinity (4.17) The marine environment is characterized by salinity. Salinity is due largely to sodium and chlorine, which make up 86 percent of sea salt. Although sea salt has a constant composition, salinity varies through the oceans. It is affected by evaporation, precipitation, and movement and mixing of water masses of different salinities. In arid and semiarid lands, little rainfall to leach and carry materials away and high evaporation rates result in an accumulation of salt in the soil. This accumulation creates a saline environment.

Study Questions

1. What is the fate of the sun's energy when it reaches Earth's atmosphere?

2. What is the fate of visible light in the plant canopy and in water?

3. What is photosynthetically active radiation?

4. What is leaf area index (LAI)? How does it relate to light penetration in the canopy?

5. What is the circadian rhythm, and how does it relate to the light environment in which organisms live?

6. Distinguish between heat and temperature.

7. Explain why seasonal stratification of temperature and oxygen takes place in deep ponds and lakes.

8. What characterizes the epilimnion, metalimnion, and hypolimnion?

9. How does heat move between organisms and their environment?

10. How does the physical structure of water influence specific heat, latent heat, viscosity, and surface tension?

11. What is osmosis? Osmotic pressure? Osmotic potential? How do they relate to the moisture environment in which organisms live?

12. Describe the water cycle.

13. What is pH? What does it tell about an organism's environment?

14. What causes tides? What are spring tides and neap tides?

Tree roots penetrate the rocky substrate, breaking it down into finer particles, which are then exposed to weathering.

Soil

OBJECTIVES

On completion of this chapter, you should be able to:

* Define soil and its general features.
* List the five major factors in soil development, and explain how soils develop.
* Describe the soil profile and its horizons.
* Describe humus.
* Explain the importance of cation exchange in the soil.
* List the distinguishing physical characteristics of soil.
* Relate climate and vegetation to the major soil orders.
* Describe the features of soil as an environment for life.
* State the causes and consequences of soil erosion.

Soil is the foundation upon which all terrestrial life and much freshwater aquatic life depend. It is the medium in which plants are rooted, a reservoir of minerals for plants upon which, in turn, animal life depends. It is the site of decomposition of organic matter and the staging area for the return of nutrients in the mineral cycle (see Chapter 25). Roots extend into a considerable portion of the soil, to which they tie the vegetation and from which they extract water and minerals in solution for photosynthesis (see Chapter 6) and other biogeochemical processes. Vegetation, in turn, influences the development of soil, its chemical and physical properties, and its organic matter content. Thus soil acts as a pathway between the organic and mineral worlds, a pathway easily altered or destroyed by human interference.

5.1 Soil Is Not Easily Defined

As familiar as it is, soil is difficult to define. One definition says that soil is a natural product formed and synthesized by the weathering of rocks and the action of living organisms. Another states that soil is a collection of natural bodies of earth, composed of mineral and organic matter and capable of supporting plant growth. Indeed, one eminent soil scientist, Hans Jenny, a pioneer of modern soil studies, will not give an exact definition of soil. In his book *The Soil Resource,* he writes:

> Popularly, soil is the stratum below the vegetation and above hard rock, but questions come quickly to mind. Many soils are bare of plants, temporarily or permanently, or they may be at the bottom of a pond growing cattails. Soil may be shallow or deep, but how deep? Soil may be stony, but surveyors (soil) exclude the larger stones. Most analyses pertain to fine earth only. Some pretend that soil in a flowerpot is not a soil, but soil material. It is embarrassing not to be able to agree on what soil is. In this pedologists are not alone. Biologists cannot agree on a definition of life and philosophers on philosophy.

Of one fact we are sure. Soil is not just an abiotic environment for plants. It is teeming with life—billions of minute and not so minute animals, bacteria, and fungi. The interaction between the biotic and the abiotic makes the soil a living system.

Soil scientists recognize soil as a three-dimensional unit or body, possessing length, width, and depth. A three-dimensional soil body large enough that we can study its physical and chemical properties is called a **pedon**. The pedon is the basic unit in the study of soils.

5.2 Soil Formation Involves Five Interdependent Factors

Hans Jenny proposed that any individual soil or soil property results from the interaction of five interdependent soil-forming factors: parent material, climate, biotic factors, topography, and time.

Parent material is the unconsolidated mass from which soils form. It is derived from parent rock or from transported material. Rocks are residual or in-place parent material. They may be igneous, sedimentary, or metamorphic. Igneous rocks are formed by the cooling of volcanic flows, surface or subterranean. The properties of these rocks depend on the rate and temperature at which they form. Sedimentary rocks are formed by the deposition of mineral particles (sediments). The properties of sedimentary rocks depend on the type of sediment from which they are formed. Some sediments are of biological origin; for example, shells of ocean invertebrates may fall to the sea floor. Metamorphic rocks are either igneous or sedimentary rocks that have been altered by heat and the pressure of overlying rock. The composition of all these rocks largely determines the chemical composition of the soil.

Wind, water, glaciers, and gravity transport other parent materials. Because of the diversity of materials, transported soils are commonly more fertile than soils derived from in-place parent materials.

Climate influences the development of soil. Temperature and rainfall, which vary with elevation and latitude, govern the rate of weathering of rocks, decomposition of minerals and organic matter, and leaching and movement of weathered material. Further, climate influences the plant and animal life in a region, both of which are important in soil development.

Biotic factors—plants, animals, bacteria, and fungi—all contribute to the formation of soil. Vegetation is largely responsible for the organic matter in the soil and the color of the surface layer. It reduces erosion, and it influences the nutrient content of the soil. For example, forests store most of their organic matter on the surface, whereas in grasslands most of the organic matter added to the soil comes from the deep fibrous root systems. Plant roots through physical and chemical processes have a major role in soil formation. Animals, bacteria, and fungi decompose organic matter, mix it with mineral matter, and aid in the aeration of soil material and the percolation of water through it.

Topography, the contour of the land, influences both the intensity of radiant energy impinging on the soil material and the amount of water that enters the soil. Both relate directly to the weathering process. More water runs off and less enters the soil on steep slopes than on level land. Therefore, soil is less developed on steep slopes, and layers are often indistinct and shallow. Water draining from slopes enters the soil on low and flat land, where the subsoil may be wet and grayish in color. Steep slopes are subject to soil erosion and soil creep—the downslope movement of soil material, which accumulates on lower slopes and lowlands.

Time is a crucial element in soil formation. The weathering of rock material; the accumulation, decomposition, and mineralization of organic material; the loss of minerals from the upper surface; gains in minerals and clay in the lower layers; and layer differentiation all require considerable time. The formation of well-developed soils may require 2000 to 20,000 years. Soil differentiation from parent material, however, may take place within 30 years. Certain acid soils in humid regions develop in 2000 years because the leaching process is speeded by acidic materials. Parent materials heavy in texture, such as clays, require a much longer time to develop into "climax" soils, because they impede downward percolation of water. Soils develop more slowly in dry regions than in humid ones. Soils on steep slopes often remain young regardless of geological age because rapid erosion removes soil nearly as fast as it forms. Floodplain soils age little through time because of the continuous accumulation of new materials. Young soils are not as deeply weathered as and may be more fertile than old soils because they have not been exposed to leaching as long. Most old soils tend to be infertile because of long-time leaching of nutrients without replacement by fresh material.

5.3 Formation of Soil Begins with Mechanical and Chemical Weathering

The formation of soil begins with the weathering of rocks and their minerals. Weathering includes both the mechanical destruction of rock materials into smaller particles and the chemical modification of primary minerals into new secondary minerals. These secondary minerals include a variety of clay minerals that may weather into still other kinds of clay minerals.

Mechanical weathering comes about through the interaction of several forces. Exposed to the combined action of water, wind, and temperature, rock surfaces flake and peel away. Water seeps into crevices, freezes, expands, and cracks the rock into smaller pieces. Growing roots of trees split rock apart.

Eventually, rocks break down into loose material. This material may remain in place, but more often than not, much of it is lifted, sorted, and carried away. Material transported by wind is known as **loess,** and that transported by water as alluvial deposits. Material transported by glacial ice is **till.** In a few places, soil materials come from accumulated organic matter, called **peat.** Materials remaining in place are called residual. This mantle of unconsolidated material is the **regolith.** It may consist of slightly weathered material with fresh primary minerals, or it may be intensely weathered and consist of highly resistant minerals such as quartz.

Accompanying this mechanical weathering and continuing long afterward is **chemical weathering** brought about by the activities of soil organisms such as lichens and mosses, the acids they produce, and the continual addition of organic matter to mineral matter. Rainwater falling upon and filtering through the accumulating organic matter picks up acids and minerals in solution and sets up a chain of complex chemical reactions in the regolith. This chemical activity decomposes primary minerals. Easily weathered, these minerals, particularly the aluminosilicates, convert to secondary minerals, particularly clays. Because iron is especially reactive with water and oxygen, iron-bearing minerals are prone to rapid decomposition. Iron may remain oxidized in the red ferric state, or it may be reduced to the gray ferrous state. The clay particles produced are shifted and rearranged within the mass by percolating water and on the surface by runoff, wind, or ice. Percolating water carries calcium, potassium, and other mineral elements, soluble salts, and carbonates deeper into the soil material or even into streams, rivers, and the sea. The greater the rainfall, the more water moves down through the soil material. These localized chemical and physical processes in the parent material result in the development of layers or horizons in the soil, giving individual soils their distinctive profiles.

Because of variations in slope, climate, and native vegetation, many different soils can develop from the same parent materials. The thickness of parent material, the kind of rock from which it was formed, and the degree of weathering affect fertility and water relations of the soil.

5.4 Living Organisms Influence Soil Formation

Plants and animals have a pronounced influence on soil development. In time, plants colonize the weathered material. Plant roots penetrate and further break down the parent material. They extract nutrients up from its depths and add them to the surface. In doing so, plants recapture minerals carried deep into the soil by weathering processes. Through photosynthesis, plants capture the Sun's energy and add a portion of it in the form of organic carbon to the soil, largely through litterfall and plant death. This energy source of plant debris enables bacteria, fungi, earthworms, and other soil organisms to colonize the area.

The decomposition of organic matter turns organic compounds into inorganic nutrients. Invertebrate animals in the soil—millipedes, centipedes, earthworms, mites, springtails, grasshoppers, and others—consume fresh material and leave partially decomposed products in their excreta. Microorganisms further reduce this material into water-soluble nitrogenous compounds and carbohydrates and an accumulation of insoluble waxes, resins, and lignins. Residual material makes up **humus**, which is eventually mineralized into inorganic compounds.

Humus is difficult to define and not fully understood. It is a dark, organic material made up of many complex compounds. It varies depending upon the vegetation from which it derives. Its decomposition proceeds slowly. The equilibrium between the formation of new humus and the decomposition of old determines the amount of humus in the soil.

Humus that develops under the cover of mixed and deciduous forests on fresh and moist soils is inseparable from the upper layer of mineral soil. A wide diversity of soil organisms fragment and decompose plant material into humic substances and mix it with the upper layer of mineral soil. This humic material gives forest topsoil its dark color.

In dry or moist acidic habitats, especially heathland and coniferous forests, the humus rests on top of the mineral soil as a matted or compacted deposit. An accumulation of litter slowly decomposes and remains unmixed with mineral soil. The reason is that soil organisms capable of living in the acidic conditions cannot mix humic materials with mineral soil. The main decomposer organisms there are acid-producing fungi.

The most prevalent type of humus is the one that is well mixed with mineral matter in the soil. Small arthropods, particularly springtails and mites, transform plant residues into feces and tiny fragments. The droppings, plant fragments, and mineral particles all form a loose, netlike structure held together by chains of small droppings that are easily separated.

5.5 The Soil Body Has Horizontal Layers or Horizons

A fresh cut along a road bank or an excavation tells something about a soil. A close-up or even cursory look reveals bands and blotches of color. If you look closer, even handling the material, you discover changes in texture and structure. Any vertical cut through a body of soil or a pedon is the **soil profile** (Figure 5.1). The apparent layers are called the **horizons**. Each horizon has a characteristic set of features, particularly color, that distinguishes it from other horizons. Each horizon has its own thickness, texture, structure, consistency, porosity, chemistry, and composition.

Generalized, soils have six major horizons: O, A, E, B, C, and R. In some soils the horizons are quite distinct. In other soils the horizons form a continuum, with no clear-cut boundary between one horizon and another. However, soils do not necessarily have all six horizons. Most often they do not.

The O horizon is the surface layer, formed or forming above the mineral layer. It consists of fresh or partially decomposed organic material that has not been mixed into mineral soil. It is usually absent in cultivated soils. This layer and the upper part of the next horizon, A, constitute the zone of maximum biological activity. Both are subject to the greatest changes in temperature and moisture, contain the most organic carbon, and are the sites where most or all decomposition takes place. The O layer fluctuates seasonally. In temperate regions, it is thickest in the fall, when new leaf litter accumulates on the surface. It is thinnest in the summer after decomposition has taken place.

The A horizon is the upper layer of mineral soil with a high content of organic matter. It is characterized by an accumulation of organic matter and by the loss of some clay, inorganic minerals, and soluble matter. The E horizon (once labeled the A_2 horizon) is the zone of maximum leaching (eluviation, thus the label E). The downward movement of water and of suspended and dissolved materials alters its chemistry and structure. The E horizon has a granular, platy, or crumblike structure, often with a whitish color. Some soils possess both an A and an E horizon. Other soils lack one or the other, depending on weathering and the incorporation of organic matter with mineral soil.

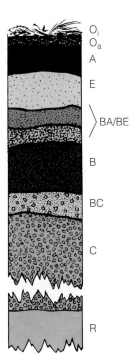

FIGURE 5.1 A generalized profile of the soil. Rarely does any one soil possess all of the horizons shown. Horizon designations are explained in the text.

Color is one of the most useful characteristics for the identification of soil. It has little direct influence on the function of a soil, but when considered with other properties, it can tell a good deal about the soil. In temperate regions, brownish-black and dark brown colors, especially in the A horizon, generally indicate considerable organic matter. The B horizon of well-drained soils may range anywhere from pale brown to reddish and yellowish colors. Dark brown and blackish colors, especially in the lower horizons, indicate poor drainage. It does not always follow, however, that dark-colored soils are high in organic matter. Soils of volcanic origin, for example, are dark from their parent material of basaltic rocks. In warm temperate and tropical regions, dark clays may have less than 3 percent organic matter. Red and yellow soils derive their colors from the presence of iron oxides, the bright colors indicating good drainage and good aeration. The intensity of red and yellow colors increases from cool regions to the equator. Other red soils obtain their color from parent material, such as red lava rock, and not from soil-forming processes. Well-drained yellowish sands are white sands that contain a small amount of organic matter and such coloring material as iron oxide. Quartz, kaolin, carbonates of calcium and magnesium, gypsum, and various compounds of ferrous iron give whitish and grayish colors to the soil. Grayish colors indicate permanently saturated soils in which iron is in the ferrous form. Imperfectly and poorly drained soils are mottled with blotches of various shades of yellow-brown and gray. The colors of soils are determined by the use of standardized color charts, notably the Munsell color charts.

Soil texture is the proportion of different sized soil particles (Figure 5.2). Texture is partly inherited from parent material and partly a result of the soil-forming process.

Particles are classified on the basis of size into gravel, sand, silt, and clay. Gravel consists of particles larger than 2.0 mm. They are not part of the fine fraction of soil. Sand ranges from 0.05 to 2.0 mm, is easy to see, and feels gritty. Silt consists of particles from 0.002 to 0.05 mm in diameter, which can scarcely be seen by the naked eye and feel and look like flour. Clay particles are less than 0.002 mm, too small to be seen under an ordinary microscope. Clay controls the most important properties of soils. One is plasticity—the ability to change shape when pressure is applied and retain that shape when pressure is removed. The other is cation exchange—the exchange of ions between soil particles and soil solution. A soil's texture is the percentage (by weight) of sand, silt, and clay. Based on proportions of these components, soils are divided into texture classes.

The B horizon is the zone of illuviation, a collector of leached material. It accumulates silicates, clay, iron, aluminum, and humus from the E horizon. It develops a characteristic physical structure involving blocky, columnar, or prismatic shapes (see Section 5.6). Below the B horizon in some soils there may be compact, slowly permeable layers, clay pans, or fragipans. A clay pan possesses much more clay than the horizon above it. It is very hard when dry, and stiff when wet. A fragipan is a brittle, seemingly cemented, subsurface horizon low in organic matter and clay, but high in silt or very fine sand. When dry a fragipan is hard. When moist it tends to erupt suddenly if pressure is applied. Both clay pans and fragipans interfere with root and water penetration in the soil.

The C horizon contains weathered material, either like or unlike the material from which the soil is presumed to have developed. Some active weathering takes place in this horizon, but it is little affected by soil formation. Below the C horizon is R, unweathered material or parent material.

5.6 Soils Have Certain Distinguishing Physical Characteristics

Soils are distinguished by differences in their physical and chemical properties. Physical properties include color, texture, structure, and depth. All may be highly variable from one soil to another.

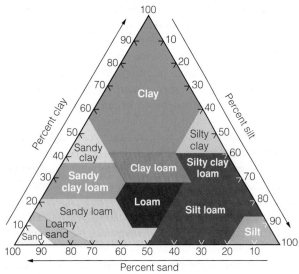

FIGURE 5.2 A soil texture chart, showing the percentages of clay (below 0.002 mm), silt (0.002 to 0.05 mm), and sand (0.05 to 2.0 mm) in the basic soil texture classes. For example, a soil with 60 percent sand, 30 percent silt, and 10 percent clay would be a sandy loam.

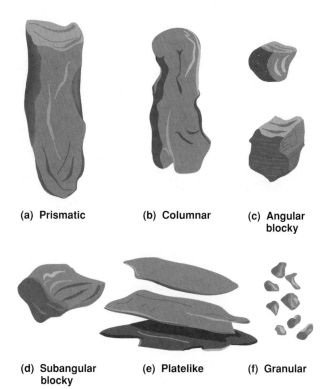

FIGURE 5.3 Some types of soil structure.

Soil texture affects pore space in the soil, which plays a major role in the movement of air and water in the soil, penetration by roots, and field capacity. In an ideal soil, particles make up 50 percent of the total volume of the soil; the other 50 percent is pore space containing soil gases. Pore space includes spaces within and between soil particles, as well as old root channels and animal burrows. Coarse-textured soils possess large pore spaces that favor rapid water infiltration, percolation, and drainage. To a point, the finer the texture, the smaller the pores, and the greater the availability of active surface for water adhesion and chemical activity. Very fine-textured or heavy soils, such as clays, easily become compacted if plowed, stirred, or walked upon. They are poorly aerated and difficult for roots to penetrate. This undesirable condition can be alleviated by the addition of organic matter.

Soil particles are held together in clusters or shapes of various sizes, called **aggregates** or **peds.** The arrangement of these aggregates is called soil structure. There are many types of soil structure, including prismatic, columnar, blocky, platelike, and granular (Figure 5.3). Structure is influenced by texture, plants growing in the soil, other organisms, and the soil's chemical status.

Soil depth varies across the landscape, depending on slope, weathering, parent materials, and vegetation. Soils developed under native grassland tend to be several meters deep. Soils developed under forests are shallow, with an A horizon of about 15 cm and a B horizon of about 60 cm. On level ground at the bottom of slopes and on alluvial plains, soils tend to be deep. Soils on ridge tops and steep slopes tend to be shallow, with bedrock close to the surface. The natural fertility and water-holding capacity of such soils are low.

5.7 Moisture–Holding Capacity Is an Essential Feature of Soils

If you dig into the surface layer of a soil about a day after a soaking rain and note the depth of water penetration, you should discover that the transition between wet surface soil and dry soil is sharp. After a heavy rain, the pore spaces fill with water. If the amount of water exceeds what the pore space can hold, we say the soil is **saturated** (Figure 5.4). Unless the soil is clay, excess moisture drains freely from the soil. If water fills all the pore spaces and is held there by internal capillary forces, the soil is at **field capacity.** Field capacity is generally expressed as the percentage of the weight or volume of soil occupied by water compared to the oven-dried weight

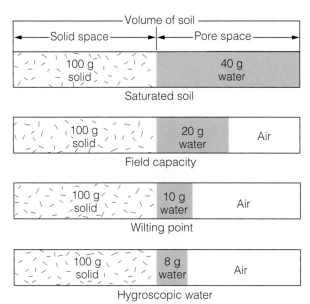

FIGURE 5.4 The proportions of solids, water, and air at the three levels and when only hygroscopic water remains. (From *Nature and Properties of Soil,* 8th edition, by Nyle C. Brady, © 1974. Reprinted by permission of Prentice-Hall, Inc., Upper Saddle River, NJ.)

of the soil at a standard temperature. The amount of water a soil holds at field capacity varies with the soil's texture—the proportion of sand, silt, and clay. Coarse, sandy soil has large pores; water drains through it quickly. Clay soils have small pores and hold considerably more water. Water held between soil particles by capillary forces is **capillary water.**

As plants and evaporation from the soil surface extract capillary water, the amount of water in the soil declines. Eventually it reaches a level at which a plant can no longer withdraw water. Soil water has now reached the **wilting point.** The amount of water retained by the soil between field capacity and wilting point (FC-WP) is the **available water capacity (AWC)** (Figure 5.4). The AWC provides an estimate of the water available for uptake by plants. Although water still remains in the soil at the wilting point—filling up to 25 percent of the pore spaces—soil particles hold it tightly, making it difficult to extract. If the soil dries even further, the only moisture remaining adheres tightly to soil particles as a thin film known as **hygroscopic water.** This film of water is unavailable to plants.

Both the field capacity and wilting point of a soil are heavily influenced by soil texture, and to a lesser extent by clay minerals, organic content, stoniness, and soil structure. Particle size of the soil directly influences the pore space and surface area onto which water adheres. Sand has 30 to 40 percent of

its volume in pore space, clays and loams 40 to 60 percent. As a result, fine-textured soils have a higher field capacity than sandy soil, but the increased area results in a higher value of the wilting point as well (Figure 5.4). Conversely, coarse-textured soils (sands) have a low field capacity and a low wilting point. Thus AWC is highest in intermediate clay loam soils.

The topographic position of a soil affects the movement of water both on and in the soil. Water tends to drain downslope, leaving soils on higher slopes and ridgetops relatively dry, and creating a moisture gradient from ridgetops to streams. However, after a dry period more water may be stored from rain on the upper slope than on the lower slope.

In all there are seven drainage classes: (1) *very excessively drained soils* and (2) *excessively drained soils,* in both of which plant roots are restricted to the upper layer of soil because of water deficiencies; (3) *well-drained soils,* in which plant roots can grow to a depth of 90 cm without restriction due to excess water; (4) *moderately well-drained soils,* in which plant roots can grow to a depth of 50 cm without restriction; (5) *somewhat poorly drained soils,* which restrict the growth of plant roots beyond a depth of 36 cm; (6) *poorly drained soils,* which are wet most of the time and are usually characterized by alders, willows, and sedges; and (7) *very poorly drained soils,* in which water stands on or near the surface most of the year.

Poorly drained areas support hydric soils. Hydric soils develop where flooding or ponding occurs long enough during the growing season to inhibit the diffusion of oxygen into the soil. Under these anaerobic conditions, iron in the soil reduces to the ferrous state, which gives a dull gray or bluish color to the horizon. Being soluble, some of this iron moves into more aerobic sites and precipitates out as oxidized iron, giving a mottled appearance to some hydric horizons. Some hydric soils possess a thick, dark surface layer because anaerobic conditions inhibit the decomposition of organic matter. Hydric soils are important indicators of wetlands (see Chapter 29).

The delineation of drainage classes of soils is important ecologically and economically. Somewhat poorly drained or hydric soils mark the beginning of wetland ecosystems. Historically, poorly drained soils were (and many still are) "improved" for agricultural and other uses by ditching and installing drainage tiles. Wetland conservation laws (often controversial) have slowed such damage. Drainage classes are also important considerations for the selection of construction and highway sites.

5.8   Cation Exchange Capacity Is Important to Soil Fertility

Chemical elements are dissolved in soil solution, bound up in organic matter, or adsorbed on soil particles as ions. An **ion** is a charged particle. Ions carrying a positive charge are **cations.** In soils ions are limited in their mobility (mostly cations) because they are closely held to particles of clay and humus. In aquatic systems the ions dissolve in water and obey the laws of diffusion and dilute solutions. These ions move from soil to plants, from plants to animals, and into the biogeochemical cycle (see Chapter 22).

Because of soil's unique platelike structure, clay particles control many of its important properties. The basic clay mineral consists of three elements: aluminum (Al^{3+}), silica (Si^{4+}), and oxygen (O^{2-}). The negatively charged edges and sides of the platelike clay particles, termed **micelles,** are exchange sites. An important feature of the micelles is that one element can substitute for another without changing the structure. For example Al may substitute for Si; iron (Fe_2 and Fe_3) and magnesium may substitute for Al^{3+}.

One of the most important features of clay minerals, this substitution results in net negative charges that must be balanced by positively charged cations. Exchangeable cations are loosely held on the surface of the micelles and can be replaced easily by others. The edges and sides, negatively charged, act as highly charged anions. They attract cations, water molecules, and organic substances. The total number of negatively charged exchange sites on clay and humus particles that attract positively charged cations is called the **cation exchange capacity** (CEC) (Figure 5.5). The CEC represents the net negative charges possessed by the soil. These negative charges enable a soil to prevent the leaching of its positively charged nutrient cations.

Exchange sites are occupied by such ions as calcium (Ca^{2+}), magnesium (Mg^{2+}), potassium (K^+), sodium (Na^+), and hydrogen (H^+). Some of these ions, especially Al^{3+} and H^+, cling more tenaciously to micelles than do others. Less tenacious, in descending order, are Ca^{2+}, Mg^{2+}, K^+, NH^+, and Na^+. These latter ions are more easily displaced from the exchange site, and their place may be taken by aluminum or hydrogen ions. The percentage of sites occupied by basic cations, primarily Ca^{2+}, Mg^{2+}, Na^+, and K^+, is called **percent base saturation.** Acidic soils have a low percent base saturation because they have a high number of exchangeable hydrogen ions. Soils with a high CEC are potentially fertile. Soils high in both CEC and base saturation are fertile unless they are saline or contain toxic heavy metals.

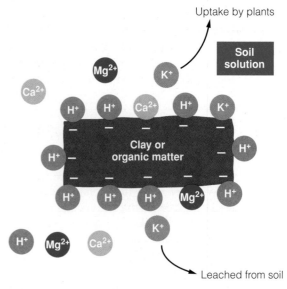

(a) Acidic soil (low pH)

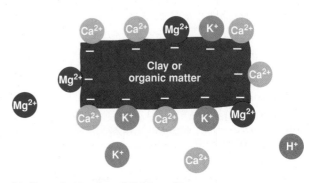

(b) Neutral to basic soil (higher pH)

FIGURE 5.5 Cation exchange in soils. (a) In acidic conditions, hydrogen (H^+) ions replace calcium (Ca^{2+}), magnesium (Mg^{2+}), and potassium (K^+) ions that are subsequently leached from the soil, impoverishing it. (b) In neutral to basic soils, calcium, magnesium, and potassium ions replace hydrogen ions.

Positively charged cations and negatively charged anions are dissolved in soil solution and occupy the exchange sites on the clay and humus particles. Cations occupying the exchange sites are in a state of dynamic equilibrium with similar cations in solution. Cations in soil solution are continuously being replaced by or exchanged with cations on the clay and humus particles in response to changes in the concentration in the soil solution. For example, the removal of certain cations from soil solution by the roots of plants reduces the concentration of those cations and enhances the release of cations from the micelles.

Hydrogen ions added by rainwater, by acids from organic matter, and by metabolic acids from roots and microorganisms increase the concentration of hydrogen ions in the soil solution and displace other cations, such as Ca^{2+}, on the micelles. As soil acidity increases, the proportion of exchangeable Al^{3+} increases and Ca^{2+}, Na^+, and other cations decrease. Such changes bring about nutrient deprivation in plants and microorganisms and also aluminum toxicity.

5.9 Climate and Vegetation Produce Different Soils

Broad regional differences in climate, vegetation, time, and degree of weathering produce major kinds of soils, called soil orders. Each order has distinctive features, summarized in Table 5.1, and its own distribution, mapped in Figure 5.6. Soil horizons vary tremendously among orders (Figure 5.7).

Two orders of soils lack B horizons—the Vertisols and Entisols. **Vertisols** are soils with a very high (30 percent) content of sticky, swelling, and shrinking-type clays. They typically develop from calcium-rich parent material in climates that feature a dry period of several months. During the dry season, the clay shrinks and the soil cracks to the surface. Materials on the surface of the soil fall into the cracks. In the wet season, water entering the cracks causes the clays to expand. The cracks close, embedding the material in the profile and causing some microbuckling on the soil surface. Such soils have limited agricultural use, mostly pasture, and create major engineering problems for road building and pipelines. They swell in response to wetting and shrink while drying. Such wetting and drying causes frequent deep cracking.

Entisols, found in a variety of climates and in areas where other soil groups dominate, are typical of mountainous and sandy soil regions. Entisols occur in areas where soil-forming processes have been interrupted or curtailed by flooding, erosion, and large-scale disturbance by human settlements and activities. They are associated with wetlands and alluvial deposits along rivers as well as areas subject to erosion and sedimentation that continually create new Entisols.

Inceptisols are mineral soils rich in weatherable materials with some subsoil development. They are associated with steeply sloping mountain lands whose slopes, subject to surface erosion, add to the accumulated materials at the base of the slopes. The major

TABLE 5.1

The Twelve Major Soil Orders

Order	Derivation and Meaning	Description
Entisol	Coined from *recent*	Dominance of soil materials; absence of distinct horizons; found on floodplains and rocky soils.
Vertisol	Latin, *verto,* "inverted"	Dark clay soils that exhibit wide, deep cracks when dry.
Inceptisol	Latin, *inceptum,* "beginning"	Texture finer than loam sand; little translocation of clay; often shallow; moderate development of horizons.
Aridisol	Latin, *aridus,* arid	Dry for extended periods; low in humus; high in base content; may have carbonate, gypsum, clay horizons.
Mollisol	Latin, *mollis,* "soft"	Surface horizons dark brown to black with soft consistency; rich in bases; soils of semihumid regions.
Spodosol	Greek, *spodus,* "ashy"	Light gray, whitish E horizon on top of a black and reddish B horizon high in extractable iron and aluminum.
Alfisol	Coined from *Al* and *Fe*	Shallow penetration of humus; translocation of clay; well-developed horizons.
Ultisol	Latin, *ultimus,* "last"	Intensely leached; strong clay translocation; low base content; humid warm climate.
Oxisol	French, *oxide,* "oxidized"	Highly weathered soils; red, yellow, or gray; rich in kalolinate, iron oxides, and often humus; in tropics and subtropics.
Histosol	Greek, *histos,* "organic"	High content of organic matter; bog and muck soils.
Andisol	Japanese, *ando, an,* "black," *do,* "soil"	Developed from volcanic ejecta; not highly weathered; upper layers dark colored; low-bulk density.
Gelisol	Greek, *gelid,* "very cold"	Presence of permafrost or soil temperature of 0°C or less within 2 meters of surface.

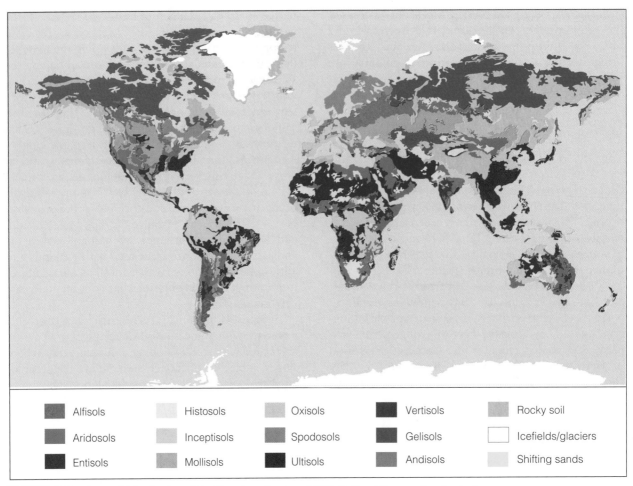

■ Alfisols	□ Histosols	■ Oxisols	■ Vertisols	■ Rocky soil
■ Aridosols	■ Inceptisols	■ Spodosols	■ Gelisols	□ Icefields/glaciers
■ Entisols	■ Mollisols	■ Ultisols	■ Andisols	■ Shifting sands

FIGURE 5.6   The world distribution of the 12 major soil orders. (USGS, National Conservation Service.)

soil-forming process associated with Inceptisols is leaching. Some Inceptisols support agriculture. Others are best devoted to forests and pastures.

Soils of arid and semiarid regions are **Aridisols.** Because vegetation is sparse, little organic matter and nitrogen accumulates in the soil. These soils are low in humus, high in base content, and often have a gypsum, clay, or carbonate horizon. Scant precipitation results in slightly weathered and slightly leached soils high in plant nutrients. The horizons are faint and thin.

Soil-forming processes in Aridisols involve calcification and salinization. Because the amount of rainfall in semiarid grassland regions generally is insufficient to remove calcium and magnesium carbonates from the profile, they are carried down only to the average depth reached by the percolating waters. Grass maintains a high calcium content in the surface soil by absorbing large quantities from the lower horizons and redepositing it on the surface as litter. Little clay is lost from the surface. The light-colored A horizon is low in organic matter, and the B horizon is characterized by an accumulation of calcium carbonate. This soil development process is **calcification.**

In other areas, soils contain excessive amounts of soluble salts, either from the parent material or from the evaporation of water that drains in from adjoining land. Infrequent rains penetrate the soil and carry the material downward to a certain depth. The water evaporates, leaving more or less cemented horizons of sodium, calcium, or magnesium salts below the surface in a soil-forming process called **salinization.** Under certain conditions, the carbonates may cement soil particles and rock fragments together below the surface to produce hard layers known as petrocalcic horizons or **caliche.** Caliche is impervious to plant roots and is difficult to excavate.

Regions subjected, past and present, to heavy volcanic activity have soils dark in color termed **Andisols.** They developed in volcanic ash, pumice, cinders, and related materials, collectively called ejecta. The rapid cooling of ejecta produces easily weathered material called volcanic glass. Ejecta is characterized by materials with a wide range of chemical composition and mineralogy that support a diversity of vegetation from desert shrub to tropical forests. The occurrence of Andisols relates to the global distribution of volcanic activity both recent and in the

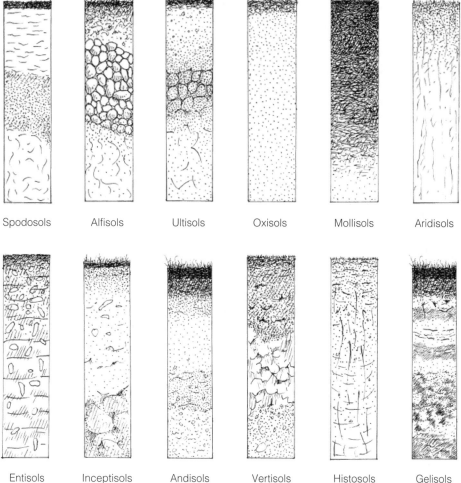

Spodosols Alfisols Ultisols Oxisols Mollisols Aridisols

Entisols Inceptisols Andisols Vertisols Histosols Gelisols

FIGURE 5.7 Profiles of the 12 major soil orders of the world.

geological past, especially during the Holocene. Easy to till and locally fertile, depending on the nature of the volcanic ejecta, Andisols have a stable, high water-holding capacity and thus are resistant to water erosion, but they are subject to wind erosion.

The subhumid-to-arid and temperate-to-tropical regions of the world—the plains and prairies of North America, the steppes of Russia, the veld and savannas of Africa, and the pampas of South America—support grassland vegetation. Dense root systems may extend many meters below the surface. Each year nearly all of the vegetative material above ground and a part of the root system go back to the soil as organic matter. Although the material decomposes rapidly, it never completely disappears before the next cycle of decay begins. Soil inhabitants mix the humus with mineral soil, creating a soil high in organic matter. These soils possess a deep, dark-colored, base-rich, and organically rich A horizon that is granular and soft. Some of these soils may have a B horizon and some lime accumulation. This soil-forming process is **melanization**. These soils are **Mollisols.**

Distinct A and B horizons characterize three other soils. All three occur in forested regions with ample rainfall but variations in temperature. As they form, bases are depleted, and insoluble iron and aluminum oxides combine in complex ways with organic acids and humus. Rainwater carries these oxides and humic materials along with clays downward into the B horizon. This leaching creates an E horizon.

This soil-forming process, called **podzolization** (from the Russian *pod*, "beneath," and *zol*, "ash"), is most intense in the highly acidic conditions of coniferous forests in cool, humid climates. It leaves a light or ash-colored A horizon. The organic (O) horizon is a layer of fermented litter on top of a layer of humus unmixed with mineral soil. Such soils are called **Spodosols.** Highly acidic, Spodosols (wood ash) are characterized by a dark-colored B horizon that forms under a gray to white E horizon. Such soils are covered mostly with coniferous and coniferous-deciduous forests. In spite of their acidity, they can support pastures, hay, and such crops as potatoes, oats, and rye.

Podzolization occurs in other soils, notably Alfisols and Ultisols. **Alfisols** are associated with humid, temperate forest regions. Like the Spodosols, these soils have accumulations of clay in the B horizon. Unlike Spodosols, Alfisols have comparatively little accumulation of organic matter on the surface. Instead organic matter tends to be well mixed by soil animals to form a darkened, organically enriched A horizon. In some soils, the division between the A and B horizons is hard to distinguish and the E horizon is missing. Because of their high base saturation and native fertility, Alfisols are very important agricultural soils.

The related **Ultisols** are characteristic of warmer climates supporting coniferous and deciduous forests. Ultisols are more intensely weathered than Alfisols and less acidic than Spodosols. They have a noticeably reddish or strong yellow color because of the release of iron from the silicates. Such reddish and yellowish soils, characteristic of the southeastern United States and tropical regions, also are in part a product of extensive leaching and some podzolization, resulting in clay accumulation in the B horizon. Added to this accumulation are clays formed by weathering in place in the B horizon. In contrast to Alfisols, the low nutrient content and high subsoil acidity of Ultisols limit agricultural productivity on them and historically have led to land abandonment. Ultisols can be productive agriculturally if fertilized, and with proper management they are valuable for timber production, especially pines.

In humid subtropical and tropical forested regions of the world where rainfall is heavy and temperatures high, weathering and leaching are much more intense, resulting in **Oxisols**. Weathering of the geologically very old substrate in these regions is almost entirely chemical, brought about by water and its dissolved substances. Because precipitation usually exceeds evaporation, the water movement is almost continuously downward. As a result, such soils are low in plant nutrients and intensely weathered to great depths. Iron and aluminum left after leaching of other minerals and bases become enriched as hydrous oxides, forming brilliant reddish colors in the E horizon. Some tropical soils, when deprived of vegetation and exposed to sun and air, tend to harden irreversibly into plinthite (formerly called laterite). Traditionally, Oxisols have supported shifting tropical cultivation, pineapples, bananas, and coffee. Recently, vast areas of tropical forests have been cleared for agriculture. With proper management, Oxisols support soybeans, wheat, corn, and pasture grasses.

The amount of water that passes through or remains in the soil determines the degree of oxidation and breakdown of soil minerals. Iron in soils where water stays near or at the surface most of the time is reduced to ferrous compounds. These compounds give a dull gray or bluish color to the horizons. The process, called **gleization,** may result in compact, structureless horizons. Gley soils are high in organic matter because vegetation produces more organic matter than can be broken down. An absence of soil microorganisms and anaerobic conditions is responsible for a low rate of humification. Gleization is the characteristic soil-formation process in cold, wet situations, especially in the tundra; but it also is common in other regions where the water table remains above the B and C horizons. We say the water table is perched.

In places with a high water table and poor drainage and in areas of high precipitation and low evaporation, organic matter only partially decomposes. These dark soils of partially decomposed organic matter are **Histosols.** Histosols are typical of peatlands. They occur in temperate regions, in the tropics, and in the arctic tundra and subarctic regions to the tropics. They include all organic soils. The main soil-forming process is the accumulation of partially decomposed organic parent material produced by plants and animals under saturated anaerobic conditions. Histosols are associated with wetlands, bogs, and other peatlands. Many Histosols have been drained for vegetable production and building sites, and mined for commercial horticultural peat, with the resulting destruction of peatland ecosystems. Sensitive to drainage, the organic matter decomposes steadily under aeration and gradually disappears, resulting in land subsidence and loss of organic matter.

The newest recognized soil order is **Gelisol.** Gelisols are characterized by permafrost—permanently frozen material underlying the solum—within 2 meters of the surface, rather than horizons. Soils included in this new order were previously scattered among other orders, notably Inceptisols, Entisols, and Histosols. Gelisols occur in the two circumpolar regions and at high elevations in mountains at lower latitudes. They have a deep active layer of accumulated organic matter supplied by tundra plants and animals, and mineral soil on top of the permafrost. During the warm season, freezing and thawing mixes the material and incorporates organic material throughout, disrupting any strong horizon development. Such activity results in patterned ground forms and **solifluction**—the slow flow of saturated soil and other unconsolidated material. Gelisols are highly fragile. Disturbance results in melting of permafrost. Because of cold temperatures, these soils cannot handle liquid, solid, or gaseous wastes.

5.10 Soil Supports Diverse and Abundant Life

The soil is a radically different environment for life than environments on and above the ground; yet the essential requirements do not differ. Like organisms

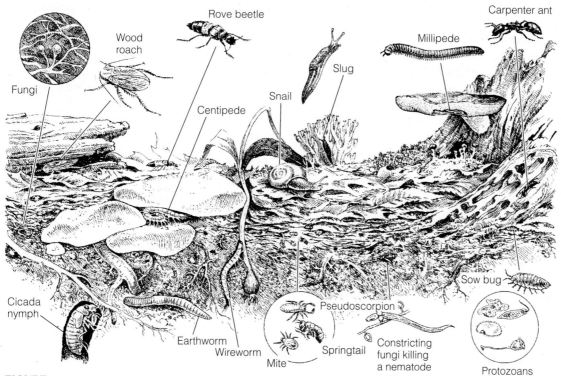

FIGURE 5.8 Life in the soil. This drawing shows only a fraction of the kinds of organisms that inhabit the soil and litter. Note the fruiting bodies of fungi, which are consumed by animals, vertebrate and invertebrate.

that live outside the soil, life in the soil requires living space, oxygen, food, and water. Without the presence and intense activity of living organisms, soil development could not proceed. Soil inhabitants from bacteria and fungi to earthworms convert inert mineral matter into a living system.

Soil possesses several outstanding characteristics as a medium for life. It is stable structurally and chemically. The soil atmosphere remains saturated or nearly so, until soil moisture drops below a critical point. Soil affords a refuge from extremes in temperature, wind, light, and dryness. These conditions allow soil fauna to make easy adjustments to unfavorable conditions. On the other hand, soil hampers the movement of animals. Except to such channeling species as earthworms, soil pore space is important. It determines the living space, humidity, and gaseous conditions of the soil environment.

Only a part of the upper soil layer is available to most soil animals as living space. Spaces within the surface litter, cavities walled off by soil aggregates, pore spaces between individual soil particles, root channels, and fissures are all potential habitats. Most soil animals are limited to pore spaces and cavities larger than themselves.

Water in the pore spaces is essential. The majority of soil organisms are active only in water. Soil water is usually present as a thin film coating the surfaces of soil particles. This film contains, among other things, bacteria, unicellular algae, protozoa, rotifers, and nematodes. The thickness and shape of the water film restrict the movement of most of these soil organisms. Many small species and immature stages of larger centipedes and millipedes are immobilized by a film of water and are unable to overcome the surface tension imprisoning them. Some soil animals, such as millipedes and centipedes, are highly susceptible to desiccation and avoid it by burrowing deeper.

When water fills pore spaces after heavy rains, conditions are disastrous for some soil inhabitants. If earthworms cannot evade flooding by digging deeper, they come to the surface, where they often die from ultraviolet radiation and desiccation or are eaten.

A diversity of life occupies these habitats (Figure 5.8). The number of species of bacteria, fungi, protists, and representatives of nearly every invertebrate phylum found in the soil is enormous. A soil zoologist found 110 species of beetles, 229 species of mites, and 46 species of snails and slugs in the soil of an Austrian deciduous beech forest.

Dominant among the soil organisms are bacteria, fungi, protozoans, and nematodes. Flagellated protozoans range from 100,000 to 1,000,000, amoebas from 50,000 to 500,000, and ciliates up to 1000 per gram of soil. Nematodes occur in the millions per square meter of soil. These organisms obtain their nourishment from the roots of living plants and from dead organic matter. Some protozoans and free-living nematodes feed selectively on bacteria and fungi.

Living within the pore spaces of the soil are the most abundant and widely distributed of all forest soil animals, the mites (Acarina) and springtails (Collembola). Together they make up over 80 percent of the animals in the soil. Flattened, they wiggle, squeeze, and digest their way through tiny caverns in the soil. They feed on fungi or search for prey in the dark interstices and pores of the organic and mineral mass.

The more numerous of the two, both in species and numbers, are the mites, tiny eight-legged arthropods from 0.1 to 2.0 mm in size. The most common mites in the soil and litter are the Orbatei. They live mostly on fungal hyphae that attack dead vegetation as well as on the sugars produced by the digestion of cellulose found in conifer needles.

Collembolae are the most widely distributed of all insects. Their common name, springtail, describes the remarkable springing organ at the posterior end, which enables them to leap great distances for their size. The springtails are small, from 0.3 to 1.0 mm in size. They consume decomposing plant material, largely for the fungal hyphae they contain.

Prominent among the larger soil fauna are the earthworms (Lumbricidae). Burrowing through the soil, they ingest soil and fresh litter and egest both mixed with intestinal secretions. Earthworms defecate aggregated castings on or near the surface of the soil or as a semiliquid in intersoil spaces along the burrow. These aggregates produce a more open structure in heavy soil and bind light soil together. In this manner, earthworms improve the soil environment for other organisms.

Feeding on the surface litter are millipedes. They eat leaves, particularly those somewhat decomposed by fungi. Lacking the enzymes necessary for the breakdown of cellulose, millipedes live on the fungi contained within the litter. The millipedes' chief contribution is the mechanical breakdown of litter, making it more vulnerable to microbial attack, especially by saprophytic fungi.

Accompanying the millipedes are snails and slugs. Among the soil invertebrates, they possess the widest range of enzymes to hydrolyze cellulose and other plant polysaccharides, possibly even the highly indigestible lignins.

Not to be ignored are termites (Isoptera), white, wingless, social insects. Except for some dipteran and beetle larvae, termites are the only larger soil inhabitants that can break down the cellulose of wood.

They do so with the aid of symbiotic protozoans living in their gut (see Chapter 16). Termites dominate the tropical soil fauna. In the tropics termites are responsible for the rapid removal of wood, dry grass, and other materials from the soil surface. In constructing their huge and complex mounds, termites move considerable amounts of soil.

Detrital-feeding organisms support predators. Small arthropods are the principal prey of spiders, beetles, pseudoscorpions, predaceous mites, and centipedes. Protozoans, rotifers, myxobacteria, and nematodes feed on bacteria and algae. Various predaceous fungi live on bacteria-feeders and algae-feeders.

5.11 Soils Have Been Severely Disturbed

Over much of the landscape, soils have been highly disturbed. Soils have been buried under fill; overturned and moved about by excavations, surface mining, and road construction; and exposed to erosion by wind and water. Upper horizons have been mixed by agricultural plowing and tillage, and compacted by heavy machinery and trampling.

Soil protected by vegetation maintains its integrity. Vegetation breaks the force of the wind and disperses raindrops, breaking their force. Rain trickles slowly through the litter, infiltrating the soil. If rainfall exceeds the soil's capacity to absorb it, the excess runs across the surface, but vegetation slows its movement.

Stripped of its protective vegetation and litter by plowing, logging, grazing, road building, and urban and suburban construction, soil is highly vulnerable to erosion—the carrying away of particles by wind and water faster than new soil can form. Depending on the soil group, new natural soil forms at the rate of 1 centimeter per 12 to 48+ years. Loss of the upper layers of humus-charged, granular, absorptive topsoil exposes the humus-deficient, less stable, less absorptive, and erodible layers beneath. If the subsoil is clay, it absorbs water so slowly that heavy rains produce a highly erosive and rapid runoff.

Soil compaction intensifies the problem. Heavy machinery, from large agricultural tractors to construction equipment, compacts soil on large areas of ground. Trampling on lawns and playing fields, concentrated use of pathways and hiking and riding trails through woods and fields, and off-road use of all-terrain vehicles compact the soil on other sites. Soil compaction occurs when any weight pushes the soil particles together and reduces the size of the pores. Greatest compaction occurs under wet and moist conditions. Moist soil particles easily slide over

(a)

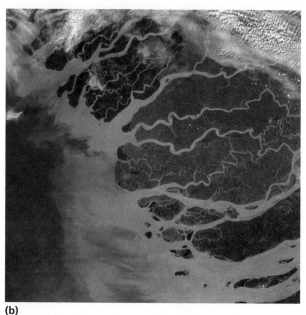
(b)

FIGURE 5.9 (a) Soil erosion from water runoff in wheatfield, Washington, U.S. (b) View from space shuttle Atlantis, 1994, of Bangladesh river delta showing erosion into ocean.

one another. Compacted soil cannot absorb water, so the water flows across the surface.

Rain falling on bare ground hammers the soil's surface, removing lightweight organic matter, breaking down soil aggregates, and forming a seal on the surface. Unable to infiltrate the soil, the water moves across the surface as runoff, carrying soil particles with it. The least conspicuous type of soil erosion is **sheet erosion**. It is a more or less even removal of soil over a field. Soil compaction can increase sheet erosion. When runoff concentrates in small channels or rills instead of moving evenly over a sloping land, its cutting force increases. **Rill erosion** channels water rapidly downslope. On areas where concentrated water cuts the same rill long enough or where runoff

concentrates in sufficient volume to cut deeply into the soil, highly destructive gullies result (Figure 5.9a). **Gully erosion** often begins in wheel ruts made by off-road vehicles in fields and forests, on logging roads and skid trails, and on livestock and hiking trails.

Bare soil, finely divided, loose, and dry, as it often is after tillage, is ripe for wind erosion. Wind picks up fine particles of dust and carries them as dust clouds. Drastic erosion occurred in the Great Plains of North America during the droughty Dust Bowl days of the 1930s. Wind erosion is increasing worldwide today, especially in arid and semiarid regions. Often, dust particles are lifted high in the atmosphere and carried for hundreds and even thousands of kilometers. Wind erosion exposes roots of plants or covers vegetation with drifting debris. In the Great Plains wind erosion exceeds water erosion.

Erosion by wind and water ruins land. Worldwide, about 12 million hectares of arable land are destroyed and abandoned annually because of soil mismanagement. Such land is usually so degraded that natural vegetation has difficulty returning. Erosion becomes progressively worse unless extreme measures are taken to restore vegetation. (See Focus on Ecology 5.1: Fighting Soil Erosion.)

Effects of erosion are felt both on and off site. Erosion of agricultural and forest lands reduces organic matter and increases clay content. It reduces water-holding capacity of the soil, intensifying drought conditions in dry weather and flooding in wet weather. Erosion degrades soil structure and reduces plant nutrients and plant rooting depths, depressing crop yields. It also reduces the diversity and abundance of soil organisms, essential to soil productivity and water infiltration. The loss of 2.5 cm of topsoil reduces corn and wheat yields by 6 percent. Costly, energy-demanding chemical fertilizers mask the ruinous effects of soil erosion on the inherent fertility of the soil. In the United States, the economic cost of soil erosion amounts to close to $27 billion annually, and the environmental costs are $17 billion.

The costs of off-site effects may be more than twice as high. Soil eroded by wind and water has to go somewhere. Erosion carries sediments into rivers, reducing light penetration and clogging navigation (Figure 5.9b). Sediment fills reservoirs and hydroelectric dams, shortening their lives and polluting the water. Wind-borne soil contributes significantly to air pollution, illness, and damage to machinery.

All forms of soil erosion destroy the integrity of ecosystems and ecological cycles. They raise the cost of food, fostering hunger and famine. For humans, the ecological consequences could well be social disorder and degradation of life. There is a true saying, "Poor soils make poor people."

FIGHTING SOIL EROSION

In the 1930s, W. C. Lowdermilk, assistant chief of the newly created Soil Conservation Service (now called the National Resources Conservation Service), wrote, "In a larger sense, a nation writes its record on the land . . . —a record that is easy to read by those who understand the simple language of the land." Past civilizations and nations have left a sorry record. The impoverished lands today—the Near East, Egypt, Syria, Lebanon, North Africa, the Mediterranean—all once supported flourishing agriculture, extensive forests, and great empires. Because of overgrazing, deforestation, and poor land use, these ancient empires based on agriculture are gone, the remains buried beneath soil eroded from the hills.

In spite of past experience, soil erosion is still one of the world's major environmental problems. Worldwide loss of topsoil amounts to 23 billion tons a year, and is increasing as human population growth forces agriculture to move into marginal, erodible land. Each year the countries of the former Soviet Union abandon an estimated one-half million hectares of cropland because of erosion. Dust storms in North Africa carry up to 100 million tons of soil annually out to the Atlantic, speeding desertification. The Ganges in India and the Yellow River in China carry billions of tons of soil to the ocean, as does the Mississippi River.

Soil erosion was severe in the United States in the 1930s, when plowing of the grasslands of the Great Plains and drought resulted in the Dust Bowl, and continuous cropping of cotton and tobacco gullied most of the South. Alarmed, President Franklin Delano Roosevelt and Congress established the Soil Conservation Service, which took massive action. Solutions included planting windbreaks in the Midwest, grassland restoration, crop rotation instead of monoculture, terracing, contour farming, and gully control. Such approaches were working until after World War II, when family farms declined and corporations with large land holdings took over agricultural production. Corporate farmers tore out windbreaks, terraces, fencerows, and soil-holding features that interfered with the use of large-scale machinery. They abandoned crop rotation for the monocultural production of corn, soybeans, and cotton. Again, soil erosion grew severe as corporate agriculture treated soil as an economic input, not as a renewable resource. Forty-four percent of U.S. cropland loses more than 11 metric tons per hectare annually through wind and water erosion. This amount is the maximum tolerable level if soil productivity is to be maintained. Twenty-three percent of cropland loses more.

Serious soil erosion is forcing agriculturists to come up with new techniques. Typically, crop production involves plowing land and cultivating crops in rows to control weeds. These expensive, time-consuming practices expose soil to severe erosion. New techniques, called conservation tillage systems, range from reduced tillage to no-tillage systems. No-tillage systems plant in the residue of the previous crop. An initial application of herbicides eliminates existing grass and weeds and the need for row cultivation. Conservation tillage significantly reduces soil erosion and the loss of soil nutrients. It increases the organic matter in soil and encourages higher populations of soil microbes and invertebrate fauna.

CHAPTER REVIEW

Summary

Soil Defined (5.1) Soil is a natural product of unconsolidated mineral and organic matter on Earth's surface. The medium in which plants grow, soil is the foundation of terrestrial ecosystems. It is the site of decomposition of organic matter and of the return of mineral elements to the nutrient cycle. It is the habitat of animal life, the anchoring medium for plants, and their source of water and nutrients.

Soil Formation (5.2–5.5) Soil results from the interaction of five factors: parent material, climate, biotic factors, topography, and time. Parent material provides the substrate from which soil develops. Climate shapes the development of soil through temperature, precipitation, and its influence on vegetation and animal life. Biotic factors—vegetation, animals, bacteria, and fungi—add organic matter and mix it with mineral matter. Topography influences the amount of water that enters the soil and rates of erosion. Time is required for full development of distinctive soils (5.2).

Soil formation begins with the weathering of rock and minerals. In mechanical weathering, water, wind, temperature, and plants break down rock. In chemical weathering, the activity of soil organisms, the acids they produce, and rainwater break down primary minerals (5.3).

Living organisms influence soil development. Plants rooted in weathering material break it down, extract nutrients from its depths, and add all-important organic material and nutrients to the surface. Soil organisms act on organic matter to produce humus. Humus is a dark-colored, noncellular, chemically complex organic material that affects soil structure and fertility (5.4).

Soils develop in layers called horizons. Five horizons are commonly recognized, although all are not necessarily present in any one soil: the O or organic layer; the A horizon, characterized by accumulation of organic matter; E, the zone of leaching of clay and mineral matter; the B horizon, in which mineral matter accumulates; and C, the underlying material, exclusive of bedrock. These horizons may divide into subhorizons (5.5).

Distinguishing Characteristics (5.6–5.8) Soils and horizons within soils differ in texture, structure, and color. Texture is determined by the proportions of soil particles of different sizes—sand, silt, and clay. Texture is important in the movement and retention of water in the soil (5.6).

The amount of water a soil can hold is one of its important characteristics. When water fills all pore spaces, the soil is saturated. When a soil holds the maximum amount of water it can retain against the force of gravity, it is at field capacity. Water held between soil particles by capillary forces is capillary water. When the moisture level is at a point at which plants cannot extract water, the soil has reached wilting point. The amount of water retained between field capacity and wilting point is available water capacity. Water held so tightly in a thin film on soil particles that it is unavailable to plants is hygroscopic water. Soils range from excessively drained soil to very poorly drained or hydric soils that mark the beginning of wetlands (5.7).

Soil particles, particularly the clay-humus complex, are important to nutrient availability and the cation exchange capacity of the soil—the number of negatively charged sites on soil particles that can attract positively charged ions. Percent base saturation is the percentage of sites occupied by ions other than hydrogen and aluminum. Soils with a high cation exchange capacity are potentially fertile (5.8).

Climate, Vegetation, and Major Soil Orders (5.9) Vegetation and climate influence soil development over large areas. Similar vegetation and climate produce soils with similar characteristics. There are 12 major soil orders. In grassland regions, calcium accumulates in the lower horizons. In arid regions, salt accumulates close to the surface. In forest regions, podzolization takes place—the leaching of calcium, magnesium, iron, and aluminum from the upper horizons and the retention of silica. In tropical regions, silica is leached and iron and aluminum oxides are retained in the upper horizon. Gleization takes place in poorly drained soils. Organic matter decomposes slowly. Iron is reduced to the ferrous state, imparting a greyish or bluish color to the horizons.

Life in the Soil (5.10) The soil provides a unique environment for diverse organisms that in turn influence soil development and structure. Bacteria, protozoans, and other microorganisms live in the thin film of water surrounding soil particles. Most microorganisms obtain their nourishment from organic matter. Larger soil invertebrates live in pore spaces and channels in the soil. They feed on fresh litter, other detrital material, bacteria, and fungi. These detritus feeders support an array of predators, from mites to spiders.

Soil Erosion (5.11) Worldwide, soils have been and are being depleted by erosion in which topsoil loss exceeds formation. Deforestation, poor agricultural practices, urbanization, road building, and other disturbances expose the soil to the erosive forces of water and wind. Soil erosion destroys natural and agricultural ecosystems and fills rivers, lakes, reservoirs, and navigation channels with silt. Wind erosion carries soil far from its source in clouds of dust and increases particulate air pollution. Soil erosion in all its forms impoverishes regions and nations, reduces food production, and causes extensive economic losses.

Study Questions

1. Why is it difficult to define soil?

2. What five major factors affect soil formation?

3. What roles do mechanical and chemical weathering play in soil formation?

4. How do living organisms develop soil?

5. What is a soil profile? What are soil horizons?

6. Distinguish among the major soil horizons.

7. How do the major soil orders reflect climate and vegetation?

8. What gives soil its textures?

9. How does cation exchange capacity affect nutrient uptake?

10. What characteristics do soils possess that make them good habitats for living organisms?

11. Discover soil texture. In a glass jar, mix several small samples of soil from your garden, a grass field, or a forest with about twice as much water. Shake vigorously and allow the sediments to settle. After 24 hours measure the depth of sand (which settles out first), silt (which settles out second on top of the sand), and clay. Divide the depth of each layer by the total depth of all the settled soil and multiply by 100 to determine the proportions of sand, silt, and clay in your soil. Compare the results with the soil texture chart (Figure 5.2).

12. From your soil conservation office or library, obtain a soil survey for your area. How do local soils relate to agricultural development, urban development, soil erosion problems, and forest distribution?

13. What is the major cause of soil erosion in your local area? Note the degree of siltation in streams, rivers, dams, and lakes after heavy rains.

14. What efforts are being made to reduce erosion in your area?

Blowin' in the Wind

"The soil is the one indestructible, immutable asset that the nation possesses. It is the one resource that cannot be exhausted; that cannot be used up."
U.S. Bureau of Soils (1909)

The following decades of exploitation and abuse on the southern plains of the United States proved this naive proclamation wrong. Soils that had taken 5000 to 10,000 years to develop were gone in a matter of years.

On the morning of April 14, 1935, the sky over the farmlands of Oklahoma was blue and clear. The life of a farm family was one of routine, and on this Sunday folks began their ritual of early morning chores before setting off to church. By mid-afternoon there were signs that the weather was about to change. The temperature began to drop, and the farm animals became restless. Suddenly a huge black cloud appeared on the horizon, approaching fast. Those traveling home from church pushed on, trying to reach home before the storm struck.

Drivers who were caught in the storm were blinded by the dust and crashed their cars into one another. Other cars came to a standstill as the static electricity caused by millions of dirt particles rubbing together shorted out ignitions. The same static electricity jammed radio broadcasts. Lost under an almost coal black sky, people caught outside searched for shelter. Some, only feet from their houses, crawled on hands and knees in search of porches or back doors in a desperate effort to escape the swirling blackness. Home was the safest place to be, but even its shelter offered only limited sanctuary. The dust was so thick that it filtered into houses. The rags stuffed under doors and windows could not hold it out.

A tidal wave of dust more than 7000 feet high had rolled over their land, engulfing everything in its path. As the storm subsided, those who ventured outside saw a transformed landscape. Drifts of soil, some more than 20 feet high, were piled up against buildings and buried farm equipment. Dead and dying livestock lined the roads. In an article written the following day, Robert Geiger, a reporter for the Associated Press, referred to the area as "the dust bowl of the continent." The area Geiger was referring to covered some 500 miles by 300 miles, over 100 million acres of land in Kansas, Colorado, Oklahoma, Texas, and New Mexico. It was home to the worst environmental disaster in American history, which set the stage for a period of social transformation that would have a profound and long-lasting impact on American culture.

Although the "black blizzard" that moved through the western plains that Sunday in April was the largest, it was only one of countless dust storms that ravaged the region between 1931 and 1939. The dust storms occurred during a period of severe drought that began in 1930. The drought alone, however, did not cause the dust storms. Although the decade of

Sand drifts on a farm near Liberal, Kansas (March 1936).

Dust storm at Lamar, Colorado (March 1936).

drought in the 1930s was the most severe on record, droughts are common in this region of the country, occurring roughly every 25 years. Rather, it was the combination of drought and misuse of the land that led to the incredible environmental devastation. In December 1936, the eight-member Great Plains Committee submitted to President Roosevelt a report entitled "The Future of the Great Plains." The committee concluded

"It was the worst environmental disaster in American history, and set the stage for a period of social transformation that would have a profound and long-lasting impact on American culture."

that the Dust Bowl was a wholly human-made disaster, produced by a history of misguided efforts to "impose upon the region a system of agriculture to which the plains were not adapted."

The central region of the United States, extending from Kansas in the east to Colorado in the west, and from the Dakotas southward to north Texas, is known as the Great Plains. It is home to a vast expanse of prairie grasslands. The annual rainfall varies over this region, declining westward from a high of 30 inches in the tallgrass prairies of Kansas to less than 18 inches in short-grass prairies that occupy the foothills of the Rocky Mountains in eastern Colorado. Much of the region that became the Dust Bowl was obtained by the United States as part of the Louisiana Purchase of

1803. Later that century, the Homestead Act of 1862 declared that any person who settled on 160 acres of the western prairie, stayed there for 5 years, made "improvements," and paid a filing fee became the rightful owner of the land. The region at this time was in the hands of cattlemen who already were overgrazing the southern plains and causing long-term damage to the grasslands. The winter of 1885–1886, the harshest in the recorded history of the region, proved to be the final and fatal blow to the cattle ranchers. With the grasslands degraded, up to 85 percent of the cattle perished, and the cattle industry went bankrupt.

Into the vacuum came farmers armed with plows, and later tractors, to do battle with the frontier. The plowing of the southern prairies removed the grassland, and with it the thick mat of roots referred to as sod—the term that gave these early farmers their nickname, "sodbusters." The continuous plowing and harvesting of wheat resulted in a constantly disturbed and exposed surface, leaving large fields bare of vegetation with straight furrows often plowed parallel to the prevailing winds. Such were the conditions on the southern plains when the "decade of drought" arrived in the early 1930s.

The black blizzards were devastating to the farmlands of the Dust Bowl, but the effects of the drought on agricultural production extended over the entire region of the Great Plains, which was America's breadbasket. In Thomas

County, Kansas, no wheat at all was harvested in 1933, 1935, 1936, and 1940. Production during the intervening years was at best only one-third that of normal years prior to the drought. Many of the farms that were not lost to the dust storms were abandoned. Throughout the Great Plains, thousands of farmers were forced to leave their land when banks foreclosed on their mortgages.

By 1934, the federal government began a series of policies aimed at battling the growing social and environmental problems on the western plains. The Taylor Grazing Act, signed by President Roosevelt in June of that year, set aside 80 million acres of federal lands for leasing to ranchers, effectively removing it from access by homesteaders. In addition, the Department of Agriculture paid farmers to retire almost 6 million acres of land, which were stabilized and leased back as grazing land.

On April 27, 1935, Congress passed the Soil Erosion Act, the first U.S. effort to establish a comprehensive, nationwide program to promote soil conservation. The act established the Soil Conservation Service in the Department of Agriculture. Under the direction of Hugh Bennett, the man largely responsible for the passage of the Soil Erosion Act, the Soil Conservation Service developed extensive conservation programs aimed at retaining topsoil and preventing irreparable damage to the land. It advocated such farming techniques as strip-cropping, terracing, crop rotation, contour plowing, and cover crops. Farmers were paid to practice farming techniques that conserved the soil. Billions of trees were planted to provide a system of long "shelter belts" one mile apart to break the force of the wind. By 1941, about 75 soil conservation districts had been organized in the states of the Great Plains to promote sound agricultural practices.

Since the Dust Bowl years, the area of land in active agricultural production has declined, while the production per acre has increased dramatically as a result of developments in plant breeding and the extensive use of chemical fertilizers. Agriculture has shifted from small family-owned farms to large-scale corporate farms—agribusiness. Throughout this transformation, much has changed, but the problem of soil erosion remains, and the consequences are no less severe. U.S. croplands now lose an average of 13 tons per hectare per year, a value that can be reduced to less than 1 ton by using more sustainable practices such as crop rotation and no-till farming. Wind and water erosion reduces the productivity of the land and affects adjacent steams and rivers as soil muddies the water and settles on the bottom, destroying habitat critical for the diversity of aquatic organisms.

To address this continuing environmental problem requires changing our concept of the essential component of the biosphere that we commonly refer to as "dirt."

Soil is the provider of nutrients formed over millennia through the combination of geological and biological processes. It is a living substrate containing a diversity of life. The ground upon which we walk is nature's past and future. It is the graveyard of past life, the place where all life is eventually broken down and transformed to the basic elements from which it was constructed, and it is also the place of birth, as those basic elements are once again used to sustain new life.

PART III

The Organism and Its Environment

Light streams through the canopy of a California coastal redwood *(Sequoia sempervirens)* grove.

Plant Adaptations to the Environment

OBJECTIVES

On completion of this chapter, you should be able to:

- Contrast primary and secondary producers.
- Describe the process of C_3 photosynthesis.
- Contrast photosynthesis with respiration.
- Discuss the relationship between water loss and CO_2 uptake in photosynthesis.
- Explain how water moves from the soil through the plant to the atmosphere.
- Explain how the process of photosynthesis responds to variations in light.
- Discuss the mechanisms used by plants to maintain thermal balance.
- Explain how carbon allocation influences plant growth.
- Contrast shade-tolerant and shade-intolerant plants.
- Discuss plant adaptations to arid environments.
- Discuss how C_4 and CAM photosynthesis are adaptations to warm and dry environments.
- Contrast plants from low-nutrient and high-nutrient environments.

All life on Earth is carbon-based. What this means is that all living creatures are made up of complex molecules built on a framework of carbon atoms. The carbon atom is able to bond readily with other carbon atoms, forming long, complex molecules. The carbon needed to construct these molecules—the building blocks of life—derives from various sources. The processes involved in acquisition and utilization of carbon are some of the most basic adaptations required for life. Humans, like all other animals, gain their carbon by consuming plant and animal materials. However, the ultimate source of carbon from which life is constructed is carbon dioxide (CO_2) in the atmosphere.

Not all living organisms can use this abundant form of carbon directly. Only one process is capable of transforming carbon in the form of CO_2 into organic molecules and living tissue. That process, carried out by green plants, algae, and some types of bacteria, is **photosynthesis.** It is essential for the maintenance of life on Earth. All other organisms derive their carbon (and most other essential nutrients) from plant and animal tissue. In general, the process by which carbon and other essential nutrients transform into part of the organism is called **assimilation.**

The acquisition and assimilation of essential nutrients and the processes associated with life—synthesis, growth, reproduction, and maintenance—require energy. Chemical energy is generated in the breakdown of carbon compounds in living cells, a process called **respiration.**

The energy that fuels photosynthesis, the process of assimilation in green plants, comes from the Sun. All other organisms in terrestrial and shallow-water systems use energy that comes directly or indirectly from photosynthesis. The source from which an organism derives its energy is one of the most basic distinctions in ecology. Organisms that derive their energy from sunlight are referred to as **autotrophs,** or **primary producers.** Organisms that derive energy from consuming plant and animal tissue, breaking down assimilated carbon compounds, are called **heterotrophs,** or **secondary producers.**

In this chapter we will examine the variety of adaptations that plants have evolved, adaptations that allow plant life to successfully survive, grow, and reproduce across virtually the entire range of environmental conditions found on Earth.

But first, let us take a closer look at the process so essential to life on Earth, or as the author John Updike so poetically phrased it, "the lone reaction that counterbalances the vast expenditures of respiration, that reverses decomposition and death."

6.1 Photosynthesis Is the Conversion of Carbon Dioxide into Simple Sugars

Photosynthesis is the process by which plants transform carbon dioxide (CO_2) into simple organic compounds. Ecologists view photosynthesis as a two-step process. The first step, referred to as light reactions, involves the absorption of sunlight (more specifically, photosynthetically active radiation or PAR; see Chapter 4, Section 4.1) by molecules of the pigment chlorophyll contained within the leaf. The energy absorbed by the chlorophyll molecule is then used to produce the high-energy compound ATP (adenosine triphosphate) as well as another compound called NADPH, which are required for the second step in photosynthesis—the conversion of CO_2 into simple sugars, or glucose ($C_6H_{12}O_6$).

In the process of photosynthesis, molecules of water (H_2O) are split. The hydrogen (H) is used in the transformation of CO_2 into glucose (carbohydrates), and oxygen (O_2) is released. Plants use one unit of water (H_2O) and produce one unit of oxygen (O_2) for every unit of CO_2 they transform into glucose. Thus, photosynthesis is responsible not only for the carbon-containing molecules we animals consume, but also for the oxygen we breathe.

Although glucose is represented as the product of photosynthesis, the initial product is a molecule that contains 3 carbon atoms; hence, the process has been called C_3 photosynthesis. These 3 carbon molecules are then used to make more complex carbon compounds, such as glucose, starches, and proteins. In turn, these compounds are utilized to produce leaves, stems, roots, flowers, and seeds.

Like many chemical reactions in nature, the process of transforming CO_2 into sugars would occur at an extremely slow rate if it were not accelerated by the action of enzymes. Enzymes are chemical compounds that speed up or catalyze chemical reactions. The enzyme that catalyzes the initial transformation of CO_2 in photosynthesis is **rubisco** (ribulose biphosphate carboxylase-oxygenase). Rubisco is the most abundant enzyme on Earth.

Some of the carbohydrates produced in photosynthesis are used in the process of cellular respiration. Respiration is the harvesting of energy from the chemical breakdown of glucose and other carbohydrates. It is an aerobic process requiring oxygen (O_2). The reaction between glucose and oxygen breaks chemical bonds in the glucose molecule, producing carbon dioxide and water. The energy generated from the breaking of these chemical bonds is stored in the form of ATP (the same chemical compound formed in the light reactions of photosynthesis).

All processes associated with life require chemical energy in the form of ATP generated by cellular respiration. Thus, all living cells carry out the process of respiration in specialized organelles called mitochondria.

Since leaves both take in carbon dioxide in the process of photosynthesis and produce carbon dioxide in the process of respiration, a simple economic approach can be used to explore the balance of these two processes. This approach is referred to as the carbon balance. If we define carbon (more precisely, CO_2) as the currency, photosynthesis represents income and respiration expenditure. The difference between the rate of carbon uptake in photosynthesis and carbon loss in respiration is the net gain, referred to as **net photosynthesis**. The rates of carbon uptake and loss, and therefore net photosynthesis, are typically measured in mol CO_2/unit leaf area (or mass)/unit time.

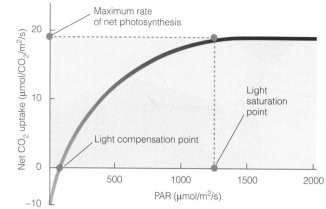

FIGURE 6.1 Response of photosynthetic activity to available light. The plant increases its rate of photosynthesis as the light level increases up to a maximum rate known as the light saturation point. After this point, any increase in PAR results in a decline in photosynthesis, or photoinhibition. The light compensation point is the light intensity at which the uptake of CO_2 for photosynthesis equals the loss of CO_2 in respiration.

6.2 The Light a Plant Receives Affects Its Photosynthetic Activity

Solar radiation provides the energy on which the light reactions for the conversion of CO_2 into glucose depend. The availability of light (PAR) to the leaf will directly influence the rate at which photosynthesis proceeds (Figure 6.1). As the amount of light declines, the rate of carbon uptake in photosynthesis will eventually decline to a level at which it equals the rate of carbon loss in respiration. At this point, the rate of net photosynthesis is zero. The light level (value of PAR) at which this occurs is called the **light compensation point**. At light levels below the compensation point, rate of carbon loss due to respiration exceeds the rate of uptake in the process of photosynthesis, and as a result, there is a net loss of carbon dioxide from the leaf to the atmosphere. This is what occurs at night.

As light levels exceed the light compensation point, photosynthetic rates increase. Eventually, the light reactions are no longer the limiting step in photosynthesis, and at this point the curve levels off (see Figure 6.1). The light level at which a further increase in light no longer results in an increase in the rate of photosynthesis is the **light saturation point**. In some plants adapted to extremely shaded environments, photosynthetic rates decline as light levels exceed sat-

uration. This negative effect of high light levels, called **photoinhibition**, can be the result of "overloading" the processes involved in the light reactions.

6.3 Photosynthesis Involves Exchanges Between the Plant and Atmosphere

The process of photosynthesis occurs in specialized cells within the leaf called **mesophyll**. For photosynthesis to take place within the mesophyll cells, CO_2 must be transported from the outside atmosphere into the leaf. In terrestrial (land) plants, CO_2 enters through openings on the surface of the leaf called stomata (Figure 6.2) through the process of diffusion.

Diffusion is the movement of a substance from areas of higher to lower concentration. Concentrations of CO_2 are often described in units of parts per million (ppm) of air. A CO_2 concentration of 355 ppm would be 355 units of CO_2 for every one million units of air. Substances flow from areas of high concentration to areas of low concentration until equilibrium is achieved—that is, until the concentrations in the two areas are equal. As long as the concentration of CO_2 in the air outside the leaf is greater than that inside the leaf, CO_2 will continue to diffuse through the stomata.

As CO_2 diffuses into the leaf through the stomata, why don't the concentrations inside and outside the leaf come into equilibrium? As CO_2 is transformed into glucose during the process of photosynthesis, the

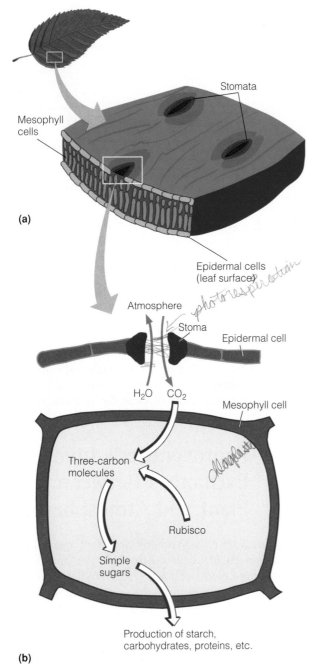

(a)

(b)

FIGURE 6.2 (a) Cross-section of a leaf showing stomata, mesophyll cells, and epidermal cells. (b) The C$_3$ pathway of photosynthesis. Carbon dioxide from the atmosphere diffuses into the leaf through the stoma to the mesophyll cells, where it is transformed into 3-carbon molecules.

concentration inside the leaf declines. As long as photosynthesis occurs, the gradient remains. If photosynthesis were to stop and the stomata were to remain open, CO$_2$ would diffuse into the leaf until the internal CO$_2$ equaled the outside concentration. But this is not how plants work.

When photosynthesis and the demand for CO$_2$ are reduced for any reason, the stomata tend to close,

reducing flow into the leaf. The reason for this closure is that stomata play a double role. As CO$_2$ diffuses into the leaf through the stomata, water vapor inside the leaf diffuses out through the same opening. This water loss through the stomata is called **transpiration.**

The rate at which water moves from inside the leaf, through the stomata, and into the surrounding outside air depends on the diffusion gradient of water vapor from inside to outside the leaf. Like CO$_2$, water vapor moves (diffuses) from areas of high concentration to areas of low concentration, from wet to dry. For all practical purposes, the air inside the leaf is saturated with water, so the outflow of water is a function of the amount of water vapor in the air—the relative humidity (see Chapter 3, Section 3.6, Figure 3.18). The drier the air (lower the relative humidity), the more the water inside the leaf will diffuse through the stomata into the outside surrounding air. The leaf must replace the water lost to the atmosphere; otherwise it will wilt and die.

As a plant loses water by transpiration through the leaves, it has to replace it by water taken up from the soil. To do so, plants need to maintain a negative pressure gradient from the soil to the atmosphere (Figure 6.3) driven by transpiration. Plants pull water from the soil (where water pressure is the highest) toward the atmosphere (where water pressure is the lowest).

As long as the relative humidity of the atmosphere is below 100 percent, there is less water in the atmosphere than in the plant, where relative humidity is effectively 100 percent. As the atmosphere becomes drier, its capacity to accept water increases (see Figure 3.18). Because of the difference in moisture outside and inside the plant, moisture escapes through the stomata and evaporates. To replace the lost water, the plant has to transport water from roots to leaves through its conductive system (xylem). As the replacement water moves up, water pressure drops below it, all the way down to the fine rootlets in contact with soil particles and pores. This tension pulls more water from the soil to the root and up through the stem to the leaf. The movement is like sucking water up a straw.

Transpiration continues while the amount of energy striking the leaf is enough to evaporate water, the stomata are open, and moisture is available in the soil. As long as plants can maintain a gradient between roots and soil, they continuously remove water from the soil. If the amount of water in the soil drops below a point where the plant can maintain the pressure gradient, stomata will close and the movement of water through the plant will cease until precipitation replenishes the supply of water in the soil.

The rate of water loss varies with daily environmental conditions, such as humidity and temperature,

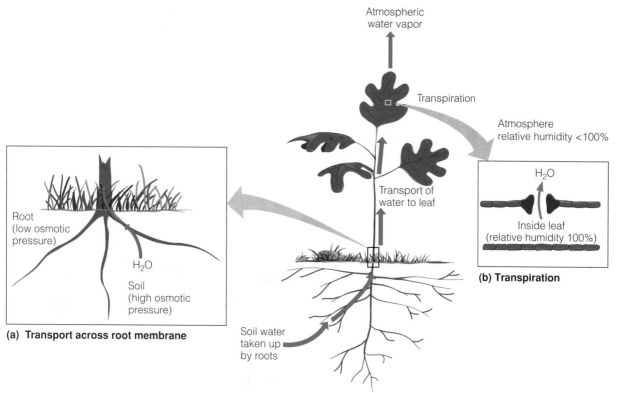

FIGURE 6.3 Transport of water along a pressure gradient from soil to leaves to air. (a) The solute concentration in the roots is higher than that in the soil, so water moves into the roots. At the other end, moisture in the atmosphere is less than that in the leaves. (b) Water moves to the atmosphere through transpiration, decreasing the osmotic pressure within the plant, which now draws more water from the soil.

and with the characteristics of plants. Opening and closing the stomata is probably the most important regulator of water loss through the plant. The trade-off between the uptake of CO_2 and the loss of water through the stomata results in a direct link between water availability in the soil and the ability of the plant to carry out photosynthesis. To carry out photosynthesis, the plant must open its stomata; but when it does, it will lose water, which it must replace. If water is scarce, the plant must balance the opening and closing of the stomata, taking up enough CO_2 while minimizing the loss of water. The ratio of carbon fixed (photosynthesis) per unit of water lost (transpiration) is called the **water-use efficiency.**

The process of CO_2 diffusion and photosynthesis in submerged aquatic plants is similar to that in terrestrial plants with one major exception. Submerged plants lack stomata. Dissolved CO_2 and bicarbonates (HCO_3) can diffuse directly across the outer cell walls of the plant. Under water, the tradeoff between carbon uptake and water loss does not exist. However, other constraints related to oxygen and salinity come into play (see Chapter 4, Sections 4.15 to 4.17).

6.4 Plant Temperatures Reflect Their Energy Balance with the Surrounding Environment

The rates of both carbon uptake in photosynthesis and the loss of carbon in respiration are dependent on the prevailing temperature. Both photosynthesis and respiration increase with temperature, and the resulting rate of net photosynthesis reflects the balance between these two processes (Figure 6.4). Extreme heat, however, damages enzymes and proteins, breaking down the photosynthetic process, resulting in a marked decline in net photosynthesis. It is the temperature of the leaf, not the air, that controls the rate of these two processes, and leaf temperature is a function of the exchange of energy (radiation) between the leaf and its surrounding environment.

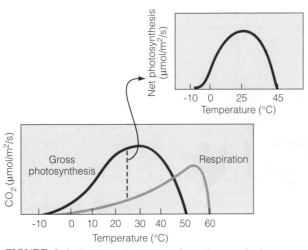

FIGURE 6.4 The general response of net photosynthetic rate to temperature. Here the optimal temperature is between 20°C and 30°C.

Plants absorb both shortwave (solar) and long-wave (thermal) radiation (see Chapter 4, Section 4.1). Plants reflect some of this solar radiation and emit long-wave radiation back to the atmosphere. The difference between the radiation a plant receives and that which it reflects and emits back to the surrounding environment is the net energy balance of the plant (R_n). The net energy balance of a plant is analogous to the concept of the energy balance of Earth presented in Chapter 3 (Section 3.1, Figure 3.2). Of the net radiation absorbed by the plant, some is used in metabolic processes and stored in chemical bonds—namely, in the processes of photosynthesis and respiration. This quantity is quite small, typically less than 5 percent of R_n. The remaining energy heats the leaves and the surrounding air. On a clear, sunny day, the amount of energy plants absorb can raise internal leaf temperatures well above ambient (air or water temperature). Internal leaf temperatures may go beyond the optimum for photosynthesis and possibly beyond critical levels (see Figure 6.4).

To maintain internal temperatures within the range of tolerance (positive net photosynthesis; see Figure 6.4), plants must dissipate heat to the surrounding environment. Terrestrial plants lose absorbed heat by convection and evaporation, and aquatic plants primarily by convection (see Chapter 4, Section 4.13). Convective loss depends upon the difference between the temperature of the leaf and the surrounding air (or water). If the temperature of the leaf is higher than that of the surrounding air, the leaf loses heat to the air moving over it. Evaporation occurs during transpiration. As plants transpire water from their leaves to the surrounding atmosphere through the stomata, the leaves lose energy and their temperature declines.

The size and shape of their leaves influence the ability of plants to lose heat through convection. Deeply lobed leaves, like those of some oaks, and small, compound leaves, like those of the black locust *(Robinia pseudoacacia)*, lose heat more effectively than broad, unlobed leaves. They expose more surface area per volume of leaf to the air for exchange of heat. The ability of terrestrial plants to dissipate heat by evaporation is dependent on the rate of transpiration, which is influenced by both the relative humidity of the air and the availability of water to the plant (Section 6.2).

6.5 Carbon Gained in Photosynthesis Is Allocated to the Production of Plant Tissues

Thus far, our discussion of plant carbon balance has focused on net photosynthesis, the balance between the processes of photosynthesis and respiration in the leaves of green plants. However, plants are not composed only of leaves; they also have roots and supportive tissues such as stems. Keeping with our simple economic model of income and expenditure, the net uptake of carbon by the whole plant (net income) will be the difference between the uptake of carbon in photosynthesis minus the loss of carbon in respiration (Figure 6.5a). The total carbon uptake or gain per unit time (income) will be the product of the average rate of carbon uptake in photosynthesis per unit of leaf area (photosynthetic surface) multiplied by the total surface area of leaves. Since all living cells respire, the total loss of carbon in respiration per unit time (expenditure) will be a function of the total mass of living tissue—that is, the sum of leaf, stem, and root tissues. The net carbon gain for the whole plant per unit time is the difference between these two values (carbon gain and carbon loss). This net carbon gain is then allocated to a variety of processes. Some of the carbon will be used in maintenance, and the rest in the synthesis of new tissues in plant growth and reproduction.

How the net carbon gain is allocated will have a major influence on the survival, growth, and reproduction of the plant. The acquisition of essential resources necessary to support photosynthesis and growth involves different plant tissues. Leaf tissue is the photosynthetic surface, providing access to the essential resources of light and CO_2. The root tissue provides access to below-ground resources such as

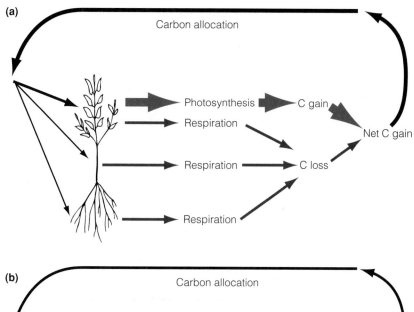

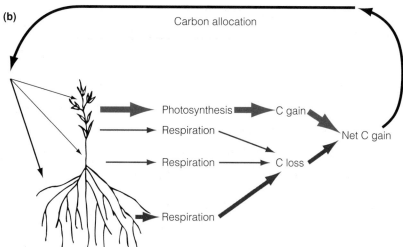

FIGURE 6.5 The net carbon gain of a plant is the difference between carbon gain in photosynthesis and carbon loss in respiration. Since leaves (or, more generally, photosynthetic tissues) are responsible for carbon gain, while all plant tissues respire (leaves, stem, and roots), the net carbon gain will be directly influenced by the pattern of carbon allocation to the production of different plant tissues. (a) Increased allocation to leaves will increase carbon gain (photosynthesis) relative to carbon loss (respiration) and therefore increase net carbon gain. (b) Increased allocation to roots will have the opposite effect, decreasing carbon gain relative to carbon loss. The result is a lower net carbon gain by the plant.

water and nutrients in the soil. Stem tissue provides vertical support, elevating leaves above the ground and increasing access to light by reducing the chance of being shaded by taller plants (see Chapter 4, Section 4.2). It also provides the conductive tissue necessary to move water and nutrients from the roots to other parts of the plant. As we discuss in the following sections, the availability of these essential resources for plant growth influences the allocation of carbon to the production of various tissues.

Under ideal conditions (no resource limitations), the allocation of carbon to the further production of leaf tissue will promote the fastest growth. Increased allocation to leaf tissue increases the photosynthetic surface, which increases the rate of carbon uptake as well as carbon loss due to respiration (both income and expenditures). Allocation to all other tissues, such as stem and root, increases the respiration rate but does not directly increase the capacity for carbon uptake through photosynthesis (expenditure only). The consequence is the reduction of net carbon gain by the plant (Figure 6.5b). However, the allocation of carbon to the production of stem and root tissue is

essential for acquiring the key resources necessary to maintain photosynthesis and growth. As these resources become scarce, it becomes increasingly necessary to allocate carbon to the production of these tissues at the expense of producing leaves. The implications of these shifts in patterns of carbon allocation will be addressed in the following sections.

6.6 Constraints Imposed by the Physical Environment Have Resulted in a Wide Array of Plant Adaptations

In Part II (Chapters 3–5), we explored variation in the physical environment over Earth's surface: the salinity, depth, and flow of water; spatial and temporal patterns in climate (precipitation and temperature); variations in geology and soils. In all but the most extreme of these environments, autotrophs—green plants—harness the energy of the Sun to fuel the conversion of carbon dioxide into glucose in the process of photosynthesis. To survive, grow, and reproduce, plants must maintain a positive carbon balance, converting enough carbon dioxide into glucose to more than offset the expenses of respiration (photosynthesis > respiration). To accomplish this, a plant must acquire the essential resources of light, carbon dioxide, water, and mineral nutrients, and tolerate other features of the environment that have a direct influence on basic plant processes, such as temperature, salinity, and pH. Although often discussed and even studied as though they are independent of each other, the adaptations exhibited by plants to these features of the environment are not independent, for reasons relating to both the physical environment and the plants themselves.

Many features of the physical environment that have a direct influence on plant processes are themselves interdependent. For example, the light, temperature, and moisture environment are all linked through a variety of physical processes discussed in Chapters 3 and 4. The amount of solar radiation not only influences the availability of visible light (photosynthetically active radiation) required for photosynthesis, but also has a direct influence on the ambient temperature (both air and water). In addition, the air temperature has a direct influence on the relative humidity, a key feature influencing the rate of transpiration and demand for water from the soil. For this reason, we see a correlation, or interdependence in the adaptations of plants to variations in these environmental factors. Plants adapted to sunny environments must be able to deal with the higher demand for water associated with higher temperatures and lower relative humidity and have characteristics such as smaller leaves and increased production of roots.

In other cases, there are tradeoffs in the ability of plants to adapt to limitations imposed by multiple environmental factors, particularly resources. One of the most important of these tradeoffs is in the acquisition of above- and below-ground resources. Allocation of carbon to the production of leaves and stems provides increased access to the resources of light and carbon dioxide, but at the expense of allocation of carbon to the production of roots. Likewise, allocation to the production of roots increases access to water and soil nutrients, but limits the production of leaves and therefore the rate of carbon gain in photosynthesis. The set of characteristics (adaptations) that allow a plant to successfully survive, grow, and reproduce under one set of environmental conditions inevitably limits its ability to do equally well under different environmental conditions. We explore the consequences of this simple premise in the following sections.

6.7 Species of Plants Are Adapted to Either High or Low Light

Although the amount of solar radiation reaching Earth's surface varies diurnally, seasonally, and geographically (see Chapter 3, Section 3.1), one of the major factors influencing the amount of light (PAR) a plant receives is the presence of other plants (see Chapter 4, Section 4.2, and Quantifying Ecology 4.1). Although the amount of light that reaches an individual plant varies continuously as a function of the area of leaves above it, plants live in one of two qualitatively different light environments—sun or shade—depending on whether they are overtopped by other plants (see Focus on Ecology 6.1: Shaded by Water). Plants have evolved a range of physiological and morphological adaptations that allow individuals to survive, grow, and reproduce in these two different light environments.

The relationship between the availability of light and the rate of photosynthesis varies among plants (Figure 6.6). Plants adapted to shaded environments tend to have a lower light compensation point, a lower light saturation point, and a lower maximum rate of photosynthesis than plants adapted to high light environments.

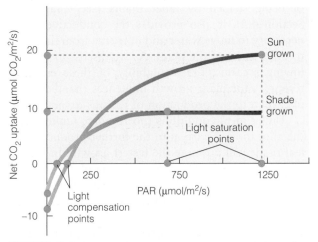

FIGURE 6.6 Patterns of photosynthetic response to light availability for shade-tolerant (sun-adapted) and shade-intolerant (shade-adapted) plants. Shade-tolerant plants have a lower light compensation point and a lower light saturation point than do shade-intolerant plants.

SHADED BY WATER

Because of the rapid attenuation of light with water depth (see Chapter 4, Figure 4.3), most aquatic plants and algae live in the equivalent of a shaded environment. Aquatic plants exhibit the same light response shown in Figure 6.1. However, they more commonly experience inhibition of photosynthesis when exposed to higher light levels in the water column. The photosynthetic rate of phytoplankton, free-floating aquatic plants, is highest at intermediate levels and decreases as the light intensity either increases or decreases (Figure A). On the other hand, the rate of respiration does not change much with depth. This means that as the phytoplankton go deeper in the water column, the rate of net photosynthesis declines as the light intensity decreases, until at some point the rate of photosynthesis is equal to the rate of respiration. This zone is called the **compensation depth,** and corresponds to the depth at which the availability of light is equal to the light compensation point discussed earlier (see Figure 6.1). Different species of phytoplankton have different photosynthetic responses to light (see Figure A), and will therefore have different compensation depths. The depth at which most aquatic plants and algae grow corresponds to the light intensity that they find most favorable for photosynthesis. Differences in photosynthetic

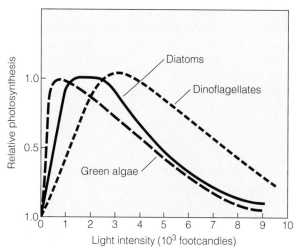

FIGURE A Response of photosynthesis to available light for three groups of phytoplankton. (From T. R. Parsons, M. Takahashi, and B. Hargrave. *Biological oceanographic processes,* 3rd edition, Pergamon Press, 1984.)

response are further influenced by seasonal variations in water temperature and light, and by adaptations for different wavelengths of visible light.

These differences relate in part to lower concentrations of the photosynthetic enzyme rubisco (Section 6.1) found in shade-grown plants. Plants must expend a large amount of energy and nutrients to produce rubisco. In shaded environments, low light, not the availability of rubisco to catalyze the fixation of CO_2, limits the rate at which photosynthesis can proceed. The plant produces less rubisco as a result. (In contrast, production of chlorophyll, the light-harvesting pigment in the leaves, often increases.) The reduced energy cost of producing rubisco and other compounds involved in photosynthesis functions to lower the rate of respiration. Since the light compensation point is the light level (PAR) necessary to maintain photosynthesis (income) at a rate that exactly offsets the expenditure of respiration (net photosynthesis = 0), the lower rate of respiration is offset by a lower rate of photosynthesis, a rate requiring less light. The result is a lower light compensation point. However, this same reduction in enzyme concentrations limits the maximum rate at which photosynthesis can occur when light is abun-

dant (high PAR). Thus, it lowers the light saturation point and maximum rate of photosynthesis.

This shift in the characteristics of the light response curve is a tradeoff for the plant. The decreased enzyme concentrations and other reductions in metabolic costs enable the plant to decrease respiration rates and maintain photosynthesis under reduced light levels. This reduced metabolic activity, however, limits the rates of photosynthesis when light levels are high. Conversely, when a plant maintains a high level of enzyme concentrations, the cost of respiration is high. However, the plant can maintain high levels of photosynthesis when more light is available. The high costs of respiration do not allow the plant to continue to maintain a positive net photosynthetic rate under low light conditions.

In addition to the changes in photosynthesis, leaves grown under low light conditions often exhibit a different morphology (size and shape). In general, leaves grown under reduced light conditions are larger (in surface area) and thinner than those grown under high light levels. The increased surface area

LEAF MORPHOLOGY AND LIGHT

The light under which a plant grows influences the morphology (size and shape) of its leaves. We find this effect both among individuals of the same species growing under different light conditions and among leaves within the same plant. As Figure A shows, the leaves of a single red oak tree *(Quercus rubra)* vary in size and shape from the top to the bottom of the tree. Leaves at the top of the tree receive higher levels of solar radiation and experience higher temperatures than leaves at the bottom. Upper leaves are smaller and more lobed. The lobes increase the surface area of the leaf in contact with the air. This design lets more heat dissipate through convection.

Leaves from the bottom of the canopy are larger, thinner, and much less lobed. Thinner leaves allow for a greater surface area of leaf per unit (weight) of carbon and other nutrients needed to construct the leaf. The amount of CO_2 that a plant can take up for photosynthesis and use for growth and maintenance depends upon both the rate of photosynthesis per unit of leaf area and the total leaf area over which photosynthesis occurs. The

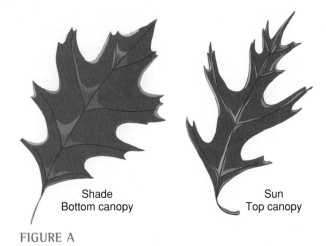

Shade
Bottom canopy

Sun
Top canopy

FIGURE A

larger, thinner leaves partially compensate for the lower rates of photosynthesis per unit leaf area at lower levels of light. They increase the surface area for capturing light— the limiting resource (see Figure A).

aids in the capture of light (see Focus on Ecology 6.2: Leaf Morphology and Light). In addition to producing broader, thinner leaves, plants grown under shaded conditions allocate a greater proportion of their net carbon gain to leaf production and less to the roots (Figure 6.7). Just as with the changes in leaf morphology (thinner, broader leaves), this shift in allocation from roots to leaves increases the surface area for the capture of the limiting resource of light. Given the lower rate of net photosynthesis associated with low light conditions, the increase in leaf area (photosynthetic surface area) is essential for the plant to maintain a positive carbon balance (income greater than expenditure). Should the plant carbon balance become negative, the plant will need to draw on carbon reserves—carbohydrates stored in various plant tissues. Plants cannot maintain this condition for long, and eventually they will die.

The patterns in photosynthetic and morphological response to variations in the light environment described above can occur among plants of the same species grown under different light conditions (as with Figure 6.7), and even among leaves on the same plant with different exposures to light (as with Focus

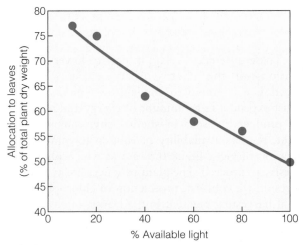

FIGURE 6.7 Changes in allocation to leaves for broad-leaved peppermint *(Eucalyptus dives)* seedlings grown under different light environments in the greenhouse. Allocation to leaves is expressed as a percentage of the total dry weight of the plant at harvest. Each point represents the average response of 5 seedlings. Light availability is expressed as a percentage of full sunlight. Levels of shading were controlled by shade-cloth of varying density. The increased allocation to leaves, together with the production of thinner leaves, acts to increase the photosynthetic surface for the capture of light. (Smith et al. 2002.)

on Ecology 6.2). Such change in the physiology or form of an organism in response to changes in environmental conditions is called **acclimatization**. These differences, however, are most pronounced between species of plants adapted to high and low light environments. These adaptations represent genetic differences in the potential response of the plant species to the light environment. Plant species adapted to high light environments are called **shade-intolerant** species or sun-adapted species. Plant species adapted to low light environments are called **shade-tolerant** species or shade-adapted species.

The difference between shade-tolerant and shade-intolerant species in their net carbon gain (growth rate) and survival under different light environments is illustrated in the work of Caroline Augspurger of the University of Illinois. She conducted a series of experiments designed to examine the influence of light availability on seedling survival and growth for a variety of tree species that inhabit the tropical rain forests of Panama. Augspurger grew tree seedlings planted in pots under two light environments—sunlight and shade. These two treatments mimic the conditions found either under the shaded environment of a continuous forest canopy, or in openings or gaps in the canopy caused by the death of large trees. The findings show a marked difference between shade-tolerant and shade-intolerant species in their patterns of seedling survival and growth when grown under high and low light conditions (Figure 6.8). Shade-tolerant species show little difference in survival and growth rates under sunlight and shade conditions. In contrast, both survival and growth rates of shade-intolerant species were dramatically reduced under shade condi-

tions. These observed differences are a direct result of the difference in the adaptations relating to photosynthesis and carbon allocation discussed earlier.

The dichotomy in adaptations between shade-tolerant and shade-intolerant species reflects a trade-off between characteristics that enable a species to maintain high rates of net photosynthesis and growth under high light conditions and the ability to continue survival and growth under low light conditions. The changes in biochemistry, physiology, leaf morphology, and carbon allocation exhibited by shade-tolerant species enable them to reduce the amount of light required to survive and grow. However, these same characteristics limit their ability to maintain high rates of net photosynthesis and growth when light levels are high. In contrast, plants adapted to high light environments (shade-intolerant species) can maintain high rates of net photosynthesis and growth under high light conditions, but at the expense of continuing photosynthesis, growth, and survival under shaded conditions.

6.8 The Link Between Water Demand and Temperature Influences Plant Adaptations

As with the light environment, terrestrial plants have evolved a range of adaptations in response to variations in precipitation and soil moisture. As we saw in

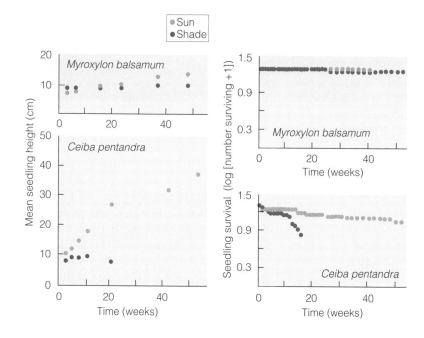

FIGURE 6.8 Seedling survival and growth over a period of one year for seedlings of two tree species on Barro-Colorado Island, Panama, grown under sun and shade conditions. *Ceiba pentandra* is a shade-intolerant species; *Myroxylon balsamum* is shade-tolerant. (From Augspurger 1982.)

WHEN WET CAN BE DRY: SALINE HABITATS

Plants of salt marshes and other saline habitats grow in a physiologically dry environment. Salinity limits the amount of water they can absorb (see Chapter 4). Known as **halophytes,** they take in water that contains high levels of solutes. Characteristically, they accumulate high levels of ions within their cells, especially in the leaves. The high solute concentration, which may equal or exceed that of seawater, allows halophytes to maintain a high cell water content in the face of a low external osmotic potential (see Chapter 4, Section 4.6).

Taking up water heavy in sodium and chloride, some halophytes dilute it with water they have stored in their tissues. Some plants have salt-secreting glands that deposit excess salt on the leaves to be washed away by rain. Others remove salts mechanically at the root membranes.

Plants of saline deserts encounter worse problems. Not only do they grow in salty soil; they must also endure dry conditions. A number of desert plants allow only certain ions to pass across root membranes and keep others out. That selectivity allows these plants to absorb essential nutrients from the soil and maintain osmotic pressure.

Halophytes, however, vary in their degree of tolerance to salt. Some plants, such as salt marsh hay grass *(Spartina patens),* grow best at low salinities. Others, such as salt marsh cordgrass *(S. alternifolia),* do best at moderate levels of salinity. A few, such as glasswort *(Salicornia* spp.), tolerate high salinities (Figure A).

FIGURE A *Salicornia,* a succulent annual or perennial depending on the species, is a halophyte, able to exist only in saline environments. It is edible and sometimes referred to as "marsh asparagus."

the earlier discussion of transpiration, however, the demand for water is linked to temperature. Recall from Chapter 3 (Section 3.6) that the amount of moisture that can be held in a volume of air increases with temperature. As a result, the amount of water required by the plant to offset losses from transpiration will likewise increase with temperature.

When the atmosphere or soil is dry (see Focus on Ecology 6.3: When Wet Can Be Dry), plants respond by partially closing the stomata and opening them for shorter periods of time. In the early period of water stress, a plant closes its stomata during the hottest part of the day. It resumes normal activity in the afternoon. As water becomes scarcer, the plant opens its

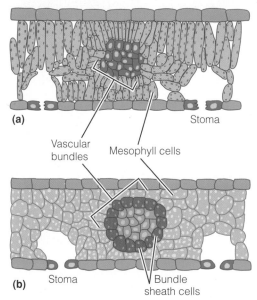

(a)

Vascular
bundles   Mesophyll cells

Stoma

(b)   Stoma   Bundle
sheath cells

FIGURE 6.9 Comparison of leaf anatomy: (a) a C_3 plant leaf; (b) a C_4 plant leaf. In the C_4 plant, bundle sheath cells form a ring around the vascular bundle.

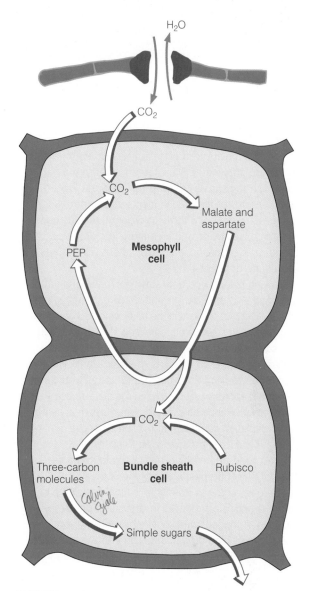

H_2O

CO_2

CO_2

PEP   **Mesophyll cell**   Malate and aspartate

CO_2

Three-carbon molecules   **Bundle sheath cell**   Rubisco

Calvin cycle

Simple sugars

FIGURE 6.10 The C_4 pathway of photosynthesis. Different reactions take place in the mesophyll and bundle sheath cells. Compare to the C_3 pathway (Figure 6.2).

stomata only in the cooler, more humid conditions of morning. This closure reduces the loss of water through transpiration, but it also reduces CO_2 diffusion into the leaf and the dissipation of heat through evaporative cooling. As a result, the rate of photosynthesis declines, and leaf temperatures may rise.

In tropical regions with distinct wet and dry seasons (see Chapter 3, Section 3.7), some species of trees and shrubs have evolved the characteristic of dropping their leaves at the onset of the dry season. These plants are termed drought deciduous. In these species, leaf senescence occurs at the onset of the dry season, and new leaves are grown just prior to the onset of rains (see Focus on Ecology 6.4: Too Much of a Good Thing).

Some species of plants, referred to as C_4 and CAM plants, have evolved a modified form of photosynthesis that functions to increase water-use efficiency in warm, dry environments. The modification involves an additional step in the conversion (fixation) of CO_2 into glucose.

In C_3 plants, the capture of light energy and the transformation of CO_2 into glucose occur in the mesophyll cells. The products of photosynthesis move into the vascular bundles, part of the plant's transport system, where they can be transported to other parts of the plant. In contrast, plants possessing the C_4 pathway have a leaf anatomy different from that of C_3 plants (Figure 6.9). C_4 plants have vascular bundles surrounded by distinctive **bundle sheath cells.** A layer of mesophyll cells concentrates around each of the bundle sheaths. C_4 plants divide photo-

synthesis between the two types of cells, the mesophyll and the bundle sheath cells.

The first step takes place in the mesophyll cells. Unlike C_3 plants, where the enzyme rubisco catalyzes the transformation of CO_2 into 3-carbon molecules, in C_4 plants another enzyme called **PEP** (phosphoenolpyruvate carboxylase) catalyzes the fixation of CO_2 into 4-carbon acids, malate and aspartate. Then malate and aspartate are transported to the bundle sheaths (Figure 6.10). There, enzymes break down the acids to form CO_2, reversing the process that is carried out in the mesophyll cells. In the bundle sheath cells, the CO_2 is transformed into sugars using the same enzyme that is used by C_3 plants—rubisco.

TOO MUCH OF A GOOD THING: PLANTS' ADAPTATIONS TO FLOODING

Too much water can place as much stress on plants as too little water. Plant species differ in their ability to deal with the stresses of flooding and waterlogged soils. Symptoms of flooding in plants intolerant of such conditions are similar to those of drought, including closing stomata, yellowing and premature loss of leaves, wilting, and rapid reduction in photosynthesis. The causes, however, are different.

Growing plants need both sufficient water and a rapid gas exchange (carbon dioxide and oxygen) with their environment. Much of this exchange takes place between the roots and the air spaces within the soil. When water fills soil pores, such gas exchange cannot take place. The plants, experiencing depressed O_2 levels, in effect asphyxiate; they drown. Roots depend upon gaseous O_2 to carry on aerobic respiration. Lacking oxygen, flooded roots are forced to shift to an alternative anaerobic metabolism. Further, anaerobic (oxygen-depleted) conditions in the soil allow chemical reactions that produce substances toxic to plants.

In response to these anaerobic conditions, some species of plants accumulate ethylene in their roots. Ethylene gas, a growth hormone, is highly insoluble in water. It is normally produced in small amounts in the roots. Under flooded conditions, ethylene diffusion from the roots slows, and oxygen diffusion into the roots virtually stops. Ethylene then accumulates to high levels in the root. Ethylene stimulates cells in the root to self-destruct and separate to form interconnected gas-filled chambers called **aerenchyma** (Figure A). These chambers, typical of aquatic plants, allow some exchange of gases between submerged and better-aerated roots.

In other plants, especially woody ones, the original roots, deprived of oxygen, die. In their place, adventitious roots (roots that arise in positions where roots normally would not grow) emerge from the submerged part of the stem. Replacing the functions of the original roots, they spread horizontally along the soil surface where oxygen is available. Some plants found in poorly drained soils—for example, red maple (*Acer rubrum*) and white pine (*Pinus strobus*)—develop shallow, horizontal root systems to cope with flooding. These shallow root systems make them highly susceptible to drought and windthrow.

Some woody species can grow on permanently flooded sites. Among them are bald cypress (*Taxodium distichum*), mangroves, willows, and water tupelo (*Nyssa aquatica*). Bald cypresses growing on sites with fluctuating water tables develop knees or **pneumatophores,** specialized growths of the root system (Figure B). These growths may be beneficial, although they are not necessary for survival. Pneumatophores on mangrove trees help with gas exchange and provide oxygen to roots during tidal cycles (see Chapter 29).

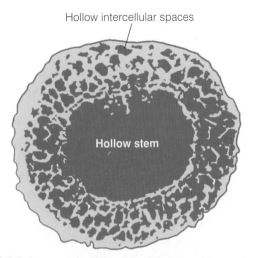

FIGURE A Aerenchyma tissue (cross section) with numerous intercellular spaces carries oxygen to the roots of aquatic plants.

FIGURE B Pneumatophores, or "knees," are typical features of a bald cypress swamp.

One of the advantages of C_4 photosynthesis is its effective use of CO_2. Due to the limitations of diffusion, the CO_2 concentration within the mesophyll will not go above the concentration of the outside air. This limits the CO_2 available to drive the chemical reactions of photosynthesis. However, carbon dioxide in the bundle sheath cells can reach concentrations 6 times greater than concentrations in the outside air. Under most conditions, the rate of photosynthesis is limited by the concentration of CO_2 inside the leaf. Therefore, the maximum rate of photosynthesis is generally greater in C_4 plants.

To understand the adaptive advantage of the C_4 pathway, we must go back to the tradeoff in terrestrial plants between the uptake of CO_2 and the loss of water through the stomata. Due to the higher photosynthetic rate, C_4 plants exhibit greater water-use efficiency. That is, for a given degree of stomatal opening and water loss, C_4 plants typically fix more carbon. This increased water-use efficiency can be a great advantage in hot, dry climates where water is a major factor limiting plant growth. However, it comes at a price. The C_4 pathway has a higher energy expenditure because of the need to produce the extra enzyme, PEP.

In the hot deserts of the world, environmental conditions are even more severe. Solar radiation is high, and water is scarce. To counteract these conditions, a small group of desert plants, mostly succulents in the families Cactaceae (cacti), Euphorbiaceae, and Crassulaceae, use a third type of photosynthetic pathway. It is the Crassulacean acid metabolism, CAM for short. The **CAM** pathway is similar to the C_4 pathway in that CO_2 is first transformed into the 4-carbon acid malate using the enzyme PEP. The malate is later turned back into CO_2 and then transformed into glucose using the enzyme rubisco. Unlike C_4 plants, however, in which these two steps are physically separate (in mesophyll and bundle sheath cells), both steps occur in the mesophyll cells, but at separate times (Figure 6.11).

CAM plants open their stomata at night, taking up CO_2 and converting it to malate using PEP. During the day, the plant closes its stomata and reconverts the malate into CO_2, which it fixes into glucose. By opening their stomata at night when temperatures are lowest and relative humidity is highest, CAM plants conserve water. Although CAM plants have high water-use efficiency, the pathway is an inefficient use of energy. Under the extremely hot and dry conditions of the desert, however, this strategy allows the continuation of photosynthesis.

Plants may also respond to a decrease in available soil water by increasing the allocation of carbon (see

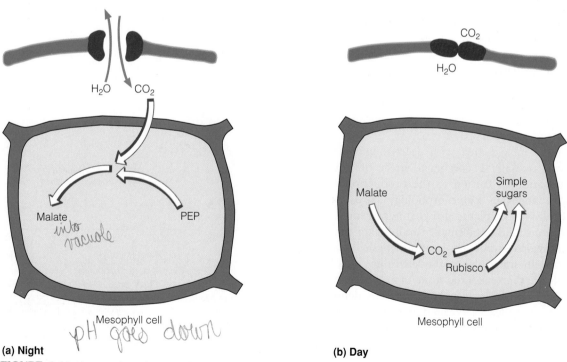

(a) Night   **(b) Day**

FIGURE 6.11 Photosynthesis in CAM plants. (a) At night, the stomata open; the plant loses water through transpiration; and CO_2 diffuses into the leaf. CO_2 is stored as malate in the mesophyll, to be used in photosynthesis by day. (b) During the day, when stomata are closed, the stored CO_2 is refixed in the mesophyll cells using light energy.

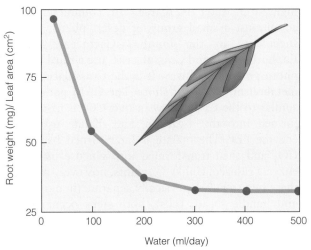

FIGURE 6.12 Relationship between plant water availability and the ratio of root mass (mg) to leaf area (cm²) for broadleaved peppermint *(Eucalyptus dives)* seedlings grown in the greenhouse. Each point on the graph represents the average value for plants grown under the corresponding water treatment. As water availability decreases, plants allocate more carbon to the production of roots relative to leaves. This increased allocation to roots increases the surface area of roots for the uptake of water, while the decline in leaf area decreases water loss through transpiration. (Smith et al. 2002.)

Figure 6.5) to the production of roots, while decreasing the production of leaves (Figure 6.12). The increased production of roots allows the plant to explore a larger volume of soil from which to extract water, while the reduction in leaf area decreases the amount of solar radiation the plant intercepts and the surface area that is losing water through transpiration. The combined effect is to increase the uptake of water per unit leaf area while reducing the total amount of water that is lost to the atmosphere through transpiration.

The decline in leaf area with decreasing water availability is actually a combined effect of both reduced allocation of carbon to the production of leaves and changes in leaf morphology (size and shape). The leaves of plants grown under reduced water conditions tend to be smaller and thicker than those of individuals growing in more mesic (wet) environments. In some plants, the leaves are small; the cell walls are thickened; the stomata are tiny; and the vascular system for transporting water is dense. Some species have leaves covered with hairs that scatter incoming solar radiation, while others have leaves coated with waxes and resins that reflect light and reduce its absorption. All these structural features reduce the amount of energy striking the leaf and thus the loss of water through transpiration.

Although an increase in root production at the expense of leaf area functions to reduce water loss (and therefore demand), it also decreases carbon gain (photosynthesis) relative to carbon loss (respiration), reducing the net carbon gain and growth rate of the plant.

You can notice these changes in plant morphology if you grow plants of the same species under different soil-water conditions, as shown in Figure 6.12.

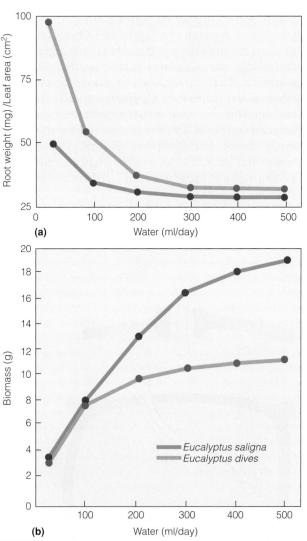

FIGURE 6.13 Comparison of patterns of carbon allocation and growth rate for two species of *Eucalyptus* along an experimental gradient of water availability. (a) Although both species exhibit the same patterns of response to declining water availability, the xeric species, *E. dives*, exhibits a consistently higher ratio of root mass (mg) to leaf area (cm²) than the mesic species, *E. saligna*, across the water gradient. (b) *E. saligna's* growth rate (biomass gain over period of experiment) continues to increase with increasing water availability. *E. dives* reaches maximum growth rate at intermediate water treatments. (Smith et al. 2002.)

They are most pronounced, however, between contrasting plant species adapted to wet and dry environments (Figure 6.13).

6.9 Plants Acclimate to the Prevailing Temperatures of Their Environment

The range of temperatures over which plants can maintain a positive rate of net photosynthesis (see Figure 6.4) vary both among species and among populations of a species growing in different temperature environments. In general, the temperature response of plants is closely related to the temperature of the environment in which they are found (Figure 6.14). This pattern suggests a high degree of acclimation (see Focus on Ecology 6.5: Degree-Days). Nevertheless, there are inherent differences in the temperature responses of species characteristic of different environments. These differences are most pronounced between plants using the C_3 and C_4 photosynthetic pathways (see Section 6.8). C_4 plants inhabit warmer, drier environments and exhibit higher optimal temperatures for photosynthesis (generally between 30°C and 40°C) than do C_3 plants (Figure 6.15).

Plants that inhabit seasonally cold environments, where temperatures drop below freezing for periods of time, have evolved a number of adaptations for survival. The leaves of some plant species can tolerate subzero (< 0°C) temperatures if the temperature decreases slowly, allowing ice to form in the cell wall. The effect is dehydration, which the plant can

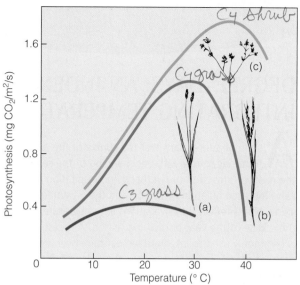

FIGURE 6.15 The effect of change in leaf temperature on the photosynthetic rates of C_3 and C_4 plants. (a) A C_3 plant, the north temperate grass *Sesleria caerulea*, exhibits a decline in the rate of photosynthesis as the temperature of the leaf increases. (b) A C_4 north temperate grass, *Spartina anglica*. (c) A C_4 shrub of the North American hot desert, Arizona honeysweet, *Tidestromia oblongifolia*. The C_4 species increase their rate of photosynthesis as the temperature of the leaf increases, up to a certain point. (Bjorkman 1973.)

reverse when the temperature rises. If the temperature falls too rapidly for dehydration to take place, ice crystals form within the cell. The crystals may puncture cell membranes. When the plant thaws, the cell contents spill out, giving the plant a cooked, wilted appearance.

The ability to tolerate extreme cold, referred to as frost hardening, is a genetically controlled characteristic that varies among species and among separated populations of the same species. In seasonally changing environments, plants develop frost hardening through the fall and achieve maximum hardening in winter. Plants acquire frost hardiness—the turning of cold-sensitive cells into hardy ones—through the formation or addition of protective compounds in the cells. They synthesize and distribute substances, such as sugars, amino acids, and other compounds, which function as antifreeze, lowering the temperature at which freezing occurs. Once growth starts in spring, plants lose this tolerance quickly and are susceptible to frost damage in late spring.

Producing the protective compounds that allow leaves to survive freezing temperatures requires a significant expenditure of energy and nutrients. Some species avoid these costs by shedding their leaves prior to the onset of the cold season. These plants are termed winter deciduous, and their leaves senesce during the fall. They are replaced during the

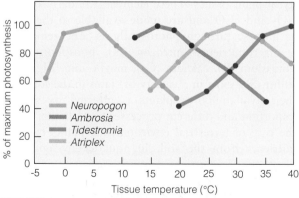

FIGURE 6.14 Relationship between net photosynthesis and temperature for a variety of terrestrial plant species from dissimilar thermal habitats: *Neuropogon acromelanus* (Arctic lichen), *Ambrosia chamissonis* (cool, coastal dune plant), *Atriplex hymenelytra* (evergreen desert shrub), and *Tidestromia oblongifolia* (summer-active desert perennial). (From Mooney et al. 1976.)

DEGREE-DAYS: AN INDEX FOR INTEGRATING TEMPERATURE OVER TIME

As temperatures vary over the course of the day and from day to day with the changing of the seasons, the relative rates of photosynthesis and respiration likewise vary. The accumulation of carbon and the growth of individual plants reflect these variations in temperature over the course of the year or growing season. A measure that is commonly used to integrate temperature over a single growing season to plant growth is the index of **degree-days.** The index of degree-days is calculated as the sum of the departures in temperatures above some **minimum base temperature** (Figure A). The base temperature typically reflects the temperature at which net photosynthesis is at or approaching zero (see Figure 6.4). The integration of mean daily temperatures above the base temperature therefore reflects photosynthetic activity and carbon accumulation as it relates to temperature. Values of degree-days have been correlated with growth rate for species in both managed (agricultural crops and forest plantations) and natural ecosystems.

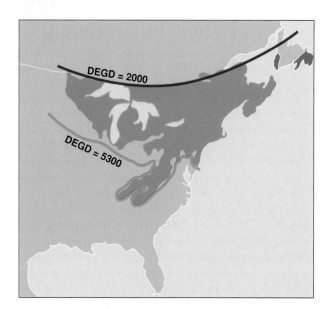

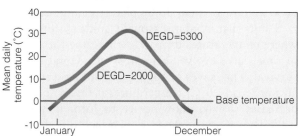

FIGURE A Mean daily temperature profiles and growing degree-days for two locations in North America using a base temperature of 0°C. Growing degree-days (DEGD) is the sum of the departures of temperatures above some minimum base temperature.

spring, when conditions are once again favorable for photosynthesis.

6.10 Plants Exhibit Adaptations to Variations in Nutrient Availability

Plants require 16 elements to carry out their metabolic processes and to synthesize new tissues. Thus, the availability of nutrients has many direct effects on plant survival, growth, and reproduction. Of the macronutrients (see Chapter 4, Section 4.14), carbon (C), hydrogen (H), and oxygen (O) form the major-ity of plant tissues. These elements are derived from CO_2 and H_2O and are made available to the plant as glucose through photosynthesis. The remaining 6 macronutrients—nitrogen (N), phosphorus (P), potassium (K), calcium (Ca), magnesium (Mg), and sulfur (S)—exist in a variety of states in the soil, and their availability to plants is affected by several important and different processes (see Chapter 5). In the case of terrestrial environments, plants take up nutrients from the soil. In aquatic environments, plants take up nutrients from the substrate or directly from the water.

The best example of the direct link between nutrient availability and plant performance involves nitrogen. Nitrogen plays a major role in photosynthesis. In Section 6.1, we examined two important compounds in photosynthesis—the enzyme rubisco and the pigment chlorophyll. Rubisco catalyzes the

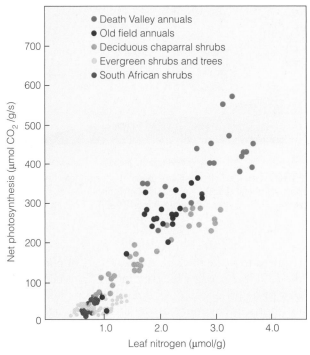

FIGURE 6.16 Influence of leaf nitrogen concentrations on maximum observed rates of net photosynthesis for a variety of species from differing habitats. (Adapted from Field and Mooney 1983.)

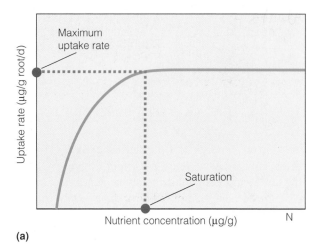

(a)

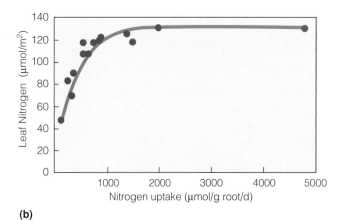

(b)

FIGURE 6.17 (a) Uptake of nitrogen by plant roots increases with concentration in soil until the plant arrives at maximum uptake. (b) Influence of root nitrogen uptake on leaf nitrogen concentrations. (Adapted from Woodward and Smith 1994.)

transformation of carbon dioxide into simple sugars, and chlorophyll absorbs light energy. Nitrogen is a component of both compounds; the plant requires nitrogen to make them. In fact, over 50 percent of the nitrogen content of a leaf is in some way involved directly with the process of photosynthesis, with much of it tied up in these two compounds. As a result, the (maximum) rate of photosynthesis for a species is correlated with the nitrogen content of its leaves (Figure 6.16).

The uptake of a nutrient depends upon both supply and demand. Figure 6.17a illustrates the typical relationship between the uptake of a nutrient and its concentration in soil. Note that the uptake rate increases with the concentration until some maximum rate. No further increase occurs above this concentration because the plant meets its demands. In the case of nitrogen, low concentrations in the soil or water mean low uptake rates. A lower uptake rate decreases the concentrations of nitrogen in the leaf (Figure 6.17b) and, consequently, the concentrations of rubisco and chlorophyll. Therefore, lack of nitrogen limits the growth of plants. A similar pattern holds for other essential nutrients.

We have seen that geology, climate, and biological activity alter the availability of nutrients in the soil (Chapter 5). As a consequence, some environments are rich in nutrients, while others are poor. How do plants from low-nutrient environments succeed?

Because plants require nutrients for the synthesis of new tissue, a plant's rate of growth influences its demand for a nutrient. In turn, the plant's uptake rate of the nutrient also influences growth. This relationship may seem circular, but the important point is that not all plants have the same inherent (maximum potential) rate of growth. In Section 6.7 (Figure 6.6), we saw how shade-tolerant plants have an inherently lower rate of photosynthesis and growth than shade-intolerant plants, even under high light conditions. This lower rate of photosynthesis and growth means a lower demand for resources, including nutrients. The same pattern of reduced photosynthesis occurs among plants that are characteristic of low-nutrient environments. Figure 6.18 shows the growth responses of two grass species when soil is enriched with nitrogen. The species that naturally grows in a high-nitrogen environment keeps increasing its rate of growth with increasing nitrogen. The species native to a low-nitrogen environment reaches its maximum rate of growth at low to medium nitrogen availability. It does not respond to further additions of nitrogen.

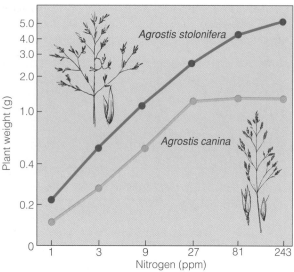

FIGURE 6.18 Growth responses of two species of grass—carpet bent grass *(Agrostis stolonifera),* found in high-nutrient environments, and velvet bent grass *(Agrostis canina),* found in low-nutrient environments—to the addition of different levels of nitrogen fertilizer. Note that *A. canina* responds to nitrogen fertilizer up to a certain level only. (Bradshaw et al. 1974.)

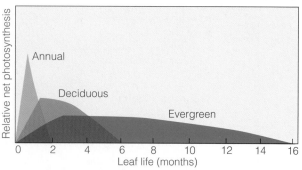

FIGURE 6.19 Relationship between the leaf longevity and photosynthetic capacity for annual plants, deciduous trees, and evergreen trees. (Mooney and Gulmon 1982.)

Some plant ecologists suggest that a low natural growth rate is an adaptation to a low-nutrient environment. One advantage of slower growth is that the plant can avoid stress under low-nutrient conditions. A slow-growing plant can still maintain optimal rates of photosynthesis and other processes critical for growth under low-nutrient availability. In contrast, a plant with an inherently high rate of growth will show signs of stress.

A second adaptation to low-nutrient environments is leaf longevity (Figure 6.19). The production of a leaf has a cost to the plant. This cost can be defined in terms of the carbon and other nutrients required to make the leaf. At a low rate of photosynthesis, a leaf needs a longer time to "pay back" the cost of its production. As a result, plants inhabiting low-nutrient environments tend to have longer-lived leaves. A good example is the dominance of pine species on poor, sandy soils in the coastal region of the southeastern United States. In contrast to deciduous tree species, which shed their leaves every year, pines have needles that live for up to 3 years.

Like water, nutrients are a below-ground resource of terrestrial plants. Their ability to exploit this resource is related to the amount of root mass. One means by which plants growing in low-nutrient environments compensate is to increase the production of roots. This increase is one cause of their low growth rates. Just as was the case with water limitation, carbon is allocated to the production of roots at the cost of the production of leaves. The reduced leaf area reduces the rate of carbon uptake in photosynthesis relative to the rate of carbon loss in respiration. The result is a lower net carbon gain and growth rate by the plant.

CHAPTER REVIEW

Summary

All life on Earth is carbon-based. The ultimate source of carbon is CO_2 in the atmosphere. Fixed by plants in photosynthesis, it is transferred directly or indirectly to all consumers. Carbon becomes part of organisms through assimilation. Assimilation requires energy derived from breakdown of carbon compounds in respiration. Because plants derive their energy from the Sun, they are called primary producers. Organisms that derive their energy directly or indirectly from plants are called secondary producers.

Photosynthesis and Respiration (6.1–6.2) Photosynthesis harnesses light energy from the Sun to convert CO_2 and H_2O into glucose. A nitrogen-based enzyme, rubisco, catalyzes the transformation of CO_2 into sugar. The first product of the reaction is a 3-carbon compound. For this reason, this photosynthetic pathway is called C_3 photosynthesis. Cellular respiration releases energy from carbohydrates to yield energy, H_2O, and CO_2. The energy released in this process is stored as the high energy compound ATP. Respiration occurs in the living cells of all organisms. (**6.1**).

The amount of light reaching a plant influences its photosynthetic rate. The light level at which the rate of carbon dioxide uptake in photosynthesis is equal to the rate of carbon dioxide loss due to respiration is called the light compensation point. The light level at which a further increase in light no longer produces an increase in the rate of photosynthesis is the light saturation point (6.2).

CO$_2$ Uptake and Water Loss (6.3) Photosynthesis involves two key physical processes: diffusion and transpiration. Diffusion is the movement of molecules from more-dense to less-dense concentration. CO$_2$ diffuses from the atmosphere to the leaf through leaf pores or stomata. As photosynthesis slows down during the day and demand for CO$_2$ lessens, stomata close to reduce loss of water to the atmosphere. Water loss through the leaf is called transpiration. The amount of water lost depends on the humidity. Water lost through transpiration must be replaced by water taken up from the soil.

Water moves from the soil into the roots, up through the stem and leaves, and out to the atmosphere. Pressure differences along a water gradient move water along this route. Plants draw water from the soil, where the water pressure is the highest, and release it to the atmosphere, where it is the lowest. Because of the lower amount of water in the atmosphere, water moves out of the leaves through the stomata in transpiration. Loss of water through the leaves reduces water pressure in the roots, so more water moves from the soil through the plant. This process continues as long as water is available in the soil. This loss of water by transpiration creates moisture conservation problems for plants. Plants need to open their stomata to take in CO$_2$, but they can conserve water only by closing the stomata (6.3).

Plant Energy Balance (6.4) Leaf temperatures affect both photosynthesis and respiration. Plants have optimal temperatures for photosynthesis, beyond which photosynthesis declines. Respiration increases with temperature.

The internal temperature of all plant parts is influenced by heat gained from and lost to the environment. Plants absorb long-wave and shortwave radiation. They reflect some of it back to the environment. The difference between the amount reflected and the amount absorbed is the plant's net radiation balance. The plant uses some of the absorbed radiation in photosynthesis. The remainder must be either stored as heat in the plant and surrounding air, or dissipated through the processes of evaporation (transpiration) and convection.

Net Carbon Gain and Carbon Allocation (6.5) The net carbon gain (per unit time) of a plant is the difference between carbon uptake in photosynthesis and carbon loss through respiration. The net carbon gain is then allocated to a variety of plant processes, including the production of new tissues. Since, in general, only leaves (photosynthetic tissues) are able to photosynthesize, yet all living tissues respire, the net carbon gain (and subsequently the growth) of a plant is influenced by the patterns of carbon allocation.

Interdependence of Plant Adaptations (6.6) Plants have evolved a wide range of adaptations to variations in environmental conditions. The adaptations exhibited by plants to these features of the environment are not independent for reasons relating to both the physical environment and the plants themselves.

Plant Adaptations to High and Low Light (6.7) Plants are either sun plants (shade-intolerant) or shade plants (shade-tolerant). Each group is characterized by adaptations to a certain light regime. Shade-adapted plants have low photosynthetic, respiratory, metabolic, and growth rates. Sun plants generally have higher photosynthetic, respiratory, and growth rates, but have lower survival rates under shaded conditions. A plant's reaction to shade is reflected in its leaves. The leaves of some species of plants change structure and shape in response to light conditions. Leaves in sun plants tend to be smaller, lobed, and thick. Shade-plant leaves tend to be larger and thinner.

Plant Response to Water Limitation (6.8) Plants of warm environments use a C$_4$ pathway of photosynthesis. It involves two steps and is made possible by leaf anatomy that differs from C$_3$ plants. C$_4$ plants have vascular bundles surrounded by chlorophyll-rich bundle sheath cells. C$_4$ plants store carbon in the form of malate and aspartate in the mesophyll cells. They transfer these acids to the bundle sheath cells, where they release the CO$_2$ they contain. Photosynthesis now follows the C$_3$ pathway. Thus, C$_4$ plants are able to store CO$_2$ as malate and carry on photosynthesis even if stomata close. C$_4$ plants have a high water-use efficiency (the amount of carbon fixed per unit of water transpired).

Succulent desert plants, such as cacti, have a third type of photosynthetic pathway, called CAM. CAM plants open their stomata to take in CO$_2$ at night, when the humidity is high. They convert CO$_2$ to a 4-carbon compound, malate. During the day, CAM plants close their stomata, convert malate back to CO$_2$, and follow the C$_3$ photosynthetic pathway.

Adaptations to Heat and Cold (6.9) Plants exhibit a variety of adaptations to both extremely cold and hot environments. Cold tolerance is mostly genetic, varying among species. Plants acquire frost hardiness through the formation or addition of protective compounds in the cell, which function as antifreeze. The ability to tolerate high air temperatures is related to plant moisture balance.

Plant Adaptations to Nutrient Availability (6.10) Terrestrial plants take up nutrients from soil through the roots. As roots deplete nearby nutrients, diffusion of water and nutrients through the soil replaces them. Plants continuously produce fine roots to forage for nutrients in the soil. Availability of nutrients has a direct effect on a plant's survival, growth, and reproduction. Nitrogen is important because rubisco and chlorophyll are nitrogen-based compounds essential to photosynthesis. Uptake of nitrogen and other nutrients depends on availability and demand. Plants with high nutrient demands grow poorly in low-nutrient environments. Plants with lower demands survive and grow, if slowly, in low-nutrient environments. Plants adapted to low-nutrient environments exhibit lower rates of growth and increased longevity of leaves. The lower nutrient concentration in their tissues means lower quality food for decomposers.

Study Questions

1. What does it mean to say that life on Earth is carbon-based?

2. Distinguish between photosynthesis and assimilation. How are they related?

3. What is respiration? Is it the reverse of photosynthesis?

4. Distinguish between primary producers and secondary producers.

5. Describe the essential features of photosynthesis.

6. What are ATP and rubisco? What is the relationship between them?

7. How do diffusion and transpiration enter into photosynthesis?

8. How do C_3, C_4, and CAM plants differ in their photosynthetic processes?

9. How does the process of photosynthesis respond to varying levels of light?

10. Contrast light saturation point and light compensation point.

11. Contrast shade-tolerant and shade-intolerant plants.

12. How does light influence the size and shape of leaves?

13. How do plants build up a tolerance to or adjust to cold?

14. What moves water through plants from soil to the atmosphere?

15. In what ways do plants cope with water limitations?

16. What are the advantages and disadvantages of closing stomata?

17. What adaptations enable land plants to live in an arid environment?

18. Contrast plants adapted to low-nutrient and high-nutrient environments.

Colorful decomposers such as honey mushrooms *(Armillaria mellea)* reside on the forest floor throughout much of continental North America.

Decomposers and Decomposition

OBJECTIVES

On completion of this chapter, you should be able to:

- Define decomposition and discuss the variety of processes involved.
- Contrast aerobic and anaerobic respiration.
- Distinguish among different groups of decomposer organisms based on function and size.
- Describe how different types of carbon compounds differ in their quality as an energy source for decomposers.
- Describe how the lignin concentration of plant litter influences the rate of decomposition.
- Contrast the differences in the decomposition of plant and animal tissues.
- Describe how climate influences microbial activity and the rate of decomposition.
- Contrast nutrient mineralization and immobilization.
- Define net mineralization rate.
- Explain how the nutrient content of plant litter influences the rate of net mineralization.
- Describe the decomposition of DOM and POM in open-water aquatic ecosystems.

Decomposition is the breakdown of chemical bonds formed during the construction of plant and animal tissue. Whereas photosynthesis involves the incorporation of solar energy, carbon dioxide, water, and inorganic nutrients into organic compounds (living matter), decomposition involves the release of energy originally fixed by photosynthesis, carbon dioxide, and water and ultimately the conversion of organic compounds into inorganic nutrients.

7.1 Decomposition Involves a Variety of Processes

Decomposition is a complex of many processes, including leaching, fragmentation, change in physical and chemical structure, ingestion, and excretion of waste products. These processes are accomplished by a variety of decomposer organisms. All heterotrophs to some degree function as decomposers. As they digest food, they break down organic matter, alter it structurally and chemically, and release it partially in the form of waste products. However, what we typically refer to as decomposers are organisms that feed on dead organic matter or detritus. This group is composed of **microbial decomposers,** a group made up primarily of bacteria and fungi, and **detritivores,** animals that feed on dead material including dung.

Decomposition of dead plant and animal matter moves through several stages, from deposition to the final breakdown into inorganic nutrients. Early stages of decomposition involve initial **leaching,** the loss of soluble sugars and other compounds that are dissolved and carried away by water, and **fragmentation,** the reduction of organic matter into smaller particles, either physically or chemically. Both of these abiotic processes result in the loss of mass and changes in the chemical composition of the detritus. As decomposition proceeds, further fragmentation and leaching take place.

Decomposers, like all heterotrophs, derive their energy and the majority of their nutrients from the consumption of organic compounds. Energy is derived through the oxidation of carbohydrates (such as glucose) in the process of respiration (see Chapter 6, Section 6.1). As organic matter is degraded, it is converted into smaller and simpler products. An essential function of decomposers is the release of inorganic nutrients as they break down organic compounds during digestion. The release of organically bound nutrients into an inorganic form is called **mineralization.** At the same time, decomposer organisms use these nutrients for their own growth. The incorporation of mineral nutrients into an organic form is known as **immobilization.**

7.2 A Range of Organisms Is Involved in Decomposition

The innumerable organisms involved in decomposition are categorized into several major groups based on their size and function (Figure 7.1). Organisms most commonly associated with the process of decomposition are the **microflora,** comprised of the bacteria and fungi. Bacteria may be aerobic, requiring oxygen for metabolism (respiration); or they may be anaerobic, able to carry on their metabolic functions without oxygen by using inorganic compounds. This type of respiration by anaerobic bacteria, commonly found in the mud and sediments of aquatic habitats and in the rumen of ungulate herbivores (Chapter 8), is fermentation.

Fermentation, which converts sugars to organic acids and alcohols, is a less efficient means of breaking down organic matter than aerobic respiration (Chapter 6). Many decomposer bacteria are **facultative anaerobes.** These bacteria use oxygen for respiration when it is available; but in its absence they can shift to fermentation, utilizing inorganic compounds as an oxidant. Other bacteria are **obligate anaerobes** that can exist only in environments devoid of oxygen.

Bacteria are the dominant decomposers of dead animal matter, whereas fungi are the major decomposers of plant material. Fungi extend their hyphae into the organic material to withdraw nutrients. Fungi range in type from species that feed on highly soluble, organic compounds, such as glucose, to more complex hyphal fungi that invade tissues with their hyphae (Figure 7.1a).

Bacteria and fungi secrete enzymes into plant and animal tissues to break down the complex organic compounds. Some of the resulting products are then absorbed as food. After one group has exploited the material to the extent of its ability, a different group of bacteria and fungi able to utilize the remaining material moves in. Thus, a succession of microflora occurs in the decomposition of organic matter until the material is finally reduced to inorganic nutrients.

Decomposition is aided by the fragmentation of leaves, twigs, and other dead organic matter (detritus) by invertebrate detritivores. These organisms fall into four major groups as classified by body width (Figure 7.2): (1) microfauna and microflora include protozoans and nematodes inhabiting the water in soil pores; (2) mesofauna, whose body width falls

FIGURE 7.1 (a) Fungi and bacteria are major decomposers of plant and animal tissues. (b) Mites are among the most abundant of small detritivores. (c) Earthworms and millipedes are large detritivores in terrestrial ecosystems, while (d) mollusks and crabs play a similar role in aquatic ecosystems.

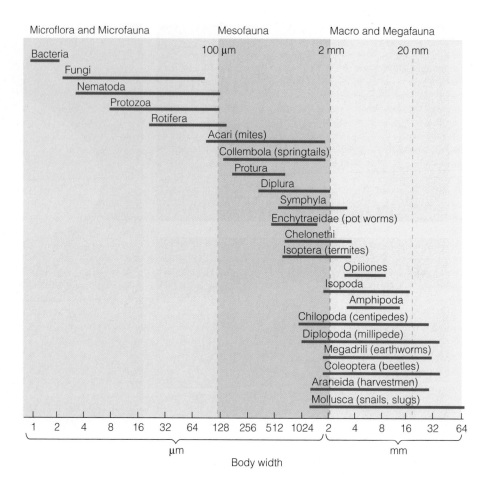

Microflora and Microfauna   Mesofauna   Macro and Megafauna

FIGURE 7.2 A general classification of soil fauna based on body width or diameter is a more functional classification relative to litter breakdown in decomposition. Bacteria and fungi fall into the category of microflora. Protozoans and nemotodes fall into the category of microfauna. (From Swift et al. 1979.)

between 100 µm and 2 mm, includes mites (Figure 7.1b), potworms, and springtails that live in soil air spaces; and (3) macrofauna and (4) megafauna, represented by millipedes, earthworms (Figure 7.1c), and snails in terrestrial habitats, and by annelid worms, smaller crustaceans, especially amphipods and isopods, and mollusks and crabs in aquatic habitats (Figure 7.1d). Earthworms and snails dominate the megafauna over 20 mm. The macrofauna and megafauna can burrow into the soil or substrate to create their own space, and megafauna, such as earthworms, have major influences on soil structure (Chapter 5). These detritivores feed on plant and animal remains and on fecal material.

Energy and nutrients incorporated into bacterial and fungal biomass do not go unexploited in the decomposer world. Feeding on bacteria and fungi are the **microbivores.** Making up this group are protozoans such as amoebas, springtails (Collembola), nematodes, larval forms of beetles (Coleoptera), and mites (Acari). Smaller forms feed only on bacteria and fungal hyphae. Because larger forms feed on both microflora and detritus, members of this group are often difficult to separate from detritivores.

7.3   The Types of Carbon Compounds in Plant Tissues Influence Their Food Quality

Dead organic matter provides both an energy (carbon) and nutrient (e.g., nitrogen) source for microbial decomposers. Dead plant tissue, referred to as litter, is by far the largest source of organic matter for decomposer organisms. Characteristics that influence the quality of plant litter as an energy source relate directly to the types and quantities of carbon compounds present—namely the types of chemical bonds present, and the size and three-dimensional structure of the molecules in which these bonds are formed. Not all carbon compounds are of equal quality as an energy source to microbial decomposers. Glucose and other simple sugars, the first products of photosynthesis, are very high-quality sources of carbon. These molecules are physically small. The breakage of their chemical bonds yields much more energy than required to synthesize the enzymes needed to

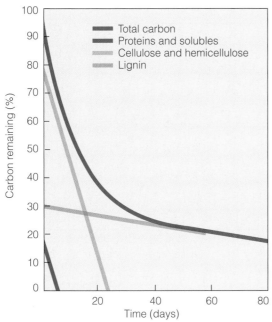

FIGURE 7.3 Variation in the rates of decay (mass loss) of different classes of carbon compounds in an experiment examining the decomposition of straw on the soil surface. At any time, the sum of the three classes of carbon compounds is equal to the value for total carbon. (Swift, Heal, and Anderson 1979.)

break them. Cellulose and hemicellulose are the main constituents of cell walls. These compounds are more complex in structure and therefore more difficult to decompose than simple carbohydrates. They are of moderate quality as a substrate for microbial decay. The much larger lignin molecules are among the most complex and variable carbon compounds in nature. There is no precise chemical description of lignin; rather, it represents a class of compounds. These compounds possess very large molecules intricately folded into complex three-dimensional structures that effectively shield much of the internal structure from attack by enzyme systems. As such, lignins, major components of wood, are among the slowest components of plant tissue to decompose.

The variation in rates of consumption of different carbon compounds are revealed in an experiment that examined the rate at which carbon was consumed during the decomposition of straw placed on a soil surface (Figure 7.3). The total carbon content of the straw, expressed as a percentage of the original mass, declined exponentially over the period of the 80-day study. However, when the total carbon was partitioned into various classes of carbon compounds, the rates at which these compounds decomposed varied widely. Proteins, simple sugars, and other soluble compounds made up some 15 percent of the original total carbon content. These compounds decomposed very quickly, disappearing completely within the first few days of the experiment. Cellulose and hemicellu-

lose made up some 60 percent of the original carbon content. Although these compounds decomposed more slowly than the proteins and simple sugars, by 3 weeks into the experiment they had been completely broken down. The third category of carbon compounds examined, the lignins, made up some 20 percent of the total original carbon. These compounds were broken down very slowly over the course of the experiment, with the vast majority of lignins remaining intact by day 80. As decomposition proceeded during the experiment, the quality of the carbon resource declined. High- and intermediate-quality carbon compounds declined at a relatively rapid rate. Thus, the proportion of total carbon remaining as lignin compounds continually increased with time. The increasing component of lignin lowered the overall quality of the remaining litter as an energy source for microbial decomposers.

Lignin compounds are of such low quality as a source of energy that they yield almost no net gain of energy to microbes during decomposition. Bacteria do not degrade lignins; they are broken down by only a single group of fungi, the basidiomycetes.

7.4 Litter Quality Influences the Rate of Decomposition

Data from a litterbag study (see Focus on Ecology 7.1: Litterbag Experiments) designed to examine the decomposition of senescent leaves from three tree species in central Virginia are presented in Figure 7.4. The graph expresses the disappearance of leaf litter as the percent of the original mass of leaves that remains at various times over the period of a year. As the bacteria and fungi consume the leaves, less mass remains. Note that the rate of decomposition for the leaf litter of these tree species is different. Sixty percent of the red maple leaves have been consumed by the end of the first year (40 percent of original mass remaining), as compared to 50 percent for white oak and only 30 percent for sycamore leaves. The difference in the rate of decomposition for these three species is a direct result of their carbon composition. The freshly fallen leaves of red maple have a lignin content of 11.7 percent, as compared to 17.7 percent for white oak and 36.4 percent for sycamore leaves. There is an inverse relationship between the rate of decomposition for plant litter and its lignin content at the initiation of decomposition. This inverse relationship between decomposition rate and the carbon quality of plant litter as a source of energy for decomposers has been reported for a wide range of

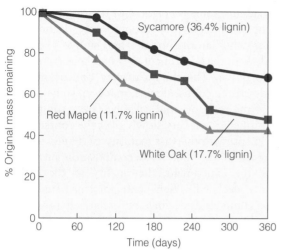

FIGURE 7.4 Results of a litterbag experiment in central Virginia designed to examine the decomposition of fallen leaves from red maple, white oak, and sycamore trees. Decomposition is expressed as the percentage of the original mass remaining at different times over the first year of the experiment. Note the inverse relationship between the rate of disappearance and the initial lignin content of the leaves for the three different species. (Smith 2002.)

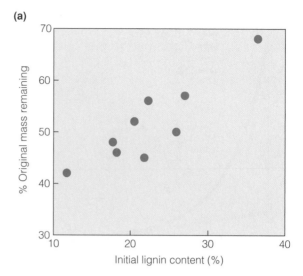

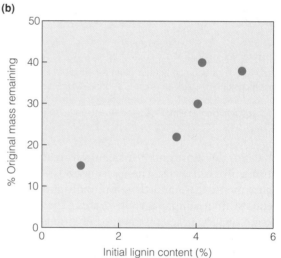

FIGURE 7.5 Relationship between initial lignin content of litter material and rate of decomposition for a variety of plant litters in both (a) terrestrial (Smith 2002) and (b) aquatic environments (Klap et al. 1999).

plant species inhabiting both terrestrial and aquatic environments (Figure 7.5).

The carbon quality of plant litter also has a direct influence on the feeding habits of the larger detritivores (megafauna). In a study conducted in the forests of eastern North America, earthworms were found to have a pronounced preference for species low in lignin content such as aspen, white ash, and basswood. They do not eat red oak leaves, high in lignins, at all. In a European study, earthworms preferred the dead leaves of elm, ash, and birch; ate sparingly of oak and beech, high in lignins; and did not touch pine or spruce needles. Millipedes likewise show a species preference related to lignin content.

Carbon quality of plant litters can have a particularly important influence on decomposition in coastal marine environments. Phytoplankton have a low lignin concentration and therefore decompose rather quickly. However, vascular plants, such as sea grasses, marsh grasses, and reeds that inhabit estuarine and marsh ecosystems have lignin concentrations similar to those of terrestrial plants. The decomposition of these plant litters is dependent on the oxygen content of the water. In the mud and sediments of aquatic habitats where oxygen levels can be extremely low, anaerobic bacteria carry out most of the decomposition (see Section 7.1). The absence of fungi, which require oxygen for respiration, hinders the decomposition of lignin compounds, therefore slowing the overall rate of decomposition (Figure 7.6).

7.5 Decomposition of Animal Matter Is More Direct

Decomposition of animal matter is more direct than decomposition of plant material. The chemical breakdown of flesh does not require all of the specialized enzymes needed to digest plant matter. That flesh not consumed by scavengers such as crows, vultures, and foxes is decomposed largely by bacteria rather than fungi and by certain arthropods such as blowflies (Calliphoridae) (Figure 7.7).

In summer and fall, when temperatures are high, microbial activity and colonization of dead animal

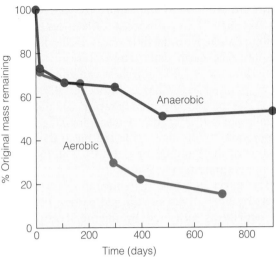

FIGURE 7.6 Results of a litterbag experiment designed to examine the decomposition of *Spartina alternifolia* litter exposed to aerobic (litterbags on the marsh surface) and anaerobic (buried 5–10 cm below the marsh surface) conditions. (Redrawn from I. Valiela 1984, *Marine Ecological Processes,* p. 310.)

FIGURE 7.8 The dung beetle is exceedingly abundant in both tropical and temperate regions. Thousands of individuals and dozens of species may be attracted to a single dropping.

tissues by blowflies is intense. Blowfly maggots emerging from eggs laid in the tissues can consume a small mammal carcass in 7 to 8 days. Between bacteria and maggots, 70 percent of the organic material of the small mammal carcass is consumed, leaving the remaining 30 percent, largely hair and bone, behind. Due to low temperatures and reduced microbial activity, decomposition during the winter and spring is restricted. As a result, the carcass can become mummified through reduced pH and other chemical changes in the tissues and is eventually fragmented and scattered.

FIGURE 7.7 Green Bottle Fly and maggots emerging from eggs laid in dead fish tissues play a major role in the decomposition of animal matter.

Unlike the bodies of dead animals, most fecal matter represents an already highly decomposed substrate. However, the dung of large herbivores still contains an abundance of partially digested organic matter that provides a rich resource for specialized detritivores in addition to earthworms, bacteria, and fungi. Among these specialized detritivores are species of flies that lay their eggs in dung, upon which the larvae will feed. The most notable of the detritivores that feed on animal dung are the dung beetles (Figure 7.8) (Scarabaeinae, Aphodiinae, and Geotrupinae). The eggs, larvae, and pupae of many species of the genus *Aphodius* develop within the dung pat. Other dung beetles, known as tumblebugs, form a mass of dung into a ball in which they lay their eggs, roll it a distance, dig a hole, and bury it as a food supply for the larvae. Aphodiinae dung beetles tunnel and form dung balls underground. The earth-boring dung beetles, Geotrupinae, spend most of their lives in deep burrows, usually beneath carrion or dung. The female lays her eggs in a plug of dung at the end of the burrow.

7.6 Decomposition Is Also Influenced by Physical Environment

In addition to the quality of the dead organic matter as a food source, the physical environment also has a direct effect on both macrodecomposers and microdecomposers and, therefore, on the rate of

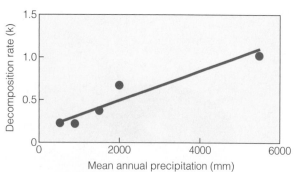

FIGURE 7.9 Relationship between rate of decomposition and annual precipitation for dead leaves of the ohia tree *(Metrosideros polymorpha)* at a range of sites on the island of Hawaii. The rate of decomposition (value of *k* on the *y* axis) at each site is calculated as the slope of the simple linear regression between the natural log of percent of original mass remaining and time (years). Data at each site were collected from litterbag studies over a period of 2 years. (Austin and Vitousek 2000.)

decomposition. Both temperature and moisture greatly influence microbial activity. Low temperatures reduce or inhibit microbial activity, as do dry conditions. The optimum environment for microbes is warm and moist. As a result, rates of decomposition are highest in warm, wet climates (Figure 7.9). Alternate wetting and drying and continuous dry spells tend to reduce both the activity and populations of microflora.

This effect of climate on the decomposition of red maple leaves at three sites in eastern North America (New Hampshire, West Virginia, and Virginia) can be seen in Figure 7.10. Although the lignin content of red maple leaves does not differ significantly at the sites, the decomposition rate increases as you move

southward from New Hampshire to West Virginia and Virginia. These observed differences can be attributed directly to the differences in the climate at the sites. The mean daily temperature at the New Hampshire site is 7.2°C and the mean annual potential evaporation (see Chapter 4, Section 4.9) is 621 mm. The mean daily temperature at the West Virginia site is 12.2°C and the mean annual potential evaporation is 720 mm, while the mean daily temperature at the Virginia site is 14.4°C and the mean potential evaporation is 806 mm.

The direct influence of temperature on decomposers results in a distinct diurnal pattern of microbial activity as measured by microbial respiration from the soil (Figure 7.11). The daily pattern of temperature is closely paralleled by the release of CO_2 from the soil due to the respiration of microbial decomposers.

7.7 Decomposers Transform Nutrients into a Usable Form

The nutrient quality of dead organic material varies. The macronutrient nitrogen can serve as an example. Most dead leaf material, such as the leaves that fall from deciduous trees during the autumn in the temperate zone, has a nitrogen content in the range of 0.5 to 1.5 percent (of total weight). The higher the nitrogen content of the dead leaf, the higher the nutrient value for the microbes and fungi that feed upon the leaf. As microbial decomposers—bacteria

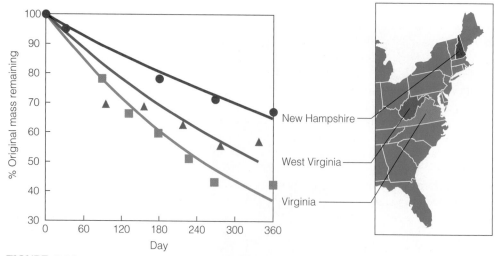

FIGURE 7.10 Decomposition of red maple litter at three sites in eastern North America. Mass loss through time was estimated using litterbag experiments at each site. The decline in decomposition rate from north to south is a direct result of changes in climate, primarily temperature. (Data from Melillo et al. 1982 for New Hampshire; Mudrick et al. 1994 for West Virginia; Smith 2002 for Virginia.)

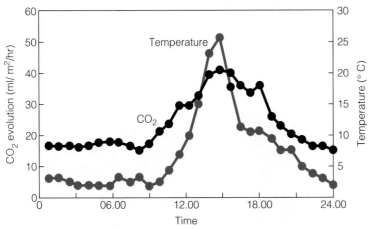

FIGURE 7.11 Diurnal changes in air temperature and decomposition in a temperate deciduous forest. Decomposition rate is measured indirectly as the release (evolution) of CO_2 from decomposing litter on the forest floor. The release of CO_2 is a measure of the respiration of decomposer organisms. (Whitkamp and Frank 1969.)

and fungi—break down dead organic material, they transform nitrogen and other nutrients tied up in organic compounds into an inorganic (or mineral) form to release energy for their metabolism. This process is called mineralization. The same decomposers that are responsible for mineralization also require nitrogen for their own growth, and therefore reuse some of the nitrogen they have produced. In doing so, they incorporate the mineral nitrogen into an organic form, a process known as immobilization. Since both of these processes, mineralization and immobilization, are taking place as decomposer organisms are consuming the litter, the supply rate of mineral nutrients to the soil during the process of decomposition, the **net mineralization rate,** is the difference between the rates of mineralization and immobilization.

Changes in the nitrogen (and other nutrient) content of litter during decomposition are typically examined concurrently with changes in mass and carbon content using litterbags (see Focus on Ecology 7.1: Litterbag Experiments). As with litter mass (see Figure 7.4) and carbon content (see Figure 7.3), the nitrogen content of the remaining litter can be expressed as a percentage of the nitrogen content of the original litter mass. Changes in the nitrogen content of plant litter during decomposition typically conform to three stages (Figure 7.12). Initially, the amount of nitrogen in the litter declines as water-soluble compounds are leached from the litter. This stage can be very short and, in terrestrial environments, is dependent on the soil moisture environment. Following the initial period of leaching, nitrogen content increases as microbial decomposers immobilize nitrogen from outside the litter. For this reason, nitrogen concentrations can rise above 100 percent, actually exceeding the initial nitrogen con-

tent of the litter material. As decomposition proceeds, and carbon quality declines (due to a higher proportional fraction of lignin; see Figure 7.3), the rate of mineralization exceeds that of immobilization, and there is a net release of nitrogen to the soil (or water).

The pattern presented in Figure 7.12 is idealized. The actual pattern of nitrogen dynamics during decomposition depends on the initial nutrient content of the litter material. If the nitrogen content of the litter material is high, mineralization may exceed the rate of immobilization from the onset of decomposition, and nitrogen concentration of the litter will not increase above the initial concentration (Figure 7.13).

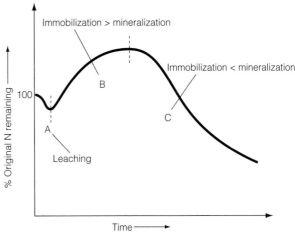

FIGURE 7.12 Idealized graph showing the change in nitrogen content of plant litter during decomposition. The initial phase (A) corresponds to the leaching of soluble compounds. Nitrogen content then increases above initial concentrations (phase B) as the rate of immobilization exceeds the rate of mineralization. As decomposition proceeds, the rate of nitrogen mineralization exceeds that of immobilization, and there is a net release of nitrogen from the litter (phase C).

LITTERBAG STUDIES

Ecologists study the process of decomposition by designing experiments that follow the decay of dead plant and animal tissues through time. Ecologists use litterbags to examine the decomposition of dead plant tissues (called plant litter). Litterbags are mesh bags constructed of synthetic material that does not readily decompose (Figure A). The holes in the bag must be large enough to allow decomposer organisms to enter and feed on the litter, but small enough not to allow decomposing plant material to fall out of the bag. Ecologists most often use mesh bags with openings of 1 to 2 mm for plant leaf litter.

A fixed amount of litter material is placed in each bag. In the experiments shown in Figure 7.4 for each of the three tree species, 30 litterbags were filled with 10 grams of leaf litter each. These bags were buried in the litter layer of the forest. At six intervals over the course of a year, 5 bags were collected for each species, and their contents were dried and weighed in the laboratory.

Note from Figure 7.4 that the mass of litter remaining in the bags decreased continuously as time progressed. This decrease shows that decomposer organisms were consuming the litter. The carbon was lost to the atmosphere as CO_2 as a result of respiration by decomposer organisms.

FIGURE A

This same experimental approach is used to examine nutrient release from decomposing litter. By analyzing the nutrient content of the remaining litter at any sample period, the uptake from and release of nutrients to the soil by microbial decomposers can be monitored (see Figure 7.13).

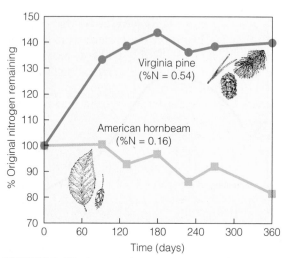

FIGURE 7.13 Change in the nitrogen content of leaf litter from two tree species inhabiting the forests of central Virginia—American hornbeam and Virginia pine. Nitrogen content is expressed as the percentage of the original mass of nitrogen remaining at different times over the first year of the experiment. All data are from litterbag experiments. Note the difference between the two species in the initial nitrogen content of the leaf litter and the subsequent rates of immobilization. (Smith 2002.)

Although the discussion and examples presented above have focused on nitrogen, the same pattern of immobilization and mineralization as a function of litter nutrient content applies to all essential nutrients (Figure 7.14). As with nitrogen, the exact pattern of dynamics during decomposition is a function of the nutrient content of the litter and the demand for the nutrient by the microbial population.

7.8 Decomposition in Aquatic Environments

Decomposition in aquatic ecosystems follows a pattern similar to that in terrestrial ecosystems, but with some major differences influenced by the watery environment. As in terrestrial environments, decomposition involves leaching, fragmentation, colonization of detrital particles by bacteria and fungi, and consumption by detritivores and microbivores. In coastal environments, permanently submerged

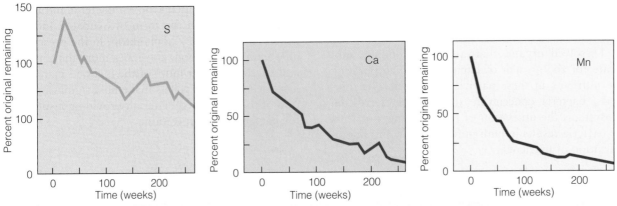

FIGURE 7.14 Patterns of immobilization and mineralization for sulfur (S), calcium (Ca), and manganese (Mn) in decomposing needles of Scots pine. Results are from a litterbag experiment over a period of 5 years. (Data from Staff and Berg 1982; redrawn from Aber and Melillo 1991.)

plant litters decompose more rapidly than do those on the marsh surface because they are more accessible to detritivores and because the stable physical environment is more favorable to microbial decomposers (Figure 7.15).

In flowing-water ecosystems, aquatic fungi colonize leaves, twigs, and other particulate matter. One group of aquatic arthropods, called **shredders,** fragment the organic particles and in the process also eat bacteria and fungi on the surface of the litter (see Chapter 27). Downstream, another group of invertebrates, the filtering and gathering **collectors,** filter from the water fine particles and fecal material left by the shredders. **Grazers** and **scrapers** feed on algae, bacteria, fungi, and organic matter collected on rocks

and large debris (all discussed in Chapter 28). Algae take up nutrients and dissolved organic matter from the water.

In still, open water of ponds and lakes and in the ocean, dead organisms and other organic material, called particulate organic matter (POM), drift toward the bottom. On its way, POM is constantly ingested, digested, and mineralized until much of the organic matter settles on the bottom in the form of humic compounds (see discussion of humus formation in Chapter 5). How much depends in part on the depth of the water through which the particulate matter falls. In shallow water, much of it may arrive in relatively large packages to be further fragmented and digested by bottom-dwelling detritivores such as crabs, snails, and mollusks.

Bacteria work on the bottom or benthic organic matter. Bacteria living on the surface can carry on aerobic respiration, but within a few centimeters below the surface of the sediment, the oxygen supply is exhausted. Under this anoxic condition, a variety of bacteria capable of anaerobic respiration take over decomposition (see Section 7.2).

Aerobic and anaerobic decomposition by benthic bacteria form only a part of the decomposition process. Dissolved organic matter (DOM) in the water column also provides a source of fixed carbon for decomposition. Major sources of DOM are the free-floating macroalgae, phytoplankton, and zooplankton inhabiting the open water. Phytoplankton and other algae excrete quantities of organic matter at certain stages of their life cycles, particularly during rapid growth and reproduction. During photosynthesis the marine alga, *Fucus vesticulosus*, produces an exudate high in carbon content. The exuded matter goes into solution rather than to bacterial decomposers. In fact, 30 percent of the nitrogen contained in the bodies of zooplankton is lost directly to the

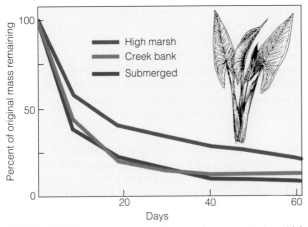

FIGURE 7.15 Decomposition of leaves of arrow arum in a tidal freshwater marsh. Decomposition is measured as the percent of original mass remaining in litterbags under three conditions: irregularly flooded high marsh exposed to alternate periods of wetting and drying, creek bed flooded two times daily (tidal), and permanently submerged. Note that the litter that is consistently wet has the highest rate of decomposition. (From Odum and Haywood 1978.)

water. Upon death, their bodies break up and dissolve within 15 to 30 minutes, too rapidly for any bacterial action to occur.

Dissolved organic matter then becomes a substrate for the growth of bacteria. By incorporating the nutrients in these particles into their own biomass, bacteria concentrate them. As in terrestrial ecosystems the utilization of these organic nutrients by bacteria results in both mineralization and immobilization of nutrients.

Ciliates and zooplankton eat bacteria and in turn excrete nutrients in the form of exudates and fecal pellets in the water. Zooplankton, in the presence of an abundance of food, consumes more than it needs and excretes half or more of the ingested material as fecal pellets. These pellets make up a significant fraction of the suspended material, providing a substrate for bacterial decomposition.

CHAPTER REVIEW

Summary

Decomposition Processes (7.1) Decomposition is the breakdown of chemical bonds formed during the construction of plant and animal tissues. Decomposition involves an array of processes including leaching, fragmentation, digestion, and excretion. An essential function of decomposers is the release of organically bound nutrients into an inorganic form.

Decomposer Organisms (7.2) The wide variety of organisms involved in the process of decomposition is classified into groups based on both function and size. The microflora, comprising the bacteria and fungi, are the group most commonly associated with decomposition. Bacteria can be grouped as either aerobic or anaerobic based on their requirement of oxygen to carry out respiration. Invertebrate detritivores are grouped into four major categories based on body size. Microbivores feed on bacteria and fungi.

Carbon Quality (7.3) Microbial decomposers utilize carbon compounds contained in dead organic matter as a source of energy. Various carbon compounds differ in their quality as energy sources for decomposers. Glucose and other simple sugars are easily broken down and provide a high-quality source of carbon. Cellulose and hemicellulose, the main constituents of cell walls, are intermediate in quality, while the lignins are of very low quality, and therefore decompose the slowest.

Litter Quality and Decomposition Rate (7.4) The quality of dead organic matter as a food source for decomposers is influenced by both the types of carbon compounds present and the nutrient content. Because of the low quality of lignins as an energy source for decomposers, there is an inverse relationship between the lignin content of plant litter and the rate at which it decomposes.

Decomposition of Animal Matter (7.5) The decomposition of animal matter is more direct than that of plant tissues, in that it does not require the array of specialized enzymes needed to digest plant matter. The breakdown of dead animal matter involves an array of organisms including scavengers, a variety of detritivores including blowflies and earthworms, as well as bacteria and fungi.

Influence of Physical Environment (7.6) The physical environment has a direct influence on both macrodecomposers and microdecomposers. Both temperature and moisture greatly influence microbial activity. Low temperatures and moisture inhibit microbial activity. As a result, the highest rates of decomposition are under warm, wet conditions. The influence of temperature and moisture on decomposer activity results in geographic variation in rates of decomposition that relate directly to climate.

Nutrient Mineralization (7.7) The nutrient quality of dead organic matter varies as a function of its nutrient content. As microbial decomposers break down dead organic matter, they transform nutrients tied up in organic compounds into an inorganic form. This process is called mineralization. The same organisms responsible for mineralization reuse some of the nutrients that they have produced, incorporating the inorganic nutrients into an organic form. This process is called immobilization. The difference between the rates of mineralization and immobilization is the net mineralization rate, which represents the net release of nutrients to the soil or water during decomposition. The relative rates of mineralization and immobilization during decomposition are related to the nutrient content of the dead organic matter being consumed.

Aquatic Environments (7.8) Decomposition in aquatic ecosystems varies as a function of water depth and flow rate. In flowing waters (streams and rivers) a variety of specialized detritivores are involved in the breakdown of plant litter imported from adjacent terrestrial ecosystems. In open-water environments, dead organisms and other organic matter, called particulate organic matter (POM), drift downward to the bottom. On its way, POM is constantly being ingested, digested, and mineralized until much of the organic matter is in the form of humic compounds by the time it reaches the bottom sediments. Bacteria decompose organic matter on the bottom sediments, utilizing both aerobic and anaerobic respiration dependent on the supply of oxygen. Dissolved organic matter (DOM) in the water column also provides a source of carbon for decomposers.

Study Questions

1. Define decomposition.

2. Name three processes involved in breakdown of dead organic matter.

3. Distinguish between aerobic and anaerobic respiration.

4. How does the type of carbon compound present in dead organic matter influence its quality as an energy source for decomposers?

5. How does litter quality influence the rate of decomposition?

6. How does lignin concentration influence the decomposition of plant litter?

7. How does climate influence the rate of decomposition?

8. Contrast the processes of mineralization and immobilization.

9. How does the rate of net mineralization typically vary through time during the process of decomposition? Why?

10. How does the nutrient content of plant litter influence the rate of net mineralization?

11. Contrast POM and DOM.

12. How does the oxygen content of water influence the process of decomposition in bottom sediments?

A West African dwarf crocodile (*Osteolammus tetraspis*) basks on a log at the edge of a rain forest river.

CHAPTER 8

Animal Adaptations to the Environment

OBJECTIVES

On completion of this chapter, you should be able to:

- Distinguish among poikilothermy, homeothermy, and heterothermy.
- Explain how animal behavior, structure, and metabolism maintain body temperature.
- Distinguish among hibernation, aestivation, and torpor.
- Tell how hyperthermia can help some mammals tolerate heat.
- Describe physiology of cold tolerance in some animals.
- Explain countercurrent circulation and its adaptive value.
- Tell how animals respond to environmentally induced moisture stress.
- Discuss the role of light in the daily and seasonal cycles of animals.
- Explain how the biological clock functions as a timekeeper.
- Relate the relationship between nutrients and the growth and reproduction of animals.
- Show the relationship between critical daylength and seaonality.
- Describe the relation of circadian rhythms to the biological clock.
- Describe the role of biological clocks in the rhythms of intertidal organisms.

137

All green plants, whether the smallest of violets or the giant sequoia trees of the western United States, derive their energy from the same process—photosynthesis. The story is quite different for animals. Because heterotrophic organisms derive their energy, and the majority of their nutrients, from the consumption of organic compounds contained in plants and animals, they are faced with literally hundreds of thousands of different types of potential food items, packaged as the diversity of plant and animal species inhabiting Earth. For this reason alone, the adaptations of animals to the environment is a much more complex and diverse topic than the adaptation of plants presented in Chapter 6. However, there are a number of key processes that are common to all animals: the acquisition and digestion of food, the maintenance of body temperature, water balance, and the adaptation to systematic variation in the light and temperature—the diurnal and seasonal cycles. In addition, there are a number of fundamentally different constraints imposed by aquatic and terrestrial environments. It is these topics that form the basis of this chapter.

8.1 Animals Have Various Ways of Acquiring Nutrients

Carbon fixed by plants in the process of photosynthesis, directly or indirectly, provides the nutritional resources of animals. Whatever the means by which they obtain their energy from organic carbon compounds contained in a wide array of potential food items, the ultimate source of these organic compounds is plants. Because plants and animals have different chemical compositions, the problem facing animals is the conversion of this plant tissue to animal tissue. Animals are high in fat and proteins, which they use as structural building blocks. Plants are low in proteins and high in carbohydrates, much in the form of cellulose and lignins in cell walls (see Chapter 7, Section 7.3). Nitrogen is a major constituent of protein. In plants the carbon to nitrogen ratio is about 40 to 1. In mammals the ratio is about 14 to 1. The task of converting cellulose and a limited supply of plant protein into animal tissue falls to the plant eaters. The diversity of potential sources of energy in the form of plant and animal tissues requires an equally diverse array of physiological, morphological, and behavioral characteristics enabling animals to acquire and assimilate these resources (Figure 8.1).

There are many ways to classify organisms based on the resources they utilize and the means by which they exploit them. The most general of these classifications is the division based on how animals use plant and animal tissues as sources of food. Organisms that rely on plant tissues as food are called **herbivores.** Organisms that feed on animal tissues are classified as **carnivores.** Organisms that feed on both plant and animal tissues are called **omnivores.** And organisms that feed on dead plant and animal matter, called detritus, are detrital feeders or detritivores (see Chapter 7). Each of these four feeding groups has characteristic adaptations that allow it to exploit the different diet.

Herbivory

Herbivores can be categorized by the type of plant material they eat. Grazers feed on leafy material, especially grasses. Browsers feed on mostly woody material. Granivores feed on seeds, and frugivores eat fruit. Others types of herbivorous animals such as avian sapsuckers (*Sphyrapicus* spp.) and sucking insects such as aphids feed on plant sap; and hummingbirds, butterflies, moths, and ants feed on plant nectar.

Grazing and browsing herbivores, with some exceptions, live on diets high in cellulose. In doing so, they face several dietary problems. Their diets are rich in carbon, but low in protein. Most of the carbohydrates are locked in indigestible cellulose, and the proteins exist in chemical compounds. Lacking the enzymes needed to digest cellulose, herbivores depend on specialized bacteria and protozoa living mutually (see Chapter 16) in their digestive tracts. These bacteria and protozoans digest cellulose and proteins, and they synthesize fatty acids, amino acids, proteins, and vitamins. For most vertebrates, bacteria and protozoans concentrate in the foregut or hindgut (Figure 8.2b), where they carry on anaerobic fermentation. In herbivorous insects, bacteria and protozoans inhabit the hindgut. Some species of cellulose-consuming wood beetles and wasps depend on fungi. These insects carry fungal spores with them externally when they invade new wood.

Ruminants, such as cattle and deer, are exemplary cases of herbivores anatomically specialized for the digestion of cellulose. They possess a highly complex digestive system consisting of a four-compartment stomach—the rumen (from which the group gets its name), reticulum, omasum, and abomasum or true stomach (Figure 8.2d)—and a long intestine. The rumen and reticulum, inhabited by anaerobic bacteria and protozoans, function as fermentation vats. These microbes break down the cellulose into

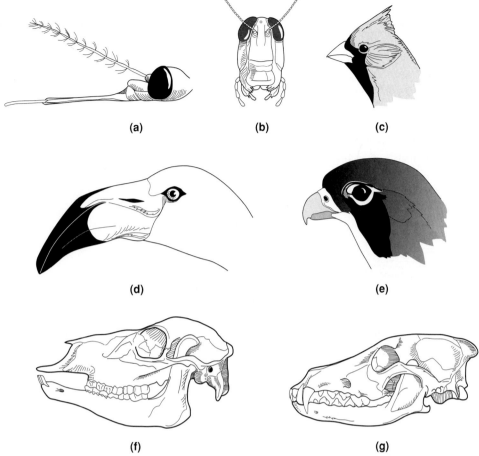

FIGURE 8.1 Mouthparts reflect how organisms obtain their food. (a) Piercing mouthparts of a mosquito. (b) Chewing mouthparts of a grasshopper. (c) The strong conical bill of a seed-eating bird. (d) The straining bill of a flamingo. (e) Tearing beak of a hawk. (f) The grinding molars of a herbivore, a deer. (g) The canine and shearing teeth of a carnivorous mammal, the coyote.

usable nutrients. Ruminants have highly developed salivary glands that excrete substances that allow the regulation of acidity (pH) and chemistry in the rumen.

As ruminants graze, they chew their food hurriedly. The food material descends to the rumen and reticulum, where it is softened to a pulp by the addition of water, kneaded by muscular action, and fermented by bacteria. At leisure the animals regurgitate the food, chew it more thoroughly to reduce plant particle size, making it more accessible to microbes, and swallow it again. The mass again enters the rumen. Finer material moves into the reticulum. Contractions force it into the third compartment, or omasum, where the material is further digested and finally forced into the abomasum, or true glandular stomach.

The digestive process carried on by the microorganisms in the rumen produces fatty acids. These acids rapidly absorb through the wall of the rumen into the bloodstream, providing the ruminant with a major source of food energy. Part of the material in the rumen converts to methane, which is expelled from the body, and part is converted into compounds that can be used directly as food energy. The ruminant digests many of the microbial cells in the abomasum to recapture still more of the energy and nutrients. Further bacterial action breaks down complex carbohydrates into sugars. In addition to carrying on fermentation, the bacteria also synthesize B-complex vitamins and amino acids.

Grazing marsupials, such as kangaroos, also possess a ruminant type of digestive system. Although the stomach is not divided into four compartments, it has regions analogous to the rumen, reticulum, abomasum, and omasum.

Most of the digestion in ruminants takes place in the foregut. Among nonruminants, such as rabbits and horses, digestion takes place less efficiently in the hindgut. Nonruminant vertebrate herbivores, such as

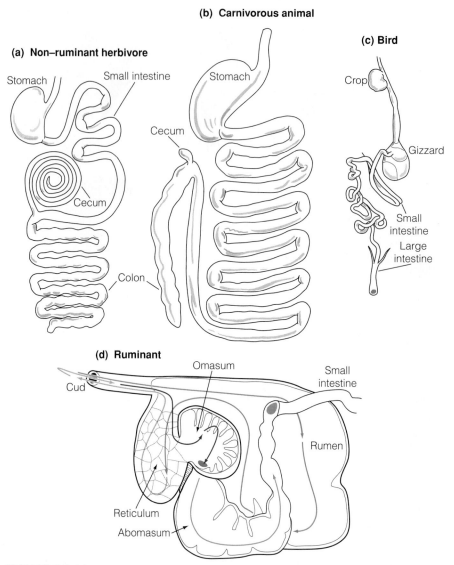

FIGURE 8.2 (a) Digestive tract of a nonruminant herbivore characterized by a long intestine and well-developed caecum. (b) The relatively simple digestive tract of a carnivorous mammal that also depicts a generalized digestive tract consisting of the esophagus, stomach (collectively the foregut), small intestine, and a small caecum and large intestine (collectively the hind gut). (c) The digestive tract of a bird with a crop, the anterior or simple stomach, and the posterior stomach or gizzard. (d) The ruminant stomach. The four-compartment stomach consists of the rumen (R), reticulum (Re), omasum (O), and abomasum (A). Food enters the rumen and the reticulum. The ruminant regurgitates fermented material (cud) and rechews it. Finer material enters the reticulum, and then the omasum and abomasum. Coarser material reenters the rumen for further fermentation.

horses, have simple stomachs, long intestinal tracts that slow the passage of food through the gut, and a well-developed caecum, a pouch attached to the colon of the intestine, where fermentation takes place (Figure 8.2a).

Lagomorphs—rabbits, hares, and pikas—resort to a form of **coprophagy,** the ingestion of fecal material for further extraction of nutrients. Part of ingested plant material enters the caecum (see Figure 8.2a), and part enters the intestine to form dry pellets. In the caecum, the ingested material is processed by microorganisms and is expelled into the large intestine and then out of the body as soft, green, moist pellets. The lagomorphs reingest the soft pellets, which are much higher in protein and lower in crude fiber than the hard fecal pellets. The amount of feces recycled by coprophagy ranges from 50 to 80 percent. The reingestion is important because the pellets, functioning as "external rumen," provide bacterially synthesized B vitamins and ensure more complete digestion of dry material and better utilization of protein. Coprophagy is widespread among the detritus-feeding animals, such as wood-eating beetles and millipedes that ingest pellets after they have been enriched by microbial activity.

Seed-eating birds—gallinaceous (chickenlike) birds, pigeons, doves, and many species of song birds—have a pouch in the esophagus called the crop (Figure 8.2c). It is a reservoir for food that passes on to a two-part stomach. The anterior glandular stomach secretes enzymes to begin digestion. The food then passes to the posterior muscular stomach, the gizzard, that functions as a powerful grinding organ. Birds assist the grinding action of the gizzard by swallowing small pebbles and gravel, or grit.

Among marine fish, herbivorous species are small and typically inhabit coral reefs. Characterized by high diversity (many different species), they make up about 25 to 40 percent of the fish biomass about the reefs. These herbivorous fish feed on algal growth that, unlike the food of terrestrial herbivores, lacks cellulose and other structural carbon compounds that are more difficult to digest. They gain access to the nutrients inside the algal cells by means of one or more of four basic types of digestive mechanisms. In some fish with thin-walled stomachs, low stomach pH weakens algal cell walls and allows digestive enzymes to come in contact with algal nutrients. Fish that possess gizzardlike stomachs ingest inorganic material that mechanically breaks down algal cells to release nutrients. Some reef fish possess specialized jaws that shred or grind algal material before it reaches the intestine. Other fish depend on microbial fermentation in the hindgut to assist in the breakdown of algal cells. These four types are not mutually exclusive. Some marine herbivores may combine low stomach pH or grinding and shredding with microbial fermentation in the hindgut.

Carnivory

Herbivores are the energy source for carnivores, the flesh eaters. Organisms that feed directly on the herbivores are termed first-level carnivores or second-level consumers. First-level carnivores represent an energy source for the second-level carnivores. Still higher categories of carnivorous animals feeding on secondary carnivores exist in some communities. As the feeding level of carnivores increases, their numbers decrease because of limited energy sources.

Unlike herbivores, carnivores are not faced with the digestion of cellulose or the quality of food. Because little difference exists in the chemical composition between the flesh of prey and the flesh of predators, there is no problem in digestion and assimilation of nutrients from their prey. Their major problem is obtaining a sufficient quantity of food.

Lacking the need to digest complex cellulose compounds, carnivores have short intestines and simple stomachs (Figure 8.2c). In mammalian carnivores, the stomach is little more than an expanded hollow tube with muscular walls. It stores and mixes foods, adding mucus, enzymes, and hydrochloric acid to speed digestion. In carnivorous birds such as hawks and owls, the gizzard is little more than an extendable pocket with reduced muscles in which digestion, started in the anterior stomach, continues. In hawks and owls, the gizzard acts as a barrier against hair, bones, and feathers that these birds regurgitate and expel as pellets by the way of the mouth.

Omnivory

Omnivory includes animals that feed on both plants and animals. The food habits of many omnivores vary with the seasons, stages in the life cycle, and their size and growth. The red fox *(Vulpes vulpes)*, for example, feeds on berries, apples, cherries, acorns, grasses, grasshoppers, crickets, beetles, and small rodents (Figure 8.3). The black bear *(Ursus americanus)* feeds heavily on vegetation—buds, leaves, nuts, berries, tree bark—supplemented with bees, beetles, crickets, ants, fish, and small to medium-sized mammals.

8.2 Animals Have Various Nutritional Needs

Animals require mineral elements and 20 amino acids, 14 of which are essential ones that cannot be synthesized by the body and have to be supplied by the diet. These nutritional needs differ little among vertebrates and invertebrates. Insects, for example, have the same dietary requirements as vertebrates, although they need more potassium, phosphorus, and magnesium than vertebrates and less calcium, sodium, and chlorine. The ultimate source of most of these nutrients is plants. For this reason, the quantity and quality of plants affect the nutrition of herbivorous consumers. When the amount of food is insufficient, consumers may suffer from acute malnutrition, leave the area, or starve. In other situations, the quantity of food may be sufficient to allay hunger, but its low quality affects reproduction, health, and longevity.

The highest-quality plant food for herbivores, vertebrate and invertebrate, is high in nitrogen in the form of protein. As the nitrogen content of their food increases, their assimilation of plant material improves, increasing growth, reproductive success,

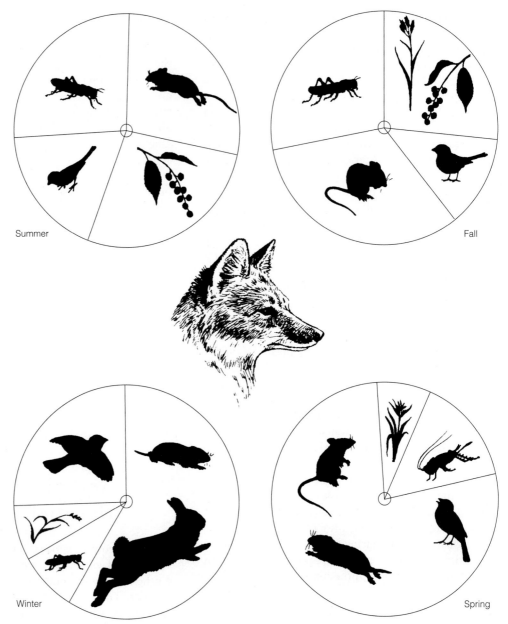

FIGURE 8.3 The red fox is an example of an omnivore. It feeds on both the herbivorous and the carnivorous levels. Its diet varies with the seasons. Note the prominence of fruits and insects in summer and of rodents and rabbits in spring and fall.

and survival. Nitrogen is concentrated in the growing tips, new leaves, and buds of plants. Its content declines as leaves and twigs mature and become senescent. Herbivores have adapted to this period of new growth. Herbivorous insect larvae are most abundant early in the growing season and these complete their growth before the leaves mature. Many vertebrate herbivores, such as deer, give birth to their young at the start of the growing season, when the most protein-rich plant foods will be available for their growing young.

Although food selection is strongly influenced by availability and season, herbivores, both vertebrate and invertebrate, do show some preference for the most nitrogen-rich plants, which they probably detect by taste and odor. For example, beaver show a strong preference for willows and aspen, two species that are high in nitrogen content. Chemical receptors in the nose and mouth of deer encourage or discourage consumption of certain foods. During drought, nitrogen-based compounds are concentrated in certain plants, making them more attractive and vulnerable to herbivorous insects. However, preference for certain plants means little if they are unavailable. Food selection by herbivores is an interaction among quality, preference, and availability.

The need for quality foods differs among herbivores. Ruminant animals, as already pointed out, can subsist on rougher or lower-quality plant materials because bacteria in the rumen can synthesize such requirements as vitamin B_1 and certain amino acids from simple nitrogen-based compounds. Therefore, the caloric content and the nutrients in a certain food might not reflect its real nutritive value for the ruminant. Nonruminant herbivores require more complex proteins. Seed-eating herbivores exploit the concentration of nutrients in the seeds. Such animals are not likely to have dietary problems. Among the carnivores, quantity is more important than quality. Carnivores rarely have a dietary problem because they consume animals that have resynthesized and stored protein and other nutrients from plants in their tissues.

FIGURE 8.4 A mineral lick used by white-tailed deer.

8.3 Mineral Availability Affects Animals' Growth and Reproduction

Mineral availability also appears to influence the abundance and fitness of some animals. One essential nutrient that has received attention is sodium, often one of the least available nutrients in terrestrial ecosystems. In areas of sodium deficiency in the soil, herbivorous animals face an inadequate supply of sodium in their diets. The problem has been noted in Australian herbivores such as kangaroos, in African elephants *(Loxodonta africana)*, in rodents, in white-tailed deer, and in moose *(Alces alces)*.

Sodium deficiency can influence the distribution, behavior, and physiology of mammals, especially herbivores. The spatial distribution of elephants across the Wankie National Park in central Africa appears to be closely correlated with the sodium content of drinking water. The most elephants are found at water holes with the highest sodium content. Three herbivorous mammals—the European rabbit *(Oryctolagus cuniculus)*, the moose, and the white-tailed deer—experience sodium deficiencies in parts of their range. In sodium-deficient areas in southwestern Australia, the European rabbit builds up reserves of sodium in its tissues during the non-breeding season. These reserves become exhausted near the end of the breeding season. During the breeding season, the rabbits selectively graze on sodium-rich plants to the point of depleting their populations.

Ruminants face severe mineral deficiencies in spring. Attracted by the flush of new growth, deer,

bighorn sheep *(Ovis canadensis)*, mountain goats *(Oreamnos americanus)*, elk *(Cervus elaphus)*, and domestic cattle and sheep feed on new succulent growth of grass, but with high physiological costs. Vegetation is much higher in potassium relative to calcium and magnesium in the spring than during the rest of the year. This high intake of potassium stimulates the secretion of aldosterone, the principal hormone that promotes retention of sodium by the kidney. Although aldosterone stimulates the retention of sodium, it also facilitates the excretion of potassium and magnesium. Because concentrations of magnesium in soft tissues and skeletal stores are low in herbivores, these animals experience magnesium deficiency. This deficiency results in a rapid onset of diarrhea and often muscle spasms (tetany). The deficiency comes late in gestation for females and at the beginning of antler growth for male deer and elk, a time when mineral demands are high.

To counteract this mineral imbalance in the spring, large herbivores seek mineral licks, places in the landscape where animals concentrate to satisfy their mineral needs by eating mineral-rich soil (Figure 8.4). Although sodium chloride is associated with mineral licks, animal physiologists hypothesize that it is not sodium the animals seek but magnesium, and in the case of bighorn sheep, mountain goats, and elk, calcium as well.

The size of deer, their antler development, and their reproductive success all relate to nutrition. Other factors being equal, only deer obtaining high-quality foods grow large antlers. Deer on diets low in calcium, phosphorus, and protein show stunted growth, and bucks develop only thin spike antlers. Reproductive success of does is highest where food is abundant and nutritious (Figure 8.5).

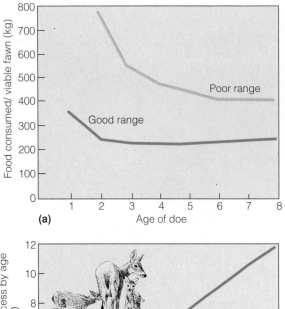

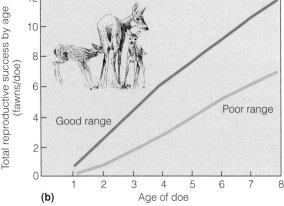

FIGURE 8.5 Differences in the reproductive success of female white-tailed deer on good and poor ranges in New York State. (a) Food consumed per viable fawn on poor range (Adirondack Mountains) was much greater than food consumed on good range (western New York). (b) Reproductive success was considerably greater on good range than on poor range.

8.4 Animals Require Oxygen to Release Energy Contained in Food

Animals obtain their energy from the organic compounds in the food they eat. They release this energy primarily through aerobic respiration, which requires oxygen (Chapter 6, Section 6.6). Oxygen is easily available in the atmosphere for terrestrial animals. But for aquatic animals, oxygen is limiting and its acquisition problematic (see Chapter 4, Section 4.4).

Differences in the means of oxygen acquisition between terrestrial and aquatic animals reflect its availability in the two environments. Small terrestrial organisms take in oxygen by diffusion across the body surface. Insects possess tracheal tubes that open to the outside through openings or spiracles on the body

wall (Figure 8.6b). The tracheal tubes carry oxygen directly to the body cells. Most terrestrial animals have some form of lungs. Structurally, lungs have innumerable small sacs that that increase surface area across which oxygen readily diffuses into the bloodstream. Amphibians take in oxygen through a combination of lungs and vascularized skin. Exceptions are the lungless salamanders that live in a moist environment and take in oxygen directly through the skin.

In addition to lungs, birds have accessory air sacs that act as bellows that keep the air flowing through the lungs as they inhale and exhale (Figure 8.6a). Airflow is one way only, forming a continuous circuit through the interconnected system, whether the bird is inhaling or exhaling. During inhalation, most of the air bypasses the lungs and enters the posterior air sac. That air then passes through the lungs to the anterior air sac. At the same time, the posterior air sac draws in more air.

In aquatic environments, organisms have to take in oxygen from the water or gain oxygen from the air by some means. Marine mammals such as whales and dolphins come to the surface to expel CO_2 and take in air to the lungs. Some aquatic insects rise to the surface to fill the tracheal system with air. Others, like the diving beetles, carry a bubble of air with them when submerged. Held beneath the wings, the air bubble contacts the spiracles of the abdomen.

Fish, the major aquatic vertebrates, pump water through their mouth. The water passes through slits in the pharynx, flows over gills, and exits back of the gill covers (Figure 8.6d). The close contact with and the rapid flow of water over the gills allows for exchanges of oxygen and carbon dioxide between water and the gills. Water passing over the gills flows in a direction opposite to that of blood, setting up a countercurrent exchange. As the blood flows through capillaries, it becomes more and more loaded with oxygen. It also encounters water more and more concentrated with oxygen because water is just beginning to pass over the gills. As water continues its flow, it encounters blood with lower oxygen concentration, aiding the uptake of oxygen through the process of diffusion (8.6c).

8.5 Regulation of Internal Conditions Involves Homeostasis and Feedback

In an ever-changing physical environment, organisms must maintain a fairly constant internal environment within the narrow limits required by their cells, organs, and enzyme systems. They need some means

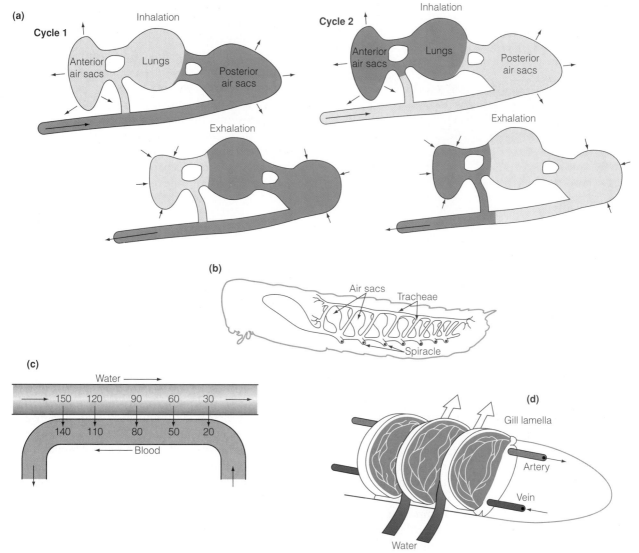

(a) Cycle 1

Inhalation

Anterior air sacs

Lungs

Posterior air sacs

Exhalation

Cycle 2

Inhalation

Anterior air sacs

Lungs

Posterior air sacs

Exhalation

(b)

Air sacs

Tracheae

Spiracle

(c)

Water →

→ 150   120   90   60   30 →

140   110   80   50   20

← Blood

(d)

Gill lamella

Artery

Vein

Water

FIGURE 8.6 Respiratory systems. (a) Gas exchange in the bird lung requires two cycles involving both inhalation and exhalation. During the first inhalation, most of the air flows past the lungs into a posterior air sac. That air passes through the lungs upon exhalation, and the next inhalation ends up in the anterior air sac. At the same time, the posterior sacs draw in more air. This flow pattern allows oxygenated blood to leave the lungs with the highest possible amount of oxygen. (b) The tracheal system and the spiracles of an insect (grasshopper). Airs enters the tracheal tubes through spiracles, openings on the body wall. (c) Water flows across the lamellae in a direction opposite to the blood flow. The blood enters the gill low in oxygen. As it flows through the lamellae, it picks up more and more oxygen from the water. The water flowing in the opposite direction gradually loses more and more of its oxygen. (d) Fish obtain oxygen from water by the way of gills. Gill filaments have flattened plates called lamellae. Blood flowing through capillaries within the lamellae picks up oxygen from the water through a countercurrent exchange.

of regulating their internal environment relative to internal conditions including body temperature, water balance, pH, and the amounts of salts in fluids and tissues. For example, the human body must maintain internal temperatures within a narrow range around 37°C. An increase or decrease of only a few degrees from this value or point could prove fatal. The maintenance of a relatively constant internal environment in a varying external environment is called **homeostasis** (Figure 8.7).

Whatever the processes involved in regulating an organism's internal environment, homeostasis involves negative feedback, meaning that when a system deviates from the normal or desired state, mechanisms function to restore the system to that state. The thermostat that controls the temperature in your home is

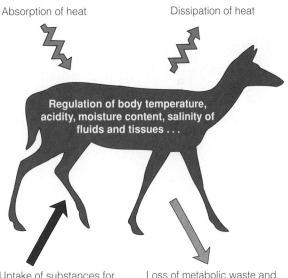

Absorption of heat   Dissipation of heat

Regulation of body temperature, acidity, moisture content, salinity of fluids and tissues . . .

Uptake of substances for cellular chemical reactions

Loss of metabolic waste and excessive uptake

FIGURE 8.7 Homeostasis calls for two-way exchange between the internal and external environments.

an example of a negative feedback system. If we wish the temperature of the room to be 20°C (68°F), we set that point on the thermostat. When the temperature of the room air falls below that point, a temperature-sensitive device within the thermostat trips the switch that turns on the furnace. When the room temperature reaches the set point, the thermostat responds by shutting off the furnace. Should the thermostat fail to function properly and not shut the furnace off, then the furnace continues to heat, the temperature continues to rise, and the furnace ultimately overheats, causing either a fire or a mechanical breakdown.

A key difference between mechanical and living systems is that in living systems, the set point is not firmly fixed as it often is in mechanical systems. Instead, organisms have a limited range of tolerances, called **homeostatic plateaus.** Homeostatic systems work within maximum and minimum values by using negative feedback to regulate activity above or below the set point. If the system deviates from that set point, a negative feedback response ensues—a control mechanism inhibits any strong movement away from the set point. Among animals, the control of homeostasis is both physiological and behavioral.

An example is temperature regulation in humans (Figure 8.8). The normal temperature, or set point, for humans is 37°C. When the temperature of the environment rises, sensory mechanisms in the skin detect the change. They send a message to the brain, which automatically relays the message to receptors that increase blood flow to the skin, induce sweating, and stimulate behavioral responses. Water excreted through the skin evaporates, cooling the body. When the environmental temperature falls below a certain

point, another reaction takes place. This time it reduces blood flow and causes shivering, an involuntary muscular exercise producing more heat.

If the environmental temperature becomes extreme, the homeostatic system breaks down. When it gets too warm, the body cannot lose heat fast enough to maintain normal temperature. Metabolism speeds up, further raising body temperature, until death results from heatstroke. If the environmental temperature drops too low, metabolic processes slow down, further decreasing body temperature, until death by freezing ensues.

8.6 Animals Exchange Energy with Their Surrounding Environment

Animals differ significantly from plants in their thermal relations with the environment. Animals can produce heat by metabolism, and their mobility allows them to seek out or escape heat and cold. The structure of their bodies influences their exchange of heat with the environment.

Consider a thermal model of an animal body (Figure 8.9). The interior or core of the body must be regulated within a defined range of temperature. In contrast, the temperature of the environment surrounding the animal's body varies. However, it is not the air or water temperature per se that an organism senses, but rather, the temperature at a thin layer of air called the **boundary layer** that lies at the surface, just above and within hair, feathers, and scales. Therefore, body surface temperature differs from both the air (or water) and the core body temperatures. Separating the body core from the body surface are layers of muscle tissue and fat, across which the temperature gradually changes from the core temperature to the body surface temperature. This layer of insulation influences the organism's **conductivity,** that is, the ability to exchange heat with the surrounding environment.

To maintain the core body temperature, the animal has to balance gains and losses of heat to the environment. It does so by changes in metabolic rate and by conduction. The core area exchanges heat, produced by metabolism and stored in the body, with the surface area by conduction. Influencing this exchange are the thickness and conductivity of fat and the movement of blood near the surface. The surface layer exchanges heat with the environment by convection, conduction, radiation, and evaporation, all influenced by the characteristics of skin and body covering.

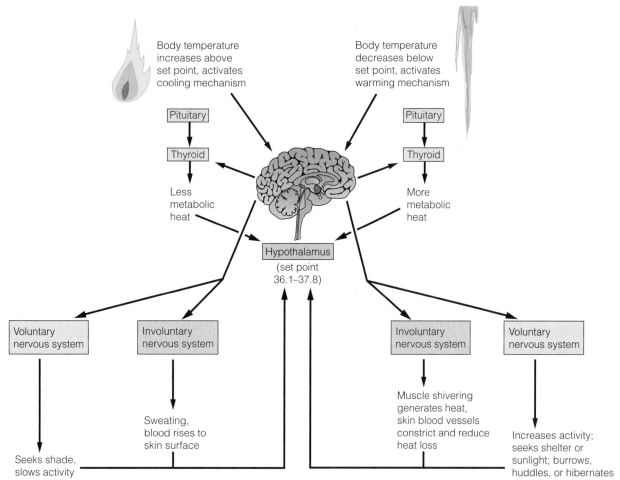

Body temperature increases above set point, activates cooling mechanism

Body temperature decreases below set point, activates warming mechanism

Pituitary → Thyroid → Less metabolic heat

Pituitary → Thyroid → More metabolic heat

Hypothalamus
(set point 36.1–37.8)

Voluntary nervous system

Involuntary nervous system

Involuntary nervous system

Voluntary nervous system

Sweating, blood rises to skin surface

Muscle shivering generates heat, skin blood vessels constrict and reduce heat loss

Seeks shade, slows activity

Increases activity; seeks shelter or sunlight; burrows, huddles, or hibernates

FIGURE 8.8 Thermoregulation is an example of homeostasis in action. The hypothalamus in your brain receives feedback or senses the temperature of blood arriving from the body core. If body core temperature rises, it responds accordingly in two ways, activating the autonomic (or involuntary) and voluntary nervous systems and the endocrine system.

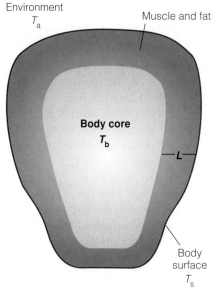

Environment
T_a

Muscle and fat

Body core
T_b

L

Body surface
T_s

FIGURE 8.9 Temperatures in an animal body. Body core temperature is T_b, the environmental temperature is T_e, the surface temperature is T_s, and L is the thickness of the outer layer of the body.

Physiology and environment heavily influence how animals confront thermal stress. Because air has a lower specific heat and absorbs less solar radiation than water, terrestrial animals face more radical and dangerous changes in their thermal environment than aquatic animals. Incoming solar radiation can produce lethal heat. The loss of radiant heat to the air, especially at night, can result in deadly cold. Aquatic animals live in a more stable energy environment (see Chapter 4), but they have a lower tolerance for temperature changes.

8.7 Animals Fall into Three Groups Relative to Temperature Regulation

To regulate temperature, some groups of animals generate heat metabolically. This internal heat production is **endothermy**, meaning "heat from within."

The result is **homeothermy** (from *homeo,* "the same"), maintenance of a fairly constant temperature independent of external temperatures. Another group of animals acquires heat primarily from the external environment. Gaining heat from the environment is **ectothermy,** meaning "heat from without." Unlike endothermy, ectothermy results in a variable body temperature. This means of maintaining body temperature is **poikilothermy** (from *poikilos,* "manifold" or "variegated").

Birds and mammals are notable **homeotherms.** They are popularly called warm-blooded. Fish, amphibians, reptiles, insects, and other invertebrates are **poikilotherms.** They are often called "cold-blooded" because they can be cool to the touch. A third group regulates body temperature by endothermy at some times and ectothermy at other times. These animals are **heterotherms** (from *hetero,* "different"). Heterotherms employ both endothermy and ectothermy, depending upon environmental situations and metabolic needs. Bats, bees, and hummingbirds belong to this group.

The terms *homeotherm* and *endotherm* are often used synonymously, as are *poikilotherm* and *ectotherm,* but there is a difference. *Ectotherm* and *endotherm* emphasize the mechanisms that determine body temperature. The other two terms, *homeotherm* and *poikilotherm,* represent the nature of body temperature (either constant or variable).

8.8 Poikilotherms Depend on Environmental Temperatures

Poikilotherms, such as amphibians, reptiles, and insects, gain heat easily from the environment and lose it just as fast. Environmental sources of heat control the rates of metabolism and activity among most poikilotherms. Rising temperatures increase the rate of enzymatic activity, which controls metabolism and respiration (Figure 8.10). For every 10°C rise in temperature, the rate of metabolism in poikilotherms approximately doubles. They become active only when the temperature is sufficiently warm. Conversely, when ambient temperatures fall, metabolic activity declines, and poikilotherms become sluggish.

Poikilotherms have an upper and lower thermal limit that they can tolerate. Most terrestrial poikilotherms can maintain a relatively constant daytime body temperature by behavioral means, such as seeking sunlight or shade. Lizards and snakes, for example, may vary their body temperature by no more than 4°C to 5°C (Figure 8.11) and amphibians by

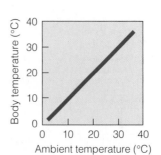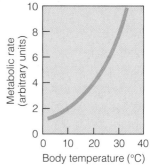

FIGURE 8.10 Relationship among body temperature, resting metabolic rate, and ambient temperature in poikilotherms. (a) Body temperature is a function of ambient temperature. (b) Resting metabolism is a function of body temperature. (After Hill and Wyse 1989:83.)

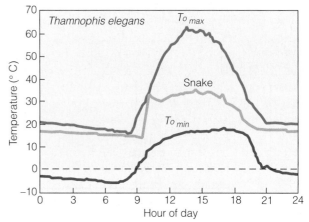

FIGURE 8.11 Daily temperature variation in the western terrestrial garter snake *Thamnophis elegans,* within its operative temperature range (T_o). Note that the snake maintains a fairly constant temperature during the day. (After Peterson et al. 1993.)

10°C when active. The range of body temperatures at which poikilotherms carry out their daily activities is the **operative temperature range.**

Poikilotherms have a low metabolic rate and a high ability to exchange heat between body and environment (high conductivity). During normal activities, poikilotherms carry out aerobic respiration. Under stress and while pursuing prey, the inability to supply sufficient oxygen to the body requires that much of their energy production is from anaerobic respiration, in which oxygen is not used. This process depletes stored energy and accumulates lactic acid in the muscles. (Anaerobic respiration takes place in the muscles of marathon runners and other athletes, causing leg cramps.) Anaerobic respiration metabolism limits poikilotherms to short bursts of activity. It results in rapid physical exhaustion, often within 3 to 5 minutes.

Aquatic poikilotherms, completely immersed, do not maintain any appreciable difference between

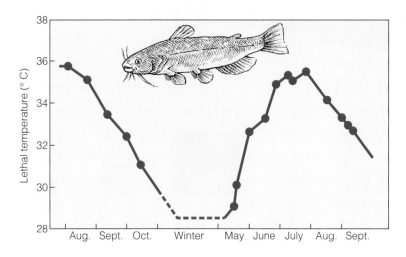

FIGURE 8.12 A diagrammatic representation of acclimatization in bullhead catfish. Tolerance for warmer or colder water shifts as temperatures gradually increase or decrease. Exposure to higher temperatures would be lethal when the organism is acclimatized to a colder temperature. (Fry 1947.)

their body temperature and the surrounding water. Aquatic poikilotherms are poorly insulated. Any heat produced in the muscles moves to the blood and on to the gills and skin, where heat transfers to the surrounding water by convection. Exceptions are sharks and tunas that possess a rete, a blood circulation system that allows them to keep internal temperatures higher than external ones (see Section 8.16). Because seasonal water temperatures are relatively stable, fish and aquatic invertebrates maintain a fairly constant temperature within any given season. They adjust seasonally to changing temperatures by acclimatization (Figure 8.12). They undergo physiological changes over a period of time. Poikilotherms have an upper and lower limit of tolerance to temperature that varies with the species. If they live at the upper end of their tolerable thermal range, poikilotherms will adjust their physiology at the expense of being able to tolerate the lower range. Similarly, during periods of cold, the animals shift to a lower temperature range that would have been lethal before. Because water temperature changes slowly through the year, aquatic poikilotherms can make adjustments slowly. Fish are highly sensitive to rapid change in environmental temperatures. If they are subjected to a sudden temperature change, they will die of thermal shock.

8.9 Homeotherms Escape the Thermal Restraints of the Environment

Homeothermic birds and mammals meet the thermal constraints of the environment by being endothermic. They maintain body temperature by oxidizing glucose and other energy-rich molecules in the process of respiration. They regulate the gradient between body and air or water temperatures by seasonal changes in insulation (the type and thickness of fur, structure of feathers, and layer of fat). They rely on evaporative cooling and on increasing or decreasing metabolic heat production.

Maintenance of a high body temperature is associated with specific enzyme systems that operate optimally within a high temperature range, with a set point about 40°C. Homeotherms generally have a high metabolic rate and a low thermal conductance. Because efficient cardiovascular and respiratory systems bring oxygen to their tissues, homeotherms can maintain a high level of aerobic energy production through aerobic respiration. Thus, they can sustain a high level of physical activity for long periods. Independent of external temperatures, homeotherms can exploit a wider range of thermal environments. They can generate energy rapidly when the situation demands, escaping from predators or pursuing prey.

8.10 Endothermy and Ectothermy Involve Tradeoffs

The two alternative approaches to the regulation of body temperature in animals, endothermy and ectothermy, are a prime example of the tradeoffs involved in the adaptations of organisms to their environment. Each strategy has advantages and disadvantages that enable the organisms to excel under different environmental conditions. For example, homeothermy allows animals to remain active regardless of environmental temperatures, whereas environmental temperatures largely dictate the activity of poikilotherms.

BODY SIZE AND METABOLISM

One of the most important features of an animal that influences its ability to regulate body temperature is size. A body exchanges heat with the external environment (either air or water) in proportion to the surface area exposed. In contrast, it is the entire body mass (or volume) that is being heated. Cold-blooded organisms (ectotherms) absorb heat across their surface, but must absorb sufficient energy to heat the entire body mass (volume). Therefore, the ratio of the surface area to the volume is a key factor in controlling the uptake of heat and the maintenance of body temperature. As the size of an organism increases, the ratio of the surface area to volume decreases (Figure A). Because the organism has to absorb sufficient energy across its surface to warm the entire body mass, the amount of energy and/or the period of time required to raise body temperature likewise increases. For this reason, ectothermy imposes a constraint on maximum body size for cold-blooded animals.

The constraint that size imposes on warm-blooded animals (homeotherms) is opposite that of cold-blooded animals. For homeotherms, it is the body mass (or volume) that produces heat through respiration, while heat is lost to the surrounding environment across its surface. The smaller the organism, the larger the ratio of surface area to volume and, therefore, the greater the relative heat loss to the surrounding environment. To maintain a constant body temperature, the heat loss must be offset by increased metabolic activity (respiration). Thus, small homeotherms consume more food per unit of body weight than large ones. Small shrews (*Sorex* spp.), for example, ranging in weight from 2 to 29 grams, daily require an amount of food (wet weight) equivalent to their own body weight. It is as if a 150-pound human needed

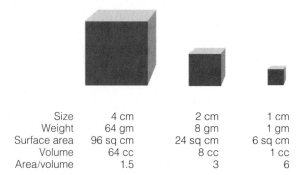

Size	4 cm	2 cm	1 cm
Weight	64 gm	8 gm	1 gm
Surface area	96 sq cm	24 sq cm	6 sq cm
Volume	64 cc	8 cc	1 cc
Area/volume	1.5	3	6

FIGURE A These three cubes—with sides of 4, 2, and 1 cm—point out the relationship between surface area and volume. A small object has more surface area in proportion to its volume than a large object of similar shape.

150 pounds of food daily to stay alive. Therefore, small animals are forced to spend most of their time seeking and eating food. The weight-specific metabolic rate (respiration rate per gram of body weight) of small endotherms rises so rapidly that, below a certain size, they could not meet their energy demands. On the average, 2 grams is about as small as an endotherm can be and still maintain a metabolic heat balance, although this minimum constraint is dependent on the thermal environment. Some shrews undergo daily torpor to reduce their metabolic needs. Because of the conflicting metabolic demands of body temperature and growth, most young birds and mammals are born in an altricial state—blind, naked, and helpless—and begin life as ectotherms. They depend upon the body heat of their parents to maintain their body temperature. That allows young animals to allocate most of their energy to growth.

However, the freedom of activity enjoyed by homeotherms comes at a high energy cost. To generate heat through respiration, homeotherms must take in calories (food). And of the food energy that is assimilated, a minimum goes to growth.

The metabolic heat that is produced by homeothermy can be lost to the surrounding environment through either conduction or convection. This heat must be replaced by additional heat generated through respiration. As a result, metabolic costs weigh heavily against smaller homeotherms and place a lower limit on body size (see Focus on Ecology 8.1). In contrast, ectotherms can allocate more of their energy intake to

biomass production than to metabolic needs. Not needing to burn calories to provide metabolic heat, ectotherms require fewer calories (food) per gram of body weight. Because they do not depend upon internally generated body heat, ectotherms can curtail metabolic activity in times of food and water shortage and temperature extremes. Their low energy demands enable some terrestrial poikilotherms to colonize areas of limited food and water.

Because ectotherms do not have the problem of metabolic heat loss, they are not restricted to any minimum body size or definite shape. Such characteristics enable poikilotherms to exploit resources and habitats unavailable to homeotherms. On the other hand, the same metabolic restrictions seem to impose an upper body-size limit (see Focus on Ecology 8.1), and restrict the distribution of the larger poikilotherms to the warmer, aseasonal regions of the subtropics and tropics. For example, the large reptiles, such as alligators, crocodiles, iguanas, komodo dragons, anacondas, and pythons are all restricted to warm tropical environments. This fact fuels the debate over whether the large dinosaurs were ectothermic or endothermic, which is based largely on the limitations imposed by ectothermy on large body size.

8.11 Heterotherms Take On Characteristics of Ectotherms and Endotherms

Species that sometimes regulate their body temperature and sometimes do not are called temporal heterotherms. At different stages of their daily and seasonal cycle or in certain situations, these animals take on characteristics of endotherms or ectotherms. They can undergo rapid, drastic, repeated changes in body temperature.

Insects are ectothermic and poikilothermic; yet in the adult stage, most species of flying insects are heterothermic. When flying, they have high rates of metabolism, with heat production as great as or greater than homeotherms. They reach this high metabolic state in a simpler fashion than do homeotherms because they are not constrained by the uptake and transport of oxygen through the lungs and vascular system. Insects take in oxygen by demand through openings in the body wall and transport it throughout the body in a tracheal system (see Section 8.4).

Temperature is critical to the flight of insects. Most cannot fly if the temperature of the body muscles is below 30°C; nor can they fly if muscle temperature is over 44°C. This constraint means that an insect has to warm up before it can take off, and it has to get rid of excess heat in flight. With wings beating up to 200 times per second, flying insects can produce a prodigious amount of heat.

Some insects, such as butterflies and dragonflies, warm up by orienting their bodies and spreading their wings to the sun. Most warm up by shivering their flight muscles in the thorax. Moths and butterflies vibrate their wings to raise thoracic temperatures above ambient. Bumblebees pump their abdomens without any external wing movements. They do not maintain any physiological set point, and they cool down to ambient temperatures when not in flight.

8.12 Torpor Helps Some Animals Conserve Energy

To reduce metabolic costs during periods of inactivity, a number of small homeothermic animals become heterothermic and enter into torpor daily. Daily **torpor** is the dropping of body temperature to approximately ambient temperature for a part of each day, regardless of season.

A number of birds, such as hummingbirds (Trochilidae) and poorwills *(Phalaenoptilus nuttallii)*, and small mammals, such as bats, pocket mice, kangaroo, and white-footed mice, experience daily torpor. Such daily torpor seems to have evolved as a means of reducing energy demands over that part of the day in which the animals are inactive. Nocturnal mammals, such as bats, go into torpor by day; and diurnal animals, such as hummingbirds, go into torpor by night. As the animal goes into torpor, its body temperature falls steeply. With the relaxation of homeothermic responses, the body temperature declines to within a few degrees of ambient. Arousal returns the body temperature to normal rapidly as the animal renews its metabolic heat production.

To escape the rigors of long, cold winters, many terrestrial poikilotherms and a few heterothermic mammals go into a long seasonal torpor, called **hibernation.** Hibernation is characterized by the cessation of activity. Hibernating poikilotherms experience such physiological changes as decreased blood sugar, increased liver glycogen, altered concentration of blood hemoglobin, altered carbon dioxide and oxygen content in the blood, altered muscle tone, and darkened skin.

Hibernating homeotherms become heterotherms and invoke controlled hypothermia (reduction of body temperature). They relax homeothermic regulation and allow the body temperature to approach ambient temperature. Heart rate, respiration, and total metabolism fall, and body temperature sinks below 10°C. Associated with hibernation are high levels of CO_2 in the body and acid in the blood. This state, called acidosis, lowers the threshold for shivering and reduces the metabolic rate. Hibernating homeotherms, however, are able to rewarm spontaneously using only metabolically generated heat.

Among homeotherms, entrance into hibernation is a controlled process difficult to generalize from one species to another. Some hibernators, such as the groundhog *(Marmota monax),* feed heavily in late summer to store large fat reserves, from which they will draw energy during hibernation. Others, like the chipmunk *(Tamias striatus),* lay up a store of food instead. All hibernators, however, convert to a means of metabolic regulation different from that of the active state. Most hibernators rouse periodically and then drop back into torpor. The chipmunk, with its large store of seeds, spends much less time in torpor than species that store large amounts of fat.

Although popularly said to hibernate, black, grizzly, and female polar bears do not. Instead, they enter a unique winter sleep from which they easily rouse. They do not enter extreme hypothermia. The bears do not eat, drink, urinate, or defecate and females give birth to and nurse young during their sleep; yet they maintain a metabolism that is near normal. To do so, the bears recycle urea, normally excreted in urine, through the bloodstream. The urea is degraded into amino acids that are reincorporated in plasma proteins.

Hibernation provides selective advantages to small homeotherms. For them, maintaining a high body temperature during periods of cold is too costly. It is far less expensive to reduce metabolism and allow the body temperature to drop. Doing so eliminates the need to seek scarce food resources to keep warm.

8.13 Poikilotherms Exploit Microclimates to Regulate Temperature

To maintain a tolerable and fairly constant body temperature during active periods, terrestrial and amphibious poikilotherms resort to behavioral means. They seek out appropriate microclimates (Figure 8.13). Insects such as butterflies, moths, bees, dragonflies, and damselflies bask in the sun to raise their body temperature to the level necessary to become highly active. When they become too warm, these animals seek the shade. Salamanders, restricted to moist, shaded habitats, lack this option, but semiterrestrial frogs, such as bullfrogs and green frogs, exert considerable control over their body temperature. By basking in the sun, frogs can raise their body temperature as much as 10°C above ambient temperature. Because of associated evaporative water losses, such amphibians must either be near or partially submerged in water. By changing position or location or by seeking a warmer or cooler substrate, amphibians can maintain body temperatures within a narrow range.

Reptiles have their own response to temperature. Most are terrestrial and exposed to widely fluctuating temperatures. The simplest way for a reptile to raise body temperature is to bask in the sun. Snakes, for example, heat up rapidly in the morning sun (Figure 8.14). When they reach the preferred temperature, the animals move on to their daily activities, retreating to the shade to cool when necessary. In this manner, they maintain a stable body temperature during the day. In the evening, the reptile experiences a slow cooling. Its body temperature at night depends upon its location.

Lizards raise and lower their bodies and change body shape to increase or decrease the conduction of heat between themselves and the rocks or soil on which they rest. They also seek sunlight or shade or

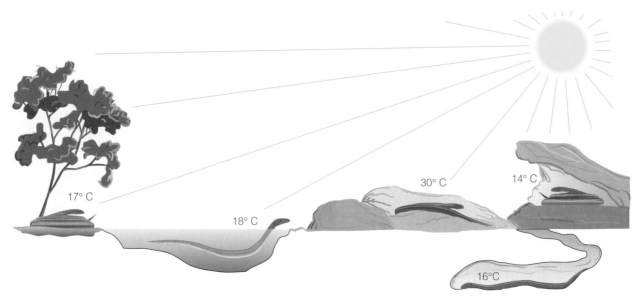

FIGURE 8.13 Microclimates a snake typically uses to regulate body temperature during the summer. (Adapted from Pearson et al. 1993.)

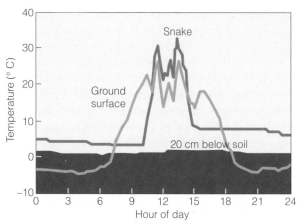

FIGURE 8.14 A snake warms up quickly in the morning, achieves a temperature plateau during the day, and cools off slowly in the evening to its nighttime temperature, which depends upon ambient temperature (T_o). (After Peterson et al. 1993.)

burrow into the soil to adjust their temperatures. Desert beetles, locusts, and scorpions exhibit similar behavior. They raise their legs to reduce contact between their body and the ground, minimizing conduction and increasing convection by exposing body surfaces to the wind. Thus, body temperatures of poikilotherms do not necessarily follow the general ambient temperature.

8.14 Insulation Reduces Heat Exchange

To regulate the exchange of heat between the body and the environment, homeotherms and certain poikilotherms use some form of insulation—a covering of fur, feathers, or body fat. For mammals, fur is a major barrier to heat flow, but its insulation value varies with thickness, which is greater on large mammals than on small ones. Small mammals are limited in the amount of fur they can carry because a thick coat could reduce their ability to move. Mammals change the thickness of their fur with the season. Aquatic mammals, especially of Arctic regions, and such Arctic and Antarctic birds as auklets and penguins have a heavy layer of fat beneath the skin. Birds reduce heat loss by fluffing the feathers and drawing the feet into them, making the body a round, feathered ball. Some Arctic birds, such as ptarmigan, have feathered feet, unlike most birds that have scaled feet that function to lose heat.

Although the major function of insulation is to keep body heat in, it also keeps heat out. In a hot environment, an animal has to either rid itself of excess body heat or prevent heat from being absorbed in the first place. One means is to reflect solar radiation from light-colored fur or feathers. Conversely, many birds of desert regions have dark feathers that absorb heat and then radiate it off again. Another means is to grow a heavy coat of fur that heat does not penetrate. Large mammals of the desert, notably the camel, employ this method. The outer layers of hair absorb heat and return it to the environment.

A number of insects, notably moths, bees, and bumblebees, have a dense, furlike coat over the thoracic region, which serves to retain the high temperature of flight muscles during flight. The long, soft hairs of caterpillars, together with changes in body posture, act as insulation, reducing convective heat exchange.

When insulation fails, many animals resort to shivering, a form of involuntary muscular activity that increases heat production. Many species of small mammals increase heat production without shivering by burning highly vascular brown fat with a high rate of oxygen consumption. Brown fat, found about the head, neck, thorax, and major blood vessels, occurs mainly in two groups of mammals: cold-acclimated adults, such as ground squirrels, and hibernators, such as bats and groundhogs. But human infants also have brown fat between their shoulder blades and the nape of the neck.

8.15 Evaporative Cooling in Animals Is Important

Many birds and mammals, and even wasps and hornets, employ evaporative cooling to reduce the body heat load. Birds and mammals lose some heat by evaporation of moisture from the skin. They accelerate evaporative cooling by sweating and panting. Only certain mammals have sweat glands, particularly horses and humans. Panting in mammals and gular fluttering in birds increase the movement of air over moist surfaces in the mouth and pharynx. Many mammals, such as pigs, wallow in water and wet mud to cool down.

Paper wasps maintain a rather constant temperature in their paper nest by fanning their wings. However, because of its gray color, the nest warms rapidly in the sunlight. If the nest temperature becomes excessive, over 35°C, wasps gather water from nearby sources, carry it to the nest, and fan their wings to speed evaporative cooling.

8.16 Some Animals Use Unique Physiological Means for Thermal Balance

Storing body heat does not seem like a sound option to maintain thermal balance in the body because of an animal's limited tolerance for heat. However, certain mammals, especially the camel, oryx, and some gazelles, do just that. The camel, for example, stores body heat by day and dissipates it by night, especially when water is limited. Its temperature can fluctuate from 34°C in the morning to 41°C by late afternoon. By storing body heat, these animals of dry habitats reduce the need for evaporative cooling and thus reduce water loss.

Many poikilothermic animals of temperate and Arctic regions withstand long periods of below-freezing temperatures in winter through supercooling and developing a resistance to freezing. **Supercooling** of body fluids takes place when the body temperature falls below the freezing point without actually freezing. The presence of certain solutes in the body that function to lower the freezing point of water (see Chapter 4) influences the amount of supercooling that can take place. Some Arctic marine fish, certain insects of temperate and cold climates, and reptiles exposed to occasional cold nights employ supercooling by increasing solutes, notably glycerol, in body fluids. Glycerol protects against freezing damage, increasing the degree of supercooling. Wood frogs *(Rana sylvatica),* spring peepers *(Hyla crucifer),* and gray tree frogs *(H. versicolor)* can successfully overwinter just beneath the leaf litter because they accumulate glycerol in their body fluids.

Some intertidal invertebrates of high latitudes and certain aquatic insects survive the cold by freezing and then thawing out when the temperature moderates. In some species, more than 90 percent of the body fluids may freeze, and the remaining fluids contain highly concentrated solutes. Ice forms outside the shrunken cells, and muscles and organs are distorted. After thawing, they quickly regain normal shape.

8.17 Countercurrent Circulation Conserves or Reduces Body Heat

To conserve heat in a cold environment and to cool vital parts of the body under heat stress, a number of animals have evolved countercurrent heat exchangers

(Figure 8.15). For example, the porpoise (*Phocaena* spp.), swimming in cold Arctic waters, is well insulated with blubber. It could experience an excessive loss of body heat, however, through its uninsulated flukes and flippers. The porpoise maintains its body core temperature by exchanging heat between arterial (coming from the lungs) and venous (returning to the lungs) blood in these structures (Figure 8.16). Veins completely surround arteries carrying warm blood from the heart to the extremities. Warm arterial blood loses its heat to the cool venous blood returning to the body core. As a result, little body heat passes to the environment. Blood entering the flippers cools, while blood returning to the deep body warms. In warm waters, where the animals need to get rid of excessive body heat, blood bypasses the heat exchangers. Venous blood returns unwarmed through veins close to the skin's surface to cool the body core. Such vascular arrangements are common in the legs of mammals and birds and the tails of rodents, especially the beaver.

Many animals have arteries and veins divided into small, parallel, intermingling vessels that form a discrete vascular bundle or net known as a rete. In a rete, the principle is the same as in the blood vessels of the porpoise's flippers. Blood flows in opposite directions, and heat exchange takes place.

Countercurrent heat exchange can also keep heat out. The oryx *(Oryx besia),* an African desert antelope exposed to high daytime temperatures, can experience elevated body temperatures yet keep the highly heat-sensitive brain cool by a rete in the head. The external carotid artery passes through a cavernous sinus filled with venous blood cooled by evaporation from the moist mucous membranes of the nasal passages (Figure 8.17). Arterial blood passing through the cavernous sinus cools on the way to the

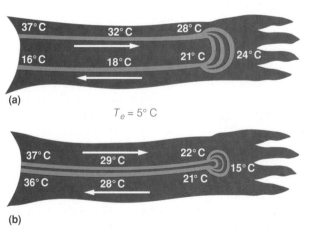

FIGURE 8.15 A model of countercurrent flow in the limb of a mammal, showing hypothetical temperature changes in the blood (a) in the absence and (b) in the presence of countercurrent heat exchange.

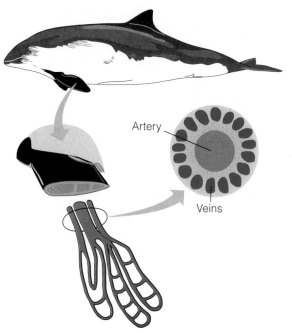

FIGURE 8.16 The porpoise and its relatives, the whales, use flippers and flukes as a temperature-regulating device. Several veins in the appendages surround the arteries. Venous blood returning to the body core is warmed through heat transfer, retaining body heat. (Adapted from Schmidt-Nielsen 1977.)

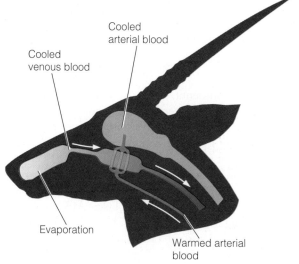

FIGURE 8.17 A desert gazelle can keep a cool head in spite of a high body core temperature by means of a rete. Arterial blood passes in small arteries through a pool of venous blood cooled by evaporation as it drains from the nasal region.

brain, reducing the temperature of the brain 2°C to 3°C lower than that of the body core.

Countercurrent heat exchangers are not restricted to homeotherms. Certain poikilotherms that assume some degree of endothermism employ the same mechanism. The swift, highly predaceous tuna (*Thunnus* spp.) and the mackerel shark *(Isurus tigris)* possess a rete in a band of dark muscle tissue used for sustained swimming effort. Metabolic heat produced in the muscle warms the venous blood, which gives up heat to the adjoining newly oxygenated blood returning from the gills. Such a countercurrent heat exchange increases the power of the muscles because warm muscles contract and relax more rapidly. Sharks and tuna maintain fairly constant body temperatures, regardless of water temperatures.

8.18 Animals Have Ways of Maintaining Water Balance

Animals have a more complex and more energy-expensive problem than plants in maintaining water balance. All animals, however, possess a more or less universal mechanism, the excretory system. The system is simple in some animals and complex in others.

Remember that osmotic pressure moves water through cell membranes from the side of greater water concentration to the side of lesser water concentration. Aquatic organisms living in freshwater have a higher salt concentration in their bodies than in the surrounding water. Their problem is to prevent uptake or to rid themselves of excess water. Freshwater fish maintain osmotic balance by absorbing and retaining salts in special cells in the body and by producing copious amounts of watery urine. Amphibians balance the loss of salts through the skin by absorbing ions directly from the water and transporting them across the skin and gill membranes. They store water from the kidneys in the bladder. If circumstances demand it, they can reabsorb the water through the bladder wall.

Terrestrial animals have three major means of gaining water and solutes: directly by drinking and by eating, and indirectly by producing metabolic water in the process of respiration (see Chapter 6, Section 6.1). They lose water and solutes through urine, feces, evaporation from the skin, and in the moist air that is exhaled. Some birds and reptiles have a salt gland and a cloaca, a common receptacle for the digestive, urinary, and reproductive tracts. They reabsorb water from the cloaca back into the body proper. Mammals possess kidneys capable of producing urine with high osmotic pressure and ion concentrations.

8.19 Animals of Arid Environments Conserve Water

In arid environments, animals, like plants, face a severe problem of water balance. They can solve the problem in one of two ways: either by evading the drought or by avoiding its effects. Animals of semiarid and desert regions may evade drought by leaving the area during the dry season, moving to areas where permanent water is available. That is the strategy employed by many of the large African ungulates. During hot, dry periods the spadefoot toad *(Scaphiopus couchi)* of the southern deserts of the United States remains below ground in a state of dormancy, called aestivation, and emerges when the rains return. Some invertebrates that inhabit ponds that dry up in summer, such as the flatworm *Phagocytes vernalis,* develop hardened casings in which they remain for the dry period. Other aquatic or semiaquatic animals retreat deep into the soil until they reach the level of groundwater. Many insects undergo diapause, a stage of arrested development in their life cycle.

Other animals remain active during the dry season but reduce respiratory water loss. Some small desert rodents lower the temperature of the air they breathe out. Moist air from the lungs passes over cooled nasal membranes, leaving condensed water on the walls. As the rodent inhales, the warm, dry air is humidified and cooled by this water.

An African desert ungulate, the oryx *(Oryx besia),* reduces daytime losses of moisture by becoming hyperthermic (increasing body temperature). A substantial rise in daytime body temperature reduces the need for evaporation. The oryx also reduces water losses by suppressing sweating and by panting only at very high temperatures. Further, the oryx reduces its metabolic rate; by lowering the internal production of heat, the animal reduces the need for evaporative cooling. By night, the oryx reduces evaporation across the skin by lowering its body temperature (Figure 8.18). With slower, more efficient breathing, the amount of water vapor it exhales is lower.

There are other approaches to the problem. Some small desert mammals reduce water loss by remaining in burrows by day and emerging by night. Many desert mammals, from the kangaroo to camels, produce highly concentrated urine and dry feces, and extract water from the food they eat. In addition,

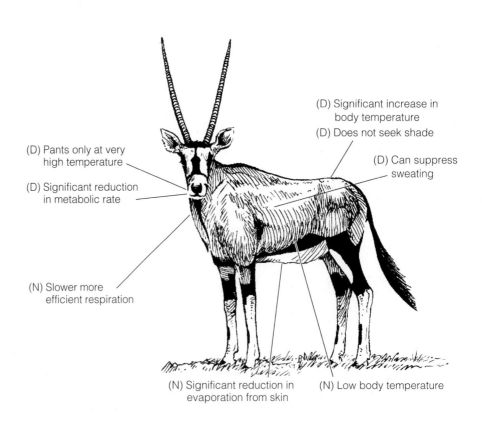

(D) Significant increase in body temperature
(D) Does not seek shade
(D) Can suppress sweating
(D) Pants only at very high temperature
(D) Significant reduction in metabolic rate
(N) Slower more efficient respiration
(N) Significant reduction in evaporation from skin
(N) Low body temperature

FIGURE 8.18 The physiological adaptation to aridity and heat of an African ungulate, the oryx. D = day; N = night.

some desert mammals can tolerate a certain degree of dehydration. Desert rabbits may withstand water losses of up to 50 percent and camels up to 27 percent of their body weight.

8.20 Animals of Saline Environments Have Special Problems

Animals living in salty environments face problems opposite to those in freshwater. These organisms have to retain their body water. When the concentration of salts is greater outside the body than within, organisms tend to dehydrate. Osmosis (see Chapter 4, Section 4.6) draws water out of the body into the surrounding environment. In marine and brackish environments, organisms have to inhibit the loss of water by osmosis through the body wall and prevent an accumulation of salts in the body.

There are many solutions to this problem. Invertebrates get around it by possessing body fluids that have the same osmotic pressure as seawater. Marine bony (teleost) fish absorb saltwater into the gut. They secrete magnesium and calcium through the kidneys and pass these ions off as a partially crystalline paste. In general, fish excrete sodium and chloride, major ions in seawater, by pumping the ions across special membranes in the gills. This pumping process is one type of active transport, moving salts against the concentration gradient. This type of transport comes at a high energy cost. Sharks and rays retain a sufficient amount of urea to maintain a slightly higher concentration of salt in the body than in surrounding seawater. Birds of the open sea can consume seawater because they possess special salt-secreting nasal glands. Gulls, petrels, and other seabirds excrete fluids in excess of 5 percent salt from these glands. Petrels forcibly eject the fluids through the nostrils; other species drip the fluids out of the internal or external nares. In marine mammals, the kidney is the main route for the elimination of salt. Porpoises have highly developed kidneys to eliminate salt loads rapidly.

Marine vertebrates in the Arctic and Antarctic have special problems. As seawater freezes, it becomes colder and saltier. In response most of these organisms increase the solute concentrations in body fluids to lower body temperature. As discussed previously, some species of fish in the Antarctic possess substances in their blood that function as antifreeze and enable the animals to survive in temperatures below the freezing point of their blood.

8.21 Daily and Seasonal Light and Dark Cycles Influence Animal Activity

The major influence of light on animals is its role in timing daily and seasonal activities including feeding, food storage, reproduction, and migratory movements. Animals react to changing light through the response of their biological clocks. (See Focus on Ecology 8.2: How Biological Clocks Measure Time.) The circadian rhythm and its sensitivity to light and dark are the major mechanisms that operate the biological clock—that timekeeper of physical and physiological activity in living things. Where is such a clock located in living things? Its position must expose it to its time-setter, light. In single-celled protists and plants, the clock appears to be located in individual cells. Light acts directly on photosensitive chemicals that activate cellular pathways. In multicellular animals, however, the clock is within the nervous system.

Skillful surgical procedures have allowed circadian physiologists to discover the location of the physiological clock in some mammals, birds, and insects. In most insects studied, the photoreceptors—located in cells at the base of the compound eyes—are connected by axons to the clock, which is located in the optic lobe of the brain or in the tissues between the optic lobes. In birds and reptiles, the clock is located in the pineal gland, functioning as a third eye resting close to the surface of the brain. In mammals, including humans, the clock is in two clumps of neurons (suprachiasmatic nuclei) just above the optic chiasm. The optic chiasm is the place where the optic nerves from the eyes intersect. Operation of the clock in mammals involves a special hormone, **melatonin**, produced by the pineal gland, that serves to measure time. More melatonin is produced in the dark than in the light. The amount produced is a measure of changing daylength.

8.22 Circadian Rhythms Have Adaptive Value

How and why circadian rhythms and biological clocks function is the domain of the physiologist. Ecologists are more interested in the adaptive value of biological clocks. One adaptive value is that the biological clock provides the organism with a time-dependent mechanism. It enables the organism to prepare for periodic changes in the environment

HOW BIOLOGICAL CLOCKS MEASURE TIME

How do biological clocks measure time? This question has been studied for years. In 1960 E. Bünning proposed a basic model—a single 24-hour oscillation sensitive to light. (Figure A). Although dated, this model is helpful. The cycle or time-measuring process begins with the onset of light or dawn. The first half (12 hours) of the cycle requires light; the second half requires darkness.

When light extends into the dark period, it triggers long-day (or short-night) response. And when the dark period extends into the light period it triggers a short-day (or long-night) response. The second model has two oscillations. Dawn regulates one oscillation and dusk the other. There are more complex models of the clock, but these basic models underlie most of them.

Given the complexity of the timing of various responses in an organism, the question arises: Does one master clock drive the system, or are other clocks involved? It appears that the biological clock is a hierarchy of clocks (Figure B). Subordinate clocks coupled to the master clock control rhythms of physiology and behavior. These clocks or pacemakers may be groups of cells within organs, where they have specific timekeeping functions.

When light enters the eye, the signal goes to the master clock. In turn, the master clock sends signals to the other clocks. One clock may take longer to reset than another with a different period. Therefore, abrupt resetting of the master clock upsets its synchronization with the other clocks. These transient disturbances in the timing of various clocks put the physiological rhythms of the body out of phase. For these reasons, you feel jet lag when you fly across several time zones. Workers on night shift suffer effects similar to jet lag.

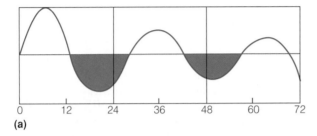
(a)

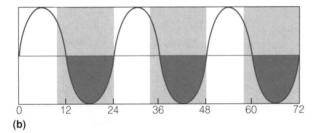
(b)

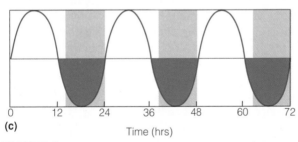
(c)

Time (hrs)

FIGURE A The Bünning model. One daily oscillation of the clock causes an alternation of half-cycles with different sensitivities to light. The dark blue portion represents the dark-sensitive part of the cycle, the white area the light-sensitive portion. The light blue areas represent periods of darkness or night. (a) The free-running clock in continuous light (or continuous dark) tends to drift out of phase with the 24-hour photoperiod. (b) Short-day conditions allow darkness to fall within the light-sensitive half-cycle. (c) In the long day, light falls during the dark-sensitive half-cycle. (Adapted from Bünning 1960.)

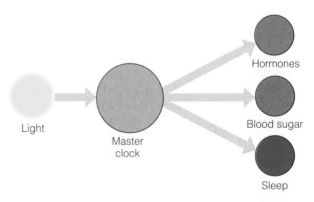

Light

Master clock

Hormones

Blood sugar

Sleep

Subservient clocks

FIGURE B The hierarchy of clocks. The master clock, entrained to environmental changes in light, resets other clocks that control physiological rhythms. (Adapted from Johnson and Hastings 1986.)

ahead of time. Circadian rhythms help organisms with physical aspects of the environment other than light or dark. For example, a rise in humidity and a drop in temperature accompany the transition from light to dark. Wood lice, centipedes, and millipedes, which lose water rapidly in dry air, spend the day in the darkness and damp under stones, logs, and leaves. At dusk they emerge, when the humidity of the air is more favorable. These animals show an increased tendency to escape from light as the length of time they spend in darkness increases. On the other hand, their intensity of response to low humidity decreases with darkness. Thus, these invertebrates come out at night into places too dry for them during the day, and they quickly retreat to their dark hiding places as light comes.

The circadian rhythms of many organisms relate to biotic aspects of their environment. Predators such as insectivorous bats must match their feeding activity to the activity rhythm of their prey. Moths and bees must seek nectar when flowers are open. Flowers must open when insects that pollinate them are flying. The circadian clock lets insects, reptiles, and birds orient themselves by the position of the sun. Organisms make the most economical use of energy when they adapt to the periodicity of their environment.

8.23 Critical Daylengths Trigger Seasonal Responses

In the middle and upper latitudes of the Northern and Southern Hemispheres, the daily periods of light and dark lengthen and shorten with the seasons (see Section 3.2). The activities of animals are geared to the changing seasonal rhythms of night and day. The flying squirrel, for example, starts its daily activity with nightfall, regardless of the season. As the short days of winter turn to the longer days of spring, the squirrel begins its activity a little later each day (Figure 8.19).

Most animals of temperate regions have reproductive periods that closely follow the changing daylengths of the seasons. For most birds, the height of the breeding season is the lengthening days of spring; for deer, the mating season is the shortening days of fall.

The signal for these responses is critical daylength (see Section 4.3). Many organisms possess both long-day and short-day responses. Because the same duration of dark and light occurs two times a year, in spring and fall, the organisms could get their signals

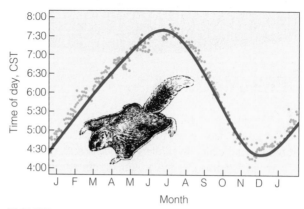

FIGURE 8.19 Seasonal variation in the time of day when flying squirrels become active. (Adapted from DeCoursey 1960.)

mixed. For them, the distinguishing cue is the direction from which the critical daylength comes. In one situation, the critical daylength is reached as long days move into short, and at the other time, as short days move into long.

Diapause, a stage of arrested growth over winter in insects of the temperate regions, is controlled by photoperiod. The time measurement in such insects is precise, usually between 12 and 13 hours of light. A quarter-hour difference in the light period can determine whether an insect goes into diapause or not. The shortening days of late summer and fall forecast the coming of winter and call for diapause. The lengthening days of late winter and early spring are signals for the insect to resume development, pupate, emerge as an adult, and reproduce.

Increasing daylength induces spring migratory behavior, stimulates gonadal development, and brings on the reproductive cycle in birds. After the breeding season, the gonads of birds regress spontaneously. During this time, light cannot induce gonadal activity. The short days of early fall hasten the termination of this period. The progressively shorter days of winter then become stimulatory. The lengthening days of early spring bring the birds back into the reproductive stage.

In mammals, photoperiod influences activity such as food storage and reproduction, too (see Focus on Ecology 8.3: Deer Antlers). Consider, for example, such seasonal breeders as sheep and deer. Melatonin initiates their reproductive cycle. More melatonin is produced when it is dark, so these animals receive a higher concentration of melatonin as the days become shorter in the fall. This increase in melatonin reduces the sensitivity of the pituitary gland to negative feedback effects of hormones from the ovaries and testes. Lacking this feedback, the anterior pituitary releases pulses of another hormone (called luteinizing hormone) that stimulates

DEER ANTLERS

The familiar antlers you see on male deer—bucks and stags—in fall are not the same antlers they wore the fall before. They lose their antlers in early winter; they gain new antlers the next fall. The acquisition of new antlers and the associated reproductive cycle are controlled by the influence of daylength on the pituitary gland, located on the floor of the brain.

The lengthening days of spring stimulate the pituitary to increase secretion of growth hormones and prolactin, a hormone associated with lactation in females. In males, these hormones stimulate the growth of antlers in the spring and early summer (Figure A). Fur-covered skin, called velvet, carries blood vessels and nerves to the growing antlers. During the shortening days of late summer, growth hormones and prolactin decrease. Under the influence of melatonin, testosterone secretion increases in the enlarging testes. The presence of testosterone inhibits the action of growth-stimulating hormones. Antler growth

ceases, and deer thrash and rub their antlers against vegetation to remove the shedding velvet.

By the onset of the season of sexual activity, called the rut, the antlers have become hardened and polished. The useful life of the newly acquired antlers, however, is short. In the shortening days of winter, pituitary stimulation of the testes declines, and testosterone, which maintains the connection between the dead bone of the antlers and the live frontal bone, diminishes. This decline of testosterone causes a loss of calcium at the point of connection between the antler and the frontal bone, and the antlers drop. In the lengthening days of spring, the cycle of antler growth and sexual resurgence begins anew.

Normally the deer is in velvet about one-third of the year. When the duration of that year is changed artificially by altering the daylength, deer may replace antlers as often as two, three, or four times a year, or only once every other year.

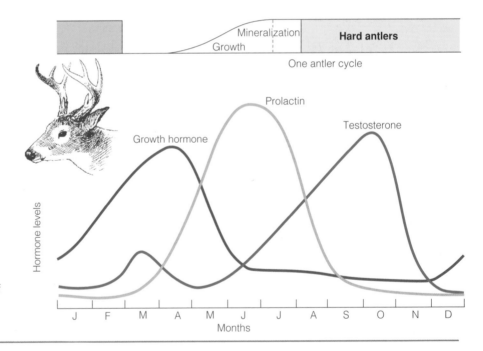

FIGURE A The seasonal course of hormonal levels in the white-tailed deer and its relationship to antler growth.

growth of ova in the ovaries and sperm production in the testes.

Activities of animals through the year reflect this seasonal response to changing daylength. The reproductive cycle of the white-tailed deer (*Odocoileus*

virginianus; Figure 8.20), for example, begins in fall, and the young are born in spring when the highest-quality food for lactating mother and young is available. In tropical Central America, home of numerous species of fruit-eating (frugivorous) bats,

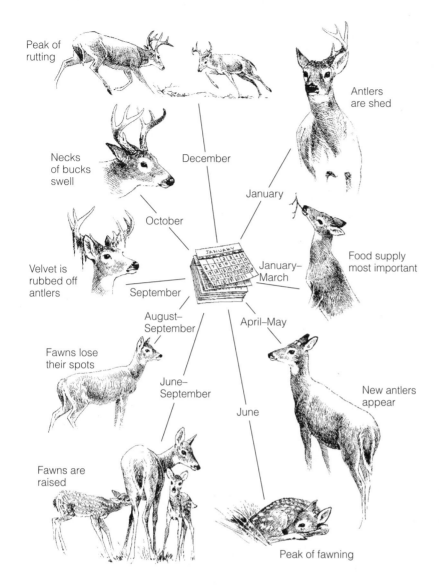

Peak of rutting

Necks of bucks swell

December

October

Velvet is rubbed off antlers

September

August–September

Fawns lose their spots

June–September

Fawns are raised

June

Peak of fawning

January

Antlers are shed

January–March

Food supply most important

April–May

New antlers appear

FIGURE 8.20 Seasonal reproductive cycle of the white-tailed deer. The cycle is attuned to the decreasing daylength of fall, when the breeding season begins, and to the lengthening days of spring, when antler growth begins.

the reproductive periods track the seasonal production of food. The birth periods of frugivorous bats coincide with the peak period of fruiting. Young are born when both females and young will have adequate food. Insects and other arthropods reach their greatest biomass early in the rainy season in the Costa Rican forests. At this time, the insectivorous bats give birth to their young.

8.24 Activity Rhythms of Intertidal Organisms Follow Tidal Cycles

Along the intertidal marshes, fiddler crabs (*Uca* spp.; the name refers to the enormously enlarged claw of the male, which he waves incessantly) swarm across the exposed mud of salt marshes and mangrove swamps at low tide. As high tide inundates the marsh, fiddler crabs retreat to their burrows, where they await the next low tide. Other intertidal organisms—from diatoms, green algae, sand beach crustaceans, and salt marsh periwinkles to intertidal fish such as blennies and cottids—also obey both daily and tidal cycles.

Fiddler crabs brought into the laboratory and held under constant temperature and light devoid of tidal cues exhibit the same tidal rhythm in their activity as they show back in the marsh (Figure 8.21). This tidal rhythm mimics the ebb and flow of tides every 12.4 hours, one-half of the lunar day of 24.8 hours, the interval between successive moonrises. Under the same constant conditions, fiddler crabs exhibit a circadian rhythm of color changes, turning dark by day and light by night.

Is the clock in this case unimodal, with a 12.4-hour cycle, or is it bimodal, with a 24.8-hour cycle,

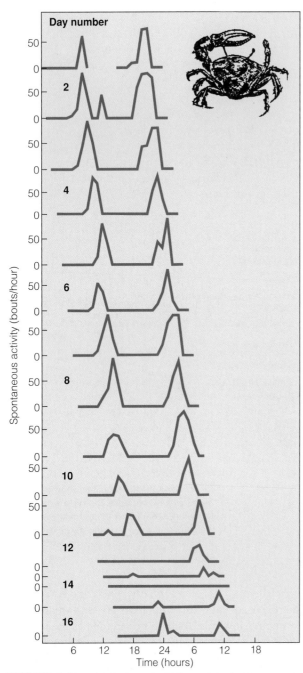

FIGURE 8.21 Tidal rhythm of a fiddler crab in the laboratory in constant light at a constant temperature of 22°C for 16 days. Because the lunar day is 51 minutes longer than the solar day, the tides occur 51 minutes later each solar day; thus peaks of activity appear to move to the right. (Adapted from Palmer 1990.)

close to the period of the circadian clock? Does one clock keep a solar-day rhythm of approximately 24 hours and another clock keep a lunar-day rhythm of 24.8 hours? J. D. Palmer and his associates at the University of Massachusetts, Amherst, did experiments to find out. The evidence suggests that one

solar-day clock synchronizes daily activities, while two strongly coupled lunar-day clocks synchronize tidal activity. Each lunar-day clock drives its own tidal peak. If one clock quits running in the absence of environmental cues, the other one still runs. This feature enables tidal organisms to synchronize their activities in a variable tidal environment. Day-night cycles reset solar-day rhythms, and tidal changes reset tidal rhythms. Even at the cellular level, organisms do not depend on one clock, any more than most of us keep a single clock at home. Organisms have built-in redundancies. Such redundancies enable the various clocks to run at different speeds, governing different processes with slightly differing periods.

8.25 Buoyancy Aids Aquatic Organisms to Stay Afloat

Important adaptations of aquatic animals are those that serve to keep them suspended in the water. The first order of business in the aquatic environment is to stay afloat. Most aquatic animals inhabiting the oceans have densities very close to that of seawater. Since living tissues are generally more dense (heavier) than water, the fact that larger animals are able to maintain buoyancy means that they must have lower density areas of their bodies that counter the higher density of most tissues.

Most fish have a gas or swim bladder (Figure 8.22b), which typically accounts for 5 to 10 percent of total body volume. Most can control the degree of buoyancy by regulating the amount of gas in the bladder. Lungs in air-breathing animals sustain neutral buoyancy.

Air is not the adaptive approach to staying afloat. Some marine animals such as the squid maintain neutral buoyancy by replacing heavy chemical ions in the body fluids with lighter ones. Squids have body cavities in which lighter ammonium ions replace the heavier sodium ions. As a result, an equal amount of body fluid is less dense than the same volume of seawater. Another mechanism is increased storage of lipids (fats and oils). Lipids are less dense than seawater. Large amounts of lipids are present in fishes that lack air bladders (such as sharks, mackerels, bluefish, and bonito). These lipids can be deposited in muscles, internal organs, and the body cavity (Figure 8.22a). In marine mammals, lipids are typically deposited as a layer of fat just below the skin (blubber) (Figure 8.22c). Blubber not only aids

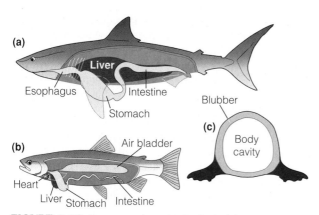

FIGURE 8.22 Buoyancy adaptation in shark, fish, and seal.
(a) Large fat-filled liver of a shark. (b) Gas bladder in a fish.
(c) Blubber surrounding the body of a seal.

in buoyancy, but also functions as insulation and energy storage.

The maintenance of neutral buoyancy in open-water environments releases organisms from many of the structural constraints imposed by gravity in terrestrial environments. Organisms such as the jellyfish, squid, and octopus quickly lose their graceful forms when removed from the water. Beached whales die from suffocation when they are no longer able to support their body weight through neutral buoyancy. It is no coincidence that the largest vertebrate and invertebrate organisms on Earth inhabit the oceans.

CHAPTER REVIEW

Summary

Acquisition of Nutrients by Animals (8.1–8.3) Major groups groups of animals are herbivores that convert plant biomass into animal tissue, carnivores that feed herbivores, omnivores that feed on both plants and animals, and detritivores that feed on dead organic matter (8.1). Directly or indirectly, animals get their nutrients from plants. Low concentration of nutrients in plants can have adverse effects on the growth, development, and reproduction of plant-eating animals. Herbivores convert plant tissue to animal tissue. Among plant eaters, the quality of food, especially its protein content and digestibility, is critical. Carnivores consume nutrients already synthesized from plants and converted into animal flesh. Their major problem is securing a sufficient quantity of food (8.2). Three essential nutrients that influence the distribution, behavior, growth, and reproduction of grazing animals are sodium, calcium, and magnesium. Grazers seek these nutrients from mineral licks and from the vegetation they eat (8.3).

Oxygen Uptake (8.4) Animals generate energy from the breakdown of organic compounds primarily through aerobic respiration, which requires oxygen. Differences in the means of oxygen acquisition between terrestrial and aquatic animals reflect its availability in the two environments. Most terrestrial animals have some form of lungs, whereas most aquatic animals utilize gills for the transfer of gases between the body and the surrounding water.

Regulation of Internal Conditions (8.5) To confront daily and seasonal environmental changes, organisms must maintain some equilibrium of their internal environment with the external one. The maintenance of a relatively constant internal environment in a variable external environment is called homeostasis. Homeostasis involves negative feedback responses. Through various sensory mechanisms,

an organism responds physiologically or behaviorally to maintain an optimal internal environment relative to its external environment. Doing so requires an exchange between the internal and external environments. This exchange involves negative and positive feedback.

Energy Exchange with Environment (8.6) Animals maintain a fairly constant internal body temperature, known as the body core temperature. They use behavioral and physiological means to maintain their heat balance in a variable environment. Layers of muscle fat and surface insulation of scales, feathers, and fur insulate the animal body core against environmental temperature changes. Terrestrial animals face a more changeable and often more threatening thermal environment than do aquatic animals.

Poikilotherms, Homeotherms, and Heterotherms (8.7–8.11) Animals fall into three major groups: poikilotherms, homeotherms, and heterotherms. Poikilotherms, so named because they have variable body temperatures influenced by ambient temperatures, are ectothermic. They depend largely on the environment as a source of heat. Animals that depend upon internally produced heat to maintain body temperatures are endothermic. Because they maintain a rather constant body temperature independent of the environment, they are called homeotherms. Many animals are heterotherms. They function either as endotherms or ectotherms (8.7).

Poikilotherms gain heat from and lose heat to the environment. Poikilotherms have low metabolic rates and high thermal conductance. Environmental temperatures control their rates of metabolism. Poikilotherms are active only when environmental temperatures are warm and are sluggish when temperatures are cool. They have, however, upper and lower limits of tolerable temperatures. Most aquatic poikilotherms do not maintain any appreciable difference between body temperature and water temperature (8.8).

Homeotherms maintain high internal body temperature by oxidizing glucose and other energy-rich molecules. They have high metabolic rates and low thermal conductance (8.9).

The two approaches to maintaining body temperature, ectothermy and endothermy, involve tradeoffs. Unlike poikilotherms, homeotherms are able to remain active regardless of environmental temperatures. For homeotherms, a high rate of aerobic metabolism comes at a high energy cost. This cost places a lower limit on body size. Because of the low metabolic cost of ectothermy, poikilotherms can curtail metabolic activity in times of food and water shortage and temperature extremes. Their low energy demands enable some terrestrial poikilotherms to colonize areas of limited food and water (8.10).

Depending upon environmental and physiological conditions, heterotherms take on the characteristics of endotherms or ectotherms. Some normally homeothermic animals become ectothermic and drop their body temperature under certain environmental conditions. Many poikilotherms, notably the insects, need to increase their metabolic rate to generate heat before they can take to flight. Most accomplish this feat by vibrating their wings or wing muscles or by basking in the sun. After flight, their body temperatures drop to ambient temperatures (8.11).

Special Animal Responses to the Thermal Environment (8.12–8.15) During environmental extremes, some animals enter a state of torpor to reduce the high energy costs of staying warm or cool. They slow their metabolism, heartbeat, and respiration, and lower their body temperature. Birds such as hummingbirds and mammals such as bats undergo daily torpor, the equivalent of deep sleep, without the extensive metabolic changes of seasonal torpor. Seasonal torpor over winter is called hibernation. Hibernation involves a whole rearrangement of metabolic activity to run at a very low level. Heartbeat, breathing, and body temperature are all greatly reduced (8.12).

Both poikilotherms and homeotherms resort to behavioral means to regulate body temperature. Mostly they exploit variable microclimates. Poikilotherms move into warm, sunny places to warm and seek shaded places to cool. Many amphibians move in and out of water. Insects and desert reptiles raise and lower their bodies to reduce or increase conductance from the ground or convective cooling. Desert animals resort to shade or spend the heat of day in underground burrows (8.13).

Body insulation of fat, fur, feathers, scales, and furlike covering on many insects reduces heat loss from the body. A few desert mammals employ heavy fur to keep out desert heat and cold. When insulation fails, many homeotherms resort to shivering and to nonshivering thermogenesis, the burning of brown fat (8.14). For homeotherms, evaporative cooling by sweating, panting, and wallowing in mud and water is an important way of dissipating body heat (8.15).

Physiological reponses to thermal environment (8.16–8.17) Some desert mammals use hyperthermia to reduce the difference between body and environmental temperatures. They store up body heat by day, then unload it to the cool desert air by night. Hyperthermia reduces the need for evaporative cooling and thus conserves water. Some cold-tolerant poikilotherms use supercooling, the synthesis of glycerol in body fluids, to resist freezing in winter. Supercooling takes place when the body temperature falls below freezing without freezing body fluids. Some intertidal invertebrates survive the cold by freezing, then thawing with warmer temperatures (8.16).

Many homeotherms and heterotherms employ counter-current circulation, the exchange of body heat between arterial and venous blood reaching the extremities. This exchange reduces heat loss through body parts or cools blood flowing to such vital organs as the brain (8.17).

Animal Responses to Water Problems (8.18–8.20) Animals possess excretory systems from simple to complex to maintain an osmotic balance with the environment. Aquatic animals need to prevent the uptake of or rid themselves of excess water. Terrestrial animals gain water by drinking, eating, and producing metabolic water. They lose water through urine, feces, respiration, and evaporation (8.18).

Animals of arid regions may reduce water loss by becoming nocturnal, producing highly concentrated urine and feces, becoming hyperthermic during the day, using only metabolic water, and tolerating dehydration (8.19).

Animal responses to saline environments are complex. Most animals maintain a water balance by means of an excretory system. Many marine invertebrates maintain the same osmotic pressure as seawater in their body cells. Fish secrete excess salt and other ions through kidneys or across gill membranes. Birds and some reptiles have salt-excreting glands (8.20).

Daily and Seasonal Changes in Daylength Influence Animal Activity (8.21–8.23) Living organisms, except bacteria, have an innate rhythm of activity and inactivity. This rhythm is free-running under constant conditions with an oscillation that deviates slightly from 24 hours. For that reason it is called a circadian rhythm. Under natural conditions the circadian rhythm is set to the 24-hour day by external time cues, notably light and dark (day and night). This setting synchronizes the activity of plants and animals with the environment. The onset and cessation of activity depend upon whether the organisms are light-active or dark-active.

Circadian rhythms operate the biological clocks of organisms. The biological clock is in the cells of plants and in the brain of multicelled animals. Animals produce more of a special hormone, melatonin, in the dark than in the light. Thus, melatonin becomes a device for measuring daylength (8.21). Circadian rhythms enable organisms to anticipate daily changes in the environment (8.22).

Seasonal changes in activity are based on daylength. Lengthening days of spring and the shortening days of fall stimulate migration in animals, and reproduction and food storage as well. These seasonal rhythms bring living organisms into a reproductive state at the time of year when the probability of survival of offspring is the highest. The rhythms synchronize within a population such activities as mating and migration, dormancy and flowering (8.23).

Tidal Cycles (8.24) Intertidal organisms are under the influence of two environmental rhythms, daylength and tidal cycles of 12.4 hours. Intertidal organisms appear to have two lunar-day clocks that set tidal rhythms, and one solar-day clock that sets circadian rhythms.

Buoyancy (8.25) Important adaptations of aquatic animals are those that serve to keep them suspended in the water. Aquatic animals use a wide variety of mechanisms to maintain buoyancy, including gas-filled bladders, lungs, and lipid deposits.

Study Questions

1. What morphological and physiological traits enable herbivores to digest plant material?

2. Why is the quality of food more important than the quantity of food for plant-eating animals?

3. Why must organisms maintain a fairly constant internal environment?

4. What is homeostasis, and how does it relate to survival in a variable environment?

5. Contrast negative feedback and positive feedback.

6. Distinguish between poikilothermy and homeothermy. What are the advantages and disadvantages of each?

7. What are the relationships among body size, metabolic rate, and temperature regulation?

8. What is acclimatization, and how does it function among poikilotherms, particularly fish?

9. Speculate on how homeothermy might have evolved.

10. What behaviors help poikilotherms maintain a fairly constant body temperature during their season of activity?

11. How can insulation maintain an animal's thermal integrity?

12. How do homeotherms respond when ambient temperatures fall below body temperature? Poikilotherms?

13. How do homeotherms use evaporative cooling?

14. Explain the value of hyperthermia in some desert mammals.

15. How does supercooling enable some insects, amphibians, and fish to survive freezing conditions?

16. How does countercurrent circulation work, and why is it important?

17. Distinguish between hibernation and torpor. Why is the black bear not a true hibernator?

18. Consider a population of fish that live below a power plant discharging heated water. The plant shuts down for three days in the winter. What effect would that have on the fish?

19. Bats hibernate in winter, becoming ectothermic for 18 days before giving birth to their young in late spring and during lactation. What is the advantage of such a physiological change? Consider the bat's summer habitat, energy demands, and daily activity patterns.

20. What is the biological clock, and where is it located?

21. What conditions must a biological clock fulfill to function as a timekeeper?

22. Discuss four uses for a biological clock.

23. Explain diapause in insects.

24. How does daylength influence the seasonal activity of plants and animals?

25. What is the adaptive value of seasonal synchronization of animal activity?

26. Why do intertidal organisms need two kinds of clocks?

St. Patrick and the Absence of Snakes in Ireland

There are no snakes in Ireland. Popular legend has it that in the fifth century Patrick, a monk who immigrated from Britain and later became his adopted country's patron saint, drove the snakes of Ireland into the sea to their destruction (Figure A). Although Patrick, whose feast day we celebrate each March 17, contributed much to the cultural development of the Irish, he is not responsible for that country's lack of snakes. The reasons for their absence lie elsewhere and include the interplay organisms have with their physical environment.

Snakes, like all reptiles and amphibians, are ectotherms (cold-blooded). As is true of all animals, snakes need to maintain certain body temperatures for the natural processes of muscle and nerve activity, as well as digestion, to take place. Being ectotherms, snakes are more or less "thermal prisoners" of their surroundings: their body temperature is dictated by that of their immediate external environment. The range of environmental temperatures within which snakes are active is narrow, from about 10°C to 40°C, with an optimum "operating" temperature of about 30°C. Compared to the range of atmospheric temperatures that exist on the surface of the earth (–110°C to 60°C), snakes' temperature tolerances are quite restricted. Although snakes depend on the external environment to provide the needed energy to maintain body temperatures within the range necessary for life processes, they are not passive prisoners of the constant variations in their thermal environments. Snakes and other terrestrial ectotherms can manipulate heat exchange between their bodies and their environments by a combination of behavioral and physiological processes. For instance, a snake can control its absorption of solar radiation—and thereby its body temperature—by altering the color of its absorptive surface (the skin) or changing the orientation of its body relative to the Sun.

Many reptiles can change their color by dispersion or contraction of dark pigments, such as melanin, in their skin. Because dark skin substantially increases the amount of solar energy that is absorbed, many individuals living in the cooler parts of a species' range are darker than their same-species counterparts in warmer climates (Figure B). Additionally, many snakes of the temperate regions can change their color to accommodate seasonal changes in the amount of solar radiation. Some snakes capitalize on this mechanism for increasing solar radiation by having dark skin on their heads that they expose to the Sun before other parts. Warming the brain and the sensory organs such as the eyes and the tongue first enhances a snake's ability to detect both danger and food. Finally, pregnant females of some species are darker than males and nonpregnant females, presumably to maintain warmer-than-normal body temperatures that are thought to accelerate embryonic development.

As noted, a second way that snakes control their absorption of solar radiation is by increasing or decreasing the amount of body area exposed to radiation. By orienting its body to lie at right angles to the direction of the Sun, and by spreading and flattening to increase its body's surface area, a snake can attain temperatures much higher than the surrounding air. When a snake's body has

FIGURE A St. Patrick, who, legend has it, drove the snakes from Ireland.

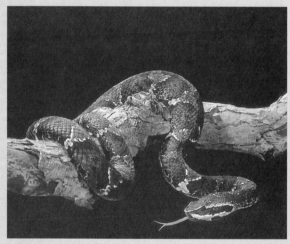

FIGURE B Light and dark phases of the Mexican moccasin (*Agkistrodon bilineatus*).

reached a suitable temperature, it avoids further heating by lightening its skin color, changing its posture to one more parallel to the Sun's rays, and eventually moving into shade or underground where heat absorption is reduced. Furthermore, the temperature of the substrate that the snake is in contact with is also important because a cool snake can crawl on a warm rock or other surface and absorb its heat.

The behavioral and physiological means by which snakes regulate body temperature permit them to occupy a surprisingly large part of Earth's land surface. Snakes occur in all continents except Antarctica.

However, the vast majority of snake species are found within the tropical and subtropical regions (at least three-fourths of the snake species existing on Earth are found between 22°N and 22°S latitude). Species diversity of reptiles decreases rapidly from the equator toward the higher latitudes. In North America, the number of lizard species is highest in the warm desert regions of the Southwest and declines continuously as you move northward (Figure C). The same pattern of species diversity is evident in Europe, where the number of reptile species declines markedly as you move from the warmer Mediterranean coast toward northern Europe and the British Isles. Although one species, the European adder *(Vipera berus)*, is found above the Arctic Circle in Scandinavia, its unusually northern distribution comes at a cost. The adder has a very limited period of activity, often only 3 to 4 months a year. In addition, the species may take up to 4 years to attain sexual maturity, and females may breed only once every 3 or 4 years, using the intervening period to build up

the fat reserves necessary to produce offspring.

The progressive decline in solar radiation and temperatures from the tropics to the poles not only reduces the abundance and diversity of reptiles but also has a direct influence on their body size. The reason for these patterns is that heat exchange occurs across the surface of a body, but warming usually occurs throughout the entire body's mass or volume. Large bodies, because of their low surface area to volume ratio, take longer to warm than smaller ones. This physical reality results in upper limits to the size of snakes (and other reptiles) depending on the distance of the snake's habitat from the tropics. All of the large snakes, such as the anaconda and python, are found within the tropical and subtropical regions. Other large reptiles—such as the iguanas, monitor lizards (which include the goanna of Australia and the komodo dragon of Indonesia), and the crocodilians (alligators, caimans, and crocodiles)—are likewise limited in their distribution to the warm, aseasonal environments of the subtropics and tropics. The maximum body size for ectotherms declines as you move north and south from the equator. This pattern is the exact opposite of that observed for endotherms (warm-blooded animals), where average body size increases from the tropics to the poles,* a pattern referred to as Bergman's rule. The environmental

*The reason for this is large bodies lose less heat than small bodies to the environment because they expose less surface area per unit volume. This is an important consideration for animals that generate their own internal heat. (Note that ectotherms tend to be limited by how much heat they can extract from their environment and endotherms by how much heat they lose to the environment.)

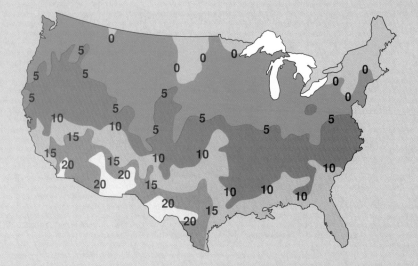

FIGURE C Patterns of lizard species diversity (number of species) in the contiguous United States.

constraint on the upper limit of body size for ectotherms is at the very heart of the current debate over whether the dinosaurs were cold- or warm-blooded. The fossil record places the dinosaurs well into the northern latitudes, with specimens found in Alaska and Siberia.

As for snakes, the fossil record reveals that the diversity of reptiles and amphibians (terrestrial ectotherms) has always been low in northern Europe. During the Pleistocene era (from 1.6 million to 10,000 years ago), a time when the temperature of Earth was considerably lower than that now observed, much of the Northern Hemisphere was covered by glaciers. Although certain refugia (small pockets of land not covered by ice) existed, the massive ice sheet virtually obliterated all life in northern Europe, including Ireland and Britain. As temperatures rose and the glaciers retreated during the Holocene (10,000 years to present), the region was recolonized by plants and animals, but the cool temperatures of northern Europe proved inhospitable to most snakes.

Nevertheless, the island of Britain, similar in climate to Ireland, ended up with three species of snakes that survive there to this day. Two enjoy an extensive distribution, but the third is found only in a small area of southern England.

Three species of snakes is a paltry sum, but nonetheless it's more than Ireland has. Why the difference? After the retreat of the continental glaciers at the end of the last ice age, Britain had a land bridge connecting it with the mainland of Europe, whereas Ireland did not. The limited diversity of snakes on the mainland meant that the pool of species available to recolonize both Britain and Ireland was small in the first place. However, Ireland's lack of a land bridge with continental Europe coupled with the limited dispersal abilities of most snake species have to date prohibited their successful recolonization there.

In fact, although devoid of snakes, Ireland does have three species of terrestrial ectotherms. Two species of amphibian (smooth newt, *Triturus vulgaris;* common frog, *Rana temporaria*) and one reptile, the small viviparous lizard *Lacerta vivipara,* were able to reach Ireland's shores despite the absence of a land bridge. So, given the

> "Snakes and other terrestrial ectotherms can manipulate heat exchange by a combination of behavioral and physiological processes."

physiological constraints imposed by temperature on the distribution of snakes together with the recent (in geological terms) climate history of the region, even if St. Patrick were to have run the snakes out of Ireland, it would have been a rather small task.

PART IV

Populations

CHAPTER 9

Wildebeest *(Connochaetes taurinus)* graze on the savanna.

Properties of Populations

OBJECTIVES

On completion of this chapter, you should be able to:

- Define population density, crude density, and ecological density.
- Summarize the types of population distribution.
- Discuss the problems and methods of determining the density of a population.
- Describe the age structure of a population.
- Explain the significance of different types of age pyramids.
- Summarize ways in which the ages of plants and animals can be determined.
- Compare the age structure of plants with that of animals.

As an individual, how do you perceive the world? Most of us regard a friend, a neighborhood maple tree, a daisy in a field, a squirrel in the park, or a bluebird nesting in the backyard as an individual. Rarely do we consider each as a part of a larger unit, a population. Although the term **population** has many different meanings and uses, for biologists and ecologists it has a very specific definition. A population is a group of individuals of the same species inhabiting a given area. This definition has two very important features. By requiring that individuals be of the same species, population is defined as a genetic unit. In the case of sexually reproducing organisms, it suggests the potential for interbreeding among members of the population. Secondly, population is a spatial concept, requiring a spatial delineation.

Because they are an aggregate of individuals, populations have unique features. They have characteristics such as density and distribution in time and space. Individuals are born, grow, and then die. These same processes are represented as rates in a population: birthrate, growth rate, and death rate. Similarly, individuals have specific characteristics relating to sex and age, whereas populations have age and sex structures that describe the aggregate.

9.1 Populations May Be Unitary or Modular

A population is considered to be a group of individuals, but what constitutes an individual? For most of us, defining an individual would seem to be no problem. We are individuals, and so are dogs, cats, spiders, insects, fish, and so on throughout much of the animal kingdom. What defines us as individuals is our unitary nature. Form, development, growth, and longevity of unitary organisms are predictable and determinate from conception on. There is no question about recognizing an individual. Within the animal kingdom, however, there are exceptions. How can you define individuals among marine and aquatic invertebrates such as corals, sponges, bryzoans, and colonial organisms (Figure 9.1) that grow by repeated production of modules. They are examples in the animal world in which the definition of an individual becomes blurred.

Defining an individual among plants becomes even more problematic. Consider a tree that has grown from a seed. It appears as an individual, but in reality it consists of populations of leaves, buds, and twigs. As it grows, it adds new stems, new buds, new leaves, twigs, branches, and, in time, fruits and seeds. Thus it adds more and more units or modules to its vertical structure, supported by dead woody tissue from previous modules. These modules of smaller units have their own demography: birthrates, death rates, and growth rates. The births and deaths of these separate modules determine not only the rate of growth but also the form the individual tree will assume. Growth is by repeated production of modules to produce a branching structure. The precise program of development is indeterminate and unpredictable.

A tree or a shrub grown from a seed is an individual with its own genetic characteristics (see Chapter 2). Once established, some species of trees—such as black locust *(Robinia pseudoacacia)* and aspen *(Populus tremuloides)*—shrubs, and numerous perennial herbaceous plants grow root extensions that send up new shoots or suckers that may remain attached to root extensions or break off to live independently (Figure 9.2). These new "individuals" or

(a)

(b)

FIGURE 9.1 Examples of clonal animal species: (a) corals and (b) sponges.

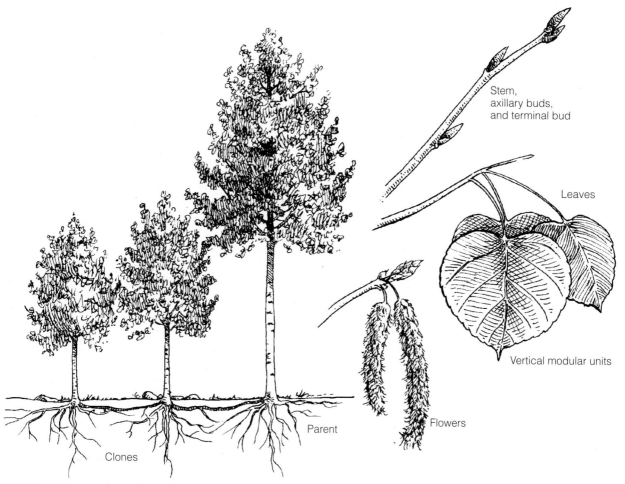

FIGURE 9.2 Horizontal and vertical modular growth in an aspen tree *(Populus tremuloides)*. Vertical growth involves modules of leaves, buds, and associated stems. Horizontal growth involves modules of roots and root buds, which give rise to clones. These clones are various ages, with the youngest individuals forming the leading edge of growth away from the parent.

clones may cover a considerable area and appear to be individuals. The individual tree or plant produced by sexual reproduction and thus arising from a zygote is a genetic individual or **genet**. Individual modules produced asexually by the genet are **ramets**. The ramets may remain physically linked to the parent genet or they may separate and exist as individuals (Figure 9.3). These individuals can produce seeds and their own lateral extensions or ramets.

Whether living independently or physically linked to the original individual, all ramets possess the same genetic constitution as the original product of sexual reproduction. Thus, by producing ramets, the genet can cover a relatively large area and considerably extend its life. Some modules die, others live, and new ones appear. Although the existing ramets may be very young and the original genetic individual may have died, the genet itself still lives through its ramets and can become quite old. The genet itself is dead only when all of its component modules are

dead. Technically, all the ramets arising from the original genetic individual should be considered as one individual in a population, but of course this can hardly be done. For this reason in the study of plant populations, ramets are typically counted as and function as individual members of the population.

9.2 The Distribution of Organisms Reflects Environmental Variation

The distribution and abundance of organisms can be defined at many levels. **Distribution** is the area in which it occurs. **Abundance** refers to its numbers or population size. Both distribution and abundance are

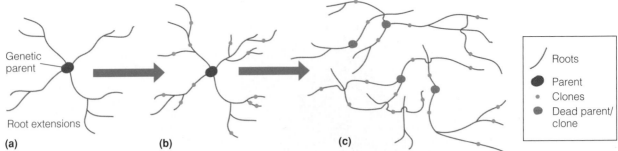

FIGURE 9.3 Connections between the original parent and its clones are lost in time, and some clones lead independent existences. (a) Genetic parent. (b) Clonal growth arising from root extensions. (c) Genetic parent is dead. Clonal growth arising from root extensions of the parent forms groups of new individual stems. Roots of some clones have separated from the parental and some clonal root extensions to establish new independent groups. However, all clones of the parent, independent or not, are genetically identical.

influenced by the occurrence of suitable habitat and tolerances for various environmental conditions.

The red maple *(Acer rubrum)* is the most widespread of all deciduous trees of eastern North America (Figure 9.4). Its northern limit coincides with the area in southeastern Canada where minimum winter temperatures drop to −40°C. Its southern limit is the Gulf Coast and southern Florida. Dry conditions halt its westward range. Within this geographic range, it grows under a wide variety of soil types, soil moisture, acidity, and elevations, from wooded swamps to dry ridges. Thus, the red maple exhibits a high degree of tolerance to temperature and other environmental conditions. In turn, this high degree of tolerance allows a widespread geographic distribution.

Minimum and maximum temperature tolerances define the limits of the distribution of many species. Although conditions close to the limits of tolerance for a species may be sufficient to maintain survival, growth, and reproduction, their values will be much below those that occur under more optimal environmental conditions. The nearer conditions approach

FIGURE 9.4 Red maple *(Acer rubrum)*, one of the most abundant and widespread trees in eastern North America, thrives on a wider range of soil types, texture, moisture, acidity, and elevation than any other forest species in North America. The northern extent of its range coincides with the −40°C minimum winter temperature in southeastern Canada.

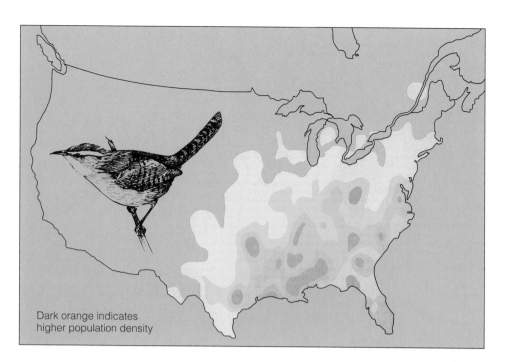

Dark orange indicates
higher population density

FIGURE 9.5 The distribution and abundance of the Carolina wren *(Thryothorus ludovicianus)* are strongly associated with temperature. The birds are least abundant about the periphery of their range. (After Root 1988.)

the minimum and maximum tolerances of the organism, the fewer the individuals. We would expect the abundance of a species to increase as we move toward optimal environmental conditions. Consider another example, the Carolina wren *(Thryothorus ludovicianus)*, a nonmigratory bird of the southeastern United States sensitive to cold winter temperatures (Figure 9.5). It is absent where the average minimum January temperature drops below –12°C. It occurs regularly only where the average minimum winter temperature is –7°C or greater. It is most abundant in its optimal range, where the average minimum January temperature is over –4°C. Note how the abundance of the species decreases as it extends from the optimum area in the southeast to the northern and western edges of its range. The western edge of the wren's range appears to be limited by another aspect of its physical environment, moisture. The wren's western distribution ends where annual rainfall is less than 52 cm.

An organism responds to a variety of environmental factors, and only when all of them are within the range of tolerance can it inhabit a location. The actual location or place where an organism lives is called its habitat. Because habitat describes a location, we can define it at many levels. Consider the horned lark *(Eremophila alpestris)*, a bird of open grassland (Figure 9.6). Although it has a continent-wide distribution, it is found primarily in the prairie grasslands of the Midwest and avoids both coniferous and deciduous forests. Even within this region, its distribution is influenced by specific tolerances. Because it requires a landscape of open, short grasses,

it is much less abundant in taller, lightly grazed grassland. The availability of space in which to establish territories limits colonization by individuals within the short grassland. Within each bird's territory, the area that it defends and utilizes to meet its needs (see Chapter 11, Section 11.10), not all parts, are usable for occupancy and specific activities.

9.3 Density Affects Individuals in a Population

A key attribute of populations is density. **Density** is the number of individuals per unit of space—so many per square kilometer, per hectare, or per square meter. It is a more specific term than *abundance* in that it is an enumeration, an actual count of individuals within a defined unit of area.

Think of the change in human numbers from the concentrated masses of the city to the sparse populations of the countryside. When you drive in city traffic or on lightly traveled country roads, you are aware of population density. You respond to that density in some way. You may seek out crowds, or you may escape to the quiet of home. High densities may frustrate or stimulate you.

Individuals in natural populations are also affected by density. Trees in crowded stands may grow more slowly, and some may succumb to a lack of water, nutrients, and light that are unequally shared. Scarce

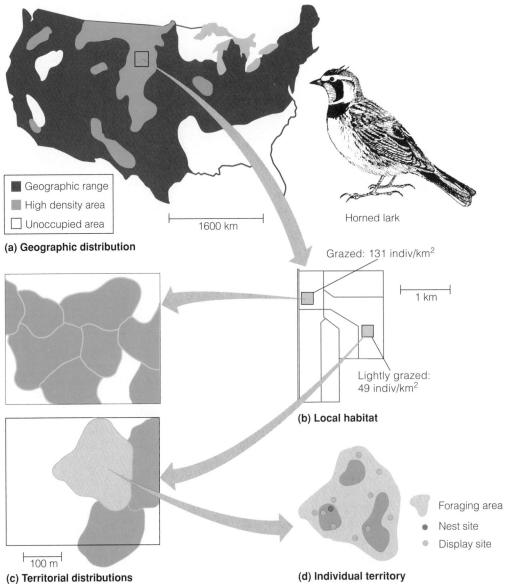

(a) Geographic distribution

■ Geographic range
■ High density area
□ Unoccupied area

1600 km

Horned lark

Grazed: 131 indiv/km^2

1 km

Lightly grazed:
49 indiv/km^2

(b) Local habitat

100 m

(c) Territorial distributions

Foraging area
● Nest site
● Display site

(d) Individual territory

FIGURE 9.6 Population distribution on different scales for the horned lark *(Eremophila alpestris)*. (a) Although the horned lark is distributed over most of the continental United States, abundance (density) varies across its geographic range. (b) On a local scale, the distribution of the bird is influenced by available grassland. (c) Within a given area of grassland, the distribution of individual birds is influenced by territorial behavior. (d) Within each territory, the bird allocates space to different activities based on the structure of the habitat. (After Wiens 1973.)

food may be denied to smaller or less aggressive mammals in a population. Some birds may deny others access to nest sites when not enough sites exist to meet the demand. Having too few individuals in a population may reduce the chances of finding a mate or inhibit behavior essential to the welfare of the population. Low population density may raise an individual's risk of succumbing to predation. Affecting the welfare of individuals in all these ways, density in part controls a population's birthrates, death rates, and growth.

9.4 Density Is Difficult to Define

However important it may be, density is an elusive characteristic, difficult to define and to determine. Density measured simply as the number of individuals per unit of space is referred to as **crude density**. The trouble with this measure is that individuals do not occupy all the space within a unit because not all of it is suitable habitat. A biologist might estimate the number of deer living in a square kilometer. The deer, however, might avoid half the area because of human habitation, land use, and lack of cover and food. Goldenrods inhabiting old fields grow in scattered clumps because of soil conditions and competition from other old-field plants.

No matter how uniform an area may appear, it is usually patchy because of small-scale differences in light, moisture, temperature, exposure, or other physical conditions. Each organism occupies only those areas that can meet its requirements. To account for patchiness, density should refer to the number of individuals per unit of available living space. That measure would be **ecological density**. For example, in a study of bobwhite quail *(Colinus virginianus)* in Wisconsin, biologists expressed density in terms of the number of birds per mile of hedgerow, the birds' preferred habitat, rather than as birds per acre. Ecological densities are rarely estimated, because determining what portion of a habitat represents living space is difficult.

9.5 Dispersion of Individuals Influences Population Density

How organisms are distributed over space has an important bearing on density. Individuals of a population may be distributed randomly, uniformly, or in clumps (Figure 9.7). Individuals are distributed randomly if the position of each is independent of the others'.

By contrast, individuals distributed uniformly are more or less evenly spaced. In the animal world, uniform dispersion usually results from some form of competition, such as territoriality (see Chapter 11, Section 11.10). Uniform distribution happens among plants when severe competition exists for crown or root space, as among some forest trees, or for moisture, as among desert plants.

The most common dispersion type is clumped dispersion, in which individuals occur in groups. Clumping results from a variety of factors, including spatial variation in habitat availability and social behavior. The distribution of human populations is clumped or aggregated because of social behavior, economics, and geography. There are various degrees and types of clumping. Groups may be randomly or nonrandomly distributed over an area. Aggregations may range from small groups to a single centralized group. If environmental conditions encourage it, populations may be concentrated in long bands or

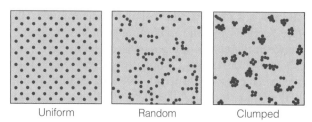

Uniform   Random   Clumped

FIGURE 9.7 Patterns of dispersion: random, uniform, and clumped.

strips along some feature of the landscape such as a river, leaving the rest of the area unoccupied.

Individuals usually are not distributed as one large population across the landscape. Rather they exist as separate or spatially disjunct populations distributed in patches across the landscape. If the patches are sufficiently close that individuals can move freely among them, then such populations, though separate, behave as a single continuous population. However, in many landscapes the habitat is so fragmented that the species exist as distinct, partially isolated subpopulations, each possessing its own population dynamics. Linking these subpopulations are movements of individuals among patches. Such separated populations interconnected by the movement of individuals are called **metapopulations**. Each subpopulation has its own birthrate, death rate, and probability of extinction. Each habitat patch vacated by local extinction has its own probability of being recolonized by individuals from other subpopulations. Such local recolonization after extinction allows metapopulations to persist. The metapopulation concept is highly applicable to the problems of habitat fragmentation and the conservation of species (see Chapter 20).

9.6 Dispersal of Individuals Influences Local Abundance

At some stage in their lives, most organisms are to some degree mobile. The movement of individuals has a direct influence on their local abundance and density. The movement of individuals in space is referred to as **dispersal**, although the term *dispersal* most often refers to the more specific movement of individuals away from each other. When individuals move out of a given population, it is referred to as **emigration**. When an individual moves from another location into a given population, it is called **immigration**. The movement of individuals among populations of a species is a key process in linking metapopulations (see Section 9.5) and maintaining the flow of genes between these subpopulations (see Chapter 2).

Many organisms, especially plants, depend on passive means of dispersal involving gravity, wind, water, and animals. The distance these organisms travel depends on the agents of dispersal. Seeds of most plants fall near the parent, and their density falls off quickly with distance (Figure 9.8). Heavier seeds, such as those of oaks, have a much shorter dispersal range than the lighter wind-carried seeds of maples, birch, milkweed, and dandelions. Some plants, such as cherries and viburnums, depend on active carriers such as particular birds and mammals to disperse their seeds by eating the fruits and carrying the seeds to some distant point. These seeds pass through the digestive tract of the animal and are deposited in their feces. Other plants possess seeds armed with spines and hooks that catch on the fur of mammals, the feathers of birds, and the clothing of humans.

For mobile animals, dispersal is active, but many depend on a passive means of transport, such as wind and moving water. Wind carries the young of some species of spiders, larval gypsy moths, and cysts of brine shrimp. In streams, the larval forms of some invertebrates disperse downstream in the current to suitable habitats. In the oceans, the dispersal of many organisms is tied to the movement of currents and tides.

Dispersal among mobile animals may involve young and adults, males and females; there is no hard-and-fast rule about who disperses. Young are the major dispersers among birds. Among rodents, such as deer mice *(Peromyscus maniculatus)* and meadow voles *(Microtus pennsylvanicus)*, subadult males and females make up most of the dispersing individuals. Crowding, temperature change, quality and abundance of food, and photoperiod all have been implicated in stimulating dispersal in various animal species.

Often, the individuals that are dispersing are seeking vacant habitat to occupy. As a result, the distance they travel will depend in part on the density of surrounding populations and the availability of suitable unoccupied areas.

Unlike the one-way movement of animals in the processes of emigration and immigration, **migration** is a round-trip. The repeated return trips may be daily or seasonal. Zooplankton in the oceans, for example, move down to lower depths by day and move up to the surface by night. Their movement appears to be a response to light intensity. Bats leave their daytime roosting places in caves and trees, travel to their feeding grounds, and return by daybreak. Other migrations are seasonal, either short range or long range. Earthworms annually make a vertical migration

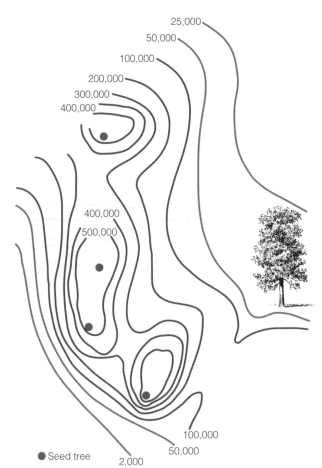

FIGURE 9.8 Pattern of annual seedfall of yellow poplar *(Liriodendron tulipifera)*. Lines define areas of equal density of seeds. With this wind-dispersed species, seedfall drops off rapidly away from the parent trees. (After Engle 1960.)

deeper into the soil to spend the winter below the freezing depths and move back to the upper soil when it warms in spring. Elk move down from their high mountain summer ranges to lowland winter ranges. On a larger scale, caribou move from the summer calving range in the boreal forests to the arctic tundra for the winter, where lichens comprise their major food source. Gray whales *(Eschrichtius robustus)* move down from the food-rich arctic waters in summer to their warm wintering waters of the California coast, where they give birth to young (Figure 9.9). Similarly, humpbacked whales *(Megaptera novaeangliae)* migrate from northern oceans to the central Pacific off the Hawaiian Islands. Perhaps the most familiar of all are long-range and short-range migration of waterfowl, shorebirds, and neotropical migrants in spring to their nesting grounds and in fall to their wintering grounds.

Another type of migration involves only one return trip. Such migrations occur among Pacific

FIGURE 9.9 Migratory pathways of two vertebrates. (a) Ring-necked ducks *(Aythya collaris)* breeding in the northeast migrate in a corridor along the coast to wintering grounds in South Carolina and Florida. (b) The gray whale *(Eschrichtius robustus)* summers in the Arctic and Bering seas and winters in the Gulf of California and the waters off Baja California.

salmon (*Oncorhynchus* spp.) spawned in freshwater streams. The young hatch and grow in the headwaters of freshwater coastal streams and rivers and travel downstream and out to sea, where they reach sexual maturity. At this stage, they return to the home stream to spawn (reproduce) and then die.

9.7 Determining Density Requires Sampling

To determine whether the distribution of individuals in a population is random, uniform, or clumped is difficult, requiring careful sampling techniques. Just as difficult is the determination of true population density. Absolute density can be determined by a complete count or by sampling. Except for unusual habitat situations, such as antelope living on an open

plain or waterfowl concentrated in a marsh, direct counts are extremely difficult or impossible. Even if possible, direct counts may be too time-consuming or expensive. For these reasons, ecologists resort to sampling to gather information on population density. A method of sampling used widely in the study of populations of plants and sessile (attached) animals involves quadrats or sampling units. Researchers divide the area of study into subunits, in which they count animals or plants of concern in a prescribed manner, usually counting individuals in only a subset or sample of the subunits. From the data they determine the mean density of the unit sampled. Multiplying the mean value by the total area provides an estimate of population size. They can use statistical procedures to calculate confidence limits of the estimate that suggest the upper and lower limits of population size within which the true population density lies. The accuracy of sampling can be influenced by the manner in which individuals are distributed over the landscape (dispersion). The density figure can also be influenced by the choice of boundaries or sample units. If a population is clumped, concentrated into small areas, and the population density is described in terms of individuals per square kilometer, you receive a false impression (Figure 9.10).

For mobile populations, animal ecologists use other sampling methods. Widely used are various mark and recapture techniques. (See Focus on Ecology 9.1: Capture-Recapture Sampling.)

For work with most animals, ecologists find that a measure of relative density or abundance is sufficient. Methods involve observations that relate to the presence of organisms rather than direct counts of individuals. Techniques include counts of vocalizations such as recording the number of drumming ruffed grouse heard along a trail, counts of animal scat seen along a length of road traveled, or the number of tracks of opossums crossing a certain dusty road. If these observations have some relatively constant relationship to total population size, you can then convert the data to the number of individuals seen per kilometer or heard per hour. Such counts, called **indexes of abundance,** cannot function alone as estimates of actual density. However, a series of such index figures collected from the same area over a period of years depicts trends in abundance. Counts obtained from different areas during the same year provide a comparison of abundance between different habitats. Most population data on birds and mammals are based on indexes of relative abundance rather than on direct counts.

9.8 Populations Have Age Structures

Unless each generation reproduces and dies in a single season, not overlapping the next generation (such as annual plants and many insects), the population will have an **age structure,** the number of individuals in different age classes. Because reproduction is restricted to certain age classes and mortality is most prominent in others, the relative proportions in which the age groups occur bear on how quickly or slowly populations grow.

Populations can be divided into three ecologically important age classes or stages: prereproductive, reproductive, and postreproductive. We might divide humans into young people, working adults, and senior citizens. How long individuals remain in each stage depends largely on the life history of the organism (see Chapter 12). Among annual species, the length of the prereproductive stage has little influence

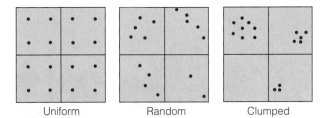
FIGURE 9.10 The difficulty of sampling. Each area contains a population of 16 individuals. We divide each area into four sampling units and choose one at random. Our estimates will be quite different, depending upon which unit we select. For the uniform population, any sampling unit gives a correct estimate. For the random population, the estimates are 20, 20, 16, and 8. For the clumped population, the estimates are 32, 20, 0, and 12.

on the rate of population growth. In organisms with variable generation times, the length of the prereproductive period has a pronounced effect on the population's rate of growth. Short-lived organisms often increase rapidly, with a short span between generations. Long-lived organisms, such as elephants and whales, increase slowly and have a long span between generations.

9.9 Determining Age Structure Requires Finding Individual Ages

The age structure of a population is the proportion of the various age classes at a given time. To determine age structure, we need some means of obtaining the ages of members of a population. For humans, this task is not a problem, but it is with wild populations. Age data for wild animals can be obtained in a number of ways, the method varying with the species. The most accurate, but most difficult, method is to mark young individuals in a population and follow their survival through time. This method requires both a large number of marked individuals and a lot of time. For this reason biologists may use other, less accurate methods. These methods include examining a representative sample of carcasses of individuals to determine the ages at death. A biologist might look for the wear and replacement of teeth in deer and other ungulates, growth rings in the cementum of the teeth of carnivores and ungulates, or annual growth rings in the horns of mountain sheep. Among birds, observations of plumage changes and wear in both living and dead individuals can separate juveniles from subadults (in some species) and adults.

Studies of the age structure of plant populations are few. The major reason is the difficulty of determining

CAPTURE-RECAPTURE SAMPLING

Capturing, marking, and recapturing animals from fish to mammals is the most widely used technique to estimate animal populations. There are many variations of this technique, ranging from single mark–single recapture to multiple capture–multiple recapture. Books are devoted to various methods of application and statistical analysis. Nevertheless, the basic concept is simple.

Capture-recapture or mark-recapture methods are based on trapping, marking, and releasing a known number of marked animals into the population. After an appropriate period of time (approximately one week for rabbits and mice), individuals are again captured from the population. Some of the individuals caught in this second period will be carrying marks, and others will not. An estimate of the population is computed from the ratio of marked to unmarked individuals in the sample. The ratio supposedly reflects the ratio of marked to unmarked animals in the population.

The simplest method, the single mark–single recapture, is known as the Lincoln index or Petersen index of relative population size. The basic model is:

$$N : M :: n : R$$

or

$$N = nM/R$$

where M is the number marked in the precensus period, R is the number of marked animals trapped in the census period, n is the total number of animals trapped in the census period, and N is the population estimate.

As an example, suppose that in a precensus period, biologists capture and tag 39 rabbits. During the census period they capture 15 tagged rabbits and 19 unmarked ones, a total of 34. The following ratio results:

$$N : 39 :: 34 : 15$$

or

$$N = nM/R = (34)(39)/15 = 88$$

The estimate of 88 rabbits is exactly that, an estimate. To determine within what range the population lies, biologists use a statistical procedure to estimate confidence limits. In this case, the confidence limits state that the chances are 95 out of 100 that the population of rabbits lies between 64 and 114 animals. Such results are typical in population studies of animals.

The capture-recapture method involves a number of assumptions:

- All individuals in a population have an equal chance of being captured. None are trap-shy and none are trap-happy.

- The ratio of marked to unmarked animals remains the same from the time of capture to the time of recapture.

- Marked individuals, once released, redistribute themselves throughout the population with respect to unmarked ones, as they were before capture.

- Marked animals do not lose their marks.

- The population is closed. No emigration or immigration takes place during the sampling period.

- No mortality or reproduction occurs during the sampling period.

A number of conditions influence the probability of capture. They include time, behavioral responses to capture, mortality patterns, and the bias of sex and age on movement. Males and younger individuals are more susceptible to capture than others in the population.

---

the age of plants and whether the plants are genetic individuals or asexual modules or clones.

Foresters have tried to use age structure as a guide to timber management. They employ size (diameter of the trunk at breast height or dbh) as an indicator of age on the assumption that diameter increases with age. The greater the diameter, the older the tree. Such assumptions, foresters discovered, were valid for dominant canopy trees; but with their growth suppressed by lack of light, moisture, or nutrients, understory trees and even so-called seedlings and saplings add little to their diameters. Although their diameters suggest youth, small trees often are the same age as large individuals in the canopy.

You can determine the approximate ages of trees in which growth is seasonal by counting annual growth rings (Figure 9.11). Attempts to age non-woody plants have not been successful. The most accurate method of determining the age structure of short-lived herbaceous plants is to mark individual seedlings and follow them through their lifetimes.

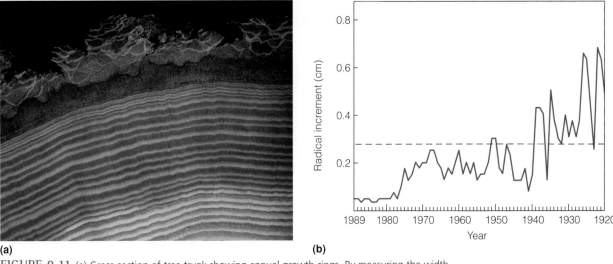

(a)　　　　　　　　　　　　　　　　　　**(b)**

FIGURE 9.11 (a) Cross-section of tree trunk showing annual growth rings. By measuring the width of each ring, a pattern of radial growth through time can be established. (b) Example of time series of radial increments for American beech tree in central Virginia. Dashed line is overall average for the tree over time.

9.10 Age Pyramids Portray Age Structure

Age pyramids (Figure 9.12) are snapshots of the age structure of a population at some period in time. Providing a picture of the relative sizes of different age groups in the population, age pyramids, with proper care, can help us judge the status of a popu-

lation. A population growing in density possesses a preponderance of young organisms. Constant or declining populations do not.

Age pyramids for plants take on a different configuration. Because dominant overstory trees can inhibit the production of seedlings and growth and survival of juvenile trees, the distribution of age classes is often highly skewed (Figure 9.13). One or two age classes dominate the site until they die or are

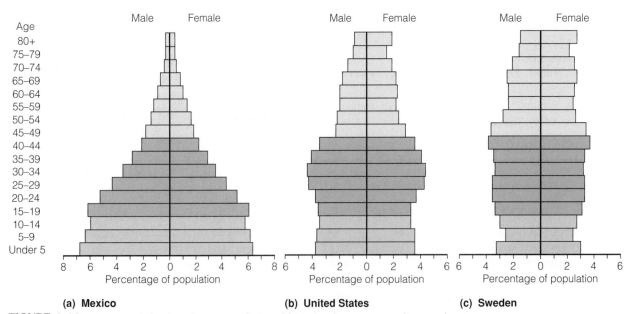

(a) Mexico　　　　　　　　**(b) United States**　　　　　　　　**(c) Sweden**

FIGURE 9.12 Age pyramids for three human populations. (a) Mexico shows an expanding population. A broad base of young will enter the reproductive age classes. (b) The age pyramid for the United States has a less tapered shape. The youngest age classes are no longer the largest. This pyramid reflects birth control, immigration, and declining fertility, which in time will further alter the age structure. (c) The age pyramid for Sweden is characteristic of a population that is approaching zero growth.

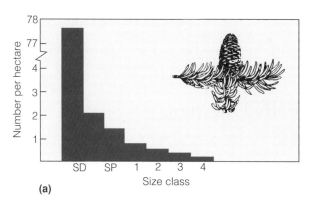

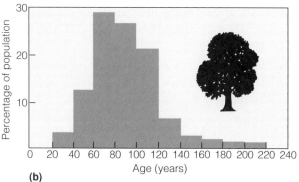

(a)

(b)

FIGURE 9.13 Size and age distributions for two plant populations. (a) A balsam fir *(Abies balsamea)* stand on Lake Superior in Ontario. Size classes are seedlings (SD), 1 to 3 cm diameter at breast height (dbh); saplings (SP) greater than 3 cm and less than or equal to 9 cm; and trees grouped into 8 cm diameter classes thereafter. This stand is young, with most of the population in younger age classes. (b) An oak *(Quercus)* forest in Sussex, England, is dominated by trees in the 60- to 120-year age classes. There has been no recruitment of new trees (younger age classes) for 20 years.

removed, allowing young age classes access to resources such as light, water, and nutrients, allowing them to grow and develop.

9.11 Sex Ratios in Populations May Shift with Age

Populations of sexually reproducing organisms in theory tend toward a 1:1 sex ratio (the proportion of males to females). The primary sex ratio (the ratio at conception) also tends to be 1:1. This statement may not be universally true, and it is, of course, difficult to confirm.

In most mammalian populations, including humans, the secondary sex ratio (the ratio at birth) is often weighted toward males, but the population shifts toward females in the older age groups. Generally, males have a shorter life span than females. The shorter life expectancy of males can be a result of both physiological and behavioral factors. For example, in many animal species, rivalries among males occur for dominant positions in social hierarchies, or for the acquisition of mates. Among birds there is a tendency for males to outnumber females because of increased mortality of nesting females, which are more susceptible to predation and attack.

CHAPTER REVIEW

Summary

Characteristics of Populations (9.1) A population is a group of individuals of the same species living in a defined area. Populations are characterized by age structure, density, distribution in time and space, birthrate, death rate, and growth rate. Most animal populations are made up of unitary individuals with a definitive growth form and longevity. Plant populations are more complex. Individual plants consist of modules: buds, leaves, twigs, flowers, and roots. Plant populations may consist of sexually produced parent plants and asexually produced stems arising from roots.

Population Distribution and Density (9.2–9.5) Distribution and abundance of organisms depend on both environmental variation and tolerances. Presence and absence define distribution, and the number of organisms defines abundance (9.2). Density, the size of a population in relation to the area it occupies, affects individuals' access to

living space, food, shelter, and mates (9.3). Because landscapes are not homogeneous, not all of the area is suitable habitat. The number of organisms in available living space is the true or ecological density (9.4).

Individuals within a population are distributed in space. If the location of each individual is independent of the others, the individuals are spaced randomly; if they are evenly spaced, the distribution is uniform. In most cases, individuals are grouped together in clumped or aggregated distributions. These groups of individuals can function as subpopulations. Movements of individuals link subpopulations into metapopulations (9.5).

Dispersal of Individuals (9.6) At some stage of their life cycle, most individuals are mobile. The dispersal of individuals has a direct impact on population density and distribution. For some organisms, such as plants, dispersal is passive and dependent on a variety of dispersal mechanisms. For mobile organisms, dispersal can occur for a variety of reasons, including the search for mates and unoccupied habitat. For some species, dispersal is a systematic process of movement between areas, a process called migration.

Sampling Populations (9.7) Determination of density and dispersion requires careful sampling and appropriate statistical analysis of the data. Plant researchers often use sample plots. Animal researchers use capture-recapture techniques or determine relative abundance by count surveys.

Age Structure (9.8–9.10) Individuals making up the population may be divided into three ecological periods: prereproductive, reproductive, and postreproductive. The distribution of individuals among age groups influences the birthrate, death rate, and growth rate. A large number of young about to enter the reproductive age suggests a potentially increasing population, whereas a high proportion of individuals in the postreproductive age classes suggests zero growth or a declining population (9.8).

To determine the age structure of a population, you must find out the ages of its members. Whereas obtaining data on age in human populations is fairly simple, determining the ages of animals and plants requires special techniques. It is difficult to determine age for plants. Although size is often used as a criterion, it can be misleading (9.9).

Age pyramids compare the percentages of the population in different age groups. Pyramids with a broad base of young suggest growing populations. Pyramids with a narrow base of young and even ratios suggest a declining or aging population. Between these two are many variations. Viewed over time, age pyramids depict changing dynamics of the population (9.10).

Sex Ratios (9.11) Populations have a sex ratio that tends to be 1:1 at conception and birth but often shifts toward females later. Sex ratios have an important influence on reproductive rates and social interactions.

Study Questions

1. Distinguish between ecological density and crude density.

2. Compare random, uniform, and clumped dispersion. What are some underlying reasons for each type of dispersion?

3. Why is it hard to determine the density of a population? Do these same problems extend to a census of human populations?

4. Of what significance are the ratios of individuals in the three general age periods?

5. What is stable age distribution?

6. Why does age structure differ between animal and plant populations?

7. How does the modular nature of plant populations add complexity to the study of age structure in plants?

8. Age pyramids, which represent the past of a population, are often used to predict its future. Think of drawbacks to this practice. What can age pyramids tell us?

CHAPTER 10

Family of warthogs travel across the Masai Mara Game Reserve in Kenya.

Population Growth

OBJECTIVES

On completion of this chapter, you should be able to:

- Express mortality as the probability of dying or surviving.
- Construct a life table.
- Explain the derivation and meaning of life expectancy.
- Distinguish among several kinds of life tables.
- Plot mortality and survivorship curves.
- Distinguish among different types of survivorship curves.
- Explain how net reproductive rates are determined.
- Distinguish between exponential and logistic growth.
- Interpret net reproductive rate as a measure of population growth.
- Explain the relationship of carrying capacity to logistic growth.
- Discuss cycles and irregular fluctuations in populations.
- Explain why populations become extinct.

The statistical study of group properties of populations—size, age and sex structure, and changes within them—is called **demography.** Populations grow through the processes of birth and immigration and decline through death and emigration. How the density of a population changes through time will reflect the balance between these inputs and outputs. Individuals come and go, but the collective, the population, will persist and grow as long as the number of individuals entering the population exceeds the number leaving.

10.1 Probability of Surviving Is Age-Specific

An excerpt from a standard rate table for a major life insurance company is shown in Table 10.1. The table provides the annual payment for a $100,000 life insurance policy for nonsmoking males of various ages. If you are a 25-year-old male, your annual payment is $125. However, if you are 50 years of age, your payment increases to $276. By age 70, the annual premium is $1671. How do insurance companies arrive at these rates? The rate schedule, in fact, represents the probability of death (or survival) for an individual of that age. In general, the older an individual gets, the greater the probability of mortality, and hence the greater the cost of life insurance. Unlike an individual, whose death is a singular and discrete event, mortality in a population is expressed as a **mortality rate,** or proportion of individuals dying during a specified time interval. To calculate mortality rate, represented by the letter q, we divide the number of individuals who died during a defined interval of time (such as a year), represented by the letter d, by the number alive at the beginning of the time period (defined by the letter N). The formula is $q = d/N$. The value q can also be used to express the probability of death (the probability that a given individual in the population will die during that time interval).

The complement of the probability of dying is the probability of survival (s), the number of survivors (individuals alive) at the end of the time period divided by the number alive at the beginning. Because the number of survivors is more important to a population than the number dying, mortality is better expressed either as the probability of surviving or as **life expectancy,** e, the average number of years into the future that an individual in the population is expected to live.

TABLE 10.1

Annual Premiums for $100,000 Life Insurance Policy, Nonsmoking Male (Annual Renewal)

Age	Annual Premium
25	$ 125
30	130
35	133
40	172
45	200
50	276
55	397
60	578
65	1012
70	1671

10.2 A Life Table Gives a Systematic Picture of Age-Specific Mortality and Survival

Although mortality and survival rates can be calculated for the population as a whole, as suggested by the age-specific insurance rates in Table 10.1, the probability of mortality changes with age. To obtain a clear and systematic picture of mortality and survival, we can construct a **life table.** The life table is simply an age-specific accounting of mortality. The technique was first developed by students of human populations, and is the basis for evaluating age-specific mortality rates by life insurance companies.

The life table consists of a series of columns headed by standard notations, each of which describes certain age-specific mortality patterns within a population (Table 10.2). It begins with a **cohort,** a group of individuals born in the same period of time (column n_x). In the example shown in Table 10.2, the initial cohort is made up of 530 individuals. These individuals are followed through time until all have died. The column labeled x represents the age classes or categories; in this example, it is expressed in years. For example, of the original 530 individuals born (0–1 years old), only 159 survived into the next age class (1–2 years old). Generally, the actual number of individuals in each age class is converted to a proportion of individuals at the beginning time

TABLE 10.2

Gray Squirrel Life Table

x	n_x	l_x	d_x	q_x	L_x	T_x	e_x
0–1	530	1.0	0.7	0.7	0.650	1.090	1.09
1–2	159	0.3	0.15	0.5	0.225	0.440	1.47
2–3	80	0.15	0.06	0.4	0.120	0.215	1.43
3–4	48	0.09	0.05	0.55	0.065	0.095	1.06
4–5	21	0.04	0.03	0.75	0.025	0.030	0.75
5–6	5	0.01	0.01	1.0	0.005	0.005	0.50

(initial cohort size). For example, in Table 10.2 the cohort size at birth of 530 converts to the proportionate value of 1.0 (530/530 = 1.0). Cohort size at age 1–2 is 159, and therefore the proportionate size is 159/530 or 0.3. These values represent the proportion of the original cohort that survive to any given age class x. Therefore, these data provide the probability at birth of surviving to age x (column l_x). (See Quantifying Ecology 10.1: Constructing a Life Table). The difference between the value of l_x for any age class and the next older (l_{x+1}) is the proportion of the original cohort that has died during that time interval (Column d_x). In Table 10.2, for example, 0.7 (or 70 percent) of the original individuals in the cohort died during their first year of life. During the second year, 0.15 (or 15 percent) of the original 530 individuals died. The proportion of individuals that died during any given time interval divided by the proportion alive at the beginning of that interval provides an **age-specific mortality rate** (q_x). This value is obtained by dividing the value of d_x by the corresponding value of l_x for any given age class.

At one time, data for life tables could be obtained only for laboratory animals and humans. As population sampling methods and age determination techniques became more refined, biologists began to construct life tables for wild animals and plants.

There are two basic kinds of life tables. One is the cohort or dynamic life table. It records the fate of a group of individuals, all born in a single short period of time, from birth to death—for example, a group of individuals born in the year 1955. A modification of the dynamic life table is the dynamic composite life table. This approach constructs a cohort from individuals born over several time periods instead of just one. For example, you might follow the fate of individuals born in 1955, 1956, and 1957. The second type is the time-specific life table. It is constructed by sampling the population in some manner to obtain a distribution of age classes during a single time period.

Although much easier to construct, this type of life table requires a number of critical assumptions. First, you assume that you sampled each age class in proportion to its numbers in the population. Secondly, you must assume that growth rate is constant, that is, the age structure is stable (see Section 10.7).

Most life tables have been constructed for long-lived vertebrate species having overlapping generations (such as humans). Many animal species, especially insects, live through only one breeding season. Because their generations do not overlap, all individuals belong to the same age class. We obtain the survivorship values, l_x, by observing a natural population several times over its annual season, estimating the size of the population at each time. For many insects, the l_x value can be obtained by estimating the number surviving from egg to adult. If records are kept of weather, abundance of predators and parasites, and the occurrence of disease, death from various causes can also be estimated.

Table 10.3 represents the fate of a cohort from a single gypsy moth egg mass. The age interval or x column indicates the life history stages, which are of

TABLE 10.3

Life Table of a Sparse Gypsy Moth Population in Northeastern Connecticut

x	n_x	l_x	d_x	q_x
Eggs	450	1.000	0.300	0.300
Instars I–III	315	0.700	0.573	0.819
Instars IV–VII	57	0.127	0.074	0.582
Prepupae	24	0.053	0.002	0.038
Pupae	23	0.051	0.0015	0.029
Adults	16	0.036	0.036	1.000

Source: Adapted from R. W. Campbell 1969.

Constructing a Life Table

There is no better way to understand a life table than to construct one. The life table consists of a series of columns relating to age-specific estimates of mortality and survival. The columns include x, the units of age; l_x, the proportion of organisms in a cohort that survive to age x, $x + 1$, and so on; and d_x, the proportion of a cohort that dies during the age interval x to $x + 1$. Dividing the proportion of individuals that died during age x by the proportion alive at the beginning of age x (l_x) provides q_x, the age-specific mortality rate.

The next two columns are L_x, the (proportionate) average years lived by all individuals in each age category, and T_x, the (proportionate) number of time units left for all individuals to live from age x onward. These two columns provide the data needed to calculate e_x, the life expectancy at the end of each age interval. The values for L_x are obtained by summing the values for l_x and l_{x+1} and dividing the sum by 2. T_x is calculated by summing all values of L_x from the bottom of the table upward to x. Life expectancy is obtained by dividing T_x for a particular age class x by the l_x value for that age class.

Consider, as an example, Table 10.1, the life table for the gray squirrel. Column n_x is the actual number of individuals alive for each age class. The data were obtained by following a cohort of squirrels marked as nestlings. The n_x values are converted to proportional values relative to the initial cohort size, in this case by dividing the number in each age category by 530: 530/530 = 1.000; 159/530 = 0.3; and so.

Obtain the d_x column by subtracting the number of survivors at $x + 1$ from the survivors at previous age x: 1.0 − 0.3 = 0.7, the mortality at age 0–1; 0.3 − 0.15 = 0.15, mortality at age 1–2, and so on.

Divide the mortality (d_x) for age x by the value of l_x for the same age interval to obtain the value q_x for the same age class: 0.7/1.0 = 0.7 for age class 0–1; 0.15/0.3 = 0.5 for age class 1–2; and so on.

Obtain the values for the L_x column by adding the survivorship of x and $x + 1$ and dividing by 2. For age class 0–1, (1.0 + 0.3)/2 = 0.65; for age class 2–3, (0.3 + 0.15)/2 = 0.225; and so on.

Calculate the T_x column by adding the values in the L_x column upward: for age class 4–5, 0.005 + 0.025 = 0.3; for age class 2–3, 0.005 + 0.025 + 0.065 + 0.12 = 0.215.

Obtain life expectancy, e_x, for each age class by dividing the T_x value by the l_x value. For age class 0–1, life expectancy is 1.09/1.0 = 1.09; for age class 4–5, life expectancy is 0.03/0.04 = 0.75.

In other words, an individual gray squirrel's life expectancy at birth is 1.09 years. If the individual lives to 4 years, on the average it can be expected to live another 0.75 years.

unequal duration. The l_x column indicates the number of survivors at each stage. The d_x column gives a breakdown of deaths in each stage. In this population, dispersal and predation account for most of the losses. Note that life expectancy is not calculated because there is none. All adults in the population will die in late summer.

10.3 Plant Life Tables Are More Complex

Mortality and survivorship in plants are not easily summarized into life tables. To begin with, age is difficult to determine. Mortality of individuals usually stimulates the growth of survivors, increasing the production of buds, leaves, and stems. Seedlings make up a large numerical proportion of individuals but an extremely small portion of the total living mass (biomass) of a plant population. Further, it is difficult to separate and even identify individuals. The parent plant may die, yet resprout, producing numerous ramets. The plant demographer has to deal with mortality (and birth) on two levels, the individual and the clones (Chapter 9, Section 9.1).

In plant demography the life table is most useful in studying three areas: (1) seedling mortality and survival; (2) population dynamics of perennial plants marked as seedlings; and (3) life cycles of annual plants. An example of the third type is Table 10.4. The time of seed formation is the initial point in the life cycle. The l_x column indicates the proportion of plants alive at the beginning of each stage and the d_x column the proportion dying.

Another approach to the life table of plants is the yield table developed by foresters (Table 10.5). The yield table considers age classes and the density

TABLE 10.4

Life Table for a Natural Population of *Sedum smallii*

x	l_x	d_x	q_x
Seed produced	1.000	0.16	0.160
Available	0.840	0.630	0.750
Germinated	0.210	0.177	0.843
Established	0.033	0.009	0.273
Rosettes	0.024	0.010	0.417
Mature plants	0.014	0.014	1.000

Data from Sharitz and McCormick 1973.

TABLE 10.5

Yield Tables for Douglas Fir*

Age (Years)	Trees per ha	Average dbh (cm)[†]	Basal Area (m²)[‡]
20	1427	14	23
30	875	23	35
40	600	30	46
50	440	38	53
60	345	46	58
70	282	53	62
80	242	59	66
90	210	65	69
100	187	70	72
110	172	75	75
120	157	80	77
130	147	83	79
140	137	87	81
150	127	91	83
160	120	94	85

*Fully stocked with site index of 200
[†]dbh—diameter at breast height (1.35 m)
[‡]Basal area is the cross-sectional area of tree at dbh = $(0.5*dbh)^2*\pi$

of trees in each age class, with additional columns giving information about the average size of individuals (diameters and basal area or cross-sectional area). Yield tables chart the mortality of trees by the reduced number of individuals in each age class. However, as the numbers decline, basal area and biomass increase, for both individuals and the population as a whole. Mortality may not indicate a declining population but a maturing one. Like life tables,

yield tables are not constant for a species. We can construct them for different environmental conditions, such as soils and moisture availability. The conditions are called site classes, referring to the environmental conditions at a particular site or location.

10.4 Life Tables Provide Data for Mortality and Survivorship Curves

We can graphically display data from the life table in two ways: a mortality curve based on the q_x column, and a survivorship curve based on the l_x column. A mortality curve plots mortality rates in terms of q_x against age. Mortality curves for the life tables presented in Table 10.2 (gray squirrel) and Table 10.4 (*Sedum smallii*) are shown in Figure 10.1. For the gray squirrel cohort, the curve consists of two parts: a juvenile phase, in which the rate of mortality is

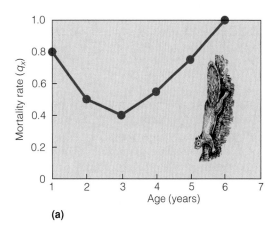

(a)

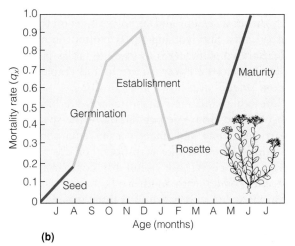

(b)

FIGURE 10.1 Examples of mortality curves for (a) gray squirrel population, based on Table 10.2, and (b) *Sedum smallii,* based on Table 10.4.

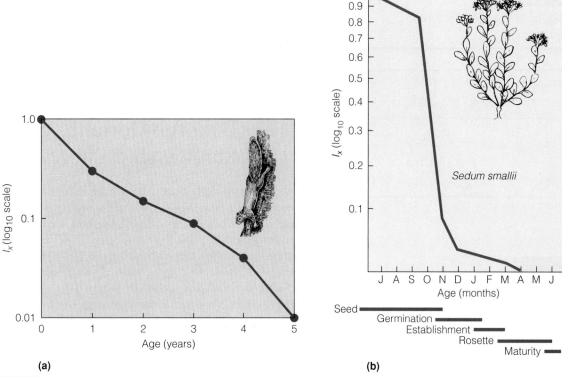

FIGURE 10.2 Survivorship curve for (a) gray squirrel, based on Table 10.2, and (b) *Sedum smallii*, based on Table 10.4.

high; and a postjuvenile phase, in which the rate decreases with age until mortality reaches some low point, after which it increases again. For plants, the mortality curve may assume a number of patterns, depending on whether the plant is annual or perennial and how we express the age structure. Mortality rates for the *Sedum* population (Figure 10.1b) are high, declining once seedlings are established.

Survivorship curves plot the l_x from the life table against time or age class *(x)*. The time interval is on the horizontal *(x)* axis and survivorship is on the vertical *(y)* axis. Survivorship *(l_x)* is plotted on a $\log_{10}$ scale. Survivorship curves for the life tables presented in Table 10.2 (gray squirrel) and Table 10.4 *(Sedum smallii)* are shown in Figure 10.2.

Life tables and survivorship curves are based on data obtained from one population of the species at a particular time and under certain environmental conditions. They are like snapshots. For this reason, survivorship curves are useful for comparing one time, area, or sex with another.

Survivorship curves fall into three general idealized types (Figure 10.3). When individuals tend to live out their physiological life span, survival rate is high throughout the life span followed by heavy mortality at the end. With this type of survivorship pattern, the curve is strongly convex, or Type I. Such a

curve is typical of humans and other mammals and has also been observed for some plant species. If survival rates do not vary with age, the survivorship curve will be straight, or Type II. Such a curve is characteristic of adult birds, rodents, and reptiles, as well as many perennial plants. If mortality rates are extremely high in early life—as in oysters, fish, many invertebrates, and many plant species, including

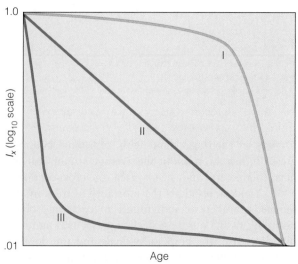

FIGURE 10.3 The three basic types of survivorship curves.

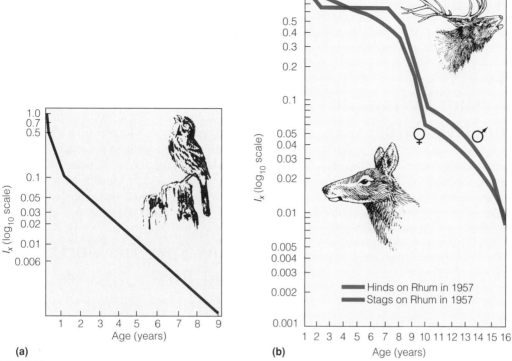

FIGURE 10.4 (a) Survivorship curve for the song sparrow *(Melospiza melodia)*. The curve is typical of birds. After a period of high juvenile mortality (Type III), the curve becomes linear (Type II). (b) Comparison of survivorship curves for male and female red deer *(Cervus elaphus)*.

most trees—the curve is concave, called Type III. These generalized survivorship curves are idealized models to which survivorship of a species can be compared (Figure 10.4). Most survivorship curves generally show Type I, Type II, and Type III at different times in a species' life history.

10.5 Natality Is Age-Specific

Birthrates or natality rates are usually expressed as births per 1000 population per unit of time. This figure is obtained by dividing the number of births over some period of time by the estimated population size at the beginning of the time period, and multiplying the resulting number by 1000. This figure is the **crude birthrate.**

This estimate of birthrate can be improved by taking two important factors into account. Only females within the population give birth. Secondly, the birthrate of females generally varies with age. Therefore, a better way of expressing birthrate is the number of births per female of age x. If we arbitrarily divide females of reproductive age into age classes and tabulate the average number of births for each age class, we get an **age-specific schedule of births.** Because population increase is a net function of reproduction by females, the age-specific birth sched-

ule can be further modified by determining only the mean number of females born in each age group, b_x (Table 10.6). The sum of the b_x values across all of the age classes provides an estimate of the average number of female offspring born to a female over her lifetime, termed the gross reproductive rate. In the example of the squirrel population presented in Table 10.6, the gross reproductive rate is 10. However, this value assumes that a female survives to the maximum age class (5–6). What we really need is a measure of reproductive rate that incorporates not only the age-specific birthrate, but also the probability of a

TABLE 10.6

Gray Squirrel Fecundity Table

x	l_x	b_x	$l_x b_x$
0–1	1.0	0.0	0.00
1–2	0.3	2.0	0.60
2–3	0.15	3.0	0.45
3–4	0.09	3.0	0.27
4–5	0.04	2.0	0.08
5–6	0.01	0.0	0.00
Σ		10.0	1.40

female's surviving to any specific age class. This estimate of reproductive rate is called the **net reproductive rate, R_0**.

10.6 Natality and Survivorship Determine Reproductive Rates

How are net reproductive rates determined? We can use the gray squirrel population described in Table 10.2 as the basis for the construction of a fecundity or fertility table (Table 10.6). The fecundity table uses the survivorship column, l_x, from the life table together with the column representing the mean number of females born to females in each age group (b_x) described above. At age 0–1, females produce no young; therefore their b_x value is 0. The b_x value for females in age class 1–2 is 2.0. For age 2–3 and 3–4, the b_x values increase to 3.0, then decline to 2.0 in age class 4–5. The value of b_x for age class 5–6 is 0.0. Although b_x may initially increase with age, survivorship (l_x) in each age class declines. To adjust for mortality, multiply the b_x values by the corresponding l_x, or survivorship, values. The resulting value, $l_x b_x$, gives the mean number of females born in each age group adjusted for survivorship.

Thus, for age class 1–2, the b_x value is 2.0; but when adjusted for survival l_x, the value drops to 0.6. For age 2–3, the b_x is 3.0 but $l_x b_x$ drops to 0.45, reflecting poor survival of adult females. The values of $l_x b_x$ are summed over all ages at which reproduction occurs. The result represents the average number of females that will be left during a lifetime by a newborn female, designated as R_0. If the R_0 value is 1, females will on average replace themselves in the population. If the R_0 value is less than 1, the females are not replacing themselves. If the value is much over 1, females are more than replacing themselves. For the gray squirrel, an R_0 value of 1.4 suggests a growing population of females. Note the significant difference between the gross and net reproductive rates (10 and 1.4 respectively). The difference reflects the fact that only a small proportion of the females born survive to the maximum age and produce 10 female offspring.

Natality in plants, like mortality, is perplexing because plants reproduce both sexually and asexually. If you consider only the genetic individual, then natality is restricted to sexual reproduction. There are two separate populations, seeds and seedlings, and two separate processes, the production of seeds and the germination of seeds. Except for annuals and biennials, which have one reproductive effort resulting in the death of the parent plant (see Chapter 12), seed production by individual plants is hard to estimate. Woody plants and other perennials, even within a population, vary in longevity, in seed production over the years, and in the ability of seeds to germinate.

Before germination, seeds usually undergo a varying period of dormancy, often before they can sprout. The seeds of some plants remain dormant for years, buried in soil or mud as a seed bank, until they are exposed to conditions conducive to germination. After germination, the seedling is subject to mortality. Thus, the plant population at all times consists of two parts, one growing and producing seeds, and the other stored as seeds in a dormant state.

10.7 Age-Specific Mortality and Birthrates Are Useful to Project Population Growth

Age-specific mortality rates (q_x) from the life table together with the age-specific birthrates (b_x) from the fecundity table can be combined to project changes in the population into the future. To simplify the process, the values for age-specific mortality are converted to age-specific survival. If q_x is the proportion of individuals alive at the beginning of an age class that die before reaching the next age class, then $1 - q_x$ is the proportion that survive to the next age class (Table 10.7); designated as s_x. With age-specific values of s_x and b_x we can project the growth of a population by constructing a **population projection table.**

We will illustrate the construction of a population projection table by using data from Table 10.6 and a hypothetical population of 10 female squirrels, all

TABLE 10.7

Age-Specific Survival and Birthrates for Squirrel Population

x	l_x	q_x	s_x	b_x
0–1	1.0	0.7	0.3	0.0
1–2	0.3	0.5	0.5	2.0
2–3	0.15	0.4	0.6	3.0
3–4	0.09	0.55	0.45	3.0
4–5	0.04	0.75	0.25	2.0
5–6	0.01	1.0	0.0	0.0

TABLE 10.8

Population Projection Table, Squirrel Population

Age	Year 0	1	2	3	4	5	6	7	8	9	10
0	20	27	34.1	40.71	48.21	58.37	70.31	84.8	101.86	122.88	148.06
1	10	6	8.1	10.23	12.05	14.46	17.51	21.0	25.44	30.56	36.86
2	0	5	3.0	4.05	5.1	6.03	7.23	8.7	10.50	12.72	15.28
3	0	0	3.0	1.8	2.43	3.06	3.62	4.4	5.22	6.30	7.63
4	0	0	0	1.35	0.81	1.09	1.38	1.6	1.94	2.35	2.83
5	0	0	0	0	0.33	0.20	0.27	0.35	0.40	0.49	0.59
Total	30	38	48.2	58.14	68.93	83.21	100.32	120.85	145.36	175.30	211.25
Lambda	λ	1.27	1.27	1.21	1.19	1.21	1.20	1.20	1.20	1.20	1.20

age 1, introduced into an unoccupied oak forest. The year of establishment will be designated as year 0. These 10 females in the year of introduction (year 0) give birth to 40 young, half of which, 20, will be female. Because females form the reproductive units of the population, we follow only the females in the construction of the table (Table 10.8). The total population of the female squirrels for the year 0 is 30 squirrels, 10 at age 1 and 20 at age 0. Not all of these squirrels will survive into the next year. The survival of these two age groups is obtained by multiplying the number of each by the s_x value. Because the s_x of the females of age 1 is 0.5, we know that 5 individuals ($10 \times 0.5 = 5$) survive to year 1 and enter age class 2. The s_x value of age 0 is 0.3, so that only 6 of the 20 in this age class in year 0 survive ($20 \times 0.3 = 6$) to year 1 and enter age class 1. In year 1, we now have 6 one-year-olds and the 5 two-year-olds, with both age classes reproducing. The b_x value of the 6 one-year-olds is 2.0, so they produce 12 offspring. The 5 two-year-olds have a b_x value of 3.0, so they produce 15 offspring. Together the two age classes produce 27 young for year 1, which now make up age class 0. The total population for year 1 (one year into the future) is 38. Survivorship and fecundity are determined in a similar manner for each succeeding year. Survivorship is tabulated year by year diagonally down the table to the right through the years, while new individuals are being added each year to age class 0.

From such a population projection table, we can calculate **age distribution** for each successive year (see Chapter 9, Section 9.8), the proportion of individuals in the various age classes for any one year, by dividing the number in each age class by the total population size for that year. Comparing the age dis-

tribution of the squirrel population in year 3 with that of the population in year 7, we observe that the population attains an unchanging or **stable age distribution** by year 7 (Table 10.9). From that year on, the proportions of each age group in the population and the rate of growth remain the same year after year, even though the population is steadily increasing.

The population projection table demonstrates an important concept of population growth. The rate of population increase is a function of the age-specific rates of survival (s_x) and birth (b_x). The constant rate of increase of the population from year to year and the stable age distribution are results of survival rates and fecundities for each age class that are constant through time.

By dividing the total number of individuals in year $t + 1$ by the total number of individuals in the previous year t, one can arrive at the **finite multiplication rate**, λ (**lambda**), for each time period. Given a stable age distribution in which lambda does not vary, lambda can be used as a multiplier to project population size (N) some time in the future:

$$N_t = N_0 \lambda^t$$

For our squirrel population, we can multiply the population size, 30, at time (year) 0 by $\lambda = 1.20$, the value derived from the population projection table, to obtain a population size of 36 for year 1. If we again multiply 36 by 1.20, or the initial population size 30 by λ^2 (1.20^2), we get a population size of 43 for year 2; and if we multiply the population at $N_0 = 30$ by λ^{10}, we arrive at a projected population size of 186 for year 10. These population sizes do not correspond to the population sizes early in the population projection table because λ fluctuates above and below the eventual value attained at stable age

TABLE 10.9

Approximation of Stable Age Distribution, Squirrel Population

| Age | \ Proportion in Each Age Class for Year | | | | | | | | | | |
	0	1	2	3	4	5	6	7	8	9	10
0	.67	.71	.71	.71	.69	.70	.70	.70	.70	.70	.70
1	.33	.16	.17	.17	.20	.17	.17	.18	.18	.18	.18
2		.13	.06	.07	.06	.07	.07	.07	.07	.07	.07
3			.06	.03	.03	.04	.04	.03	.03	.03	.03
4				.02	.01	.01	.01	.01	.01	.01	.01
5					.01	.01	.01	.01	.01	.01	.01

distribution. Only after the population achieves a stable age distribution does the λ value of 1.20 project future population size.

10.8 Exponential Growth Is Like Compound Interest

The equation $N_t = N_0 \lambda^t$ describes a population that grows exponentially, like compound interest. Such growth can occur when λ is greater than 1, the environment remains constant, and resources are in excess of demand. Some hypothetical examples of exponential growth are in Figure 10.5. Notice how the shapes of the curves vary with the value of λ. The closer λ comes to 1.0, the slower the growth. The population with value of λ = 1.05 is barely replacing itself, and the population with the value below 1.0 is declining. These curves suggest several features of population growth; it is influenced by life history features, such as age at the beginning of reproduction, the number of young produced, survival of young, and length of the reproductive period. A population may increase at an exponential rate until it overshoots the ability of the environment to support it. Then the population declines sharply from starvation, disease, or emigration.

The J-shaped or exponential curve is characteristic of some invertebrate and vertebrate populations introduced in a new or sparsely inhabited (low population size) environment. An example of an exponential growth curve is the rise and decline of the reindeer herd introduced on St. Paul, one of the

Pribilof Islands, Alaska (Figure 10.6). Introduced on St. Paul in 1910, reindeer expanded rapidly from 4 males and 22 females to a herd of 2000 in only 30 years. So severely did the reindeer overgraze their range that the herd plummeted to 8 animals in 1950. The decline produced a curve typical of a population that exceeds the resources of the environment. Growth stops abruptly and declines sharply in the face of environmental deterioration. From a low point, the population may recover to undergo another phase of exponential growth; it may decline to extinction; or it may recover and fluctuate around a level far below the original high point.

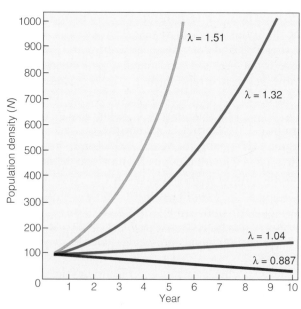

FIGURE 10.5 Exponential growth for four hypothetical populations with different values of λ.

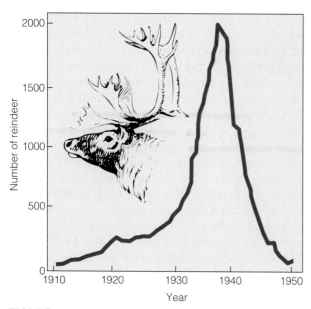

FIGURE 10.6 Exponential growth of the St. Paul reindeer *(Rangifer tarandus)* herd and its subsequent decline. (Scheffer 1951.)

10.9 Rate of Increase Is Used in Population Studies

Derived from the population projection table, lambda (λ) integrates estimates of both birthrates and death rates derived from the life table analysis. It is a discrete parameter of population increase, because it is a ratio of population sizes at the beginning and end of a discrete time unit. In the case of the gray squirrel population (Tables 10.2 and 10.7), yearly age classes are used. Therefore, lambda is an annual estimate of population growth. For a variety of reasons, ecologists often want to express change as a continuous process. For example, some populations reproduce through the year, rather than in discrete breeding seasons. Other populations possess overlapping generations whose members may give birth or die during the same time interval. In these cases, it is more usual to express population growth as an instantaneous rate of change in births and deaths. To do so, the finite rate of increase can be converted to the per capita rate of increase that measures instantaneous population growth.

The rate of increase is obtained by taking the natural log of λ. Thus, for the gray squirrel, $r = ln\ 1.2 = 0.18$. Lambda, then, is often expressed as e^r where e is the base of the natural logarithm, 2.71828 ($e^r = 2.71828^{0.18} = 1.2 = \lambda$). As with λ, the determination of r by this method only approximates the rate of increase. Determining the maximum value of r for a population, r_{max}, referred to as the intrinsic rate of increase, requires a more complex calculation.

Although the jump from lambda (λ) to r as a measure of population growth rate may seem unnecessarily complicated, the use of r has certain advantages for population ecologists. It is easier to compare the growth rates of different populations using r rather than λ. Consider the declining population with $\lambda = 0.887$. If the population were increasing at the same rate, λ would be 1.127. It is hard to see that $\lambda = 0.887$ is the inverse of $\lambda = 1.127$. It is easy to see the connection when these values of lambda are expressed as $r = 0.120$ and $r = -0.120$. Thus, r allows a direct comparison of rates. In addition, r allows us to compute the doubling time of a population.

Doubling time is the time required for a population to double its size. If $N_t/N_0 = 2$, then $e^{rt} = 2$; $rt = ln\ 2 = 0.6931$. Therefore doubling time $t = 0.6931/r$. The doubling time for the gray squirrel population, 0.6931/0.18, is 3.9 or approximately 4 years. Demographers use the same method to determine the doubling time of human populations.

Now, with some of the mystery removed from r, we will use the term from this point on. To present exponential growth in terms of r, you would use the equation

$$N_t = N_0 e^{rt}$$

where e^{rt} has been substituted for λ^t.

Perhaps the most important reason that population ecologists use r rather than λ as an estimate of population growth is mathematical analysis. Since r is an instantaneous estimate of population growth (rather than based on some discrete fixed-time unit such as years), the above equation for exponential growth can be rewritten as the following differential equation:

$$\frac{dN}{dt} = rN$$

In this form, *dN/dt* represents the instantaneous change in population size, not the population size per se. As we will see in the following sections, presenting the exponential model of population growth in this manner allows ecologists to expand the model to incorporate a number of factors that can directly influence population growth.

10.10 Population Growth Is Limited by the Environment

In the real world, the environment is not constant and resources are limited. As the density of a population increases, demand for resources increases. If the rate of consumption exceeds the rate at which resources can be resupplied, the resource base will shrink. Shrinking resources and the potential for an unequal distribution of those resources will result in increased mortality, decreased fecundity, or both. As a result, population growth declines with increasing density, eventually reaching a level at which population growth will cease. That level is called **carrying capacity**, or K. The carrying capacity is defined as the maximum sustainable population size for a given environment. It is a function of the supply of resources (e.g., food, water, space, etc.).

By adding the concept of carrying capacity, K, to the exponential model of population growth, we can account for the effects of density that slows population growth:

$$\frac{dN}{dt} = rN \left(\frac{K-N}{K} \right)$$

The new model, referred to as the logistic model, effectively has two components: the original exponential term (rN) and a second term ((K − N)/K), which functions to reduce population growth as the population size approaches the carrying capacity. When the population density (N) is low relative to the carrying capacity (K), the term (K − N)/K is close to 1.0, and population growth follows the exponential model (rN). However, as the population grows and N approaches K, the term (K − N)/K approaches zero, slowing population growth (Figure 10.7). Should the population density exceed K, population growth becomes negative and population density declines back toward carrying capacity.

As an illustration of logistic growth we can return to the example of the gray squirrel population (Table 10.2). Let us assume a carrying capacity of 200 individuals. Given the value of r = 0.18 calculated from the life table and an initial population size of 30 individuals, the predicted pattern of population growth under both the exponential and logistic models of population growth are shown in Figure 10.8.

The logistic growth curve is theoretical, a mathematical model of how populations might grow under favorable conditions. Although natural populations might appear to grow logistically, they rarely do. For example, Figure 10.9 considers the growth of a human population in Monroe County, West Virginia. The county has always had a stable economic base of agriculture and small industry. It was settled in the early 1700s and was well established in 1800, when the first United States census was taken. The population grew most rapidly between 1800 and 1850, so those years provided the data to estimate r. The population reached 13,200 in 1900 and has since fluctuated around that number. The rate of increase r was calculated as 0.074 and K was set at 13,200. Although the actual growth curve of the population mimics the logistic, the calculated logistic growth curve rose much more steeply than the actual growth curve and predicted that the population would reach K around 1870, 30 years before it actually did.

The reasons for nonconformity are straightforward. The age structure was not stable; birthrates and death rates varied from census period to census period; immigration and emigration were common to the population. The most surprising feature of the population is its stability after reaching K.

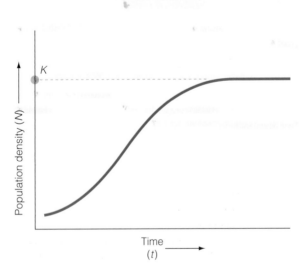

FIGURE 10.7 Logistic growth curve showing carrying capacity, K.

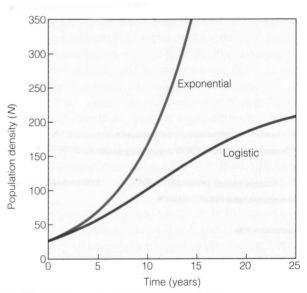

FIGURE 10.8 Exponential and logistic growth curve for the gray squirrel population from Tables 10.2 and 10.7; r = 0.18; K = 200, and N_0 = 30.

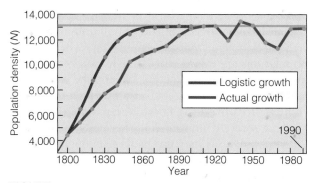

FIGURE 10.9 Actual and logistically predicted growth of population in Monroe County, West Virginia.

10.11 Populations Fluctuate Around Upper and Lower Limits

The logistic model suggests that populations function as systems, regulated by positive and negative feedback. Positive feedback promotes growth (as illustrated by the exponential curve); negative feedback of resource limitation slows it. As the population approaches carrying capacity, it theoretically responds instantaneously, with the growth rate declining to zero as the density approaches the maximum sustainable population size.

Rarely does such feedback work as smoothly in practice as the equation suggests. Often adjustments lag, and available resources may be sufficient to allow the population to overshoot the carrying capacity. Unable to sustain itself, the population then drops to some point below carrying capacity, but not before it has altered resource availability for future generations. Recall the example of reindeer introduced to the Pribilof Islands, Alaska (see Figure 10.6). The density of the previous generation and the depletion of resources, especially food, result in a time lag into population recovery.

Time lags make a population fluctuate, sometimes widely. Populations may be influenced by some powerful outside force such as weather or by some chaotic or random changes inherent in the population. A population may fluctuate about the equilibrium level, K, rising and falling between upper and lower limits (Figure 10.10). Such fluctuations are **stable limit cycles**. Some populations oscillate between high and low points in a manner more regular than we would expect to occur by chance. Such fluctuations are **population cycles**.

The two most common oscillation intervals in animal populations are 9- to 10-year cycles, typical

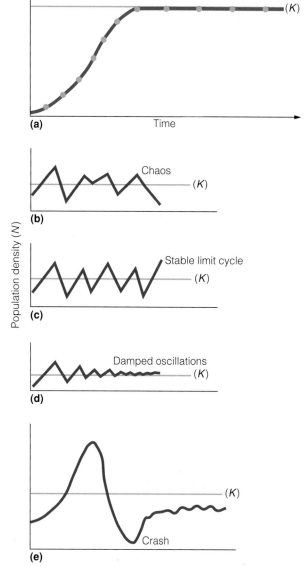

FIGURE 10.10 Logistic growth curve and examples of fluctuations around K.

of the snowshoe hare (Figure 10.11a) and 3- to 4-year cycles, typical of lemmings (Figure 10.11b). These cyclic fluctuations are confined largely to simpler ecosystems, such as northern coniferous forests and tundra. Cycles in the snowshoe hare involve an interaction among the hare and its overwinter food supply (mostly small aspen twigs). A growing population of hares reduces the ability of plants to recover from being browsed. Once decreased plant growth triggers an overwinter food shortage, a weakened hare population declines from increased mortality. Now the hare population is low and the vegetation recovers, stimulating a resurgence of the hare and initiating another cycle.

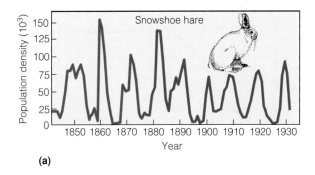

(a)

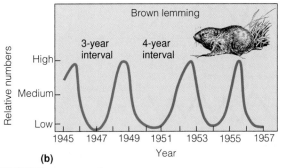

(b)

FIGURE 10.11 (a) The 9- to 10-year cycle of the showshoe hare *(Lepus americanus)* in Canada. (MacCulich 1937.) (b) The 4-year cycle of the brown lemming *(Lemmus sibiricus)* near Barrow, Alaska. (Pitelka 1957.)

10.12 Populations Can Decline Toward Extinction

When deaths exceed births and emigration exceeds immigration, populations decline. R_0 becomes less than 1.0; r becomes negative. Unless the population can reverse the trend, it may become so low that it declines toward extinction (see Figure 10.6).

The extinct heath hen *(Tympanuchus cupido cupido)* is a classic example. Formerly abundant in New England, this eastern form of the prairie chicken was reduced by habitat destruction and excessive hunting to remnant population on Martha's Vineyard off Massachusetts and in the pine barrens of New Jersey. By 1880 it was restricted to Martha's Vineyard. Two hundred birds made up the total population in 1890. Conservation measures increased the population to 2000 by 1917; but then fire, gales, cold weather, and excessive predation by goshawks reduced the population to 50. The number of birds rose slightly by 1920, then declined to extinction in 1925. The last bird died in captivity in 1932.

The case of the heath hen points out the major causes of extinction, hastened by human interfer-

ence. The overriding cause is habitat destruction. Loss of habitat forces remnant populations to exist in small, fragmented patches of habitat. These small populations are highly vulnerable to chance environmental catastrophes and predation. The fewer the animals, the greater the individual's chances of succumbing to a predator. Loss of even a few individuals can severely impair the population's viability. Small populations may not be large enough to stimulate the social behavior necessary for successful reproductive activity. Heath hens were communal or lek breeders (see Chapter 12, Section 12.6). The species required a minimal number of displaying males to attract females and stimulate reproductive activity. The population was too small to carry off successful reproduction.

Extinction is a natural process, albeit a selective one. Species differ in their probability of extinction depending on their characteristics as well as on random factors. Some of the qualities of a species that favor a high rate of extinction are large body size, small or restricted geographic range, habitat specialization, lack of genetic variability to cope with a changing environment, and inability to switch to alternate food sources in the face of a declining resource base. However, the recent extinctions and rapid declines of populations are unnatural, the result of ever-increasing pressures of human population (see the Ecological Application after Chapter 12: Cheating Nature).

10.13 Past Extinctions Have Been Clustered in Time

Extinctions are not spread evenly across Earth's history (Figure 10.12). Most extinctions are clustered in geologically brief periods of time (less than several tens of millions of years). One mass extinction occurred in the late Permian period, 225 million years ago, when 90 percent of the shallow-water marine invertebrates disappeared. Another occurred in the Cretaceous period, 65 to 125 million years ago, when the dinosaurs vanished. An asteroid striking Earth, interrupting oceanic circulation, altering the climate, and causing volcanic and mountain-building activity is believed to have caused that extinction event.

One of the great extinctions of mammalian life took place during the Pleistocene, when such species as the woolly mammoth, giant deer, mastodon, and giant sloth vanished from Earth. Some students of the Quaternary (that geological period from the end of the Tertiary period to the present time) believe that climate changes as ice sheets advanced and

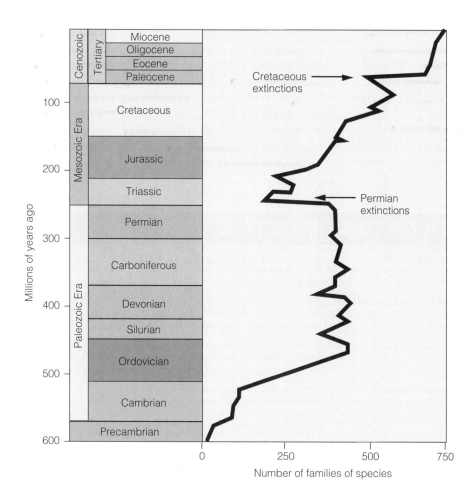

FIGURE 10.12 The geological time scale and mass extinctions in the history of life. The fossil record profiles mass extinctions during geological times. The most recent mass extinction, which occurred during the Cretaceous, wiped out more than half of all species, including all the dinosaurs. The Permian mass extinction event resulted in the loss of 96 percent of all marine species, and perhaps as much as 50 percent of the total species on Earth.

retreated caused the extinctions. Others argue that Pleistocene hunters overkilled large mammals, especially in North America, as human populations swept through North and South America between 11,550 and 10,000 years ago. Perhaps the large grazing herbivores could not withstand the combined predatory pressure of humans and other large carnivores. The greatest number of present-day extinctions have taken place since A.D. 1600. Humans have caused well over 75 percent of these extinctions through habitat destruction, introduction of predators and parasites, and exploitative hunting and fishing.

10.14 Extinction of Species Begins Locally

We usually think of extinction as taking place simultaneously over the full range of the species. Actually, it begins with isolated local extinctions when local conditions deteriorate or habitat disappears. Eventually, one local extinction after another adds up to total extinction.

The most important cause of extinction is habitat destruction or alteration, which is a local phenomenon. Cutting and clearing a forest, draining and filling a wetland, converting grassland to cropland, constructing highways and industrial complexes, and building cities and suburbs, with new housing and malls, greatly reduce available habitat for most species. When a habitat disappears, its unique plant and animal life also disappears.

Because of the rapidity of habitat destruction, no evolutionary time exists for a species to adapt to changed conditions. Forced to leave, dispossessed animals usually find the remaining habitats filled and face competition from others of their own kind or from different species. Restricted to marginal habitats, animals may persist for a while as nonreproducing members of a population or succumb to predation or starvation. As the habitat becomes more and more fragmented, the affected animal populations are fragmented into small isolated populations out of contact with others of the same species. As a result, isolated populations have less genetic variation, making them less adaptable to environmental change, a topic explored in Chapter 2. The survival of local populations often depends heavily on immigration of new individuals. As distance between local populations increases and as the size of local populations declines, immigration becomes impossible. When the local population falls below some minimum level, it may become extinct simply through random fluctuations in reproductive success.

Much the same situation exists with plants. They, too, face habitat destruction brought about by agriculture, mining, and urban and suburban development. Such activities result in the mass elimination of whole populations, many of them restricted to certain habitats. Unlike animals, most plants have poor dispersal abilities and cannot escape to favorable habitats, nor can they adapt quickly to changed environmental conditions. Added to habitat destruction is the invasion of habitats by alien (nonindigenous) species that humans introduce, which take over and crowd out native plant species.

CHAPTER REVIEW

Summary

Mortality, Survivorship, and Life Tables (10.1–10.4) Mortality, often concentrated in the young and the old, is the greatest reducer of populations. Mortality is measured by dividing the number dying in a given time period by the number alive at the beginning of the period (10.1). Mortality and its complement, survivorship, are best analyzed by means of a life table, an age-specific summary of mortality. Life tables for animals can be derived from data on mortality and survivorship (10.2). Life tables for plants are more difficult to develop because plants have modular growth and two life-cycle stages, germination and seedling development (10.3). From the life table we derive both mortality curves and survivorship curves. They are useful for comparing demographic trends within a population and among populations under different environmental conditions, and for comparing survivorship among various species. In general, mortality curves assume a J shape. Survivorship curves fall into three major types: Type I, in which individuals tend to live out their physiological life span; Type II, in which mortality and thus survivorship are constant through all ages; and Type III, in which survival of young is low. Survivorship curves follow similar patterns in both plants and animals (10.4).

Natality and Fecundity Tables (10.5–10.6) Birth is the greatest influence on population increase. Like mortality rate, birthrate is age-specific. Certain age classes contribute more to the population than others (10.5). Natality in plants is complicated because they have both vegetative (asexual) and sexual reproduction. Seed production and germination are true natality, because they produce new genetic individuals. Modular populations of clones—leaves, buds, and twigs—exhibit their own "birth" rates. The fecundity table provides data on the gross reproduction of each age class, b_x, and survivorship, l_x, of each age class, the sum of the products of which gives the net reproductive rate, R_0 (10.6).

Reproductive Rates and Population Growth (10.7–10.11) Age-specific estimates of survival and birth rates from the fecundity table can be used to project changes in population density. The procedure involves using the age-specific sur-

vival rates to move individuals into the next age class, and age-specific birthrates to project recruitment into the population. The resulting population projection table provides future estimates of both population density and age structure. Estimates of changes in population density can be used to calculate lambda (λ), a discrete estimate of population growth rate (10.7). In an unlimited environment, populations expand geometrically or exponentially. Exponential growth, described by a J-shaped curve, is like compound interest. Such growth may occur when a population is introduced into an unoccupied habitat (10.8). Instantaneous rate of increase can be expressed as the rate of increase in which growth is considered continuous rather than discrete. This approach is most useful in studies of populations with overlapping generations or that reproduce throughout the year. Both lambda and the rate of increase are useful in comparing population growth under different environmental conditions (10.9). Because resources are limited, geometric growth cannot be sustained indefinitely. Population growth eventually slows and arrives at some point of equilibrium with the carrying capacity (K) (10.10).

Natural populations rarely achieve a stable level. They fluctuate within upper and lower limits around a mean. Some fluctuations have peaks and lows that occur more regularly than we would expect by chance. The two most common intervals are 3 to 4 years, as in lemmings, and 9 to 10 years, as in snowshoe hares (10.11).

Extinction of Populations (10.12–10.14) A population may decline to extinction (10.12). Past extinctions have been clustered in time. Extinction is a natural process. Over long geological periods of time, old species disappear and new ones evolve (10.13). Current extinctions, however, are not brought about by natural processes but by human activity. At the present time, habitat fragmentation and destruction and overexploitation of populations are accelerating extinctions at an alarming rate (10.14).

Study Questions

1. Explain the meanings of mortality, natality, survivorship, and fecundity. How are they related?

2. What is a life table? What information do you need to construct one?

3. What are the advantages and weaknesses of a life table in a study of population dynamics?

4. What is the difference between a mortality curve and a survivorship curve? From what columns of the life table are they derived?

5. Why are natality and mortality more difficult to study in plants than in animals?

6. What are the differences between exponential and logistic population growth? How are they related?

7. How does R, net reproductive rate, relate to λ, the finite rate of increase, and r, the rate of increase? When do you use each one?

8. What is the difference between the exponential and logistic models of population growth?

9. How do you determine the carrying capacity of a habitat? What causes fluctuations around carrying capacity?

10. What are the causes of extinction? Relate them to some currently endangered species, such as the whooping crane, California condor, manatee, spotted owl, or desert tortoise.

11. How do the concepts in this chapter apply to the worldwide problem of human population growth? Review the population growth rates of some less-developed countries. For information, consult the *Population Bulletin* published by the Population Reference Bureau.

A subdominant individual assumes a submissive pose when confronted by the dominant wolf in the pack.

Intraspecific Population Regulation

OBJECTIVES

On completion of this chapter, you should be able to:

- Define competition.
- Relate the concept of carrying capacity with density-dependent population regulation.
- Distinguish between scramble and contest competition, and between interference and exploitation.
- Discuss the effects of intraspecific competition on growth, mortality, and fecundity.
- Explain the significance of dispersal.
- Describe a social hierarchy.
- Explain how social dominance may regulate populations.
- Define territoriality and explain its possible role in population regulation.
- Discuss how plants can establish spatial zones of influence.
- Explain how density-independent forces affect density-dependent responses.

No population continues to grow indefinitely. Even those exhibiting exponential growth eventually confront the limits of the environment. As the density of a population changes, interactions occur among members of the population that tend to regulate its size. These interactions include a wide variety of mechanisms relating to physiological, morphological, and behavioral adaptations.

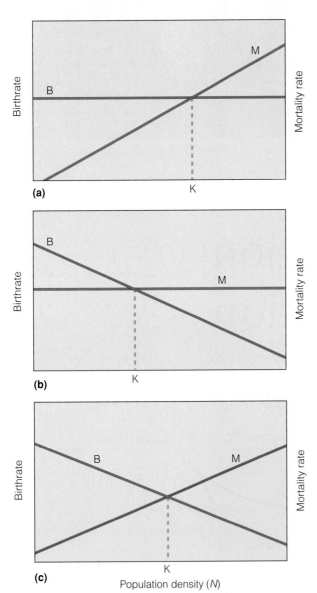

FIGURE 11.1 Regulation of density in three situations. (a) Birthrate *(B)* is independent of population density, as indicated by the horizontal line. Only the death rate *(M)* increases with density. At *K*, equilibrium is maintained by increasing mortality. (b) The situation is reversed. Mortality is independent, but birth declines with density. At *K*, a decreasing birthrate maintains equilibrium. (c) Full density-dependent regulation. Both birthrate and mortality are density-dependent. Fluctuations in either one hold the population at or near *K*.

11.1 Population Regulation Involves Density Dependence

The concept of carrying capacity introduced in Chapter 10 suggests a negative feedback between population increase and resources available in the environment. As population density increases, the per capita availability of resources declines. The decline in per capita resources eventually reaches some critical level at which it functions to regulate population growth. Implicit in this model of population regulation is **density dependence**. Density-dependent effects influence a population in proportion to its size. They function by slowing the rate of increase. In the case of the logistic growth model, this is done in a mathematically and biologically simplistic manner through the inclusion of a carrying capacity *(K)* in the term *(K − N)/K*. In reality, density-dependent mechanisms act by influencing the rates of birth and death (Figure 11.1). When population density is low, resources are not limiting, birthrate is high relative to death rate, and the population increases. Conversely, when population density is high, resources become limiting; birthrate may decline, mortality may rise, or both.

Mechanisms of density-dependent population regulation may include factors other than the direct effects of resource availability. For example, the spread of disease and parasites can be influenced by population density (see Chapter 16).

Other factors that can directly influence rates of birth and death function independently of population density. If some environmental factor such as adverse weather conditions affects the population without regard to the number of individuals, or if the proportion of individuals affected is the same at any density, then the influence is referred to as **density-independent**.

11.2 When Resources Are Limited, Competition Results

Implicit in the concept of carrying capacity is competition among individuals for essential resources. **Competition** among individuals takes place when a resource is in short supply relative to the number seeking it. Competition among individuals of the same species is referred to as **intraspecific competition**. As long as the availability of resources does not impede the ability of individuals to survive, grow, and reproduce, no competition exists. When resources are insufficient to satisfy all individuals, the means by which they are allocated has a marked influence on the welfare of the population.

When resources are limited, a population may exhibit one of two responses: scramble competition or contest competition. **Scramble competition** occurs when growth and reproduction are depressed equally across individuals in a population as the intensity of competition increases. **Contest competition** takes place when some individuals claim enough resources while denying others a share. Generally, under the stress of limited resources, a species will exhibit only one type of competition. Some are scramble species and others are contest species. One species may practice both types at different stages in the life cycle. For example, the larval stages of some insects endure

scramble competition until the population declines, and the adult stages face contest competition.

The outcomes of scramble and contest competition vary. In its extreme, scramble competition can lead to all individuals receiving insufficient resources for survival and reproduction, resulting in local extinction. In contest competition, only a fraction of the population suffers—the unsuccessful individuals. The survival, growth, and reproduction of individuals that successfully compete for the limited resources function to sustain the population.

In many cases, competing individuals do not directly interact with one another. Instead, individuals respond to the level of resource availability that is depressed by the presence and consumption of other individuals in the population. For example, large herbivores such as zebras grazing on the savannas of Africa may influence each other not through direct interactions, but by reducing the amount of grass available as food. Similarly, as a tree in the forest takes up water through its roots, it decreases the remaining amount of water in the soil for other trees. In these cases, competition is described in terms of **exploitation.**

In other situations, however, individuals interact directly with each other, preventing others from occupying a habitat or accessing resources within it. In this case, competition is termed **interference.**

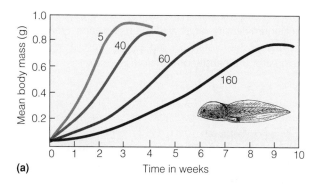

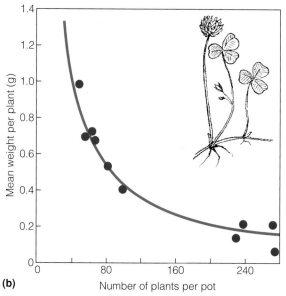

(b)
Number of plants per pot

FIGURE 11.2 The effect of population density on the growth of individuals. (a) The growth rate of the tadpole *Rana tigrina* declines swiftly as density increases from 5 to 160 individuals confined in the same space (Dash and Hota 1980). (b) The growth rate and subsequent weight of white clover (*Trifolium repens*) plants declines markedly with increasing density of individuals planted in the pot (Chatworthy 1960).

11.3 Intraspecific Competition Affects Growth and Development

Because the intensity of intraspecific competition is density-dependent, it increases gradually, at first affecting just the quality of life. Later it affects individual survival and reproduction. (See Focus on Ecology 11.1: The African Buffalo.)

As population density increases toward a point at which resources are insufficient, individuals in scramble competition reduce their intake of food. That diet slows the rate of growth and inhibits reproduction. Examples of this inverse relationship between density and rate of body growth may be found among populations of ectothermic (cold-blooded) vertebrates (Figure 11.2a). Tadpoles reared experimentally at high densities experienced slower growth, required a longer time to change from tadpoles to frogs, and had a lower probability of completing this transformation. Those that did reach threshold size were smaller than those living in less

THE AFRICAN BUFFALO

The African buffalo *(Syncerus caffer)*, studied in detail on the Serengeti in East Africa by A. R. C. Sinclair, is a large bovine ungulate of the African savanna. Its food is grass, preferably the protein-rich leaves. In 1894, rinderpest, a measleslike disease of cattle, spread from domestic cattle to the African buffalo and wildebeest *(Connochaetes taurinus)*. The disease decimated the herds in the early 1900s. Veterinarians eliminated rinderpest in cattle, which reduced its incidence in wild bovines. The disease continued to cause high mortality among juvenile buffalo as late as 1964. Released from heavy juvenile mortality, the African buffalo expanded dramatically, reaching an apparent equilibrium density in the 1970s (Figure A).

The critical time of year for the buffalo population is the dry season, when food may become scarce. Rainfall determines the productivity of grass. The greater the rainfall, the more vigorously the grass grows, increasing the amount of forage available in the dry season. Equilibrium density varies with the mean annual rainfall. The greater the rainfall, the more grass is available, and the greater the density of buffalo (Figure B).

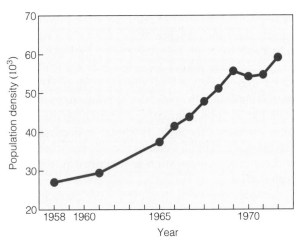

FIGURE A Change in population of African buffalo in Serengeti National Park. (After Sinclair 1977.)

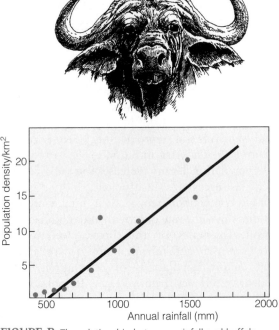

FIGURE B The relationship between rainfall and buffalo abundance.

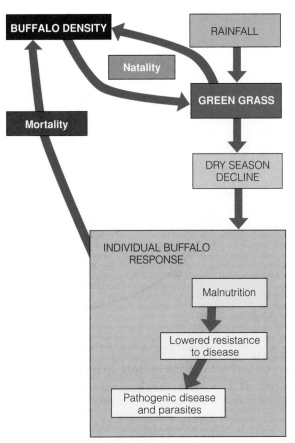

FIGURE C Population regulation in the African buffalo.

During the wet season on the African savanna, food is abundant, but during the dry season, the quality of food declines as the grasses dry. Buffalo become more selective, seeking green leaves and moving to the moist riverine habitat. They break into smaller units and use different areas. As the dry season progresses, buffalo become less selective, eating dry leaves and stems they otherwise would reject. As food quantity and quality decline, scramble competition becomes keener. Individuals in any one area reduce the food available to neighboring animals. The more buffalo present, the less food is available for each individual. Eventually the animals use up their fat reserves.

Undernourished and lacking the protein intake necessary to maintain their immunity to disease and the para-sites they normally harbor, adults become ill. Older animals are the most vulnerable. The number dying depends upon the rapidity with which adults use up their energy reserves before the arrival of the rainy season and new growth. If the next season sees more rainfall and that rainfall extends sporadically into the dry season, the mortality of adults the following year declines (Figure C).

Thus mortality of adults, influenced by rainfall and poor resources, determines a variable equilibrium density and regulates the population. By contrast, juvenile mortality appears to be density independent. One or several environmental variables cause fluctuations in the population. Density-dependent adult mortality compensates for the disturbances and dampens fluctuations.

---

dense populations. Fish living in overstocked ponds exhibit a similar response to density. Bluegills *(Lepomis macrochirus)*, for example, normally grow to the size of a dessert plate; but in overstocked and underharvested farm ponds, many do not grow beyond the size of a silver dollar and never reproduce.

A similar response of reduced growth rate under conditions of resource competition has been observed experimentally in plant populations. White clover plants *(Trifolium repens)* grown in pots with varying densities of individuals (Figure 11.2b) clearly show an inverse relationship between growth rate and population density. The mean weight of individual plants declines with increasing density of individuals in the pot. This decline is a direct consequence of resource limitation. At low densities, all individuals are able to acquire sufficient resources to meet demands for growth. As the density is increased (more individuals planted per pot), demand exceeds the supply of resources in the pot, and both growth rate and plant size decline.

In plant populations, the reduction in per capita resource availability that results from intraspecific competition can influence patterns of carbon allocation and morphology (form) as well as growth rate (Figure 11.3). Tree seedlings of broad-leaf peppermint *(Eucalyptus dives)* were grown in a greenhouse under varying levels of water availability (ml water per pot per day). Seedlings were grown either individually (1 individual per pot) or in populations of

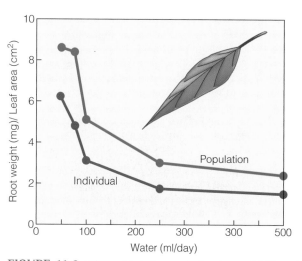

FIGURE 11.3 Relationship between plant water availability and the ratio of root mass (mg) to leaf area (cm²) for broad-leaf peppermint *(Eucalyptus dives)* seedlings grown in the greenhouse. Plants were grown under one of two treatments: (1) single individual per pot, and (2) population of 16 individuals per pot. As water availability decreases (less water provided to the pot per day), plants in both treatments allocate more carbon to the production of roots relative to leaves. However, plants grown under competition (populations of 16 individuals) consistently allocate more carbon to roots than corresponding individuals grown under the same water supply. These results show that competition functions to reduce water availability per plant and that plants respond accordingly to decreased water supply. (Smith unpublished.)

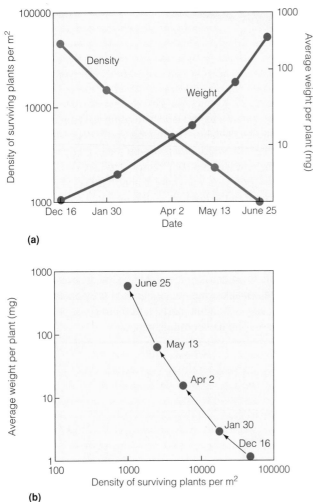

(a)

(b)

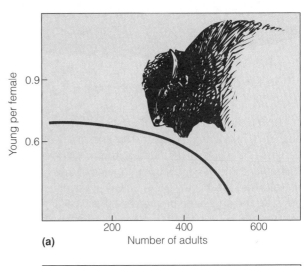

(a)

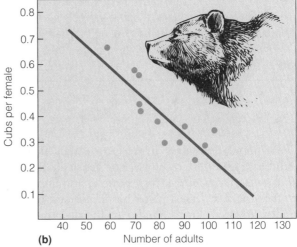

(b)

FIGURE 11.4 (a) Changes in the number of surviving individuals and average plant size (weight in mg) through time for an experimental population of horseweed *(Erigeron canadensis)*. (b) Data from (a) replotted to show the relationship between population density and average plant weight. Competition results in mortality, which in turn increases the per capita availability of resources, resulting in increased growth of the survivors. (Yoda 1963.)

FIGURE 11.5 Linear and nonlinear density-dependent change in large mammal populations. (a) The birthrate of the American bison begins to decline sharply after the population reaches a certain density. (b) The grizzly bear shows a constant decrease in young as the population increases. (Figure 14.3b from D. R. McCullough, "Population Dynamics of the Yellowstone Grizzly," in *Dynamics of Large Mammal Populations,* C. W. Fowler and T. D. Smith, eds. (New York: John Wiley & Sons, 1981), p. 177. © 1981 John Wiley & Sons, Inc. Reprinted by permission.)

16 individuals per pot. As expected, reduced water availability resulted in an increase in the production of roots relative to leaves (see Chapter 6, Section 6.8). However, this pattern is much more pronounced for individuals grown in populations, since competition results in plants receiving only a fraction of the total water supplied. Competition for water results in a functionally drier environment for the individual plants in the population.

In addition to reduced growth rate, a common response of plants to high population density is reduced survival. Mortality functions to increase resource availability for the remaining individuals, allowing for increased growth (see Focus on Ecology 11.2: Thinning the Ranks). In an experiment aimed at exploring this relationship, K. Yoda planted seeds

of horseweed *(Erigeron canadensis)* at a density of 100,000 seeds per square meter. As the seedlings grew, competition for the limited resources ensued (Figure 11.4a). The number of seedlings surviving declined within months to a density of approximately 1000 individuals. The death of individuals increased the per capita resource availability, and the average size of the surviving individuals increased as population density declined. The inverse relationship between population density and average plant size during the experiment can be seen more easily in Figure 11.4b. This progressive decline in density and increase in biomass of remaining individuals in a single species population is known as self-thinning.

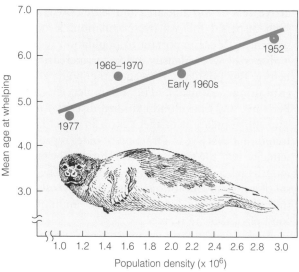

FIGURE 11.6 The mean age of sexual maturity for female harp seals *(Phoca groenlandica)* (and other marine and terrestrial animals) is related not so much to age as to weight. Seals arrive at sexual maturity when they reach 87 percent of average adult body weight. Seals attain this weight at an earlier age when population density is low. (From Lett et al. 1981:144.)

11.4 Intraspecific Competition Can Reduce Reproduction

In addition to having a direct influence on the survival and growth of individuals, competition within a population can function to reduce fecundity. The timing of the response depends upon the nature of the population. Among large mammals with a long life span and low reproduction, regulating mechanisms often do not function until the population approaches carrying capacity. At that point, density mechanisms set in and tend to overcompensate. The birthrate of bison *(Bison bison)* shows such a response (Figure 11.5a), declining sharply once the population has reached a high density. However, the birthrates of other large mammals, such as the grizzly bear *(Ursus arctos),* appear to be affected across a much wider range of population density (Figure 11.5b), with birthrate declining linearly with population density. The mechanisms by which competition influences reproductive rate can vary with species. In populations of harp seals, the onset of sexual maturity in females is related to body weight. Reduced growth rates (weight gain) under high population densities increases the mean age at which females become reproductive (Figure 11.6).

Density-dependent controls on fecundity are also observed in plant populations. The amount of grain

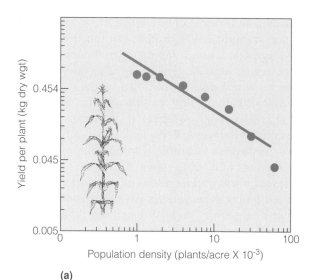

(a)

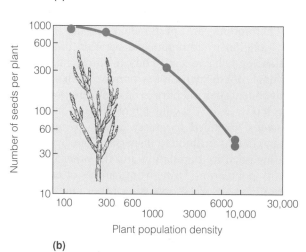

(b)

FIGURE 11.7 Two examples of density-dependent effects on fecundity in plants. (a) Grain (seed) production of individual corn plants declines with increasing density of plants per acre. (Fery and Janick 1971.) (b) Number of seeds produced per plant declines with increasing population density in the annual salt marsh herb *Salicornia europaea.* (Watkinson and Davy 1985.)

produced by individual corn plants is dramatically reduced when planted at higher densities (Figure 11.7a). Similarly, the number of seeds produced per plant declines with increasing density of individuals in populations of the annual herb *Salicornia europa* (Figure 11.7b) inhabiting coastal salt marshes.

11.5 High Density Is Stressful to Individuals

As a population reaches a high density, individual living space becomes restricted. Often, aggressive contacts among individuals increase. One hypothesis of population regulation in animals is that increased crowding and social contact cause stress. Such stress triggers hormonal changes that can function to suppress growth, curtail reproductive functions, and delay sexual activity. They may also suppress the immune system and break down white blood cells, increasing vulnerability to disease. In mammals, social stress among pregnant females may increase mortality of the young in the fetal stage (unborn) and cause inadequate lactation, stunting the growth and development of nursing young. Thus, stress results in decreased births and increased infant mortality. Such population-regulating effects have been confirmed in confined laboratory populations of several species of mice and to a lesser degree in enclosed wild populations of woodchucks *(Marmota monax)* and rabbits *(Oryctolagus cuniculus)*. Evidence of effects of stress in free-ranging wild animals, however, is difficult to obtain.

Pheromones are perfumelike chemical signals that function to communicate between individuals, influencing behavior and body function much like a hormone does. Pheromones present in the urine of adult rodents can encourage or inhibit reproduction. One study involved wild female house mice *(Mus musculus)* confined to grassy areas surrounded by roadways that prohibited dispersal. One group lived in a high-density population and the other in a low-density population. Urine from females of each population was absorbed onto filter paper and placed in laboratory cages with juvenile test females. Exposing juvenile females to urine from high-density populations delayed puberty, whereas exposing females to urine from low-density populations did not. The results suggest that pheromones contained in the urine of adult females in high-density populations function to delay puberty and help slow further population growth.

11.6 Dispersal May or May Not Be Density-Dependent

Instead of coping with stress, some animals disperse, leaving the population to seek vacant habitats. Although dispersal is most apparent when population density is high, it goes on all the time. Some individuals leave the parent population whether it is crowded or not. There is no hard-and-fast rule about who disperses.

When a lack of resources forces some individuals to disperse, the ones to go are usually subadults driven out by adult aggression. The odds are that such individuals will perish, although a few may arrive at some suitable area and settle down. Because such dispersal follows overpopulation, it does not prevent it. More important to population regulation is dispersal when density is low or increasing, well before the population reaches the point of over-exploiting resources. For example, Dominique Berteaux and Stan Boutin studied dispersal in a red squirrel population in the Yukon in Canada. They found that every year a fraction of older reproductive females left their home areas during the summer months when food availability was high. The dispersal of adults functions to increase the survival of their juvenile offspring during the winter months when food resources are scarce.

Some dispersing individuals, especially juveniles, can maximize their probability of survival and reproduction only if they leave their birthplace. When intraspecific competition at home is intense, dispersers can relocate in habitats where resources are more accessible, breeding sites more available, and competition less. Further, the disperser reduces the risk of inbreeding (see Chapter 19). At the same time, dispersers incur risks that come with living in unfamiliar terrain.

Dispersers, according to recent studies, travel no farther than necessary. How far they go depends on the density of surrounding populations and the availability of suitable habitat. Individuals may travel in a straight line from their birthplace, or they may make a number of exploratory forays before leaving, and then occupy the first uncontested site they locate (Figure 11.8). How well they fare in their new location depends upon the quality of habitat.

Dispersal often involves a **source habitat** in which reproduction exceeds mortality and a **sink habitat** in which mortality exceeds reproduction. Potential sinks often are empty or even marginal or unsuitable habitat areas where conditions such as high predation rates, poor nesting sites, and lack of protective cover increase mortality and preclude successful reproduction. Even though some sink habitats support large populations, the species would disappear from the area without continual immigration. Because a species may actually be more abundant in a sink than in the source habitat, the sink area may appear to be optimal habitat based solely on population density. In reality, the habitat may be luring dispersers into danger or reproductive failure.

Such population sinks can be distinguished from optimal habitats only with some knowledge of the reproduction and survival of the species within it. Such knowledge is critical in the conservation of species. Because of population abundance, we may mistakenly select a sink habitat as a wildlife refuge. Such a decision could lead both sink and source populations to local extinction.

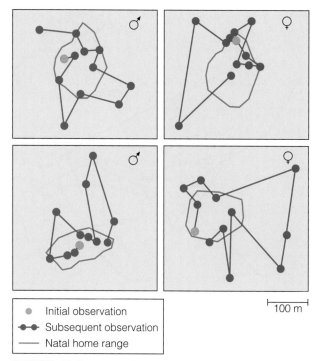

- ● Initial observation
- ●—●— Subsequent observation
- —— Natal home range

`|—— 100 m ——|`

FIGURE 11.8 Dispersal forays made by young radio-collared red squirrels *(Tamiasciurus hudsonicus)* in Assiniboine, Alberta, Canada. (Larsen and Boutin 1994.)

Does dispersal actually regulate a population? Although dispersal is positively correlated with population density, no relationship that can be generalized exists between the proportion of the population leaving and its increase or decrease. Dispersal may not function as a regulatory mechanism, but it contributes strongly to population expansion and aids in the persistence of local populations.

11.7 Social Behavior May Limit Populations

Intraspecific competition can express itself in social behavior, the degree to which individuals of the same species tolerate one another. Social behavior appears to be a mechanism that limits the number of animals living in a particular habitat, having access to a common food supply, and engaging in reproductive activities.

To prove that social behavior limits populations in a density-dependent fashion, population ecologists have to show that (1) a substantial portion of the population consists of surplus animals that do not breed because they either die or attempt to breed and fail; (2) such individuals are prevented from

breeding by dominant individuals; (3) nonbreeding individuals are capable of breeding if dominant individuals are removed; and (4) breeding animals are not completely using the available food and space.

11.8 Social Dominance Can Affect Reproduction and Survival

Many species of animals live in groups with some kind of social organization. This organization is based on aggressiveness, intolerance, and the dominance of one individual over another. Two opposing forces are at work. One is mutual attraction of individuals; the other is a negative reaction against crowding, the need for personal space.

Each individual occupies a position in the group based on dominance and submissiveness. In its simplest form, the group includes an alpha individual dominant over all others, a beta individual dominant over others except the alpha, and an omega individual subordinate to all others. Individuals settle social rank by fighting, bluffing, and threatening at initial encounters between any given pair of individuals or at a series of such encounters. Once individuals establish social rank, they maintain it by habitual subordination of those in lower positions. Threats and occasional punishment handed out by those of higher rank reinforce this relationship. Such organization stabilizes and formalizes intraspecific competitive relationships and resolves disputes with a minimum of fighting and wasted energy.

Social dominance plays a role in population regulation when it affects reproduction and survival in a density-dependent manner. An example is the wolf. Wolves live in small groups of 6 to 12 or more individuals, called packs. The pack is an extended kin group consisting of a mated pair, one or more juveniles from the previous year who do not become sexually mature until the second year, and several related nonbreeding adults.

The pack has two social hierarchies, one headed by an alpha female and the other headed by an alpha male, the leader of the pack, to whom all other members defer. Below the alpha male is the beta male, closely related, often a full brother, who has to defend his position against pressure from males below him in the social hierarchy.

Mating within the pack is rigidly controlled. The alpha male (occasionally the beta male) mates with the alpha female. She prevents lower-ranking females from mating with the alpha and other males, while the alpha male inhibits other males from mating with her. Therefore, each pack has one reproducing pair and one litter of pups each year. They are reared cooperatively by all members of the pack.

The size of packs, which is heavily influenced by the availability of food, governs the level of the wolf population in a region. Priority for food goes to the producing pair. At high pack density and decreased availability of food, individuals may be expelled or leave the pack. Unless they have the opportunity to settle successfully in a new area and form a pack, they may not survive. Thus, at high wolf densities, mortality increases and birthrates decline. When the population of wolves is low, sexually mature males and females leave the pack, settle in unoccupied habitat, and establish their own packs that have an alpha (reproducing) female. In this case, nearly every sexually mature female reproduces, and the wolf population increases. At very low densities, however, females may have difficulty locating males with whom to establish a pack and so fail to reproduce or even survive.

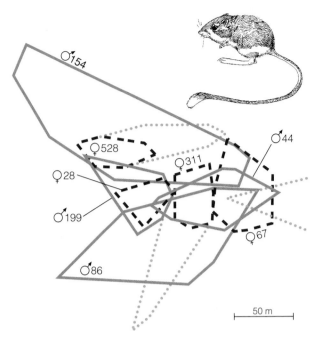

FIGURE 11.9 Home ranges of eight adult Merriam's kangaroo rats *(Dipodomys merriami)* in Arizona. The home ranges of males are indicated by solid lines, females by dashed lines. Dotted lines show excursions outside the usual home ranges. Note the overlap of all home ranges forming the greater home range. The males' ranges are larger; females' ranges lie within the home ranges of males. (Jones 1989.)

11.9 Social Interactions Influence Activities and Home Ranges

Social interactions among individuals influence the movement, distribution, and reproductive activity of animals over an area. The space that an animal normally uses during a year is its **home range.** The two sexes may have the same or different home ranges (Figure 11.9), which may overlap. Although the home range is not defended, aggressive interactions may influence the movements of individuals within another's home range. Some species, however, defend a core area of the home range against others. If the animal defends any part of its home range, we define that part as a **territory,** a defended area. If the animal defends its entire home range, its home range and territory are the same.

Home range is highly variable, even within a species. Seldom is a home area rigid in its use, its size, or its establishment. The home range may be compact and continuous, or it may be broken into discontinuous parts reached by trails. Irregularities in spatial and temporal distribution of food produce corresponding irregularities in home range and frequency of visitation.

Overall size of the home range varies with the available food resources, mode of food gathering, body size, and metabolic needs. Among mammal species, the home range size is related to body weight (Figure 11.10). In general, carnivores require a larg-

er home range than herbivorous and omnivorous animals of the same size. Males and adults have larger home ranges than females and subadults. Weight alone is often sufficient to account for differences within a species without invoking any mechanisms relating to competitive interactions.

Food supply can also directly influence home range size. The more concentrated the food supply, the smaller the home range. Home ranges of some mammals and the center of activity within them appear to be in a constant state of flux, determined by such variables as the location of food and the reproductive condition of neighboring males and females.

The degree of aggressive behavior among individuals may limit the size of a home range. In many mammal species, dominant individuals hold the largest home ranges, with some overlap, while subdominant individuals occupy home ranges within those of dominant individuals. The dominant animal is able to range more freely over a larger area and have access to more resources than subdominant individuals. Usually, a dominant male can control highly desirable locations, especially in relation to food and females, and force subdominant animals to move less freely in the area.

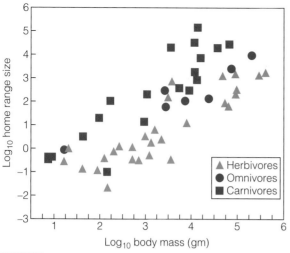

FIGURE 11.10 Relationship between the size of home range and body weight of North American mammals. (Harastad and Bunnell 1979.)

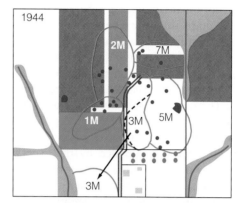

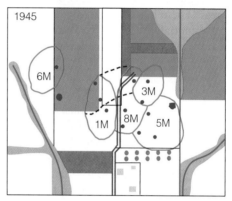

FIGURE 11.11 Territories of the grasshopper sparrow *(Ammodramus savannarum)* determined by observations of male banded birds, indicated by 1M, 2M, etc. Dots indicate song perches. Note how they are distributed near the territorial boundaries. Shaded areas represent crop field. Dashed lines indicate territory shifts prior to second nesting. The return of the same males to nearly the same territory the second year is termed philopatry. (Smith 1963.)

Restriction of activities to a home area confers certain advantages. Animals become familiar with the area and the location of food, shelter, and escape cover. The animal is more secure than it would be if it were forced into unfamiliar areas.

11.10 Territoriality May Regulate Populations

Territoriality is a situation in which an animal defends an exclusive area not shared with rivals (Figure 11.11). Defense involves well-defined behavioral patterns—song and call, intimidation displays such as spreading wings and tail in birds and baring fangs in mammals, attack and chase, and marking with scents that evoke escape and avoidance in rivals. As a result, territorial individuals tend to occur in more or less regular patterns of distribution (see Section 9.5).

Why should a cardinal or a wolf pack defend a territory? The proximate reasons vary. For some, the reason is the acquisition and protection of food, a nesting site, or a mating area, or attraction of a mate. The ultimate reason is an increased probability of survival and reproductive success. By defending a territory, the individual secures sole access to an area of habitat and the resources it contains. If space is limited, territoriality may force others into suboptimal habitats, reducing their reproductive success. At the same time, the individual increases the proportion of its own offspring in the population.

Defending a territory is costly in energy and time and can interfere with other essential activities such as feeding, courtship, mating, and rearing young. Because the defense of a territory has both benefits and costs, in some situations a territory is a net benefit to the individual, and in other situations it is not. To be worthwhile, a territory must meet its owner's needs, and the benefits of defending the territory need to outweigh the energetic costs of its defense. If resources are unpredictable or patchy, or resources are scarce, it may be advantageous not to defend any space.

A feature of territory is size. As the size of a territory increases, the cost of defense increases. In general, territory size tends to be no larger than required. That size may vary from year to year (see Figure 11.11) and from locality to locality, depending upon resources and the number of animals seeking space.

For some animals, birds in particular, it is not the size of the territory that counts, but rather its quality. Some males, perhaps the most aggressive, claim the best territories, usually measured by features of vegetation that relate to food availability or make them superior nesting sites. Less successful males occupy suboptimal territories, and some males secure no territories at all. The most successful males are assured of a mate, whereas male birds settled on a poor territory may be unable to attract mates.

The total area available divided by the average size of the territory determines the number of territorial owners a habitat can support. When the available area is filled, owners evict excess animals or deny them access. These individuals make up a floating reserve of potential breeders. The existence of such a reserve of potentially breeding adults has been described for a number of bird species, including the red grouse *(Lagopus lagopus)* in Scotland, the Australian magpie *(Gymnorhina tibicen)*, Cassin's auklet *(Ptychoramphus aleuticus)* of Alaska, and the white-crowned sparrow *(Zonotrichia leucophrys)* of California. Studies of a banded (marked for identification) population of white-crowned sparrows indicated a surplus of potential breeding individuals. In fact, 24 percent of the individuals holding territories had been floaters (no territory) for a period ranging from 2 to 5 years prior to acquiring a territory. Floaters quickly replaced territory holders that disappeared during the breeding season.

Few data exist on the social organization of the floaters. Floaters may form flocks with a dominance hierarchy on areas not occupied by territory holders, as do red grouse and Australian magpies. They may live singly outside occupied territories, as white-crowned sparrows do; or they may spend much time on the breeding territories of others, accepted as nonrivals, as do the rufous-collared sparrows *(Zonotrichia capensis)* of Costa Rica.

Territoriality can function as a mechanism of population regulation. If all pairs that settle on an area get a territory, territoriality only spaces a population and does not regulate it. By contrast, if territorial size has a lower limit, the number of pairs that can settle on an area is limited, and individuals that fail to do so have to leave. In such a situation, territoriality might regulate the population, but only if an excess of nonterritorial males and females of reproductive age occurs, as is the case for the red grouse and white-crowned and rufous-collared sparrows mentioned above. Then reproduction is limited by territoriality, and we have density-dependent population regulation.

The size of home ranges and territories is of more than academic interest. It is an indicator of carrying capacity and of how much habitat is necessary to support viable populations. Animals with small home ranges or territories can be contained in smaller areas of contiguous habitat than can large mammals and predatory birds, the populations of which may require thousands of square kilometers. Management of such expansive habitats is a major ecological and economic problem.

11.11 Plants Capture and "Defend" Space

Plants are not territorial in the same sense that animals are territorial; but plants can capture and hold onto space. This phenomenon in the plant world is analogous to territoriality in animals, especially if you accept an alternative definition of territoriality—individual organisms spaced out more than we would expect from a random occupancy of suitable habitat.

Plants from dandelions to trees do capture a certain amount of space and exclude individuals of their own and other species. When a dandelion plant spreads its rosettes of leaves on the ground, it eliminates all other plants from the area it covers. Plants also establish zones of resource depletion associated with their canopy (leaves) and root systems. Taller individuals intercept light (see Chapter 4, Section 4.2), shading the ground below and limiting successful establishment to species that can tolerate the reduced availability of light (see Chapter 6, Section 6.7). Likewise, the uptake of water and nutrients from the soil limits availability to individuals with overlapping rooting zones. Because of their longevity, some plants, especially trees, occupy space for a long time, preventing invasion by individuals of the same or other species. Plants successful in capturing space increase their fitness at the expense of others.

Plants also capture and maintain space by the release of organic toxins. Various chemical compounds released by roots and leaves accumulate in the soil. These compounds inhibit the germination of seeds and the growth of other plants, both herbaceous and woody (see Focus on Ecology 14.1: A Chemical Solution to Competition in Plants), reducing competition for light, nutrients, and space.

THINNING THE RANKS

When you buy seeds to plant in your garden, whether they are flowers or vegetables, there are instructions on the back of the package for planting. Included will be a recommended spacing: the space you should allow between seeds as you plant them in the soil. The recommended spacing is related to the size the plants will achieve, assuring that each individual has adequate space and resources to grow to its potential. Should the seeds be planted too closely, crowding will occur as the plants grow, and individuals will reduce the availability of resources to neighboring plants, reducing their growth (see Figures 11.3, 11.4, and 11.7). For this reason, it is the practice of gardeners to regularly space their plantings to reduce competition and allow the individuals to grow vigorously. The same is true for tree plantations.

Foresters plant trees in rows that form a checkerboard pattern when viewed from above. Each individual is equally spaced from its adjacent neighbors. The spacing allows individuals sole access to water and nutrients in the soil and reduces the potential for one individual to shade its neighbors. If the trees are planted too close, competi-

tion among them reduces the growth of each. However, if the trees are planted too far apart, the forester is wasting valuable land for producing wood.

Foresters often extend the practice of controlling the spacing of individuals to natural stands of young trees. Although the individual trees are the same age, variation among individuals occurs, and some will grow faster than others. As individuals begin to crowd each other, competition for space in the crown (canopy) takes place. The smaller the crown, the less light each tree intercepts, and the slower they grow. The forester does not let nature determine the winner. Foresters remove competing, less vigorous individuals in a process called "stand thinning." Foresters keep the largest, most vigorous trees, which they call crop trees. They are the ones with the highest potential for future growth. Foresters remove enough of the surrounding trees to allow 10 feet spacing between the crop tree's crown and the crowns of surrounding trees. Such thinning allows the crop tree to expand its crown, increasing its photosynthetic capacity and thus its potential growth.

11.12 Density-Independent Influences Can Be Strong

We have seen that population growth and fecundity are heavily influenced by density-dependent responses. However, there are other, often overriding influences on population growth that do not relate to density. These influences are termed density-independent. Strictly speaking, density-independent factors do not regulate population growth, because regulation implies feedback. Density-independent influences, however, can have considerable impact on birthrates and death rates, at times limiting population growth (Figure 11.12). They can mask or at times even eliminate density-dependent regulation.

A cold spring may kill the flowers of oak trees, causing a failure of the acorn crop. Because of the failure, squirrels may suffer widespread starvation the following winter. Although the proximate cause

of starvation is the density of squirrels and the low food supply, weather was the ultimate cause of the decline. In general, annual and seasonal changes in the environment tend to cause irregular population fluctuations. Conditions beyond an organism's limits of tolerance can have disastrous effects. They can affect growth, maturation, reproduction, survival, and movement. They can even lead to the extinction of local populations.

The influence of weather is irregular and unpredictable. It functions largely by influencing the availability of food. Pronounced changes in population growth often correlate directly with variations in moisture and temperature. For example, outbreaks of spruce budworm (*Choristoneura fumiferana*) are usually preceded by 5 or 6 years of low rainfall and drought. Outbreaks end when wet weather returns. Such density-independent effects can take place on a local scale where topography and microclimate influence the fortunes of local populations.

In desert regions, a direct relationship exists between precipitation and rate of increase in certain

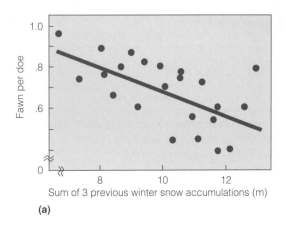

(a)

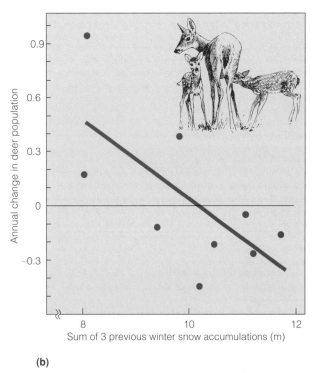

(b)

FIGURE 11.12 The relationship between the sum of the previous 3 winter monthly snow accumulations in northeastern Minnesota and the population of white-tailed deer. (a) Fecundity (fawn to doe ratio). (b) Percent annual change in next winter's population. (From Mech et al. 1987.)

rodents and birds. Merriam's kangaroo rat *(Dipodomys merriami)* occupies lower elevations in the Mojave Desert. The kangaroo rat has the physiological capacity to conserve water and survive long periods of aridity. However, it does require the prevailing patterns of seasonal moisture availability to be sufficient to stimulate the growth of herbaceous desert plants in fall and winter. The kangaroo rat becomes reproductively active in January and February when plant growth, stimulated by fall rains, is green and succulent. Herbaceous plants provide a source of water, vitamins, and food for pregnant and lactating females. If rainfall is scanty, annual plants fail to develop, and reproduction by kangaroo rats is low. This close relationship between population dynamics and seasonal rainfall and success of winter annuals is also apparent in other rodents and birds occupying similar desert habitats.

CHAPTER REVIEW

Summary

Density-Dependent Regulation (11.1) Populations do not increase indefinitely. As resources become less available to an increasing number of individuals, birthrates decrease, mortality increases, and population growth slows. If the population declines, mortality decreases, births increase, and population growth speeds up. Between positive and negative feedback, the population arrives at regulation.

Competition for Resources (11.2–11.4) Intraspecific competition occurs when resources are in short supply. Competition can take two forms: scramble and contest. In scramble competition, growth and reproduction are depressed equally across individuals as competition increases. In contest competition, dominant individuals claim sufficient resources for growth and reproduction. Others produce no offspring or perish. Competition can involve interference among individuals or indirect interactions via exploitation of resources (11.2). Competition for scarce resources can decrease or retard growth and development. Up to a point, plants respond to competition by modifying form and size. After that they experience density-dependent mortality (11.3). High population density and competition can also function to delay reproduction in animals and reduce fecundity in both plants and animals (11.4).

Responses to Competition (11.5–11.6) The stress of crowding may cause delayed reproduction, abnormal behavior, increased adrenal activity, reduced ability to resist disease and parasitic infections in animals, and reduced growth and production of seeds in plants (**11.5**). Stress might also lead animals to disperse. Dispersal is a constant phenomenon in populations at presaturation levels. Individuals seem to be genetically programmed to disperse. Many end up in submarginal habitats, but some succeed in new or unfilled habitats. A population sink may keep drawing immigrants but fail to sustain them. At saturation levels, dispersal is a response to overcrowding, and the dispersers are surplus to the population. There is no strong indication that dispersal regulates populations, but it does help them expand (**11.6**).

Social Behavior (11.7–11.8) Intraspecific competition may be expressed through social behavior. The degree of tolerance can limit the number of animals in an area and some animals' access to essential resources (**11.7**). A social hierarchy is based on dominance. Dominant individuals secure most of the resources. Shortages are borne by subdominant individuals. Such social dominance may be a mechanism of population regulation (**11.8**).

Home Range and Territoriality (11.9–11.11) Social interactions influence the distribution and movement of animals. The area an animal normally covers in its life cycle is its home range, which can overlap the home ranges of others. The size of a home range is influenced by body size, metabolic needs, food resources, and interactions within a species. Dominant individuals can strongly influence movements and home range size of subdominant individuals. If the animal or a group of animals defends a part or all of its home range as its exclusive area, it exhibits territoriality. The defended area is its territory (**11.9**). Animals defend territories by songs, calls, displays, chemical scents, and fighting. Owners can afford a territory only if benefits exceed the costs of defense. Territoriality is a form of contest competition in which a portion of the population is excluded from reproduction. These nonreproducing individuals act as a floating reserve of potential breeders, available to replace losses of territory holders. In such a manner, territoriality can act as a population-regulating mechanism (**11.10**).

Plants are not territorial in the same sense as animals, but they do hold on to space, excluding other individuals of the same or smaller size. Plants capture and hold space by intercepting light, moisture, and nutrients or by releasing organic toxins (**11.11**).

Density-Independent Effects (11.12) Density-independent influences, such as weather, affect but do not regulate populations. They can reduce populations to the point of local extinction. Their effects, however, do not vary with population density. Regulation implies feedback.

Study Questions

1. What is competition? How are scramble and contest competition important in population regulation?

2. How might density-dependent effects influence growth and size of individuals?

3. How might stress influence population growth in mammals? How do plants respond to crowding?

4. How is dominance expressed in a social hierarchy? Does it regulate population size?

5. Distinguish between home range and territory.

6. How can territorial behavior affect reproductive success?

7. How can territoriality function in population regulation? What is the role of the floater?

8. Comment on the remark, "If this woods is cut, the animals will just move elsewhere." What is the probable fate of the displaced individuals, and why?

9. Look up the home ranges of such large mammals as the elephant, rhinoceros, grizzly bear, and others. Discuss the problem of maintaining viable populations on such limited areas as national parks.

10. What are some problems associated with the release of animals into unfilled or partially filled habitats?

Two bull elephant seals *(Mirounga angustirostris)* clash for possession of a section of beach and its attendant females.

CHAPTER 12

Life History Patterns

OBJECTIVES

On completion of this chapter, you should be able to:

- Discuss the tradeoffs and consequences of sexual and asexual reproduction.
- Discuss modes of sexual reproduction.
- Define a mating system, and distinguish among monogamy, polygamy, polygyny, and polyandry.
- Explain the two types of sexual selection.
- Discuss the importance of energy allocation to reproduction.
- Compare semelparity and iteroparity.
- Discuss the relationship between the number of young and parental investment.
- Explain the relationship among size, age, and fecundity.
- Contrast *r*-selection and *K*-selection; and R, C, and S life history strategies in plants.
- Discuss the importance of habitat selection in reproductive success.

An organism's life history is its lifetime pattern of growth, development, and reproduction. Life histories combine a rich array of adaptations relating to physiology, morphology, and behavior. They involve adaptations to the prevailing physical environment, such as those discussed in Part III (The Organism and Its Environment). Perhaps most importantly, life histories involve adaptations relating to the organism's interactions with other organisms—the biological environment. Without question, the most important of these life history characteristics are those adaptations relating to reproduction. The true measure of an organism's reproductive success is fitness (see Chapter 2). Differential reproduction of individuals and the process of natural selection drive evolution. Achieving fitness involves, among other things, survivorship and fecundity; modes of reproduction; age at reproduction; number of eggs, young, or seeds produced; parental care; size; and time to maturity. As with adaptation to the physical environment (Part III), adaptations relating to reproduction involve tradeoffs imposed by constraints of physiology, energetics, and the prevailing physical environment—the organisms' habitat.

FIGURE 12.1 The freshwater hydra reproduces asexually by budding.

12.1 Reproduction May Be Sexual or Asexual

In Chapter 2 we explored how genetic variation among individuals within a population arises from the shuffling of genes and chromosomes in sexual reproduction. In sexual reproduction, two individuals produce haploid (one-half the normal number of chromosomes) gametes—egg and sperm—that combine to form a diploid cell or zygote that has a full complement of chromosomes. Because the possible number of gene recombinations is enormous, recombination is an immediate and major source of genetic variability among offspring. However, not all reproduction is sexual. Many organisms reproduce asexually. Asexual reproduction produces offspring without the involvement of egg and sperm. It takes many forms but in all cases the new individuals are genetically the same as the parent. Strawberry plants spread by runners. The one-celled *Paramecium* reproduces by dividing in two. Hydras, coelenterates that live in fresh water (Figure 12.1), reproduce by budding; the bud pinches off as a new individual. In spring, wingless female aphids emerge from overwintering eggs and give birth to wingless females without fertilization, a process called **parthenogenesis** (Gk *parthenos* = virgin, L. *genesis* = to be born).

Organisms that rely heavily on asexual reproduction revert on occasion to sexual reproduction. Many of these reversions to sexual reproduction are induced by an environmental change at some time in their life cycle. During warmer parts of the year, hydras turn to sexual reproduction to produce overwintering eggs from which young hydras emerge to mature and reproduce asexually. After giving birth to several generations of wingless females, aphids produce a generation of winged females. These winged females migrate to different food plants, become established, and reproduce parthenogenetically. Later in the summer, these same females move back to the original food plants and give birth to true sexual forms, winged males and females that lay eggs rather than give birth to young. After mating, each female lays one or more overwintering eggs. Because the males produce sperm with the X chromosome only, the eggs that hatch in spring will produce wingless females that give birth to female young.

Each form of reproduction, asexual and sexual, has its tradeoffs. The ability to survive, grow, and reproduce indicates that an organism is well adapted to the prevailing environmental conditions. Asexual reproduction produces offspring that are genetically identical to the parent and are therefore well adapted to the local environment. Because all individuals are capable of reproducing, asexual reproduction results in a potential for high population growth. However, the cost of asexual reproduction is the loss of genetic recombination that increases variation among offspring. Low genetic variability among individuals in the population means that the population responds more uniformly to a change in environmental conditions than does a sexually reproducing population. If a change in environmental conditions is detrimental, the effect on the population can be catastrophic.

In contrast, the mixing of genes and chromosomes that occurs in sexual reproduction produces genetic variability to the degree that each individual in the population is genetically unique. This genetic variability produces a broader range of potential responses to the environment, increasing the probability that some individuals will survive environmental changes. But this variability comes at a cost. Each individual can contribute only one-half of its genes to the next generation. It requires organ systems that, aside from reproduction, have no direct relationship to an individual's survival. Production of gametes (egg and sperm), courtship activities, and mating are energetically expensive. The expense of reproduction is not shared equally by both sexes. The production of eggs by females is energetically much more expensive than the production of sperm by males. As we shall examine in the following sections, this difference in energetic investment in reproduction between males and females has important implications in the evolution of life history characteristics.

12.2 Sexual Reproduction Takes a Variety of Forms

Sexual reproduction takes a variety of forms. The most familiar involves separate male and female individuals. It is common to most animals. Plants with that characteristic are called **dioecious** (Greek *di*, "two," and *oikos*, "home"); examples are holly (*Ilex* spp.) trees and stinging nettle (*Urtica* spp.) (Figure 12.2a).

In some species, individual organisms possess both male and female organs. They are **hermaphroditic** (Greek *hermaphroditos*, from a son of Hermes and Aphrodite who became joined in one body with a nymph while bathing). In plants, individuals can be hermaphroditic by possessing bisexual flowers with both male organs (stamens) and female organs (ovaries), such as lilies and buttercups (Figure 12.2b). Such flowers are termed perfect. Asynchronous timing of the maturation of pollen and ovules reduces the chances of self-fertilization. Other plants are

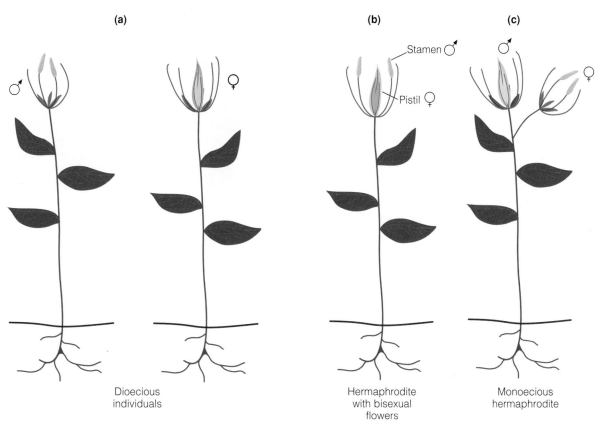

FIGURE 12.2 Floral structure in (a) dioecious plants (separate male and female individuals), (b) a hermaphroditic plant possessing bisexual flowers, and (c) a monoecious plant possessing separate male and female flowers.

FIGURE 12.3 Hermaphroditic earthworms mating.

FIGURE 12.4 Parrotfishes (Scaridae) that inhabit coral reefs exhibit sex change. When a large dominant male mating with a harem of females is removed (by a predator or experimenter), within days, the largest female in the harem becomes a dominant male and takes over the missing male's function.

monoecious (Greek *mon*, "one," and *oikos*, "house"). They possess separate male and female flowers on the same plant, as do birch (*Betula* spp.) and hemlock (*Tsuga* spp.) trees (Figure 12.2c). Such flowers are called imperfect. This strategy of sexual reproduction can be an advantage in the process of colonization. A single self-fertilized hermaphroditic plant can colonize a new habitat and reproduce, establishing a new population, as do self-fertilizing annual weeds that colonize disturbed sites.

Among animals, hermaphroditic individuals possess the sexual organs of both males and females (both testes and ovaries), a condition common in invertebrates such as earthworms (Figure 12.3). In these species, referred to as **simultaneous hermaphrodites,** the male organ of one individual is mated with the female organ of the other, and vice versa. The result is that a population of hermaphroditic individuals is able to produce two times as many offspring as a population of unisexual individuals.

Other species are **sequential hermaphrodites.** Animals—such as some mollusks and echinoderms—and some plants may be males first during one part of their life cycle and females in another part. Some fish may be females first, then males. Sex change usually takes place as individuals mature or grow larger. A change in the sex ratio of the population stimulates sex change among some animals. Removing individuals of the other sex initiates sex reversal among some species of marine fish (Figure 12.4). Among some coral reef fish, removal of females from a social group stimulates males to change sex and become females. In other species, removal of males stimulates a one-to-one replacement of males by sex-reversing females. Among the mollusks, the Gastropoda (snails and slugs) and Bivalvia (clams and mussels) have sex-

FIGURE 12.5 The jack-in-the-pulpit becomes asexual, male, or female depending on energy reserves. The plant gets its name from the flower stem or spadix enclosed in a hoodlike sheath. This fruiting plant is the female stage.

changing species. Almost all of these species change from male to female.

Plants also can undergo sex change. One such plant is jack-in-the-pulpit *(Arisaema triphyllum),* a clonal herbaceous plant found in woodlands (Figure 12.5). Jack-in-the-pulpit may produce male flowers one year, an asexual vegetative shoot the next, and female flowers the next. Over its life span, a jack-in-the-pulpit may produce both sexes as well as an asexual vegetative shoot, but in no particular sequence. Usually an asexual stage follows a sex change. Sex change in jack-in-the-pulpit appears to be triggered by the large energetic cost of producing female flowers. If the plant is to survive, it cannot produce female flowers in successive years.

12.3 Mating Strategies Take Several Forms

On a brushy rise of ground at the edge of a forest, a pair of red foxes has dug a deep burrow. Inside are the female and her litter of pups. Outside at the burrow entrance are scattered bits of fur and bones, the leftovers of meals carried to the den by the male for his mate and pups. Back in the woods, a female deer has hidden a young dappled fawn in a patch of ferns on the forest floor. The fawn's father is nowhere near and has no knowledge of this offspring's existence. His only interaction with the mother was during the previous fall, when she was one of a number of females with which he mated during a short period of several days.

The fox and the deer represent two extremes in what is referred to as the **mating system,** the pattern of mating between males and females in a population. The structure of mating systems ranges from **monogamy** which involves the formation of a lasting pair bond between one male and one female, to **promiscuity,** in which males and females mate with one or many of the opposite sex and form no pair bond.

Monogamy is most prevalent among birds and rare among mammals, with the exception of several carnivores such as foxes and weasels, and a few herbivores such as the beaver, muskrat, and prairie vole. It exists mostly among species in which cooperation by both parents is needed to raise the young successfully. Most species of birds are seasonally monogamous (during the breeding season), because most young birds are helpless at hatching and need food, warmth, and protection. The avian mother is no better suited to provide these needs than the father. Instead of seeking other mates, the male can increase his fitness more by continuing his investment in the young. Without him, the young carrying his genes may not survive. Among mammals, the situation is different. The females lactate (produce milk), providing food for the young. Males often can contribute little or nothing to the survival of the young, so it is to their advantage to mate with as many females as possible. Among the exceptions are the foxes, wolves, and other canids among which the male provides for the female and young and defends the territory. Both males and females regurgitate food for the weaning young.

Monogamy, however, has another side. Among numerous species of monogamous birds, the female or male may "cheat" by engaging in extra-pair copulations while maintaining the reproductive relationship with the primary mate and caring for the young. By engaging in extra-pair relationships, the female may increase her fitness by rearing young sired by two or more males. The male increases his fitness by having his offspring produced by several females.

Polygamy is the acquisition by an individual of two or more mates, none of which is mated to other individuals. It can involve one male and several females or one female and several males. A pair bond exists between the individual and each mate. The individual having multiple mates (be it male or female) is generally not involved in caring for the young. Freed from parental duty, the individual can devote more time and energy to competition for more mates and resources. The more unevenly such critical resources as food or quality habitat are distributed, the greater the opportunity for a successful individual to control the resource and several mates.

The number of the opposite sex an individual can monopolize depends upon the degree of synchrony in sexual receptivity. For example, if females in the population are sexually active for only a brief period, as with the white-tailed deer, the number a male can monopolize is limited. However, if females are receptive over a long period of time, as with elk, the size of a harem a male can control depends upon the availability of females and the number of mates the male has the ability to defend.

Environmental and behavioral conditions result in various types of polygamy. In **polygyny,** an individual male pairs with two or more females. In **polyandry,** an individual female pairs with two or more males. Polyandry is interesting because it is the exception rather than the rule. This system is best developed in jacanas, phalaropes, and sandpipers. The female competes for and defends resources essential for the male. In addition, females compete for available males. As in polygyny, this mating system depends upon the distribution and defensibility of resources, especially quality habitat. The female produces multiple clutches of eggs, each with a different male. After the female lays a clutch, the male begins incubation and becomes sexually inactive.

The nature and evolution of male-female relationships are influenced by environmental conditions, especially the availability and distribution of resources and the ability of individuals to control access to resources. If the male has no role in the feeding and protection of young and defends no resources available to them, the female gains no advantage by remaining with him; and likewise, the male gains no increase in fitness by remaining with the female. If the habitat is sufficiently uniform so that little difference exists in the quality of territories held by individuals, selection would favor monogamy because female fitness in all habitats (territories) would be nearly the same. However, if the habitat is diverse, with some parts more productive than others, competition may be intense, and some males will settle on poorer territories. Under such conditions, a female may find it more advantageous to join another female in the territory of the male defending a rich resource than to settle alone with a male on a poorer territory. Selection under those conditions will favor a polygamous relationship, even though the male may provide little aid in feeding the young.

12.4 Acquisition of a Mate Involves Sexual Selection

The flamboyant plumage of the peacock (Figure 12.6) presented a troubling problem for Charles Darwin. Its tail feathers are big and clumsy, and require a considerable allocation of energy to grow. In addition, they are very conspicuous and present a hindrance when a peacock is trying to escape predators. In the theory of natural selection (see Chapter 2), what could account for the peacock's tail? Of what possible benefit could it be?

In his book *The Descent of Man and Selection in Relation to Sex*, published in 1871, Darwin observed that the elaborate and often outlandish plumage of birds as well as the horns, antlers, and large size of polygamous males seemed incompatible with natural

FIGURE 12.7 This bull elk is bugling a challenge to other males in a contest to control a harem.

selection. To explain why males and females of the same species often differ greatly in body size, ornamentation, and color (referred to as **sexual dimorphism**), Darwin developed a theory of **sexual selection.** He proposed two processes to account for these differences between the sexes: intrasexual selection and intersexual selection.

Intrasexual selection involves male-to-male (or female-to-female, in the case of polyandry) competition for the opportunity to mate. It leads to exaggerated secondary sexual characteristics, such as large size, aggressiveness, and organs of threat such as antlers and horns (Figure 12.7) that aid in competition for access to mates.

Intersexual selection involves the differential attractiveness of individuals of one sex to another. Intersexual selection leads to characteristics in males such as bright or elaborate plumage used in sexual displays, as well as the elaboration of some of the same characteristics related to intrasexual selection (such as horns and antlers). There is intense rivalry among males for female attention. In the end, the female determines the winner, selecting an individual as a mate. But do such characteristics as bright coloration, elaborate plumage, and size really influence the selection of males by females of the species?

Marion Petrie of the University of Newcastle, England, conducted a number of experiments to examine intersexual selection in peacocks *(Pavo cristatus)*. She measured the tail length of male peacocks chosen by females as mates over the breeding season. Her results show that females selected males with larger tail feathers. She then removed tail feathers from a group of large-tailed males and found that reduced tail size led to a reduction in mating

FIGURE 12.6 Male peacock in courtship display.

success. In the case of the peacock's tail, size does matter.

But tail size in and of itself is not what is important; it is what tail size implies about the individual. The large, colorful, and conspicuous tail makes the male more vulnerable to predation, or in many other ways reduces the male's probability of survival. The fact that a male can carry these handicaps and survive is an honest appraisal of his health, strength, and genetic superiority. Females showing preference for males with large tail feathers will produce offspring that will carry genes for high viability. Thus, the driving force behind the evolution of exaggerated secondary sexual characteristics in males is selection by females. In fact, the offspring of female peacocks paired with males having large tail feathers had higher rates of survival and growth than the offspring of those paired with males having short tails. A similar mechanism may be at work in the selection of male birds with bright plumage. One hypothesis proposes that only healthy males can develop bright plumage. There is evidence from some species that males with low parasitic infection have the brightest plumage. Females selecting males based on differences in the brightness of plumage are in fact selecting males that are the most disease resistant, therefore increasing their fitness.

12.5 Females May Acquire Mates Based on Resources

A female exhibits two major approaches in choosing a mate. In the sexual selection discussed above, the female selects for characteristics such as exaggerated plumage or displays that are indirectly related to the health and quality of the male as a mate. The second approach is that the female can base her choice on resources such as habitat or food that will improve her fitness.

For monogamous females, the criterion for mate selection appears to be acquisition of a resource, usually a defended high-quality habitat or territory (see Section 11.10). Does the female select the male and accept the territory that goes with him, or does she select the territory and accept the male that goes with it? There is some evidence from laboratory and field studies that female songbirds base their choice, in part, on the variety of the male's song. In aviary studies, female great tits (Parus major) were more receptive of males with more varied or elaborate songs. In the field, male sedge warblers (Acrocephalus schoenobaenus) with more complex song repertoires

appear to hold the higher-quality territories, so the more complex song may convey that fact to the female. None of these studies, however, determined the fitness of females attracted to these males.

Among polygamous species, the question becomes more complex. In those cases in which females acquire a resource along with the male, the situation is similar to that of monogamous relationships. Among polygamous birds, for example, the females show strong preference for males with high-quality territories. On territories with superior nesting cover and an abundance of food, females can attain reproductive success even if they share the territory and the male with other females.

When polygamous males offer no resources, it would seem that the female has limited information upon which to act. She might select a winner from among males that best others in combat—as in bighorn sheep, elk, and seals. She might select mates by intensity of courtship display or some morphological feature that may reflect a male's genetic superiority and vitality.

Among some polygamous species, it is hard to see female choice at work. In elk and seals, a dominant male takes charge of a harem of females. (Figure 12.7). Nevertheless, the females express some choice. Protestations by female elephant seals over the attention of a dominant male may attract other large males nearby, who attempt to replace the male. Such behavior ensures that females will mate only with the highest-ranking male.

In other situations, females seem clearly in control. Males put on an intense display for them. Such advertising can be costly for courting polygamous males. Because of conspicuous behavior and inattention, they may be subject to intense predation.

Outstanding examples of female choice appear in lek species. These animals aggregate into groups on communal courtship grounds called leks or arenas. Males on the lek defend small territories that hold no resources and advertise their presence by colorful vocal and visual displays. Females visit the leks of displaying males, select a male, mate, and move on. Although few species engage in this type of mating system, it is widespread in the animal world, from insects to frogs to birds and mammals. Males defend small, clustered mating territories, whereas females have large overlapping ranges that the males cannot economically defend. Leks provide an unusual opportunity for females to choose a mate among the displaying males. Congregating about dominant males with the most effective displays, subdominant and satellite males may steal mating opportunities. A majority of matings on the lek, however, is done by a small percentage of the males in a dominance hierarchy formed in the absence of females.

12.6 Mating in a Small Population Can Result in Inbreeding

Populations, however widespread, consist of rather tight-knit local groups or demes. When such populations are small, the choice of mates can be limited, creating the potential for inbreeding. **Inbreeding** is mating between relatives. With inbreeding, mates on average are more closely related than they would be if they had been chosen at random from the population. Inbreeding occurs because populations are small or the potential mates that are genetically related are in close proximity.

The extreme form of inbreeding is self-fertilization, which occurs among some plants. With each generation of inbreeding, the homozygotes AA and aa breed true (see Chapter 2, Section 2.3). Offspring from the heterozygotes Aa will be one-half heterozygous Aa, one-quarter homozygous AA, and one-quarter homozygous aa. The same situation will occur in the next generation. Add these new homozygotes to those already in the population, and eventually self-fertilized populations become exclusively homozygous AA and aa. The more usual situation is close inbreeding, usually between brother and sister, parent and offspring, or first cousins, all individuals who share a number of like genes. Thus, inbreeding increases homozygosity and decreases heterozygosity (see Chapter 2).

Inbreeding can be detrimental. Rare, recessive, deleterious genes become expressed. They can cause decreased fertility, loss of vigor, reduced fitness, reduced pollen and seed fertility in plants, and even death. These consequences are referred to as **inbreeding depression.**

Of course, not all inbreeding is bad. Occasional inbreeding will fix rare alleles that otherwise might be lost. Animal and plant breeders use inbreeding to select for certain desirable genes (see the Ecological Application after Chapter 2: Taking the Uncertainty out of Sex.)

Except for self-fertilizing plants, close inbreeding in nature is rare. It amounts to less than 2 percent for natural populations for which there are data. Certain safeguards in nature reduce inbreeding. They include spatial separation or differences between sexes in the dispersal of young. One sex stays behind; the other leaves. Monogamous mating habits reduce inbreeding in birds. Female mammals tend to stay near their birthplace, whereas young males leave or are driven away. Among prairie dogs *(Cynomys ludovicianus)*, young males leave the family group before breeding, and adult males may move to new breeding groups when their daughters mature. Although dispersal does reduce inbreeding, sex-biased dispersal may have evolved for other reasons, such as enhanced reproductive success.

Kin recognition may also reduce close inbreeding. Because of their close association in early life, siblings in some species, such as prairie dogs, recognize one another over time. Females mate with unrelated males, leave the group if a related male returns, or fail to come into estrus (reproductive receptivity) if their father is in the group. All these safeguards break down if the population is small and highly isolated.

Outcrossing or **outbreeding**—mating between unrelated individuals—can mitigate the adverse effects of inbreeding. Although it adds genetic diversity, outbreeding can carry its own problems. Long-distance dispersal can introduce individuals to a population that may not be well adapted to local conditions. Offspring that are parented by individuals of two different local populations may be poorly adapted to the local environment of either parent. For example, bobwhite quail from the warm southern United States were introduced to the northern United States to augment small populations of cold-adapted northern bobwhite quail. Poorly adapted to the cold weather, the "hybrid" offspring succumbed to winter, with the eventual loss of the affected quail population. Such maladaptation of the offspring is referred to as **outbreeding depression.**

12.7 Organisms Budget Time and Energy to Reproduction

Organisms spend their energy, like income in a household or a business, to meet many needs. Some energy must go to growth, to maintenance, to acquiring food, to defending territory, and to escaping predators. Energy must also be allocated to reproduction. The time and energy allocated to reproduction make up an organism's **reproductive effort.**

The more energy an organism spends on reproduction, the less it can allocate for growth and maintenance. For example, reproducing females of the terrestrial isopod *Armadillidium vulgare* have a lower rate of growth than nonreproducing females. Nonreproducing females devote as much energy to growth as reproducing females devote to both growth and reproduction.

The amount of energy organisms invest in reproduction varies. The investment includes not only the production of offspring, but also the costs of care and nourishment. Herbaceous perennials invest between 15 and 20 percent of annual net production to reproduction, including vegetative propagation. Wild annuals that reproduce only once expend 15 to 30 percent; most grain crops, 25 to 30 percent; and corn and barley, 35 to 40 percent. The lizard *Lacerta vivipara* invests 7 to 9 percent of its annual energy assimilation to reproduction. The female Allegheny Mountain salamander *Desmognathus ochophaeus* spends 48 percent of its annual energy budget on reproduction, including energy stored in eggs and energy costs of brooding.

How often can an organism afford to reproduce? One approach is to initially invest all energy into growth, development, and energy storage, followed by one massive reproductive effort, and death. In this strategy, an organism sacrifices future prospects by expending all its energy in one suicidal act of reproduction. This mode of reproduction is called **semelparity.**

Semelparity is employed by most insects and other invertebrates, by some species of fish, notably salmon, and by many plants. It is common among annuals, biennials, and some species of bamboos. Many semelparous plants, such as ragweed, are small, short-lived, and found in ephemeral or disturbed habitats. For them, it would not pay to hold out for future reproduction, for the chances are slim. They gain their maximum fitness by expending all their energies in one bout of reproduction.

Other semelparous organisms, however, are long-lived and delay reproduction. Mayflies (Ephemeroptera) may spend several years as larvae before emerging from the surface of the water for an adult life of several days devoted to reproduction. Periodical cicadas spend 13 to 17 years below ground before they emerge as adults to stage a single outstanding exhibition of reproduction. Some species of bamboo delay flowering for 100 to 120 years, produce one massive crop of seeds, and die. The Hawaiian silverswords (*Argyroxiphium* spp.) live 7 to 30 years before flowering and dying. In general, species that evolve semelparity have to increase fitness enough to compensate for the loss of repeated reproduction.

Organisms that produce fewer young at one time and repeat reproduction throughout their lifetime are called **iteroparous.** Iteroparous organisms include most vertebrates, perennial herbaceous plants, shrubs, and trees. For an iteroparous organism the problem is timing reproduction—early in life or later. Whatever the choice, it involves tradeoffs. Early reproduction means less growth, earlier maturity, reduced survivorship, and reduced potential for future reproduction. Later reproduction means increased growth, later maturity, and increased survivorship, but less time for further reproduction. In effect, if an organism is to make a maximum contribution to future generations, it has to balance the profits of immediate reproduction against future prospects, including the cost to fecundity (total offspring produced) and its own survival.

12.8 Parental Investment Depends upon the Number and Size of Young

In theory, a given allocation to reproduction can potentially produce many small young or fewer large ones. The number of offspring affects the parental investment each receives. If the parent produces a large number of young, it can afford only minimal investment in each one. In such cases, animals provide no parental care, and plants store little food energy in seeds (Figure 12.8). Such organisms usually inhabit disturbed sites, unpredictable environments, or places such as the open ocean where opportunities for parental care are difficult at best. By dividing energy for reproduction among as many young as possible, these parents increase the chances that some young will successfully settle somewhere and reproduce in the future.

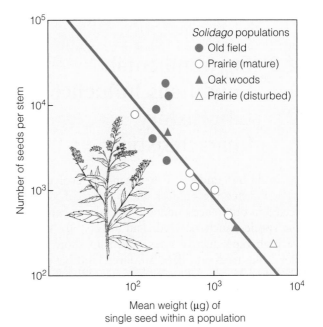

FIGURE 12.8 Inverse relationship between mean weight of individual seeds and the number of seeds produced per stem for populations of goldenrod *(Solidago)* in a variety of habitats. (Werner and Platt 1976.)

Parents that produce few young are able to expend more energy on each individual. The amount of energy will vary with the number of young, their size, and their maturity at birth. Some organisms expend less energy during incubation. The young are born or hatched in a helpless condition and require considerable parental care. These animals, such as young mice or nestling robins, are **altricial.** Other animals have longer incubation or gestation, so the young are born in an advanced stage of development. They are able to move about and forage for themselves shortly after birth. Such young are called **precocial.** Examples are gallinaceous birds, such as chickens and turkeys, and ungulate mammals, such as cows and deer.

The degree of parental care varies widely. Some species of fish, such as cod *(Gadus morhua)*, lay millions of floating eggs that drift freely in the ocean with no parental care. Other species, such as bass, lay eggs in the hundreds and provide some degree of parental care. Among amphibians, parental care is most prevalent among tropical toads and frogs and some species of salamanders. Among reptiles, crocodiles are an exception. They actively defend the nest and young for a considerable time. Invertebrates exhibit parental care to varying degrees. Octopus; crustaceans such as lobsters, crayfish, and shrimp; and certain amphipods such as millipedes, brood and defend eggs. Parental care developed best among the social insects: bees, wasps, ants, and termites. Social insects perform all functions of parental care, including feeding, defense, heating and cooling, and sanitation.

12.9 Environmental Conditions Influence Life History Characteristics

The conditions under which organisms evolve and adapt influence the number and the size of young, reflected in differences among reproductive patterns. Some species, such as weeds and insects, are small, have high reproductive rates, and lead short lives. Others, like trees and deer, are large, and have low reproductive rates and long lives.

Among plants, short-lived annuals have numerous small seeds. High production of numerous offspring ensures that some will survive and germinate the next growing season. Optimum seed size and number for perennial plants relate to dispersal ability, colonizing ability, and the need to escape predation. Plants that colonize disturbed environments that are widely distributed or available for only a short period of time may have small wind-blown seeds that carry great distances. Because they invest minimal energy in each seed, these plants can afford heavy losses of seeds. Plants subject to heavy predation may have small seeds that provide a less attractive source of food. Again an abundance of seeds is an insurance that some will escape predation (see Chapter 15).

Plants associated with more stable environments may produce fewer and larger seeds, with a large store of energy that the seedling can use to become established. Such plants may invest a considerable amount of energy in toxins or heavy seed coats to reduce predation.

The plant ecologist J. Phillip Grime of Sheffield, England, developed a life history classification for plants based on three primary strategies *(R, C,* and *S)* that relate plant adaptations to different habitats (Figure 12.9). Species exhibiting the *R,* or ruderal, strategy rapidly colonize disturbed sites but are small in stature and short-lived. Allocation of resources is primarily to reproduction, with characteristics allowing for a wide dispersal of seeds to newly disturbed sites. Predictable habitats with abundant resources favor species that allocate resources to growth, favoring resource acquisition and competitive ability (*C* species). Habitats where resources are limited favor stress-tolerant species (*S*-species) that allocate resources to maintenance. These three strategies form the end points of a triangular classification system that allows for intermediate strategies, depending upon such environmental factors as resource availability and frequency of disturbance.

The population ecologists R. MacArthur (late) of the University of Pennsylvania, E. O. Wilson of Harvard University, and later E. Pianka of the University of Texas arrived at a different approach and called these types *r* species and *K* species. Their theory of *r*- and *K*-selection predicts that species in different environments will differ in life history traits such as size, fecundity, age at first reproduction, number of reproductive events during a lifetime, and total life span. Species popularly known as *r*-strategists are typically short-lived. They have high reproductive rates at low population densities, rapid development, small body size, large number of offspring (but with low survival), and minimal parental care. They make use of temporary habitats. Many inhabit unstable or unpredictable environments that can cause catastrophic mortality independent of population density. For these species, environmental resources are rarely limiting. They exploit non-competitive situations. Adaptable *r*-strategists, such as weedy species, have means of wide dispersal, are good colonizers, and respond rapidly to disturbance.

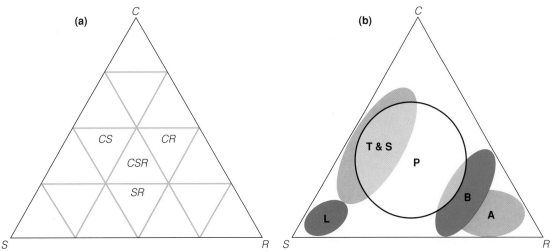

FIGURE 12.9 Grime's model of life history variation in plants based on three primary strategies: ruderals (R), competitive (C), and stress-tolerant (S). (a) These primary strategies define the three points of the triangle. Intermediate strategies are defined by combinations of the three (e.g., CS, CR, CSR, and SR). (b) Grime's assessment of life history strategies of most trees and shrubs (T&S), lichens (L), biennial herbs (B), perennial herbs (P), and annual herbs (A). (Grime 1977.)

K-strategists are competitive species with stable populations of long-lived individuals. They have a slower growth rate at low populations, but they maintain that growth rate at high densities. *K*-strategists can cope with physical and biotic pressures. They possess both delayed and repeated reproduction, and have a larger body size and slower development. They produce few seeds, eggs, or young. Among animals, parents care for the young; among plants, seeds possess stored food that gives the seedlings a strong start. *K*-strategists' mortality relates more to density than to unpredictable environmental conditions. They are specialists, efficient users of a particular environment, but their populations are at or near carrying capacity and are resource-limited. These qualities, combined with their lack of means of wide dispersal, make *K*-strategists poor colonizers of new and empty habitats.

Although it is tempting to classify organisms as either *r*-strategists or *K*-strategists, it is difficult to force species into such a classification. At some time in its life history, the same species may exhibit *r* traits or *K* traits. Meadow mice living in environments where dispersal can easily take place exhibit characteristics of an *r* species, whereas those with no place to disperse assume the characteristics of a *K* species. White-tailed deer exhibit *r* traits when a population spreads into new habitat or is greatly reduced by hunting. At high densities, however, the population exhibits *K* traits, with low overall recruitment of young. The concept of *r* species and *K* species is most useful to compare organisms of the same type, individuals within a population, or populations within a species.

12.10 Food Supply Affects the Production of Young

Within a given region, production of young may reflect the abundance of food. In environments where the availability of food resources is highly variable through time, the number of offspring that can physiologically be produced may be greater than the number that can be provided for during certain years. Under these circumstances, it may be necessary to reduce the number of young. In many species of birds, asynchronous hatching and **siblicide**, the killing of one sibling by another, function to reduce the number of offspring.

In asynchronous hatching, the young are of several ages. The older siblings beg more vigorously for food, forcing the harried parents to ignore the calls of the younger, smaller sibling, which perishes. For example, the common grackle *(Quiscalus quiscula)* begins incubation before the entire clutch of 5 eggs has been laid. The eggs laid last are heavier, and the young from them grow fast. However, if food is scarce, the parents fail to feed these late offspring because of more vigorous begging by the larger siblings. The last-hatched young then die of starvation. Thus, asynchronous hatching favors the early-hatched young, ensuring the survival of some siblings under adverse conditions.

In other situations, the older or more vigorous young simply kill their weaker siblings. A number of

FIGURE 12.10 Example of siblicide in which the larger of two offspring kills the smaller sibling. Masked booby parents collaborate by ignoring the battle between their offspring.

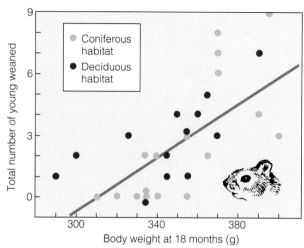

FIGURE 12.11 Lifetime reproductive success of the European red squirrel is correlated to body weight in the first winter as an adult. (Wauters and Dhondf 1989.)

birds, including raptors, herons, egrets, gannets, boobies, and skuas, practice siblicide. The parents normally lay 2 eggs, possibly to insure against infertility of a single egg. The larger of the 2 hatchlings kills the smaller sibling or runt, and the parents redirect all resources to the surviving chick (Figure 12.10). These birds are not alone in siblicidal tendencies. The females of some parasitic wasps lay 2 or more eggs in a host, and the larvae fight each other until only one survives.

12.11 Fecundity Depends upon Age and Size

For many species, the number of offspring produced varies with the age and size of the parent. This relationship is especially strong in plants and ectothermic (cold-blooded) animals. Among plants, perennials delay flowering until they have attained a sufficiently large size (and leaf area) to support seed production. Many biennials in poor environments also delay flowering beyond the usual two-year life span until environmental conditions become more favorable. Annuals show no relationship between size and the percentage of energy devoted to reproductive output; as a result, size differences among annuals result in differences in the number of seeds produced. Small plants produce fewer seeds, even though the plants may be contributing the same share of energy to reproduction as larger plants.

Similar patterns exist among poikilothermic (cold-blooded) animals. Fecundity in fish increases with size, which increases with age. Because early fecundity reduces both growth and later reproductive success, fish obtain a selective advantage by delaying sexual maturation until they grow larger. Gizzard shad *(Dorosoma cepedianum)* reproducing at 2 years of age produce about 59,000 eggs. Those delaying reproduction until the third year produce about 379,000 eggs. Among the gizzard shad, only about 15 percent spawn at 2 years of age and about 80 percent at 3 years. The number of eggs produced by loggerhead sea turtles *(Caretta caretta)* is constrained by the female's egg-carrying capacity, which is related to body size.

An apparent relationship also exists between body size and fecundity among endotherms (warm-blooded animals). Heavier females are more successful in reproduction, and more of their young survive. For example, the body weight of female European red squirrels *(Sciurus vulgaris)* in Belgian forests is strongly correlated with lifetime reproductive success (Figure 12.11). Few squirrels weighing less than 300 grams reproduced.

12.12 Reproductive Effort May Vary with Latitude

Birds in temperate regions have larger clutch sizes (the number of eggs produced) than do those in the tropics (Figure 12.12), and mammals at higher latitudes have larger litters than those at lower latitudes. Lizards exhibit a similar pattern. Those living at

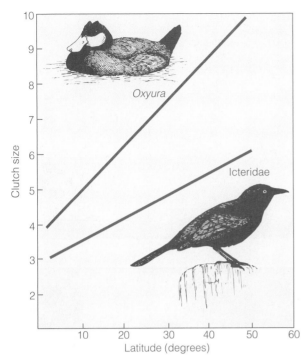

FIGURE 12.12 The relationship between clutch size and latitude in birds. Represented are the subfamily Icteridae (blackbirds, orioles, and meadowlarks) in North and South America and the worldwide genus *Oxyura* (ruddy and masked ducks) of the subfamily Anatinae. (Cody 1966.)

lower latitudes have smaller clutches, have higher reproductive success, reproduce at an earlier age, and experience higher adult mortality than those living at higher latitudes.

Insects, too, such as the milkweed beetle (*Oncopeltus* spp.), exhibit a latitudinal pattern in reproductive effort. Temperate and tropical milkweed beetles have a similar duration of the egg stage, egg survivorship, developmental rate, and age at sexual maturity. Although the clutch sizes are the same, the temperate species lay a larger number of eggs because they lay more clutches. Total egg production of the tropical species is only 60 percent of that of the temperate species.

Plants also follow the general principle of allocation on a latitudinal basis. For example, three species of cattail grow in North and Central America. The common cattail *(Typha latifolia)* grows under a broad range of climates, with a distribution extending from the Arctic Circle to the equator. The narrow-leaved cattail *(T. angustifolia)* is restricted to the northern latitudes. The southern cattail *(T. domingensis)* grows only in the southern latitudes. These three cattails show a climatic gradient in the allocation of energy to reproduction. *T. angustifolia* and the northern populations of *T. latifolia* grow earlier and faster than *T. domingensis*, and produce a greater number of

rhizomes (ramets). The southern populations produce larger rhizomes.

Why might this pattern of geographic variation in reproductive allocation occur? David Lack, an English ornithologist, proposed that clutch size in birds has evolved to equal the largest number of young the parents can feed. Thus, clutch size is an adaptation to food supply. Temperate species, he argued, have larger clutches because increasing daylength during the spring provides a longer time to forage for food to support large broods. In the tropics, where daylength is roughly 12 hours, foraging time is more limited.

An American ecologist, Martin Cody, modified Lack's ideas by proposing that clutch size results from different allocations of energy to egg production, avoidance of predators, and competition. In temperate regions, periodic local climatic catastrophes (such as a very harsh winter or summer drought) can hold a population below the level that the resources could support. Organisms respond with a higher rate of increase and larger clutches. In tropical regions, with predictable climates and increased probability of survival, there is no need for extra young.

A third hypothesis, proposed by N. P. Ashmole, states that clutch size varies in direct proportion to the seasonal variation in resources, especially food. Population density is regulated primarily by mortality in winter, when resources are scarce. Greater winter mortality means more food for the survivors during the breeding season. This abundance is reflected in larger clutches. Thus, geographical variation in mean clutch size and the size of the breeding population is inversely related to winter food supply.

Although reproductive output among organisms does appear to be greater at higher latitudes, the number of comparable species for which there are data is too small to confirm the hypotheses. Many more studies along a latitudinal gradient are needed.

12.13 Habitat Selection Influences Reproductive Success

Reproductive success depends heavily on choice of habitat. Settling on a less-than-optimal habitat can result in reproductive failure. The process in which organisms actively choose a specific location to inhabit is called **habitat selection.** Given the importance of habitat selection on an organism's fitness, how are they able to assess the quality of an area in which they settle? What do they seek in a living place? Such questions have been intriguing ecologists for many years. Understanding the relationship between habitat selection, reproductive success, and population dynamics has become particularly important in light of the loss of habitats for many species as a result of human impacts on the landscape (see Chapter 10, Section 10.14, and Ecological Application: Asteroids, Bulldozers, and Biodiveristy.).

Habitat selection has been most widely studied in birds, particularly those species that defend breeding territories (see Chapter 11, Section 11.10). Territories can be delineated, and features of the habitat can be described and contrasted with the surrounding environment. Of particular importance is the ability to contrast those areas that have been chosen as habitats with adjacent areas that have not. Using this approach, a wide variety of studies have demonstrated a strong correlation between structural features of vegetation and its selection as habitat. These studies suggest that habitat selection most likely involves a hierarchical approach. Birds appear to initially assess the general features of the landscape—the type of terrain; presence of lakes, ponds, streams, and wetlands; gross features of the vegetation such as open grassland, shrubby areas, and types and extent of forests. Once in a broad general area, the birds respond to more specific features, such as the structural configuration of vegetation, particularly the presence or absence of various vertical layers such as shrubs, small trees, tall canopy, and degree of patchiness (Figure 12.13). Frances James, an avian ecologist at Florida State University, coined the term "niche gestalt" to describe the vegetation profile associated with the breeding territory of a particular species.

In addition to the physical structure of the vegetation, the actual plant species present can also be important. Certain species of plants might produce preferred food items, such as seeds and fruits, or influence the type and quantity of insects available as prey for insectivorous birds.

The structural features of the vegetation that define its suitability for a given species may be related to a variety of specific needs, such as food, cover, and nesting sites. The lack of song perches may prevent some birds from colonizing an otherwise suitable habitat. An adequate nesting site is another requirement. Animals require sufficient shelter to protect themselves and young against enemies and adverse weather. Cavity-nesting animals require suitable dead trees or other structures in which they can construct cavities. In areas where such sites are absent, populations of birds and squirrels can be increased dramatically by providing nest boxes and den boxes.

(a)   Yellowthroat

(b)   Hooded warbler

(c)   Ovenbird

FIGURE 12.13 The vegetation structure that characterizes the habitat of three neotropical warblers: (a) Yellowthroat *(Geothlypis trichas),* a bird of shrubby margins of woodland and wetlands and brushy fields. (b) Hooded warbler *(Wilsonia citrina),* a bird of small forest openings. (c) Ovenbird *(Seiurus aurocapillus),* an inhabitant of deciduous or mixed conifer-deciduous forests with open forest floor. The labels 9–10 m, 18–30 m, and 15–30 m, refer to height of vegetation. (James 1971.)

Habitat selection is a common behavioral characteristic of a wide variety of vertebrates other than birds; fish, amphibians, reptiles, and mammals furnish numerous examples. Garter snakes (*Thamnophis elegans*) living along the shores of Eagle Lake in the sagebrush-ponderosa pine country of northeastern California select rocks of intermediate thickness (20 to 30 cm) over thinner and thicker rocks as their retreat sites. Shelter under thin rocks becomes lethally hot; shelter under thicker rocks does not allow the snakes to warm to their preferred range of body temperature (T_b; see Chapter 8). Under the rocks of intermediate thickness, snakes are able to achieve and maintain their preferred body temperature for a long period. Insects, too, cue in on habitat features. Thomas Whitham studied habitat selection by the gall-forming aphid *Pemphigus*, which parasitizes the narrow-leaf cottonwood (*Populus angustifolia*). He found that aphids select the largest leaves to colonize and discriminate against small leaves. Beyond that, they select the best positions on the leaf. Occupying this particular habitat, which provides the best food source, produces individuals with the highest fitness.

Even though a given habitat may provide suitable cues, it still may not be selected. The presence or absence of others of the same species may influence individuals to choose or avoid a particular site. In social or colonial species like herring gulls, an animal will choose a site only if others of the same species are already there. On the other hand, the presence of predators and human activity may discourage a species from occupying an otherwise suitable habitat.

Most species exhibit some flexibility in habitat selection. Otherwise, these animals would not settle in what appears to us as a less-suitable habitat or colonize new habitats. Often, individuals are forced to make this choice. Available habitats range from optimal to marginal; the optimal habitats, like good seats at a concert, fill up fast. The marginal habitats go next, and latecomers or subdominant individuals are left with the poor habitats where they may have little chance of reproducing successfully.

Do plants select habitats, and if so how? Plants can hardly get up and move about to find a suitable site. Plants, like animals, fare better in certain habitat types, characterized by such environmental factors as light, moisture, nutrients, and presence of herbivores. The only recourse plants have in habitat selection is through the evolution of dispersal strategies that influence the probability that a seed will arrive at a place suitable for germination and seedling survival. Habitat selection involves the ability of plants to disperse with the aid of wind, water, or animal agents to preferred patches of habitat, which more often than not involves an element of chance.

12.14 Life History Characteristics Drive Population Dynamics

An organism's life history—its lifetime pattern of growth, development, and reproduction—directly influences its demography—the age structure, sex ratio, and patterns of population growth. In Chapter 10 we examined how population growth was influenced by age-specific patterns of survival and reproduction. As we have seen in this chapter, these patterns of survival and fecundity are key features of the species' life history. The method of reproduction (sexual or asexual), mating system, reproductive allocation, and number of offspring produced for a given allocation of energy all interact to influence age-specific patterns of fecundity. The reproductive allocation and degree of parental care will be reflected in patterns of survivorship (Chapter 10, Section 10.6). The production of a large number of offspring with little parental care will result in high rates of juvenile mortality (Type III survivorship curve, Figure 10.4) relative to species producing fewer young and investing in their care and nourishment. Life history characteristics are not independent of each other; they form an interrelated set of characteristics that are the product of natural selection, often reflecting constraints imposed by the prevailing environmental conditions. And as we will see in the following chapters, a species' life history characteristics will also influence the nature of its interactions with other species.

Summary

Asexual and Sexual Reproduction (12.1–12.3) The ability of an organism to leave behind reproducing offspring is its fitness. Organisms that contribute the most offspring to the next generation are the fittest. Reproduction can be asexual or sexual. Asexual reproduction or cloning results in new individuals that are genetically the same as the parent. Sexual reproduction combines egg and sperm in a diploid cell or zygote. Sexual reproduction produces genetic variability among offspring (**12.1**).

Sexual reproduction takes a variety of forms. Plants with separate males and females are called dioecious. An organism possessing both male and female sex organs is hermaphroditic. Plant hermaphrodites have bisexual flowers or, if they are monoecious, separate male and female flowers on the same individual. Some plants and animals change gender (**12.2**). Mating systems include two basic types—monogamy and polygamy. In polygyny, the male acquires more than one female; in polyandry, the female acquires more than one male. The potential for competitive mating and sexual selection is higher in polygamy than in monogamy (**12.3**).

Sexual Selection (12.4–12.5) Selection of a proper mate is essential if an organism is to contribute to the next generation. An important component of mating strategy is sexual selection. In general, males compete with males for the opportunity to mate with females, but females finally choose mates. Sexual selection favors traits that enhance mating success, even if they handicap the male by making him more vulnerable to predation. Male competition is intrasexual selection, while intersexual selection involves the differential attractivemenss of individuals of one sex to another. By choosing the best males, females ensure their own fitness (**12.4**). Females may also choose mates based on the acquisition of resources, usually a defended territory or habitat. By choosing a male with a high-quality territory, the female may increase her fitness (**12.5**).

Inbreeding (12.6) Small population size may result in inbreeding, the mating of related individuals. Inbreeding functions to increase homozygosity in a population. The increase in homozygosity can increase the expression of rare, deleterious genes, a consequence referred to as inbreeding depression. Certain safeguards have evolved to reduce inbreeding, including kin recognition and dispersal of young.

Energy Investment in Reproduction (12.7–12.8) To maximize fitness, an organism has to balance immediate reproductive efforts against future prospects. One alternative, semelparity, is to invest maximum energy in a single reproductive effort. The other alternative, iteroparity, is to allocate less energy each time to repeated reproductive efforts (**12.7**). The number of young produced relates to the parental investment. The amount of time and energy parents allot to reproduction is reproductive effort. Organisms that produce a large number of offspring have a minimal investment in each offspring. They can afford to send a large number into the world with a chance that a few will survive. By so doing, they increase parental fitness but decrease the fitness of the young. Organisms that produce few young invest considerably more in each one. Such organisms increase the fitness of the young at the expense of the fitness of the parents (**12.8**).

Environmental Influences (12.9–12.12) Organisms living in variable or ephemeral environments or facing heavy predation produce numerous offspring, ensuring that some will survive. A large number of young is characteristic of annual plants, short-lived mammals, insects, and semelparous species. Having few young is characteristic of long-lived species. Iteroparous species may adjust the number of young in response to environmental conditions and the availability of resources (**12.9**). Production of young often reflects the availability of food. In times of food scarcity, parents may fail to feed some offspring. In other situations, vigorous young kill their weaker siblings (**12.10**).

A direct relationship between size and fecundity exists among plants and ectotherms. The larger the size, the more young are produced. Among endotherms, too, heavier females are more successful in reproduction (**12.11**).

In general, clutch and litter sizes increase from the tropics to the poles. This gradient may reflect length of daylight, which influences foraging time, or the more stable climate in the tropics (**12.12**).

Habitat Selection (12.13) Reproductive success depends heavily on the choice of habitats. Habitat selection is partly genetic and partly psychological. Most studies of habitat selection have focused on birds that defend breeding territories. Results suggest that habitat selection involves a hierarchical approach, with the initial selection based on general features of the landscape; within this area, individuals respond to specific features of the vegetation or habitat.

Life History Characteristics and Population Dynamics (12.14) An organism's life history characteristics influence its demography—age structure, sex ratio, and population dynamics. Life history characteristics are not independent of each other; they form an interrelated set of characteristics that are the product of natural selection, often reflecting constraints imposed by prevailing environmental conditions.

Study Questions

1. What is fitness?
2. Distinguish between sexual and asexual reproduction.
3. Contrast dioecious, monoecious, and hermaphroditism.
4. What are some advantages of hermaphroditism?

5. What is a mating system? Distinguish among monogamy, polygamy, polygyny, and polyandry.

6. What is sexual selection? In what ways does it differ for males and females?

7. Distinguish between intrasexual selection and intersexual selection.

8. What is the driving force in the evolution of secondary sexual characteristics in males?

9. What is involved in reproductive effort?

10. How does energy allocation affect reproduction and growth?

11. What are semelparity and iteroparity, and what conditions favor each?

12. What are the advantages of early and late reproduction?

13. Under what conditions should parents have many young or few young?

14. Compare investments in atricial and precocial young.

15. How might resource availability affect reproduction?

16. Explain the relationship between size and fecundity.

17. What is the difference between r-selected and K-selected organisms? Between R, C, and S plant strategies?

Cheating Nature

By the beginning of the 19th century, England was reaping both the benefits and the consequences of the Industrial Revolution. Between 1800 and 1880, the population of London swelled from 1 million to 4.5 million, and the environment of London became intolerable. Streets were covered with manure from thousands of horse-drawn carriages; raw sewage flowed into the gutters and emptied into the Thames River, which provided drinking water to much of the population. In the ever-increasing population of London, particularly the poor, Thomas Malthus, a political economist and historian, saw disaster lurking. Malthus believed that unless actions were taken to reduce the rate of growth, the human population was on an inevitable course of self-destruction. In *An Essay of the Principle of Population,* first published in 1798, Malthus discussed the tendency for populations to grow geometrically. The inevitable consequence, he believed, was that populations would eventually expand beyond the capacity of the environment to support their number. Malthus saw a rising mortality rate—death—as the inevitable consequence and principle control on population growth.

Two hundred years later, on October 12, 1999, the United Nations declared that the human population had reached the milestone of 6 billion. Over the preceding two centuries, human population had grown geometrically, increasing over sixfold (Figure A), thus proving Malthus correct on one front. The human population, however, had not exceeded the capacity of the environment to support its numbers. How could this be? The answer lay in the unique human ability to cheat nature by modifying both the environment and the manner in which we interact with it, and in doing so, redefining the constraints that it imposes upon our species.

The incredible rise in population size over the past 200 years masks significant changes that occurred during the previous millennia. By plotting the same information using a logarithmic scale ($\log_{10}$), a much richer picture of human history is exposed (Figure B). The history of human population growth encompasses three great technological-cultural phases: hunter-gatherer, agricultural, and industrial or modern. During these three phases, each of which was shorter than its predecessor, population has increased, with growth rates eventually slowing as the limits of human growth were approached. These limits were imposed by climate, food, space, energy, and disease. In each of these phases, humans redefined their relationship with the environment, eventually breaking down the equilibrium between population and resources, and setting the stage for a new period of growth.

During the first phase of human history, dependence on plants and animals for energy was the major constraint on human population growth. The hunter-gatherer societies that existed prior to 10,000 BP consisted of small, autonomous bands of a few hundred individuals. Humans were dependent on the productivity and abundance of plants and animals that made up natural ecosystems and on their ability to extract and utilize those natural resources.

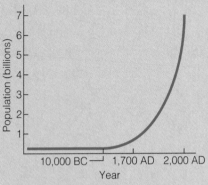

FIGURE A Human population growth.

People were nomadic, tracking their resource in space and time, and vulnerable to environmental change.

The Neolithic period (8000–5000 B.C.) saw the development of agriculture—the cultivation of plants and the domestication of animals. During this period, humans underwent a major shift in the manner in which they interacted with the environment, redefining the relationship between humans and nature. The adoption of agriculture led to the development of permanent villages, the division of labor, and the rise of a new social structure. But perhaps most importantly, it greatly increased the quantity and predictability of food resources, effectively increasing the human carrying capacity. The ceiling imposed by the productivity of natural ecosystems on hunter-gatherer societies had been lifted, and the human population, influenced by changes in both mortality and fertility, increased steadily by several orders of magnitude. Although the transition from hunter-gatherer to agricultural society greatly eased the constraints imposed by the environment on the human carrying capacity, the continued

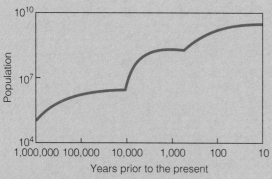

FIGURE B Cycles of demographic growth.

dependence on plants and animals as the sole sources of energy still set an upper limit on productivity. That constraint was to change with the Industrial Revolution.

Beginning in the mid-18th century, a transformation began, first in England and then throughout Europe and America; the transformation continues today in less-industrialized regions of the world. The mechanical energy of animal and human labor upon which the human population relied was

"The human species has redefined its own life history, with reproduction based less on biological factors and more dependent on the control of fertility by individuals. . . ."

replaced by machines fueled by coal, a much more concentrated form of energy. Goods traditionally made at home or in small shops and so-called cottage industries began to be manufactured in factories. The result was a shift from a rural, agricultural-based economy to an urban, industrial-based economy. Improvements occurred on many fronts, including nutrition, personal hygiene, waste disposal and sanitation, medicine, trade and

industrial production, migration and urbanization, and the expansion of agricultural lands. Once again, the human species had redefined its relationship with nature, by tapping a new source of energy that was the accumulated product of hundreds of millions of years (see the Ecological Application after Chapter 30: How a Lack of Mushrooms Helped Power the Industrial Revolution). As with the advent of agriculture, the ceiling to population growth imposed by the environment was once again lifted. Only this time, the rate of population expansion would be far greater. The cause was a fall in mortality rates as a result of better nutrition, personal hygiene, waste disposal and sanitation, and advances in the control of infectious disease. The greatest decline was in infant mortality. As a result of the decline in mortality rate, the annual growth rate of the European population rose fourfold between 1750 and 1850.

The rise in the population growth rate eventually peaked and began to decline during the second half of the 20th century. The decline was a result of falling birthrates as a result of social transformations during the Industrial Revolution. In the new urban-industrialized society, the "cost" of childbearing increased. In contrast to agricultural society, children in

the industrial-based economy became wage earners and producers at a much later age, and they required a much greater investment in terms of material needs, health care, and education. Perhaps more important for the newly emerging working class, the care of children also deprived their mothers of employment opportunities. Although we often associate the reduction of birthrates with the invention of the birth control pill in the mid 20th century, fertility control has a long history. In the last quarter of the 18th century, the practice of limiting family size spread from being a common practice of royalty to the peasantry; and by the end of the 19th century, it had spread throughout Europe and New England.

The period of demographic transition experienced in Europe was characterized by several phases involving historical changes in the rates of mortality and birth, and consequently the rate of population increase (Figure C). Mortality declined first, resulting in an increase in the population growth rate. This was followed by a decline in fertility. As the birthrate declined, the growth rate declined, eventually approaching a new equilibrium of zero population growth. This same pattern of demographic transition that has characterized Western countries over the past two centuries is currently underway in the developing nations of the world. The first wave of reduced fertility rate in the Western industrialized nations during the mid-20th century was not enough to put the brakes on global population growth. Only when the birthrates began to decline in South America and East and Southeast Asia (including China and India) did the global rate of population growth peak and then begin to decline. By 1900, the

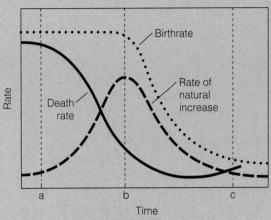

FIGURE C Demographic transition model.

global average life expectancy still stood at 30 years, virtually unchanged from five centuries prior, but by 1950 it had risen to 47, and by 1998 it was 67.

There has been a revolution in public health since World War II, and nowhere has it had more of an impact than in the developing countries. Between 1930 and 1990, the population of developing countries more than tripled. In the first three decades following World War II, life expectancy in these countries rose from 32 to 50 years, an increase of over 50 percent. Between 1965 and 1975, the world's population growth rate reached its maximum (2.1 percent)

and has been dropping since then as a result of declining fertility (birthrate). What does the future hold? Global growth rate has fallen to 1.26 percent, and it is expected to fall to 0.46 percent by 2050. Regardless of this decline, an increase in the global population of some 3.3 billion persons over the next 50 years is predicted because of sheer momentum. This is due in large part to the combined effect of the current population size and the age structure, more specifically the large number of young people who will enter their reproductive years over the upcoming decades. But as the history of the human population has shown us, predicting the future is difficult at best. Two centuries ago, Thomas Malthus saw a rising mortality rate as the inevitable consequence of and control on population growth. What emerged, however, was the adjustment of population by means

of a check on reproduction. The human species has redefined its own life history, with reproduction based less on biological factors and more on the control of fertility by individuals—a possibility that Malthus did not foresee.

Philosophers and scientists throughout the ages have debated the characteristic that distinguishes humans from all other creatures. Perhaps the answer is our unique ability to explore the relationship between cause and effect and in so doing, to extrapolate, which enables us to see future consequences of our current actions. In Celtic mythology, *taisch* was the Gaelic word for second sight, the ability to see the future or distant events. It can be both a blessing and a curse, but with it comes a burden of responsibility. Virtually all of the current environmental concerns, from the loss of biodiversity to global climate change, are linked to the growing human population. Science may well provide us with the ability to see the future consequences of our actions, but it is up to us to determine what those actions will be, and whether we will once again redefine our relationship with nature.

PART V

Communities

Forest and pond—an illustration of the contrasts between terrestrial and aquatic communities.

Community Structure

OBJECTIVES

On completion of this chapter, you should be able to:

* Define both community and guild.
* Discuss the various types of species interactions that occur within communities.
* Explain how species dominance influences community structure.
* Define the concept of species diversity.
* Define food chain and food web.
* Explain how vertical structure of vegetation influences the terrestrial community.
* Discuss how the physical environment influences the structure of aquatic communities.
* Discuss spatial variation in community structure.
* Discuss problems associated with delineating communities.
* Define succession.
* Contrast primary and secondary succession.
* Discuss the successional changes in vegetation following the retreat of the Pleistocene ice sheet.
* Contrast the organismal with the individualistic concept of the community.

As you walk through a forest or swim along a coral reef, you see a collection of individuals of different species: plants and animals that form local populations. Sharing environments and habitats, these species of plants and animals interact in various ways. The species that occupy a given area, interacting either directly or indirectly, are called a **community**. This definition embraces the concept of community in its broadest sense. It is a spatial concept: the collective of species that occupy a place that has a defined boundary. Because ecologists generally do not study the entire community, the term *community* is often used in a more restrictive sense, referring to a subset of the species, such as the plant, bird (avian), small mammal, or fish community. This use of the term *community* suggests relatedness or similarity among the members in their taxonomy, response to the environment, or utilization of resources. Within the community, some species carry out similar functions or exploit the same resource. These groups are called **guilds.** For example, birds feeding mostly on insects make up an insect-feeding guild.

Although ecologists classify communities in different ways, all communities have certain characteristics that define their biological and physical structures. These characteristics vary in both space and time.

13.1 Species Within the Community Have Positive, Negative, or Zero Effects on Each Other

Living in close association, species may interact. They may compete for a shared resource, such as food, light, space, or moisture. One may depend upon the other as a source of food. They may provide mutual aid, or they may have no direct effect on each other at all.

If we designate the positive effect of one species on another as +, a detrimental effect as –, and no effect as 0, we can express the different ways in which populations of two species interact (Table 13.1). When neither of the two populations affects the other, the relationship is (00), or neutral. If the two populations mutually benefit, the interaction is (++), or positive, and the relationship is **mutualism.** If the relationship is not essential for the survival of either population, it is nonobligatory mutualism. If the relationship is essential for the survival of one or both populations, it is obligatory mutualism

TABLE 13.1

Population Interactions, Two-Species Systems

	Response	
Type of Interaction	A	B
Neutral	0	0
Mutualism	+	+
Commensalism	+	0
Competition	–	–
Amensalism	–	0
Predation	+	–
Parasitism	+	–
Parasitoidism	+	–

(Chapter 16). When one species maintains or provides a condition that is necessary for the welfare of another but does not affect its own well-being, the relationship (+0) is **commensalism.** For example, the trunk or limb of a tree provides the substrate on which an epiphytic orchid grows. The arrangement benefits the orchid, which gets nutrients from the air and moisture from aerial roots, while the tree is unaffected.

When the relationship is detrimental to the populations of both species (––), the interaction is **competition.** In some situations, the interaction is (–0). One species reduces or adversely affects the population of another, but the affected species has no influence in return. This relationship is **amensalism,** considered by many ecologists as a form of asymmetric competition.

Relationships in which one species benefits at the expense of the other (+–) are predation, parasitism, and parasitoidism. **Predation** is when one organism eats another, typically killing the prey (Chapter 15). Predation always has a negative effect on the individual prey. In **parasitism,** one organism feeds on the other, rarely killing it outright. The two, parasite and host, live together for some time. The host survives, although its fitness is reduced. **Parasitoidism** is like predation in that it kills the host eventually. Parasitoids, which include certain wasps and flies, lay eggs in or on the body of the host. When the eggs hatch, the larvae feed upon the host. By the time the larvae reach the pupal stage, the host has succumbed (Chapter 16). We will examine each of these types of species interactions in the following chapters and explore their role in shaping communities.

Some Measures of Dominance

1. Dominance =
$$\frac{\text{basal area or aerial coverage, species A}}{\text{area sampled}}$$

2. Relative dominance =
$$\frac{\text{basal area or coverage, species A}}{\text{total basal area or coverage, all species}}$$

3. Relative density =
$$\frac{\text{total individuals, species A}}{\text{total individuals, all species}}$$

4. Frequency =
$$\frac{\text{plots or points where species A occurs}}{\text{total number of sample plots or points}}$$

5. Relative frequency =
$$\frac{\text{frequency value, species A}}{\text{total frequency value, all species}}$$

6. Importance value = relative frequency + relative dominance + relative density

All the above results may be multiplied by 100.

7. Simpson's index of dominance:

$$\text{dominance} = \frac{\sum n_i(n_i - 1)}{N(N - 1)}$$

where N is total number of individuals of all species and n_i is the total number of individuals of species A

13.2 Communities Vary in Biological Structure

The mix of species, including both their number and relative abundance, defines the biological structure of a community. A community can be composed of a few common species; or it can have a wide variety of species, some common with high population density, but most rare with low population density. When a single or few species predominate within a community, these organisms are **dominants.**

There is no single criterion for characterizing a species as being dominant (see Quantifying Ecology 13.1: Some Measures of Dominance). The dominants in a community may be the most numerous, possess the highest biomass, preempt the most space, make the largest contribution to energy flow or nutrient cycling (Chapter 21), or by some other means control or influence the rest of the community.

Some ecologists ascribe the dominant role to those organisms that are greatest in number, but abundance alone is not sufficient. A species of plant, for example, can be widely distributed but exert little influence on the community as a whole. In the forest, the small or understory trees can be numerically superior; yet the community is controlled by a few large trees that overshadow the smaller ones. In such a situation, the dominant organisms are not those with the largest numbers but those that have the greatest biomass or that preempt most of the canopy space and thus control the distribution of light.

In other cases, a scarce organism dominates by its activity. The predatory starfish *Pisaster*, for example, preys on several associated species and reduces competitive interactions among them, so they are able to coexist. If the predator is removed, a number of the prey species disappear, and one becomes dominant. In effect, this predator controls the nature of the community and must be regarded as the dominant species. Such species are often called **keystone species.**

The dominant species may not be the most essential species in the community from the standpoint of energy flow or nutrient cycling, although this is often the case. Dominant species achieve their status at the expense of other species in the community. For example, the American chestnut tree *(Castania dentata)* was a dominant component of the oak-chestnut forests in eastern North America until early in the twentieth century. At that time, its populations were decimated by the chestnut blight introduced from Asia. Since that time, a variety of species, including oaks, hickories, and yellow-poplar, have taken over the position of the chestnut in the forest.

The concept of species dominance is context dependent. A predator species that maintains one mix of species in the community may not be defined as dominant if the criterion is the cycling of nutrients or density. We must define our criteria when characterizing a species as being dominant.

13.3 Number and Relative Abundance Define Species Diversity

Among the array of species that make up the community, a few are abundant, but most are rare. You can discover this characteristic by counting all the individuals of each species in a number of sample plots within a community and determining what

TABLE 13.2

Structure of One Mature Deciduous Forest in West Virginia

Species	Number	Percentage of Stand
Yellow-poplar (Liriodendron tulipifera)	76	29.7
White oak (Quercus alba)	36	14.1
Black oak (Quercus velutina)	17	6.6
Sugar maple (Acer saccharum)	14	5.4
Red maple (Acer rubrum)	14	5.4
American beech (Fagus grandifolia)	13	5.1
Sassafras (Sassafras albidum)	12	4.7
Red oak (Quercus rubra)	12	4.7
Mockernut hickory (Carya tomentosa)	11	4.3
Black cherry (Prunus serotina)	11	4.3
Slippery elm (Ulmus rubra)	10	3.9
Shagbark hickory (Carya ovata)	7	2.7
Bitternut hickory (Carya cordiformis)	5	2.0
Pignut hickory (Carya glabra)	3	1.2
Flowering dogwood (Cornus florida)	3	1.2
White ash (Fraxinus americana)	2	.8
Hornbeam (Carpinus carolinia)	2	.8
Cucumber magnolia (Magnolia acuminata)	2	.8
American elm (Ulmus americana)	1	.39
Black walnut (Juglans nigra)	1	.39
Black maple (Acer nigra)	1	.39
Black locust (Robinia pseudoacacia)	1	.39
Sourwood (Oxydendrum arboreum)	1	.39
Tree of heaven (Ailanthus altissima)	1	.39
	256	100.00

TABLE 13.3

Stucture of a Second Deciduous Forest in West Virginia

Species	Number	Percentage of Stand
Yellow-poplar (Liriodendron tulipifera)	122	44.5
Sassafras (Sassafras albidum)	107	39.0
Black cherry (Prunus serotina)	12	4.4
Cucumber magnolia (Magnolia acuminata)	11	4.0
Red maple (Acer rubrum)	10	3.6
Red oak (Quercus rubra)	8	2.9
Butternut (Juglans cinerea)	1	.4
Shagback hickory (Carya ovata)	1	.4
American beech (Fagus grandifolia)	1	.4
Sugar maple (Acer saccharum)	1	.4
	274	100.0

percentage each contributes to the total number of individuals of all species. This measure is known as **relative abundance.**

As an example, samples representing the species composition of two forest communities are presented in Tables 13.2 and 13.3. The sample from the first forest consists of 24 species of trees over 10 cm dbh (diameter of a tree measured at 1.5 meters above the ground). Two species, yellow-poplar and white oak, make up nearly 44 percent of the stand. The 4 next most abundant trees—black oak, sugar maple, red maple, and American beech—each make up a little over 5 percent of the stand. Nine species range from 1.2 to 4.7 percent, while the 9 remaining species as a group represent about 5 percent of the stand. The second forest presents a somewhat different picture. This community consists of 10 species, of which 2—yellow-poplar and sassafras—make up almost 84 percent of the total stand density. Both forest communities illustrate the pattern of a few common species associated with many rare ones.

These two forest communities differ in their **species diversity.** Species diversity relates to both the number of species, **species richness,** and how individuals

Indexes of Diversity

The Shannon index of diversity is one of a number of diversity indexes. Based on information theory, it measures the degree of uncertainty. If diversity is low, then the certainty of picking a particular species at random is high. If diversity is high, then it is difficult to predict the identity of a randomly picked individual. High diversity means high uncertainty.

Another common index is Simpson's. It takes a different approach—the number of times we would have to take pairs of individuals at random to find a pair of the same species. This index of diversity is the inverse of Simpson's dominance index (see Quantifying Ecology 13.1):

$$\text{diversity} = \frac{N(N-1)}{\sum n_i(n_i - 1)}$$

or

$$1 - \frac{\sum n_i(n_i - 1)}{N(N-1)}$$

In a collection of species, high dominance means low diversity.

The Shannon and Simpson indexes take into consideration both the richness and evenness of species. A much simpler index of diversity that does not take evenness into account is Margalef's:

$$\text{diversity} = \frac{(S-1)}{\ln N}$$

where S is the number of species and N is the number of individuals. Such an index does not express the differences among communities having the same S and N, so it is much less useful.

To quantify species diversity, several indexes have been proposed (see Quantifying Ecology 13.2: Indexes of Diversity). The most widely used is the Shannon index, which ecologists have adapted from communication or information theory. The formula for the Shannon index is

$$H = -\sum_{i=1}^{S} (p_i)(\ln p_i)$$

where H is the measure of species diversity, S is the number of species, p_i is the proportion of individuals in the total sample belonging to the ith species, and $\ln p_i$ is the natural logarithm of p_i. The index takes into consideration the number as well as the relative abundance of species. The woodland described in Table 13.2 has a diversity index of 2.49; the one described in Table 13.3 has a diversity index of 1.30.

The two components, species richness and species evenness, can be separated. Species richness is simply the number of species. In the first woodland, that number is 24, and in the other, 10. To determine evenness, you first have to calculate H_{max}, what H would be if all the species in the community had an equal number of individuals:

$$H_{max} = \ln S$$

where S is the species richness. For the first woodland H_{max} is 3.18, and for the second it is 2.30. Now we can calculate the evenness (J):

$$J = H/H_{max}$$

The evenness of the first woodland is 0.78, and that of the second is 0.57. The first woodland has a more even distribution of individuals among species than does the second.

13.4 Food Webs Describe Species Interactions

Perhaps the most fundamental process in nature is the acquisition of food—the energy and nutrients required for assimilation. The species interactions outlined in Table 13.1—predation, parasitism, competition, and mutualism—are all involved in the acquisition of food resources. For this reason, ecologists studying the structure of communities often focus on the feeding relationships of the component species—the manner in which species interact in the process of acquiring food resources.

The most abstract representation of feeding relationships within a community is the food chain. A **food chain** is a descriptive diagram—a series of arrows, each pointing from one species to another for which it is a source of food. For example, grasshoppers eat grass; clay-colored sparrows eat

are apportioned among the species, **species evenness.** The first forest community has both greater species richness and greater species evenness than the second.

Species richness and species evenness are useful in measuring species diversity. A community that contains a few individuals of many species is said to have a higher diversity than a community having the same number of individuals with most of them belonging to a few species. For example, a community with 10 species of 10 individuals each has a higher diversity than another community with 10 species but with the 100 individuals apportioned 91, 1, 1, 1, 1, 1, 1, 1, 1, 1.

FIGURE 13.1 A food web for a prairie grassland community in the Midwest. Arrows flow from food (prey species) to consumer (predator species).

grasshoppers; and marsh hawks prey on the sparrows. We write this relationship as follows:

grass → grasshopper → sparrow → hawk

Feeding relationships in nature, however, are not simple, straight-line food chains. Rather they involve numerous food chains meshed into a complex **food web** with links leading from primary producers and detritus through an array of consumers: detritivores, herbivores, carnivores, and omnivores (Figure 13.1). Such food webs are highly interwoven, with linkages representing a wide variety of species interactions.

Although any two species are linked by only a single arrow representing the relationship between predator (the consumer) and prey (the consumed), the structure of communities cannot be understood solely in terms of the direct interactions between species. For example, a predator may reduce competition between two prey species by controlling their population sizes below carrying capacity. An analysis of the mechanisms controlling community structure must include these indirect effects represented by the structure of the food web, a topic we will explore in Chapter 17.

13.5 Communities Have Defining Physical Structure

Communities are characterized not only by the mix of species and the interactions among them—the biological structure—but also by physical features. The physical structure of the community reflects abiotic factors, such as the depth and flow of water in aquatic environments. It also reflects biotic factors, such as the spatial arrangement of organisms. In a forest, for example, the size and height of the trees and the density and dispersion of their populations define the physical attributes of the community.

The form and structure of terrestrial communities largely reflect the vegetation. The plants may be tall or short, evergreen or deciduous, herbaceous or woody. Such characteristics can describe growth forms. Thus, we might speak of shrubs, trees, and herbs and further subdivide the categories into needle-leaf evergreens, broadleaf evergreens, broadleaf deciduous trees, thorn trees and shrubs, dwarf shrubs, ferns, grasses, forbs, mosses, and lichens. Ecologists often classify and name terrestrial communities based on the dominant plant growth forms

and their associated physical structure: forests, woodlands, shrublands, or grassland communities (see Focus on Ecology 13.1: Plant Life Form Classification and Chapter 24).

In aquatic environments, physical structure defined by the dominant organisms is also used to classify and name communities. Kelp forests, seagrass meadows, and coral reefs are examples. The physical structure of aquatic communities is more often, however, defined by features of the physical or abiotic environment, such as water depth, flow rate, or salinity.

13.6 Communities Have a Characteristic Vertical Stratification

Each community has a distinctive **vertical structure** (Figure 13.2). On land, vertical structure is determined largely by the life form of the plants—their size, branching, and leaves—that in turn influences and is influenced by the vertical gradient of light. The vertical structure of the plant community provides the physical framework in which many forms of

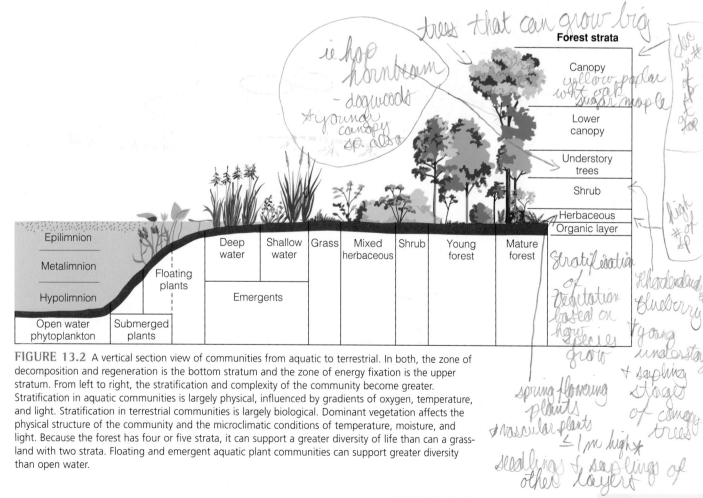

FIGURE 13.2 A vertical section view of communities from aquatic to terrestrial. In both, the zone of decomposition and regeneration is the bottom stratum and the zone of energy fixation is the upper stratum. From left to right, the stratification and complexity of the community become greater. Stratification in aquatic communities is largely physical, influenced by gradients of oxygen, temperature, and light. Stratification in terrestrial communities is largely biological. Dominant vegetation affects the physical structure of the community and the microclimatic conditions of temperature, moisture, and light. Because the forest has four or five strata, it can support a greater diversity of life than can a grassland with two strata. Floating and emergent aquatic plant communities can support greater diversity than open water.

PLANT LIFE FORM CLASSIFICATION

In 1903, the Danish botanist Christen Raunkiaer developed a system of plant classification that provides a useful alternative to the traditional system based on a plants' growth form (tree, shrub, grass, etc.). Raunkier classified plant life by the relation of the embryonic or meristemic tissues that remain inactive over the winter or prolonged dry periods—the **perennating tissues**—to their height above ground. Such perennating tissue includes buds, bulbs, tubers, roots, and seeds. Raunkiaer recognized six principal life forms, which are summarized

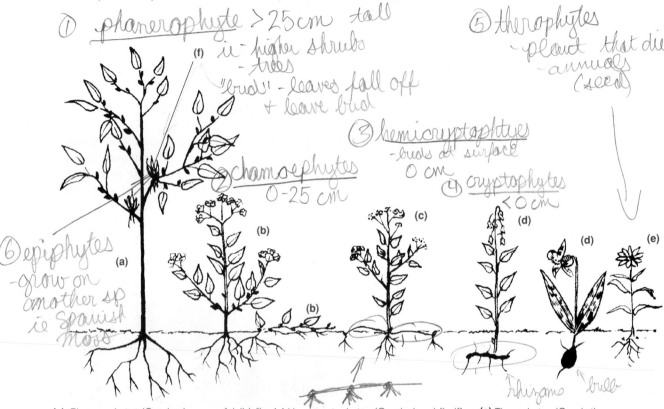

Handwritten annotations:
① phanerophyte >25cm tall
ie higher shrubs
- trees
"bud" - leaves fall off
+ leave bud
⑤ therophytes
- plant that die
- annuals (seed)
③ hemicryptophytes
- buds at surface
0 cm
④ cryptophytes
<0 cm
② chamaephytes
0-25 cm
⑥ epiphytes
- grow on another sp
ie spanish moss
rhizome   bulb

(a) Phanerophytes (Greek *phaneros*, "visible"). Perennial buds carried well up in the air and exposed to varying climatic conditions. Trees and shrubs over 25 cm; typical of moist, warm environments.

(b) Chamaephytes (Greek *chamia*, "on the ground"). Perennial shoots or buds on the surface of the ground to about 25 cm above the surface. Buds receive protein from fallen leaves and snow cover. Typical of cool, dry climates.

(c) Hemicryptophytes (Greek, *hemi*, "half" and *kryptos*, "hidden"). Perennial buds at the surface of the ground, where they are protected by soil and leaves. Many plants have rosette leaves. Characteristic of cold, moist climates.

(d) Cryptophytes (Greek, *kryptos*, "hidden"). Perennial buds buried in the ground on a bulb or rhizome, where they are protected from freezing and drying. Typical of cold, moist climates.

(e) Therophytes (Greek *theros*, "summer"). Annuals, with complete life cycle from seed to seed in one season. Plants survive unfavorable periods as seeds. Typical of deserts and grasslands.

(f) Epiphytes (Greek *epi*, "upon"). Plants growing on other plants; roots up in the air.

FIGURE A Raunkiaer's life forms. The parts of the plants that die back during winter or prolonged dry periods are unshaded; the persistent plant parts with buds (or seeds in the case of therophytes) are dark.

animal life are adapted to live. A well-developed forest ecosystem, for example, has several layers of vegetation. From top to bottom, they are the *canopy,* the *understory,* the *shrub layer,* the *herb* or *ground layer,* and the *forest floor.* We could continue down into the root layer and soil strata.

The upper layer, the **canopy,** is the primary site of energy fixation through photosynthesis. The structure of the canopy has a major influence on the rest of the forest. If it is fairly open, considerable sunlight will reach the lower layers. If ample water and nutrients are available, a well-developed understory and

in Figure A. All of the species within a community can be grouped into these six classes. For example, trees and shrubs greater than 25 cm in height are classified as *phanerophytes* because their leaf-producing buds are elevated above the ground on stems. In contrast, grasses are classified as *cryptophytes* because their above-ground tissues die back in winter or during prolonged dry periods. A community with a high percentage of perennating tissue well above ground is characteristic of warmer, wetter climates. A community where most of the plants are *crypto-* *phytes* and *hemicryptophytes* is characteristic of colder or drier environments. The perennating tissue from which new leaves grow is underground.

When the species within a community are classified into life forms and each life form is expressed as a percentage, we get a life form spectrum that reflects the plants' adaptations to the environment, particularly the climate (Figure B). Such a system of classification provides a standard means of describing the structure of a community for purposes of comparison.

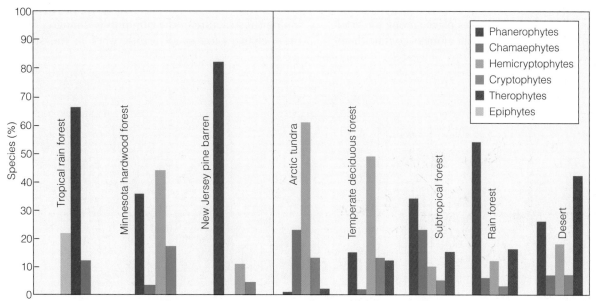

FIGURE B Left: Life form spectra of a tropical rain forest (Richards 1996), a temperate forest in Minnesota (Buell and Wilbur 1948), and a New Jersey pine barren (Sterns and Buell 1951). Note the absence of hemicryptophytes, chamaephytes, and therophytes in the tropical rain forest and the prominence of epiphytes. The pine barrens are dominated by phanerophytes. Right: Life form spectra of major ecosystems. Note the importance of hemicryptophytes in the Arctic tundra and temperate deciduous forest, and the importance of phanerophytes and therophytes in the desert ecosystem.

shrub strata will form. If the canopy is dense and closed, light levels are low, and the understory and shrub layers will be poorly developed.

In the forests of the eastern United States, the **understory** consists of tall shrubs such as witch hobble *(Viburnum alnifolium)*, understory trees such as dogwood *(Cornus* spp.) and hornbeam *(Carpinus caroliniana)*, and younger trees, some of which are the same species as those in the canopy. The nature of the **herb layer** depends on the soil moisture and nutrient conditions, the slope position, the density of

the canopy and understory, and the exposure of the slope, all of which vary from place to place throughout the forest. The final layer, the **forest floor,** is the site where the important process of decomposition takes place and where decaying organic matter releases nutrients for reuse by the forest plants (Chapter 7).

Aquatic ecosystems such as lakes and oceans have strata determined largely by light penetration. They have distinctive profiles of temperature and oxygen (see Chapter 4). In the summer, well-stratified lakes have a layer of well-mixed water, the **epilimnion;** a second layer, the **metalimnion,** which is characterized by a **thermocline** (a steep and rapid decline in temperature); the **hypolimnion,** a deep, cold layer of dense water at about 4°C (40°F), often low in oxygen; and a layer of bottom sediments. In addition, two other structural layers are recognized, based on light penetration: an upper zone, the **photic zone,** where the availability of light supports photosynthesis primarily by phytoplankton, and in deeper waters, the **aphotic zone,** or area without light. The bottom zone where decomposition is most active is referred to as the **benthic zone.**

Each vertical layer in the community is inhabited by characteristic organisms. In addition to the vertical distribution of plant life described above, various types of consumers and decomposers occupy all levels of the community, although decomposers are typically found in greater abundance in the forest floor (soil surface) and sediment (benthic) layers. Although considerable interchange takes place among the vertical strata, many highly mobile animals confine themselves to only a few layers (Figure 13.3). The species occupying a given vertical layer may change during the day or season. Such changes reflect daily and seasonal variations in the physical environment, such as humidity, temperature, light, and oxygen concentrations in the water; shifts in the abundance of essential resources such as food; or different requirements of organisms for the completion of their life cycles. For example, zooplankton migrate vertically in the water column over the course of the day in response to varying light (see Chapter 9, Section 9.6).

13.7 Spatial Change in Community Structure Is Zonation

As you move across the landscape, the physical and biological structure of the community changes. Often these changes are small, subtle ones in the species

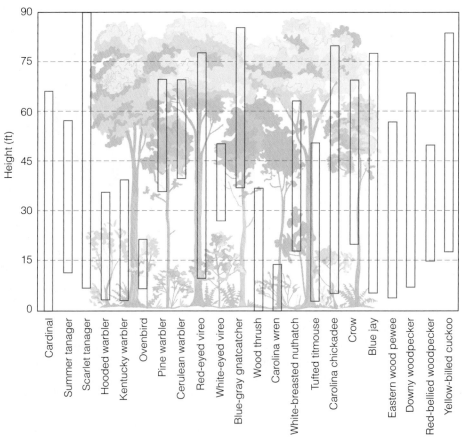

FIGURE 13.3 Vertical distribution of bird species within the forest community on Walker Branch watershed. Height range represented by colored bars is based on the total observations of birds during the breeding season regardless of activity. (After Anderson and Shugart 1974.)

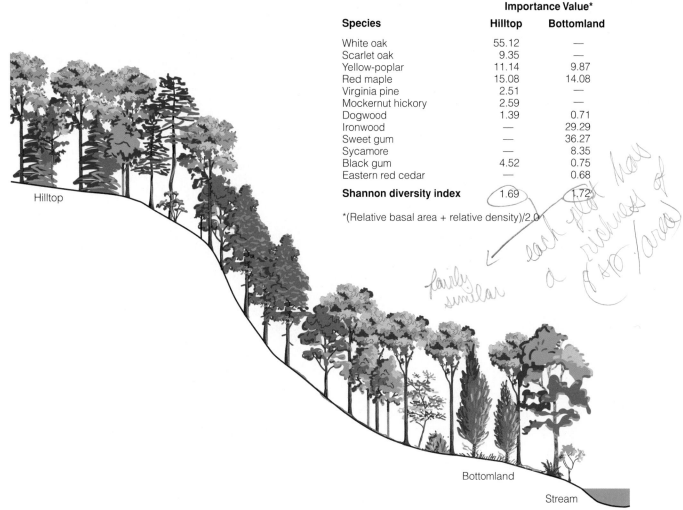

Species	Importance Value*	
	Hilltop	Bottomland
White oak	55.12	—
Scarlet oak	9.35	—
Yellow-poplar	11.14	9.87
Red maple	15.08	14.08
Virginia pine	2.51	—
Mockernut hickory	2.59	—
Dogwood	1.39	0.71
Ironwood	—	29.29
Sweet gum	—	36.27
Sycamore	—	8.35
Black gum	4.52	0.75
Eastern red cedar	—	0.68
Shannon diversity index	1.69	1.72

*(Relative basal area + relative density)/2.0

[handwritten annotations: "fairly similar", "each plot has a richness of (#sp./area)"]

FIGURE 13.4 Changes in species composition of forest stands along a topographic gradient in Fluvanna County, Virginia. The table summarizes stand composition as sampled at two points: on the hilltop and along the stream.

composition or height of the vegetation. However, as you travel farther, these changes become more pronounced. For example, central Virginia just east of the Blue Ridge Mountains is a landscape of rolling hills. The area is a mosaic of forest and field. As you walk through a forest, the physical structure of the community—the canopy, understory, shrub layer, and forest floor—appears much the same. However, the biological structure, the mix of species that compose the community, may change quite dramatically. As you move from a hilltop to the bottomland along a stream, the mix of trees changes from oaks (*Quercus* spp.) and hickory (*Carya* spp.) to species associated with much wetter environments, such as sycamore *(Platanus occidentalis)*, hornbeam *(Carpinus caroliniana)*, and sweetgum *(Liquidambar styraciflua)* (Figure 13.4). In addition to changes in the vegetation, the animal species—insects, birds,

and small mammals—that occupy the forest likewise change. These changes in the physical and biological structure of communities as you move across the landscape are referred to as **zonation.**

Patterns of spatial variation in community structure or zonation are common to all environments, aquatic and terrestrial. Figure 13.5 provides an example of zonation in a salt marsh. Notice the variations in the physical and biological structure of the communities as you move from the shore through the marsh to the upland. The dominant plant growth forms in the marsh are grasses and sedges. These growth forms give way to shrubs and trees as you move to dry land and the depth of the water table increases. Within the zone dominated by grasses and sedges, the dominant species change as you move back from the tidal areas. These differences result from a variety of environmental changes across a

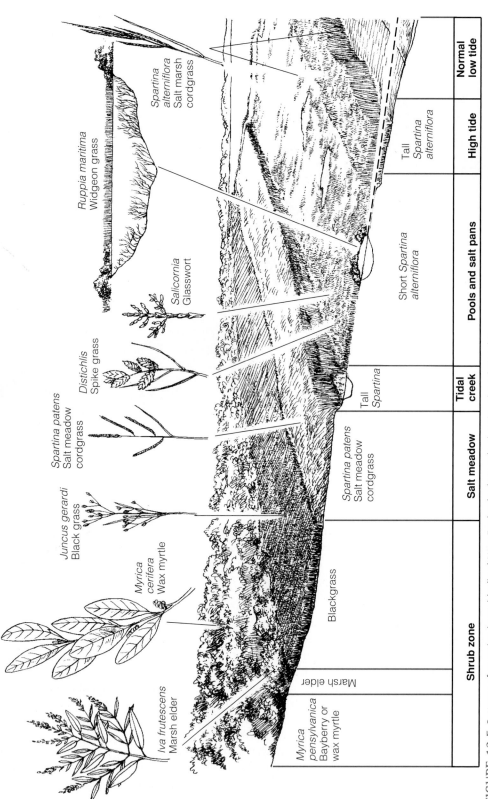

FIGURE 13.5 Patterns of zonation in an idealized New England salt marsh, showing the relationship of plant distribution to microtopography and tidal submergence.

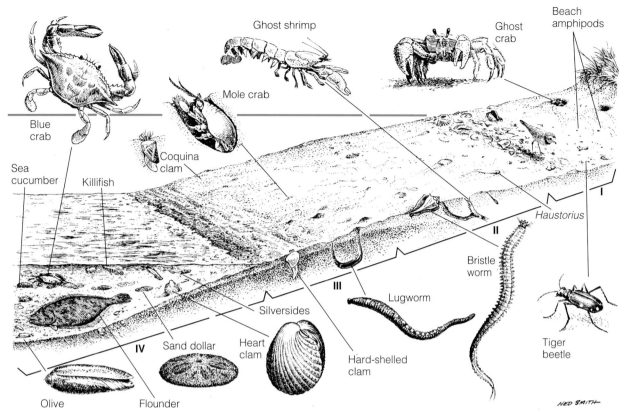

FIGURE 13.6 Life on a sandy ocean beach along the mid-Atlantic coast. The distribution of organisms changes along a gradient from land to sea. I, supratidal zone: ghost crabs and sand fleas; II, flat beach zone: ghost shrimp, bristle worms, clams; III, intratidal zone: clams, lugworms, mole crabs; IV, subtidal zone. The dashed line indicates high tide.

spatial gradient, including microtopography, water depth, sediment oxygenation, and salinity. The changes are marked by distinct plant communities that are defined by changes in dominant plants and in structural features such as height, density, and dispersion.

The intertidal zone of a sandy beach provides an example in which the pattern of zonation is dominated by heterotrophic organisms rather than autotrophs (Figure 13.6).

13.8 Defining Boundaries Between Communities Is Often Difficult

As previously noted, the community is a spatial concept: the species that occupy a given area. Ecologists typically distinguish between communities or community types based on observable differences in their physical and biological structure: different species assemblages that are characteristic of different phys-

ical environments. How different must two adjacent areas be before we call them separate communities? This is not a simple question. Consider the forest in Figure 13.4. Given the difference in species composition from hilltop to bottomland, most ecologists would define these two areas as different vegetation communities; but as you walk between them, the distinction may not seem so straightforward. If the transition between the two communities is abrupt, there may not be a problem in defining the community boundaries. However, the species composition and patterns of dominance may shift gradually. In this case, the boundary is not as clear.

Ecologists have a variety of sampling and statistical techniques to delineate and classify communities. Generally, all employ some measure of community similarity or difference (see Quantifying Ecology 13.3: Community Similarity). Although it is easy to describe the similarities and differences between two areas in terms of species composition and structure, the actual classification of areas into distinct groups of communities involves a degree of subjectivity, and it often depends on the objectives of the particular study.

Community Similarity

A number of methods are available for measuring similarity of communities. The one most often recommended is Morisita's index, based on Simpson's index of dominance. However, here are two simpler approaches: Sorensen's *coefficient of community* and *percent similarity.*

To find the **coefficient of community,** apply the equation

$$CC = \frac{2c}{s_1 + s_2}$$

where *c* is the number of species common to both communities and s_1 and s_2 are the number of species in communities 1 and 2.

For the woodland examples:

$$s_1 = 24 \text{ species}$$
$$s_2 = 10 \text{ species}$$
$$c = 9 \text{ species}$$
$$CC = \frac{2(9)}{24 + 10} = \frac{18}{34} = .529$$

Coefficient of community does not consider the relative abundance of species. It is most useful when the major interest is the presence or absence of species.

To calculate **percent similarity** (PS), first tabulate species abundance in each community as a percentage. Then add the lowest percentage for each species that the communities have in common. For the two woodlands, 16 species are exclusive to one community or the other. The lowest percentage for those 16 species is 0, so they need not be added in.

$$PS = 29.7 + 4.7 + 4.3 + 0.8 + 3.6$$
$$+ 2.9 + 0.4 + 0.4 + 0.4 = 47.2$$

Percent similarity does consider relative abundance of species in each community.

13.9 Classification of Communities Is Scale-Dependent

The forest in Figure 13.4 was small. As we consider ever larger areas, differences in community structure, both physical and biological, increase. An example is the pattern of forest zonation in the Great Smoky Mountains National Park (Figure 13.7). The zonation is a complex pattern related to elevation, slope position, and exposure. Note that the description of the forest communities in the park contains few species names. Names like hemlock forest are not meant to suggest a lack of species diversity; they are just a shorthand method of naming communities for the dominant species. Each community could be described by a complete list of species, their population sizes, and their contributions to the total biomass. However, such lengthy descriptions are unnecessary to communicate the major changes in the structure of communities across the landscape. In fact, as we expand the area of interest to include the entire eastern United States, the nomenclature for classifying forest communities becomes even broader. In Figure 13.8, which is a broad-scale description of forest zonation in the eastern United States developed by Lucy Braun, all of Great Smoky Mountains National Park shown in Figure 13.7 (located in southeastern Tennessee and northwestern North Carolina) is described as a single forest community type—oak-chestnut, a type that extends from New York to Georgia.

These large-scale examples of zonation make an important point to which we return when we examine the processes responsible for spatial changes in community structure: our very definition of community is a spatial concept. Like the biological definition of population (Chapter 9), the definition of community refers to a spatial unit that occupies a given area. In a sense, the distinction among communities is arbitrary, based on the criteria for classification. As we shall see, the methods used in delineating communities as discrete spatial units has led to problems in understanding the processes responsible for patterns of zonation.

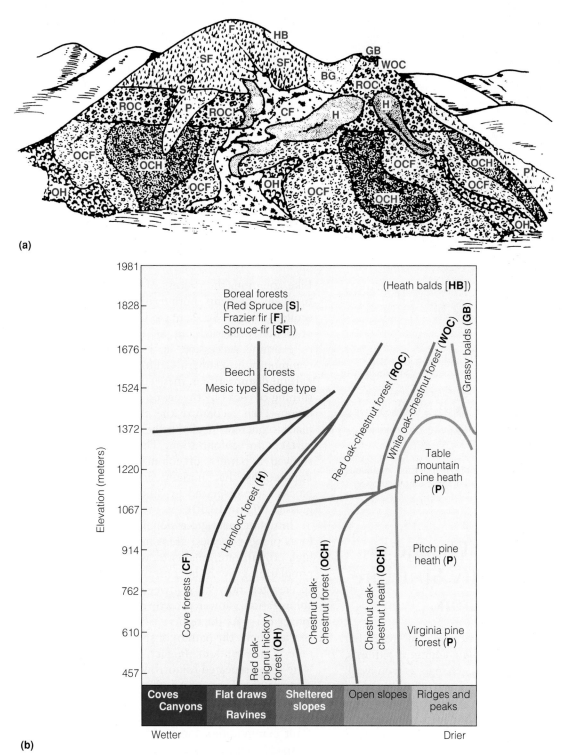

(a)

(b)

Wetter   Drier

FIGURE 13.7 Two descriptions of forest communities in Great Smoky Mountains National Park.
(a) Topographic distribution of vegetation types on an idealized west-facing mountain and valley.
(b) Idealized arrangement of community types according to elevation and aspect. (Whittaker 1954.)

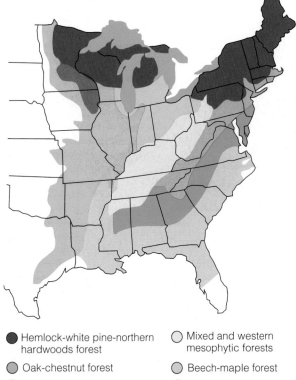

- ● Hemlock-white pine-northern hardwoods forest
- ○ Mixed and western mesophytic forests
- ● Oak-chestnut forest
- ● Beech-maple forest
- ● Oak-pine forest
- ● Maple-basswood forest
- ○ Southeastern evergreen-southern mixed hardwood forest
- ● Oak-hickory forest

FIGURE 13.8 The large-scale distribution of eastern deciduous forest communities in the eastern United States is defined by 8 regions. (Braun 1950.)

13.10 Temporal Change in Community Structure Is Succession

Community structure varies not only in space, but also in time. We have seen how community patterns, both physical and biological, change as you walk from hilltop to stream. Now suppose that you stand in one position, such as in a field adjacent to the forest, and observe the area as time passes. You return to the field each summer to stand in this same spot and look around. In the field that once was covered only in grasses and other herbaceous vegetation, you begin to see the arrival of woody plants, shrubs, and small trees. The density of woody vegetation increases until the field becomes a forest (Figure 13.9). The forest is initially dominated by fast-growing, shade-intolerant species such as quaking aspen. Other deciduous tree species such as red maple become established in the understory, eventually making their way to the canopy as the aspen trees die. Even these

species eventually are replaced by slower-growing, shade-tolerant tree species such as oaks, hickories, and sugar maple.

The process that you have observed, the gradual and seemingly directional change in the structure of the community through time from field to forest, is **succession.** Succession in its most general definition is the change in community structure through time. In contrast to zonation, succession refers to a given point in space—a single location.

13.11 Succession Is Common to All Environments

Like zonation, the process of succession is common to all environments, both terrestrial and aquatic. The ecologist Wayne Sousa of the University of California at Berkeley carried out an interesting experiment to examine succession in a rocky intertidal algal community in southern California. A major form of natural disturbance in these communities is the overturning of rocks by the action of waves. Algae populations then recolonize the cleared surfaces. Sousa placed concrete blocks in the water to provide new surfaces for colonization. The results of his study showed a pattern of colonization and extinction. Populations that initially colonized the concrete blocks were displaced by other species as time progressed (Figure 13.10).

Initial or early successional species (often referred to as pioneer species) are usually characterized by a high growth rate, small size, wide dispersal, and fast population growth. In contrast, the late successional species generally have lower rates of dispersal and colonization, slower growth rates, larger sizes, and longer lives. As the terms *early* and *late successional species* imply, the pattern of species replacement over time is not random. In fact, if Sousa's experiment were to be repeated tomorrow, we would expect the patterns of colonization and extinction in the successional sequence to be somewhat similar.

Other patterns of succession occur in terrestrial plant communities. Figure 13.11 shows the pattern of species replacement following a clear-cut at the Hubbard Brook Experimental Forest in New Hampshire. Prior to forest clearing, the understory was dominated by seedlings and saplings of beech and sugar maple. Large individuals of these two tree species dominated the canopy, and the seedlings represented successful reproduction of the parent trees. Following the removal of the larger trees in 1970, the beech and maple seedlings declined and were soon replaced by raspberry thickets and seedlings of

FIGURE 13.9 Successional changes over 55 years in a western Pennsylvania field. (a) In 1942 it was moderately grazed. (b) The same area 21 years later in 1963. (c) In 1972 quaking aspen has claimed some of the ground. (d) A view of the rail fence in the right background of (a). (e) In 1963 the rail fence has rotted and white pine and aspen grow in the area. (f) In 1997 the field has been completely claimed by a young stand of aspen and maple.

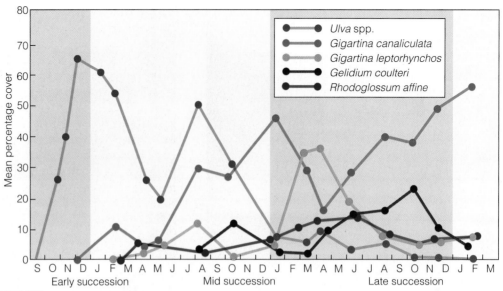

FIGURE 13.10 Mean percentage of five algal species that colonized concrete blocks introduced into the intertidal zone in September 1974. Note the change in species dominance over time. (Sousa 1979.)

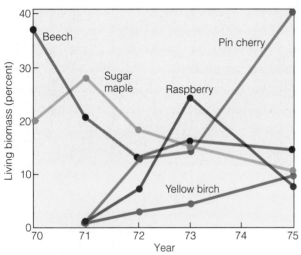

FIGURE 13.11 Changes in the relative abundance (percentage of living biomass) of woody species in the Hubbard Brook Experimental Forest after a clear-cut in 1970. (From F. Bormann and G. E. Likens, *Pattern and Process in a Forested Ecosystem*, p. 116, Fig. 4.4b. New York: Springer-Verlag, 1979. Copyright © 1979 Springer-Verlag. Used by permission.)

sun-adapted (shade-intolerant), fast-growing, early successional tree species such as pin cherry and yellow birch. After many years, these species eventually will be replaced by the late successional species of beech and sugar maple that had previously dominated the site.

The observed similarity in patterns of species colonization and extinction through time for sites across a wide range of environmental conditions, both terrestrial and aquatic, suggests to ecologists a common mechanism or mechanisms influencing the process of succession. The fact that community structure does not vary randomly in either space (zonation) or time (succession), but exhibits repeatable, often predictable patterns, has been the motivating force for research throughout the history of ecology. As we will see in Chapter 17, the search for causation has not been a simple task. It remains one of the major areas of ecological study.

13.12 Succession Is Either Primary or Secondary

The two studies above point out the similar nature of successional dynamics in very different environments. They also exemplify two different types of succession—primary and secondary. **Primary succession** occurs on a site previously unoccupied by a community—a newly exposed surface such as sand, bare rock, or the cement blocks in the rocky intertidal environment.

In contrast, **secondary succession** occurs on previously occupied sites following disturbance. Disturbance is any process that results in the removal (either partial or complete) of the existing community (Chapter 19). In the Hubbard Brook example, the disturbance did not remove all individuals. In such cases, the amount (density and biomass) and composition of the surviving community have a major influence on successional dynamics.

Primary succession occurs on newly exposed land surfaces created by geological processes and in newly formed bodies of water. The island of Hawaii is known for its active volcanoes. Lava flows on Hawaii represent newly formed landscapes on which primary succession occurs (Figure 13.12). The previous plant communities in these areas have been destroyed by fire and heat, and the existing soils have been covered by a new layer of igneous rock.

Another example of primary succession is the colonization of newly exposed glacial till (rock material deposited as glacier melts) in the area of Glacier Bay National Park, Alaska. Over the past 200 years, the glacier that once covered the entire region of the bay has been retreating (Figure 13.13). As the glacier retreats, the newly exposed landscape is initially

FIGURE 13.12 Trees and ferns colonize a lava flow on the island of Hawaii.

(b)

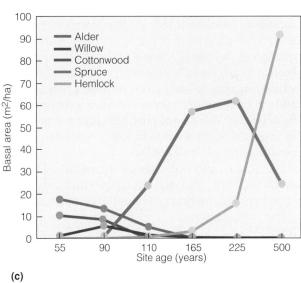

(a)

(c)

FIGURE 13.13 (a) The Glacier Bay fjord complex in southeastern Alaska, showing the rate of ice recession since 1760. As the ice retreats, it leaves moraines along the edge of the bay in which primary succession occurs. (b) Primary succession along riverine environments of Glacier Bay National Park, Alaska. (c) Changes in community composition with stand age for sites at Glacier Bay. (Hobbie 1994)

colonized by species such as alder (*Alnus* spp.) and cottonwood (*Populus* spp.) (Figure 13.13c). Eventually these early successional species are replaced by spruce (*Picea* spp.) and hemlock, and the resulting forest resembles the surrounding landscape.

13.13 Succession Involves Heterotrophic Species

Although our discussion and examples of succession have thus far focused on temporal changes in the autotrophic component of the community (plant succession), changes in the heterotrophic component of the community also occur. A well-studied example is succession in the heterotrophic communities involved in decomposition. Dead trees, animal carcasses, and droppings form substrates on which communities of organisms involved in decomposition exist. Within these communities, groups of plants and animals succeed each other in a process of colonization and replacement. In these instances, succession is characterized by early dominance of fungi and invertebrates that feed on dead organic matter. Available energy and nutrients are most abundant in the early stages of succession, and decline steadily as succession proceeds.

When a windstorm uproots or breaks a tree, the fallen tree becomes the stage for succession of plant and animal colonists until the log becomes part of the forest soil (Figure 13.14). The newly fallen tree, its bark and wood intact, is a ready source of shelter and nutrients. The first to exploit this resource are bark beetles and wood-boring beetles that drill through the bark, feed on the inner bark and the cambium, reducing it to frass (feces) and fragments, and tunnel galleries in which to lay eggs. Both adults and larvae drill more tunnels as they feed. Ambrosia beetles tunnel into the sapwood, creating galleries in which fungi grow. The tunnels provide a passageway, and the frass and softened wood form a substrate for bacteria. Loosened bark provides cover for predatory insects that are soon to follow: centipedes, mites, pseudoscorpions, and beetles.

As decay proceeds, the softened wood holds more moisture, but the accessible nutrients have been depleted, leaving behind more complex, decay-resistant compounds. The pioneering arthropods leave for other logs. Fungi able to break down cellulose and lignin remain (see Chapter 7). Moss and lichens find the softened wood an ideal habitat. Plant seedlings, too, take root on the softened logs, and their roots penetrate the heartwood, providing a pathway for fungal growth into the depths of the log.

FIGURE 13.14 A succession of plant and animal species are involved in the decomposition of fallen logs.

Eventually the log is broken into light brown, soft, blocky pieces, and the bark and sapwood are gone. At this advanced stage of decay, the log provides the greatest array of microhabitats and the highest species diversity. Invertebrates of many kinds find shelter in the openings and passages; salamanders and mice dig tunnels and move into the rotten wood. Fungi and other microorganisms abound, and numerous species of mites feed on decomposed wood and fungi. At last the log crumbles into a red-brown mulchlike mound of lignin materials resistant to decay, its nutrients and energy largely depleted. The tree is incorporated into the soil.

As plant succession advances, animal life changes, too (Figure 13.15). Each successional stage has its own distinctive fauna. Because animal life is often influenced more by structural characteristics than by species composition (see Chapter 12, Section 12.13), successional stages of animal life may not correspond to the stages identified by plant ecologists.

During plant succession, animals can quickly lose their habitat by vegetation change. In eastern North America, grasslands and old fields support meadowlarks, meadow mice, and grasshoppers. When woody plants—both young trees and shrubs—invade, a new structural element appears. Grassland

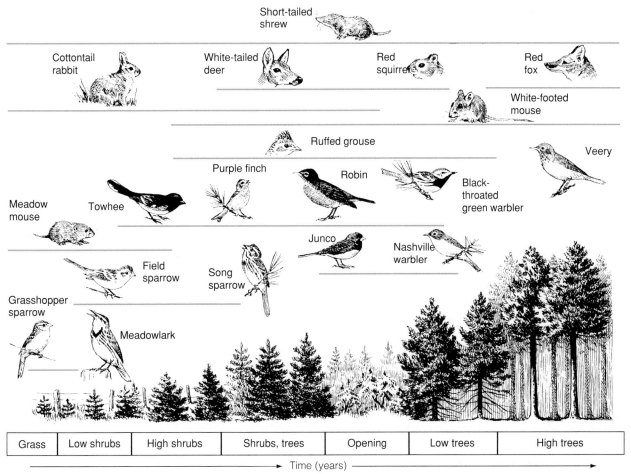

| Grass | Low shrubs | High shrubs | Shrubs, trees | Opening | Low trees | High trees |

Time (years)

FIGURE 13.15 Changes in the composition of animal species inhabiting various stages of plant succession, from old-field to conifer forest, in central New York. Species appear or disappear as vegetation density and height change. (Smith 1960.)

animals soon disappear, and shrubland animals take over. Towhees, catbirds, and goldfinches claim the thickets, and meadow mice give way to white-footed mice. As woody plant growth proceeds and the canopy closes, species of the shrubland decline, replaced by birds and insects of the forest canopy. As succession proceeds, the vertical structure becomes more complex. New species appear, such as tree squirrels, woodpeckers, and such birds of the forest understory as the hooded warblers and ovenbirds.

13.14 Community Structure Changes over Geological Time

Since its inception some 4.6 billion years ago, Earth has changed profoundly. Land masses emerged and broke into continents. Mountains formed and eroded, seas rose and fell, and ice sheets advanced to cover large expanses of the Northern and Southern Hemispheres and then retreated. All these changes affected the climate and other environmental conditions from one region of Earth to another. Many species of plants and animals evolved, disappeared, and were replaced by others. As environmental conditions changed, so did the distribution and abundance of plant and animal species.

Records of plants and animals composing past communities lie buried as fossils: bones, insect exoskeletons, plant parts, and pollen grains. The study of the distribution and abundance of ancient organisms and their relationship to the environment is paleoecology. The key to present-day distributions of animals and plants can often be found in paleoecological studies. For example, paleoecologists have reconstructed the distribution of plants in eastern North America following the last glacial maximum of the Pleistocene.

The Pleistocene was an epoch of great climatic fluctuations throughout the world. At least four times in North America and three times in Europe ice sheets advanced and retreated. With each movement, the biota retreated and advanced again with a somewhat different mix of species.

Each glacial period was followed by an interglacial period. The climatic oscillations in each interglacial period had two major stages, cold and temperate. During the cold stage, tundralike vegetation dominated the landscape. As the glaciers retreated, light-demanding forest trees such as birch and pine advanced. Then, as the soil improved and the climate warmed, these trees were replaced by more shade-tolerant species such as oak and ash. As the next glacial period began to develop, species such as firs and spruces took over the forest. Now both climate and soil began to deteriorate. Heaths began to dominate the vegetation, and forest species disappeared.

The last great ice sheet, the Laurentian, reached its maximum advance about 20,000 BP to 18,000 BP during the Wisconsin glaciation stage in North America. Canada was under ice. A narrow belt of tundra about 60 to 100 km wide bordered the edge of the ice sheet and probably extended southward into the high Appalachians. A few relict examples of this tundra persist today. Boreal forest, dominated by spruce and jack pine, covered most of the eastern and central United States as far as western Kansas.

As the climate warmed and the ice sheet retreated northward, plant species invaded the glaciated areas. The maps in Figure 13.16 reflect the advances of four major tree genera in eastern North America following the retreat of the ice sheet. Margaret Davis developed these maps from patterns of pollen deposition in sediment cores taken from lakes in eastern North America. By examining the presence and quantity of pollen deposited in sediment layers and dating the sediments with radiocarbon, she was able to obtain a picture of the spatial and temporal dynamics of tree communities over the past 18,000 years.

These analyses identify plants at the level of genus rather than species because, in many cases, we cannot identify species from pollen grains. Note that the rates at which different genera and probably species expanded their distribution northward with the retreat of the glacier are markedly different. The differences in the rates of range expansion are most likely due to the differences in the temperature responses of the species, the distances and rates at which seeds can disperse, and interactions among species. The implication is that, over the past 18,000 years, the distribution and abundance of species and the subsequent structure of forest communities in eastern North America have changed dramatically (Figure 13.17).

13.15 Two Contrasting Views of the Community

In the beginning of this chapter, we defined the community as group of species (populations) that occupy a given area, interacting either directly or indirectly. Interactions can have both a positive and a negative influence on species populations. How important are these interactions in determining community structure? This question led to a major debate in ecology in the first half of the twentieth century, a debate that still influences our views of the community.

When you walk through most forests, you see a variety of plant and animal species—a community. If you walk far enough, the dominant plant and animal species will change (see Figure 13.4). As you move from hilltop to valley, the structure of the community will differ. But what if you continue your walk over the next hilltop and into the adjacent valley? You will most likely notice that although the communities on the hilltop and valley are quite distinct, the communities on the two hilltops or valleys are quite similar. As a botanist might put it, they exhibit relatively consistent floristic composition. At the International Botanical Congress of 1910, botanists adopted the term **association** to describe this phenomenon. An association is a type of community with (1) relatively consistent species composition, (2) a uniform, general appearance (physiognomy), and (3) a distribution that is characteristic of a particular habitat, such as the hilltop or valley. Whenever the particular habitat or set of environmental conditions repeats itself in a given region, the same group of species occurs.

Some scientists of the time thought that association implied processes that might be responsible for structuring communities. The logic was that if clusters or groups of species repeatedly associate together, that is indirect evidence for either positive or neutral interactions among them. Such evidence favors a view of communities as integrated units. One of the leading proponents of this thinking was the Nebraskan botanist Frederick Clements. Clements developed what has become known as the **organismic concept of communities**. Clements likened associations to organisms, with each species representing an interacting, integrated component of the whole. The process of succession was viewed as the development of the organism.

As depicted in Figure 13.18a, in Clements' view the species in an association have similar distributional

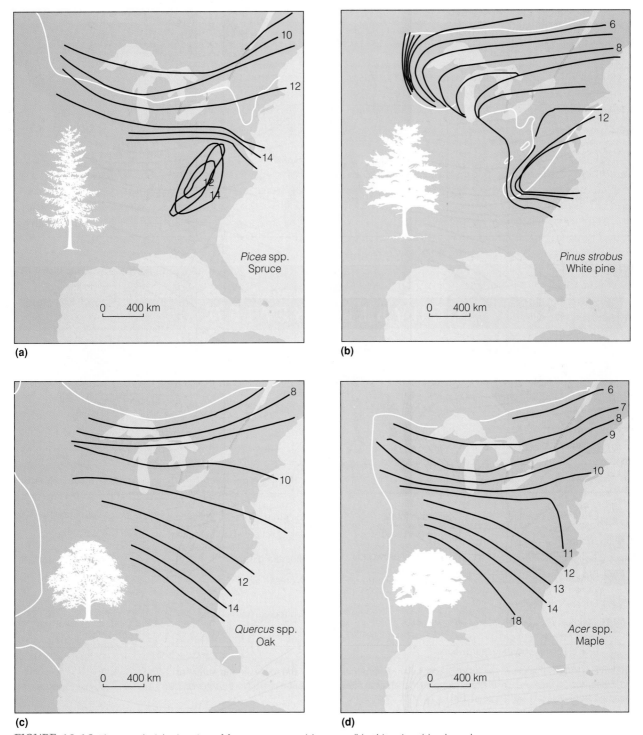

FIGURE 13.16 The postglacial migration of four tree genera: (a) spruce, (b) white pine, (c) oak, and (d) maple. The dark lines represent the leading edges of the northward-expanding populations. The white lines indicate the boundaries of the present-day ranges. The numbers are thousands of years before the present. (Davis 1981.)

limits along the environmental gradient, and many of them rise to maximum abundance at the same point. The transitions between adjacent communities (or associations) are narrow, with few species in common. This view of the community suggests a common evolutionary history and similar fundamental responses and tolerances (see Chapter 2: Focus on Ecology 2.3) for the component species.

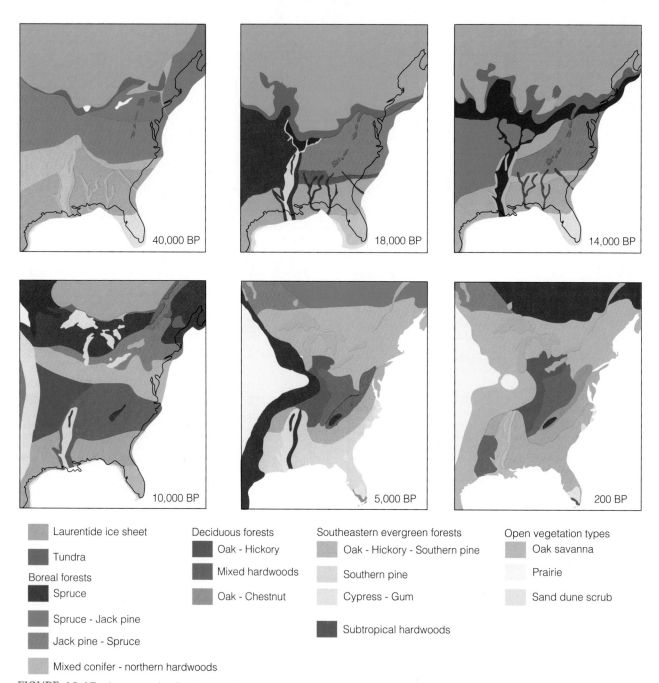

Laurentide ice sheet

Tundra

Boreal forests
Spruce

Spruce - Jack pine

Jack pine - Spruce

Mixed conifer - northern hardwoods

Deciduous forests
Oak - Hickory

Mixed hardwoods

Oak - Chestnut

Southeastern evergreen forests
Oak - Hickory - Southern pine

Southern pine

Cypress - Gum

Subtropical hardwoods

Open vegetation types
Oak savanna

Prairie

Sand dune scrub

FIGURE 13.17 Changes in the distribution of plant communities during and following the retreat of the Wisconsin ice sheet, reconstructed from pollen analysis at sites throughout eastern North America. (Adapted from Delcourt and Delcourt 1981.)

Mutualism and coevolution play an important role in the evolution of species making up the association. The community has evolved as an integrated whole; the species interactions are the "glue" that holds it together.

In contrast to Clements' organismal view of communities was the botanist H. A. Gleason's view of community. Gleason stressed the individualistic nature of species distribution. His view became known as the **individualistic continuum concept.** The continuum concept states that the relationship among coexisting species (species within a community) is a result of similarities in their requirements and tolerances, not a result of strong interactions or common evolutionary history. In fact, Gleason concluded that changes in species abundance along environmental

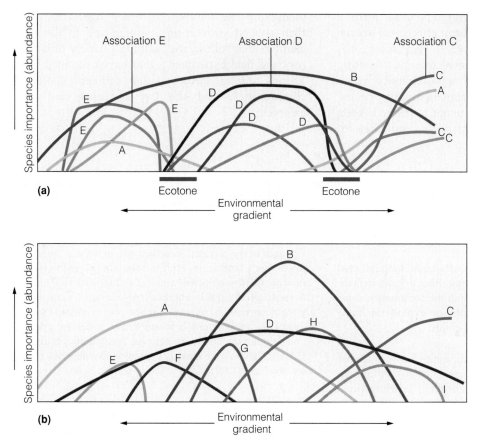

FIGURE 13.18 Two models of community. (a) The organismal or discrete view of communities proposed by Clements. Clusters of species (Cs, Ds, and Es) show similar distribution limits and peaks in abundance. Each cluster defines an association. A few species (for example, A) have sufficiently broad ranges of tolerance that they occur in adjacent associations, but in low numbers. A few other species (for example, B) are ubiquitous. (b) The individualistic or continuum view of communities proposed by Gleason. Clusters of species do not exist. Peaks of abundance of dominant species, such as A, B, and C, are merely arbitrary segments along a continuum.

gradients occur so gradually that it is not practical to divide the vegetation (species) into associations. In contrast to Clements' view, Gleason's view was that species distributions along environmental gradients do not form clusters, but represent the independent responses of species. Transitions are gradual and difficult to identify (Figure 13.18b). What we refer to as the community is merely the group of species found to coexist under any particular set of environmental conditions.

The major difference between these two views is the importance of interactions, evolutionary and current, in the structuring of communities. Although it is tempting to choose between these views, current thinking involves elements of both perspectives, as we will see.

13.16 Understanding Community Structure Requires Understanding Species Interactions

The field of ecology has a rich history of cataloguing the distribution and abundance of species and of mapping communities, an area of study known as natural history. Science, however, is more than cataloguing nature; it involves the search for mechanisms, the underlying processes that give rise to the patterns observed in nature. For ecological communities, the major focus of this search is the study of species interactions and how they vary across environmental gradients (changing environmental conditions). In most ecological communities, these interactions form a complex web of direct and indirect effects (see discussion of food webs in Section 13.4). As a result, it is virtually impossible for a scientist to simultaneously consider all components of a community when exploring the nature of interactions among species. As a compromise, scientists attempt to define subsets of the community that present likely candidates for identifying the nature of species' interactions and their role in structuring communities. In the three chapters that follow, we examine a variety of laboratory and field experiments that have examined interactions among species within different ecological communities, and the key concepts that have emerged.

CHAPTER REVIEW

Summary

Community Defined The group of species (populations) that occupy a given area, interacting either directly or indirectly, is called a community. Within the community, some species carry out similar functions or exploit the same resource. These groups are called guilds.

Species Relationships (13.1) Relationships between species can be positive (+), negative (−), or neutral (0). In mutualism (++), populations benefit each other; in competition (−−), both populations are affected adversely; in commensalism (+0), one population benefits and the other is unaffected; and in predation and parasitism (+−), one population benefits and the other is harmed.

Biological Structure (13.2–13.4) When a single or a few species predominate within a community, they are referred to as dominants. The dominants may be the most numerous, possess the greatest biomass, preempt the most space, or make the largest contribution to energy flow (13.2). Communities may be characterized by species diversity. Species diversity involves two components: species richness, the number of species in a community; and species evenness, how individuals are apportioned among the species (13.3). Feeding relationships can be graphically represented as a food chain—a series of arrows, each pointing from one species to another for which it is a source of food. Within a community, numerous food chains mesh into a complex food web with links leading from primary producers through an array of consumers (13.4).

Physical Structure (13.5–13.6) Communities are also characterized by physical structure. In terrestrial communities, the structure is largely defined by the vegetation. Vertical structure on land reflects the life forms of plants. In aquatic communities, it is largely defined by physical features such as light, temperature, and oxygen profiles (13.5). All communities possess an autotrophic and a heterotrophic layer. The autotrophic layer carries out photosynthesis. The heterotrophic layer uses carbon stored by the autotrophs as a food source. Vertical layering provides the physical structure in which many forms of animal life live (13.6).

Spatial Variation in Community Structure (13.7–13.9) Changes in the physical structure and biological communities across a landscape result in zonation. Zonation is common to all environments, aquatic and terrestrial. Zonation is most pronounced where sharp changes occur in the physical environment, as in aquatic communities (13.7). In most cases, transitions between communities are gradual, and defining the boundary between communities is difficult (13.8). The way we classify a community depends on the scale we use (13.9).

Temporal Variation in Community Structure (13.10–13.13) With time, natural communities change. This gradual sequential change in the relative abundance of species in a community is succession. Opportunistic, early successional species yield to late successional species (13.10). Succession occurs in all environments. The similarity of successional patterns in different environments suggests a common set of processes (13.11). Primary succession begins on sites devoid of or unchanged by organisms. Secondary succession begins on sites where organisms are already present (13.12).

Changes in the heterotrophic component of the community also occur during succession. The succession of organisms involved in the decomposition of fallen logs in a forest provide an example. Successional changes in vegetation affect the nature and diversity of animal life. Certain sets of species are associated with the structure of vegetation found during each successional stage (13.13).

Long-Term Changes (13.14) The present pattern of vegetational distribution reflects the glacial events of the Pleistocene. Plants retreated and advanced with the movements of the ice sheets. The rates and distances of their advances are reflected in the present-day ranges of species and the distribution of plant communities.

Concept of the Community (13.15) Historically, there have been two contrasting concepts of the community. The organismal concept views the community as a unit, an association of species, in which each species is a component of the integrated whole. The individualistic concept views the co-occurrence of species as a result of similarities in requirements and tolerances.

Understanding Species Interactions (13.16) A major focus of community ecology is the study of species interactions and how they vary across environmental gradients (changing environment conditions).

Study Questions

1. Define community.

2. How should we measure species dominance?

3. Distinguish between species richness and species evenness.

4. Diversity is greatest in tropical rain forests. What does this fact tell you about dominance and species distribution there?

5. What is a food web? Contrast direct and indirect interactions within a food web.

6. Contrast the vertical stratification of an aquatic community with that of a terrestrial community.

7. If two communities have the same values for species diversity, do they necessarily have a high percent similarity value? Explain your answer.

8. Contrast zonation and succession. Provide examples of each.

9. Discuss some problems that arise in delineating and classifying communities.

10. Distinguish between primary succession and secondary succession.

11. What is the relationship of plant succession to animal habitats?

12. Locate an area with which you were familiar years ago, and compare the vegetation then and now. What changes have taken place? What brought them about?

13. Distinguish between the organismal and individualistic concepts of the community.

Spotted hyenas *(Crocuta crocuta)*, jackals *(Canus aureus),* and vultures vie for the carcass of a wildebeest.

Interspecific Competition

OBJECTIVES

On completion of this chapter, you should be able to:

- Describe the types of interactions between competing species.
- Describe the relationship between the logistic growth model and the Lotka-Volterra model of competition.
- Describe four theoretical outcomes of interspecific competition.
- Explain how potentially competing species may coexist.
- Discuss the role of nonresource environmental factors in competition.
- Explain how environmental variation can influence competition among species.
- Explain resource partitioning.
- Relate the concept of the niche to interspecific competition.
- Discuss the difficulty in determining the role of competition in structuring communities.

In Chapter 3 of *The Origin of Species*, Charles Darwin wrote: "as more individuals are produced than can possibly survive, there must in every case be a struggle for existence, either one individual with another of the same species, or with the individuals of distinct species, or with the physical conditions of life." The concept of interspecific competition is one of the cornerstones of evolutionary ecology. Darwin based his idea of natural selection on competition, the "struggle for existence." Because it is advantageous for individuals to avoid this struggle, competition has been regarded as the major force behind species divergence and specialization. Nevertheless, the role of interspecific competition in structuring communities is controversial in ecology and is often difficult to detect and quantify.

14.1 Interspecific Competition Involves Two or More Species

The relationship in which the populations of both species are affected adversely (−−) is interspecific competition. In interspecific competition, as in intraspecific competition, individuals seek a common resource in short supply, but in interspecific competition the individuals are of two or more species. Both kinds of competition may take place simultaneously. Gray squirrels, for example, compete among themselves for acorns during a year when the production of acorns by oak trees is low. At the same time, white-footed mice, white-tailed deer, wild turkey, and blue jays vie for the same crop. Because of competition, one or more of these species may broaden the base of their foraging efforts. Populations of various species may be forced to turn away from acorns to food that is less in demand.

Like intraspecific competition, interspecific competition takes two forms: exploitation and interference (see Chapter 11, Section 11.2). As an alternative to this simple dichotomous classification of competitive interactions, Thomas Schoener of the University of California at Davis proposed that six different types of interactions are sufficient to account for most instances of interspecific competition: (1) consumption, (2) preemption, (3) overgrowth, (4) chemical interaction, (5) territorial, and (6) encounter.

Consumption competition occurs when individuals of one species inhibit individuals of another through the consumption of a shared resource, such as the competition among various animal species for acorns described above. Preemptive competition occurs primarily among sessile organisms, such as

barnacles, where the occupation by one individual precludes the establishment (occupation) by others. Overgrowth competition occurs when one organism literally grows over another (with or without physical contact), inhibiting access to some essential resource. An example of this interaction is when a taller plant shades those individuals below, reducing available light, as seen in Chapter 4. In chemical interactions, chemical growth inhibitors or toxins released by an individual inhibit or kill other species. Allelopathy in plants (see Focus on Ecology 14.1: A Chemical Solution to Competition in Plants), in which chemicals produced by some plants inhibit germination and establishment of other species, provides an example of this type of species interaction. Territorial competition results from the behavioral exclusion of others from a specific space that is defended as a territory (see Chapter 11, Section 11.10), while encounter competition results when nonterritorial encounters between individuals result in a negative effect on one or both of the participant species. Various species of scavengers fighting over the carcass of a dead animal provide an example of the latter type of interaction (see photograph on page 269).

14.2 There Are Four Possible Outcomes of Interspecific Competition

In the early part of the twentieth century, two mathematicians, the American Alfred Lotka and the Italian Vittora Volterra, independently arrived at mathematical expressions to describe the relationship between two species using the same resource. Both began with the logistic equation for population growth that we developed in Chapter 10:

$$\frac{dN}{dt} = rN \left(\frac{K - N}{K} \right)$$

Next, both modified the logistic equation for each species by adding to it a term to take into account the competitive effect of one species on the population growth of the other. For species 1 this term is αN_2. N_2 is the population density of species 2 and α is the competition coefficient that quantifies the per capita effect of species 2 on species 1. Similarly, for species 2, the term is βN_1, where β is the per capita competition coefficient that quantifies the per capita effect of species 1 on species 2. The competition coefficients can be thought of as factors for converting an individual of one species into the equivalent number of

A CHEMICAL SOLUTION TO COMPETITION IN PLANTS

A particular form of interference competition among plants is **allelopathy,** the production and release of chemical substances by one species that inhibit the growth of other species. These substances range from acids and bases to simple organic compounds that reduce competition for nutrients, light, and space. Produced in profusion in natural communities as secondary substances, most compounds remains innocuous. A few, however, influence community structure. For example, broom sedge (*Andropogon virginicus*) produces chemicals that inhibit the invasion of old fields by shrubs and thus maintains its dominance. Bracken fern *(Pteridium aquilinum),* the most widely distributed vascular plant in the world, produces plant poisons that accumulate in the upper surface of the soil. These phytotoxins kill the germinating seeds of many plants, especially conifers, and reduce the growth of seedlings. These allelopathic effects, along with the heavy, smothering overwinter accumulation of dead fronds, allow bracken ferns to dominate large areas of ground. Likewise, the black walnut *(Juglans nigra)* of the eastern North American deciduous forest is antagonistic to many plants.

In desert shrub communities, a number of shrubs (*Larrea, Franseria,* and others) release toxic compounds to the soil through rainwater. Under laboratory conditions, at least, these substances inhibit germination and growth of seedlings of annual herbs. In southern California, two shrubs—sagebrush *(Artemisia)* and sage *(Salvia)*—that commonly invade annual grasslands release aromatic compounds called terpenes (camphor is one such example) to the air. These terpenes are adsorbed from the atmosphere onto soil particles. In certain clay soils, these terpenes accumulate during the dry season in quantities sufficient to inhibit the germination and growth of herbaceous seedlings. As a result, invading patches of shrubs are surrounded by belts devoid of herbs and by wider belts in which the growth of grassland plants is reduced. Allelopathy may not be the only reason for belts devoid of vegetation. Studies of plant-animal interactions suggest that although plants do produce toxins, the absence of plants may result from grazing by hares and consumption of seeds by rodents, birds, and ants.

---

individuals of the competing species based on their shared use of the resources that define the carrying capacities. An individual of species 1 is equal to β individuals of species 2. Likewise, an individual of species 2 is equivalent to α individuals of species 1. These terms, in effect, convert the population density of the one species into the equivalent density of the other. Now we have a pair of equations, which consider both intraspecific and interspecific competition.

Species 1: $\dfrac{dN_1}{dt} = r_1 N_1 \left(\dfrac{K_1 - N_1 - \alpha N_2}{K_1} \right)$

Species 2: $\dfrac{dN_2}{dt} = r_2 N_2 \left(\dfrac{K_2 - N_2 - \beta N_1}{K_2} \right)$

As you can see, in the absence of interspecific competition—either α or $N_2 = 0$ in equation 1 and β or $N_1 = 0$ in equation 2—the population of each species grows logistically to equilibrium at K, or carrying capacity. In the presence of competition, the picture changes.

For example, the carrying capacity for species 1 is K_1, and as N_1 approaches K_1 the population growth

(dN_1/dt) approaches zero. However, species 2 is also vying for the limited resource that determines K_1, so we must consider the impact of species 2. Since α is the per capita effect of species 2 on species 1, the total effect of species 2 on species 1 is αN_2. We must consider the effects of both species in calculating population growth. As the combined population effect $(N_1 + \alpha N_2)$ approaches K_1, the growth rate of species 1 will approach zero. The greater the density of the competing species (N_2), the greater the influence on reducing the growth rate (dN_1/dt) of species 1.

Depending upon the combination of values for the Ks and for α and β, the Lotka-Volterra equations predict four different outcomes. Figure 14.1 graphically depicts all the possible outcomes. In two situations, one species wins out over the other. In one case, species 1 inhibits further increase in species 2 while continuing to increase. In this case, species 2 is driven to extinction. In the other case, species 2 inhibits further increase in species 1 while continuing to increase, and species 1 disappears. In the third situation, each species, when abundant, inhibits the growth of the other species more than it inhibits

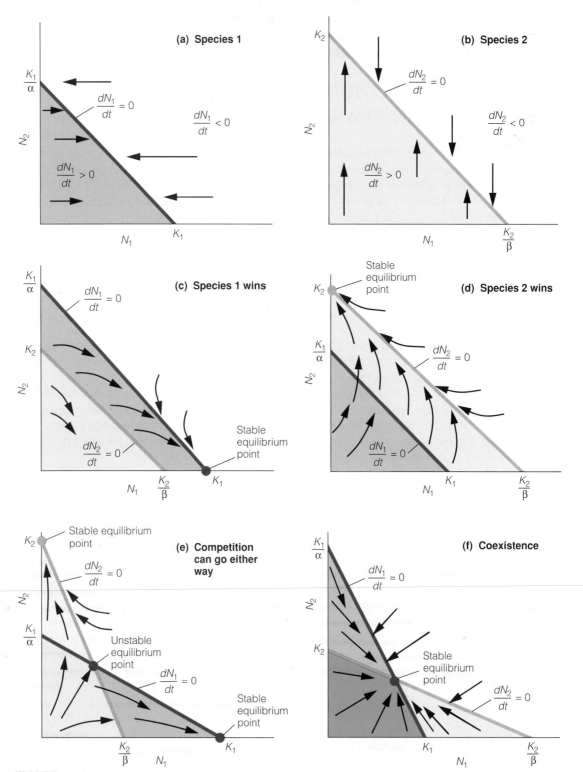

FIGURE 14.1 The Lotka-Volterra model of competition between two species. (a–b) In the absence of competition, populations of species 1 and 2 come to equilibrium. In the shaded area below the line, the population growth is positive and the population increases (as indicated by the arrows), whereas above the line population decreases. (c) The isocline of species 1 falls outside the isocline of species 2. Species 1 always wins, leading to the extinction of species 2. (d) The situation is the reverse of (c). (e) Each species inhibits the growth of the other more than its own growth. The more abundant species often wins. (f) Each species inhibits the growth of its own population more than that of the other by intraspecific competition. The species coexist.

its own growth. Eventually, one or the other species wins, driving the other to extinction. The outcome depends upon the initial population densities of the two species (which species is most abundant). In the fourth situation, neither population can achieve a density capable of eliminating the other. Each species inhibits its own population growth more than that of the other species.

In each graph the x axis represents the population size of species 1 and the y axis the population size of species 2. Two lines are plotted on each graph, one representing each of the two species. The diagonal line for species 1 represents the combined population densities of species 1 and 2 that equal K_1 and therefore $dN/dt = 0$. The line for species 2 represents the combined population densities of species 1 and 2 that equal K_2. For any point on the species 1 line, $N_1 + N_2 = K_1$. When $N_1 = K_1$, then N_2 must be zero. Since α is the per capita effect of species 2 on species 1, the population density of species 2 that is exactly equivalent to the carrying capacity of species 1 ($\alpha N_2 = K_1$) will be $N_2 = K_1/\alpha$. Therefore, when $N_2 = K_1/\alpha$, then N_1 must be equal to zero. For the line describing species 2, the values where the line crosses the axes will be K_2 and $N_1 = K_2/\beta$, the population density of species 1 that is equal to the carrying capacity of species 2.

The diagonal lines are called **zero growth isoclines**. The zero growth isoclines for species 1 and 2 are shown in Figures 14.1a and 14.1b respectively. In the space below the line, or isocline, for species 1 (Figure 14.1a), combinations of N_1 and N_2 are below carrying capacity; therefore, the population is increasing (dN_1/dt greater than zero). This increase is represented by the arrows that are parallel to the x axis and point in the direction of increasing values of N_1. In the space above the isocline, combinations of N_1 and N_2 are greater than carrying capacity; therefore, the population is declining (dN_1/dt is negative). In this region the arrows point in the direction of decreasing population size. Figure 14.1b depicts the analogous situation for species 2.

Figures 14.1c–f depict the possible outcomes when the two isoclines are combined. In Figure 14.1c, the isocline of species 1 is parallel to, and lies outside, the isocline of species 2. In this case, even when the population of species 2 is at its carrying capacity (K_2), its density cannot stop the population of species 1 from increasing ($K_2 < K_1/\alpha$). As species 1 continues to increase, species 2 eventually becomes extinct. In Figure 14.1d, the situation is reversed. Species 2 wins, leading to the exclusion of species 1.

In Figures 14.1e and 14.1f, the isoclines cross, but the outcomes of competition are quite different. In Figure 14.1e, note that along the x axis, the value of K_2/β is less than the value of K_1. Recall that the value

$N_1 = K_2/\beta$ is the density of species 1 exactly equal to the carrying capacity of species 2 (K_2). Because K_2/β is less than K_1, species 1 can achieve population densities that would exceed the density required to drive the population of species 2 to extinction. Likewise, because K_1/α is less than K_2 along the y axis, species 2 can achieve population densities high enough to drive species 1 to extinction. Which species "wins" the competition depends upon the initial populations of the two species.

In Figure 14.1f, the result of competition differs in that neither species can exclude the other. The result is coexistence. Note that in this case, K_1 is less than K_2/β. The population of species 1 can never reach a density sufficient to eliminate species 2. For this to happen, the population of species 1 would have to reach $N_1 = K_2/\beta$. Likewise, K_2 is less than K_1/α, so species 2 can never achieve a high enough population density to eliminate species 1. Intraspecific competition (K) inhibits the growth of each population more than interspecific competition inhibits the growth of the other species.

14.3 Laboratory Experiments Support the Lotka–Volterra Equations

The theoretical Lotka-Volterra equations stimulated studies of competition in the laboratory, where under controlled conditions the outcome is more easily determined than in the field. One of the first to study the Lotka-Volterra competition model experimentally was the Russian biologist G. F. Gause. He used two species of *Paramecium*, *P. aurelia* and *P. caudatum*. *P. aurelia* has a higher rate of increase than *P. caudatum* and can tolerate a higher population density. When he introduced both into one tube containing a fixed amount of bacterial food, *P. caudatum* died out (Figure 14.2). In another experiment, Gause reared the loser, *P. caudatum*, with another species, *P. bursaria*. These two species coexisted because *P. caudatum* fed on bacteria suspended in solution, whereas *P. bursaria* confined its feeding to bacteria at the bottom of the tubes. Each used food unavailable to the other.

In the 1940s and 1950s, Thomas Park at the University of Chicago conducted a number of classic competition experiments with laboratory populations of flour beetles. He found that the outcome of competition between *Tribolium castaneum* and *T. confusum* depended upon environmental temperature,

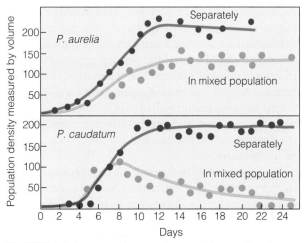

FIGURE 14.2 Competition experiments with two ciliated protozoans, *Paramecium aurelia* and *P. caudatum,* grown separately and in a mixed culture. In a mixed culture, *P. aurelia* outcompetes *P. caudatum,* and the result is competitive exclusion. (Gause 1934.)

humidity, and fluctuations in the total number of eggs, larvae, pupae, and adults. Often the outcome of competition was not determined until many generations had passed.

In a much later study, David Tilman of the University of Minnesota and his associates grew laboratory populations of two species of diatoms, *Asterionella formosa* and *Synedra ulna*. Both species require silica for the formation of cell walls. The researchers monitored not only population growth and decline but also the level of silica in the water. When grown alone in a liquid medium to which silica was continually added, both species kept silica at a low level. When grown together, *S. ulna* took silica to a level too low for *A. formosa* to survive and reproduce (Figure 14.3). By reducing resource availability, *Synedra* drove *Asterionella* to extinction.

In both of these laboratory experiments, the patterns were similar. The two species grew exponentially at low densities. As their densities increased, their population growth declined. Each began to influence the other. Under these conditions, individuals of one species had a greater fitness than those of the other.

14.4 Studies Support the Competitive Exclusion Principle

In three of the four situations predicted by the Lotka-Volterra equations, one species drives the other to extinction. The results of the laboratory studies

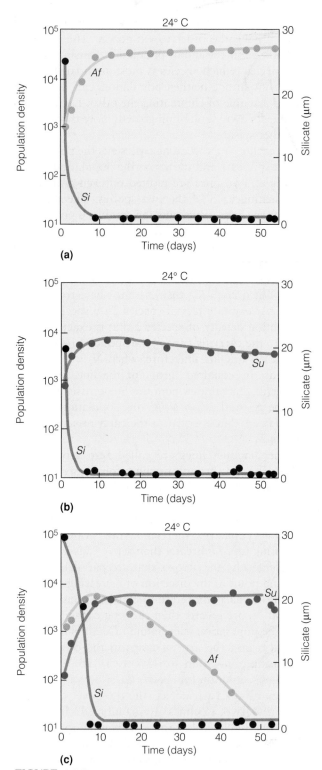

FIGURE 14.3 Competition between two species of diatom, *Asterionella formosa* (Af) and *Synedra ulna* (Su), for silica. (a–b) Grown alone in a culture flask, both species reach a stable population that keeps silica (Si) at a constant low level. *Synedra* draws silica lower. (c) When the two are grown together, *Synedra* reduces silica to a point at which the other species dies out. (Tilman et al. 1981.)

presented above tend to support the mathematical models. These and other observations have led to the concept called Gause's principle (named after the Russian biologist G. F. Gause), although the idea was far from original with him. More recently called the **competitive exclusion principle,** it states that complete competitors cannot coexist. What the term *complete competitors* means is that the two species (noninterbreeding populations) possess exactly the same ecological requirements and live in exactly the same place. Under this set of conditions, if population A increases the least bit faster than population B, then A eventually will occupy the area completely and B will become extinct.

Competitive exclusion, then, invokes more than competition for a limited resource. The competitive exclusion principle involves assumptions about both the species involved and the environment in which they exist. First, it assumes that the competitors have exactly the same resource requirement and that both competitors remain genetically unchanged (natural selection does not function to reduce competition). Secondly, it assumes that immigrants from other areas with different environmental conditions do not move into the population of the losing species. The competitive exclusion principle also assumes that environmental conditions remain constant. Such conditions rarely if ever exist. Because they are not ecologically identical, species can compete for an essential resource without meeting the criteria for complete competitors. The idea of competitive exclusion, however, has stimulated a more critical look at competitive relationships in natural situations, including studies to determine what ecological conditions are necessary for coexistence of species that share a common resource base. Important factors include influence of nonresource environmental constraints on competitive ability, spatial and temporal variations in resources, competition for multiple resources, and resource partitioning. We shall examine each topic and how it functions to influence the nature of competition in the following sections.

14.5 Competition Is Influenced by Nonresource Factors

Interspecific competition involves individuals of two or more species vying for the same limited resource. The relative abilities of the different species to compete for the shared resource are influenced by a wide variety of factors. For example, in the early stages of secondary succession in an old-field community, a wide variety of plant species colonize the site and compete for essential resources, including space. The species that have high rates of photosynthesis and allocate carbon to height growth and the production of leaves will overgrow and shade other species, preempting space and gaining access to the essential resource of light. At first, it would seem that the set of characteristics (adaptation) allowing fast growth under high light conditions (see Chapter 6, Section 6.7) would be the decisive factor determining the superior competitor(s). Field experiments have shown this to be true, with shade-intolerant, fast-growing species quickly dominating these environments. However, features of the environment other than light also have a direct influence on the germination, establishment, and growth rates of species and the outcome of competition. Environmental factors such as temperature, soil or water pH, relative humidity, and salinity directly influence physiological processes related to plant growth and reproduction, but are not consumable resources over which species compete.

Fakhri Bazzaz of Harvard University examined the responses of a variety of annual plants that dominate during the early stages of old-field succession to variations in air temperature. Species differed significantly in the range of temperatures over which maximum rates of seed germination occurred (Figure 14.4). These differences among species in germination rates have a direct influence on patterns of seedling establishment, subsequent competition for resources, and the structure of old-field communities during the early stages of succession. Differences in the temperature responses of these species can result in year-to-year variation in community composition within an old field as a result of differences in temperatures during the early growing season. Similar results have been observed in salt marsh communities, where variations in salinity (both spatial and temporal) can shift the relative competitive ability of plant species based on their salt tolerance.

14.6 Temporal Variation in the Environment Influences Competitive Interactions

One species excludes another when it exploits a shared limiting resource more effectively, resulting in an increase in its population at the expense of its competitor. However, as with the example of the annual plant species studied by Fakhri Bazzaz, each

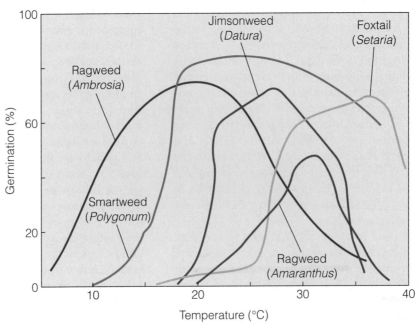

FIGURE 14.4 Patterns of seed germination of five annual plant species along a gradient of temperature. These species dominate the early stages of secondary succession in field communities of the midwestern United States. (Bazzaz 1996.)

species does best under particular conditions (Figure 14.4). Low temperature may favor one competitor; if temperatures rise over time, the advantage may shift to another species. When environmental conditions vary through time, the competitive advantages also change. As a result, no one species will reach sufficient density to displace its competitors. In this manner, environmental variation allows competitors to coexist, where under constant conditions, one would exclude the other.

A similar example is provided by the work of Peter Dye in the grasslands of southern Africa. He examined annual variations in the biological structure of a savanna community in southwest Zimbabwe. From 1971 to 1981, the dominant grass species shifted from *Urochloa mosambicensis* to *Heteropogon contortus* (Figure 14.5a). This shift in species dominance is similar to that seen in the examples of succession presented in Chapter 13 (see Figures 13.10 and 13.11); however, this observed shift in dominance is a result of yearly variations in rainfall (Figure 14.5b). Rainfall during the 1971–1972 and 1972–1973 rainy seasons was much lower than average. *Urochloa mosambicensis* can maintain higher rates of survival and growth under dry conditions than can *Heteropogon contortus*, making it a better competitor under conditions of low rainfall. With the return to higher rainfall during the remainder of the decade, *Heteropogon contortus* became the dominant grass species. Annual rainfall in this semiarid region of southern Africa is highly variable,

and fluctuations in species composition such as those shown in Figure 14.5 are a common feature of the community.

In addition to shifting the relative competitive abilities of species, variation in climate can function as a density-independent limitation on population density (see Chapter 11, Section 11.12). Periods of drought or extreme temperatures may function to depress populations below carrying capacity. If these events are frequent enough relative to the time required for the population to recover (approach carrying capacity), resources may be sufficiently abundant during the intervening periods to reduce or even eliminate competition.

14.7 Competition Occurs for Multiple Resources

In many cases, competition between species involves multiple resources, and often competition for one influences the ability of an organism to access other resources. One such example is the practice of interspecific territoriality, where competition for space influences access to food and nesting sites (see Chapter 11, Section 11.10).

A wide variety of bird species in the temperate and tropical regions exhibit interspecific territoriality. Most often, this practice involves the defense of territories against closely related species, such as the

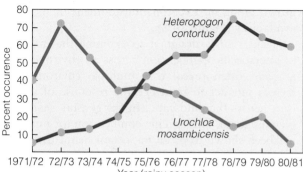

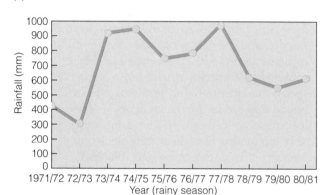

(a)

(b)

FIGURE 14.5 (a) Shift in the dominant grass species in a savanna community in southwest Zimbabwe over the period 1971–1981. (b) *Urochloa mosambicensis* was able to successfully compete under the drier conditions during the 1971–1972 and 1972–1973 rainy seasons. With the increase in rainfall beginning in the 1973–1974 season, *Heteropogon contortus* came to dominate the site. (Adapted from Dye and Spear 1982.)

gray and dusky flycatchers of the western United States. Some bird species, however, defend their territories against a much broader range of potential competitors. For example, the acorn woodpecker *(Melanerpes formicivorus)* defends territories against jays and squirrels, as well as against other species of woodpeckers. Strong interspecific territorial disputes likewise occur among brightly colored coral reef fish.

Competition among plants provides numerous examples of how competition for one resource can influence the ability of an individual to exploit other essential resources, leading to compounding effects on growth and survival. R. H Groves and J. D. Williams examined competition between populations of subterranean clover *(Trifolium subterraneum)* and skeleton weed *(Chondrilla juncea)* in a series of greenhouse experiments. Plants were grown in both monocultures (single populations) and in mixtures (two populations combined). The investigators used a unique experimental design to determine the independent effects of competition for above-ground (light) and below-ground (water and nutrients) resources. In the monocultures, plants were grown in pots, allowing for both the canopies (leaves) and roots to intermingle. In the two-species mixtures, three different approaches were used (Figure 14.6): (1) plants of both species were grown together in the pot, allowing both the canopies and roots to intermingle; (2) plants of both species were grown in the same soil but with their canopies separated; and (3) species were grown in separate soil (pots) with their canopies intermingled.

Clover was not significantly affected by the presence of skeleton weed; however, the skeleton weed was affected in all three treatments where the two populations were grown together. When the roots were allowed to intermingle, the biomass (dry weight of plant population) of skeleton weed was reduced by 35 percent of the biomass for the species when grown as a monoculture. The biomass was reduced by 53 percent when the canopies were intermingled. When both the canopies and roots were intermingled, the biomass was reduced by 69 percent, indicating an interaction in the competition for above-ground and below-ground resources. Clover plants were the superior competitor for both above- and below-ground resources, resulting in a compounding effect of competition for these two resources. This type of interaction has been seen in a variety of laboratory and field experiments. The species with the faster growth rate overtops and shades the slower-growing species, reducing growth and demand for below-ground resources. This functions to increase access to resources and further growth by the superior competitor.

Grown alone   Root competition   Shoot competition   Root and shoot competition

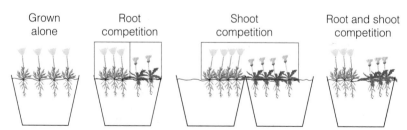

FIGURE 14.6 Experimental design used to examine above- and below-ground competition between subterranean clover and skeleton weed. Boxes enclosing above-ground plant parts (stems, leaves, and flowers) represent partitions to separate the two species. (Groves and Williams 1975.)

14.8 Relative Competitive Abilities Change along Environmental Gradients

As environmental conditions change, so will the relative competitive abilities of species. The shifts in competitive ability can result either from changes in the carrying capacities of species (values of K; see Section 14.2) related to a changing resource base, or from changes in the physical environment that interact with resource availability.

Numerous laboratory and field studies have examined the outcomes of competition among plant species across experimental gradients of resource availability. Mike Austin and colleagues at the CSIRO research laboratory in Canberra, Australia, have conducted a number of greenhouse studies to explore the changing nature of interspecific competition among plant species across experimental gradients of nutrient availability. Results of these studies show that the relative competitive abilities of the various plant species involved in the experiments change under different nutrient concentrations (Figure 14.7). Stewart Pickett and Fakhri Bazzaz found similar shifts in the competitive ability of summer annuals grown along an experimental gradient of moisture availability (Figure 14.8).

Field studies designed to examine the influence of interspecific competition across an environmental gradient often reveal that multiple environmental factors interact to influence the response of organisms across the landscape. In the coastal regions of New England, interspecific competition for nutrients has been shown to be an important factor determining the patterns of salt-marsh plant zonation (see Figure 13.5). However, the relative competitive abilities of species for nutrients is influenced by the ability of plant species to tolerate a gradient of increasing physical stress relating to waterlogging, salinity, and oxygen availability in the soil and sediments (Figure 14.9). The upper boundaries (toward shoreline) of species distribution are set by interspecific competition for nutrients, while the lower boundaries are set by the ability to tolerate the physical stress associated with increasing water depth.

Chipmunks furnish a striking example of the interaction of competition and tolerance to physical stress in determining species distribution along an environmental gradient. In this case, physiological tolerance, aggressive behavior, and restriction to habitats in which one organism has competitive advantage all play a part. On the eastern slope of the Sierra Nevada Mountains live four species of chipmunks: the alpine chipmunk *(Eutamias alpinus)*, the

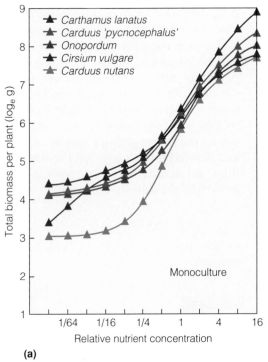

(a)

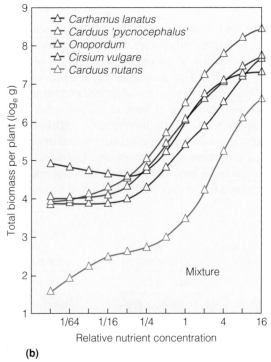

(b)

FIGURE 14.7 Response of six thistle species to an experimental gradient of nutrient availability: (a) single species populations (monocultures), and (b) mixed populations. (Data from Austin et al. 1986.)

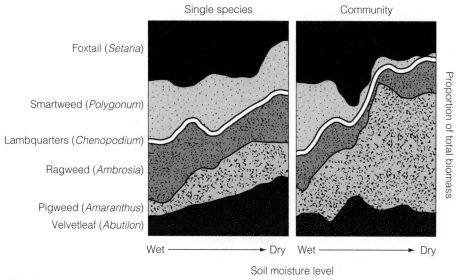

FIGURE 14.8 Differential growth response of summer annuals on a moisture gradient grown as individuals and in a community (as mixture). (Data from Pickett and Bazzaz 1978.)

lodgepole pine chipmunk *(E. speciosus),* the yellow pine chipmunk *(E. amoenus),* and the least chipmunk *(E. minimus),* at least three of which have strongly overlapping food requirements. Each species occupies a different altitudinal zone (Figure 14.10).

The line of contact is determined partly by interspecific aggression. Aggressive behavior by the dominant yellow pine chipmunk determines the upper range of the least chipmunk. Although the least chipmunk can occupy a full range of habitats from

sagebrush desert to alpine fields, it is restricted in the Sierras to sagebrush habitat. Physiologically, it is more capable of handling heat stress than the others, enabling it to inhabit extremely hot, dry sagebrush. If the yellow pine chipmunk is removed from its habitat, the least chipmunk moves into vacated open pine woods. However, if the least chipmunk is removed from the sagebrush habitat, the yellow pine chipmunk does not invade the habitat. The aggressive behavior of the lodgepole pine chipmunk in turn

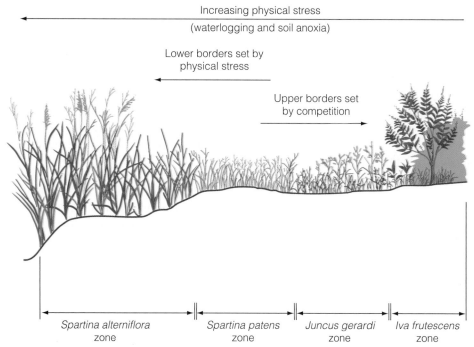

FIGURE 14.9 Zonation of the dominant perennial plant species in a New England salt-marsh community. (From Emery et al. 2001.)

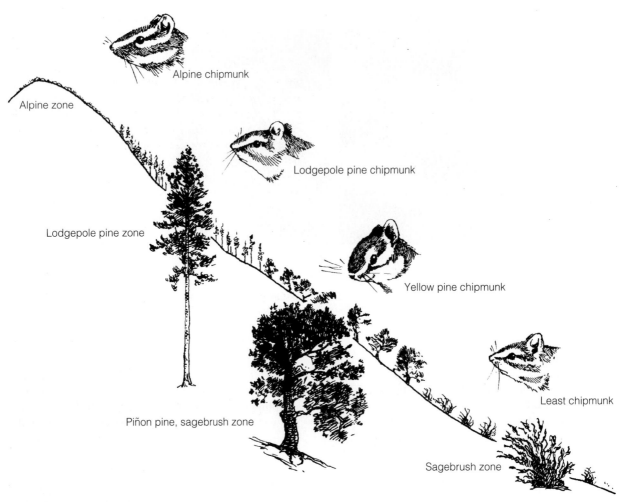

FIGURE 14.10 A transect of the Sierra Nevada Mountains in California, 38° north latitude, showing vegetation zonation and the altitudinal ranges of four species of chipmunks *(Eutamias)* on the east slope. (After Heller-Gates 1971.)

determines the upper limit of the yellow pine chipmunk. The lodgepole pine chipmunk is restricted to shaded forest habitat because it is vulnerable to heat stress. Most aggressive of the four, the lodgepole pine chipmunk also may limit the downslope range of the alpine chipmunk. Thus, the range of each chipmunk is determined both by aggressive exclusion and by its ability to survive and reproduce in a habitat hostile to the other species.

14.9 Species Coexistence Involves Partitioning Available Resources

All terrestrial plants require light, water, and essential nutrients such as nitrogen and phosphorus. Consequently, competition between various co-occurring species ought to be common. The same

should be true for the variety of insect-feeding bird species inhabiting the canopy of a forest, large mammalian herbivores feeding on the grasslands of east Africa, and predatory fish species that make the coral reef their home. How is it that these diverse arrays of potential competitors can coexist in the same community? Observations of similar species sharing the same habitat suggest that they coexist by partitioning available resources. Animals use different kinds and sizes of food, feed at different times, or forage in different areas. Plants require different proportions of nutrients or have different tolerances for light and shade. Each species exploits a portion of the resources unavailable to others, resulting in differences among co-occurring species that would not be expected purely due to chance.

Field studies provide many examples of apparent resource partitioning. One example involves three species of annual plants growing together on prairie soil abandoned one year after plowing. Each plant exploits a different part of the soil resource (Figure

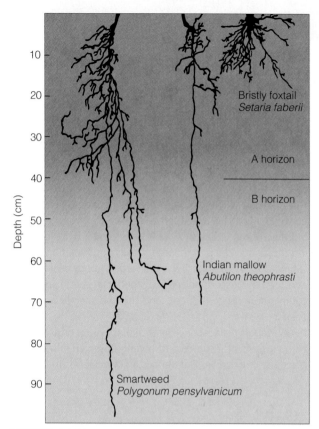

FIGURE 14.11 Partitioning of the prairie soil resource at different levels by three species of annual plants one year after disturbance.

branched taproot extending to intermediate depths, where moisture is adequate during the early part of the growing season but is less available later on. The plant is able to carry on photosynthesis at a low water availabilty (see Chapter 6, Section 6.16). The third species, smartweed *(Polygonum pensylvanicum)*, possesses a taproot that is moderately branched in the upper soil layer and develops mostly below the rooting zone of other species, where it has a continuous supply of moisture. The response of these three plant species to variation in water availability can be seen in the results of the water gradient experiment shown in Figure 14.8.

Tamar Dayan, at Tel Aviv University, examined possible resource partitioning in a group of coexisting species of wild cats inhabiting the Middle East. Dayan and colleagues found clear evidence of nonrandom differences among coexisting species in the size of the canine teeth, which are crucial to the way in which cats capture and kill their prey. For cats, canine teeth are used to kill prey, so there is a general relationship between their size and the prey species selected. The pattern observed suggests an exceptional regularity in the spacing of species along an axis defined by the size of canine teeth (Figure 14.12).

These patterns of resource partitioning have the effect of reducing competition among co-occurring species. By dividing the resource, each species reduces direct competition with the others. Resource partitioning results from specific physiological, morphological, or behavioral adaptations (see Chapter 2) that allow individuals access to essential resources and therefore influence fitness. Since these differences also function to reduce competition among co-occurring species, they are often regarded as an outcome of interspecific competition in the past. Competition is at the heart of Darwin's theory of

14.11). Bristly foxtail *(Setaria faberii)* has a fibrous, shallow root system that draws on a variable supply of moisture. It recovers rapidly from drought, takes up water rapidly after a rain, and carries on a high rate of photosynthesis even when partially wilted. Indian mallow *(Abutilon theophrasti)* has a sparse,

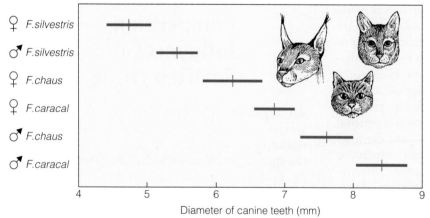

FIGURE 14.12 Size (diameter) of canine teeth for small cat species that co-occur in Israel. Note the regular pattern of differences in size between species. Size is correlated with the size of prey consumed by each. (From Dayan et al. 1990.)

GHOSTS OF COMPETITION PAST

Many ecologists have viewed the apparent partitioning of resources among co-occurring species (see Figures 14.11 and 14.12) as evidence for the influence of competition in the evolution of species characteristics. Competition is at the heart of Darwin's theory of natural selection. Characteristics that enable an organism to reduce competition will function to increase fitness, therefore influencing the evolution of characteristics related to the acquisition of resources. Consider two bird species that feed on seeds. The pattern of seed selection for both species (A and B) can be expressed as a bell-shaped curve on a graph with seed size as the *x* axis and proportion of total diet consisting of a given seed size as the *y* axis (Figure A). Imagine two species show considerable overlap in the size of seeds they select; if we assume that interspecific competition reduces the fitness of individuals of both

species, those individuals that select seeds from the tails of the distributions where overlap is minimum will encounter less competition and increased fitness. The range of seeds selected by an individual is constrained by the size and shape of the beak. If competition between the two species favors the selection of smaller seeds by species A and larger seeds by species B, natural selection will favor those individuals in population A with small beaks, while favoring those individuals with larger beaks in population B. Ultimately, the two species will diverge in both beak size and the size of seeds selected as food. Direct interspecific competition will be reduced, and the two species will coexist.

Although the scenario described above is consistent with patterns of resource partitioning observed in nature, it is difficult to prove. Differences among species may

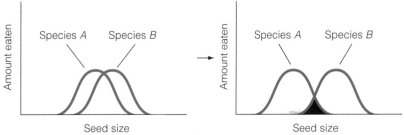

FIGURE A Competition arising from competition for intermediate size seeds may result in the two species shifting diets to reduce overlap in the range of seed sizes chosen as food.

natural selection (see Chapter 2, Section 2.1). Characteristics that enable an organism to reduce competition will function to increase fitness, therefore influencing the evolution of characteristics related to the acquisition of resources. Although this hypothesis is consistent with observed patterns of resource partitioning found in many communities, actually determining whether these differences in resource utilization evolved in response to severe competition for resources in the past is difficult, to say the least (see Focus on Ecology 14.2: Ghosts of Competition Past). Nevertheless, ecologists regard competition as instrumental in directing the evolutionary divergence of species (Chapter 2, Section 2.6).

14.10 Interspecific Competition Influences the Realized Niche

An organism free from interference from other species could use the full range of conditions and resources to which it is adapted. We call this range the species' **fundamental niche** (see Chapter 2, Focus on Ecology 2.3). Competition from other species often restricts a species to a portion of its fundamental niche. The portion of the fundamental niche that a species actually exploits in the presence of competitors is its **realized niche**. An example is provided

relate to adaptation for the ability to exploit a certain environment or range of resources independent of competition. Differences among species have evolved over a long period of time for which we have limited information about resources and potential competitors that may have influenced natural selection, leading Joseph Connell, an ecologist at the University of California at Santa Barbara, to refer to this theory as the "ghosts of competition past." Some of the strongest evidence in support of this theory is from studies that examine differences in the characteristics of (sub)populations of a species that face different competitive environments. One of the better examples involves two Darwin's finches of the Galápagos Islands, the medium ground finch *Geospiza fortis* and the small ground finch *G. fuliginosa,* both of which feed on an overlapping array of seed sizes. The medium ground finch is allopatric (lives in the absence of the other species) on Daphne Island, and the small ground finch is allopatric on Los Hermanos. Populations of the different species on these two islands have widely overlapping beak depths (Figure B). On the island of Santa Cruz, where the two species are sympatric (live together), beak depth of the medium ground finch is much larger than on Daphne Major, and the beak depth of the small ground finch is significantly smaller than on the island of Los Hermanos. In the face of competition for food, selection favored individual medium ground finches with large beaks that could effectively exploit larger seeds while favoring individual small ground finches that fed on smaller seeds. The out-

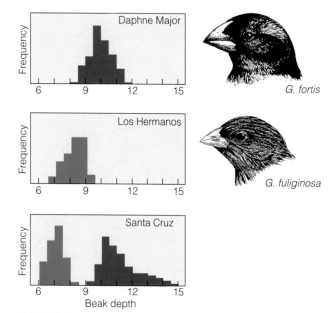

FIGURE B Apparent character displacement in beak size in populations of Galapagos finches. Both species co-occur on the island of Santa Cruz, while only *G. fortis* is found on the the island of Daphne Major and *G. fuliginosa* on the island of Los Hermanos. (Grant 1986.)

come of this competition was a shift in both beak depths and feeding niches. This shift is one of the many examples of **character displacement** in the wild.

---

by the work of J. B. Grace and R. G. Wetzel of the University of Michigan. Two species of cattail *(Typha)* occur along the shoreline of ponds in Michigan. One species, *Typha latifolia,* dominates in the shallower water, while *Typha angustifolia* occupies the deeper water farther from shore. When these two species were grown along the water depth gradient in the absence of the other species, a comparison of the results with their natural distributions reveals how competition influences their realized niche (Figure 14.13). Both species can survive in shallow waters, but only the narrow-leaved cattail, *Typha angustifolia,* can grow in water deeper than 80 cm. When the two species grow together along the same gradient of water depth, their distributions, or realized niches, change. Even though *Typha angustifolia* can grow in shallow waters (0 to 20 cm depth) and

above the shoreline (–20 to 0 cm depth), in the presence of *Typha latifolia* it is limited to depths of 20 cm or deeper. Individuals of *T. latifolia* outcompete individuals of *T. angustifolia* for the resources of nutrients, light, and space, limiting its distribution to the deeper waters. Note that the maximum abundance of *T. angustifolia* occurs in the deeper waters, where *T. latifolia* is not able to survive.

Although each of the two species of cattail has exclusive use of a subset of the range of habitats along the shoreline, the species coexist at intermediate depths. When two or more organisms use a portion of the same resource simultaneously, be it food or habitat, it is referred to as **niche overlap.** The amount of niche overlap is assumed to be proportional to the degree of competition for that resource. Niche overlap, however, does not always mean high

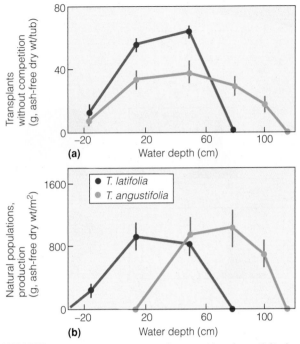

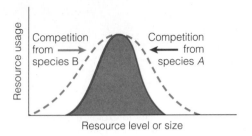

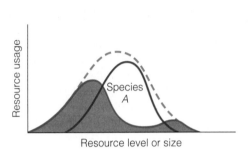

FIGURE 14.13 The distribution of two species of cattail *(Typha latifolia* and *Typha angustifolia)* along a gradient of water depth: (a) grown separately in an experiment; (b) growing together in natural populations. The responses of the two species in the absence of competition reflect their fundamental niche (physiological tolerances). The distribution of each species is altered by the presence of the other. They are forced to occupy only their realized niches. (Grace and Wetzel 1981.)

FIGURE 14.14 Two possible relationships between fundamental and realized niches for a hypothetical species C. The fundamental niche is shown as a dashed red line and the realized niche is shaded red. (a) Species A and B compete for the resource and cause the realized niche of species C to compress to the central optimum. Note that the curves are symmetrical and bell-shaped. (b) Dominant species A (blue line) forces species C out of its optimum into the peripheral part of its fundamental niche, causing a bimodal, asymmetrical realized niche shown in red. (Taken from Krebs 2001, as modified from Austin 1999).

competitive interaction. In fact, resources may not be in short supply. Extensive niche overlap may indicate that little competition exists and that resources are abundant.

As preceding examples have illustrated (for example, Figures 14.10 and 14.13), competition may force species to restrict their use of space, range of foods, or other resource-oriented activities. As a result, species do not always occupy that part of their fundamental niche that corresponds to the conditions under which they do best in terms of highest growth rate, reproduction, or fitness (Figure 14.14). Jessica Gurevitch, of the University of New York at Stony Brook, examined the role of interspecific competition on the local distribution of *Stipa neomexicana*, a C_3 perennial grass found in the semiarid grassland communities of southeastern Arizona. *Stipa* is found only on the dry ridge crests where grass cover is low, rather than in moister, low-lying areas below the ridge crests where grass cover is greater. In a series of experiments, Gurevitch removed neighboring plants from individual *Stipa* plants in ridge crest, mid-slope, and lower-slope habitats. The survival, growth, and reproduction of these plants were compared with control individuals (whose neighboring plants were

not removed). Her results clearly show that *Stipa* has a higher growth rate, produces a greater number of flowers per plant, and has higher rates of seedling survival in mid- and lower-slope habitats (Figure 14.15). Its population density in these habitats is limited by competition with more successful C_4 grass species. *Stipa* distribution (or realized niche) is limited to suboptimal habitats due to competition.

For simplicity, we represent a niche as one-dimensional (as in Figure 14.14). In reality, a species' niche includes many types of resources: food, a place to feed, cover, space, and so on. Rarely do two or more species possess exactly the same combination of requirements. Species may overlap on one gradient but not on another. The total competitive interaction may be less than that suggested by the niche overlap on one gradient alone (Figure 14.16).

The converse of a compression or shift in a species niche as a result of competition is when a species expands its niche in response to the removal of a competitor, termed **competitive release**. Competitive release may occur when a species invades an island that is free of potential competitors, moves into habitats it never occupied on a mainland, and increases its abundance. Such expansion may also

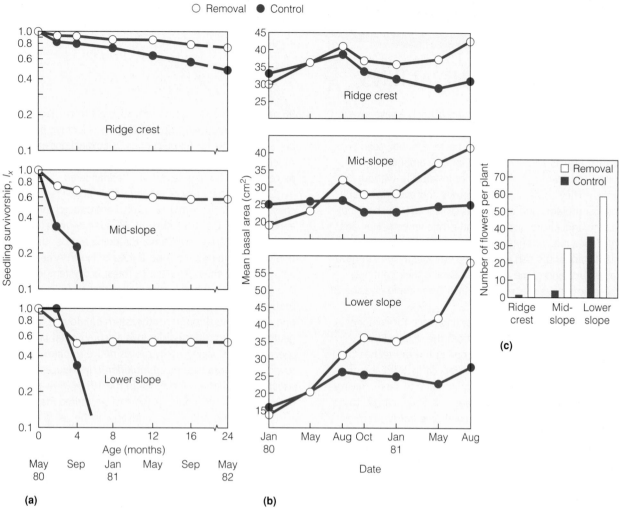

FIGURE 14.15 Response of *Stipa neomexicana* plants in three different habitats (ridge crest, mid-slope, and lower slope. Results of both treatment (neighboring plants removed) and control (neighbors *not* removed) plants are shown for (a) seedling survival, (b) mean growth rate, and (c) flowers produced per plant. Under natural conditions, the distribution of *Stipa* is restricted to the ridge crest habitats as a result of competition from other grass species. (Gurevitch 1986.)

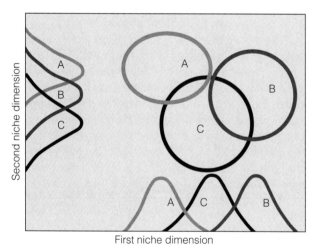

FIGURE 14.16 Niche overlap based on two environmental gradients. Species may exhibit considerable overlap on one gradient and little or none on the other. When we sum niche dimensions (in the circles), we find niche overlap reduced considerably. (Adapted from Pianka, 1978.)

follow when a competing species is removed from a community, allowing remaining species to move into microhabitats they previously could not occupy.

14.11 The Importance of Competition Is Difficult to Demonstrate in the Field

Demonstrating interspecific competition in laboratory "bottles" or the greenhouse is one thing; demonstrating competition under natural conditions in the field is another. In the field, you (1) have little control over the environment; (2) have difficulty knowing whether the populations are at or below carrying

INVASION OF THE ALIENS

Aliens—nonnative plants and animals—have invaded every continent and nearly every island on Earth. They come in all forms: viruses, bacteria, fungi, seed plants, insects and other invertebrates, amphibians, fish, birds, and mammals. Many arrive insidiously, without notice. They come hidden as viruses and bacteria inhabiting living plants and animals, as stowaways on ships' ballast, in agricultural products and exotic timbers. Humans introduce many nonnative species, including exotic garden plants, aquarium fish and plants, fish for stocking in natural waters, and game animals. Escaping their confines, these nonnative species begin their often-rapid invasion of the country.

Such invasions have been taking place for centuries. Stowaway black rats on ships from the Middle East brought bubonic plague to Europe in the fourteenth century. European explorers and settlers carried diseases to Native Americans who were not immune to them. Sailing ships carried brown rats to remote islands. Escaping to land, they multiplied and destroyed predator-naive birds and mammals. Alien plants, such as purple loosestrife and Japanese honeysuckle (Lonicera japonica), are displacing native plants. In a world shrunken by rapid transportation and widespread movement of humans, invaders have been increasing at an unprecedented rate, destroying populations of native species and altering ecosystems to their own needs.

Not all would-be invaders are successful, but those that are gain an early foothold. Their local population remains low and undetected for a long period; then they suddenly expand and move out from their beachhead. The invaders are mostly r-strategists. They reproduce rapidly, producing a large number of spores, seeds, and young that are excellent dispersers. Scattered by the wind, carried by insects, birds, and mammals, and transported on automobiles, trucks, trains, and riverboats, the invaders take advantage of disturbed sites, empty niches, and competitive release. If the abiotic environment is appropriate for the invaders, they are likely to succeed.

Once established in an area, the invaders displace native species. The Great Lakes have been invaded by 139 nonnative aquatic species that are changing ecological relationships in them. The San Francisco Bay area is occupied by 96 nonnative invertebrates, from sponges to crustaceans. Exotic fish, introduced purposefully or accidentally, have been responsible for 68 percent of fish extinctions in North America during the past 100 years and for the decline of 70 percent of the fish species listed as endangered.

Other invaders have changed whole ecosystems. The chestnut blight (Endothia parasitica), introduced through imports of the Chinese chestnut, all but exterminated the chestnut (Castanea mollissimi), the dominant deciduous tree in the eastern hardwood forests of North America, and converted the oak-chestnut forest ecosystem to oak. The gypsy moth, which accidentally escaped from a lot brought into Massachusetts for experimental silk production, has spread through northeastern oak forests, endangering oaks and other hardwoods as commercial timber species and displacing many moths native to hardwood forests. Multiflora rose (Rosa multiflora), introduced and widely planted as a living fence in the 1950s, now covers pastures and forest edges in the eastern United States.

The most notorious recent invader is the zebra mussel (Driessena polymorpha). A European species that arrived in the ballast waters of ships, it was discovered in 1988 in Lake St. Claire, Ontario, north of Lake Erie. By 1997 the zebra mussel had spread to each of the Great Lakes and the major river systems of the United States. Its exploding populations are having a detrimental effect on aquatic ecosystems. The zebra mussel attaches itself to native mussels, covering them to the point of suffocation. Feeding on phytoplankton, the zebra mussel, with its tremendous water-filtering capacity, reduces the population upon which larval fish such as the alewife and smelt depend. Their filtering clarifies water so much that it increases the growth of aquatic plants.

The zebra mussel also is causing great economic damage. Attaching themselves to any hard surface, zebra mussels clog water intake pipes of power plants, water treatment plants, and industrial facilities, costing millions of dollars for replacement and maintenance. They foul boat hulls and engine cooling systems. Hitchhiking on the hulls of barges, recreational boats, and boat trailers, and drifting downstream with the current, the zebra mussel has the potential of affecting two-thirds of the nation's waterways ecologically and economically for years to come.

capacity; and (3) lack full knowledge of the life-history requirements or the subtle differences between the species.

In the previous sections, we reviewed a variety of studies that examined the role of competition in the field. Perhaps the most common are removal experiments, where one of the potential competitors is removed, and the response of the remaining species is monitored. Although these experiments might appear straightforward, providing clear evidence of competitive influences, the removal of individuals may have direct and indirect effects on the environment that are not intended or understood by the investigators and that can influence the response of the remaining species. For example, the removal of (neighboring) plants from a location may increase light reaching the soil surface, soil temperatures, and evaporation. The result may be reduced soil moisture and increased rates of decomposition, influencing the abundance of below-ground resources. The sometimes "hidden treatment effects" can hinder the interpretation of experimental results.

Despite the difficulty in determining the influence of competition on community structure, since Darwin (1859), ecologists have considered interspecific competition, especially competitive exclusion, as the cornerstone of community structure. As we have seen in the preceding examples, interspecific competition has been analyzed in a wide variety of plants and animals over the past 50 years, and several leading ecologists have reviewed these studies to explore the prevalence of competition in nature and the strength of its effects. The results of these reviews suggest, if they do not prove with certainty, that competition exists between certain species and within guilds of species. But do these studies demonstrate that interspecific competition exerts a strong influence on community organization?

The answer to this question depends on many things, including the definition of community. As we stated in the introduction to Chapter 13, ecologists use the term *community* in a variety of ways. If the community in question is limited to members of a particular group of species that exploit resources in a similar fashion, such as the guild of ground-foraging birds, competition may be shown to be important in controlling their relative abundances. But does competition within a guild of ground-foraging birds influence the structure of the entire bird community, or the community in its fullest sense—the entire group of plants and animals inhabiting the forest? Determining the effects of interspecific competition on shaping the whole community is difficult at best. The clearest examples of competition's influencing community structure are the effects of introductions or invasions of exotic species (see Focus on Ecology 14.3: Invasion of the Aliens). We shall return to the role of competition in structuring communities later, after we have examined the role of other types of population interactions that influence the distribution and abundance of species within the community.

CHAPTER REVIEW

Summary

Models of Interspecific Competition (14.1–14.2) In interspecific competition, individuals of two or more species share a resource in short supply, reducing the fitness of both. As with intraspecific competition, competition between species can involve either exploitation or interference. Six different types of interactions are sufficient to account for most instances of interspecific competition: (1) consumption, (2) preemption, (3) overgrowth, (4) chemical interaction, (5) territorial, and (6) encounter (14.1).

The Lotka-Volterra equations describe four possible outcomes of interspecific competition. Species 1 may win over species 2; species 2 may win over species 1. Both of these outcomes represent competitive exclusion. The other two outcomes involve coexistence. One is unstable equilibrium, in which the species that was most abundant at the outset usually wins. A final possible outcome is stable equilibrium, in which two species coexist, but at a lower population level than if each existed in the absence of the other (14.2).

Competitive Exclusion (14.3–14.4) Laboratory experiments with species interactions support the Lotka-Volterra model (14.3). Results of the experiments led to formulation of the competitive exclusion principle—two species with exactly the same ecological requirements cannot coexist. This principle has stimulated critical examinations of competitive relationships outside the laboratory, especially how species coexist and how resources are partitioned (14.4).

Coexistence (14.5–14.6) Environmental factors such as temperature, soil or water pH, relative humidity, and salinity are features of the environment that directly influence physiological processes related to growth and reproduction, but are not consumable resources over which species compete. By differentially influencing species within a community, these nonresource factors can influence the outcome of competition (14.5). Environmental variability may give each species a temporary advantage, allowing the coexistence of competitors, where under constant conditions, one would exclude the other (14.6).

Multiple Factors (14.7–14.8) In many cases, competition between species involves multiple resources, and often competition for one influences the ability of an organism to access other resources (14.7).

As environmental conditions change, so will the relative competitive ability of species. The shifts in competitive ability can result either from changes in the carrying capacities related to a changing resource base, or from changes in the physical environment that interact with resource availability. Natural environmental gradients often involve the covariation of multiple factors, both resources and nonresource or stress factors, such as salinity, temperature, and water depth (14.8).

Resource Partitioning and the Niche (14.9–14.10) Many species that share the same habitat coexist by partitioning available resources (14.9). Closely associated with interspecific competition is the concept of the niche. Basically, a niche is the functional role of an organism in the community. That role might be constrained by interspecific competition. In the absence of competition, an organism occupies its fundamental niche. In the presence of interspecific competition, the fundamental niche is reduced to a realized niche, the conditions under which an organism actually exists. When two different organisms use a portion of the same resource, such as food, their niches overlap. Overlap may or may not indicate competitive interaction. A species compresses or shifts its niche when competition forces it to restrict its type of food or habitat. In some cases, the realized niche may not correspond to the set of conditions that are optimal for the species. In the absence of competition, the species experiences competitive release, expanding its niche (14.10).

Difficulty in Proving Competition (14.11) Due to the difficulty of establishing competition in the field and the limited ability to explore the array of direct and indirect interactions that occur among the diversity of species inhabiting natural communities, determining the effects of interspecific competition on shaping the whole community is difficult at best.

Study Questions

1. What is interspecific competition, and what types of interactions does it involve?

2. Describe four outcomes of competition based on the Lotka-Volterra competition equations.

3. What is competitive exclusion? What is the problem with the concept?

4. How can environmental factors that are not resources influence competition between species?

5. How can temporal variation in environmental conditions influence competition?

6. What is allelopathy? Why is it a form of competition?

7. Provide an example of competition for multiple resources.

8. How can competition among species vary across environmental gradients?

9. What is resource partitioning? What are some weaknesses of the concept?

10. Define niche. Distinguish between a fundamental niche and a realized niche.

11. What is niche overlap?

12. What is character displacement?

13. Why are island flora and fauna so vulnerable to invasion by nonnative species? Discuss in terms of interspecific competition and competitive exclusion. Use the Hawaiian Islands as an example.

Brown bears *(Ursus arctos)* and gulls prey on migrating salmon in an Alaskan river.

Predation

OBJECTIVES

On completion of this chapter, you should be able to:

- Define predation and distinguish among its forms.
- Sketch the Lotka-Volterra model of predation and discuss the links between predator and prey populations.
- Explain functional response and the three forms it can take.
- Explain numerical response and the processes involved.
- Define search image and switching and their importance to predation.
- Discuss the coevolution of predator defenses.
- Explain optimal foraging.
- Explain the differences between quantitative and qualitative plant defenses.
- Describe plant-herbivore and herbivore-carnivore systems.

When the poet Alfred Lord Tennyson wrote "Tho' Nature, red in tooth and claw," he was no doubt seeking to evoke savage images of predation. Although the very term *predator* brings to mind images of lions on the savannas of Africa or the great white shark cruising coastal waters, predation is defined more generally as the consumption of all or part of one living organism by another. Although all heterotrophic organisms derive their energy from the consumption of organic matter, predators are distinguished from scavengers and decomposers (see Chapter 7) in that they feed on living organisms. As such, they function as agents of mortality with the potential to regulate prey populations. Likewise, being the food resource, prey population has the potential to influence the growth rate of the predator population. These interactions between predator and prey species can not only have consequences on community structure, but can also serve as agents of natural selection, influencing the evolution of both predator and prey.

15.1 Predation Takes a Variety of Forms

Most of us associate predation with a hawk taking a mouse or a wolf killing a deer, a form of predation called **carnivory**. That is a narrow view. A fly laying its eggs on a caterpillar to develop there at the expense of its victim is exhibiting a form of predation called **parasitoidism**. The parasitoid attacks the host (the prey) indirectly by laying its eggs on the host's body. When the eggs hatch, the larvae feed on the host, slowly killing it. In **parasitism**, too, a predator lives on or within the host (see Chapter 16). The parasite feeds on its host but seldom kills it outright (which, in effect, would destroy both its habitat and its food). A deer feeding on shrubs and grass and a mouse eating a seed are practicing a form of predation called **herbivory**. Seed consumption is outright predation because the embryonic plant is killed. Grazing on plants without killing them can be considered a form of parasitism. When they kill the plant outright, grazers become predators of plants. A special form of predation is **cannibalism**, in which the predator and the prey are the same species.

Predation is more than a transfer of energy. It is a direct and often complex interaction of two or more species, the eater and the eaten. The numbers of some predators may depend upon the abundance of their prey. The population of the prey may be controlled by its predator.

15.2 Mathematical Models Describe the Basics of Predation

In the 1920s, Lotka and Volterra turned their attention from competition (see Chapter 14, Section 14.2) to the effects of predation on population growth. Independently, they proposed mathematical statements to express the relationship between predator and prey populations. They provided one equation for the prey population and another for the predator population.

The population growth equation for the prey population consists of two components: the exponential model of population growth presented in Chapter 10 *(dN/dt = rN)* and a mortality term that represents the removal of the prey from the population by the predator:

$$dN_{prey}/dt = rN_{prey} - CN_{pred}N_{prey}$$

where $CN_{pred}N_{prey}$ is mortality of prey due to the predator, which depends upon the number of prey (N_{prey}), the number of predators (N_{pred}), and the average per capita capture rate of prey by predators *(C)*.

The equation for the predator population is

$$dN_{pred}/dt = B(CN_{pred}N_{prey}) - DN_{pred}$$

where *B* is the efficiency of converting prey consumed ($CN_{pred}N_{prey}$) to the production of new predators (reproduction), and *D* is the death rate of predators. The Lotka-Volterra equations assume that the prey population grows exponentially and that reproduction in the predator population depends upon the number of prey consumed.

Recall that, as with the Lotka-Volterra model of competition presented in Chapter 14, these equations represent the growth rate of the prey and predator populations. To understand how these two populations interact we can use the same graphical approach used to examine the outcomes of interspecific competition in Chapter 14 (Section 14.2). In the absence of predators (or very low predator density), the prey population will grow exponentially ($dN_{prey}/dt = rN_{prey}$). As the predator population increases, prey mortality will increase until eventually the rate of mortality due to predation ($CN_{pred}N_{prey}$) is equal to the inherent growth rate of the prey population (rN_{prey}) and the net population growth for the prey species will be zero ($dN_{prey}/dt = 0$). We can solve for the predator population density at which this occurs:

$$CN_{pred}N_{prey} = rN_{prey}$$
$$CN_{pred} = r$$
$$N_{pred} = r/C$$

Simply put, the growth rate of the prey population is zero when the density of predators is equal to the per capita growth rate of the prey population divided by the per capita capture rate of the predators. If the predator density exceeds this value, the growth rate of the prey becomes negative, and the population density declines. The mortality due to predation ($CN_{pred}N_{prey}$) is greater than the inherent growth rate of the prey population (rN_{prey}).

Likewise we can examine the influence of prey density on the growth rate of the predator population. The growth rate of the predator population will be zero ($dN_{pred}/dt = 0$) when the rate of predator increase (resulting from the consumption of prey) is equal to the rate of mortality:

$$B(CN_{pred}N_{prey}) = DN_{pred}$$
$$BCN_{prey} = D$$
$$N_{prey} = D/BC$$

The growth rate of the predator population is zero when the density of prey is equal to the mortality rate of the predator population divided by the product of the per capita rates of capture and conversion efficiency of captured prey into new predators (reproduction).

We can examine these results graphically (Figure 15.1a) using two axes that represent the population densities of the prey (*x* axis) and predator (*y* axis). Having defined the density of prey at which the growth of the predator population is equal to zero ($N_{prey} = D/BC$) and the density of predators at which the growth rate of the prey density is zero ($N_{pred} = r/C$), we can draw the corresponding zero isoclines (Chapter 14, see Section 14.2). If values of prey density are to the right of the predator isocline, the predator population increases; if to the left of the isocline, predator density declines, likewise for the prey isocline. For values of predator density below the isocline, the prey growth rate is positive. For values above the isocline, the prey density declines. By combining the two isoclines (Figure 15.1b), changes in the growth rates of predator and prey populations can be examined for any combination of population densities.

These paired equations, when solved, show that the two populations rise and fall in oscillations (Figure 15.1c). The reason for oscillation is that as the predator population increases, it consumes a progressively larger number of prey, until the prey population begins to decline. The declining prey population no longer supports the large predator population. The predators now face a food shortage, and many of them starve or fail to reproduce. The predator population declines sharply to a point where the

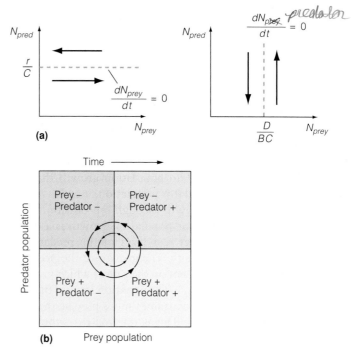

(a)

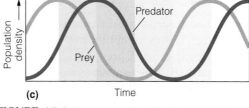

(b)

(c)

FIGURE 15.1 Patterns predicted by the Lotka-Volterra model of predator-prey interaction. (a) The zero isoclines ($dN/dt = 0$) for the prey and predator populations. Note that the zero isocline for the prey population is defined by a fixed density of predators and the predator isocline by a fixed population density of prey. The arrows show the direction of change in population density relative to the isoclines. (b) The combined zero isoclines provide a means of examining the combined population trajectories of the predator and prey populations. The arrows represent the combined population trajectory. A minus sign indicates population decline, and a plus sign population increase. This trajectory shows the cyclic nature of the predator-prey interaction. (c) By plotting the implied changes in density for both the predator and prey populations through time, it can be seen that the two populations continuously cycle out of phase with each other, with the density of predators lagging behind that of the prey. The shaded regions in the background of the graph relate to the corresponding regions in (b).

reproduction of prey more than balances its losses through predation. The prey increases, eventually followed by an increase in the population of predators. The cycle may continue indefinitely. The prey is never quite destroyed; the predator never completely dies out.

15.3 Model Suggests Mutual Population Regulation

The Lotka-Volterra model of predator-prey interactions suggests a mutual regulation of predator and prey populations. In the equations presented above, the growth of predator and prey populations are linked by the single term relating to the consumption of prey: $CN_{pred}N_{prey}$. For the prey population, this term serves to regulate population growth through mortality. In the predator population, it serves to regulate population growth through reproduction. The regulation of predator population growth is a direct result of two distinct responses of the predator to changes in prey density. The growth of the predator population is dependent on the rate at which prey are captured ($CN_{pred}N_{prey}$). The equation implies that the individual predator consumes more prey as the prey population increases. This is referred to as the **functional response**. Secondly, this increased consumption of prey results in an increase in predator reproduction [$B(CN_{pred}N_{prey})$], referred to as the **numerical response**.

This model of predator-prey interaction has been widely criticized for overemphasizing the mutual regulation of predator and prey populations. The continuing appeal of these equations to population ecologists, however, lies in the straightforward mathematical descriptions and in the oscillatory behavior that seems to occur in predator-prey systems. Perhaps its greatest value is in stimulating a more critical look at predator-prey interactions in natural communities, including those conditions that influence the control of prey populations by predators. A variety of factors have emerged including the availability of cover (refuges) for the prey, the increasing difficulty of locating prey as it becomes scarcer, choice among multiple prey species, and evolutionary changes in predator and prey characteristics (coevolution). We shall examine each of these topics and how it influences the nature of predator-prey interactions in the following sections.

15.4 Functional Responses Relate Prey Taken to Prey Density

The English entomologist M. E. Solomon introduced the idea of functional response in 1949. A decade later, the ecologist C. S. Holling explored the concept in more detail. The basis of a functional response is

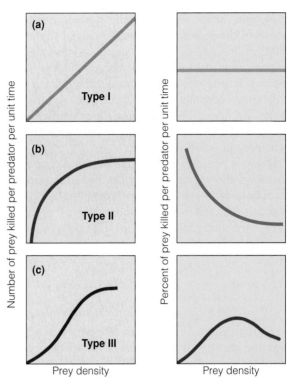

FIGURE 15.2 Three types of functional response curves, which relate the per capita rate of predation to prey density. (a) Type I. The number of prey taken per predator increases linearly as prey density increases. Expressed as a proportion of the prey density, the rate of predation is constant—independent of prey density. (b) Type II. The predation rate rises at a decreasing rate to a maximum level. Expressed as a proportion of prey density, the rate of predation declines as the prey population grows. (c) Type III. The rate of predation is low at first, then increases in a sigmoid fashion approaching an asymptote. Plotted as a proportion of prey density, the rate of predation is low at low prey density, rising to a maximum before declining as the rate of predation reaches its maximum.

that a predator will take more prey as the density of the prey increases. How a predator's rate of consumption responds to changes in the population density of prey is a key factor influencing the ability of a predator to regulate the prey population. Holling classified functional responses into three types (Figure 15.2).

In a Type I response the number of prey taken per predator increases in a linear fashion (Figure 15.2a). This is the response assumed in the Lotka-Volterra model presented in Section 15.2, where the capture rate is described by the constant C. This functional response implies that mortality in the prey species is density independent. Observations of European kestrel *(Falco tinnunculus)* predation on *Microtus* vole populations in northern Finland suggest a Type I functional response (Figure 15.3a).

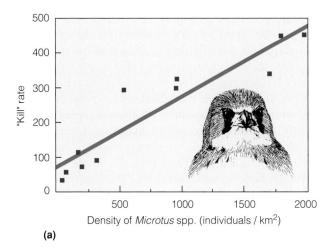

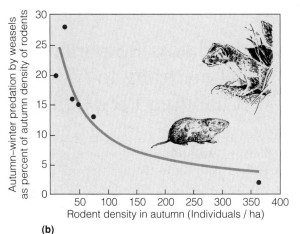

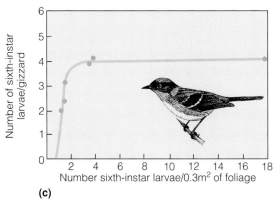

FIGURE 15.3 Three examples of functional response curves relating the rate of predation to prey density. (a) Type I: Functional response of a pair of European kestrels to *Microtis* vole density during the breeding season (after Korpimaki and Norrdahl 1991). (b) Type 2: Functional response of weasels preying on rodents in the deciduous forests of Bialowieza National Park, Poland. Note predation expressed as percentage. See Figure 15.2. (c) Type 3: Functional response of bay-breasted warblers feeding on spruce budworm larvae. (Mook et al. 1963.)

In a Type II response the capture rate increases in a decelerating fashion only up to some maximum rate that is attained at some high prey density (Figure 15.2b). This functional response implies a declining mortality rate with increasing prey density (Figure 15.3b). Two factors potentially interact to result in the capture rate's reaching some maximum, rather than continuing to increase indefinitely (as in Type I). First, predators may become satiated, at which time the rate of feeding is limited by the time required for digestion and assimilation of food. Secondly, as the predator captures more prey, the time it spends handling and eating the prey functions to decrease the time spent searching and hunting for additional prey.

The Type III response resembles Type II in having an upper limit to the capture rate; however, the number of prey taken is low at first, increasing in an S-shaped (sigmoid) fashion, as the capture rate approaches the maximum value (Figures 15.2c and 15.3c). A Type III functional response can potentially regulate a prey population because the capture rate varies with prey density. At low densities, the capture rate is negligible, but as prey density increases (as indicated by the upward sweep of the curve), predatory pressure increases in a density-dependent fashion.

A number of factors may result in a Type III response. The availability of cover (refuge) from which to escape from predators may be an important factor. If the habitat provides only a limited number of hiding places, it will serve to protect most of the prey population at low density, increasing the susceptibility of individuals as the population grows. Secondly, the predator may switch to alternative sources of food when the prey are scarce.

15.5 Predators Develop a Search Image for Prey

Another reason for the sigmoidal shape of the Type III functional response curve may be the predator's **search image**, an idea first proposed by the animal behaviorist L. Tinbergen. When a new prey species appears in the area, its risk of becoming selected as food by a predator is low. The predator has not yet acquired a search image—a way to recognize it as a potential food item. Once the predator has captured an individual, it may find it a desirable food. Having identified the species as a desirable prey, it has an easier time locating others of the same kind. The more adept the predator becomes at securing a particular prey item, the more intensely it concentrates on it. In time, the number of this particular prey species becomes so reduced or its population becomes so dispersed that encounters between it and the predator lessen. The search image for that prey item begins to wane, and the predator may turn its attention to another prey species.

15.6 Predators May Turn to Alternate, More Abundant Prey

Although a predator may have a strong preference for a certain prey, in most cases it can turn to another, more abundant prey species that provides more profitable hunting. If rodents, for example, are more abundant than rabbits and quail, foxes and hawks will concentrate on rodents.

Ecologists call the act of turning to more abundant alternate prey **switching** (Figure 15.4). In switching, the predator concentrates a disproportionate amount of feeding on the more abundant species and pays little attention to the less abundant species. As the relative abundance of the second prey species increases, the predator turns its attention to it.

At what point in prey abundance a predator switches depends considerably on the predator's food preference. A predator may hunt longer and harder for a palatable species before it turns to a more abundant, less palatable alternate prey. Conversely, the predator may turn from the less desirable species at a much higher level of abundance than it would from a more palatable species.

In spite of switching, some predators deliberately seek out certain prey, no matter how scarce. Predators in a California grassland show a distinct preference for meadow voles (*Microtus californicus*) over harvest mice (*Reithrodontomys megalopis*) and other rodents even though alternate prey are more abundant. An abundance of alternate prey apparently enables carnivores to maintain a sufficiently high population and to obtain enough energy to allow continued predatory pressure on the preferred species. Among herbivores, deer show a pronounced preference for a certain species of woody plants. Meadow mice (*Microtus pennsylvanicus*) often concentrate on seeds of preferred grasses, even though the seeds of other species are more abundant.

15.7 Predators Respond Numerically to Changing Prey Density

As the density of prey increases, the predator growth rate should respond positively. A numerical response of predators can occur through reproduction by predators (as suggested by the conversion factor *B* in the Lotka-Volterra equation for predators), or through the emigration or movement of predators into areas of high prey density. The latter is referred to as an aggregative response (Figure 15.5). The response of predators to aggregate in areas of high prey density is a critical feature in determining the ability of a predator population to regulate prey

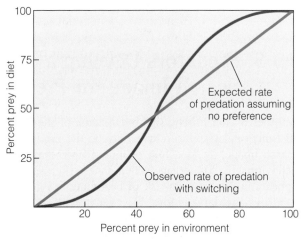

FIGURE 15.4 A model of prey switching. The straight line represents the expected rate of predation assuming no preference by the predator. The prey are eaten in a fixed proportion to their availability (density). The curved line represents the change in predation rate observed in the case of prey switching. At low densities, the proportion of the prey species in the diet of the predator is less than expected based on chance (based on its proportional availability to other prey species). Over this range of abundance, the predator is selecting alternative prey species. When prey density is high, the predator takes more of them than expected. Switching occurs at the point where the lines cross. The habit of prey switching results in a Type III functional response between predator and the prey species.

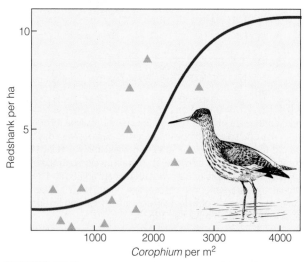

FIGURE 15.5 Aggregative response in the redshank (*Tringa totanus*). The curve plots the density of the redshank in relation to the average density of its arthropod prey (*Corophium* spp.). (Hassel and May 1974.)

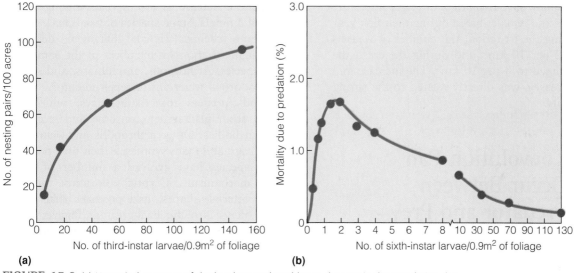

(a)

(b)

FIGURE 15.6 (a) Numerical response of the bay-breasted warbler to changes in the population density of spruce budworms in New Brunswick, Canada. (b) The predicted pattern of mortality rate estimated by combining the numerical response shown in (a) with the functional response of predation rate to prey density shown in Figure 15.3c. (Mook et al. 1963.)

density. The reason for the importance of aggregative response, or emigration, is that the population of most predators grows slowly in comparison to that of their prey. The numerical response of the bay-breasted warbler provides an example. Recall from Figure 15.3c that this bird exhibits a Type III functional response; the per capita rate of predation increases with increasing prey density. As the density of prey, spruce budworm, increases, the per capita rate of consumption by individual bay-breasted warblers increases up to some maximum capture rate (see Figure 15.3c). In addition, the density of birds increases with increasing prey density (Figure 15.6a). This increase in density is a result of an increase in the number of nesting pairs occupying the area (breeding territories per hectare). The net result is that the percent of prey organisms eaten per unit time by the entire predator population increases with the increase in prey density (Figure 15.6b). The combined functional and numerical responses by the bay-breasted warbler population interact to regulate the prey (spruce budworm) population at low population levels.

Other numerical responses by predator populations involve an increase in reproductive effort. An example is the response of a weasel *(Mustela nivalis)* population to the density of two rodents, the bank vole *(Clethrionomys glareolus)* and the yellow-necked mouse *(Apodemus flavicollis)*, in Biatowieza National Park in eastern Poland in the early 1990s. During this time, the rodents experienced a two-year irruption in population size brought about by a heavy crop of oak, hornbeam, and maple seeds. The abundance of food stimulated the rodents to breed throughout the winter. The long-term average popu-

lation density was between 28 and 74 animals per hectare. During the irruption, the rodent population reached 300 per hectare, and then declined precipitously to 8 per hectare (Figure 15.7).

The weasel population followed the fortunes of the rodent population. At normal rodent densities, the winter weasel density ranged between 5 to 27 per 10 km², declining by early spring to 0 to 19. Following reproduction, the midsummer density rose to 42 to 47 per 10 km². Because reproduction usually requires a certain minimal time, a lag typically exists between an increase of a prey population and a numerical response by a predator population. No time lag, however, exists between increased rodent reproduction and weasels' reproductive response. Weasels breed in the spring, and with an abundance

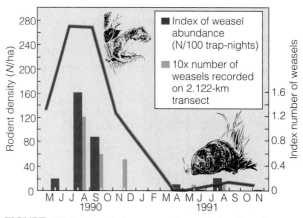

FIGURE 15.7 Numerical response of weasels (predators) to an irruption and crash in the populations of forest rodents (prey). Brown bars represent weasel data from live trapping; orange bars represent data from captures, visual observations, and radio tracking. (Jedrzejewski et al. 1995.)

of food they may have two litters or larger litter size. Young males and females breed during their first year of life. During the irruption, the number of weasels grew to 102 per 10 km^2, and during the crash the number declined to 8 per 10 km^2. The increase and decline in weasels was directly related to the spring rodent density.

15.8 Coevolution Can Occur Between Predator and Prey

By acting as agents of mortality, predators exert a selective pressure on prey species. That is to say, any characteristic that enables individual prey to avoid being detected and captured by a predator will increase its fitness. Natural selection should function to produce "smarter," more evasive prey (fans of the Road Runner® cartoon should already understand this concept). However, failure to capture prey results in reduced reproduction and increased mortality of predators. Therefore, natural selection should also function to produce "smarter," more skilled predators. As prey species evolve characteristics to avoid being caught, predators evolve more effective means to capture them. To survive as a species, the prey must present a moving target that the predator can never catch. This view of the *coevolution* between predator and prey led the evolutionary biologist Van Vallen to propose the Red Queen hypothesis. In Lewis Carroll's *Through the Looking Glass and What Alice Found There,* there is a scene in the Garden of Living Flowers, a world in which everything is continuously moving. Alice is surprised to see that no matter how fast she moves, the world around her remains motionless, to which the Red Queen responds "Now, *here,* you see, it takes all the running you can do, to keep in the same place." So it is with prey species. To avoid extinction at the hands of predators, they must evolve means of avoidance; they must keep moving just to stay where they are.

15.9 Prey Have Evolved Defenses Against Predators

Animal species have evolved a wide range of characteristics that function to avoid detection, selection, and capture by predators. These characteristics are collectively referred to as predator defenses.

Chemical defense is widespread among many groups of animals. Some species of fish release alarm pheromones (chemical signals) that, when detected, induce flight reactions in members of the same and related species. Arthropods, amphibians, and snakes employ odorous secretions to repel predators. Many arthropods possess toxic substances, which are acquired from plants they consume and stored in their own bodies. Other arthropods and venomous snakes, frogs, and toads synthesize their own poisons.

Prey species have evolved a number of other defense mechanisms. **Cryptic coloration** includes colors, patterns, shapes, and postures that allow prey to blend into the background (Figure 15.8). Some animals are protectively colored, blending into the background of their normal environment. Such protective coloration is common among fish, reptiles, and many ground-nesting birds. Object resemblance is common among insects. For example, walking sticks (Phasmatidae) resemble twigs, and katydids (Pseudophyllinae) resemble leaves (Figure 15.9). Some animals possess eyespot markings, which intimidate potential predators, attract their attention away from the animal, or delude them into attacking a less vulnerable part of the body. Associated with cryptic coloration is **flashing coloration**. Certain butterflies, grasshoppers, birds, and ungulates, such as the white-tailed deer, display extremely visible color patches when disturbed and put to flight. The flashing coloration may distract and disorient predators; or, as in the case of the white-tailed deer, it may serve as a signal to promote group cohesion when confronted by a predator

FIGURE 15.8 Cryptic coloration in the flounder (*Paralichthys*), which inhabits the shallow coastal waters of eastern North America. Most flounders can change color and pattern rapidly to match that of the bottom sediments, allowing them to avoid detection by both predators and potential prey.

FIGURE 15.9 The walking stick (Phasmatidae), which feeds on the leaves of deciduous trees, strongly resembles a twig.

FIGURE 15.10 The white-tailed deer of North America received its name from the white tail-patch that serves as both an alarm and distraction as it flees predators.

(Figure 15.10). When the animal comes to rest, the bright or white colors vanish, and the animal disappears into its surroundings.

Animals that are toxic to predators or utilize other chemical defenses often possess **warning coloration,** bold colors with patterns that serve as warning to would-be predators. The black-and-white stripes of the skunk, the bright orange of the monarch butterfly, and the yellow-and-black coloration of many bees and wasps and some snakes serve notice of danger to their predators (Figure 15.11). All their predators, however, must have an unpleasant experience with the prey before they learn to associate the color pattern with unpalatability or pain.

Some animals living in same habitats with inedible species sometimes evolve a coloration that resembles or mimics the warning coloration of the toxic species (Figure 15.12). This type of **mimicry** is called Batesian mimicry after the English naturalist H. E. Bates, who described it from observation of tropical butterflies. The mimic, an edible species, resembles the inedible species, called the model. Once the predator has learned to avoid the model, it avoids the mimic also. In this manner, natural selection reinforces the characteristic of the mimic species that resembles that of the model species.

Some animals employ **protective armor** for defense. Clams, armadillos, turtles, and numerous beetles all withdraw into their armor coats or shells when danger approaches. Porcupines, echidnas, and hedgehogs possess quills (modified hairs) that discourage predators.

Still other animals utilize **behavioral defenses.** One is to give an alarm call when a predator is sighted. Because high-pitched alarm calls are not species specific, a wide range of nearby animals recognizes them. Alarm calls often bring in numbers of potential prey that mob the predator. Other behavioral defenses include distraction displays, which are most common among birds. They direct the attention of the predator away from the nest or young.

For some prey, living in groups is the simplest form of defense. Predators are less likely to attack a concentrated group of individuals. By maintaining a tight, cohesive group, prey make it difficult for any

(a)

(b)

FIGURE 15.11 The bright and distinctive coloration patterns of the (a) monarch butterfly *(Dannus plexippus)* and the (b) poison-dart frog *(Phyllobates bicolor)* serve as a warning to potential predators of their toxicity.

predator to obtain a victim (Figure 15.13). Sudden, explosive group flight can confuse a predator, which is unable to decide which individual to follow.

A more subtle form of defense is the timing of reproduction so that most of the offspring are produced in a short period of time. Then prey are so abundant that the predator can take only a fraction of them, allowing a percentage of the young to escape and grow to a less-vulnerable size. Such is the strategy called predator satiation employed by large ungulates such as the caribou *(Rangifer tarandus)* in the Arctic and the wildebeest *(Connochaetes taurinus)* and buffalo in Africa. The 13-year and 17-year appearances of the periodical cicadas *(Magicicada* spp.) function in much the same manner. By appearing suddenly in enormous numbers, the cicadas quickly satiate predators.

15.10 Predators Have Evolved Efficient Hunting Tactics

As prey have evolved ways of avoiding predators, predators have evolved better ways of hunting. Predators have three general methods of hunting: ambush, stalking, and pursuit. Ambush hunting means lying in wait for prey to come along. This method is typical of some frogs, alligators, crocodiles, and lizards and certain insects. Although ambush hunting has a low frequency of success, it requires minimal energy. Stalking, typical of herons and some cats, is a deliberate form of hunting with a quick attack. The predator's search time may be

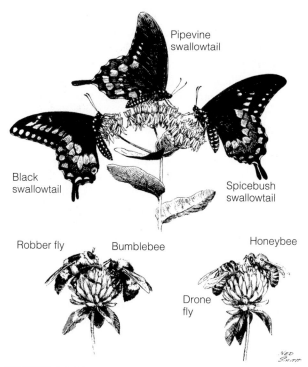

FIGURE 15.12 Mimicry in insects. The hover, or drone fly *(Eristalis tenax)*, deceives predatory birds by mimicking the honeybee *(Apis mellifera)*, which has a stinger. The robber fly *(Laphria* spp.) illustrates aggressive mimicry. It is the mimic of the bumblebee *(Megabombus pennsylvanicus)*, on which it preys.

great, but pursuit time is minimal. Pursuit hunting, typical of many hawks, lions, wolves, and insectivorous bats, involves minimal search time because the predator usually knows the location of the prey, but pursuit time is usually great. Stalkers spend more time and energy encountering prey. Pursuers spend more time capturing and handling prey.

Predators, like their prey, may use cryptic coloration to blend into the background or break up their outlines (Figure 15.14). Predators use deception by resembling the prey. Robber flies *(Mallophora bomboides)* mimic bumblebees, their prey (see Figure 15.12). The female of certain species of fireflies imitates the mating flashes of other species, attracting males of those species, which she promptly kills and eats. Predators may also employ chemical poisons, as do venomous snakes, scorpions, and spiders. They may form a group to attack large prey, as lions and wolves do.

15.11 Animals Use a Foraging Strategy

Have you ever watched a robin hopping across the lawn? It flies in, lands, hops for a short distance as if sizing up the situation, and stops. Then it moves on deliberately in a series of irregular paths across the grass. The bird pauses every few feet, stares ahead as if in deep concentration, or cocks its head toward the ground. It either moves on or crouches low as if to brace itself. It pecks quickly in the ground and pulls out an earthworm. On occasion, the earthworm pulls back and wins the tug-of-war. The robin does not push the action. It lets go and hops to another spot to repeat the food-seeking activity.

The behavior of the robin has four parts. First, the robin has to decide where to hunt for food. Once

FIGURE 15.13 Musk ox *(Obibos moschatus)* form a circle, each facing outward, to present a combined defense when threatened by predators.

FIGURE 15.14 The alligator snapping turtle uses a combination of cryptic coloration and mimicry to avoid detection and attract prey. While lying motionless on the bottom with its mouth wide open, it wiggles its worm-shaped tongue to attract and ambush potential prey. (See front center of lower jaw.)

on the lawn, it has to search for food items. Having located some potential food, the robin has to decide whether to pursue it. If it begins pursuit by pecking, the robin has to attempt a capture, in which it might or might not be successful. By capturing an earthworm, the robin earns some units of energy. In these activities, the robin needs to gain more energy than it expends.

The problem facing the robin and all other animals is securing sufficient energy to maintain itself, feed its young, and store fat for migration or winter, without spending too much time doing so. To the robin, time is energy. To spend too much time securing energy without a sufficient return for the effort results in a type of personal bankruptcy. If the robin fails to find suitable and sufficient food in the area of the lawn it is searching, it leaves to search elsewhere. If successful, the bird probably will return until the spot is no longer an economical place to feed.

The means animals employ to secure food are their **foraging strategy**. Of practical interest to ecologists is the optimal foraging strategy against which the actual foraging strategy can be measured. An optimal foraging strategy is one that provides the maximum net rate of energy gain (energy gain per unit of time or effort). A foraging strategy involves two separate but related components. One is optimal diet; the other is optimal foraging efficiency.

When the robin searches the lawn for food, it should, by optimal foraging theory, select only those items that provide the greatest energy return for the energy expended. That places an upper and lower limit on the size of food items accepted. If a food item is too large, it requires too much time to handle; if it is too small, it does not deliver enough energy to cover the costs of capture. The robin should reject or ignore less valuable items, such as small beetles and caterpillars, and give preference to small and medium-sized earthworms. These more valuable food items would be classified as preferred foods, ones taken out of proportion to their availability relative to all other food items.

The robin, according to optimality theory, should take the most valuable food items first. When it depletes those items, it should turn its attention to the next most valuable food items. Eventually, the animal should expand its diet to include the poorer foods when the discovery of high-value foods falls below a certain rate.

Field studies provide some insight into how animals forage under natural conditions. N. Davies studied the feeding behavior of the pied wagtail (*Motacilla alba*) and the yellow wagtail (*Motacilla flava*) in a pasture near Oxford, England. The birds fed on various dung flies and beetles attracted to cattle droppings. Prey included large, medium, and small flies and small beetles. The wagtails showed a decided preference for medium-sized prey (Figure 15.15a). The size of prey selected corresponded to the optimum-sized prey the birds could handle profitably (Figure 15.15b). The birds ignored the small sizes. Easy-to-handle small prey did not return sufficient energy, and large sizes required too much time and effort.

15.12 Foragers Seek Productive Food Patches

The robin lives in a patchy environment made up of clumps of trees, shrubby thickets or plantings, and open lawns. The robin could forage in all these patches of vegetation, eating all food items it came across, but it does not. Rather, the bird concentrates its foraging activities in patches of open lawn. Moreover, the robin may restrict its searching to certain parts of the lawn, where the environment is more favorable for earthworms, until the earthworms are depleted or have moved deeper into the soil in response to a drying surface. Then the robin has to turn its attention to other patches and other food, such as ground beetles, flies, sow bugs, snails, and millipedes.

The robin has been following the rules of optimal foraging: (1) concentrate on the most productive

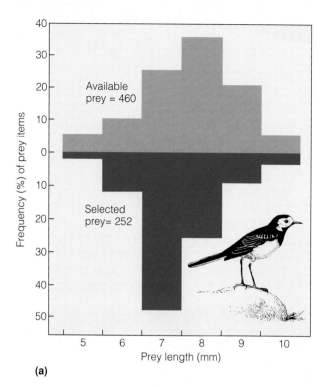

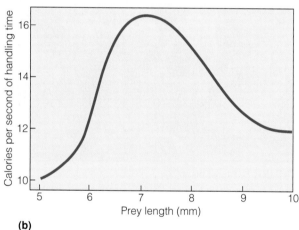

(b)

FIGURE 15.15 (a) Pied wagtails show a definite preference for medium-sized prey, which are taken in disproportionate amounts compared to the frequencies of prey available. (b) The prey size chosen by pied wagtails is the optimal size, providing maximum energy per handling time. (Davies 1977.)

patches; (2) stay with those patches until their profitability falls to a level equal to the average for the foraging area as a whole; (3) leave the patch once it has been reduced to the level of average productivity; and (4) ignore patches of low productivity.

Risk of predation affects foraging activities and sites for many animals, notably birds and fish. This risk makes their foraging activities somewhat less than optimal. They may avoid food-rich patches and opt for poorer, safer patches.

Being opportunists, animals will take some less-than-optimal food items (by ecologists' standards) upon discovery. They may quit foraging in a patch before food items drop to some minimal level. They do not necessarily pass up profitable patches because the patches do not meet some theoretical expectation. Animals quickly learn where food is and where food is not. They will stay with a patch as long as the rate of replenishment exceeds the rate of depletion. Some animals are highly restricted in their choices. Sedentary animals, such as corals and barnacles that are filter feeders, have to take what food flows past them. Others have severely restricted foraging patterns that limit their choices of food.

In spite of its weakness, the concept of optimal foraging is valuable. It provides a way to compare how animals actually forage, select, and harvest their food.

15.13 Herbivores Prey on Plants

Although the term *predator* is typically associated with animals that feed on other animals, herbivory is a form of predation in which animals prey on plants. (See Focus on Ecology 15.1: Carnivorous Plants.) Herbivory represents a special type of predation because herbivores typically do not kill the plants on which they feed. Because the ultimate source of food energy for all heterotrophs is carbon fixed by plants in the process of photosynthesis (see Chapter 6), plant-herbivore interactions represent a key component of all communities.

If you measure the amount of biomass actually eaten by herbivores, it may be small, perhaps 6 to 10 percent of total plant biomass present in a forest community, or as high as 30 to 50 percent in grassland communities. In years of major insect outbreaks, however, or in the presence of an overabundance of large herbivores, consumption is considerably higher (Figure 15.16). Consumption, though, is not necessarily the best measure of the importance of herbivory within a community. Grazing on plants can have a more subtle impact on both plants and herbivores.

Removal of plant tissue—leaf, bark, stems, roots, and sap—affects a plant's ability to survive, even though the plant may not be killed outright. Loss of foliage and subsequent loss of roots decrease plant biomass, reduce the vigor of the plant, place it at a competitive disadvantage with surrounding vegetation, and lower its reproductive effort. That is especially true in the juvenile stage, when the plant is

CARNIVOROUS PLANTS

Terrestrial and semiaquatic plants gain many of their nutrients from the soil. Some plants, however, that live in acidic, nutrient-poor communities—heaths, bogs, swamps, freshwater marshes, and impoverished soil in forest openings—supplement their nutrition by feeding directly on animals. Out of 250,000 or more species of plants, 400 resort to carnivory as a source of nitrogen and phosphorus.

Carnivorous plants trap their prey, which may vary from protozoans, small crustaceans, mosquito larvae, and minute water insects to small tadpoles, large insects, and small amphibians. Carnivorous plants fall into two groups. Active trappers employ rapid plant movements to open trap doors or to close traps. Passive trappers employ pitfalls or sticky adhesive traps.

Two examples of active trappers are the Venus flytrap (*Dionaea muscipula*), restricted to coastal plain of North and South Carolina, and the ubiquitous aquatic and semi-aquatic bladderworts (*Utricularia* spp.). The Venus flytrap (Figure A) employs clam-shaped, hinged leaves. Around their unattached edges are numerous guard hairs and minute nectar glands that attract insects. On the surface

FIGURE A Venus flytrap

FIGURE B Pitcher plant

most vulnerable and least competitive with surrounding vegetation.

A plant may be able to compensate for the loss of leaves by increasing photosynthesis in the remaining leaves. However, it may be adversely affected by the loss of nutrients, depending on the age of the tissues removed. Young leaves are dependent structures, importers and consumers of nutrients drawn from reserves in roots and other plant tissues. Grazing herbivores, both vertebrate and invertebrate, often concentrate on younger leaves and shoots because they are lower in structural carbon compounds, such as lignins, that are difficult to digest and provide little if any energy (see Chapter 7, Section 7.3). By selectively feeding on younger tissues, grazers remove considerable quantities of nutrients from the plant.

Plants respond to defoliation with a flush of new growth that drains nutrients from reserves that otherwise would have gone into growth and reproduc-

tion. If defoliation of trees is complete (see Figure 15.16a), as often happens during an outbreak of gypsy moths (*Porthetria dispar*) or fall cankerworms (*Alsophila pometaria*), leaves that regrow in their place are often quite different in form. The leaves are often smaller, and the total canopy (area of leaves) may be reduced by as much as 30 to 60 percent. In addition, the plant uses stored reserves to maintain living tissue until new leaves form, reducing reserves that it will require later. Regrown twigs and tissues are often immature at the onset of cold weather, reducing their ability to tolerate winter temperatures. Such weakened trees are more vulnerable to insects and disease. Defoliation kills coniferous species.

Browsing animals such as deer, rabbits, and mice selectively feed on the soft, nutrient-rich growing tips (apical meristems) of woody plants, often killing the plants or changing their growth form. Bark-of each half are three small trigger hairs and a covering of minute digestive glands. An insect attracted to the brightly colored leaf touches the trigger hairs, causing the trap to close quickly. The bladderwort, as its name suggests, possesses small, elastic, flattened bladders with the entrance sealed by a flap of cells. When prey touch the tactile cells on the flap, the trap door opens. The bladder walls spring apart, causing a sucking motion that sweeps a current of water into the bladder. Then the door closes, trapping the prey.

FIGURE C Sundew

Pitcher plants (*Sarracenia* spp.) and the various sundews (*Drosera* spp.) are examples of passive trappers. Pitcher plants (Figure B) employ pitfalls. The leaves are modified into pitcherlike or funnel-like traps that arise from underground stems. Bright coloration and secretions of nectar along the hood and the rolled-up lip attract insects. When they alight and move down the leaf, the insects are unable to back up against the stiff, downward-directed hairs. They fall into watery fluid containing digestive enzymes secreted by the leaf. The sundews (Figure C) attract insects to sticky leaves by color, scent, and glistening droplets of adhesive. Sundews have two types of glands on the leaf surface that secrete adhesive droplets. Long-stalked glands on the edge of the leaf ensnare the insect. Shorter-stalked glands in the middle secrete digestive juices. The long-stalked glands slowly bend into the center of the leaf, securing the prey to the digestive area of the leaf.

The digestive glands of carnivorous plants vary considerably but share some features. The outer secretory cells of the leaf epidermis are specialized for enzyme synthesis. The capture of prey induces an osmotically driven outflow of fluid from the vascular system that flushes out digestive enzymes stored in the glands. After digestion is complete, the leaf reabsorbs the secretion pool, with its nitrogen-rich and phosphorus-rich products of digestion, through its cell walls. The plant moves the fluids through the vascular system to other parts of the plant.

burrowing insects feed on the growing cambium tissues of the bole (tree trunk), disrupting the transport of water and nutrients, and making the tree susceptible to fungal pathogens.

Some herbivores, such as aphids, do not consume tissue directly but tap plant juices instead, especially in new growth and young leaves. Sap-sucking insects can decrease growth rates and biomass of woody plants by 25 percent.

Grasses have their meristems, the source of new growth, close to the ground. As a result, grazers first eat the older rather than the more expensive younger tissue. Therefore, grasses are well adapted to grazing, and, up to a point, most benefit from it (see Focus on Ecology 15.2: Grazing the West). The photosynthetic rate of leaves declines with leaf age. Grazing stimulates production by removing older tissue functioning at a lower rate of photosynthesis, increasing the light availability to underlying young leaves. Some

grasses can maintain their vigor only under the pressure of grazing, even though defoliation reduces sexual reproduction. Not all grasses, however, are grazing-tolerant. Species with vulnerable meristems or storage organs (see Chapter 6) can be quickly eradicated under heavy grazing.

15.14 Plants Defend Themselves from Herbivores

Most plants are sessile; they cannot move. Thus, avoiding predation requires adaptations that discourage being selected by herbivores. The array of characteristics that plants employ to deter herbivores include both chemical and structural defenses.

(a)

(b)

FIGURE 15.16 Examples of the impact of high rates of herbivory. (a) Intense predation on oaks by gypsy moths in the forests of eastern North America, and (b) contrast between heavily grazed grassland in Konza Prairie reserve and an adjacent fenced area where large herbivores have been excluded.

For herbivores, often the quality of food rather than the quantity is the constraint on food supply. Because of the complex digestive process needed to break down plant cellulose and convert plant tissue into animal flesh, high-quality forage rich in nitrogen is necessary (see Chapter 8). Without that richness, herbivores can starve to death on a full stomach. Low-quality foods are tough, woody, fibrous, and indigestible. High-quality foods are young, soft, and green, or they are storage organs such as roots, tubers, and seeds. Most food is low quality, and herbivores that have to live on such resources suffer high mortality or reproductive failure.

Many plant species contain chemicals that reduce the availability of their proteins to herbivores. Oaks and other species contain tannins in vacuoles (pockets) in their leaves, which bind with proteins and inhibit their digestion. Lignins and other complex structural compounds are difficult to impossible for most herbivores to digest, making the nitrogen and other essential nutrients bound in these compounds unavailable to the herbivore. These types of compounds are referred to as **quantitative inhibitors** in that that they reduce digestibility and thus potential energy gain from food. These chemicals, mostly tannins and resins, are concentrated near the surface tissues in leaves, bark, and seeds. They function to lower palatability and reduce the ability of microorganisms to break them down in herbivore digestive systems (see Chapter 8, Sections 8.1 and 8.2).

A second group of chemical defenses are termed **qualitative inhibitors**—toxic substances such as cyanogenic compounds (cyanide) and alkaloids such as nicotine, caffeine, cocaine, morphine, and mescaline that interfere with metabolism. Plants can synthesize these substances quickly at little cost. They are effective at low concentrations and readily transported to the site of attack. They can be shuttled about the plant, from growing tips to leaves to stems, roots, and seeds. They can also be transferred from seed to seedling. These substances function as poisons, decreasing the impact of herbivory by reducing damage to the plant.

Although the qualitative inhibitors function to protect against most herbivores, some specialized herbivores have developed ways of breaching these chemical defenses. Some insects can absorb or metabolically detoxify the chemical substances. They even store the plant poisons to use them in their own defense, as the larvae of monarch butterflies do, or in the production of pheromones (chemical signals). Some beetles and certain caterpillars cut circular trenches in leaves before feeding, stopping the flow of chemical defenses to the leaf area on which they will feed.

Plants may also employ structural defenses—hairy leaves, thorns, and spines—that can function to discourage feeding by herbivores (Figure 15.17).

A number of hypotheses have been put forward to explain why plants differ in the types of defenses they have evolved to avoid herbivores. A feature common to all of these hypotheses is the tradeoff between the costs and benefits of defense. The cost of defense in the diversion of energy and nutrients from other needs must be offset by the benefits of avoiding

GRAZING THE WEST

Grazing on public western rangeland is a highly debated topic in the popular press and among environmentalists, range managers, and grassland ecologists. Environmentalists are exerting pressure to eliminate grazing by sheep and cattle on rangelands. They say that grazing is destroying these ecosystems. On the other side are ranchers and livestock interests who want to maintain or increase grazing on public rangeland.

The issue revolves about the idea that grasses, when mowed or grazed, respond to the loss of leaves by increased growth, called overcompensation. (Lawn grasses overcompensate for mowing—that is why you have to mow often, especially early in the growing season.) Most experimental evidence for overcompensation comes from studies of how individual species of grass on highly fertile sites respond to defoliation. Other evidence comes from community-level studies of African grasslands grazed by a diversity of highly mobile native herbivores. Studies at both levels suggest that moderate grazing does promote the productivity of many species of grass above a level that prevails in the absence of grazing. Further, moderate grazing under productive conditions promotes the diversity of plants, whereas in the absence of grazing, certain species dominate and others disappear.

Some range managers and ranchers misinterpret or misunderstand the nature of the studies. They have turned the concept that grazing may benefit plants and increase grassland productivity into a range policy advocating heavy grazing of grassland.

Because evidence for overcompensation comes from highly productive monocultural grassland systems, many grassland ecologists and range managers question whether this concept can be broadly applied to arid and semiarid rangelands. Overcompensation may not occur in these less-productive rangelands, where resources are limiting and crowding of plants does not occur. A number of studies of rangeland systems and semiarid grassland plants show that grazing often harms plants.

Some grassland ecologists point out that grazing affects several levels: individual plants, populations, and communities. Populations of various grassland species have direct and indirect responses to grazing. Direct responses to defoliation involve survival, fecundity, and growth. Indirect responses involve changes in competition between plants and in microclimatic conditions around the plant. Species that respond positively to grazing will disappear if grazing ceases. Plants that are less desirable to grazers can become dominant if palatable species are overgrazed. Types of herbivores also influence the response of plants. Sheep, which graze close to the ground, have a different effect than cattle, which remove leaves higher off the ground. At the ecosystem level, grazing can change the structure and plant composition of rangelands, redistribute nutrients through droppings, and cause erosion. Overgrazing of arid and semiarid rangeland can cause woody vegetation to replace grasses.

From this debate we learn caution. We must not lightly extrapolate findings gained from the study of one species or one community to other species and other communities. Results obtained from monocultural ecosystems seldom apply to natural ecosystems that are ecologically complex. The theory of overcompensation is most applicable to monocultural fertile domestic grasslands, not to arid and semiarid systems. It applies to systems that evolved under the influence of less abundant, migratory, large herbivores, not high-density, sedentary herds of cattle and sheep.

The debate also points to the need for better communication between ecologists and people who use the land. Ecologists should explicitly indicate the limitations of their findings and suggest how those findings might properly apply to management. We need better melding of agricultural research and ecological research in those areas where the findings of both impinge on the management of natural ecosystems.

---

predation. Quantitative defenses are costly, but are long-lived. In contrast, the poisons that form the basis of qualitative defenses are less expensive to produce but are shorter-lived, needing to be replenished from time to time. As a consequence, slower-growing, longer-lived plants often exhibit quanitative defenses. Conversely, faster-growing, shorter-lived plant species are more likely to evolve qualitative defenses.

15.15 Plants, Herbivores, and Carnivores Interact

The concept of a food chain was introduced in Chapter 13 (see Section 13.4). It is a representation of feeding relationships within a community—a series of arrows linking predator (the consumer) and

FIGURE 15.17 Thorns of the acacia tree serve as protection against predators.

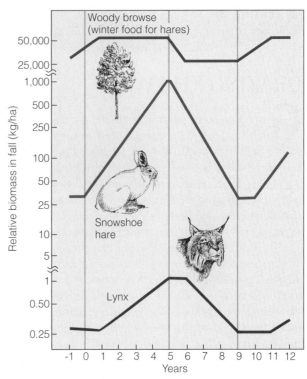

FIGURE 15.18 The three-way interaction of woody vegetation, snowshoe hare, and lynx. Note the time lag between the cycles of the three populations. (Adapted from Keith et al. 1974.)

prey (the consumed). In our discussion thus far, we have discussed herbivory on plants and carnivory on animals as two separate topics, linked only by the common theme of predation. However, they are linked in a very important way as levels on the food chain. Plants are consumed by herbivores, which in turn are consumed by carnivores. As such, we cannot really understand a herbivore-carnivore system without understanding plants and their herbivores; nor can we understand plant-herbivore relations without understanding predator-herbivore relationships. All three—plants, herbivores, and carnivores—are interrelated. Ecologists are beginning to understand these three-way relationships.

A classic case is the three-level interaction of plants, the snowshoe hare *(Lepus americanus),* and its predators—lynx *(Felis lynx),* coyote *(Canis latrans),* and horned owl *(Bubo virginianus)* (Figure 15.18). The snowshoe hare inhabits the high-latitude forests of North America. In winter, it feeds on the buds of conifers and twigs of aspen, alder, and willows, which are termed browse. Browse consists largely of smaller stems and young growth rich in nutrients. The hare-vegetation interaction becomes critical when the amount of essential browse falls below that needed to support the population over winter (approximately 300 grams per individual per day). Excessive browsing during periods when the hare population is high reduces future woody growth, bringing on a food shortage.

The shortage and poor quality of food lead to malnutrition, parasite infections, and chronic stress. Those conditions and low winter temperatures weaken the hares, reducing reproduction and making them extremely vulnerable to predation. Intense predation causes a rapid decline in the number of hares.

Now facing their own shortage of food, the predators fail to reproduce, and populations decline. Meanwhile, released from the pressures of browsing by hares, plant growth rebounds. As time passes, with the growing abundance of winter food together with the decline in predatory pressure, the hare population starts to recover, beginning another cycle. Thus, an interaction between predators and food supply (plants) produce the hare cycle, and in turn the hare cycle impacts the population dynamics of its predators.

15.16 Does Predation Influence Community Structure?

As we have seen in the preceding sections, the relationship between a predator and its prey is influenced by an array of specific behavioral, morphological, and physiological adaptations, making it difficult to generalize about the influence of predation on prey populations. Nonetheless, numerous laboratory and field studies offer convincing evidence that predators can significantly alter prey abundance.

Whereas the influence of competition on community structure is somewhat obscure, the influence of predation is more demonstrable. Because all heterotrophs derive their energy and nutrients from the consumption of other organisms, the influence of predation can more readily be noticed throughout a community. As we shall see in Chapter 17 when we return to the topic of food webs, the direct influence of predation on the population density of prey species can have the additional impact of influencing the interactions among prey species, particularly on competitive relationships.

CHAPTER REVIEW

Summary

Forms of Predation (15.1) Predation is defined generally as the consumption of all or part of one living organism by another. Forms of predation include carnivory, parasitoidism, cannibalism, and herbivory.

Models of Predation (15.2–15.7) Interactions between predator and prey have been described in mathematical models by Lotka and Volterra. The models predict oscillations of predator and prey populations (15.2), suggesting a mutual regulation of predator and prey populations (15.3).

The interaction between predator and prey involves a functional response and a numerical response. A functional response is one in which the number of prey taken increases with the density of prey. There are three types of functional responses. In Type I, the number of prey affected increases in a linear fashion. In Type II, the number of prey affected increases at a decreasing rate toward a maximum value. In Type III, the number of prey taken increases in a sigmoidal fashion as the density of prey increases.

Only Type III is important as a population-regulating mechanism (15.4). A Type III functional response may involve a search image. Predators develop a search image for a particular prey as that item becomes more abundant (15.5). The Type III response implies that predators have a choice. Switching takes place when a predator turns to alternate, more abundant prey, taking it in a disproportionate amount (15.6).

A numerical response is the increase of predators with an increased food supply. Numerical response may involve an aggregative response, the influx of predators to a food-rich area. More importantly, a numerical response involves a change in the rate of growth of a predator population through changes in fecundity (15.7).

Prey Defense, Predator Offense (15.8–15.10) As prey species evolve characteristics to avoid being caught, predators evolve more effective means to capture them (15.8). Chemical defense in animals usually takes the form of distasteful or toxic secretions that repel, warn, or inhibit would-be attackers. Cryptic coloration and behavioral patterns enable the prey to escape detection. Warning coloration declares that the prey is distasteful or disagreeable. Some palatable species mimic unpalatable species for protection. Armor and aggressive use of weapons defend some prey. Alarms and distraction displays help others. Another form of defense is predator satiation. Prey species reproduce so many young at once that predators can take only a fraction of them (15.9).

Predators have evolved strategies of their own. Different methods of hunting include ambush, stalking, and pursuit. Predators also employ cryptic coloration to hide and aggressive mimicry in which they mimic the appearance of the prey (15.10).

Foraging Strategy (15.11–15.12) Central to the study of predation is the concept of optimal foraging. This strategy gives the predator a maximum rate of energy gain. There is a breakeven point above which foraging is profitable and below which it is not. An optimal diet includes the most efficient size of prey for handling and net energy return (15.11). Optimal foraging concentrates activity in the most profitable patches. When the predator depletes the food to the average profitability of the area as a whole, it abandons that patch for one more profitable. Because of risk of predators, social structure, and predator preference, foraging may be less than optimal (15.12).

Herbivory (15.13–15.14) The amount of plant biomass actually eaten by herbivores varies among communities. Plants respond to defoliation with a flush of new growth, which draws down nutrient reserves. Such drawdown can weaken plants, especially woody ones, making them more vulnerable to insects and disease. Moderate grazing may stimulate leaf growth in grasses up to a point. By removing older leaves less active in photosynthesis, grazing stimulates the growth of new leaves (15.13). Plants affect herbivores by denying them palatable or digestible food or by producing toxic substances that interfere with growth and reproduction. Certain herbivore specialists are able to breach the chemical defenses. They detoxify the secretions, block their flow, or sequester them in their own tissues as a defense against predators (15.14).

Vegetation-Herbivore-Carnivore Systems (15.15) Plant-herbivore and herbivore-carnivore systems are closely related. An example of a three-level feeding interaction is the cycle of vegetation, hares, and their predators. Malnourished hares fall quickly to predators. Recovery of hares follows recovery of plants and decline in predators.

Community Structure (15.16) Because all heterotrophs derive their energy and nutrients from the consumption of other organisms, the influence of predation can be readily noticed throughout a community.

Study Questions

1. Define predation, carnivory, herbivory, and cannibalism.

2. What are the basic features and assumptions of the Lotka-Volterra equations?

3. What is a functional response in predation?

4. Distinguish among Type I, Type II, and Type III functional responses. Which may function to regulate prey populations, and why?

5. What is a search image? Switching? Relate them to Type III functional responses.

6. Define numerical response, and discuss the processes involved.

7. What defenses do prey use against predators?

8. Contrast ambush with pursuit and stalking. Evaluate costs and benefits.

9. Relate the methods of hunting to several defensive and offensive strategies, such as cryptic coloration.

10. What is an optimal foraging strategy? What is an optimal diet?

11. How is foraging influenced by a patchy environment?

12. How do plants defend themselves against herbivores? Include two approaches to chemical defense.

13. What effects do herbivores have on plants?

14. How do vegetation, herbivores, and carnivores relate in the scheme of predation?

The little tent-building bat *(Uroderma bilobatum)* of the neotropical rain forest is an important seed disperser of tropical trees.

CHAPTER 16

Parasitism and Mutualism

OBJECTIVES

On completion of this chapter, you should be able to:

- Define parasitism and describe the types of parasites.

- Discuss the modes of transmission.

- Describe the coevolutionary relationship between host and parasite.

- Discuss the potential array of host responses to parasite infection.

- Discuss parasitism as a form of population regulation for host species.

- Define mutualism and discuss it in the context of coevolution.

- Discuss the role of mutualism in nutrition, defense, reproduction, and dispersal.

- Contrast obligate and facultative forms of mutualism.

- Discuss the general process of facilitation and how it relates to mutualism.

In our discussion of predator-prey interactions in Chapter 15, we examined the concept of coevolution, the evolution of one species in response to interaction with another species. Prey have evolved means of defense against predators, and predators have evolved ways to breach those defenses: an evolutionary "game" of adaptation and counter-adaptation. The process of coevolution is even more evident in the interactions between parasites and their hosts. The parasite lives on or in the host organism for some period of its life, a relationship referred to as **symbiosis** (Greek *sym*, "together," and *bios*, "life" or "mode of life"). Symbiosis, as defined by Lynn Margulis, is the "intimate and protracted association between two or more organisms of different species." In the case of the parasitic relationship, the host organism is the habitat as well as the source of nourishment for the parasite. For the parasite, this is an **obligatory relationship**; the parasite requires the host organism for its survival and reproduction. In response to parasites, host species have evolved a variety of defense mechanisms to minimize the negative impact of the parasite's presence. In the situation where adaptations have countered the negative impacts, the relationship may be termed **commensalism**: a symbiotic relationship in which one partner benefits without significantly affecting the other. Coevolution may even take a symbiotic relationship one step farther, providing an advantage to both species involved: a relationship known as **mutualism**.

16.1 Parasites Draw Resources from Host Organisms

Parasitism is a relationship in which two organisms live together (symbiosis), one deriving its nourishment at the expense of the other. Parasites, strictly speaking, draw nourishment from the tissues of the organism on which they live, referred to as the **host**. Typically, parasites do not kill their hosts as predators generally do, although the host may die from secondary infection or suffer stunted growth, emaciation, or sterility.

The general category of parasites consists of a wide range of organisms, including viruses, bacteria, protists, fungi, plants, and an array of invertebrates, among them arthropods. A heavy load of parasites is termed an **infection**, and the outcome of an infection is a **disease**.

Parasites are distinguished by size. Ecologically, parasites may be classified as **microparasites** and **macroparasites**. Microparasites include viruses, bacteria, and protozoans. They are characterized by small size and a short generation time. They develop and multiply rapidly within the host and are the class of parasites that we typically associate with the term *disease*. The duration of the infection is generally short relative to the expected life span of the host. Transmission from host to host is most often direct, although other species may serve as carriers or vectors.

Macroparasites are relatively large. Examples include flatworms, acanthocephalans, roundworms, flukes, lice, fleas, ticks, fungi, rusts, smuts, dodders, broomrape, and mistletoe. Macroparasites have a comparatively long generation time, and typically do not complete an entire life cycle in a single host organism. They may spread by direct transmission from host to host or by indirect transmission, involving intermediate hosts and carriers.

Although macroparasites and microparasites, also called pathogens, are extremely important in interspecific relations, only within the past decade or so have ecologists begun to appreciate their role in population dynamics and community structure. Parasites have dramatic effects when they are introduced into host populations that have not evolved defenses against them. In these cases, diseases sweep through and decimate the population.

16.2 Hosts Provide Diverse Habitats for Parasites

Hosts are the habitats for parasites, and the diverse array of parasites that have evolved exploit every conceivable habitat on and within their hosts. Parasites that live on the skin within the protective cover of feathers and hair are **ectoparasites**. Others, known as **endoparasites**, live within the host. Some burrow beneath the skin. They live in the bloodstream, heart, brain, digestive tract, liver, spleen, mucosal lining of the stomach, spinal cord, nasal tracts, lungs, gonads, bladder, pancreas, eyes, gills of fish, muscle tissue, or other sites. Parasites of insects live on the legs, on the upper and lower body surfaces, and even on the mouthparts.

Plant parasites, too, divide up the habitat. Some live on the roots and stems; others penetrate the roots and bark to live in the woody tissue beneath. Some live at the root collar, commonly called a crown, where the plants emerge from the soil. Others live within the leaves, on young leaves, on mature leaves, or on flowers, pollen, or fruits.

A major problem for parasites, especially parasites of animals, is gaining access to and escaping from the host. Parasites can enter and exit host animals through a variety of pathways including the mouth, nasal passages, skin, rectum, and urogenital system; traveling to their point of infection through the pulmonary, circulatory, or digestive systems.

For parasites, host organisms are like islands that eventually disappear (die). Dependent upon the host as habitat, to survive and multiply, parasites have to escape from one host and locate another, something they cannot do at will. Endoparasites can escape only during a larval stage of their development, known as the *infective stage*, when they must make contact with the next host. The process of transmission from one host to another can occur by either direct or indirect means and can involve adaptations by parasites to virtually all aspects of feeding, social, and mating behaviors in host species.

16.3 Direct Transmission Can Occur Between Host Organisms

Direct transmission is the transfer of a parasite from one host to another without the involvement of an intermediate organism. The transmission can occur by direct contact with a carrier, or the parasite can be dispersed from one host to another through the air, water, or other substrate. Typically, microparasites are transmitted directly, as is the case for the influenza (airborne) and smallpox (direct contact) viruses, and the variety of bacterial and viral parasites associated with sexually transmitted diseases.

Many important macroparasites of animals and plants move from infected to uninfected hosts by direct transmission as well. Among internal parasites, the roundworms *(Ascaris)* live in the digestive tracts of mammals. Female roundworms lay thousands of eggs in the gut of the host that are expelled with the feces, where they are dispersed to the surrounding environment (water, soil, ground vegetation). If they are swallowed by a host of the correct species, the eggs hatch in the host's intestines, and the larvae bore their way into the blood vessels and come to rest in the lungs. From here, they ascend to the mouth, usually by causing the host to cough, and are swallowed again to reach the stomach, where they mature and enter the intestines.

The most important debilitating external parasites of birds and mammals are spread by direct contact. They include lice, mites that cause mange, ticks, fleas, and botfly larvae. Many of these parasites lay their eggs directly on the host, but fleas lay their eggs and their larvae hatch in the nests and bedding of the host, from which they leap onto nearby hosts.

Some macroparasites of flowering plants also spread by direct transmission. One group is *holoparasites,* plants that lack chlorophyll and draw their water, nutrients, and carbon from the roots of host plants. Notable among them are members of the broomrape family (Orobanchaceae). Two examples are squawroot *(Conopholis Americana),* which parasitizes the roots of oaks (Figure 16.1), and beech-drops *(Epifagus virginiana),* which parasitize mostly the roots of beech trees. Seeds of these plants are dispersed locally, and, upon germination, their roots extend through the soil and attach to the roots of the host plant.

Some fungal parasites of plants spread through root grafts. For example, *Fomes annosus,* an important fungal infection of white pine *(Pinus strobus),* spreads rapidly through pure stands of the tree when roots of one tree grow onto (and become attached to) the roots of a neighbor.

16.4 Transmission Between Hosts Can Involve an Intermediate Vector

Some parasites are transmitted between hosts by an intermediate organism, or **vector.** For example, an arthropod vector, the black-legged tick *Ixodes scapularis,* is responsible for transmitting Lyme disease, the major arthropod-borne disease in the United States. Named because the first noted occurrence was at Lyme, Connecticut, in 1975, the disease is caused by a bacterial spirochete, *Borrelia burgdorferi.* It lives in the bloodstream of vertebrates, from birds and mice to deer and humans. The spirochete depends upon the tick for transmission from one host to another.

Malaria parasites infect a wide variety of vertebrate species, including humans. The four species of protozoan parasites *(Plasmodium)* that cause malaria in humans are transmitted to the bloodstream by the bite of an infected female mosquito of the genus *Anopheles* (Figure 16.2). Mosquitoes are known to transmit over 50 percent of the approximately 102 *arboviruses* (short for *arthropod-borne viruses*) that can produce disease in humans, including dengue and yellow fever.

Insect vectors are also involved in the transmission of parasites among plants. European and native elm bark beetles *(Scolytus mutisriatus* and *Hylurgopinys rufipes)* carry spores of the devastating Dutch elm disease *(Ceratocystis ulmi)* from tree to tree.

Mistletoes *(Phoradendron* spp.) belong to a group of plant parasites known as *hemiparasites*

FIGURE 16.1 Squawroot *(Conopholis americana)*, a member of the broomrape family, is a hemiparasite on the roots of oak.

(Figure 16.3). Although photosynthetic, they draw water and nutrients from their host plant. The transmission of mistletoes between host plants is linked to seed dispersal. Birds feed on the mistletoe fruits. The seeds pass through the digestive system unharmed, and are deposited on trees where the birds perch and defecate. The sticky seeds attach to limbs and send out rootlets that embrace the limb and enter the sapwood.

16.5 Transmission Can Involve Multiple Hosts and Stages

In Chapter 10 we introduced the concept of life cycle, the phases associated with the development of an organism, typically divided into juvenile (or pre-

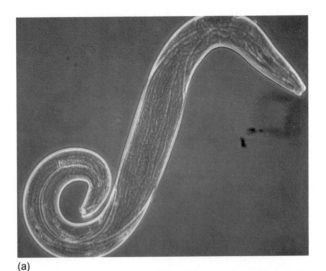

(a)

(b)

FIGURE 16.2 Malaria is a recurring infection produced in humans by protozoan parasites (a) transmitted by the bite of an infected female mosquito of the genus *Anopheles* (b). At present, over 40 percent of the world's population is at risk and over 1 million are killed each year by malaria.

FIGURE 16.3 Mistletoe (*Phoradendron* spp.) with fruits on European birch (*Betula* sp.). Mistletoes are hemiparasites. Although capable of photosynthesis, they penetrate the host tree, extracting water and nutrients.

reproductive), reproductive, and post-reproductive. Some species of parasites cannot complete their entire life cycle in a single host species. The host species in which the parasite becomes an adult and reaches maturity is referred to as the *definitive host*. All others are *intermediate hosts*, which harbor some developmental phase. Parasites may require one, two, or even three intermediate hosts. Each

stage can develop only if the parasite can be transmitted to the appropriate intermediate host. Thus, the dynamics of a parasite population is closely tied to the population dynamics, movement patterns, and interactions of the various host species.

Many parasites, both plant and animal, use this form of indirect transmission, spending different stages of the life cycle with different host species. Figure 16.4 shows the life cycle of the meningeal worm *(Parelaphostrongylus tenuis)*, a parasite of the white-tailed deer in eastern North America. Snails or slugs that live in the grass serve as the intermediate host species for the larval stage of the worm. The deer picks up the infected snail while grazing. In the deer's stomach, the larvae leave the snail, puncture the deer's stomach wall, enter the abdominal membranes, and travel via the spinal cord to reach spaces surrounding the brain. Here the worms mate and produce eggs. Eggs and larvae pass through the bloodstream to the lungs, where the larvae break into air sacs and are coughed up, swallowed, and passed out with the feces. The snails acquire the larvae as they come into contact with the deer feces on the ground. Once within the snail, the larvae continue to develop to the infective stage.

The blood fluke *(Schistosoma)* that infects humans likewise uses snails as an intermediate host. Eggs leave the human body in feces, and hatch into

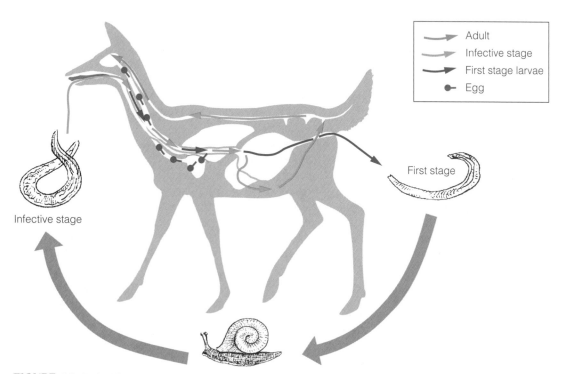

FIGURE 16.4 The life cycle of a macroparasite, the meningeal worm *Parelaphostrongylus tenuis*, which infects white-tailed deer, moose, and elk. Transmission is indirect, involving snails as intermediate host. (After Anderson 1963.)

larvae that swim freely in bodies of standing water. These larvae burrow into the body of a snail, where they reproduce asexually. The newly produced larvae (different larval stage) leave the snail, living in the water until they come into contact with a human. Upon contact, they burrow into the skin and migrate to thin-walled veins in the intestine, where the cycle begins anew.

16.6 Hosts Respond to Parasitic Invasions

Just as the coevolution of predators and prey have resulted in the adaptation of defense mechanisms by prey species, likewise host species exhibit a range of adaptations that function to minimize the impact of parasites. Some responses are defensive mechanisms, functioning to reduce parasitic invasion. Other defense mechanisms aim to combat parasitic infection once it has occurred.

Some animals respond defensively to avoid parasitism. Birds and mammals rid themselves of ectoparasites by grooming. Among birds, the major form of grooming is preening, which involves manipulating plumage with the bill and scratching with the foot. Both activities remove adults and nymphs of lice from the plumage. Deer seek dense, shaded places where they can avoid deerflies, which are common to open areas.

If infection should occur, the first line of defense involves the inflammatory response. The death or destruction (injury) of host cells stimulates the secretion of histamines (chemical alarm signals), which induce increased blood flow to the site, resulting in inflammation. This reaction brings in white blood cells and associated cells that directly attack the infection. Scabs can form on the skin, reducing points of further entry. Internal reactions can produce hardened cysts in muscle or skin that enclose and isolate the parasite. An example is the cysts that encase the roundworm *Trichinella spiralis* (Nematoda) in the muscles of pigs and bears, which causes trichinosis when ingested by humans in undercooked pork.

Plants respond to bacterial and fungal invasion by forming cysts in the roots and scabs in fruits and roots, cutting off contact of the fungus with healthy tissue. Plants react to attacks on leaf, stem, fruit, and seed by gall wasps, bees, and flies by forming abnormal growth structures unique to the particular gall insect (Figure 16.5). Gall formation exposes the larvae of some gall parasites to predation. The conspicuous, swollen knobs of the goldenrod ball gall, for example, attract the downy woodpecker *(Picoides*

(a) Pine cone gall   **(b) Globular hickory leaf gall**

(c) Oak bullet gall   **(d) Goldenrod ball gall**

FIGURE 16.5 Galls are a growth response to an alien substance in plant tissues. In this case, the presence of a parasitic egg stimulates a genetic transformation of the host's cells. (a) The pine cone gall, a bud gall on willows caused by the gall midge *Rhabdophaga strobiloides*. (b) The hickory leaf gall, induced by the gall aphid *Phylloxera canyaeglobuli*. (c) The oak bullet gall, caused by the gall wasp *Disholcaspis globulus*. (d) The goldenrod ball gall, a stem gall induced by a gallfly, *Eurosta solidagin*.

pubescens), which excavates and eats the larva within the gall.

The second line of defense is the immune response (or immune system). When a foreign object such as a virus or bacteria, termed an antigen (a contraction of *antibody-generating*), enters the bloodstream, it elicits an immune response. White cells called lymphocytes (produced by lymph glands) produce antibodies. The antibodies target the antigens present on the parasite's surface or released into the host, helping to counter their effects. These antibodies are energetically expensive to produce. They also are potentially damaging to the host's own tissues. Fortunately, the immune response does not have to kill the parasite to be effective. It only has to reduce the feeding, movements, and reproduction of the parasite to a tolerable level. The immune system is extremely specific, and it has a remarkable "memory." It can "remember" antigens it has encountered in the past, and react to them more quickly and vigorously in subsequent exposures.

The immune response, however, can be breached. Some parasites vary their antigens more or less continuously. By doing so, they are able to keep one jump ahead of the host's response. The result is a chronic infection of the parasite in the host. Antibodies specific to an infection normally are composed of proteins. If the animal suffers from poor nutrition and its protein deficiency is severe, normal production of antibodies is inhibited. Depletion of energy reserves breaks down the immune system and allows viruses or other parasites to become pathogenic. The ultimate breakdown in the immune system occurs in humans infected with the human immunodeficiency virus (HIV), the causal agent of AIDS, which is transmitted sexually or through the use of shared needles, or by infected donor blood. The virus attacks the immune system itself, exposing the host to a range of infections that prove fatal.

16.7 Parasites Can Impact Host Survival and Reproduction

Given that organisms have a limited amount of energy, it is not surprising that parasitic infections function to reduce both growth and reproduction. J. J. Schall of the University of Vermont examined the impact of malaria on the western fence lizard (Scleroporus occidentalis) inhabiting California. Clutch size is approximately 20 percent smaller in malaria-infected females compared with noninfected individuals. The cause of reduced reproduction is that infected females are less able to store fat during the summer, resulting in less energy for egg production the following spring. Parasitic infection of plants—notably grasses and sedges by endophytic fungi that grow within the leaves and stems, and by smuts and rusts that grow externally—can prevent the plants from reproducing sexually, inhibiting or aborting flowers.

Although most parasites do not kill their host organisms, increased mortality can result from a variety of indirect consequences of infection. Nestling birds with heavy tick infestations exhibit reduced growth and higher rates of mortality. Heavily parasitized animals often behave abnormally. Rabbits infected with the bacterial disease tularemia (Francisella tularensis), transmitted by the rabbit tick (Haemaphysalis leporis-palustris), are sluggish, increasing their vulnerability to predation. Pacific killifish (Fundulus parvipinnis) parasitized by trematodes display abnormal behavior such as sur- facing and jerking. This behavior calls attention to the trematodes' definitive host—fish-eating birds. By altering its intermediate host's behavior, making it more susceptible to predation, the trematode ensures the completion of its life cycle.

16.8 Parasites and Hosts May Evolve an Uneasy Truce

For parasite and host to coexist under a relationship that is hardly benign, the two have to come to an uneasy truce. The parasite gains no advantage if it kills its host. A dead host means dead parasites. Conversely, the host needs to resist this invasion by eliminating the parasites or at least minimizing their effects. The parasite and its host have to strike a balance. The host has to reach a level of immune response between beneficial and harmful consequences. It has to direct enough of its metabolic resources to minimize the cost of parasitism, yet not unduly impair its own growth and reproduction. The parasite has to achieve optimal growth and reproduction without overwhelming its host. Not to do so would be highly detrimental to both. This mutual tolerance involves host and parasite in adapting to each other to some degree.

An example of coevolutionary adaptation and counteradaptation is the parasite-host interaction of the European rabbit (Oryctolagus cuniculus) and the viral infection myxomatosis. To control the intro- duced rabbit, the Australian government introduced the viral parasite into the population. The first epi- demic of myxomatosis was fatal to 97 to 99 percent of the rabbits. The second resulted in mortality of 85 to 95 percent; the third, 40 to 60 percent. The effect on the rabbit population was less severe with each succeeding outbreak, suggesting that the two popula- tions had adjusted to each other.

In this adjustment, certain strains of the virus, intermediate in virulence, tended to replace highly virulent strains. The transmission of myxomatosis depends upon Aedes and Anopheles mosquitoes, which feed only on the blood of living animals. Rabbits infected with a less-virulent strain live longer than those infected with a more-virulent strain. Because the mosquitoes have access to the less-virulent form for a longer time, the less-virulent strain has a competitive advantage. Also involved was a passive immunity to myxomatosis conferred upon the young born to mothers that were immune. The final outcome has been the development of resistance to myxomatosis in the rabbit population.

16.9 Parasites May Regulate Host Populations

The impact of parasites on host populations depends on the mode of transmission and the density and dispersion of the host population. Microparasites, dependent for the most part on direct transmission, require a high host density to persist. For them, ideal hosts live in groups or herds. These parasites need a long-lived infective stage that does not ensure long-term immunity in the host population. Immunity reduces parasite populations, if it does not eliminate them. An example of a parasite in wild populations that does not confer long-term immunity is rabies. One that does confer immunity to animals that survive the disease is distemper.

Indirect transmission, typical of macroparasites, is more complex. Parasites with indirect transmission require a highly effective transmission stage. They do well and persist for a long time in low population densities of hosts. The longevity of each parasitic stage varies in different hosts. Longevity is high in the definitive host and much lower in the intermediate host.

Parasitism can have a debilitating effect on host populations, most evident when parasites invade a population with no evolved defenses (see Focus on Ecology 16.1: Plagues upon Us). In such cases, the spread of the disease may be virtually density-independent, reducing populations, exterminating them locally, or restricting the distribution of the host. The chestnut blight *(Endothia parasitica)*, introduced into North America from Europe, nearly exterminated the American chestnut *(Castanea dentata)* and removed it as a major component of the forests of eastern North America. Dutch elm disease, caused by a fungus *(Graphium ulmi)* spread by beetles, has nearly removed both the American elm *(Ulmus americana)* from North America and the English elm *(U. glabra)* from Great Britain. Anthracnose *(Biscula destructiva)*, a fungal disease, is decimating flowering dogwood *(Cornus florida)*, an important understory tree in the forests of eastern North America.

Rinderpest, a viral disease of domestic cattle, was introduced to East Africa in the late 19th century and subsequently decimated herds of African buffalo *(Syncerus caffer)* and wildebeest *(Connochaetes taurinus)*. Avian malaria carried by introduced mosquitoes has eliminated most native Hawaiian birds below 1000 meters, the altitude above which the mosquito cannot persist.

On the other hand, parasites may function as density-dependent regulators on host populations.

Such incidents typically occur with directly transmitted endemic (native) parasites that are maintained in the population by a small reservoir of infected carrier individuals. Outbreaks of these diseases appear to occur when the density of the host population is high, and they tend to reduce host populations sharply, resulting in population cycles of host and parasite similar to those observed for predator and prey (see Chapter 15, Section 15.2). Examples are distemper in raccoons and rabies in foxes, both significant in controlling their host populations.

In other cases, the parasite may function as a selective agent of mortality, infecting only a subset of the population. The distribution of macroparasites, especially those with indirect transmission, is highly clumped. Some individuals in the host population carry a higher load of parasites than others (Figure 16.6). These individuals are the ones that are most likely to succumb to parasite-induced mortality, suffer reduced reproductive rates, or both. Such deaths often are caused not directly by parasites but indirectly by secondary infection. Herds of bighorn sheep *(Ovis canadensis)* in western North America may be infected with up to 7 different species of lungworms (Nematoda). Highest infections occur in the spring

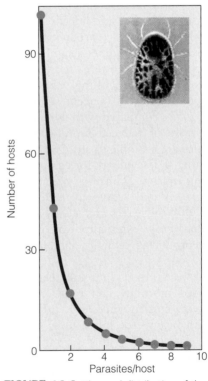

FIGURE 16.6 Clumped distribution of the tick *Ixodes trianguliceps* (Birula) on a population of the European field mouse, *Apodemus sylvaticus*. Most of the individuals in the host population carry no ticks. A few individuals carry most of the parasite load. (After Randolph 1975.)

PLAGUES UPON US

Humans have always been bedeviled by parasites, but more so in recent stages of human history. During our first 2 million years as hunter-gatherers, the most bothersome parasites were macroparasites such as roundworms, directly transmitted. Only microparasites with high transmission rates that produced no immunity could persist in such small groups of hosts.

Once humans became sedentary agriculturalists, however, and aggregated in villages, populations became large enough to sustain bacterial and viral parasites. Many of these parasites evolved from those causing diseases in domestic animals. Measles, for instance, evolved from canine distemper. Populations were too small at first to support disease continuously, without reinfection from some neighboring settlement. Once a settlement grew into a sufficiently large city, the population was dense enough to maintain a reservoir of infection. As commerce developed between cities, people and goods began to move long distances. They introduced diseases from one part of the world to another where the populations lacked immunity. Periodic epidemics swept through the cities.

A classic example of the importation, direct transmission, and rapid spread of a disease is bubonic plague. Plague is caused by a rod-shaped bacillus, *Yersinia pestis (Pasteurella pestis)*. It is transmitted directly from host to host mostly by the bite of its vector (fleas from infected rodents) and from person to person by mucous droplets spread by coughing. Infected individuals become ill in a few hours to a few days, showing symptoms of high fever and swollen lymph nodes. Death often follows within several days. Plague was called the black death because of the dark color on the faces of many dead victims.

The reservoir of the bacillus is burrowing rodents. The bacillus is most closely associated with the black rat (*Rattus rattus*), a native of central India, also the original focal point of the disease. Agile climbers, black rats easily boarded cargo ships that carried them to port cities in Asia and the Mediterranean region. Hidden in the baggage of caravans, they spread across the steppes of Asia, where undoubtedly they transferred the plague bacillus to burrowing rodents in the region. In 1331 an epidemic of the plague swept through China. Mongol armies carried the disease with them as they swept across Asia to the Mediterranean. At the siege of Caffa in 1346 on the Crimean peninsula, the Mongol army was devastated by the plague and withdrew after catapulting the victims' corpses into town. Trade resumed, and ships carried infected black rats to ports of southern Europe.

Conditions were right for the spread of disease. Europe was experiencing huge population growth; the climate was worsening; and crops were failing. By the end of December 1347, the disease that had decimated the Mongol army spread to Italy and southern France; by December 1348 it reached southern Germany and England; and by December 1350 it reached Scandinavia. Between 1348 and 1350, one-third of the European population, including entire villages, succumbed to the black death, upsetting the social, political, and economic stability of Europe. Later outbreaks occurred in 1630 in Milan, in 1665 in London, and in 1720–1721 in Marseilles. Sporadic local outbreaks occurred throughout the world until 1944, when antibiotics quickly cured the disease when diagnosed early. The plague bacillus still thrives worldwide, including North America, harbored by burrowing rodents.

Other diseases introduced into populations lacking immunity mimic the spread and devastation of the black death. Smallpox, measles, and typhus were carried to the New World by Spanish and English explorers and settlers, and spread rampant through indigenous populations of South, Central, and North America. Disease devastated the Aztecs in Peru and nearly exterminated the Native American population of New England, allowing uncontested settlement of the region by the English. In more recent times, a massive flu epidemic spread worldwide in 1918, killing 21 million people, including 500,000 people in the United States. Flu still remains a threat because of its high mutation rate. Strains evolve faster than resistance to the disease develops. As a result, flu returns in waves of different types.

Because many of the old plagues, like measles, have been contained by vaccinations and other health measures, many of us have become complacent about disease. Nevertheless, new diseases—Ebola in Zaire and AIDS everywhere—warn us that plagues are still with us. New mutant forms of old diseases such as tuberculosis, once considered conquered; a massive increase in world population; a changing global climate; and rapid transcontinental movement of people and goods set the stage for future plagues.

when the lambs are born. Heavy lungworm infections in the young bring about a secondary infection, pneumonia, that kills the lambs. Such infections can sharply reduce mountain sheep populations by reducing reproductive success.

16.10 Parasitism Can Evolve into a Positive Relationship

Parasites and their hosts live together in a symbiotic relationship in which the parasite derives its benefit (habitat and food resources) at the expense of the host organism. At some stage in the coevolution of the host and parasite, the relationship may become beneficial to both. For example, a host tolerant of parasitic infection may begin to exploit the relationship. In time, the two species become dependent on each other. Such a relationship is called mutualism.

Mutualism is a relationship between members of two species that benefits both. Out of this relationship, individuals of both species enhance their survival, growth, or reproduction. Evidence suggests that this interaction is more a reciprocal exploitation than a cooperative effort between individuals.

Mutualism may be symbiotic or nonsymbiotic. In symbiotic mutualism, individuals coexist and their relationship is obligatory. At least one member of the pair becomes totally dependent on the other. At the extreme, the two interacting organisms function as one individual (see Focus on Ecology 16.2: Permanent House Guests). In nonsymbiotic mutualism, the two organisms do not physically coexist yet depend upon each other for some essential function. Mutualistic associations are involved in a variety of processes related to the acquisition of energy and nutrients, protection and defense, reproduction and dispersal.

16.11 Symbiotic Mutualisms Are Involved in the Transfer of Nutrients

The digestive system of herbivores is inhabited by a diverse community of mutualistic organisms that play a critical role in the digestion of plant materials. The chambers of a ruminant's stomach (see Chapter 8, Section 8.1) contain large populations of bacteria and protozoa that carry out the process of fermenta-

tion. The inhabitants of the rumen are primarily anaerobic, adapted to this peculiar environment. Ruminants are perhaps the best studied, but not the only example of the role of mutualism in animal nutrition. The stomachs of virtually all herbivorous mammals and some species of birds and lizards rely on microbial flora to digest cellulose in plant tissues.

Mutualistic interactions are also involved in the uptake of nutrients by plants. Nitrogen is an essential constituent of protein, a building block of all living material. Although nitrogen is the most abundant constituent of the atmosphere, approximately 79 percent in its gaseous state, it is unavailable to most life. It must first be converted into a chemically usable form. One group of organisms that can utilize gaseous nitrogen (N_2) is the nitrogen-fixing bacteria of the genus *Rhizobium*. *Rhizobium* bacteria are widely distributed in the soil, where they can grow and multiply. But in this free-living state, they do not fix nitrogen. Legumes—a group of plant species that include clover, beans, and peas—attract the bacteria through the release of exudates and enzymes from the roots. *Rhizobium* bacteria enter the root hairs, where they multiply and increase in size. This invasion and growth results in swollen, infected root hair cells, which form root nodules (Figure 16.7). The bacteria receive carbon and other resources from the host plant, while the bacteria contribute the fixed nitrogen to the plant, allowing the plant to function and grow independent of the availability of mineral (inorganic) nitrogen in the soil (see Chapter 6).

Some forms of mutualism are so permanent and obligatory that the distinction between the two interacting populations becomes blurred. A good example is mycorrhizae, a mutualistic relationship between plant roots and fungi. The fungi assist the plant with the uptake of nutrients from the soil. In return, the plant provides the fungi with carbon, a source of energy.

Endomycorrhizae are common to many trees of temperate and tropical forests. Mycelia—masses of fine fungal filaments in the soil—infect the tree roots. They penetrate the cells of the host to form a finely bunched network called an arbuscle (Figure 16.8a). The mycelia act as extended roots for the plant, but do not change the shape or structure of the roots. They draw in nitrogen and phosphorus at distances beyond those reached by the roots and root hairs. Another form, *ectomycorrhizae*, produce shortened, thickened roots that look like coral (Figure 16.8b). The threads of the fungi work between the root cells. Outside the root, they develop into a network that functions as extended roots.

Mycorrhizae are especially important in nutrient-poor soils. They aid in the decomposition of litter

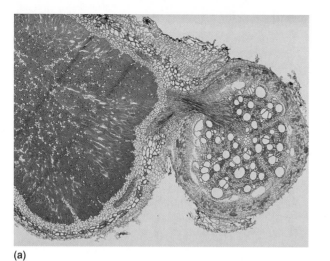

(a)

(b)

FIGURE 16.7 Nitrogen-fixing *rhizobium* bacteria (a) infect the roots forming nodules (b). The bacteria derive carbon from the plant and in return provide the plant with nitrogen.

and the uptake of nutrients, especially nitrogen and phosphorus, from the soil into the root tissue. In addition, mycorrhizae reduce susceptibility of their hosts to disease by providing a physical barrier and stimulating the roots to produce chemical inhibitory substances. Without ectomycorrhizae, certain groups of trees, especially conifers, oaks, and birches, could not become established, survive, and grow. In fact, many mycorrhizae are specific to certain trees, such as pines and Douglas fir.

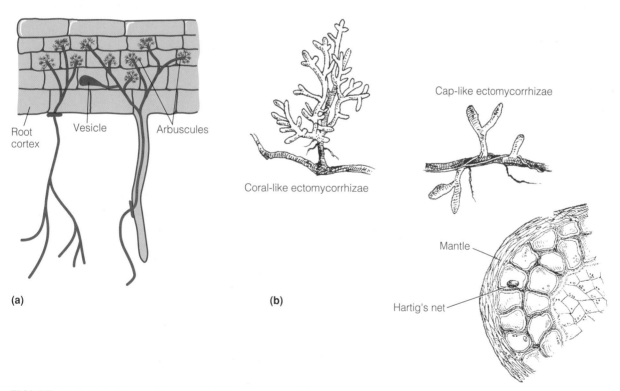

(a)

Root cortex

Vesicle

Arbuscules

(b)

Coral-like ectomycorrhizae

Cap-like ectomycorrhizae

Mantle

Hartig's net

FIGURE 16.8 (a) Endomycorrhizae grow within tree roots, and fungal hyphae enter the cells. (b) Ectomycorrhizae form a mantle of fungi about the tips of rootlets. Hyphae invade the tissues of rootlets between the cells. The network is called Hartig's net.

PERMANENT HOUSE GUESTS

Some forms of mutualism are so permanent and obligatory that the distinction between the two interacting organisms becomes blurred. Reef-forming corals of the tropical waters provide an example. These corals secrete an external skeleton composed of calcium carbonate. The individual coral animals, called polyps, occupy little cups or corallites in the larger skeleton that forms the reef (Figure A). These corals have small symbiotic plant cells (algae) called zooxanthellae in their tissues. Although the coral polyps are carnivores that feed on zooplankton suspended in the surrounding water, they acquire only about 10 percent of their daily energy requirement from zooplankton. They obtain the remaining 90 percent from carbon produced by the symbiotic algae through photosynthesis. Without the algae, these corals would not be able to survive and flourish in their nutrient-poor environment.

Lichens present an example of a symbiotic association in which the fusion of mutualists has made it even more difficult to distinguish the nature of the individual. A lichen (Figure B) consists of a fungus and an alga combined within a spongy body called a *thallus.* The alga supplies food to both organisms, while the fungus protects the alga from harmful light intensities, produces a substance that accelerates photosynthesis in the alga, and absorbs and retains water and nutrients for both organisms. There are about 25,000 known species of lichens, each composed of a unique combination of fungus and alga.

Perhaps the most remarkable case of endosymbiosis (one species living within another) is a theory involving the evolution of multicellular organisms. In Chapter 2, we gave the impression that you, like all living organisms, have a single set of genes arranged along chromosomes in the nucleus of each cell. We now have to confess that this is not really the case. The nucleus is one of numerous small bodies, called *organelles,* that reside within the cell. One kind of organelle, the *mitochondria,* carries out the process of respiration (Chapter 6, Section 6.1) that occurs in all living cells (Figure C). Mitochondria seem to have evolved from small, free-living cells much like bacteria (called prokaryotes). These small prokaryotes were able to use oxygen to release large amounts of energy from organic molecules in the process of cellular respiration. They may possibly have been ingested as food by the

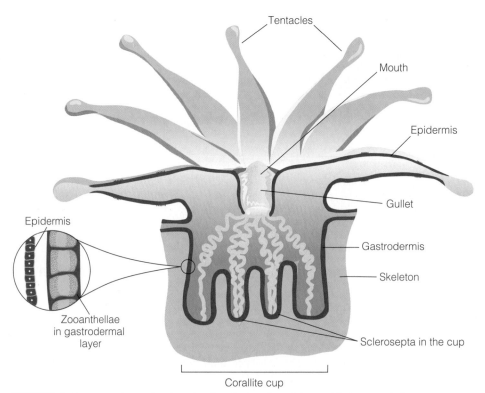

FIGURE A Anatomy of a coral polyp, showing the location of the symbiotic zooanthellae.

larger prokaryotes. If indigestible, they may have remained alive, functioning as internal parasites while continuing to perform respiration. The engulfed cells may have grown increasingly dependent on the host cells for organic molecules needed to carry out their chemical activities. Likewise, the host cells would have derived an increasing proportion of their energy from the ATP generated by the cells performing respiration. As the cells became more interdependent, they may have become inseparable, functioning as a single organism. This theory, first developed by Lynn Margulis of the University of Massachusetts, has become widely accepted and is supported by strong circumstantial evidence. For one thing, the mitochondria have their own genes, passed down separately from the genes in the nucleus of the cell. That is to say, each cell in your body has two sets of genes: one in the nucleus and the other within the mitochondria. Although the genes contained in the nucleus of your cells that define your individual characteristics are drawn equally from your mother and father, all of your mitochondrial genes (the genes contained in the mitochondria in each cell of your body) came from your mother; they are contained within the unfertilized egg. In addition to replicating their own DNA, mitochondria reproduce within the cell by a process resembling the reproduction of bacteria.

FIGURE A Lichens consist of a fungus and an alga combined within a spongy body call a thallus.

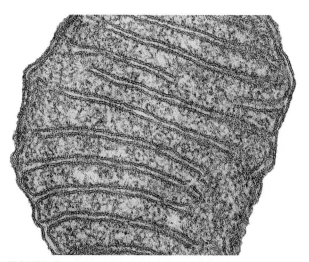

FIGURE B Mitochondria that occur in all living cells carry out the process of respiration.

16.12 Some Symbiotic Mutualisms Are Defensive

Other mutualistic associations involve defense of the host organism. A major problem for many livestock producers is the toxic effects of certain grasses, particularly perennial ryegrass and tall fescue. These grasses are infected by symbiotic endophytic fungi that live inside plant tissues (Figure 16.9). The fungi (Clavicipitaceae and Ascomycetes) produce alkaloid compounds in the tissue of the host grasses. The alkaloids, which impart a bitter taste to the grass, are toxic to grazing mammals, particularly domestic animals, and a number of insect herbivores. In mammals, the alkaloids constrict small blood vessels in the brain, causing convulsions, tremors, stupor, gangrene of the extremities, and death. At the same time, these fungi seem to stimulate plant growth and seed production. This symbiotic relationship suggests a defensive mutualism between plant and fungi. The fungi defend the host plant against grazing. In return, the plant provides food to the fungi in the form of photosynthates.

A group of Central American ant species (*Pseudomyrmex* spp.) that live in the swollen thorns

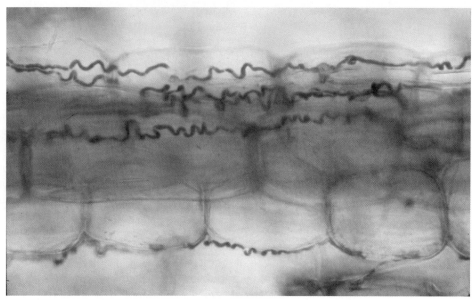

FIGURE 16.9 Endophytic fungi in a blade of fescue *(Festuca).*

of acacia (*Acacia* spp.) trees provides a second example of defensive mutualism. In addition to providing shelter, the plants provide a balanced and almost complete diet for all stages of development. In turn, the ants protect the plants from herbivores. At the least disturbance, the ants swarm out of their shelters, emitting repulsive odors and attacking the intruder until it is driven away.

16.13 Mutualisms May Be Nonsymbiotic

Many mutualistic relationships are not symbiotic. Mutualists live physically separate lives, yet depend upon each other for some essential function. Pollination in flowering plants and seed dispersal provide a variety of examples. The evolution of these mutualistic relationships is different from those that have arisen from the evolution of host-parasite relationships. In most cases, however, they have evolved from predation or some more general form of exploitation.

In plant-pollinator relationships, birds and insects came to plants to feed on pollen. In the course of this exploitation, these animals happened to carry pollen to other plants of the same species. Such plants experienced improved reproductive success and began to exploit the visitors as a means of dispersing pollen. Selection favored the development of mechanisms to maintain the relationship, such as sugar-rich nectar.

Nonsymbiotic mutualisms are generally not confined to two species. A species may form a mutualistic relationship with any number of related species. Because the benefits of such mutualisms, usually associated with seed dispersal and pollination, spread over many plants, pollinators, and seed dispersers, these mutualisms are called *diffuse.*

16.14 Mutualisms Are Often Necessary for Pollination

The goal of pollination is to transfer pollen from the anthers of one plant to the stigma of another plant of the same species (see Chapter 12, Section 12.2). Some plants simply release their pollen in the wind. This method works well and costs little when plants grow in large homogeneous stands, as grasses and pine trees do. Wind dispersal is unreliable when individuals of the same species are scattered individually or in patches across a field or forest. In these circumstances, pollen transfer depends upon insects, birds, and bats.

Nectivores, animals that feed on nectar, visit plants to exploit a source of food. While feeding, the nectivores inadvertently pick up pollen and carry it to the next plant they visit. With few exceptions, the nectivores are typically generalists, feeding on many different plant species. Because each species flowers briefly, nectivores depend on a progression of flowering plants through the season. Nectivores cannot afford to commit themselves to a single species of flower, but they do concentrate on one species while its flowers are available.

Plants are the ones that have to specialize. They have to entice certain animals by color, fragrances, and odors, dusting them with pollen, and then rewarding them with a rich source of food: sugar-rich nectar, protein-rich pollen, and fat-rich oils. Providing such rewards is expensive for plants. Nectar and oils are of no value to the plant except as an attractant for potential pollinators. They represent energy that the plant otherwise might expend in growth.

Many species of plants, such as blackberries, elderberries, cherries, and goldenrods, are generalists themselves. They flower profusely and provide a glut of nectar that attracts a diversity of pollen-carrying insects, from bees and flies to beetles. Other plants are more selective, screening their visitors to ensure some efficiency in pollen transfer. These plants may have long corollas, allowing access only to insects and hummingbirds with long tongues and bills and keeping out small insects that eat nectar but do not carry pollen. Some plants have closed petals that only large bees can pry open. Orchids, whose individuals are scattered widely through their habitats, have evolved a variety of precise mechanisms for pollen transfer and reception so that pollen is not lost when the insect visits flowers of other species (Figure 16.10).

16.15 Mutualisms Are Involved in Seed Dispersal

Plants with seeds too heavy to be dispersed by wind depend on animals to carry them some distance from the parent plant and deposit them in sites favorable for germination and seedling establishment. Some seed-dispersing animals upon which the plant depends may be seed predators as well, eating the seeds for their own nutrition. Plants depending on such animals must produce a tremendous number of seeds over their reproductive lives. They sacrifice most of them to ensure that a few will be dispersed, come to rest on a suitable site, and germinate.

For example, a very close mutualistic relationship exists between wingless-seeded pines of western North America [whitebark pine *(Pinus albicaulis)*, limber pine *(Pinus flexilis)*, southwestern white pine *(Pinus strobiformis)*, and pinyon pine *(Pinus edulis)*] and several species of jays [Clark's nutcracker *(Nucifraga columbiana)*, pinyon jay *(Gymnorhinus cyanocephalus)*, western scrub jay *(Aphelocoma californica)*, and Steller's jay *(Cyanocitta stelleri)*]. In fact, there is a close correspondence between the ranges of these pines and jays. The relationship is

FIGURE 16.10 Only males of a single wasp species pollinate the elbow orchid of southeastern Australia. A structure on the flower mimics the body of the smaller female wasp and emits an odor that imitates the pheremones (chemical signals) produced by the female. When a male wasp struggles to mate with the female decoy, pollen sticks to his body, which he then transfers to other elbow orchids.

especially close between Clark's nutcracker and the whitebark pine (Figure 16.11). Only Clark's nutcracker possesses the morphology and behavior appropriate to disperse the seeds successfully away from the tree. The bird carries as many as 50 seeds in cheek pouches up to 22 kilometers away from the source and caches them deep enough in the soil of forest and open fields to reduce predation by rodents.

Seed dispersal by ants is prevalent among a variety of herbaceous plants that inhabit the deserts of the southwestern United States, the shrublands of Australia, and the deciduous forests of eastern North America. Such plants, called *myrmecochores*, have an ant-attracting food body on the seed coat called an *elaiosome* (Figure 16.12). Appearing as shiny tissue on the seed coat, the elaiosome contains certain chemical compounds essential for the ants. The ants carry seeds to their nests, where they sever the elaiosome and eat it or feed it to their larvae. The ants discard the intact seed within abandoned galleries of the nest. The area around ant nests is richer in nitrogen and phosphorus than the surrounding soil, providing a good substrate for seedlings. Further, by removing seeds far from the parent plant, the ants significantly reduce losses to seed-eating rodents.

Plants may enclose their seeds in a nutritious fruit attractive to fruit-eating animals, the **frugivores** (Figure 16.13). Frugivores are not seed predators. They eat only the tissue surrounding the seed and, with some exceptions, do not damage the seed. Most frugivores do not depend exclusively on fruits, which are only seasonally available and deficient in proteins.

(a)

FIGURE 16.11 (a) The Clark's nutcracker *(Nucifraga columbiana),* a jay of the high mountains of western North America, eats and stores the seeds of whitebark pine *(Pinus albiculis)* in which it is perched. Similarity in the geographic range of these two species (b) reflects their interdependence.

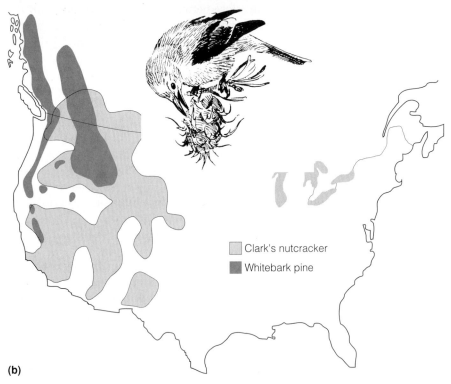

Clark's nutcracker
Whitebark pine

(b)

To use frugivorous animals as agents of dispersal, plants must attract them at the right time. Plants discourage the consumption of unripe fruit by cryptic coloration, such as green unripened fruit among green leaves, and by unpalatable texture, repellent substances, and hard outer coats. When seeds mature, plants attract fruit-eating animals by presenting attractive odors, softening the texture, increasing sugar and oils, and "flagging" their fruits with colors.

Most plants have fruits that can be exploited by a large number of animal dispersers. Such plants opt for quantity dispersal, the scattering of a large

FIGURE 16.12 Seeds of the Pacific bleeding heart have white appendages called elaiosomes. Bleeding hearts, trilliums, and several dozen other plants have these appendages on their seeds that contain oils attractive to ants. The ants carry the seeds to their nest, where the elaiosomes are removed and consumed as food, leaving the seeds unharmed.

FIGURE 16.13 The frugivorous cedar waxwing (Bombycilla cedrorum) feeds on the red berries of mountain ash (Sorbus).

number of seeds with the chance that a diversity of consumers will drop some seeds in a favorable site. Such a strategy is typical of, but not exclusive to, plants of the temperate regions, where fruit-eating birds and mammals rarely specialize in one kind of fruit and do not depend exclusively on fruit for sustenance. The fruits are usually succulent and rich in sugars and organic acids. They contain small seeds that pass through the digestive tract unharmed. To accomplish this passage, such plants have evolved seeds with hard coats resistant to digestive enzymes. Such seeds may not germinate unless they have been conditioned or scarified by passage through the digestive tract. Large numbers of small seeds may be so dispersed, but few are deposited on suitable sites.

In tropical forests, 50 to 75 percent of the tree species produce fleshy fruits whose seeds are dispersed by animals. Rarely are these frugivores obligates of the fruits on which they feed, although exceptions include a large number of tropical fruit-eating bats (see photo on page 309).

16.16 Facilitation is a Form of Mutualism

Many of the examples of mutualism that we have examined in the previous sections are obligatory (termed obligate mutualisms), which have coevolved to the point to which one or both members of the association cannot persist without the other. In other situations, the association is nonessential, but nonetheless increases the fitness of the parties involved. This type of mutualism is referred to as *facultative*. Facultative interactions, however, cover the broader range of interactions among organisms that benefit at least one of the participants and cause harm to neither. Termed **facilitation**, the situation where one species benefits from the presence or action of another is an important feature of most ecological communities. Simply by their mere presence, many species alter the local environment, often making a stressful habitat more hospitable for other species. Trees cast shade onto the forest floor, reducing light, temperature, and surface evaporation, in turn increasing relative humidity and soil moisture. Corals form reefs, increasing the structural complexity of the community and providing habitat for a wide variety of other species. Although the role of positive interactions among species within communities has been long recognized, ecological research on the role of positive interactions is still far less common than on other forms of interactions, particularly competition and predation.

16.17 Mutualism Can Influence Community Structure

Mutualism is easy to appreciate at the individual level. We grasp the interaction between an ectomycorrhizal fungus and its oak or pine host; we count the acorns dispersed by squirrels and jays, and we measure the cost of dispersal to oaks in terms of seeds consumed. Mutualism improves the growth and reproduction of the fungus, the oak, and the seed predators. But what are the consequences at the level of the population and community?

Defining population and community effects of mutualism is considerably more difficult than defining the effects of interspecific competition, predation, and parasitism. Mutualism exists at the population level only if the growth rate of species A increases with the increasing density of species B, and vice versa. For symbiotic mutualists where the relationship is obligate, the influence is straightforward. Remove species A, and the population of species B no longer exists. If ectoymycorrhizal spores fail to infect the rootlets of young pines, they will not develop. If the young pine invading a nutrient-poor field fails to acquire a mycorrhizal symbiont, it will not grow well, if at all.

Discerning the role of nonsymbiotic mutualisms, or even facultative interactions, on community structure can be more difficult. As presented in Sections 16.14 and 16.15, mutualistic relationships are common in plant reproduction, where plant species often depend on animal species for pollination, seed dispersal, or germination. Although, in some cases, the relationships between pollinators and certain flowers are so close that loss of one could result in the extinction of the other, in most cases the effects are subtler and require detailed demographic studies to determine the consequences on species fitness.

When the mutualistic interaction is diffuse, involving a number of species, as is often the case with pollination systems and seed dispersal by frugivores, the influence of specific species-species interactions are difficult to determine. In other situations, the mutualistic relationship between two species may be mediated or facilitated by a third species, in much the same manner as vector organisms and intermedi-

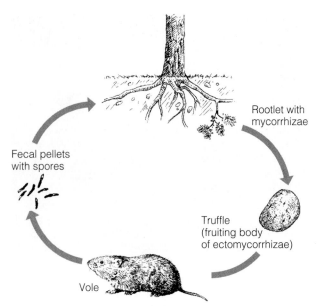

FIGURE 16.14 A mutualism involving three species and both symbiotic and nonsymbiotic interactions. Voles eat truffles, the below-ground fruiting bodies of some mycorrhizae. The spores become concentrated in fecal pellets. The voles function to disperse the spores to locations where they can infect new host plants.

ate hosts in parasite-host interactions. The mutualistic relationships among conifers, mycorrhizae, and voles in the forests of the Pacific Northwest is one such example (Figure 16.14). The conifers depend on mycorrhizal fungi associated with the root system for nutrient acquisition from the soil. In return, the mycorrhizae depend on the conifers for energy in the form of carbon (see Section 16.13). The mycorrhizae also have a mutualistic relationship with voles, which feed on the fungi and disperse the spores, infecting the root systems of other conifer trees.

Perhaps the greatest limitation in evaluating the role of mutualism in community structure is that many mutualistic relationships, if not most, arise from indirect interaction in which the affected species never come into contact, influencing each other's fitness or population growth rate indirectly through a third species or by altering the local environment (habitat modification), topics we shall revisit in the following chapter. Although it is difficult to demonstrate, mutualism may well be as significant as either competition or predation to community structure.

CHAPTER REVIEW

Summary

Coevolution is the evolution of one species in response to its interaction with another species. Many species have some evolutionary effect on each other. Prey evolve defenses, and predators improve their hunting efficiency. A closer coevolutionary relationship exists between parasites and their hosts. The parasite lives on or in the host organism for some period of its life, a relationship referred to as symbiosis.

Characteristics of Parasites (16.1) Parasitism is a situation in which two organisms live together, one deriving its nourishment at the expense of the other. Parasitic infection can result in disease. Microparasites include the viruses, bacteria, and protozoa. They are small in size, have a short generation time, multiply rapidly in the host, tend to produce immunity, and spread by direct transmission. They are usually associated with dense populations of the host. Macroparasites are relatively large in size and include parasitic worms, lice, ticks, fleas, rusts, smuts, fungi, and other forms. They have a comparatively long generation time and rarely multiply directly in the host, persist with continual reinfection, and spread by both direct and indirect transmission.

Parasite-Host Relationships (16.2–16.5) Parasites exploit every conceivable habitat in host organisms. Many are specialized to live at specific sites, such as the roots of a plant or the liver of an animal. The problem of parasites is to gain entrance into and escape from the host. The life cycles of parasites revolve about these two problems (16.2). Transmission for many species of parasites occurs directly from one host to another, either through direct physical contact or through the air, water, or other substrate (16.3). For other parasites, transmission between hosts occurs by means of other organisms, called vectors. These carriers become intermediate hosts of some developmental or infective stage of the parasite (16.4). Other species of parasites require more than one type of host. Indirect transmission takes them from definitive to intermediate to definitive host. Indirect transmission often depends on the feeding habits of the host organisms (16.5).

Host Response to Parasitism (16.6–16.8) Hosts respond to parasitic infections through behavioral changes, by inflammatory responses at the site of infection, and through subsequent activation of their immune systems (16.6). A heavy parasitic load can decrease reproduction of the host organism. Although most parasites do not kill their hosts, mortality can result from secondary factors. As a consequence, parasites can both reduce fecundity and increase mortality rates of the host population (16.7). The death of a host does not benefit a parasite. Natural selection favors less virulent forms of the parasite that can live in the host without killing it. Hosts and parasites most often develop mutual tolerance with a low-grade, widespread infection (16.8).

Population Response (16.9) Under certain conditions, parasitism can regulate a host population. When introduced to a population that has not developed defense mechanisms, parasites can spread quickly, leading to high rates of mortality and in some cases to virtual extinction of the host species.

Mutualistic Relationships (16.10) Mutualism is a positive reciprocal relationship between two species that may have evolved from predator-prey or host-parasite relationships. Symbiosis describes the relationship in which the two mutualists physically live together. Symbiotic relationships are often obligatory for one or both involved species.

Types of Mutualistic Interactions (16.11–16.15) Symbiotic mutualisms are involved in the uptake of nutrients in both plants and animals. The chambers of a ruminant's stomach contain large populations of bacteria and protozoa that carry out the process of fermentation. Some plant species have a mutualistic association with nitrogen-fixing bacteria that infect and form nodules on their roots. The plants provide the bacteria with carbon, while the bacteria provide nitrogen to the plant. Fungi form mycorrhizal associations with the root systems of plants, assisting in the uptake of nutrients. In return, they derive energy in the form of carbon from the host plant (16.11). Other mutualistic associations are associated with the defense of the host organism (16.12). Many mutualistic relationships are not symbiotic. Mutualists live physically separate lives, yet depend upon each other for some essential function (16.13). Nonsymbiotic mutualisms are involved in the pollination of many species of flowering plants. The pollinator extracts nectar from the flower, and in doing so collects and exchanges pollen with other plants of the same species. To reduce wastage of pollen, some plants possess morphological structures that permit only certain animals to reach the nectar (16.14). Mutualism is also involved in seed dispersal. Some seed-dispersing animals upon which the plant depends may be seed predators as well, eating the seeds for their own nutrition. Plants depending on such animals must produce a tremendous number of seeds to ensure that a few will be dispersed, come to rest on a suitable site, and germinate. Alternatively, plants may enclose their seeds in a nutritious fruit attractive to frugivores (fruit-eating animals). Frugivores are not seed predators. They eat only the tissue surrounding the seed and, with some exceptions, do not damage the seed (16.15).

Facilitation (16.16) When an association is nonessential but nonetheless increases the fitness of the parties involved, it is referred to as facultative. Facultative interactions cover the broader range of interactions among organisms that benefit at least one of the participants and cause harm to neither.

Community Structure (16.17) Mutualistic relationships, both direct and indirect, may integrate community structure and composition in ways that we are just beginning to appreciate and understand.

Study Questions

1. Define parasitism. Why is it more reasonable to classify parasites by size than by taxonomic groups?

2. Distinguish between a definitive host and intermediate hosts. Look up examples beyond those mentioned in the text.

3. Examine transmission of some common parasites such as tapeworms, roundworms, and the organisms responsible for malaria, Rocky Mountain spotted fever, and giardiasis. How might some of these parasitic diseases be controlled?

4. How might a patchy or clumped distribution of hosts affect the spread of parasites?

5. What is mutualism? Look up some examples of mutualism and examine them critically. Are they in fact mutualistic?

6. Distinguish among symbiosis, obligate symbiotic mutualism, and obligate nonsymbiotic mutualism.

7. Is fruit predation a chancy way of distributing seeds? Why?

8. Is mutualism reciprocal exploitation, or two species acting together for mutual benefit?

9. Explore examples of how parasitic infections have altered populations, ecosystems, and even human history. (Hint: Consider the chestnut blight's effect on North American deciduous forests; the night mosquito, bird pox, and the fate of the Hawaiian honeycreepers; rats, fleas, and the black death in Europe; rinderpest in African wild and domestic ungulates. There are many others.)

Douglas fir and western hemlock with an abundance of dead wood and decomposing logs—a setting characteristic of old-growth forests.

CHAPTER 17

Processes Shaping Communities

OBJECTIVES

On completion of this chapter, you should be able to:

- Discuss how the fundamental niche of a species constrains community structure.

- Explain how diffuse and indirect interactions occur in food webs.

- Discuss how tradeoffs between stress tolerance and competitive ability can influence community patterns along environmental gradients.

- Contrast autogenic and allogenic environmental change.

- Show how organisms can influence succession by changing environmental conditions.

- Discuss why dominance changes along a temporal gradient in succession.

- Discuss how and why patterns of diversity change over the course of succession.

- Explain how habitat links plant and animal components of the community.

The community is a group of plant and animal species that inhabit a given area. As such, understanding the biological structure of the community is a question of controls on the distribution and abundance of species. Thus far in this book, we have examined a wide variety of topics that address this broad question, from adaptations of organisms to the physical environment, the evolution of life history characteristics and their influence on population demography, and the interactions among different species. In Chapter 13 we examined the structure and dynamics of communities, exploring how patterns of species distribution and abundance vary in time and space. We described how dominance and diversity change as you move across the landscape or as time progresses. However, scientists want to do more than describe. They want to understand and ultimately predict. What processes shape these patterns of community structure? Why is the basic process of succession following disturbance consistent in different environments? Why are communities in some environments more or less diverse than others? How will communities respond to the addition or removal of a species? In this chapter, we integrate our discussion of the adaptation of organisms to the physical environment presented in Part III (Chapters 6–8) with the discussion of species interactions in the previous three chapters to search for processes that control community structure that can be generalized to a wide variety of communities.

17.1 The Fundamental Niche Constrains Community Structure

Recall the concept of the fundamental niche from Chapter 2 (Focus on Ecology 2.3: The Niche). All living organisms have a range of environmental conditions under which they can survive, grow, and reproduce. This range of environmental conditions is not the same for all organisms. The conditions under which an organism can function successfully are the consequence of a wide variety of physiological, morphological, and behavioral adaptations. As well as allowing the organism to function under a specific range of environmental conditions, these same adaptations also limit its ability to do equally well under different conditions. In Part III we explored numerous examples of this premise. Plants adapted to high light environments exhibit characteristics that preclude them from being successful under low light conditions (Chapter 6). Animals that regulate body temperature through ectothermy (cold-blooded animals) are able

to reduce energy requirements during periods of resource shortage. Dependence on external sources of energy, however, limits diurnal and seasonal periods of activity as well as their geographic distribution (Chapter 8). The number of offspring produced at any one time constrains the nature of parental care (Chapter 12). Each set of adaptations reflects a solution to a set of environmental conditions, and conversely precludes adaptation to another.

Secondly, environmental conditions vary both in time and space (Part II). This observation, when combined with the inherent differences in the fundamental niche of species, provides a starting point from which to explore the processes that structure communities. We can represent the fundamental niches of a variety of species with bell-shaped curves along some environmental gradient, such as availability of water or light for plants (Figure 17.1). The response of each species is defined in terms of abundance. Although the fundamental niches overlap, each species has limits beyond which it cannot survive. The distribution of fundamental niches along the environmental gradient represents a primary constraint on the structure of communities. For any given range of environmental conditions, only a subset of species can survive, grow, and reproduce. As the environmental conditions change from location to location, or through time at any given location, the possible distribution and abundance of species will change; and this changes the structure of the community.

This view of community is what ecologists refer to as a "null" model. It assumes that the presence and abundance of the individual species found in a given community are a result of the independent responses of the individual species to the prevailing

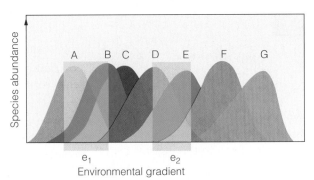

FIGURE 17.1 Fundamental niches of seven hypothetical species along an environmental gradient (e.g., moisture, temperature, or elevation) in the absence of competition from other species. The species all have bell-shaped responses to the gradient, but each has different tolerance limits defined by a minimum and maximum value along the gradient. As conditions change, for example from e_1 to e_2, the set of species that can potentially occur in the community changes. These changes can occur in either time or space.

physical environment. Interactions among species have no influence on community structure. In light of the examples we have reviewed in the previous three chapters, this assumption must seem somewhat odd. However, it is helpful as a framework against which to compare the actual patterns observed within the community. For example, this null model is the basis for comparisons in the experiments, examined in the previous chapters, in which the interactions between two species (competition, predation, parasitism, and mutualism) are explored by physically removing one species and examining the population response of the other. If the population of the remaining species does not differ from that observed in the presence of the species that has been removed, it can be assumed that the apparent interaction has no influence on their abundance within the community.

There is a great deal of evidence (noted in Chapters 14 to 16) that species interactions do influence both the presence and abundance of species within a wide variety of communities. Because the results from studies that examine the interactions of a limited group of species are removed from the context of the larger community, they most likely underestimate the importance of species interactions on the structure and dynamic of communities.

17.2 Species Interactions Are Diffuse

One reason such experiments tend to underestimate the importance of species interactions in communities is that interactions are often diffuse, involving a number of species. The work of Norma Fowler at Duke University provides an example. She examined competitive interactions within a field community by selectively removing species of plants from experimental plots and assessing the growth responses of the remaining species. Her results showed that competitive interactions within the community tended to be rather weak and diffuse. The removal of single species had little effect. The response to the removal of groups of species, however, tended to be much stronger, suggesting that individual species compete with a variety of other species for essential resources within the field. In diffuse competition, the direct interaction between any two species may be weak, making it difficult to determine the effect of any given species on another. Collectively, however, competition may be an important factor limiting the abundance of all species involved.

Diffuse interaction, where one species may be influenced by interactions with many different species, is not limited to competition. In the example

of predator-prey cycles in Chapter 15 (Section 15.15), a variety of predator species (including the lynx, coyote, and horned owl) are responsible for the periodic cycles in the snowshoe hare population. Examples of diffuse mutualisms relating to both pollination and seed dispersal are presented in Chapter 16 (Sections 16.14 and 16.15), where a single plant species may be dependent on a variety of animal species for successful reproduction.

Although food webs present only a limited view of species interactions within a community, they provide an excellent means of illustrating the diffuse nature of species interactions. Charles J. Krebs of the University of British Columbia has developed a generalized food web for the boreal forest communities of northwestern Canada (Figure 17.2). This food web contains the plant-snowshoe hare-carnivore system discussed in Chapter 15 (see Figure 15.17). The arrows point from predator to prey, and an arrow that circles back to the same box (species) represents cannibalism (great horned owl and lynx). Although the food web only presents the direct links between predator and prey, it also implies the potential for competition among predators for a shared prey resource. It serves to illustrate the diffuse nature of species interactions within this community. For example, snowshoe hares are preyed upon by 11 of the 12 predators present within the community. Any single predator species may have a limited effect on the snowshoe hare population, but the combined impact of predation functions to regulate the snowshoe hare population. This same example illustrates the diffuse nature of competition within this community. Although the 12 predator species feed upon a wide variety of prey species, snowshoe hares represent an important shared food resource for the 3 dominant predators: lynx, great horned owls, and coyotes.

17.3 Food Webs Illustrate Indirect Interactions

Food webs also illustrate a second important feature of species interactions within the community—indirect effects. Indirect interactions occur when one species does not interact with a second species directly, but influences a third species that does have a direct interaction with the second (see Focus on Ecology 17.1: The Politics of Nature). For example, in the food web presented in Figure 17.2, lynx do not have a direct interaction with white spruce; however, by reducing snowshoe hare and other herbivore populations that feed upon white spruce, lynx can have a positive effect on the white spruce population

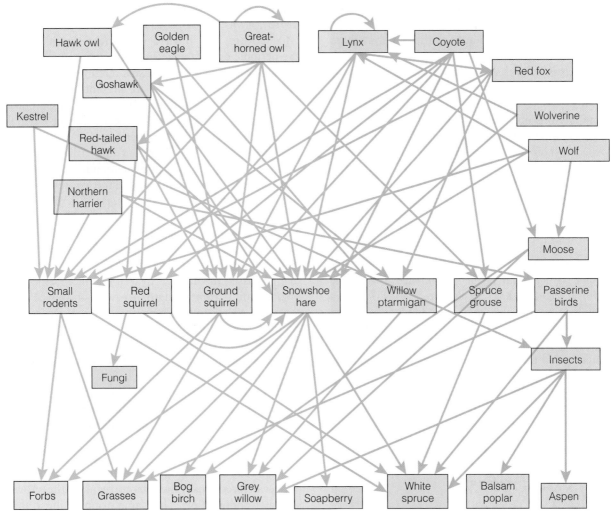

FIGURE 17.2 A generalized food web for the boreal forests of northwestern Canada. The dominant species within the community are shown in yellow. Arrows link predator with prey species. Arrows that loop back to the same species (box) represent cannibalism. (Modified from C. Krebs, *Ecology*, 5th edition, p. 464.)

(survival of seedlings and saplings). The key feature of indirect interactions is that they can propagate throughout the community.

The role of predation in shaping the structure of communities by influencing the outcome of competitive interactions among prey species provides another example of indirect effects within food webs. Robert Paine of the University of Washington was one of the first ecologists to demonstrate this point. The intertidal zone along the rocky coastline of the Pacific Northwest is home to a variety of mussels, barnacles, limpets, and chitons (all invertebrate herbivores). All of these species are preyed upon by the starfish *Pisaster* (Figure 17.3). Paine conducted an experiment in which he removed the starfish from some areas (experimental plots) while leaving other areas undisturbed for purposes of comparison (controls). Following the removal of the starfish, the

number of prey species in the experimental plots dropped from 15 at the beginning of the experiment to 8. In the absence of predation, several of the mussel and barnacle species that were superior competitors in the absence of predation excluded the other species, reducing the overall diversity of the community. This type of indirect interaction is called **keystone predation,** where the predator enhances one or more inferior competitors by reducing the abundance of the superior competitor.

One of the more interesting indirect interactions that occurs within food webs is **indirect mutualism.** An example comes from a study of subalpine ponds in Colorado by Stanley Dodson of the University of Wisconsin. It involves the relationships among two species of herbivorous species of *Daphnia* and their predators, a midge larva *(Chaoborus)* and a larval salamander *(Ambystoma).* The salamander larvae

THE POLITICS OF NATURE

The old saying that "politics and war make for strange bedfellows" refers to the observation that unusual alliances are often formed in times of conflict. The alignment of nations during the period of the Cold War (the era of political tension between the Soviet Union and the United States from 1945 to 1991) illustrates this point. Countries that had little or no direct interactions, such as Angola (southwestern Africa), Nicaragua (Central America), Ethiopia (northwest Africa), Cambodia (southeast Asia), and Bulgaria (eastern Europe), found themselves to be indirect allies by way of their association with the Soviet Union. Likewise, they became enemies of the United States because of that same association.

History is rife with examples of indirect conflict and alliances between countries as a result of common enemies or friends. Likewise, in nature, interactions between species are often the consequence of an indirect interaction with a third party (species). A predator can indirectly benefit a species with which it has no direct interaction if the predator selectively preys upon a species with which it competes. Alternatively, the predator might indirectly harm that same species if the predator reduces the population of a species with which it has a mutualistic association.

So it would seem, as with politics, in nature:

Friends of friends are friends.

Friends of enemies are enemies.

Enemies of friends are enemies.

Enemies of enemies are friends.

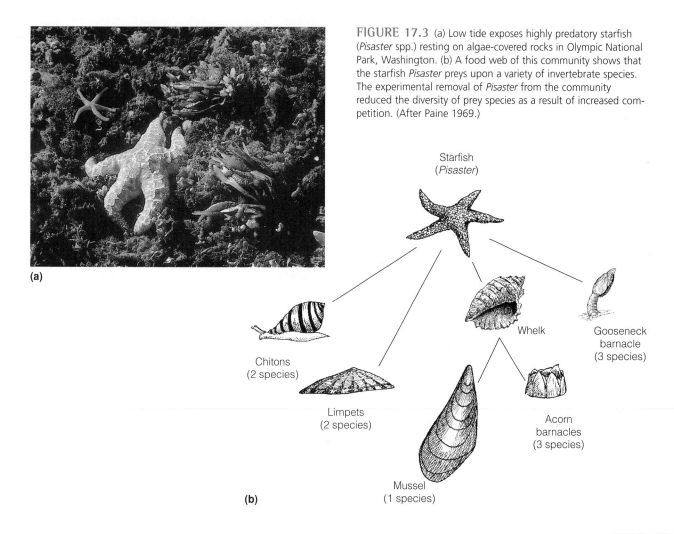

(a)

(b)

FIGURE 17.3 (a) Low tide exposes highly predatory starfish (*Pisaster* spp.) resting on algae-covered rocks in Olympic National Park, Washington. (b) A food web of this community shows that the starfish *Pisaster* preys upon a variety of invertebrate species. The experimental removal of *Pisaster* from the community reduced the diversity of prey species as a result of increased competition. (After Paine 1969.)

Starfish
(*Pisaster*)

Chitons
(2 species)

Limpets
(2 species)

Mussel
(1 species)

Whelk

Acorn
barnacles
(3 species)

Gooseneck
barnacle
(3 species)

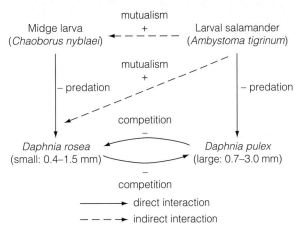

FIGURE 17.4 Diagram showing the relationship among the midge larvae *(Chaoborus),* larval salamander *(Ambystoma),* and two species of *Daphnia (D. rosea* and the larger *D. pulex)* that inhabit pond communities in the mountains of Colorado. The removal of salamander larvae from some ponds resulted in the competitive exclusion of *D. rosea* by *D. pulex,* and the local extinction of the midge population that preyed upon it. (After Dodson 1974.)

FIGURE 17.5 Moose population can have a significant impact on plant community structure along the river environments in which the moose feed.

prey on the larger of the two *Daphnia* species, while the midge larvae prey on the small species (Figure 17.4). In a study of 24 pond communities in the mountains of Colorado, Dodson found that where salamander larvae were present, the number of large *Daphnia* was low and the number of small *Daphnia* high. But in ponds in which the salamander larvae were absent, small *Daphnia* were absent and midges could not survive. The two species of *Daphnia* apparently compete for the same resources. When the salamander larvae are not present, the larger of the two *Daphnia* species can outcompete the smaller. With the salamander larvae present, predation reduces the population growth rate of the larger *Daphnia,* allowing for the coexistence of the two species. In this example, two indirect mutualistic relationships arise. The salamander larvae indirectly benefit the smaller species of *Daphnia* by reducing the population size of its competitor. Subsequently, the midge apparently depends on the presence of salamander larvae for its survival in the pond. This possible indirect mutualism can be demonstrated only under controlled experiments involving population manipulations of the four species.

The idea of indirect mutualism is highly speculative, requiring strong experimental demonstration in natural communities. It does suggest that mutualism, as well as competition and predation, can be an integrating force in the structuring of natural communities through indirect effects. There is a growing appreciation within ecology for the role of indirect effects in shaping community structure. Understanding these complex interactions is more than an academic exercise; it has direct implications for the conservation and management of natural communities. As with the example of starfish in the intertidal zone, the removal of a species from the community can have many unforeseen consequences. For example, Joel Berger of the University of Nevada and colleagues have examined how the local extinction of grizzly bears *(Ursus arctos)* and wolves *(Canis lupus)* from the southern Greater Yellowstone Ecosystem, resulting from decades of active predator control, have impacted the larger ecological community. One unforeseen consequence of the loss of these large predators is the decline of bird populations that utilize the vegetation along rivers (riverine habitat) within the region. The elimination of large predators from the community resulted in an increase in the moose population (prey species). Moose selectively feed on willow *(Salix* spp.) and other woody species that flourish along the river shorelines (Figure 17.5). The increase in moose populations had a dramatic impact on the vegetation in the riverine areas that provides habitat for a wide variety of bird species, leading to the local extinction of some populations.

17.4 Food Webs Suggest Controls of Community Structure

The wealth of experimental evidence illustrates the importance of both direct and indirect interactions on community structure. On that basis, rejecting the "null" model presented in Section 1 of this chapter

RIVETS OR REDUNDANCY?

Human activity, primarily through the destruction of habitats, is having a dramatic impact on the biological diversity of the planet. We are currently undergoing a mass extinction, with estimates of annual extinction rates reaching thousands of species (see the Ecological Application after Ch. 19: Asteroids, Bulldozers, and Biodiversity). What is the consequence of such a loss of species on the functioning of Earth's communities and ecosystems? This question is at the forefront of conservation ecology. At the heart of this question is the importance of species interactions on community structure. Will the loss of species from a community have a cascading effect, resulting in the further extinction of other species? Will it perhaps even change the nature of whole ecosystems and their ability to provide essential human services? Two contradictory models have been proposed that directly address these questions regarding the importance of biological diversity in ecological communities.

In 1981, Stanford University ecologists Paul and Anne Ehrlich proposed a hypothesis that has come to be known as the **rivet model.** The diversity of species within a community, state the Ehrlichs, is analogous to the rivets on an airplane, with each species playing a small but significant role in the functioning of the whole (eco)system. The loss of each rivet weakens the plane by a small but noticeable amount. Eventually, the combined effect results in an abrupt loss of structural integrity, and the plane will fall from the sky.

This view of the community stresses the importance of species interactions. The loss of even a single species (such as a keystone predator) could have a cascading effect throughout the community, with devastating results.

A contrary hypothesis came a decade later from ecologist Brian Walker of Australia's Commonwealth Scientific and Industrial Research Organization. Walker's hypothesis, known as the **redundancy model,** asserts that most species are superfluous. Rather than functioning like rivets, species are more like passengers on the plane, where only a few key passengers (species) are needed to keep the (eco)system functioning.

According to Walker's model, species can be classified into functional groups based on their role in the community, much like the concept of species guilds. Species within any functional group can be thought of as redundant. This view of community structure stresses the importance of functional groups in controlling community structure and function rather than individual species. The loss of any one species within a functional group may have little or no effect on the community. It is the loss of a functional group that will have a cascading effect on community structure. If one predator species is lost from the community, another species within the community will take its place. The key question becomes how much or little redundancy there is within the community.

As is often the problem with opposing hypotheses in science, observational and experimental data can be found to support both views of community structure. There is little doubt that some, often dominant, species play a critical role within communities, and their loss would have a dramatic effect on both the structure and functioning of the community. On the other hand, communities are not like a house of cards, where removing any one species has the cascading effect of sending the structure crashing to the ground. The critical task of ecologists is to determine where, between these two contradictory models, lies the truth, and to do so before the answer is merely academic.

would be justified. But given the complex of direct and indirect interactions suggested by food webs, how can we begin to understand which interactions are important in controlling community structure and which are not? Are all species interactions important? Does some smaller subset of interactions exert a dominant effect, while most have little impact beyond those species directly involved? The former would suggest that the community is like a house of cards, where removing any one species may have a cascading effect on all others. The latter suggests a more loosely connected assemblage of species.

These questions are at the forefront of conservation ecology because of the dramatic decline in biological diversity that is occurring as a result of human activity (see Focus on Ecology 17.2: Rivets or Redundancy?). Certain species within the community can exert a dominant influence on community structure, such as the predatory starfish that inhabits the rocky intertidal communities. However, the relative importance of most species on the functioning of communities is largely a mystery. One approach being used to simplify the task is to group species into **functional groups** based on criteria relating to

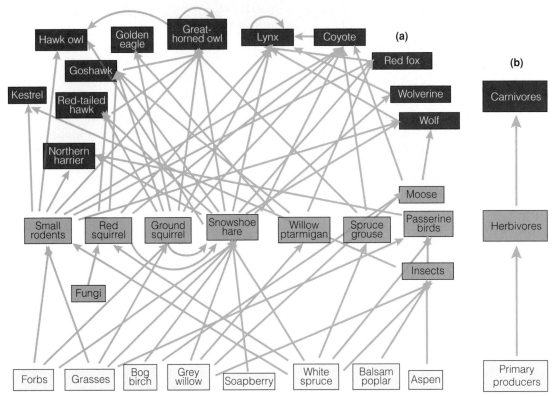

FIGURE 17.6 Aggregation of (a) species forming the food web for Canadian boreal forests presented in Figure 17.2 into (b) trophic levels (generalized feeding groups). Arrows point from prey to predator (reverse of pattern shown in Figure 17.2).

their function within the community. For example, the concept of guilds presented in Chapter 13 is a functional grouping of species based on similar functions within the community or exploitation of the same resource—grazing herbivores, pollinators, cavity-nesting birds. By aggregating species into a smaller number of functional groups, it is possible to explore the processes controlling community structure in more general terms. For example, what is the role of mammalian predators in boreal forest communities? In fact, this functional grouping of species can be seen in the food web presented in Figure 17.2, where the categories (boxes) of forbs, grasses, small rodents, insects, and passerine birds represent groups of functionally similar species.

One way in which food webs are simplified is to aggregate species into general feeding groups based on the source from which they derive food energy. In Chapter 6, we defined organisms that derive their energy from sunlight as autotrophs, or **primary producers.** Organisms that derive energy from consuming plant and animal tissue are called heterotrophs or **secondary producers** and are further subdivided into herbivores, carnivores, and omnivores based on their consumption of plant tissues, animal tissues, or both. These feeding groups are referred to as **trophic levels,** after the Greek work *trophikos,* meaning nourishment. The food web presented in Figure 17.2 has been aggregated into three trophic levels in Figure 17.6: primary producers, herbivores, and carnivores. Although this is an obvious oversimplification, using this approach raises some fundamental questions concerning the processes that control community structure.

As with food webs, the arrows in a simple food chain based on trophic levels point in the direction of energy flow—from primary producers to herbivores, and from herbivores to carnivores. The structure of food chains suggests that populations at any given trophic level are controlled (limited) by the populations in the trophic level below. This is called **bottom-up control.** Plant populations control herbivore populations, which in turn control the diversity and population densities of carnivore populations. However, as we have seen from the previous discussion of predation and food webs, **top-down control** can occur as well, when predator populations can control the diversity of prey species.

The work by Mary Power and her colleagues at the University of Oklahoma Biological Station suggests that the role of top predators (carnivores) on community structure can extend to lower trophic levels, influencing primary producers as well as herbivore populations. Power and colleagues showed that a top predator (largemouth bass, *Micropterus salmoides*) had strong indirect effects that cascaded down through the food web to influence the abundance of benthic algae in stream communities of the midwestern United States. In these stream communities, herbaceous minnows (primarily *Campostoma ananomalum*) graze on algae, and in turn largemouth bass (and two species of sunfish) feed on the minnows. During periods of low flow, isolated pools form in the streams. As part of the experiment, bass were removed from some pools, and the populations of algae and minnows were monitored. Pools with bass had low minnow populations and a luxuriant growth of algae. In contrast, pools from which the bass were removed had high minnow populations and low populations (biomass) of algae. In this example, top predators (carnivores) were shown to control the abundance of plant populations (primary producers) indirectly through their direct control on herbivores.

A famous article written by Nelson Hairston, Fred Smith, and Larry Slobodkin first introduced the concept of top-down control with the frequently quoted "the world is green" proposition. These three ecologists proposed that the world is green (plant biomass accumulates) because predators keep herbivore populations in check. Although this proposition is supported to some degree by the results of experiments like those by Mary Power and her colleagues, experimental data required to test this hypothesis are extremely limited, particularly in terrestrial ecosystems. However, the proposition continues to cause great debate within the field of community ecology.

17.5 Species Interactions along Environmental Gradients Involve Both Stress Tolerance and Competition

We have now seen that the biological structure of a community is first constrained by the environmental tolerances of the species (fundamental niche). These tolerances are then modified through both direct and indirect interactions with other species (realized niche). Competitors and predators can function to restrict a species from a community, while mutualists can function to facilitate a species' presence and abundance within the community. As we move across the landscape, variations in the physical environment will alter the nature of both these con-straints on species distribution and abundance. Species differ in their range of environmental tolerances, and the nature of species interactions change based on the environmental context.

The discussion of plant adaptations to environmental conditions in Chapter 6 provides some insight into how the competitive abilities of plant species may vary across environmental gradients of resource availability. Adaptations of plants to variations in the availability of light (Section 6.7), water (Section 6.8), and nutrients (Section 6.10) result in a general pattern of tradeoffs between the characteristics that enable a species to survive and grow under low resource availability and those that allow for high rates of photosynthesis and growth under high resource availability (Figure 17.7a). Competitive success in plants is often linked to growth rate and the acquisition of resources (see Chapter 14). Those species that have the highest growth rate and acquire most of the resources at any given point on the resource gradient have the competitive advantage there. The difference in adaptations to resource availability among the species reflected in Figure 17.7a result in each species' having a competitive advantage over some range of resource conditions. The result is a pattern of zonation along the gradient (Figure 17.7b) that reflects the changing relative competitive abilities. The ability of each species to tolerate the stress associated with resource limitation defines its lower resource boundary. Competition defines the upper boundary. Such a tradeoff in tolerance and competitive ability is most likely responsible for the patterns of zonation along gradients of soil moisture, such as the one reported by the plant ecologist Robert Whittaker (formerly of Cornell University) for the lower elevations of the central Siskiyou Mountains of Oregon and California presented in Figure 17.8.

Competition among plant species rarely involves a single resource (Section 14.7). The experiments of R. H. Groves and J. D. Williams examining competition between populations of subterranean clover and skeleton weed (see Section 14.6 and Figure 14.7) clearly show that there is an interaction between competition for both above-ground (light) and below-ground resources (water and nutrients). The differences in adaptations relating to the acquisition of these resources when they are in short supply can result in changing patterns of competitive ability along gradients where these two classes of resources covary. Allocation of carbon to the production of leaves and stems provides increased access to the resources of light, but at the expense of allocation of carbon to the production of roots. Likewise, allocation of carbon to the production of roots increases access to water and soil nutrients, but limits the production of leaves and therefore the rate of carbon gain in photosynthesis. As the availability of water (or nutrients) increases along a gradient (such as in Figure 17.8), the competitive advantage shifts from those species adapted to low availability of water (high root production) to those species that allocate carbon to the production of leaves and height growth, but require higher water availability to survive (Figure 17.9).

This framework of tradeoffs in species characteristics related to tolerance and competitive ability along resource gradients provides a powerful tool for understanding plant community structure. However, interpretation can become more complicated when dealing with environmental gradients and communities where there are interactions of resource and nonresource factors (see Chapter 14, Section 14.5).

This type of competition is nicely illustrated in the pattern of plant zonation in salt-marsh communities along the coast of New England (see Figure 14.9). Nancy Emery and her colleagues at Brown University conducted a number of field experiments to identify the factors responsible for patterns of species distribution in these coastal communities. Experiments included the addition of nutrients, the removal of neighboring plants, and reciprocal transplants—planting species in areas where they are not naturally found to occur along the gradient. The results of the experiments indicate that the patterns of zonation are an interaction between the relative competitive abilities of species for nutrients and the ability of plant species to tolerate increasing physical stress. The low marsh is dominated by *Spartina alterniflora* (smooth cordgrass), a large perennial grass with extensive rhizomes. The upper edge of *S. alterniflora* is bordered by *Spartina patens* (saltmeadow cordgrass), a perennial turf grass, which is replaced at higher elevations in the marsh by *Juncus gerardi* (black needle rush), a dense turf grass (Figure 17.10). While the low marsh experiences daily flooding by the tides, the *S. patens* and *J. gerardi* zones are inundated only during high tide cycles (see Chapter 4). These differences in the frequency and duration of tidal inundation establish a spatial gradient of increasing salinity, waterlogging, and reduced

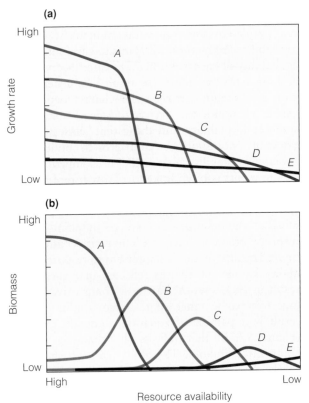

FIGURE 17.7 (a) General pattern of tradeoff between a species' ability to survive and grow under low resource availability and the maximum rate of growth achieved under high resource availability. Assuming that the superior competitor at any given point along the resource gradient (*x* axis) will be the species with the highest rate of growth, the relative competitive ability of the species changes with resource availability. (b) The result outcome of competition will be a pattern of zonation, where the lower boundaries for the species are a result of differences in their tolerances for low resource availability, and the upper bounds are set by competition. (From Smith and Huston 1989.)

FIGURE 17.8 Distribution of tree species along a soil-moisture gradient at low elevations in the central Siskiyou Mountains of Oregon and California. Data are from 50 stands sampled between 610 and 915 meters in elevation. (After Whittaker 1960.)

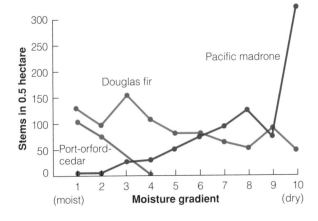

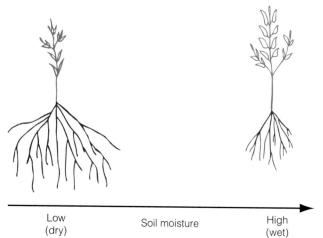

Low (dry) — Soil moisture — High (wet)

FIGURE 17.9 General trends in plant adaptations (characteristics) that serve to increase fitness along a soil moisture gradient. At low soil moisture, allocation to the production of roots at the expense of leaves aids in acquiring water and reducing transpiration, allowing the plant to survive (tolerance). As water availability increases, the overall increase in plant growth results in competition for light as some individuals overtop others. A shift in allocation to height growth (stems) and the production of leaves increases a plant's competitive ability.

oxygen levels across the marsh (see Figure 17.10). Individuals of *S. patens* and *J. gerardi* that were transplanted into lower marsh positions exhibited stunted growth and increased mortality. Thus, the lower distribution of each species is determined by its physiological tolerance to the physical stress imposed by tidal inundation (its fundamental niche). In contrast, individuals of *S. alterniflora* and *S. patens* exhibited increased growth when transplanted onto higher marsh positions where the neighboring plants had been removed. They were excluded by competition from higher marsh positions when neighboring plants were present (not removed). These results indicate that the upper distribution of each species in the marsh was limited by competition.

At first this example would seem to be a clear case of the tradeoff between adaptations for stress tolerance and competitive ability (high growth rate and resource use) as suggested in Figure 17.8. But such was not the case. The experimental addition of nutrients to the marsh changed the outcome of competition, but not in the manner that might be predicted. The addition of nutrients completely

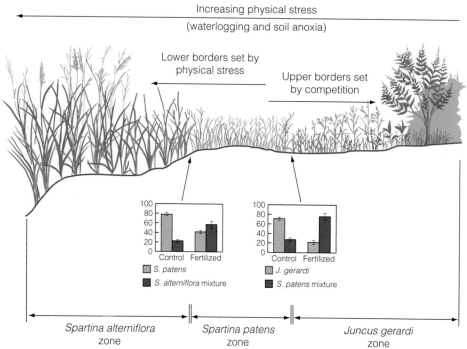

FIGURE 17.10 Patterns of plant zonation and physical stress along an elevational gradient in a tidal salt marsh. While the lower marsh experiences daily flooding by the tides, the upper zones are inundated only during the high-tide cycles. The higher water levels of the lower marsh result in lower oxygen levels in the sediments and higher salinities. The lower (elevation) boundary of each species is determined by its tolerance to physical stress, while the upper boundaries are limited by competition. The bar graphs show the shifts in percent cover by the two adjoining species within the border zones under normal (control) conditions and when fertilizer was added to increase nutrient availability. Note that the increase in nutrient availability resulted in the subordinate competitor under ambient conditions (control) becoming the dominant (superior) competitor. (Data from Emery et al. 2001, *Ecology* 82: 2471–2485.)

reversed the relative competitive abilities of the species, allowing *S. alterniflora* and *S. patens* to shift their distributions to higher marsh positions (see Figure 17.10).

J. gerardi, the dominant under ambient (low) nutrient conditions, allocates more carbon to root biomass than either species of *Spartina*. That allows *Juncus* to be the superior competitor under conditions of nutrient limitation, but limits its ability to tolerate the higher water levels of the lower marsh. In contrast, *S. alterniflora* allocates a greater proportion of carbon to above-ground tissues, producing taller tillers (stems and leaves), an advantage under the high water levels of the lower marsh. The tradeoff in allocation to below- and above-ground tissues results in the competitive hierarchy and thus patterns of zonation observed under ambient conditions. When nutrients are not limiting (nutrient addition experiments), competition for light dictates the competitive outcome among marsh plants. The greater allocation of carbon to height growth by the *Spartina* species increased their competitive ability on the upper marsh.

In the salt-marsh plant community, a tradeoff between competitive ability below-ground and the ability to tolerate the physical stress associated with the low oxygen and high salinity levels of the lower marsh appear to drive zonation patterns across the salt-marsh landscape. In this environment, the stress gradient does not correspond to the resource gradient as in Figure 17.9, allowing the characteristics for stress tolerance to enhance competitive ability under high resource availability.

Spatial variation in both resources and nonresource factors that directly influence physiological processes give rise to patterns of zonation in both terrestrial and aquatic communities. Patterns of temperature and moisture resulting from regional variations in climate (see Chapter 3) are the major determinant of regional and global patterns of vegetation distribution, and they form the basis of most vegetation classification systems (see Chapter 24, Biogeography and Biodiversity). On a more local scale, climate interacts with soils and topography to influence patterns of temperature and soil moisture (see Chapter 4). The underlying geology of an area interacts with climate to influence soil characteristics such as texture (see Chapter 5). In turn, texture has a direct effect on soil moisture-holding capacity, cation exchange, and base saturation (see Chapters 4 and 5), influencing the moisture and nutrient environment of plants. In aquatic environments, water depth, flow rate, and salinity are the major environmental gradients that directly influence the distribution and dynamics of communities.

17.6 Community Structure Changes along a Temporal, Autogenic Gradient

The changes in environmental conditions that bring about shifts in the physical and biological structure of communities across the landscape are varied. They can, however, be grouped into two general classes: allogenic and autogenic. **Autogenic** environmental change is a direct result of the organisms within the community. For example, the vertical profile of light in a forest is a direct result of the interception and reflection of solar radiation by the trees. In contrast, **allogenic** environmental change is a feature of the physical environment. Examples are the decline in average temperature with elevation in mountainous regions (Chapter 3), the decrease in temperature with depth in a lake or ocean (Chapter 4), or the changes in salinity and water depth in coastal environments.

In Chapter 13, we defined succession as changes in community structure though time—specifically, changes in species dominance. One group of species initially colonizes an area, but as time progresses they decline, to be replaced by other species (see Figure 13.5). The fact that we find this general pattern of changing species dominance as time progresses in most environments suggests a common underlying mechanism.

Plant succession has been a major focus of study since the birth of ecology as a science (see Focus on Ecology 17.3: A Succession of Ideas). Although many hypotheses and models have been put forward to explain succession, no general consensus has been achieved. One obstacle is the diversity of environments and associated communities in which succession has been studied. No single cause fits all the examples. Despite this lack of consensus, a general model of plant succession has begun to emerge.

One feature common to all plant succession is autogenic environmental change. In both primary and secondary succession, colonization alters environmental conditions. One clear example is the alteration of the light environment. Leaves reflecting and intercepting solar radiation create a vertical profile of light within a plant community (see Chapter 4, Section 4.2). As you move from the canopy to ground level, the light available to drive the processes of photosynthesis declines. In the initial period of colonization, few if any plants are present. In the case of primary succession, the newly exposed site has never been occupied. In the case of secondary succession,

A SUCCESSION OF IDEAS

The study of succession has been a focus of ecological research for over a century. The concept of succession was largely developed by the botanists E. Warming in Denmark and Henry Cowles in the United States in the early part of the 20th century. The intervening years have seen a variety of hypotheses that have attempted to address the processes that drive succession: the seemingly directional change in species composition through time.

Frederick Clements (1916, 1936) developed a theory of plant succession and community dynamics known as the monoclimax hypothesis. The community was viewed as a highly integrated superorganism, and the process of succession represented the gradual and progressive development of the community to the ultimate or climax stage. The process was seen as analogous to the development of an individual organism.

In 1954, F. Egler proposed a hypothesis he termed initial floristic composition. In Egler's view, the process of succession at any site is dependent on which species get there first. Species replacement is not an orderly process, because some species suppress or exclude others from colonizing the site. No species is competitively superior to another. The species that arrive first inhibit newcomers until they eventually die, making the site accessible to others. Succession is therefore very individualistic, dependent on the particular species that colonize the site and the order in which they arrive.

Joseph Connell and Ralph Slatyer (1977) proposed a framework for viewing succession that considers a range of species interactions and responses through succession. They offered three models. One is the facilitation model. Early successional species modify the environment so that it becomes more suitable for later successional species to invade and grow to maturity. In effect, early-stage species prepare the way for late-stage species, facilitating their success.

The inhibition model involves strong competitive interactions. No one species is completely superior to another. The site belongs to those species that get there first; in the colorful words of a Civil War general describing how armies take and hold ground, "the firstest gets the mostest." The first species to arrive holds the site against all invaders. They make the site less suitable for both early and late successional species. As long as they live and reproduce, they maintain their position. They relinquish it only when they are damaged or die, releasing space to another species. Gradually, however, species composition shifts as short-lived species give way to long-lived ones.

A third model, the tolerance model, holds that later successional species are neither inhibited nor aided by species of earlier stages. Later-stage species can invade a newly exposed site, become established, and grow to maturity independent of species that precede or follow them. They can do so because they tolerate a lower level of some resources. Such interactions lead to communities composed of those species most efficient in exploiting available resources. An example might be a highly shade-tolerant species that could invade, persist, and grow beneath the canopy because it is able to exist at a lower level of one resource—light. Ultimately, through time, one species would prevail.

Since the work of Connell and Slatyer, the search for a general model of plant succession has continued among ecologists. Although numerous hypotheses have been put forward in recent years, a general trend in thinking has emerged: a trend that focuses on how the adaptations and life-history traits of individual species influence species interactions and ultimately species distribution and abundance under changing environmental conditions.

---

plants have been killed or removed by some disturbance. Under these circumstances, the availability of light at the ground level is high, and seedlings are able to become established. As plants grow, their leaves intercept sunlight, reducing the availability of light to shorter plants. This reduction in available light will reduce rates of photosynthesis, slowing the growth of these shaded individuals. Assuming that not all plant species photosynthesize and grow at the same rate, those species of plants that can grow tall the fastest will have access to the light resource. They reduce the availability of light to the slower-growing species. This reduction in light enables the fast-growing species to outcompete the other species and dominate the site. However, in changing the availability of light below the canopy, the dominant species create an environment that is more suitable for other species that will later displace them as dominants.

Recall from Chapter 6 (Section 6.7) that not all species of plants respond to variation in available light in the same manner. Sun-adapted, shade-intolerant plants exhibit a different response to light than do shade-adapted, shade-tolerant species. Shade-intolerant species exhibit high rates of photosynthesis and growth under high-light environments. Under low-light levels, they are not able to survive (see Chapter 6, Figure 6.7). In contrast, shade-tolerant plant species exhibit much lower rates of photosynthesis and growth under high-light conditions, but are able to continue photosynthesis, growth, and survival under lower light. There is a fundamental physiological tradeoff between the characteristics that enable high rates of growth under high-light conditions and the ability to continue growth and survival under shaded conditions (also see Figure 17.7).

In the early stages of plant succession, shade-intolerant species dominate because of their high growth rates. Shade-intolerant species overtop and shade the slower-growing, shade-tolerant species. As time progresses and light levels decline below the canopy, seedlings of the shade-intolerant species cannot grow and survive in the shaded conditions. At this time, although the shade-intolerant species dominate the canopy, no new individuals are being recruited into their populations. In contrast, shade-tolerant species are able to germinate and grow under the canopy. As the shade-intolerant plants that make up the canopy die, shade-tolerant species in the understory replace them.

Figure 17.11 shows this pattern of changing population recruitment, mortality, and species composition through time for a forest community in the Piedmont region of North Carolina. Fast-growing, shade-intolerant pine species dominate in early succession. As time progresses, the number of new pine seedlings declines as the light decreases at the forest floor. Now the shade-tolerant oak and hickory species are able to establish seedlings in the shaded conditions of the understory. As the pine trees in the canopy die, the community shifts from a forest dominated by pine species to one dominated by oaks and hickories.

In this example, succession results from changing competitive ability under autogenically changing environmental conditions. The shade-intolerant species are able to dominate the early stages of succession because of their ability to grow quickly in the high-light environment. However, as autogenic changes in the light environment occur, the ability to tolerate and grow under shaded conditions enables the shade-tolerant species to rise to dominance.

Light is not the only environmental factor that changes over the course of succession. Other autogenic changes in environmental conditions influence patterns of succession. Consider the example of primary succession on newly deposited glacial sediments and lava flows (as discussed Chapter 13, Section 13.12 and Figure 13.13). Because of the absence of a well-developed soil, little nitrogen is

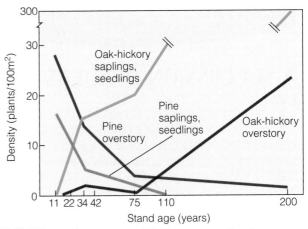

FIGURE 17.11 Dominance shift of overstory and understory (seedlings and saplings) pines, oaks, and hickories during secondary succession in the Piedmont region of North Carolina. Early successional pine species initially dominate the site. Pine seedling regeneration declines as the light decreases in the understory. Shade-tolerant oak and hickory seedlings establish themselves under the reduced light conditions. As pine trees in the overstory die, oak and hickory replace them as the dominant species in the canopy. (After Billings 1938.)

present in these newly exposed surfaces, restricting the establishment, growth, and survival of most plant species. However, those terrestrial plant species that have the mutualistic association with nitrogen-fixing *Rhizobium* bacteria are able to grow and dominate the site (see Chapter 16, Section 16.11). These plants provide a source of carbon (food) to the bacteria that inhabit their root systems. In return, the plant has access to the atmospheric nitrogen fixed by the bacteria. Alder, which colonizes the newly exposed glacial sediments in Glacier Bay, is one such plant species (see Figure 13.13c).

As individual alder shrubs shed their leaves or die, the nitrogen they contain is released to the soil by decomposition. Now other plant species can colonize the site (Figure 17.12). As nitrogen becomes increasingly available in the soil, species that do not have the added cost of mutualistic association and exhibit faster rates of growth and recruitment come to dominate the site. As in the Piedmont forest example, succession is a result of autogenic change in the environment and the relative competitive abilities of the species colonizing the site.

The exact nature of succession varies from one community to another, and it involves a variety of species interactions and responses that include competition and inhibition, differences in environmental tolerances, and facilitation. However, in all cases, the role of temporal, autogenic changes in environmental conditions and the differential response of species to those changes are key features of community dynamics.

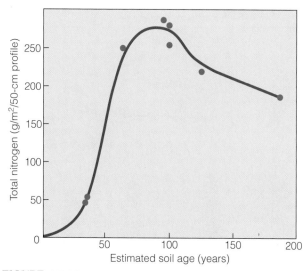

FIGURE 17.12 Changes in soil nitrogen during primary succession in Glacier Bay National Park since the retreating glacier exposed the surface for colonization by plants. Initially, the virtual absence of nitrogen limits colonization to alder, which has a mutualistic association with *Rhizobium* bacteria, allowing it access to atmospheric nitrogen. As plant litter decomposes, nitrogen is released to the soil through mineralization. With the buildup of soil organic matter and nitrogen levels, other plant species are able to colonize the site and displace alder (see Figure 13.13). (Adapted from Oosting 1942.)

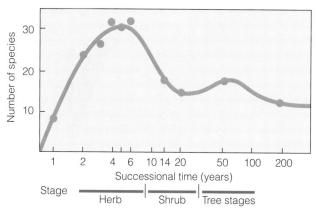

FIGURE 17.13 Changes in plant diversity during secondary succession of an oak-pine forest in Brookhaven, New York. Diversity is reported as species richness in 0.3-hectare samples. Species richness increases into the late herbaceous stages, declines into the shrub stage, increases once again into the early forest stages, and declines thereafter. The peaks in diversity correspond to the periods of transition between these stages, where species from both stages are present at the site. (Adapted from Whittaker 1975, Whittaker and Woodwell 1968.)

17.7 Species Diversity Changes During Succession

In addition to shifts in species dominance, patterns of plant species diversity change over the course of succession. Patterns of diversity through succession have been investigated by comparing sites within an area that are at different stages of succession; such groups of sites are known as chronoseres or chronosequences. Studies of secondary succession in old-field communities have shown that species diversity increases with site age (time since abandonment). The plant ecologist Robert Whittaker observed a different temporal pattern of species diversity for sites in upstate New York (Figure 17.13). Species diversity increases into the late herbaceous stages and then decreases into shrub stages. Species diversity then increases again in young forest, only to decrease as the forest ages.

The processes of species colonization and replacement drive succession. To understand the changing patterns of species richness and diversity during succession, we need to understand how these two processes vary in time. Colonization increases species richness. Species replacement typically results from competition, or an inability of the species to tol-

erate the changing environmental conditions. Species replacement acts to decrease species richness.

During the early phases of succession, diversity increases as new species colonize the site. However, as time progress, species become displaced—replaced as dominants by slower-growing, more shade-tolerant species. The peak in diversity during the middle stages of succession corresponds to the transition period, after the arrival of later successional species but prior to the decline (replacement) of early successional species. The two peaks in diversity seen in Figure 17.13 correspond to the transitions between the herbaceous- and shrub-dominated phases, when both groups of plants were present, and the transition between early and later stages of woody plant succession. Species diversity declines as shade-intolerant tree species displace the earlier successional trees and shrubs.

The rate at which displacement occurs is influenced by the growth rates of species involved in the succession. If growth rates are slow, the process will move slowly; if growth rates are fast, the process of displacement will occur more quickly. This observation led Michael Huston, an ecologist at Oak Ridge National Laboratory, to conclude that patterns of diversity through succession will vary with environmental conditions (particularly resource availability) that directly influence the rates of plant growth. By slowing the population growth rate of competitors that will eventually displace earlier successional species, the period of coexistence is extended and species diversity will remain high. This hypothesis

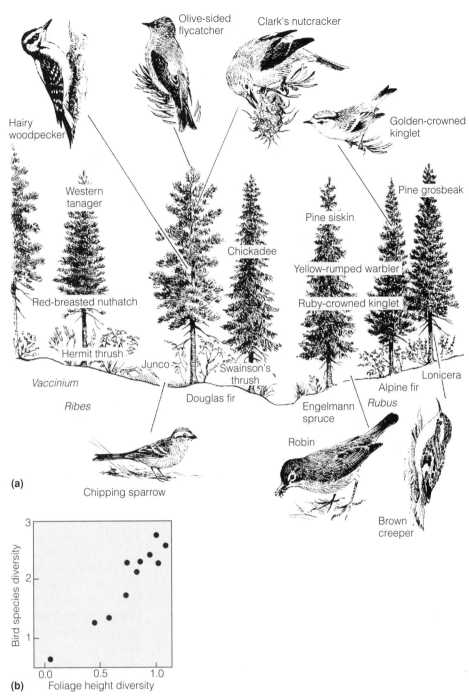

(a)

Chipping sparrow

(b)

FIGURE 17.14 Bird species become more diverse as vertical stratification increases in forest ecosystems. (a) The vertical distribution of some bird species in a spruce-fir forest in Wyoming. (Also see Figure 13.13.) (b) The relationship between bird species diversity and foliage height diversity for areas of deciduous forest in eastern North America. Foliage height diversity is a measure of the vertical structure of the forest. The greater the number of vertical layers of vegetation, the greater the diversity of bird species present in the forest. (After MacArthur and MacArthur 1961.)

has the interesting consequence of predicting the highest diversity at low to intermediate levels of resource availability by extending the period over which species coexist.

Disturbance can have an effect similar to the effect of reduced growth rates by extending the period over which species coexist. In the simplest sense, disturbance acts to reset the clock in succession. By reducing or eliminating plant populations, the site is once again colonized by early successional species, and the process of colonization and species replacement begins again. If the frequency of disturbance (defined by the time interval between disturbances; see Chapter 19) is high, then later successional species will never have the opportunity to colonize the site. Under this scenario, diversity will remain low. In the absence of disturbance, later succession species will eventually displace earlier successional species, and species diversity will decline. At an intermediate frequency of disturbance, colonization can occur, but competitive displacement is held to a minimum. The pattern of high diversity at intermediate frequencies of disturbance was proposed independently by Michael Huston and by Joseph Connell of the University of California at Santa Barbara, and is referred to as the **intermediate disturbance hypothesis**.

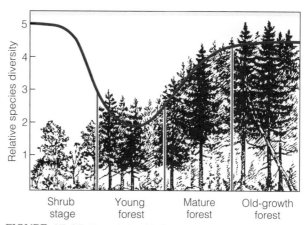

FIGURE 17.15 The relationship between successional stage and the number of mammal species present in Pacific Northwest forests.

17.8 Habitat Links the Structure of Plant and Animal Components of the Community

In Chapter 12 (Section 12.13), we explored the topic of habitat selection by animals. The structural features of the vegetation that influence habitat suitability for a given species may be related to a variety of specific needs, such as food, cover, and nesting sites. Because these needs vary among species, the structure of vegetation has a pronounced influence on the diversity of animal life within the community. A strong relationship exists between the degree of vertical stratification (foliage height diversity) and bird species diversity across a wide range of communities (Figure 17.14). Increased vertical structure means more resources and living space and a greater diversity of habitats (also see Chapter 13, Figure 13.3). Grasslands, with their two strata, support 6 or 7 species of birds, all which nest on the ground. A deciduous forest in eastern North America may support 30 or more species occupying different strata. The scarlet tanager *(Piranga olivacea)* and wood pewee *(Contopus virens)* occupy the canopy; the hooded warbler *(Wilsonia citrina)* is a forest shrub species; and the ovenbird *(Seiurus aurocapillus)* both forages and nests on the forest floor. Insects show similar stratification. Among the pine bark beetles inhabiting northeastern North America, the large turpentine beetle *(Buprestis apricans)* is restricted to the base of trees and the pine engraver beetle *(Ips*

oregoni) to the upper trunk and large branches. A third species, the small and abundant *Pityogenes hopkinsi,* lives on smaller branches in the canopy.

As shown in Chapter 13, Figure 13.15 (Section 13.13), the species composition and diversity of animal life changes as plant succession advances (Figure 17.15). Each successional stage has its own distinctive fauna. In forest succession, herbaceous and shrub stages typically support the greatest diversity of animal species. These species depend upon the early stages of plant succession for their habitat. As that stage passes away, so do the animals that occupy it. Therefore, certain species of animal life are highly dependent on disturbances that restore or maintain early stages of succession. Among such animals are bobwhite quail, cottontail rabbit, prairie warbler, and woodcock. Because of their dense canopy, lack of understory vegetation, and low degree of vertical stratification, young forest stands exhibit the lowest diversity. Diversity increases as the forest matures, with old-growth forests supporting a higher diversity than other forest stages. Old growth has more varied habitats, including dead and fallen trees and gaps in the canopy that stimulate new understory growth.

The various stages of wetland succession support their own forms of animal life (Chapter 29). Fish live in open water, and waterfowl and herons feed there. Floating aquatic vegetation supports hydras, frogs, diving beetles, gill-breathing snails, and insects colonizing the undersides of floating leaves. Emergent vegetation is occupied by nesting waterfowl and bitterns, red-winged blackbirds, muskrats, amphibians, flies and mosquitoes, mayflies, and dragonflies. Incoming woody vegetation such as alder and willow creates habitats for swamp sparrows, yellowthroats, woodcocks, and white-footed mice.

CHAPTER REVIEW

Summary

Fundamental Niche and Community Structure (17.1) The range of environmental conditions tolerated by a species defines its fundamental niche. These constraints on species' ability to survive and flourish will limit their distribution and abundance to a certain range of environmental conditions. Species differ in the range of conditions they tolerate. As environmental conditions change in both time and space, the possible distribution and abundance of species will change. This framework provides a "null" model against which to compare observed community patterns.

Food Webs and Species Interactions (17.2–17.4) Experiments that examine only two potentially interacting species tend to underestimate the importance of species interactions in communities because interactions are often diffuse, involving a number of species. In diffuse competition, the direct interaction between any two species may be weak, making it difficult to determine the effect of any given species on another. Collectively, however, competition may be an important factor limiting the abundance of all species involved. Diffuse interactions involving predation and competition can be seen in the structure of food webs (17.2). Food webs also illustrate the occurrence of indirect interactions among species within the community. Indirect interactions occur when one species does not interact with a second species directly, but influences a third species that does have a direct interaction with the second. For example, a predator may enhance one or more inferior competitors by reducing the abundance of the superior competitor on which it preys. Indirect mutualism results when one species benefits another indirectly through its interactions with others, reducing either competition or predation (17.3). To understand the role of species interactions in structuring communities, food webs are often simplified by grouping species into functional groups based on their similarity in the use of resources, or their role within the community. One such functional classification is to classify species into trophic levels based on general feeding groups (primary producers, herbivores, carnivores, etc.). The resulting food chains suggest the possibility of either bottom-up (primary producers) or top-down (top carnivores) control on community structure and function (17.4).

Species Interactions along Environmental Gradients (17.5–17.7) The biological structure of a community is first constrained by the environmental tolerances of the species (realized niche), which are then modified through both direct and indirect interactions with other species (realized niche). As we move across the landscape, variations in the physical environment will alter the nature of both these constraints on species distribution and abundance. There is a general tradeoff between a species' stress tolerance and its competitive ability along gradients of resource availability. This tradeoff can result in patterns of zonation across the landscape where variations in resource availability exist. The relationship between stress tolerance and competitive ability is more complex along gradients that include both resource and nonresource factors, such as temperature, salinity, or water depth (17.5).

Environmental changes can be autogenic or allogenic. Autogenic changes are a direct result of the activities of organisms in the community. Changes in environmental conditions independent of organisms are allogenic. Succession is the progressive change in community composition through time in response to autogenic changes in environmental conditions. One example is the changing light environment and the shift in dominance from fast-growing, shade-intolerant plants to slow-growing, shade-tolerant plants observed in terrestrial plant succession. Autogenic changes in moisture and nutrient availability likewise have a major influence on succession (17.6).

Patterns of species diversity change during the course of succession. Species colonization increases species richness, while species replacement acts to decrease the number of species present. Species diversity increases during the initial stages of succession as the site is colonized by new species. As early successional species are displaced by later arrivals, species diversity tends to decline. Peaks in diversity tend to occur during the stages of succession that correspond to the transition period, after the arrival of later successional species but prior to the decline of early successional species. Patterns of diversity during succession are influenced by both resource availability and disturbance (17.7).

Habitat Diversity and Species Diversity (17.8) The structure of vegetation has a pronounced influence on the diversity of animal life within the community. Increased vertical structure means more resources and living space and a greater diversity of habitats. As with plant species diversity, the diversity of animal life likewise changes as succession proceeds. As the structure of vegetation changes through time, the suitability of the vegetation to provide habitat for animal species will likewise change. The result is a shift in the presence and abundance of animal species within the community.

Study Questions

1. How does the fundamental niche of species constrain community structure?
2. Define diffuse competition.
3. Define indirect interaction in the context of food webs.
4. Give an example of how predation can result in indirect mutualism.
5. Contrast plant adaptations for stress tolerance and competitive ability along resources gradients.

6. How do stress tolerance and competition interact to influence patterns of zonation?

7. Contrast autogenic and allogenic environmental change, and provide examples of each.

8. How does competition for resources, especially light, influence the direction of succession?

9. In what ways can one species make the environment more favorable for another?

10. How does the structure of vegetation within a community influence the diversity of animal life?

CHAPTER 18

Netting salmon off the Alaskan coast.

Human Interactions Within Communities

OBJECTIVES

On completion of this chapter, you should be able to:

- Discuss the concept and application of sustainable yield to the exploitation of natural populations.
- Explain the potential conflict between sustainable yield and economic concerns.
- Discuss how the introduction of nonnative species can influence community structure.
- Define the terms *weed* and *pest,* and explain how they relate to human activities.
- Identify the various methods of pest control.
- Discuss the various approaches used to restore ecological communities.

As heterotrophic organisms, humans depend upon other species of plants and animals as sources of energy and nutrients—food. In fact, humans utilize a wide array of species, both plant and animal, as "natural resources" to meet a variety of needs, including clothing and shelter. Throughout history, human populations have had an impact on their local environments through the use of natural resources. Since early Grecian times, forest cutting, agricultural practices, overgrazing, and soil erosion converted the Mediterranean region into shrubland. Cutting of pine forests as far back as Roman times converted the Scottish highlands into moors.

The role of humans within ecological communities is a very broad topic, and to even touch upon the full range of issues is beyond the scope of this text. Therefore, in this chapter we restrict our discussion to the direct interaction of humans with the populations of other species. We address the more indirect interactions through the modification and destruction of habitat and alterations of the physical environment (air, soil, and waters) in following chapters (see Chapters 19, 23, and 30; Ecological Applications: Asteroids, Bulldozers, and Biodiversity and Blowin' in the Wind). Specifically, we will explore the topics of the utilization/exploitation and control of natural populations. In doing so, we will see that, like other members of the broader ecological community, humans interact directly and indirectly with a wide array of species, functioning as predator, competitor, and mutualist.

18.1 Exploitation of Natural Populations Led to Management

Humans have exploited plant and animal populations as a source of food, clothing, and shelter throughout history. The advent of agriculture some 10,000 years ago reduced the dependence of humans on natural populations. However, over 80 percent of the world's commercial catch of fish and shellfish is from the harvest of naturally occurring populations in the oceans (71 percent) and inland freshwaters (10 percent). Likewise, forest plantations account for only 10 percent of the world's total harvest of forest resources. The remaining 90 percent is harvested from native forests.

Although there are numerous historical accounts of overexploitation and population declines, not until the later 1800s was any effort made to manage fishery resources to ensure their continuance. At that time, wide fluctuations in the catches of fish in the North Sea began to have an economic impact on the commercial fishing industry. Ensuing debates raged over the cause of the decline and whether commercial harvesting was having an impact on fish populations. Many people disagreed about whether the removal of fish had an effect on reproduction. Not until a Danish fishery biologist, C. D. J. Petersen, developed a method for estimating population size based on a technique of tagging, releasing, and recapture (see Focus on Ecology 9.1) were biologists able to make some assessment of fish populations. Together with data from egg surveys and the aging of fish from commercial catches, these studies suggested that overharvesting indeed was the culprit. But the debate continued, and it was not until after World War I that the controversy was laid to rest.

During the war, fishing in the North Sea had stopped. After the war, fishermen experienced sizable increases in their catches. Fishery biologists suggested that renewed fishing would once again reduce population sizes and that catches would stabilize and eventually decline with overexploitation. Their predictions were correct, and with time, attention turned to the question of sustainable harvest.

18.2 Continued Exploitation Depends on Sustained Yield

The amount of resource harvested per unit time is called **yield**. Harvesting at a level that will ensure a similar yield over and over without forcing the population into decline is called **sustained yield**. This requires that the yield (amount removed per unit time) not exceed the rate of production (amount of resource produced per unit time). To understand the concept of sustained yield, we need to return to the topic of population growth examined in Chapter 10.

Consider a population of fish that is growing in a manner described by the logistic model of population growth presented in Chapter 10 (Section 10.10) and that is being harvested. We can incorporate the effect of fishing on the population (N) in a manner similar to that used in modeling the influence of predation on prey populations (Chapter 15, Section 15.2).

$$\frac{dN}{dt} = rN\left(1 - \frac{N}{K}\right) - qEN$$

Note that the density-dependent term $(K - N)/K$ has been expressed as the equivalent term $(1 - \frac{N}{K})$ for purposes of convenience. As with predation, the harvest rate is the product of three terms: prey density

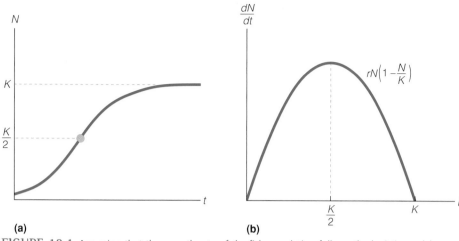

FIGURE 18.1 Assuming that the growth rate of the fish population follows the logistic model presented in Chapter 10 [$dN/dt = rN (1 - N/K)$]: (a) In the absence of fishing, the population will grow to carrying capacity, K. (b) The relationship between the rate of population growth, dN/dt, and population density, N, takes the form of a parabola, reaching a maximum value at a population density of $N = K/2$.

(N—fish population), a constant q that represents the efficiency of the predator (catchability), and a term representing the fishing effort, E. Fishing effort (E) functions as a substitute for predator density and takes into consideration a variety of factors such as the size of the fishing fleet and the length of the season. The product of catchability and effort, qE, is the fishing mortality; it has the same dimension as r and will play an important role in what follows.

In the absence of fishing, the population will grow to carrying capacity (K) (Figure 18.1a). The relationship between the rate of population growth (dN/dt—the change in N over the time interval t) and population density takes the form of a parabola (Figure 18.1b). At a population density of zero, the rate of growth is likewise zero. As population density increases, the rate of population growth increases, reaching its maximum at a population density of half the carrying capacity ($K/2$), referred to as the inflection point. As the density continues to rise above this point, the rate of population growth declines, reaching zero as the population density approaches the carrying capacity ($N = K$).

To maintain a sustainable harvest, the growth rate of the fish population must be greater than or equal to the rate of harvest. The rate of harvest is equal to the growth rate of the population when:

$$rN\left(1 - \frac{N}{K}\right) = qEN$$

Solving this equation algebraically, we see that there are two solutions:

$$N^* = K\left(1 - \frac{qE}{r}\right)$$

$$N^* = 0$$

The second solution corresponds to the extinction of the fish population, so only the first is relevant to our discussion.

The values of N^* for a fish population with constant values of r and K will vary depending on the fishing effort (qE). For a given fishing effort (value of qE), the harvest rate (qEN) will increase linearly as a function of N (Figure 18.2). For values of N above N^*, the rate of harvest (qEN) is greater than the rate of population growth [$rN(1 - \frac{N}{K})$] and the population declines. If the rate of harvest is less than the rate of population growth, the population increases. In both instances, the population returns to a density of N^*. If we continue to further increase the fishing effort (mortality), we eventually reach a point where the harvest rate exceeds the growth rate for all densities (N) and the population is driven to extinction (Figure 18.2c). The relationship between N^* and fishing effort (qE) is shown in Figure 18.3. When fishing effort is zero, the population will be at carrying capacity (K). The upper limit to fishing effort will be when the per capita rate of mortality due to fishing effort (qE) is equal to maximum per capita growth rate of the population (r). Each point along the line shown in Figure 18.3 is the value of N^* for a corresponding fishing effort (qE) at which growth rate of the fish population is equal to the rate of harvest—the condition for sustained yield.

The sustained yield, Y, at each point is therefore:

$$Y = qEN^*$$

Using this equation, the relationship between yield and effort is plotted in Figure 18.4. The yield initially rises with effort, reaching a **maximum sustained yield** (MSY). Further increase in fishing effort will lower

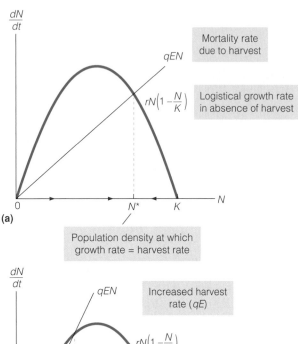

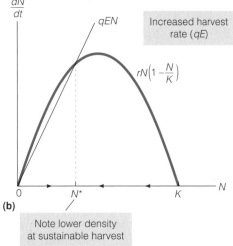

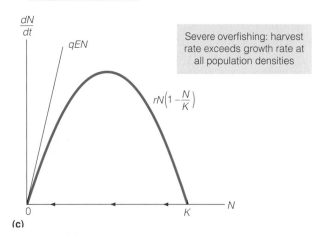

(a)

Mortality rate due to harvest

qEN

Logistical growth rate in absence of harvest

$rN\left(1 - \dfrac{N}{K}\right)$

Population density at which growth rate = harvest rate

(b)

qEN

Increased harvest rate (qE)

$rN\left(1 - \dfrac{N}{K}\right)$

Note lower density at sustainable harvest

(c)

qEN

Severe overfishing: harvest rate exceeds growth rate at all population densities

$rN\left(1 - \dfrac{N}{K}\right)$

FIGURE 18.2 (a) The relationship between the rate of population growth, dN/dt, and population density for the two components of the fish population model with harvest: logistic population growth [$rN(1 - N/K)$] and mortality due to harvest (qEN). The line describing harvest assumes a given value for fishing effort (qE). The population density at which growth rate is equal to harvest rate (sustainable harvest) is labeled as N^* on the x axis. For population densities above N^*, harvest exceeds the growth rate and the population declines. For densities below N^*, the population growth rate exceeds the rate of harvest and the density increases. (b) As the fishing effort (value of qE) increases, the corresponding population density decreases. (c) In the case of severe overfishing (high fishing effort), the rate of harvest exceeds the growth rate at all densities, and the population will be driven to extinction. (After Kot 2001.)

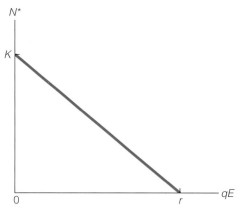

FIGURE 18.3 The relationship between fishing effort and the corresponding population density at sustainable yield, N^*. When fishing effort is very low or zero (no fishing), the population will approach carrying capacity (K). The population density at sustainable yield declines as fishing effort increases. For sustainable yield to occur, the rate of harvest cannot exceed the maximum (intrinsic) rate of population growth defined by r.

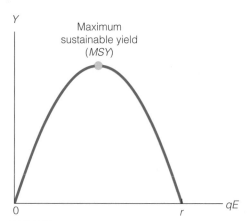

FIGURE 18.4 Relationship between fishing effort (qE) and yield (Y) under conditions of sustainable yield ($Y = qEN^*$). The relationship takes the form of a parabola. An increase in effort increases sustainable yield up to a point. Further increases in effort lower the yield as the stock (N^*) becomes increasingly overexploited and reduced. The maximum value of Y (and corresponding value of effort) is the maximum sustainable yield.

the yield as population becomes overexploited and depleted. At the maximum sustainable yield, harvest takes the population down to a level at which the remaining stock can replace the amount removed before the next harvest period.

18.3 Achieving Sustainable Yield Is Often Difficult

The ultimate goal of a managed fishery is to achieve that maximum sustainable yield. This requires information on the intrinsic growth rate of the population (*r*) and a program to monitor fish populations. Fishing effort is typically controlled through regulating the number of licenses, methods of fishing, and timing and length of the fishing season.

In effect, the concept of sustainable yield is an attempt at being a "smart predator," maintaining the prey population at a density where the production of new individuals just offsets the mortality represented by harvest. The higher the rate of population increase, the higher the rate of harvest that produces the maximum sustainable yield. Species characterized by a very high rate of population growth often lose much of their production to a high density-independent mortality, influenced by variation in the physical environment such as temperature (see Chapter 11, Section 11.12). The management objective for these species is to reduce "waste" by taking all individuals that otherwise would be lost to natural mortality. Such a population is difficult to manage. The stock can be depleted if annual patterns of reproduction are interrupted due to environmental conditions. An example is the Pacific sardine (*Sardinops sagax*). Exploitation of the Pacific sardine population in the 1940s and 1950s shifted the age structure of the population to younger age classes. Before exploitation, reproduction was distributed among the first five age classes (years). In the exploited population, close to 80 percent of reproduction was associated with the first two age classes. Two consecutive years of environmentally induced reproductive failure caused a collapse of the population from which the species never recovered (Figure 18.5).

Sustainable yield requires a detailed understanding of the population dynamics of the fish species. Recall from Chapter 10 that the intrinsic rate of population growth, *r*, is a function of the age-specific birth and mortality rates. Unfortunately, the usual approach to maximum sustained yield fails to adequately consider size and age classes, differential rates of growth among them, sex ratio, survival, reproduction, and environmental uncertainties, all data difficult to obtain. Adding to the problem is the common-property nature of the resource—because it belongs to no one, it belongs to everyone to use as each sees fit.

Perhaps the greatest problem with models of sustainable harvest is that they fail to incorporate the most important component of population exploitation—economics. Once commercial exploitation

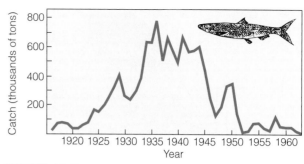

FIGURE 18.5 The annual catch of the Pacific sardine along the Pacific coast of North America. Overfishing, environmental changes, and an increase in the population of a competing fish species, the anchovy, resulted in a collapse of the population. (Murphy 1966.)

begins, the pressure is on to increase it to maintain the underlying economic infrastructure. Attempts to reduce the rate of exploitation meet strong opposition. People argue that reduction will mean unemployment and industrial bankruptcy—that, in fact, the harvest effort should increase. This argument is shortsighted. An overused resource will fail, and the livelihoods it supports will collapse, because in the long run the resource will be depleted. That fact is written in abandoned fish processing plants and rusting fishing fleets. With conservative exploitation, the resource could be maintained.

Another problem is that traditional population management, especially by fisheries, considers stocks of individual species as single biological units rather than as components of a larger ecological system. Each stock is managed to bring in a maximum economic return, overlooking the need to leave behind a certain portion to continue its ecological role as predator or prey. This attitude encourages a tremendous discard problem, euphemistically called "bypass." Employing large drift nets that encompass square kilometers of ocean, fishermen haul in not only commercial species they seek, but a range of other marine life as well, including sea turtles, dolphins, and scores of other species of fish. Fishermen bypass this unwanted catch by dumping it back into the sea. Discarded fish alone make up one-fourth of the annual marine catch. Of the 27 metric tons of fish taken in 1995 in the Pacific Northwest, 9 metric tons was bypass. The ecological effects of bypass can be enormous. Because much of the bypass consists of juvenile and undersized fish of commercial species, the practice can seriously affect the future of those fisheries. The removal or reduction of other species can alter the nature of interactions within the community (see Chapter 17, Section 17.3). Such disturbances can alter food webs of ocean ecosystems and upset the functioning of the pelagic ecosystem.

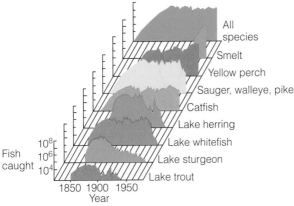

FIGURE 18.6 Annual catches of selected species of fish, and all species combined, by the Lake Erie commercial fishery since 1820. The vertical scale is logarithmic. (Reprinted with permission from H. A. Regier and W. L. Hartman, "Lake Erie's Fish Community: 150 Years of Cultural Stress," *Science* 180 (1973): 1248–1255. © 1973 American Association for the Advancement of Science.)

The history of the fishery in Lake Erie illustrates this point (Figure 18.6). Before the War of 1812, the lake, whose shores were lightly settled, held an abundance of whitefish (*Coregonus clupeaformis*), lake trout (*Salvelinus namaycush*), blue pike (*Stizostedion vitreum glaucum*), sauger (*Stizostedion canadense*), and lake herring (*Coregonus artedii*). After the war, settlement increased rapidly, and so did the exploitation of the fishery resource. A subsistence fishery grew into a thriving industry by 1820. For the next 70 years, rapid improvement in transportation, fishing boats, gear, and techniques and an expanding market increased the average rate of the catch by 20 percent a year. By 1890 this rapid growth in catch slowed as stocks became depleted. However, increased intensity of fishing, further improvement of equipment, and heavy capital input maintained the size of the catch until the late 1950s.

Early to go was the lake sturgeon (*Acipenser fulvescens*), at first netted and burned on the shore because it destroyed fishing nets. Later it was exploited as a market fish along with lake trout and whitefish. Fishing intensity grew with incentives to increase production during World War I and with the introduction of a gill net that was extremely wasteful in its catch. In 1950 the development of the nylon gill net, which could be left in the water much longer, made fishing more intense than ever. Now walleye (*Stizostedion vitreum vitreum*), blue pike, and yellow perch (*Perca flavescens*) joined the ranks of overexploited species. By 1960 the stocks of walleye and blue pike had been commercially depleted. The fishery is now dominated by yellow perch.

Adding to the stress was pollution in the lake. Industrial, agricultural, and urban wastes, toxic pol-

lutants and biocides, and runoff from shore developments wiped out phytoplankton that supported fish life. An invasion of rainbow smelt dealt the final blow. Young smelt feed on plankton and crustaceans, the older smelt on small fish. In turn, the smelt is valuable prey for adult lake trout, blue pike, walleyes, and sauger. With the decimation of those species, smelt increased rapidly and became the predators of young remnant stocks. Had the invasion of rainbow smelt (*Osmerus mordax*) been halted and had fishing been regulated, the preferred commercial species would still be available in harvestable numbers.

The history of this Great Lakes fishery is a microcosm of the history of marine fisheries. Certain stocks of fish, such as the Pacific sardine (*Sardinops sagax*), Atlantic halibut (*Hippoglossus hippoglossus*), Atlantic cod, and Peruvian anchovy, have been exploited to commercial extinction, causing ecological and economic damage. The sad plight of the whales is another example of decreased catches followed by increased hunting intensity, made possible by greater capital input and technological advances such as factory ships and fleets of hunters. Despite warnings of overharvesting, the marketplace and short-term profits dictated the take.

18.4 Sustained Yield Is Also Applied to Forestry

The goal of sustained yield in forestry is, like that of sustained yield in fisheries, to achieve a balance between net growth and harvest. Sustained yield in forestry is much more sophisticated than in fisheries simply because it is easier to inventory, measure growth and potential yields of biomass, and manipulate populations or stands of trees. To achieve this end, foresters have an array of silvicultural and harvesting techniques from clear-cutting to selection cutting.

Clear-cutting involves removal of the forest and reversion to an early stage of succession (see Figure 19.21). Unless carefully managed, clear-cut areas can be badly disturbed by erosion, impacting the subsequent recovery of the site as well as adjacent aquatic communities. In selection cutting, mature single trees or groups of trees scattered through the forest are removed. Selection cutting produces only small openings or gaps in the forest canopy. Although this form of timber harvest can minimize the scale of disturbance within the forest, it can result in changes in species composition and diversity. Regardless of the differences in approach, a number of general principles can be made regarding sustainable forestry.

Forest trees function in the manner discussed for competition in plant populations in Chapter 11

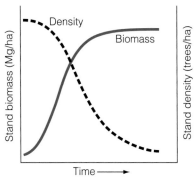

FIGURE 18.7 Hypothetical example of the relationship between stand density and average tree size in a forest undergoing secondary succession. Density declines, and most of the biomass in the forest stand is contained in a few large canopy trees.

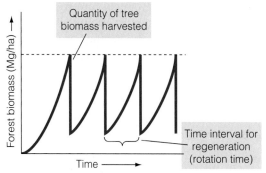

FIGURE 18.8 To achieve sustained yield, a sufficient time must be allowed between harvests (rotation time) for biomass to return to preharvest levels. Rotation time will depend on the growth rate of the species and site conditions that influence forest productivity.

(Section 11.3). The establishment of a forest begins with one or more populations of small individuals that grow and compete for the essential resources of light, water, and nutrients. As biomass in the forest increases, the density of trees decreases, while the average tree size increases (Figure 18.7; also see Figure 11.4). Eventually, most of the biomass in a forest stand is contained in a small number of large canopy trees. When these trees are harvested, a sufficient time must pass for the forest to once again repeat this process. For sustained yield, the period of time between harvests must be sufficient for the forest to regain the level of biomass at the time of the previous harvest (Figure 18.8). The period of time between harvests is called the **rotation time.**

The rotation time is dependent on a variety of factors related to the tree species, site conditions, type of management, and the intended use of the trees that are being harvested. Wood for paper products (pulp wood), fence posts, and poles are harvested from fast-growing species allowing a short rotation period (15–40 years). These species are often grown in highly managed plantations where trees can be spaced to reduce competition and fertilized to maximize growth rates. Trees harvested for timber (saw logs) require a much longer rotation period. Hardwood species used for furniture and cabinetry are typically slower growing and may have a rotation time of 80 to 120 years. Sustained forestry of these species works best on extensive areas where blocks of land can be maintained in different age classes.

A significant amount of nutrients are lost from the forest when trees are harvested and removed. The loss of nutrients in plant biomass is often compounded by further losses from soil erosion and various post-harvest management practices—particularly the use of fire (see Chapter 21, Section 21.6). The reduction of nutrients will reduce subsequent plant growth, requiring a longer rotation period for subsequent harvests, or causing a reduction in forest yield if the rotation period is maintained. Forest managers often counter the loss of nutrients by using chemical fertilizers, creating other environmental problems for adjacent aquatic ecosystems (see Chapters 21 and 23).

Sustained yield is a key concept in forestry and is practiced to some degree by large timber companies and federal and state forestry agencies. But, all too often, industrial forestry's approach to sustained yield is to grow trees as a crop rather than maintaining a forest ecosystem. Their management approach is a form of agriculture in which trees are grown as crops: clear-cut, spray herbicides, plant or seed the site to one species, clear-cut, and plant again. Clear-cutting practices in some national forests, especially in the Pacific Northwest and the Tongas National Forest in Alaska, hardly qualify as sustained yield management when below-cost timber sales are mandated by the government to meet politically determined harvesting quotas. Even more extensive clear-cutting of forests is taking place in the northern forests of Canada, especially British Columbia, and in large areas of Siberia. As the timber supply dwindles in the Pacific Northwest, the timber industry that moved west following the depletion of eastern hardwoods and pine forests of the lake states is moving east again to exploit the regrowth of eastern hardwood forests, especially the rich and diverse central hardwood forest. From Virginia and eastern Tennessee to Arkansas and Alabama, timber companies have built more than 140 highly automated chip mills that cut up trees of all sizes into chips for paper pulp and particle board. Feeding the mills requires clear-cutting 500,000 hectares annually. The growing demand for timber has boosted timber prices, stimulating more clear-cutting. The rate of harvest is wholly unsustainable. In face of growing timber demands, sustained yield management has hardly filtered down to smaller parcels of private land.

The problem of sustained yield forestry, like that of sustained yield fisheries, is its economic focus on the resource with little concern for the forest as a biological community. A carefully managed stand of trees, often reduced to one or two species, is not a forest in an ecological sense. Rarely will a naturally regenerated forest, and certainly not a planted one, ever support the diversity of life found in old-growth forests. By the time the trees reach economic or financial maturity—based on the type of rotation—they are cut again. Economic maturity is not the same as ecological maturity. For this reason, prospects of future old-growth forests are becoming as extinct as the passenger pigeon that once depended on them (see the Ecological Application after Ch. 23: Time to Rethink the Law).

18.5 Human Activities Have Benefited Some Species at the Expense of Others

Although the exploitation and management of natural populations provides over 80 percent of fish and shellfish harvested globally on an annual basis, the vast majority of human food resources are derived from agriculture—the production of crops and the grazing of livestock. Even though botanists estimate that, worldwide, there are as many as 30,000 native plant species with parts (seeds, roots, leaves, fruits, etc.) that can be consumed by humans, only 15 plant and 8 animal species produce 90 percent of our food supply. The seeds of only 3 annual grasses—wheat, rice, and corn (maize)—constitute over 80 percent of the cereal crops consumed by the world population. Although initially derived from native plant species, the varieties of the cereal crops that are cultivated today are the products of intensive selective breeding and genetic modification by plant and agricultural scientists around the world (see the Ecological Application after Ch. 2: Taking the Uncertainty out of Sex). The same is true for domestic animals used for food production.

Approximately 11 percent of Earth's ice-free land area is under cultivation. Another 25 percent is used as pastureland for grazing livestock (primarily cattle and sheep). In effect, a very small number (at most a few tens) of plant and animal species now dominate over a third of the ice-free land area of the planet. From the perspective of population interactions, humans have developed a mutualistic relationship with a small set of species that are used as a source of food, acting as agents of dispersal, providing suitable habitat, maximizing their reproduction, and greatly expanding both their geographic distribution and abundance. At the same time, the clearing of land for agriculture (both crops and pasture lands) has dramatically reduced the distribution and abundance of literally tens of thousands of other plant and animal species, in many cases leading to their extinction. The alteration and loss of habitat is the leading cause of modern extinctions (Chapter 10, Sections 10.12–10.14).

In addition to the loss of habitat resulting from the conversion of natural ecosystems into crop and pasture lands, agricultural practices also influence adjacent aquatic communities. Nutrient input from fertilizers often results in eutrophication and reduced oxygen levels (see Chapter 27), altering aquatic communities. Siltation caused by the erosion of farmlands is an insidious form of pollution, for it is widespread, it often goes unnoticed, and the damage it does is often permanent. Silt prevents the growth of aquatic plants and settles on the stream bottom, smothering larvae, mussels, and other bottom organisms and mollusks. Silty water flowing through the gravel nests of trout and salmon causes heavy mortality of eggs. Thousands of miles of trout and salmon streams have been destroyed by siltation, which, more than any other cause, limits the natural reproduction of these fish.

Although human activity and alteration of the landscape has led to the decline of many plant and animal species, other species have flourished. In eastern North America, for example, white-tailed deer have experienced an explosive population growth during the past 30 years. The clearing of forested lands has created a patchwork of forest and field, providing excellent habitat for deer populations. Deer have literally invaded suburban developments that provide both protection and a rich source of highly palatable food in the form of garden and landscape plantings.

In some cases, the creation of habitat by human activity has altered the migratory patterns of species, allowing populations to remain in an area year-round. Nonmigratory Canada geese in the eastern United States are common permanent residents of golf courses, parks, suburban housing developments—wherever grass and sufficient water are available to them. These geese descended from captive populations held as live decoys back in the 1930s and stopover migratory birds that stayed, attracted by the rich source of food.

18.6 Human-Introduced Invasive Species Significantly Alter Natural Communities

Intentionally or unintentionally, humans have functioned as agents of dispersal for countless species of plants and animals, transporting them outside their natural geographic ranges. Although many species failed to survive in their new homes, others flourish. Freed from the constraints of their native competitors, predators, and parasites, they successfully established themselves and spread. These nonnative (nonindigenous) plants and animals are referred to as **alien** or **invasive species.**

Animal invaders often cause the extinction of vulnerable native species through predation, grazing, competition, and habitat alteration. Island species suffer the most. In Hawaii, for example, during the past 200 years, 263 of the islands' native creatures disappeared; 300 are listed as endangered or threatened; and 1400 life forms are in trouble or extinct. Among the island's 111 birds, 51 are extinct and 40 are endangered. On the Pacific island of Guam, the brown tree snake (*Boiga irregularis*), a native of New Guinea, accidentally reached the island around 1950, probably aboard military equipment transported there for dismantling. The snake has eliminated 9 of 12 native bird species, 6 of 12 native lizards, and 2 of 3 native fruit bats; it even invades houses on the island and bites sleeping infants.

Plant invaders, many introduced as horticultural plants, outcompete native species and alter fire regimes, nutrient cycling, energy budgets, and hydrology. Introduction of exotics is the cause behind 95 percent of species loss and endangerment in Hawaii. Among the 1126 native flowering plants of Hawaii, 93 are extinct and 40 are endangered. On the North American continent, the ornamental perennial herb purple loosestrife (Figure 18.9a) has eliminated native wetland plants to the detriment of wetland wildlife. The Australian paperbark tree (*Maleleuca quinquenervia*) (Figure 18.9b), introduced as an ornamental plant in Florida, is rapidly displacing cypress, sawgrass, and other native species in the Florida Everglades, drawing down water, and fostering more frequent or intense fires. *Myrica faya*, a nitrogen-fixing tree, was imported from the Canary Islands to the big island of Hawaii in 1900 to halt soil erosion. The tree is now spreading over 30,000 acres of Hawaii National Park, increasing nitrogen supplies in poor volcanic soil at a rate 90 times greater than native species, paving the way for invasion of other nonindigenous plants, and displacing the native forest.

The problem of invasive species is not restricted to terrestrial environments. The Great Lakes have been invaded by 139 nonnative aquatic species that impacted native plant and animal species. The San Francisco Bay area is occupied by 96 nonnative invertebrates, from sponges to crustaceans. Exotic fish, introduced purposefully or accidentally, have been responsible for 68 percent of fish extinctions in North America during the past 100 years and for the decline of 70 percent of the fish species listed as endangered.

A classic example of the way an invasive animal species can alter community structure was the accidental introduction of peacock bass (*Cichla ocellatus*), a native of the Amazon, into Gatun Lake in the

(a)

(b)

FIGURE 18.9 Two invasive plant species that have significant negative impacts on native communities: (a) the perennial herb purple loosestrife, and (b) Australian paperbark tree has invaded the Florida Everglades resulting in severe impacts on the ecosystem.

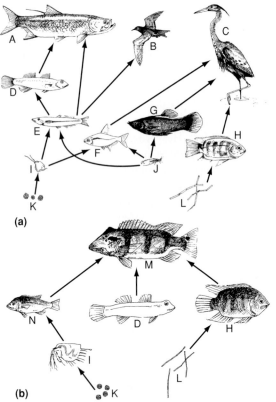

FIGURE 18.10 Generalized food webs of common Gatun Lake populations contrasting (a) regions without or before introduction of *Cichla* (peacock bass) and (b) regions with *Cichla*. Thick arrows indicate major importance of food item to predator or herbivore; thin arrows indicate minor importance. Key to species: A, *Tarpon atlanticus* (tarpon); B, *Chlidonias niger* (black tern); C, several species of herons and kingfishers; D, *Gobiomorus dormitor*; E, *Melaniris chagresi*; F, Characinidae, including four common species; G, Poecillidae, including two common species—the exclusively herbivorous *Poecilia mexicana* and the exclusively insectivorous *Gambusia nicaraguagensis*; H, *Chichlasoma maculicauda*; I, zooplankton; J, terrestrial insects; K, nannophytoplankton; L, filamentous green algae; M, adult *Cichla ocellaris*; N, young *Cichla*. (After Zaret and Paine 1973.)

Panama Canal Zone. A popular sport and food fish and a voracious predator, it escaped from impoundments. Its presence has had a devastating effect on the fish population and has profoundly affected community structure (Figure 18.10). In the lake the peacock bass feeds mostly on adult fish of the genus *Melaniris*, decreasing its population. Other predatory species that feed on *Melaniris*, such as Atlantic tarpon, black terns, and herons, have sharply declined. A complex community structure has been highly simplified. Six or eight common fish species within the community have been eliminated or seriously reduced, all by the introduction of a single top-level predator in the lake community.

Other invaders have changed whole ecosystems. The chestnut blight, introduced through the import of the Chinese chestnut, has all but exterminated the American chestnut (*Castanea mollissimi*), the dominant deciduous tree in the eastern hardwood forests of North America, and converted the oak-chestnut forest ecosystem to oak. The gypsy moth, which accidentally escaped from a population brought into Massachusetts for experimental silkworm production, has spread through northeastern oak forests, endangering oaks as commercial timber species and displacing many moths native to hardwood forests. Multiflora rose (*Rosa multiflora*), introduced and widely planted as a living fence in the 1950s, now covers pastures and forest edges in the eastern United States.

18.7 Many Species Clash with Human Interests

Humans have always had to contend with unwanted plants and animals that interfere with our well-being. We have developed general terms to describe these species, both plant and animal—weeds and pests. **Weed** is a general term describing plants growing anywhere they are not desired. Native plant species that grow in adjacent fields, for example, become weeds when they disperse into adjacent gardens, competing with crops for space, light, and nutrients, reducing yields. The most persistent and abundant weeds are plant species that are easily dispersed, allowing them to colonize newly disturbed sites. They often remain in the soil as seeds for a long time, responding quickly to conditions that favor germination and growth.

Pests are animals that humans consider undesirable, a classification that varies with time, place, and circumstances as well as individual attitudes. Mice, rats, cockroaches, fleas, mites, lice, and mosquitoes have long been unwelcome visitors, adapting to human habitation and cultural practice. Many species are vectors for disease (see Chapter 16). Mosquitoes are known to transmit over 50 percent of the approximately 102 *arboviruses* (short for *arthropod-borne viruses*) that can produce disease in humans. With the development of agriculture, large predators that threatened domestic animals, herbivores that invaded fields and gardens, and grain-feeding birds all became pests.

Among the most invasive of weeds and pests are the species that colonize agricultural crops and forests, causing significant economic loss. Extensive acreages of monoculture (single species) cropland

and forest tree plantations provide huge areas of abundant food and cover, providing conditions for economically damaging outbreaks of pest species that feed upon these plants. It is estimated that about 55 percent of the world's potential food supply is lost to pests, either before or after harvest. Agricultural pests are represented by a wide array of plant, animal, and fungal species, both native and alien.

After centuries, it is evident that only where we can isolate and destroy invading colonies can we eradicate invasive pest and weed species. Too often, they unobtrusively gain a foothold and reach a population level at which we cannot eradicate them. The best we can do is control their numbers. The idea is to reduce both alien and native pest and weed species to a level below which they will not cause economic injury. At that point, costs of control should be less than or at best equal to the net increase in the value derived from such control. Even so, many people initiate control measures when no strong economic values are involved, such as spraying insecticides and herbicides on lawns.

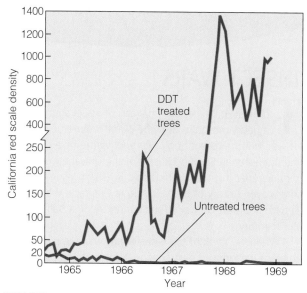

FIGURE 18.11 Experimental monthly applications of DDT on scale-infested lemon trees reduced not only the pest but also its natural enemies. The scale resurged dramatically. Natural predators and parasites kept the scale in check on unsprayed trees. (From DeBach 1974.)

18.8 A Variety of Means Are Deployed to Control Weed and Pest Species

One of the common means of controlling weeds and pests is with chemicals. The ancient Sumerians used sulfur to combat crop pests, and the Chinese as long ago as 3000 B.C. used substances derived from plants as insecticides. By the early 1800s, such chemicals as Paris green, Bordeaux mixture, and arsenic were commonly used to combat insect and fungal pests. The major chemical weapons, however, appeared after World War II with the development of organic insecticides (containing carbon) to combat insect vectors of human disease, especially in tropical areas. The success of these insecticides encouraged their rapid use in agriculture, for which the chemical industry provided an arsenal of over 500,000 biocides (see Focus on Ecology 18.1: Insect Wars).

These organic chemicals, with varying degrees of toxicity and persistence, are either synthetic (of human manufacture) or botanical (derived from plants). Major groups of synthetic pesticides include chlorinated hydrocarbons (see Chapter 23, Section 23.9), organophosphates, and carbamates. In one manner or another, they disrupt the nervous system. Botanical insecticides include pyrethrum, nicotine, and rotenone that is highly toxic to fish.

Ease of application, effectiveness in small doses, low cost, and toxicity gave chemical insecticides the appearance of a panacea. They became, as the entomologist Paul DeBach described them, ecological narcotics. Instead of solving the pest problem, the insecticides compounded it by killing natural predators as well as pests. Unchecked, the surviving pests resurged (Figure 18.11). Killing natural predators also released other insect pests that had been held in check. Now their populations increased dramatically, creating new pest problems.

A major failing of chemical pesticides is the ability of insects to evolve resistance to pesticides via natural selection (see Chapter 2). As one pesticide replaces another, the pest populations acquire a resistance to them all. By 1988, over 1600 insect pests worldwide had developed resistance to one or more pesticides. Some insects, notably houseflies, certain mosquitoes, the Colorado potato beetle, and cotton bollworms have overcome the toxic effects of every pesticide to which they have been exposed. Insect pests need only about 5 years to evolve pesticide resistance; their predators do so much more slowly.

The chemicals for weed control are organic herbicides, many of them toxic. They fall into three classes, based on their effects on plants. Contact herbicides, such as atrazine, kill foliage by interfering with photosynthesis. Absorbed by plants, systemic herbicides such as 2,4-D and 2,4,5-T overstimulate growth

INSECT WARS

To most of us, picking up a can of insect spray to zap a pesky mosquito or fly does not conjure up images of war, but the recent book *War and Nature* by environmental historian Edmund Russell of the University of Virginia does just that. In his book, Russell investigates the interwoven history of the U.S. Army's Chemical Warfare Service and pest control efforts during the first half of the 20th century.

When we think of the casualties of war, we naturally think of weapons of destruction—guns and bombs. Yet by creating troop concentrations, food shortages, exhaustion, and poor sanitation, throughout history war has resulted in the outbreak of disease epidemics, causing both public health and military problems. World War I saw some of its worst casualties from typhus. Typhus broke out in Siberia in November 1914 and spread through both troops and refugees. The epidemic killed over 150,000 people in less than 6 months. During World War II, louse-borne typhus was a serious problem for American troops on both the European and Pacific fronts. While in the early years of World War II in the Pacific, malaria caused eight to ten times more American casualties than did enemy soldiers.

As part of a program to combat the spread of typhus, the military used both aerosol bombs (compressed insecticide in metal containers) and louse powder to combat lice; both used pyrethrum as the active ingredient. Ground from the petals of several species of chrysanthemum, pyrethrum achieved prominence as "insect powder" before 1800. It was first produced in Persia (Iran), then Dalmatia (part of Austria-Hungary at the time), and France became the major exporter in the 18th and 19th centuries. Because flowers were picked by hand, the production of pyrethrum was concentrated in areas where labor was cheap. During World War II, naval warfare had virtually halted the export of pyrethrum, and the need for a substitute was urgent.

A Swiss chemist, Paul Muller, first developed DDT in the late 1930s, and although it was shown to be an effective insecticide with low toxicity to humans, it received little attention in the United States. By 1942, however, a lot had changed. The U.S. Bureau of Entomology was given reports from the Swiss company for which Muller worked, and tests of DDT were undertaken by various government agencies. They found that DDT powder works to kill lice four times longer than pyrethrum powder; sprayed on walls of buildings, DDT killed adult mosquitoes for months. As a synthetic, DDT could be manufactured in the United States. It provided the solution that the military needed, and large-scale production was begun. The army was able to start spraying DDT around the world as soon as the chemical arrived in combat theatres using the already deployed chemical warfare equipment.

Organizationally, war stimulated the creation, growth, and linkage of military and civilian institutions devoted to pest control and chemical warfare, accelerating development in both spheres. Scientifically and technologically, pest control and chemical warfare each created

hormones. The plants grow faster than they can obtain necessary nutrients and die. Soil sterilants kill microorganisms necessary for certain plants to grow.

Chemical control has its dark side. The chemicals affect not just the target species but a wide array of nontarget species as well, often eliminating them and upsetting natural food webs, especially through the suppression of natural predatory species. The surviving pests then resurge in greater numbers than ever. Some herbicides contain dioxins, which have been linked with birth defects and cancers in humans.

Some of the most serious pest species are alien species, introduced unintentionally. When an animal or a plant is successfully introduced, intentionally or not, into new habitat outside of the natural range, it often leaves its enemies behind. Freed from predation, the species can quickly become a pest or a weed.

This fact has led to a search for suitable natural enemies to introduce.

The classic success story is the case of the introduced cottony-cushion scale moth *Icerya purchasi*, a native of Australia that sucks food from the leaves and twigs of citrus fruit trees. First seen in California in 1872, within 15 years the insect threatened the citrus fruit industry. In 1887 entomologists discovered that the ladybug beetle *Rodolia cardinalis (Vadalia cardinalis)* and a parasitic fly *Cryptochaetum iceryae* controlled the scale in its native habitat. The two parasites were reared and released in California citrus groves with amazing success. By the end of 1889, the scale was no longer a problem.

Biological control, like chemical control, can backfire. The success of the cactus-feeding moth *Cactoblastus* in controlling *Opuntia* in Australia

knowledge and tools that the other used to meet its goals. During the mid 1930s, for example, the German chemist I.G. Farben developed a new class of insecticides called organophosphates. Although extremely effective, several of these compounds were lethal to humans as well as to insects. Undesirable in an insecticide, this attribute was valuable as a potential chemical weapon. One of these compounds, later named tabun, was the first organophosphate nerve gas. So in creating a new class of insecticides, Farben had also created a new class of chemical weapons.

PENICK INSECTICIDAL BASES...
Super Ammunition for the Continued Battle of the Home Front
FIGURE A

At home, DDT spraying inspired military and civilian researchers to link chemical warfare and pest control in new ways. By the mid-20th century, the American chemical industry grew in size, expertise, profitability, and status. American scientists developed new chemicals to kill insect pests. Farmers applied more kinds of insecticides in greater quantities than ever before. Using surplus military planes piloted by veterans, crop dusting delivered DDT to farm fields.

Research on the application of classes of chemical compounds once used for war as insecticides moved the battlefield to the farm, and the enemy was the insect pest—redefining chemical warfare as pest control. Chemical companies capitalized on the expertise that had been developed and the capital that had been put in place as part of the war effort. At first the rhetoric of World War II pervaded this transfer of military technology to civilian pest control (Figure A), taking advantage of the role of insecticides (specifically DDT) as a "war hero."

As the 20th century progressed, our perception of chemical warfare, both on the battlefield and farm, changed. In 1972, the U.S. government banned the domestic use of DDT, although production and distribution to other countries continues. International discussions to ban the use of chemical weapons began that same year. It would take another 25 years, however, before the banning of development and use of chemical weapons by the Chemical Weapons Convention would take effect.

encouraged its introduction to several West Indies islands to control *Opuntia* there. In time the moth made its way to Florida, where it now threatens the existence of several native *Opuntia* species. The European lady beetle *(Coccinella septempunctata)* was introduced into the United States to control the Russian wheat aphid. The species was so successful that it has replaced many native American ladybird beetles; it has become the ladybug with which most of us are familiar. The moral is that, although using alien predators as biological controls can be effective, these species possess their own inherent dangers that must be assessed before they are released, so as not to create new pests that compound the problem.

Numerous wild plants and animals have evolved their own defenses against natural enemies (see Section 16.7). One approach to pest control, then, is to breed for genetic resistance. This tool has been used successfully in crop plants, such as corn, wheat, and rice. The process involves crossing cultivars with wild relatives to capture the gene into the gene pool of the cultivated plant. The latest technique to increase plant resistance to weeds and insect pests is genetic engineering, particularly the use of recombinant DNA material to introduce the gene for the desired trait into the plant (see the Ecological Application after Ch. 2: Taking the Uncertainty out of Sex). The transfer of a single gene may confer resistance to viruses or to herbicides, or encode for production of endotoxins, poisons that inhibit the feeding activity of insect enemies.

Another technique is splicing genes of *Bacillus thuringiensis* (Bt) into crop plants, notably corn, cotton, and potatoes. This splicing creates transgenic varieties that produce stomach poisons fatal to

ENDANGERED SPECIES—ENDANGERED LEGISLATION

Endangered species are the subject of a hot political controversy. The root of the controversy is the Endangered Species Act of 1973. Administered by the National Biological Service (formerly the Fish and Wildlife Service), the act covers vertebrate animals, invertebrates, and plants. It has a number of provisions. First, it requires listing endangered and threatened species. The list now includes over 1100 species. An endangered species is one that has so few individual survivors that the species could become extinct over its natural range. There are, unfortunately, numerous examples worldwide. In the United States the whooping crane, the black-footed ferret, and the California condor are among them. Threatened species, even more numerous, are still relatively abundant over their natural range, but are declining in numbers and are likely to become endangered. Among them are species of Pacific salmon, many grassland birds, neotropical warblers, and many species of cacti.

The act also mandates the designation of critical habitat to be protected and recovery programs to increase the abundance and distribution of endangered species. Some programs have been very successful, including the restoration of the bald eagle.

The act makes it illegal for the federal government to fund any projects that would harm listed species, including habitat destruction. It makes it illegal for U.S. citizens to harm, capture, or market animals that appear on the federal endangered and threatened species list. It also regulates the importation of endangered plants and animals from foreign lands and provides for participation in international agreements regulating commerce in wild animals.

Up for reauthorization since 1992, the Endangered Species Act has vocal and powerful opponents. Because the act protects the individual organism as well as its habitat, land developers, loggers, and the mining and fishing industries see the act as an obstacle to their interests. They are exerting intense pressure to allow the use of economic factors to override the Endangered Species Act. With the Endangered Species Act out of the way, they could cut the remnant old-growth forests, drain wetlands, mine in parks, and ignore precautions to save threatened marine life, regardless of the consequences to a species. Succumbing to such pressures, congressional politicians withhold or reduce funds to slow down the listing of species and cut back protection of habitats. In spite of protests about the act, it has affected only one-tenth of 1 percent of all of the projects involving potential threats to listed species. The opponents of the act give little heed to the great contributions wild plants and animals make to the economy through natural products, recreation, and ecotourism, or to their financial value as a reservoir of genetic resources and a source of medicinal ingredients.

Unfortunately, the Endangered Species Act spotlights species rather than communities or ecosystems. The real issue is not protecting individual species such as the spotted owl, marbled murrelet, and red-cockaded woodpecker, but protecting the ecosystems of which they, associated species, and humans are a part.

---

crop-damaging pests. The use of these bioengineered seeds eliminates the need for spraying the fields with pesticides. So effective is this method of pest control that it kills all of the particular pest species involved, such as corn borer or cotton boll weevil, but with a potentially major drawback. Because the insect pests are continually exposed to the toxin, the few that survive can evolve a resistance to Bt, just as to conventional pesticides.

Genetic engineering for pest control has other inherent dangers. Plants with engineered traits could evolve new genotypes with somewhat different life histories or physiological traits. Engineered crop plants might transfer genes over relatively long distances by hybridizing with related plants that differ in their life history characteristics. Many crops such as celery, asparagus, and carrot have weedy relatives with high reproductive output and efficient seed dispersal. If these weedy relatives acquire the engineered gene, then the gene could be spread through the range of plants. This transfer would create weeds against which current herbicides would be ineffective, and insect-resistant plants could speed the co-evolution of even more insect-resistant pests.

18.9 Restoration Efforts Attempt to Undo Damage to Communities

In recent years, considerable efforts are underway to restore natural communities impacted by human activities (see Focus on Ecology 18.2: Endangered Species—Endangered Legislation), stimulating a new approach to human intervention that is termed restoration ecology. The goal of restoration ecology is to return an ecosystem to a close approximation of its conditions prior to disturbance through the application of ecological principles. It involves a continuum of approaches from the reintroduction of species and the restoration of habitats, to attempts to reestablish whole communities as functioning ecosystems.

The least intensive restoration effort involves the rejuvenation of existing communities by eliminating invasive species, replanting native species, and reintroducing natural disturbances such as short-term periodic fires in grasslands and low-intensity ground fires in pine forests. Lake restoration involves the reduction of inputs of nutrients, especially phosphorus, from the surrounding land that stimulate growth of algae, restoration of aquatic plants, and the reintroduction of fish species native to the lake. Wetland restoration may involve restoration of hydrological conditions, so that the wetland is flooded at the appropriate time of year, and replanting of aquatic plants.

More intensive restoration involves recreation of the community from scratch, best accomplished on relatively small areas. This kind of restoration involves preparation of the site, introduction of an array of appropriate native species over time, and employment of appropriate management to maintain the community, especially against the invasion of nonnative species from adjacent surrounding areas. A classic example is the re-creation of a prairie ecosystem on a 60-acre field near Madison, Wisconsin. The previous prairie had been plowed, grazed, and overgrown. The process of restoration involved destroying occupying weeds and brush, reseeding and replanting native prairie species, and burning the site once every 2 to 3 years to approximate a natural fire regime. After nearly 60 years, the plant community now resembles the original native prairie.

Much attention is being focused on the restoration of wetlands that have been lost due to the diversion of water, or drained for land development. New techniques are also being developed for the construc-

FIGURE 18.12 Black rhino captured at Umfolozi Reserve in Natal, South Africa, being released into a newly established reserve at Pilansburg, South Africa.

tion of wetlands where they did not exist before, often for the purpose of wastewater and stormwater treatment.

Other restoration efforts involve the reintroduction of species lost from the community through overexploitation or deliberate extermination, such as the reintroduction of the wolf to the Yellowstone ecosystem. Comeback of such species as the white-tailed deer, pronghorn antelope (*Antilocapra americana*), and wild turkey (*Meleagris gallopavo*) was achieved by a series of actions that allowed low populations to expand. Strict protection from hunting, followed by highly regulated hunting seasons once the populations were near recovery, were combined with the establishment of refuges and reserves by states and the federal government to protect both animals and habitat. Often, wild individuals taken from pockets of abundance were introduced to areas of scarcity and empty habitats to reestablish populations.

Restoration efforts often involve transplantation of individuals from one location to another (Figure 18.12). Restoration of some species such as whooping crane (*Grus americana*), masked bobwhite (*Colinus virginiana*), Hawaiian goose or nene (*Branta sandwicensis*), peregrine falcon (*Falco peregrinus*), and California condor (*Gymnogyps californianus*) and mammals such as the wolf and European wisent (*Bison bonasus*) have relied on the introduction of individuals from captive-bred populations. Introduction of captive-bred individuals to the wild requires prerelease and postrelease conditioning, including the acquisition of food, shelter finding, interaction with con-specifics, and fear and avoidance of humans. Despite these inherent difficulties, reintroduction programs have halted and in some cases reversed the downward spiral toward extinction.

Summary

Sustained Yield (18.1–18.4) Humans exploit natural populations for the essential resources of food, clothing and shelter, particularly in the practices of fisheries and forestry. First attempts at managing exploited populations involved fisheries (18.1). If plant and animal populations exploited for human use are to be continuously productive, the yield or amount harvested per unit time should not exceed the rate of production, the amount produced per unit time. Harvesting at a level that will ensure a similar yield over and over without forcing the population into decline is called sustained yield. The level of yield above which a population declines is maximum sustainable yield (18.2). Management of species for sustainable yield is dependent on specific life history characteristics. Sustained yield models fail to consider all aspects of population dynamics, including natural mortality and environmental uncertainty. Sustained yield management considers the resource as a single biological unit, not as part of an ecosystem. An overexploited commercial species shows warning signs, such as a sharp decrease in catch, dominance of young age classes, and failure to replace losses. Too often, economic considerations outweigh biological considerations, causing the commercial and even biological extinction of a species (18.3).

Sustained yield in forestry seeks to balance the volume of timber cut with the volume of timber growth. This requires waiting a sufficient time between harvests for the forest (tree population) to recover to preharvest levels. The period of time required for recovery, called the rotation time, is dependent on the growth rate of the species, site conditions, and the intended use of the trees that are being harvested. Economic considerations often override sustained yield. Logging of older-growth forests and management practices often change species composition and structure of forests. Plantations often replace natural regenerated forests (18.4).

Agriculture and Land Use (18.5) The vast majority of human food resources are derived from agriculture—the production of crops and the grazing of livestock. Only 15 plant and 8 animal species produce 90 percent of our food supply, with approximately 26 percent of the ice-free land area of the planet either under cultivation or used as pasture. Humans function as agents of dispersal for this small set of species, providing suitable habitat, maximizing their reproduction, and greatly expanding both their geographic distribution and abundance. At the same time, the clearing of land for agriculture has dramatically reduced the distribution and abundance of thousands of other species, in many cases leading to their extinction. Agriculture and forestry practices have also resulted in soil erosion and inputs of nutrients to adjacent aquatic ecosystems from the use of chemical fertilizers, both of which have a major impact on the biotic communities of streams and rivers. Human activity and alteration of the landscape has increased the availability of habitat for some species, increasing their populations.

Invasive Species (18.6) Humans have both intentionally and unintentionally introduced alien (nonnative) species to communities. These invasive species have outcompeted and replaced native plants and animals. Invasive predatory species can significantly alter food chains and interactions among species.

Controlling Species (18.7–18.8) Pests and weeds are animals and plants growing anywhere humans do not want them. Pests are responsible for significant agricultural losses and can function as vectors of disease. The aim of control is to hold the population below the level at which it causes economic injury (18.7).

The use of organic chemical pesticides and herbicides prompts the evolution of resistant strains and causes toxic pollution. It kills natural enemies, allowing a pest population to increase dramatically. Biological controls—natural enemies and parasites—are effective in keeping some pests below the economic injury level. Breeding for genetic resistance to pests and weeds is another important and effective tool. Transplanted genes produce plants that are resistant to viral diseases and insect attack. If engineered genes transfer to wild relatives, we risk creating highly resistant strains of weeds and pests (18.8).

Restoration (18.9) Restoration efforts attempt to restore degraded ecosystems by eliminating invasive species, planting, restoring cycles of natural disturbance, and reintroducing animal species lost from the ecosystem. Restoration of declining wildlife species in North America saved from extinction many of our most common species, such as wild turkey and deer. Current restoration efforts are difficult because of conflicting demands for land.

Study Questions

1. What relationship between yield and production is necessary for sustained yield? What is maximum sustained yield?

2. Report on the effects of modern commercial ocean fishing, especially with huge drift nets, on the fishery resource and on other ocean life.

3. What is rotation time in the management and harvest of forest resources? What factors influence the rotation time?

4. How do economic interests interfere with sustained yield practices?

5. Comment on the statement that a stand of trees is not necessarily a forest.

6. In what ways does agriculture influence patterns of species distribution and abundance in ecological communities?

7. Define the terms *pest* and *weed*.

8. Under what conditions do plants and animals become weeds and pests?

9. What are the major approaches to controlling pests and weeds? What are the ecological advantages and disadvantages of each?

10. What approaches must we take to conserve and restore threatened species? Why is habitat a key element? Why should we emphasize habitat and not individual species?

Landscape showing the contrast between highly managed agricultural fields in the foreground and native forest communities occupying the hills in the background.

Landscape Ecology

OBJECTIVES

On completion of this chapter, you should be able to:

- Identify the focus of study in landscape ecology.
- Distinguish between edge and ecotone and discuss the edge effect.
- Discuss how patch size influences species diversity.
- Relate the theories of island biogeography and metapopulations to habitat fragmentation.
- Discuss the role of corridors as mechanisms for dispersal.
- Define disturbance.
- Distinguish among the scale, frequency, and intensity of disturbances.

In Chapter 13 we defined the community as a group of plant and animal species that occupies a given area. Although, by definition, ecological communities have a spatial boundary (see Chapter 13, Section 13.8), they also have a spatial context within the larger landscape. Consider the view of the Virginia countryside in Figure 19.1. This particular landscape spreads over gradually ascending terrain where a flat coastal plain meets the slopes of the Blue Ridge Mountains to the west. It is a patchwork quilt of forest, pine plantation, fields and hedgerows, golf course greens, water, and human habitations. This patchwork of different types of land cover is called a **mosaic,** using the analogy of mosaic art in which an artist combines many small pieces of various colored material to create a larger pattern or image (Figure 19.2). The artist creates the emerging pattern, defining boundaries by using different shapes and colors of materials to construct the mosaic. In a similar fashion, the landscape mosaic is a product of the boundaries defined by changes in the physical and biological structure of the communities that form its elements. In the artist's mosaic, the elements only interact visually to present the emerging image. In the landscape mosaic, the elements (ecological communities) interact in a wide variety of ways depending on their size and spatial arrangement. The study of the causes behind the formation of these patches and the ecological consequences arising from their spatial patterns on the landscape is called **landscape ecology.**

19.1 Environmental Processes Create a Variety of Patches in the Landscape

The visually distinct communities that make up the landscape mosaic are called **patches.** A patch is a relatively homogeneous area—such as a farmer's field, a pond, or a lawn—that differs from its surroundings, both in its structure and in the composition of species inhabiting it. Patches vary in size, shape, and type and are embedded within a complex **matrix** of surrounding areas. These landscape patches are produced by a variety of environmental factors, including regional variations in geology, topography, soils, and climate. Upon this natural stage, human activity makes its mark on the broad-scale distribution of communities. For example, humans may both diversify and diminish the landscape pattern whenever they transform one community into another, such as clearing a forest to create a housing development.

Many landscape patterns that we observe today in the United States reflect early land survey methods that divided land into sections, half sections, and finally quarter sections of 160 acres. Historically, these land surveys were set on straight lines east to west (consider the Mason-Dixon line and many state boundaries), with no attempt to follow topography

FIGURE 19.1 A view of a Virginia landscape showing a mosaic of patches consisting of different types of land cover: natural forest, plantations, fields, water, and rural development.

FIGURE 19.2 Mosaic in Khirbat al Mafjar, Israel, dating from 710–750 AD. The boundaries defining the different objects in the scene (lion, tree, antelope, etc.) are created using small pieces of stone differing in shape and color.

or natural boundaries. The straight-line survey resulted in square corners where woods and fields, croplands and developments, and other landscape elements met (see Figure 19.1). This checkerboard pattern, often overlooked by the casual observer, has had a lasting impact on the landscape.

Human activity has been the dominant force determining the size and shape of landscape patches. However, natural variations in geology and soil condition as well as natural events, such as fire and grazing by herbivores, also influence the formation of patches. As a result, patches vary considerably in size and shape. They may be square, elongate, round, or convoluted, covering tens of square kilometers or only a few meters. The area, shape, and orientation of landscape patches all determine their suitability as habitats for plants and animals and influence many ecological processes, such as wind flow, dispersal of seeds, and the movement of animals. The following section looks at the special zones formed where patches intersect. These areas are often inhabited by species common to each adjacent community, resulting in some of the most unique and highly diverse communities in the landscape.

19.2 Transition Zones Offer Diverse Conditions and Habitats

Among the landscape's most conspicuous features are the edges, which mark the perimeter of each patch. Many of these edges indicate an abrupt change in physical conditions—topography, sub-strate, soil type, or microclimate—between communities. An example is a patch of wetland associated with a depression in which water accumulates. Where long-term natural features underlie adjoining vegetation, edges are usually stable and permanent, and we call them **inherent edges**. Other edges, however, may result from natural disturbances, such as fire, storms, and floods, or from human-related activities, such as timber harvesting and livestock grazing. Such temporary edges, which require continuous maintenance to keep them from reverting to their natural state, are called **induced edges**.

The place where the edge of one patch meets the edge of another is called a **border** (Figure 19.3). The two edges and the border combined make up the **boundary**. Some boundaries between landscape patches are abrupt, with a sharp visual contrast between the vegetation of adjoining patches, such as between a forest and an adjacent agricultural field. Other boundaries are less visually distinct. For example, in the area where two forest communities meet (see Chapter 13, Figure 13.4), vegetation growing in one patch blends with vegetation from the other to form a transition zone called an **ecotone**.

Certain plant and animal species colonize edge and ecotone environments. Plant species found in such areas have generally adapted to the often-unique physical conditions imposed by transition

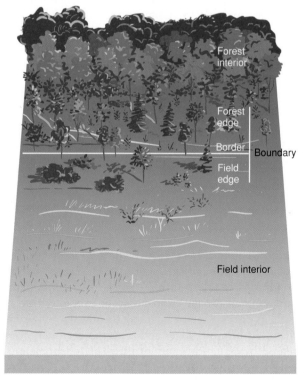

FIGURE 19.3 Spatial relationships of border, boundary, edge, and interior on a landscape. (Based on Forman 1995.)

(a)

(b)

FIGURE 19.4 (a) The ruffed grouse requires a patchy environment of forest cover with shrubs, sapling trees, and mature trees, as well as openings in the forest. (b) Photo of the indigo bunting *(Passerina cyanea),* which inhabits woodland edges, hedgerows, roadside thickets, and large gaps in forests that create edge conditions. The male requires tall, open song perches and the female a dense thicket in which to build a nest.

zones. For example, shrub and tree species that have colonized the boundary between adjacent forest and field communities differ from their counterparts growing in the understory of the adjacent forest. Edge species tend to be more shade-intolerant (sun-adapted) and better able to tolerate dry conditions caused by higher air temperatures and rates of evapotranspiration. Animal species inhabiting edge environments usually require two or more plant communities within their home range, or territory. For example, the ruffed grouse *(Bonasa umbellatus)* requires forest openings with abundant low shrubs and herbaceous plants to feed its young, a dense stand of sapling trees for nesting cover, and mature forests for winter food and cover (Figure 19.4a). Because a ruffed grouse's territory measures between 4 to 8 hectares, the whole spectrum of plant habitats must be contained within that area. Other species, such as the indigo bunting *(Passerina cyanea),* that are restricted exclusively to the edge environment, are referred to as **edge species** (Figure 19.4b).

Because ecotones blend elements from both adjacent patches, their structure and composition is often very different from either patch, and thus they offer unique habitats with relatively easy access to the adjacent communities (Figure 19.5). These diverse conditions enable transition zones to support plant and animal species from adjacent communities as well as those species adapted to an edge environment. As a result, edges and ecotones are often populated by a rich diversity of life. This phenomenon, called the **edge effect,** is influenced by the area of edge available (length and width) and by the degree of contrast between adjoining plant communities. In general, the greater the contrast between adjoining patches of vegetation, the greater the diversity of species. Therefore, an edge between forest and grassland should support a greater variety of birds and animals than an edge between a young and a mature forest.

Although edge effect may increase species diversity, it can also create problems. Abrupt edges are particularly attractive to predators. Raccoons, opossums, and foxes use edges as travel lanes. Avian predators such as crows and jays seek out and rob the nests of smaller birds inhabiting more open edge habitat. Edges further alter interaction between species by either restricting or facilitating dispersal across the landscape. For example, dense or thorny shrubs growing in the boundary of forest and field can block animals from passing through.

19.3 Patch Size and Shape Are Critical to Species Diversity

The mosaic of patches that define the landscape is constantly changing. Conspicuous among these changes is the ongoing fragmentation of large tracts of land by human development (Figure 19.6). Where humans continue localized clearing, fragmentation eventually results in a mosaic of smaller, often-isolated patches of forest, grassland, or shrubland. If fragmentation reduces patches below a critical size, the

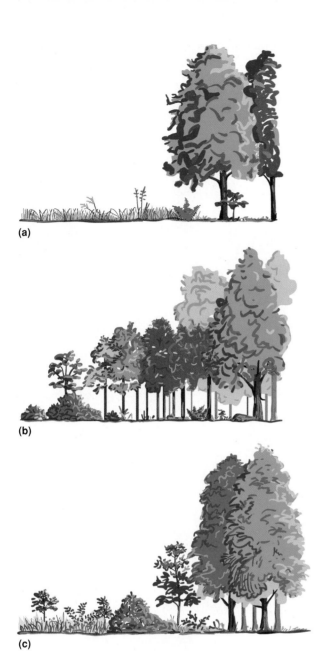

(a)

(b)

(c)

FIGURE 19.5 Contrast in edge is important in increasing species richness. A high-contrast edge (a) is more valuable to edge species than a low-contrast edge (b) because two quite different vegetation types adjoin. Low-contrast edges do not provide enough difference between vegetation communities to be of maximum value to edge species. Of greatest value is an advancing edge (c), such as woody vegetation invading an adjoining field. An advancing edge not only provides variation in height, but also, in effect, creates two edges on the site.

remaining area may not be able to support many of its original species, and local extinction occurs. As the landscape becomes more fragmented and contiguous areas of habitat for species become smaller and more isolated, the relationship between species diversity and patch size (or area) takes on a critical importance in the conservation of biological diver-sity. What size of habitat patch maintains greatest species diversity? At what size of patch do area-sensitive species disappear? Such questions have stimulated studies of how plant and animal species respond to habitat fragmentation.

Patch size has a critical influence on community structure. As a general rule, large patches of habitat contain a greater number of both individuals (population density) and species (species richness) than do small patches. Perhaps the most obvious reason that large areas support more individuals is that they typically have a greater carrying capacity for a given species. In other words, they can support a greater number of home ranges (or territories) than smaller patches can (see Chapter 10). Within the animal community, there is a general relationship between body size (weight) and the size of an animal's home range (see Figure 11.10). In addition, for a given body size, the home range of carnivores is greater than that of herbivores. Therefore, the distribution of larger species, particularly large predators such as grizzly bears, wolves, and mountain lions, will be limited to larger, contiguous patches of habitat.

The relationship between patch size and species diversity is more complex. Larger patches are more likely to contain variations in topography and soils that give rise to a greater diversity of plant life (both taxonomic and structural), which in turn will create a wide array of habitats for animal species (see Chapter 17, Section 17.8). The size and shape of patches also affect the relative abundance of edge (or perimeter) and interior environments (Figure 19.7a). Only when a patch becomes large enough that it is deeper than its edge is wide can it develop interior conditions (Figure 19.7b). For example, at one extreme, a very small patch is all edge habitat, but as it increases in size, the ratio of edge to interior decreases. Altering the shape of the patch can change this edge-to-interior relationship. For example, the long, thin habitat in Figure 19.9c is all edge. Such long, narrow patches of woodland, whose depth does not exceed the width of the edge, are edge communities regardless of the total patch area.

In contrast to edge species, other species, termed **interior species**, require environmental conditions characteristic of interior habitats, away from the abrupt changes associated with edge environments. Figure 19.8 shows the relationship between forest area (patch size) and the probability of occurrence for six different bird species of eastern North America. Edge species, such as the gray catbird (*Dumetella carolinensis*) and American robin (*Turdus migratorius*), have a high probability of occurring in small forest patches dominated by edge environments. As patch size increases, the probability of finding them

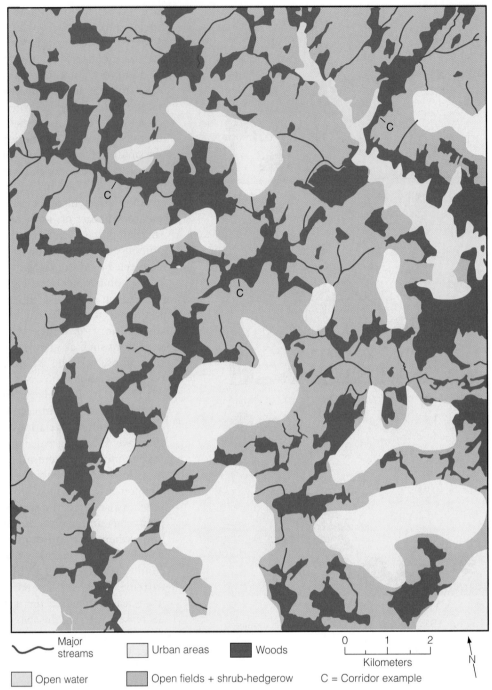

Major streams

Urban areas

Woods

Open water

Open fields + shrub-hedgerow

0   1   2
Kilometers

N

C = Corridor example

FIGURE 19.6 At the time of European settlement, this section of eastern Maryland was completely forested. Over several hundred years, land was cleared for agriculture, fragmenting the remaining forest land into isolated patches. In the past quarter century, urban and suburban developments have broken up forests and farms even more. So far a few corridors remain, connecting various forest islands. (Adapted from J. Whitcomb et al., "Effects of the Forest Fragmentation on Avifauna of the Eastern Deciduous Forest," in R. L. Burger and D. M. Sharpe, eds., *Forest Island Dynamics in Man-Dominated Landscapes,* p. 127. New York: Springer-Verlag, 1981. Copyright © 1981 Springer-Verlag. Used by permission.)

decreases. In contrast, the ovenbird *(Seiurus auro-capillus)* is a species adapted to the interior of older forest stands, which are characterized by large trees and sparse shrub cover in the understory layer. Accordingly, the probability is low that it will occur in small patches. Intermediate to these two groups are

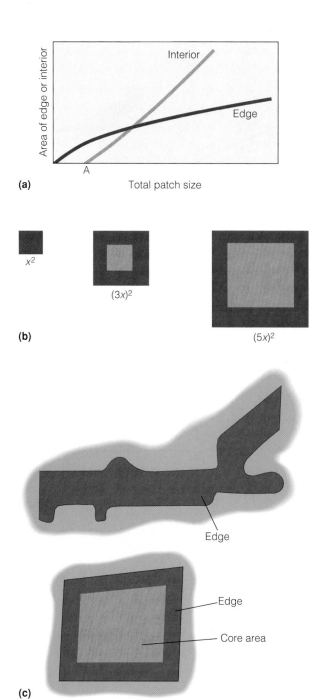

(a)

Total patch size

(b)

x^2

$(3x)^2$

$(5x)^2$

Edge

Edge

Core area

(c)

FIGURE 19.7 Relationship of habitat patch size to edge and interior conditions. All habitat patches are surrounded by edge. (a) The general relationship between patch size and area of edge and interior. Below point A, the habitat is all edge. As size increases, interior area increases and the ratio of edge to interior decreases. (b) Assuming that the depth of the edge remains constant, the ratio of edge to interior decreases as the habitat size increases. When the patch is large enough to maintain shaded, moist conditions, an interior begins to develop. (c) This relationship of size to edge holds for a square or circular habitat patch. Long, narrow woodland islands whose width does not exceed the depth of the edge would be edge communities, even though the area may be the same as that of square or circular ones. (After Temple 1986.)

area-insensitive species, such as the Carolina chickadee *(Poecile carolinensis),* that are at home in habitat patches of various sizes.

The minimum patch area needed to maintain interior species differs between plants and animals. For plants, patch size per se is less important to the persistence of a species than are environmental conditions. For many shade-tolerant plant species found in the forest interior, the minimum size is the area needed to sustain moisture and light conditions typical of the interior. That area depends in part on the ratio of edge to interior and on the nature of the surrounding edge habitat. If it is too small or too open, sunlight and wind will penetrate and dry the interior environment, eliminating herbaceous and woody species that require moister soil conditions. For example, in the northeastern United States, forest fragmentation can result in the decline of mesic (moisture-requiring) species such as sugar maple and beech, while encouraging the growth of more xeric species such as oaks.

Several studies that examine bird species in forest patches reveal a pattern of increasing species diversity with an increase in patch size, but only up to a point. R. F. Whitcomb and colleagues studied patterns of species diversity in forest patches in western New Jersey. Their findings suggest that maximum bird diversity is achieved with woodlands 24 hectares in size. Similar patterns were observed in studies investigating the species composition of bird communities in large and small forest patches (from 3 to 7620 hectares) within agricultural regions in Illinois and Ontario, Canada. With patches of intermediate size, a general pattern of maximum species diversity results from the negative correlation between edge species and the size of habitat patches combined with the positive correlation between interior species and increased area (see Figure 19.8).

These studies suggest that two or more small forest patches will support more species than an equivalent area of contiguous forest. However, smaller woodlands did not support true forest interior species, such as the pileated woodpecker or ovenbird (see Figure 19.8), which require extensive wooded areas. Therefore, estimates of species diversity do not present a complete picture of how forest fragmentation is impacting biological diversity in the landscape. The practice of establishing intermediate size patches may achieve maximum diversity, but it will also exclude species that require larger areas to survive and reproduce. Large forest tracts with a high degree of heterogeneity are required to support a range of species characteristic of both edge and interior habitats.

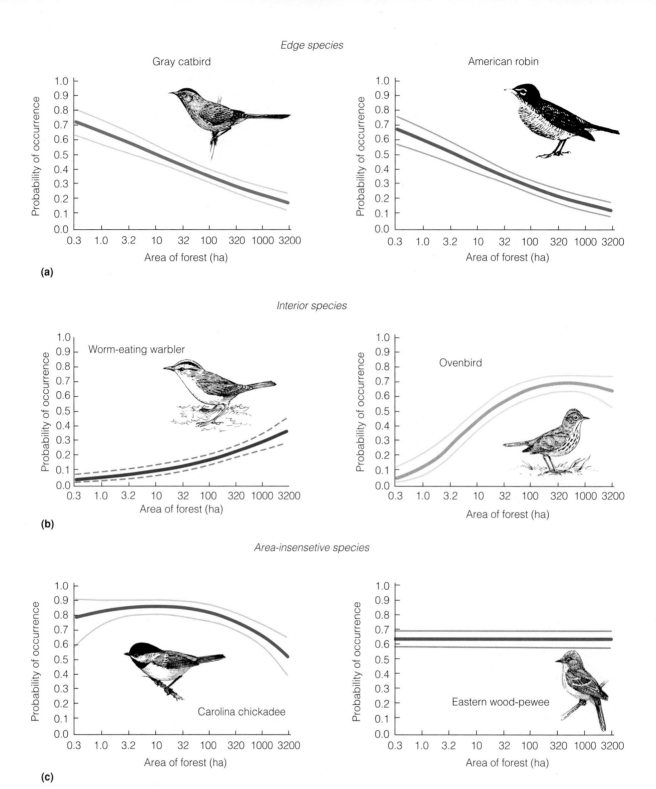

FIGURE 19.8 Difference in habitat responses between edge species, area-sensitive or interior species, and area-insensitive species. The graphs indicate the probability of detecting these species from a random point in patches of various sizes. The dashed lines indicate the 95 percent confidence intervals for the predicted probabilities. (a) The catbird and the robin are familiar edge species. As patch size increases, the probability of finding them decreases. (b) The worm-eating warbler and the ovenbird are ground-nesting birds of the forest interior. The probability of finding them in small patches is low. (c) The chickadee and wood-pewee are area-insensitive, at home in both small to large forest patches. (After Robbins et al. 1989)

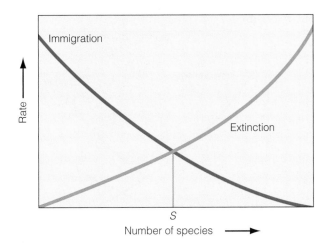

FIGURE 19.9 According to the theory of island biogeography, immigration rate declines with increasing species richness, while extinction rate increases. The balance between rates of extinction and immigration (immigration rate = extinction rate) defines the equilibrium number of species *(S)* on the island. (After MacArthur and Wilson 1967.)

19.4 The Theory of Island Biogeography Applies to Landscape Patches

The influence of area (patch size) on species richness did not escape the notice of early naturalist-explorers and biogeographers, who noted that large islands hold more species than small islands. Johann Reinhold Forster, a naturalist on Captain Cook's second voyage to the Southern Hemisphere (1772–1775), noted that the number of different species found on islands depended on the size of the island. The zoogeographer P. Darlington offered a rule of thumb: On islands, a tenfold increase in area leads to a doubling of the number of species.

The various patches, large and small, that form the vegetation patterns across the landscape suggest islands of different sizes. Some are near to each other; others are remote and isolated. A patch of forest, for example, may sit within a sea of cropland or housing developments, isolated from other forest patches on the landscape. The sizes of these patches and their distances from each other on the landscape have a pronounced influence on the nature and diversity of life they hold.

The similarity between ocean-bound islands and isolated patches of habitat led ecologists to apply the theory of **island biogeography** to the study of terrestrial landscapes. First developed in 1963 by Robert MacArthur (formerly of Princeton University) and Edward O. Wilson (Harvard University), the theory is quite simple: The number of species established on an island represents a dynamic equilibrium between the immigration of new colonizing species and the extinction of previously established ones (Figure 19.9).

Consider an uninhabited island off the mainland. The species on the mainland make up the species pool of possible colonists. Those species with the greatest ability to disperse from the mainland will be the first to colonize the island. As the number of species on the island increases, the rate of immigration of new species to the island will decline. This decline results because the more mainland species that successfully colonize the island, the fewer potentially new species for colonization that remain on the mainland (the source of immigrating species). When all the mainland species exist on the island, the rate of immigration will be zero.

If we assume that extinctions occur at random, the rate of species extinctions on the island will increase with the number of species occupying the island based purely on chance. Other factors, however, will amplify this effect. Later immigrants may be unable to establish populations because earlier arrivals have already utilized available habitats and resources. As the number of species on the island increases, competition among the species will most likely increase, causing the rate of extinction to progressively increase.

An equilibrium species richness *(S)* is achieved when the rate of immigration equals the rate of extinction (see Figure 19.9). If the number of species inhabiting the island exceeds this value, the rate of extinction is greater than the rate of immigration, resulting in a decline in species richness. If the number of species is below this value, the rate of immigration is greater than extinction, and the number of species increases. At equilibrium, the number of species residing on the island remains stable, although the composition of species may change. The rate at which one species is lost and a replacement species gained is the **turnover rate**.

The distance of the island from the mainland and the size of the island both affect the equilibrium species richness (Figure 19.10). The greater the distance between the island and the mainland, the less the likelihood that many immigrants will successfully complete the journey. The result is a decrease in the equilibrium number of species (Figure 19.10a).

On larger islands, extinction rates, which vary with area, are lower because a greater area generally contains a wider array of resources and habitats. For this reason, large islands can both support more individuals of each species and meet the needs of a

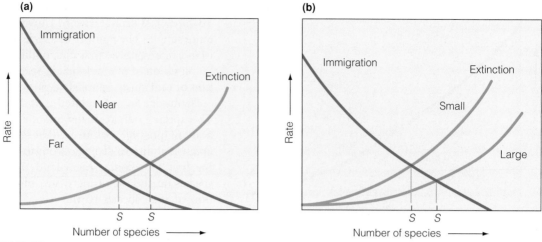

FIGURE 19.10 (a) Immigration rates are distance-related. Islands near a mainland have a higher immigration rate and associated equilibrium species richness *(S)* than do islands distant from a mainland. (b) Extinction rates relate to area, being higher on small islands than on large ones. The equilibrium number of species varies according to island size, with larger islands having a greater equilibrium species richness than smaller islands. (After MacArthur and Wilson 1967.)

wider variety of species. The lower rate of extinction on larger islands results in a higher equilibrium number of species as compared to smaller islands (Figure 19.10b).

Although the theory of island biogeography was initially applied to oceanic islands, there are many other types of "islands." Mountaintops, bogs, ponds, dunes, and areas fragmented by human land use as well as hosts and their parasites are all essentially island habitats. As Daniel Simberloff of the University of Tennessee, one of the first ecologists to experimentally test the predictions of this theory, put it: "Any patch of habitat isolated from similar habitat by different, relatively inhospitable terrain traversed only with difficulty by organisms of the habitat patch may be considered an island."

Exactly how does island theory apply to landscape patches? Landscape patches or islands differ considerably from the oceanic islands for which the theory was developed. Oceanic islands are terrestrial environments surrounded by an aquatic barrier to dispersal. They are inhabited by organisms of various species that arrived there by chance over a long period of time or that represent remnant populations that resided in the area long before the island was formed. By contrast, organisms inhabiting landscape patches are samples of populations that extend over a much wider area. These patch communities contain fewer species than are found in the larger area, and only a few individuals may represent each species. However, unlike isolated oceanic islands, terrestrial patches remain associated with other terrestrial environments, which often present fewer barriers to movement and dispersal among patches.

19.5 In Fragmented Landscapes, Corridors Permit Movement Between Isolated Patches

The fragmentation of large contiguous habitats, be they forest, prairie, or marshland, results in a mosaic of distinct patches—islands between which the movement of species is limited or even made impossible by the inhospitable nature of the intervening environment. In some situations, **corridors,** or strips of similar habitat, may connect separate patches and thereby allow greater movement between them. The habitat of a corridor is similar to that of the patches it connects, but different from the surrounding landscape in which it is set. Usually, corridors are of human origin. Some may be narrow-line corridors, such as hedgerows and lines of trees planted as windbreaks, bridges over fast-flowing streams, highway median strips, and drainage ditches (Figure 19.11a). Wider bands of vegetation called strip corridors consist of both interior and edge environments. Examples of such corridors include broad strips of woodlands left between housing developments, overgrown rights-of-way beneath power lines, and belts of vegetation along the banks of streams and rivers (Figure 19.11b).

Corridors have two functional roles in the landscape. They provide a habitat for a variety of plants and animals, and they permit travel between patches.

(a)

(b)

FIGURE 19.11 Examples of corridors: (a) hedgerow and (b) riverine vegetation.

In Europe, a long history of hedgerows in the rural landscape has encouraged the development of typical hedgerow animal and plant communities (see Figure 19.11a). In suburban and urban settings, corridors serve as stopover habitat for migrating birds and provide protective cover for prey species. Sometimes corridors have a negative impact on certain populations. For example, they offer scouting positions for predators that need to remain concealed while hunting in adjacent patches.

When corridors interconnect to form networks, they offer dispersal routes for species traveling between habitat patches. They probably function best as travel lanes for individuals moving within the bounds of their home range. By facilitating movement among different patches, corridors can encourage gene flow between subpopulations and help reestablish species in habitats that have experienced local extinctions. Corridors also act as filters by providing dispersal routes for some species but not for others. Different-sized gaps in corridors allow certain organisms to cross and restrict others—the **filter effect**. Corridors can also create avenues for spreading disease between patches.

Roads—corridors designed as dispersal routes for humans—dissect the landscape and have various effects on adjacent patches of land. Two- to four-lane high-speed roads are a major source of mortality for wildlife ranging from large mammals to tiny insects, and they effectively divide populations of many species. All types of roads in some way affect roadside vegetation. Salt spread on highways during snow removal, particulate matter from tires, and chemical pollutants from truck and automobile exhaust all affect roadside vegetation. During storms and snowmelt, water runoff carries these pollutants and debris into adjacent patches. In addition, noise from passing traffic discourages wildlife from occupying otherwise suitable habitat. Most important, perhaps, road corridors allow people access to remote areas, with often-disastrous ecological effects, as exemplified by logging roads cut through tropical rain forest. Where roads invade, people and development follow.

19.6 Metapopulations Decrease Vulnerability to Local Extinctions

The distribution of a species over the landscape is limited to areas that provide a suitable habitat (see Chapter 12). In many landscapes, the habitat is so fragmented that the species exists as distinct, partially isolated subpopulations, each possessing its own population dynamics. Each subpopulation has its own birthrate, death rate, and probability of extinction. Linking these subpopulations are individuals moving between patches. Such separated populations interconnected by the movement of individuals are called **metapopulations** (see Chapter 9, Section 9.5).

As predicted by island biogeography theory (see Section 19.4), large patches of grassland and forest support larger populations of individuals and greater diversity of species than do small patches. They also provide a source of individuals for repopulating small patches. How rapidly colonization occurs depends on the size of the patch from which the colonizers migrate, the distances between patches, and the ability of a particular species to disperse (Figure 19.12). The presence of corridors greatly aids this process of dispersal and recolonization.

As with species on small islands, populations in small patches are vulnerable to local extinction. Small population size can result in loss of genetic

FIGURE 19.12 Example of habitat patches, each with their own subpopulation, embedded within a landscape mosaic. Habitat for the Bay checkerspot butterfly *(Euphydryas editha bayensis)* metapopulation is fragmented due to both natural and human-induced factors. The shaded areas are patches of serpentine grasslands—the butterfly's habitat. The largest population occupies the Morgan Hill area. Only those patches closest to the Morgan Hill area indicated by arrows are usually occupied, suggesting that the butterfly is a poor disperser. Extinctions and recolonization of smaller patches is common. (From Harrison et al. 1988.)

diversity through inbreeding (see Chapter 12, Section 12.6) or random shifts in gene frequency called *genetic drift* (see Focus on Ecology 19.1: Genetic Drift and Effective Population Size). Both inbreeding and genetic drift increase the frequency of homozygotes (see Chapter 2, Section 2.3) in the population. Often the deleterious result is the expression of recessive genes that cause genetic disorders, decreased fertility, and even death.

The persistence of a metapopulation is the result of a complex dynamic among subpopulations. The rates of birth, death, immigration, and emigration of each subpopulation interact with the size and spatial arrangement of habitat patches to determine whether a given patch functions as a source or sink. In **source patches,** the local reproductive rate exceeds the local mortality rate, leading to a surplus of individuals that are able to colonize other patches of habitat. In **sink patches,** mortality rate exceeds reproduction rate so that the local population will go extinct if it is not regularly recolonized by new immigrants. Dependence on key source patches is a critical feature of such metapopulations. The identification of key source patches and the corridors that link with them is critical for the conservation of species.

19.7 The Impact of Disturbance Is Determined by Frequency, Intensity, and Scale

The distinct patterns of communities that we see in the landscape, as well as the plant and animal species that inhabit them, are heavily influenced by disturbances, both past and present. A **disturbance** is any relatively discrete event, such as a fire, windstorm, flood, drought, or epidemic, that disrupts community structure and function. Disturbances both create and are influenced by patterns on the landscape. For example, communities on ridge tops are more susceptible to damage from wind and ice storms, while bottomland communities along streams and rivers are more susceptible to flooding. In turn, these disturbances result in a new pattern of patches on the landscape.

Ecologists often distinguish between a particular disturbance event—a single storm or fire—and the **disturbance regime** or pattern of disturbance that characterizes a landscape over a long period of time. The disturbance regime has both spatial and temporal characteristics, which include intensity, scale, and frequency. **Intensity** is measured by the proportion of the total biomass or population of a species that the disturbance kills or eliminates. It is influenced by the magnitude of the physical force involved, such as the strength of the wind or the amount of energy released during a fire. **Scale** refers to the spatial extent of the impact of the disturbance relative to the size of the affected landscape. **Frequency** is the mean number of disturbances that occur within a particular time interval (Figure 19.13). The mean time between disturbances for a given area is the return interval. Often a disturbance's frequency is linked to its intensity and scale. For example, natural disturbances taking place on a small scale, such as the death of an individual tree in a tropical forest, occur quite frequently. By contrast, large-scale disturbances, such as fire, are rarer and occur on average only once every 50 to 200 years. In the intertidal zone of the ocean, abrasive action of waves tears away mussels and algae from rocks along the shore. In grasslands, badgers and groundhogs dig burrows and expose small patches of mineral soil. The outcome of such small-scale disturbances is the creation of a **gap,** an opening that becomes a site of localized regeneration and growth within the community. (Plant ecologist A. S. Watt first applied the term *gap* in 1947.) Within the gap,

GENETIC DRIFT AND EFFECTIVE POPULATION SIZE

The offspring produced in sexual reproduction represent only a subset of the parents' genes. If two parents have brown hair, for example, but both have the recessive gene for blond hair (heterozygous Bb, where B is the dominant gene for brown hair and b is the recessive gene for blond hair; see Chapter 2), there is a 25 percent chance that their child will be homozygous for brown hair (BB). Should they limit their family size to two children, and should both have the genes BB for hair color, the recessive gene will not be passed on to successive generations. Now suppose this same random process occurs throughout the population. In effect, there is a reduction in the genetic variation within the population—collectively known as the gene pool. The resulting random change in the frequency of alleles (alternative forms of a gene) within the population (gene pool) as a result of chance is called genetic drift. It occurs in all populations and represents one of the mechanisms of evolution. This loss of variation can result in a gene's being fixed—only one form or allele remaining (for example, the dominant gene for brown hair, B).

Although the process of genetic drift occurs in all populations, it occurs at a faster rate in small, isolated populations (or subpopulations on the landscape). Through time, some genes become fixed or homozygous for one allele. If the genes involved are maladaptive, the small population may become extinct. If they confer a selective advantage in the new environment, they will enhance the fitness of individuals within the population. Thus, chance may play an important part in evolutionary change.

But what constitutes a small population? The number of individuals in a population is generally greater than the number that is actually contributing genes to the next generation. In polygamous populations, for example, a few dominant males are responsible for all mating, so the alleles that these males happen to carry contribute disproportionately to the following generations. From a genetic point of view, the nonbreeding males in such populations might as well not exist. For this reason, the actual size of a small population or a subpopulation is of little meaning. Of greatest importance is the genetically effective population size that is passing genes to successive generations.

Without going into mathematical detail, the effective population size can be smaller than the actual population density for several reasons that include variations in the number of offspring individuals produce (such as in the case of polygamy), unequal numbers of males and females in the population, and inbreeding (mating of closely related individuals).

Human population growth, fragmentation and elimination of habitats, and poaching are reducing populations of increasing numbers of species. Many are precariously low, including the black rhinoceros (*Diceros bicorn*), Sumatran rhinoceros (*Dicerorhinus sumatrensis*), tiger (*Panthera tigris*), chimpanzee (*Pan troglodytes*), tamarins (*Leontopithecus* spp.), red-cockaded woodpecker (*Picoides borealis*), and spotted owl (*Strix occidentalis*); the list goes on and on. Already the gene pool has been dramatically reduced, and the question is what sizes of populations are needed to save the species.

The number of individuals needed to ensure the persistence of a subpopulation in a viable state for a period of several hundred years must be large enough to cope with chance variations in individual births and deaths, random series of environmental changes, genetic drift, and catastrophes. Genetic models suggest that populations with an effective population size of 100 or less and an actual size of less than 1000 are highly vulnerable to extinction. An accumulation of mildly harmful mutations that become fixed could drive such populations to extinction within 100 generations.

The number needed is called the minimum viable population. What is a minimum viable metapopulation? What is required is a sufficient number of subpopulations and suitable habitat patches along with a reserve of unoccupied patches for potential colonization. The actual population number is more elusive; it is dependent on the life history of the species (longevity, mating system, etc.) and the ability of individuals to disperse among habitat patches. However difficult to quantify, the concepts of effective population size, genetic drift, and minimum viable population size are of paramount importance in the conservation of species and maintenance of biological diversity.

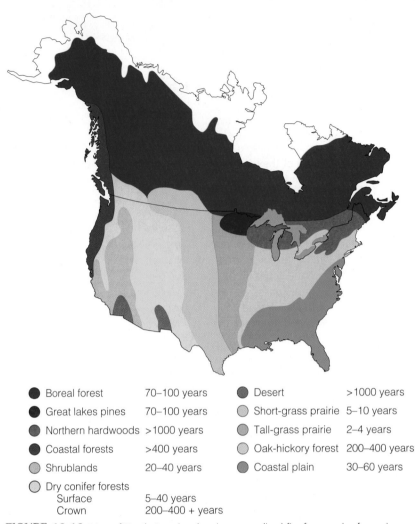

Boreal forest	70–100 years	Desert	>1000 years
Great lakes pines	70–100 years	Short-grass prairie	5–10 years
Northern hardwoods	>1000 years	Tall-grass prairie	2–4 years
Coastal forests	>400 years	Oak-hickory forest	200–400 years
Shrublands	20–40 years	Coastal plain	30–60 years
Dry conifer forests			
Surface	5–40 years		
Crown	200–400 + years		

FIGURE 19.13 Map of North America showing generalized fire frequencies for major vegetation communities. (Map from Aber and Melillo 1991; Based on data from Heinselman 1973, 1981; Gray and Schlesinger 1981; Christensen 1981; Bormann and Likens 1979.)

the physical environment often differs substantially from conditions in the surrounding area. For example, the microclimate in a gap created by the death of a canopy tree differs from that of the rest of the forest. Light and soil temperature in the gap increase, while soil moisture and relative humidity decrease. These new conditions both encourage the germination and growth of sun-adapted species and stimulate the growth of smaller trees already established in the understory. The race is on as individuals grow in height; their fate will be determined by the rate at which crowns of trees surrounding the gap can expand to fill the opening.

Community response to larger-scale disturbances such as fire, logging, or other forms of land clearing go beyond reorganizing populations that already occupy the site. Large-scale disturbances result in substantial reduction or even elimination of local populations and significantly modify the site's physical environment. In these situations, a period of colonization follows the disturbance. Some species may already be present, but dormant, as seeds of woody and herbaceous plants, roots, and rhizomes; stump sprouts; and surviving saplings. Other colonizing species are carried to the site by the wind or animals. Long-term recovery involves the process of secondary succession discussed in Chapter 13 (Section 13.12) in which species characteristic of the original community eventually replace early colonizing species.

19.8 Various Natural Processes Function as Disturbances

Many disturbances arise from natural causes, such as wind and ice storms, lightning-fires, floods, grazing animals, and insect outbreaks. While disturbance typically results in the death of organisms and loss of biomass, it can also function as a powerful force for change in the physical environment. Storm floods cut away banks, change the courses of streams and rivers, scour stream bottoms, and redeposit sediments, as well as carry away aquatic organisms. High winds and tides break down barrier dunes and allow seawater to invade the shoreline, changing the geomorphology of barrier islands (Figure 19.14). By modifying the nature of an ecosystem in ways that favor the survival of some species and eliminate others, disturbance can either reduce or foster species diversity (see Focus on Ecology 19.2: One Man's Meat).

Wind and moving water are two powerful agents of disturbance. Large, inflexible canopy trees that grow in unprotected areas like open fields are vulnerable to windthrow. So, too, are trees growing on poorly drained soil; their shallow roots are not well anchored. On rocky intertidal and subtidal shores, powerful waves overturn boulders and dislodge sessile organisms. This action clears patches of hard substrate, making them available for recolonization. Hurricanes represent large-scale wind- and water-related disturbances that have a devastating impact on ecosystems (Figure 19.15). In the Caribbean region, the frequency (average time interval between occurrences) of hurricanes with the intensity of Hurricane Hugo (winds over 166 km/hour), which struck the southeastern United States in 1989, is once every 50 to 60 years.

Fire is a major agent of disturbance, altering both the biological and physical environment (see Focus on Ecology 19.3: Fire Ecology). Losses of plants and animals can be significant, reducing both biomass and species diversity. Fire also consumes standing dead material, accumulated plant litter, and soil organic matter, and the nutrients contained are released. Fire prepares the seedbed for some species of trees by exposing mineral soil. By removing some or all of the previous vegetation, fire can lead to an increase in the availability of light, water, and nutrients to both remaining (surviving) and newly colonizing plants.

Grazing animals are a common cause of disturbances. For example, in the southwestern United States, domestic cattle droppings disperse seeds of mesquite (*Prosopis* spp.) and other woody shrubs, encouraging those species to invade already overgrazed rangeland. The disturbance of herbivores is not limited to domestic species. In parts of eastern North America where human habitation has eradicated most large predators, populations of white-tailed deer have

FIGURE 19.15 Hurricane Hugo in 1989 devastated many pine forests on the coastal plain of South Carolina.

FIGURE 19.14 Erosion of coastal dunes by storm surges creates breaks in the front dunes, resulting in areas of inundation.

ONE MAN'S MEAT

When E. B. White, paraphrasing the Roman poet Lucrecius, wrote "one man's meat is another man's poison," he was commenting that circumstances or conditions that one person might find favorable may well be intolerable for another. An individual's needs and desires are relative. Although a product of natural selection and evolution rather than culture and personal philosophy, the same can be said for disturbance. Perhaps Charles Darwin could have rephrased White's insight as "one species' disturbance is another species' adaptation." As features of the environment, the various forms of natural disturbance function as agents of natural selection (see Chapter 2). Species have adapted to disturbances such as fire, hurricanes, and floods in a variety of ways, either depending on the environmental conditions that they create or tolerating their impacts so as to survive and recover. In the extreme case, species may require the agent of disturbance in order to reproduce.

Disturbance alters the structure of the community (the collection of plant and animal species) as well as the underlying physical environment. Some species of plants and animals are adapted to the conditions that follow a disturbance (see the discussion of life history strategies in Section 12.9). Plant species typically are small in stature and short-lived, able to maintain fast growth rates under the conditions of reduced competition from other plants and high resource availability. These species generally have a large allocation of energy and resources to reproduction, producing a large number of small seeds for wide dispersal to newly disturbed sites. J. P. Grime refers to these species as *ruderals* (see Chapter 12, Section 12.9), while the limnologist G. E. Hutchinson used the more descriptive term *fugitive species*.

Many species of vertebrates also depend upon periodic disturbances to maintain their habitat, rejuvenate forage plants, and provide the proper mosaic of vegetation types. Among strongly fire-dependent species are Kirtland's warbler *(Dendroica kirtlandii)*, restricted to the jack pine forests of the lower peninsula of Michigan, and the Dartford warbler *(Sylvia sarda)* and Marmora's warbler *(S. undata)* of the Mediterranean shrublands of Sardinia. Without periodic fires to maintain their restricted habitats, these birds would face extinction. Fire produces a mosaic of shrubs, timber, and open land in the forested landscapes of eastern North America. This patchy environment is essential for such species as snowshoe hare, black bear, white-tailed deer, and ruffed grouse.

While some species exploit the conditions created by disturbance, others have evolved characteristics that serve to minimize the impact of disturbance. The structure of stem tissues in palm trees contributes to their capacity to withstand stresses resulting from high winds (and hurricanes) along coastal environments. Some tree species have thick insulated bark (many pines and oaks, western larch, giant sequoia, and redwood trees) that functions to protect the more sensitive interior tissues from being damaged by heat, allowing them to survive less intense surface fires. Some fire-adapted species of shrubs and trees respond to fire by resprouting (also called coppicing). Although fire kills the tops and foliage, new growth appears as bud sprouts and root sprouts.

Fire can also set into motion the process of regeneration for some fire-adapted species by stimulating the germination of seeds. For example, fire plays a major role in the giant sequoia forests of the western United States. Fire removes litter that has accumulated on the forest floor, providing a seedbed of soft, friable soil that favors seed germination. Fire also kills fungi in the litter and soil. The role of fire in providing conditions for enhanced seed germination and establishment is important to a wide variety of conifer species that inhabit the forests of both eastern and western North America.

Some plant species can truly be characterized as fire dependent. Jack, lodgepole, and knobcone pines are coniferous tree species that retain unripened cones for many years. Seeds remain viable within the cones until a crown fire destroys the stand. Then the heat opens the cones and releases the seeds—an event called **serotiny**—to the newly prepared seedbed.

For plant species or communities that depend on fire for regeneration or suppression of later successional species, fire dependence appears to have evolved with flammability. The plants often have fine branches, volatile resins and terpenes in foliage, and an accumulation of dead material, which increase their ability to burn. Trees of the genus *Eucalyptus* that inhabit the woodlands and forests of Australia provide an excellent example. These fire-dependent species produce leaf oils that are highly flammable. For species that depend on fire for regeneration or to create suitable habitat, fire cannot be considered a disturbance; rather, suppression of fire becomes the disturbance.

FIRE ECOLOGY

Many regions of the world are characterized by vegetation that evolved under the influence of fire. These include the grasslands of North America, the shrublands of the southwestern United States and the Mediterranean region, the African grasslands and savannas, the eucalyptus forests of Australia, the southern pinelands of the United States, and aged stands of coniferous forests of western North America. In many regions, fire is a major determinant of landscape patterns. How frequently a fire burns over a given area—its return rate—is influenced by the occurrence of droughts, accumulation and flammability of the fuel (biomass), the resulting intensity of the burn, and human interference. Prior to European settlement of North America, fires occurred about every 3 years in the grasslands of the Midwest. This amount of time was needed for sufficient fuel in the form of dead stems and leaves to accumulate. In forest ecosystems, the frequency of fires varies greatly, depending on the type of forest. Frequent light fires that burn only the surface layer may have a return interval of 1 to 25 years, whereas fires that destroy canopy trees may have a return interval of 50, 100, or even 300 years.

Type and behavior depend on the kind and amount of fuel, moisture, wind, and other meteorological conditions, season of the year, and nature of the vegetation. **Surface fire,** the most common type, feeds on the litter layer. In grasslands, it consumes dead grass and mulch, converting organic matter to ash. Usually, fire does not harm the roots, stalks, tubers, and underground buds, but it does kill most of the invading woody vegetation. In the forest,

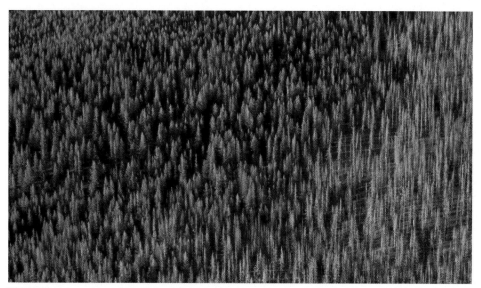

FIGURE A Crown fire resulting in a landscape mosaic of patches of burned and unburned forest.

eliminated certain trees, such as white cedar *(Thuja occidentalis)* and American yew *(Taxus canadensis)* from the forest. The African elephant *(Loxodonta africana)* has long been considered a major influence on the development of savanna communities. The combination of high density and restricted movement, limited to the confines of national parks and conservation areas, can result in the large-scale destruction of woodlands (Figure 19.16).

The beaver *(Castor canadensis)* modify many forested areas in North America and Europe. By damming streams, beavers alter the structure and dynamics of flowing water ecosystems. When the dammed streams flood lowland areas, beavers convert forested stands into wetlands. Pools behind dams become catchments for sediments. By feeding on aspen, willow, and birch, beavers maintain stands of these trees, which otherwise would be replaced by later

surface fires consume leaves, needles, debris, and humus. They kill herbaceous plants and seedlings and scorch the bases and occasionally the crowns of trees. Surface fires may kill thin-barked trees like maples by scorching the inner layers. Thick-barked trees, like oaks and pines, are better protected.

If the fuel load is high and the wind strong, surface fires may leap into the forest canopy to cause a **crown fire,** one that sweeps through the canopy of the forest. Crown fires are most prevalent in coniferous forests because of the flammability of the foliage. If the canopy is unbroken, the fire may sweep across it, and tops and branches fall to the ground to further feed the fire. A crown fire kills most above-ground vegetation through which it burns, skips, and hops, leaving patches unburned,

creating a mosaic of unburned and regenerating vegetation (Figure A). Thus fire or the lack of it creates or changes the structure and pattern of vegetation and develops a mosaic of different stand ages.

Ground fire that consumes organic matter down to the mineral substrate or bare rock is the most destructive (Figure B). It is most prevalent in areas of deep, dried-out organic matter (peat) and of extremely dry, light organic matter such as an accumulation of conifer needles. Such a fire is flameless with extremely high temperatures, and it persists until all available fuel is consumed. In spruce and pine forests, with their heavy accumulation of fine litter, a ground fire can burn down to expose rocks and mineral soil. Such fires eliminate any opportunity for that plant community to return, irreversibly changing the nature of the landscape.

FIGURE B Fires of great intensity can have a profound influence on ecosystems. After a spruce forest located on the Alleghany Plateau area of West Virginia was cut in the mid-1800s, intense ground fires burned, fueled by piles of debris from logging. Fire consumed the ground layer, exposing bedrock and mineral soil. The forest never recovered.

successional species. Thus, the actions of beavers create diverse patches—pools, open meadows, and thickets of willow and aspen—within the larger landscape.

Birds may appear to be an unlikely cause of major changes in vegetation. However, in the lowlands along the west coast of Hudson Bay, large numbers of the lesser snow goose (*Chen caerulescens caerulescens*) have altered both brackish and freshwater marshes. Snow geese grub for roots and rhizomes of graminoid plants in early spring and graze intensively on leaves of

grasses and sedges in summer. A dramatic increase in the number of geese has left large areas stripped of vegetation, resulting in the erosion of peat and the exposure of underlying glacial gravels. Since plant species that have colonized these new patches differ from the surrounding marsh vegetation, they have given rise to a new mosaic of patches on the landscape.

Outbreaks of insects such as gypsy moths and spruce budworms defoliate large areas of forest and cause the death or hinder the growth of affected

FIGURE 19.16 Consuming great quantities of woody vegetation and uprooting trees, elephants of the African savanna have an important influence on community succession. The life cycles of certain trees and the maintenance of the savanna ecosystem depend in part on the disturbance regime of elephants.

FIGURE 19.17 Most of the tall grass prairie and oak savanna of the United States has been converted to corn and soybeans.

trees. Gypsy moth infestation of a hardwood forest may result in a mortality rate of 10 to 50 percent. Spruce and fir stands infested by spruce budworms may lose up to 100 percent. The impact of forest insects is most intense in large expanses of homogeneous forest occupied by older trees.

19.9 Human Disturbance Creates Some of the Most Long-Lasting Effects

Some of the most lasting disturbances to the landscape are human-induced. Human activity has a more profound impact on ecosystems than natural disturbances do because often they are ongoing—involving continuous management of an ecosystem. One of the more permanent and radical changes in vegetation communities occurs when we remove natural communities and replace them with cultivated cropland and pasture (Figure 19.17). Many of us have the idea that agriculture is a relatively recent disturbance. In fact, prehistoric human populations were cultivating crops and converting land to pasture as long as 5000 years BP (before present). Their activities changed the

pattern of the landscape by extending or reducing the ranges of herbaceous and woody plants, allowing opportunistic weedy species to spread and altering the dominance structure in woodlands.

Another major large-scale disturbance is timber harvesting. Disturbance to the forest depends on the logging methods employed, which range from removing only selected trees to clearing entire blocks of timber. In selection cutting, foresters remove only mature single trees or selected groups of trees scattered throughout the forest. Selection cutting produces only gaps in the forest canopy (see Section 19.7). Therefore, while this form of timber harvest can minimize the scale of disturbance, it can also change the species composition and diversity of the forest.

Clear-cutting involves removing wide blocks of trees (Figure 19.18). Unless foresters manage these cleared areas carefully, erosion can badly disturb the ecosystem, impacting not only recovery of the site, but adjacent aquatic communities as well (see Chapter 21).

19.10 The Landscape Represents a Shifting Mosaic of Changing Components

Unlike the artist's mosaic shown in Figure 19.2 that has survived unchanged for centuries, the mosaic of communities that define the landscape is ever-changing.

(a)

(b)

FIGURE 19.18 (a) Block clear-cutting in a coniferous forest in the western United States. Such cutting fragments the forest. (b) Unless carefully managed, clear-cutting can cause severe disturbance to a forest ecosystem (see Chapter 21).

Through time, disturbances—large and small, frequent and infrequent—alter the biological and physical structure of communities making up the landscape, giving way to the process of succession. This view depicts the landscape as a **shifting mosaic** of patches, the overall pattern constantly changing as each patch passes through successive stages of development (Figure 19.19). The ecologists F. Herbert Borman and Gene Likens applied this concept to describe the process of succession in forested landscapes, using the term *shifting-mosaic steady state*. The term *steady state* is a statistical description of the collection of patches, the average state of the forest. In other words, the mosaic of patches shown in Figure 19.19 is not static. Each patch is continuously changing. Disturbance causes patches in the mosaic that are currently classified as late successional to revert back to early successional. Patches classified as early successional will undergo shifts in species composition, and later successional species will come to dominate. Although the mosaic is continuously changing, the average composition of the forest (average over all patches) may remain fairly constant—in a steady state. This example of a continuously changing population of patches that remains fairly constant when viewed collectively rather than individually is very similar to the concept of a stable age distribution presented in Chapter 10 (Section 10.7). In a population with a stable age distribution, the proportion of individuals in each age class remains constant even though individuals are continuously entering and leaving the population through births and deaths.

If we return to the Virginia countryside in Figure 19.1, we now see the mosaic of patches—forest, pine plantation, hedgerows, fields, golf course, pond, and human habitations—not as a static image, but as a dynamic landscape. Most of the forested lands were once fields and pastures cleared for agriculture. During the late 19th and early 20th centuries, when agriculture in the region declined, farmers aban-

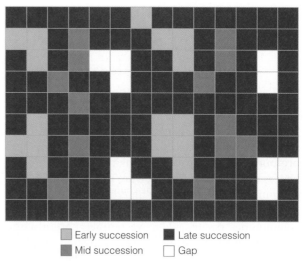

Early succession   Late succession
Mid succession   Gap

FIGURE 19.19 Representation of a forested landscape as a mosaic of patches in various stages of successional development. Although each patch is continuously changing, the average characteristic of the forest may remain relatively constant—in a steady state.

doned their fields and the land reverted to forest. The current mosaic of land cover is maintained by active processes, including many human-induced disturbances. Within these patches, communities function as islands, some bridged by corridors. Their populations are part of larger metapopulations linked by the dispersal of individuals. As time passes, the landscape will continue to change. Patterns of land use will shift boundaries; succession will alter the structure of communities; and natural disturbances such as fire and storms will form new dynamic patches within the mosaic.

CHAPTER REVIEW

Summary

Landscapes (19.1) Landscapes consist of mosaics of patches related to and interacting with each other. The study of the causes and consequences of these spatial patterns is landscape ecology. The various patches in the landscape have their own unique origin, from remnant patches of original vegetation to introduced patches requiring human maintenance. Natural disturbance and human activity determine their size and shape.

Edges and Ecotones (19.2) The place where two different communities meet is an edge, which may be abrupt or diffuse. An edge may be inherent, produced by a sharp environmental change such as a topographical feature or shift in soil type, or it may be induced, created by some form of disturbance that is limited in extent and changes through time. Often, the vegetation of one patch blends with another to form a gradual transition zone called an ecotone. Typically, ecotones have a high species richness because they support selected species of adjoining communities and also a group of opportunistic species adapted to edges.

Fragmentation, Patch Size, and Diversity (19.3). A positive relationship exists between species diversity and area. Generally, large areas support more species than small areas. The increase in species diversity with increasing patch size is related to a number of factors. Many species are area-sensitive—that is, they require large unbroken blocks of habitat. Larger areas typically encompass a greater number of habitats and therefore will support a greater array of animal species. Another feature of patch size relates to the difference between the habitats provided by edge and interior environments. In contrast to edge species, interior species require conditions found in the interior of large habitat patches, away from the abrupt changes in environmental conditions associated with edge environments.

Landscape Patches and Island Biogeography (19.4) The theory of island biogeography proposes that the number of species an island holds represents a balance between immigration and extinction. The island's distance from a mainland or source of potential immigrants influences immigration rates. Thus, islands that are most distant from a mainland would receive fewer immigrants than islands that are closer to the mainland. The area of the island influences extinction rates. Because small islands hold smaller numbers of individuals of a species and have less variation in habitat, they experience higher extinction rates than large islands. In habitat patches as in islands, large areas support more species than small areas.

Corridors (19.5) Linking patches are corridors, strips of vegetation similar to the patch but different from the surrounding matrix. Corridors act as conduits or travel lanes, function as filters or barriers, and provide dispersal routes among patches.

Patch Size Influences Populations (19.6) Habitat fragmentation and human exploitation have reduced many species to isolated or semi-isolated populations. These subpopulations inhabiting fragmented habitats form metapopulations. Maintaining them depends upon movement of individuals among habitat patches.

Disturbances and Landscape Mosaic (19.7–19.9) Disturbance, a discrete event that disrupts communities and populations, also initiates succession and creates diversity. Disturbances have spatial and temporal characteristics. Small-scale disturbances make gaps in the substrate or vegetation, creating patches of different composition or successional stage. Large-scale disturbances favor opportunistic species. Severe disturbances can replace the community altogether. Of great ecological importance are intensity, frequency, and return interval. Too-frequent disturbances can eliminate certain species (19.7). Fire is the major natural large-scale disturbance. It has both beneficial and adverse effects. Other major natural disturbance regimes include wind and animals (19.8). Major human-induced disturbances include logging, mining, agriculture, and development. These produce profound, often permanent, changes in the landscape (19.9).

Landscape as a Shifting Mosaic (19.10) The landscape is dynamic, and patches are in various stages of development and disturbance. This view of the landscape suggests a pattern of shifting mosaics. The term *shifting mosaic* refers to a community's being composed of a mosaic of patches, each in a phase of successional development. The term *steady state* is a statistical description of the collection of patches, the average state of the landscape.

Study Questions

1. What is landscape ecology?

2. What major factors are behind the basic landscape patterns? What is habitat fragmentation?

3. Distinguish between an edge and an ecotone.

4. Explain the edge effect.

5. Contrast interior species with edge species. Are all species occupying a small patch edge species?

6. How does patch size influence species diversity? Discuss possible reasons for this relationship.

7. What is island biogeography theory?

8. What aspects of island biogeography relate to patches in a landscape mosaic?

9. What types of corridors can you find in Figure 19.1? What is their origin?

10. What is disturbance? Contrast scale, frequency, and intensity as they relate to disturbance.

Asteroids, Bulldozers, and Biodiversity

In the early morning of June 30, 1908, a fireball visible for hundreds of miles exploded in the atmosphere some six miles above the headwaters of the Tunguska River in the Siberian region of Russia. Thirty-five miles from the site, people were knocked to the ground, unconscious. The detonation was heard at least 600 miles away. Light was reflected from the dust scattered high into the atmosphere from the blast, brightening the night skies throughout western Europe. Russian scientists who visited the remote region 20 years later found trees blown down in a paral-

> **"Earth is currently under-going a mass extinction on par with previous events, with an estimated annual loss of species in the tens of thousands."**

lel line more than 20 miles from the site. The cause of this destruction is unclear—scientists are still divided as to whether it was a meteorite or fragment of a comet nucleus that exploded before striking Earth's surface.

Approximately 65 million years ago, at the end of the Cretaceous period, in the region of the Yucatan Peninsula, scientists now believe that a similar, but much larger event occurred. A large meteorite struck Earth's surface, leaving a crater 180 kilometers in diameter under the waters of the Caribbean. Evidence from deep-sea

sediment cores reveals a remarkable record of the meteorite's impact and the resulting debris. The debris, blasted high into the atmosphere, may have triggered a major decline in Earth's temperature. Scientists now believe that the assault by this massive asteroid or comet was in large part responsible for the extinction of 70 percent of all species that inhabited Earth at that time, including the dinosaurs. In the period that followed, the various species that eventually would come to dominate the oceans and land surface changed dramatically from the previous inhabitants. For example, on land, mammals diversified and increased in size, a process that eventually gave rise to a particular species of primate—*Homo sapiens,* otherwise known as humans.

Paleontologists refer to the loss of species at the end of the Cretaceous as a "mass extinction event." It was not the only such event; Earth has undergone a number of mass extinctions, such as the one during the Permian (250 million years ago) when over 50 percent of Earth's species disappeared from the fossil record, including 96 percent of all marine species. The fossil record teaches us that these mass extinction events change the course of evolution, including a dramatic shift in the types of organisms that inhabit the planet.

Earth is currently undergoing a mass extinction on par with previ-

ous events, with an estimated annual loss of species in the tens of thousands. The current mass extinction event, however, is different from those of the past. The cause in not extraterrestrial, such as from a meteor or comet, or a change in sea level or climate. The destruction is due to bulldozers, agriculture, and urbanization—the destruction of habitat.

Although in North America's recent past, hunting for food and other goods has led to the extinction of many mammal and bird species, when compared to current losses, the number of species extinguished by overkill is relatively small. Hunting has led to the extermination of marine mammals such as Stellar's sea cow (extinct about 1767), the New England sea mink (about 1880), and the Caribbean monk seal (about 1952). Overkill has been found to be the main cause of virtually all 46 modern extinctions of large terrestrial mammals. Among birds, about 15 percent of the 88 modern species extinctions and 83 subspecies extinctions have been attributed to overkill. Affected species include the great auk (1844) and the passenger pigeon (1914). In some instances overkill has resulted from the often mistaken belief that a wild species was a threat to gardens or to domestic animals. Victims of this belief include the Carolina parakeet (Figure A), the only native U.S. parakeet (1914).

Other causes of a recent extinction include the introduction of predators and competitors to systems with isolated endemic species unable to cope with the

FIGURE A Carolina parakeet.

pressures imposed by the new arrivals. Predators introduced by humans have wiped out many species, particularly in island habitats that previously had been relatively free of predators. The largest number of such extinctions have been caused by rats, carried across the Pacific by Polynesians and Micronesians over the past few thousand years, and spread throughout the rest of the world by Europeans over the past 500 years. Examples include the five species of birds killed off on Lord Howe Island shortly after rats arrived with a ship in 1918 and the extinction of four species of birds, a bat, and numerous invertebrates after rats arrived on New Zealand's Big South Cape Island in 1964. Although competition can also cause extinction, interspecific competition most commonly acts to reduce the geographic range and abundance of native species, thereby making them more vulnerable to other causes of extinction. In North America, the introduction of the European house sparrow and the starling has con-

stricted the ranges of such native birds as the purple finch and the eastern bluebird, but has not yet led to their extinction.

By far the most important cause of extinction by humans in the destruction of natural habitats in order to create farms and ranches for raising crops and livestock. Such habitat alteration can begin the slide of a species to extinction long before the last vestige of its habitat is eliminated. Fragmentation of the original home ranges, reduction of breeding sites and feeding grounds, and other factors can cause population size to decline to critically low numbers well before a species habitat has been completely destroyed. The largest number of current species extinctions is occurring in the tropics, particularly the tropical rain forests.

Although accounting for only 7 percent of the total land area, tropical rain forests are home to more than 50 percent of all terrestrial plant and animal species. Over the last several decades, extensive land clearing in the tropical rain forests of Asia, Africa, and South America has greatly decreased their global distribution and resulted in the extinction of an inestimable number of plant and animal species. In the Amazon region of Brazil, the rate of forest clearing during the 1980s exceeded 1 percent per year, decreasing the extent of forest from 1 million km^2 to current estimates of just over 50,000 km^2. Forest clearing in Madagascar has resulted in the removal of over 90 percent of the original forest cover

(Figure B). This loss of habitat has resulted in the large-scale extinction of endemic species, including six genera of lemurs. Since 1960, 95 percent of the rain forest cover in Ecuador has been destroyed, resulting in an estimated loss of over 200,000 species. These are but a few examples.

The changes in land use that bring about this destruction and large-scale extinction are not limited to the tropics or to terrestrial environments. Pollution of our inland waterways, dredging and filling of coastal wetlands, and the destruction of coral reefs by pollution and siltation are having a similar effect to forest clearing on Earth's freshwater and coastal environments.

To truly understand the impact that humans are having on the diversity of species inhabiting Earth, we must place the current rates of species extinction into the context of the total biological diversity of the planet—its biodiversity. What is the present state of biodiversity on Earth? According to E. O. Wilson, Frank B. Baird Professor of Science at Harvard University, about 1.4 million living species of all kinds of organisms have been described by science. Approximately 750,000 are insects, 41,000 are vertebrates, and 250,000 are plants (vascular plants and bryophytes). The remainder consist of a complex array of invertebrates, fungi, algae, and microorganisms. Scientists who catalogue plant and animal species diversity, called systematists, say that this picture of species diversity is still very incomplete, except for a few well-studied groups such as the vertebrates and flowering plants. Recent investigations in the canopy of Peruvian rain forests suggest that the total number of species inhabiting Earth could be much higher. Such studies are identifying thousands of new species each year. (Other major habitats that

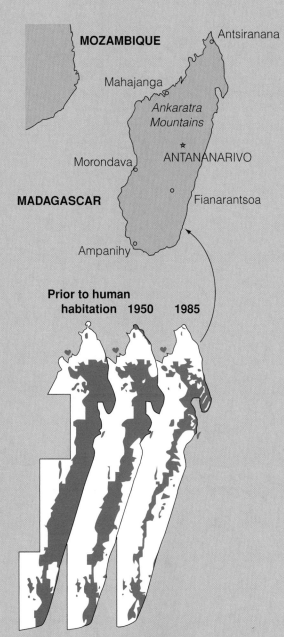

MOZAMBIQUE

Antsiranana

Mahajanga

Ankaratra Mountains

Morondava

ANTANANARIVO

MADAGASCAR

Fianarantsoa

Ampanihy

Prior to human habitation   1950   1985

FIGURE B Decline of tropical rain forest in Madagascar. Area in dark green represents forest cover.

how humans affect the biodiversity of the planet? The term *biodiversity* (*bio* = biological, *diversity* = variety) did not exist before the mid-1980s. In large part the concern is a direct product of technology. It is no coincidence that the first Earth Day celebration in the spring of 1970 came only nine months following *Apollo 11*'s historic first landing on the moon. Our first view of Earthrise from the lunar surface put the finite nature of our shared planet into perspective.

The growing use of satellite remote sensing (see Figure B) in the late 1970s and early 1980s allowed us for the first time to quantify the decline of natural ecosystems, particularly in remote regions of the globe. Once the true extent of the problem was realized, the concern moved from academic circles to society at large. By the 1990s this awareness and concern extended into the international community with the convening of the United Nations Earth Summit in Rio de Janeiro in 1992. At the Earth Summit, over 150 nations signed the Convention on Biodiversity, making the preserva-

tion of the world's biodiversity an international priority. Although there are many reasons voiced, the arguments about the importance of maintaining biodiversity can be grouped into three categories: economic, evolutionary, and moral.

The economic argument is based largely on self-interest. Many of the products we use come from organisms with which we share this planet. Obviously, the foods we eat are all derived from other organisms. Every time we buy a drug or other pharmaceutical, there is almost a 50-50 chance that we can attribute some of the essential constituents to a wild species. The value of medicinal products derived from such sources is now over $40 billion every year. (Of course, there is no way to assign value to the lives that are saved by the use of medicines obtained from natural products.) We derive rubber, solvents, and paper from trees. Cotton, flax, leather, and a host of other natural materials are used to clothe us. Modern industrial society owes a lot to Earth's genetic resources that in one way or another contribute to a number of products that better our standard of living. Today's benefits from nature's cornucopia are astonishing enough, but they represent only the tip of the iceberg. At this point, scientists have taken only a preliminary look at some 10 percent of the 250,000 plant species, a considerable number of which have already proved to be of enormous economic importance. Additionally, we have scarcely scratched the surface of the potential of the products derived from the animal kingdom. As these species are lost through extinction, so is their potential for human exploitation.

The second argument for the preservation of biodiversity is based on genetics (see Chapter 19). The

remain poorly explored include the coral reefs, the floor of the deep sea, and the soil of tropical forests and savannas.) According to Wilson, we do not know the true number of species on Earth, not even to the nearest order of magnitude. His own guess is that the absolute number falls somewhere between 5 and 30 million.

Why should we care about how many other organisms there are or

current patterns of biodiversity are a product of ecological and evolutionary processes that have acted on species that existed in the past. The processes of mutation, mixing of genetic information through sexual reproduction, and natural selection, together with the essential ingredient of time, give rise to new species. All species eventually go extinct, many leaving no trace of their past presence other than fossilized impressions buried deep in the earth. Others, however, fade into extinction after having given rise to new species. For example, it is believed that all modern birds can trace their evolutionary history to the earliest known bird, *Archaeopteryx,* which lived during the Jurassic period (fossil record dates to 145 million years ago). If *Archaeopteryx* had been driven to extinction before acting as the evolutionary seed of more modern birds, the variety of life at our backyard bird feeders would be quite different than it is today. Likewise, the mass extinction of modern-day species limits the potential evolution of species diversity in the future.

The third category of arguments in support of conserving biodiversity is philosophical. Humans are but one of millions of species inhabiting Earth, relative newcomers to the long evolutionary history of life on our planet. It is the nature of all organisms to both respond to and alter their surrounding environment. However, it is unlikely that any other species in the history of Earth has had such a dramatic impact on its environment in such a short period of time. The fundamental question facing humanity is a moral one. To what degree will we allow human activities to continue to result in the extinction of tens of thousands of species with whom we share this planet? It is on this question that the debate on the value of biodiversity will center. Arguments based on economics will fall to the wayside as technology allows us to synthesize medicines and other products, and concerns for the evolutionary future of our planet appear all too abstract when balanced against the needs of our growing human population. Science can work to identify and quantify the problem, but its solution lies outside the realm of science. It involves social, economic, and ethical issues that influence all our lives. Unlike so many problems facing society that science is called upon to solve, this is one that science can identify but that members of society—including you—must help decide how to solve.

PART VI

Ecosystems

Trillium, an early spring flower, draws on energy reserves in the roots to support early blooms. Then they send their production back into the roots to build up reserves for the following spring.

Ecosystem Productivity

OBJECTIVES

On completion of this chapter, you should be able to:

- Describe the concept of the ecosystem.
- Relate the laws of thermodynamics to ecology.
- Define the types of ecological production.
- Discuss how climate and nutrient availability influence primary productivity.
- Describe secondary production and its allocation.
- Compare assimilation and production efficiencies of ectotherms and endotherms.
- Distinguish between grazing and detrital food chains.
- Trace the flow of energy through an ecosystem using the model of food chains.
- Define trophic levels and ecological pyramids.

Up to this point, we have looked at interactions of organisms with their physical environment, interactions among individuals within the same population, and interactions between populations of different species. All of these interactions came together in the concept of the community, an assemblage of species in a given place interacting directly and indirectly with each other. The concept of the community, however, involves the biota only. Now we add the last dimension: the interaction of the biotic community with the abiotic world to give us the ecosystem.

20.1 The Ecosystem Consists of the Biotic and Abiotic

The distribution and abundance of species and the biological structure of the community vary in response to environmental conditions. However, it is equally true that the organisms themselves, in part, define the physical environment. An example is the role of autogenic changes in light and nutrient availability in driving plant succession. It is this inseparable link between the biological environment (the community) and the physical environment that led A. G. Tansley to coin the term **ecosystem** in an article in the journal *Ecology* in 1935. Tansley wrote:

> The more fundamental conception is . . . the whole *system* (in the sense of physics) including not only the organism-complex, but also the whole complex of physical factors forming what we call the environment. . . . We cannot separate them [the organisms] from their special environment with which they form one physical system. . . . It is the systems so formed which . . . [are] the basic units of nature on the face of the earth. . . . These *ecosystems*, as we may call them, are of the most various kinds and sizes.

In the concept of the ecosystem, the biological and physical components of the environment are a single interactive system.

Like the community, the ecosystem is a spatial concept; it has boundaries. Like the community, these boundaries are often difficult to define. At first examination, a pond ecosystem is clearly separate and distinct from the surrounding terrestrial environment. A closer inspection, however, reveals a less distinct boundary between aquatic and adjacent terrestrial ecosystems. Some plants along the shoreline, such as cattails, may be either partially submerged or

rooted in the surrounding land, able to tap the shallow water table with their roots. Amphibians move between the shoreline and the water. Surrounding trees drop leaves into the pond, adding to the dead organic matter that feeds the decomposer community on the pond bottom.

Regardless of these difficulties, ecosystems theoretically have spatial boundaries. Having defined the boundaries, we can view our ecosystem in the context of its surrounding environment.

The primary focus of ecosystem ecology is the exchange of energy and matter. Exchanges from the surrounding environment into the ecosystem are **inputs.** Exchanges from inside the ecosystem to the surrounding environment are **outputs.** An ecosystem with no inputs is called a **closed ecosystem;** one with inputs is an **open ecosystem.** Inputs and outputs, together with exchanges of energy and matter among components within the ecosystem, form the basis of our discussion in this and the following three chapters.

20.2 All Ecosystems Have Three Basic Components

In the simplest terms, all ecosystems, both aquatic and terrestrial, consist of three basic components—the autotrophs, the heterotrophs, and nonliving matter (see Figure 20.1). The autotrophs, or primary producers, are largely green plants. These organisms generally use the energy of the Sun in photosynthesis (Chapter 6, Section 6.2) to transform inorganic compounds into simple organic compounds.

The heterotrophs, or consumers, use the organic compounds produced by the autotrophs as a source of food. Through decomposition, heterotrophs eventually transform these complex organic compounds into simple inorganic compounds that are once again used by the primary producers. The heterotrophic component of the ecosystem is often subdivided into two subsystems: consumers and decomposers. The consumers feed largely on living tissue, and the decomposers break down dead matter into inorganic substances. No matter how we classify them, all heterotrophic organisms are consumers, and all in some way act as decomposers (see Chapters 7 and 8).

The third, or nonliving matter, component consists of the soil, sediments, particulate matter, dissolved organic matter in aquatic ecosystems, and litter in terrestrial ecosystems. All of the dead organic matter is derived from plant and consumer remains and is acted upon by the decomposers. Such dead

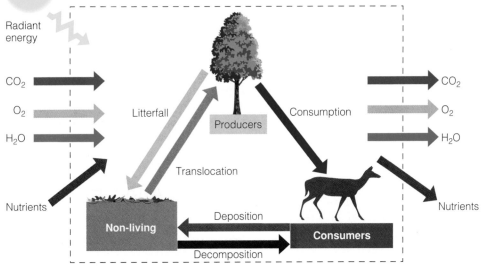

FIGURE 20.1 Schematic diagram of an ecosystem. The dashed lines represent the boundary of the system. The three major components are the producers, the consumers, and the nonliving components: inactive or dead organic matter, the soil matrix, nutrients in solution in aquatic ecosystems, sediments, and so on. The arrows indicate interactions within the ecosystem and exchanges with the external environment.

organic matter is critical to the internal cycling of nutrients in the ecosystem (Chapter 21).

The driving force of the system is the energy of the Sun. This energy, harnessed by the primary producers, flows from producers to consumers to decomposers and eventually dissipates as heat.

20.3 The Laws of Thermodynamics Govern Energy Flow

Production in ecosystems involves the fixation and transfer of energy from the Sun. Green plants fix solar energy in the process of photosynthesis (Chapter 6). The products of photosynthesis, photosynthates, accumulate as plant biomass. Nonphotosynthetic organisms convert this stored energy into heterotrophic biomass. This fixation and transfer of energy through the ecosystem is governed by the laws of thermodynamics.

Energy exists in two forms: potential and kinetic. **Potential energy** is stored energy—that is, capable of and available for performing work. **Kinetic energy** is energy in motion. It performs work at the expense of potential energy. Work is of at least two kinds: the storage of energy and the arranging or ordering of matter.

Two laws of thermodynamics govern the expenditure and storage of energy. The **first law of thermodynamics** states that energy is neither created nor destroyed. It may change form, pass from one place to another, or act upon matter in various ways. Regardless of what transfers and transformations take place, however, no gain or loss in total energy occurs. Energy is simply transferred from one form or place to another. When wood burns, the potential energy lost from the molecular bonds of the wood equals the kinetic energy released as heat. When energy is lost from the system into the surrounding environment, the reaction is **exothermic.**

On the other hand, energy may be paid into a reaction. Here, too, the first law of thermodynamics holds true. In photosynthesis, for example, the molecules of the products (simple sugars) store more energy than the reactants that combined to form the products. The extra energy that is stored in the products is acquired from the sunlight that is harnessed by the chlorophyll within the leaf (see Chapter 6). Again, there is no gain or loss in total energy. When energy from the outside is put into a system to raise it to a higher energy state, the reaction is **endothermic.**

Although the total amount of energy in any reaction, such as burning wood, does not increase or decrease, much of the potential energy degrades into a form incapable of doing further work. It ends up as heat, disorganized or randomly dispersed molecules in motion, useless for further transfer. The measure of this disorder is **entropy.**

The transfer of energy involves the **second law of thermodynamics.** It states that when energy is transferred or transformed, part of the energy assumes a form that cannot pass on any further. Entropy increases. When coal is burned in a boiler to produce steam, some of the energy creates steam, and part is dispersed as heat to the surrounding air. The same thing happens to energy in the ecosystem. As energy is transferred from one organism to another in the form of food, a large part of that energy is degraded as heat—is no longer transferable. The remainder is stored as living tissue.

At first appearance, biological systems seemingly do not conform to the second law of thermodynamics. The tendency of life is to produce order out of disorder, to decrease rather than increase entropy. The second law theoretically applies to closed systems in which no energy or matter is exchanged with the surrounding environment. With the passage of time, closed systems tend toward maximum entropy; eventually, no energy is available to do work. Living systems, however, are open systems with a constant input of energy.

20.4 Primary Production Fixes Energy

The flow of energy through a terrestrial ecosystem starts with the harnessing of sunlight by green plants. The rate at which radiant energy is converted by photosynthesis to organic compounds is referred to as **primary productivity** because it is the first and basic form of energy storage.

Gross primary productivity is the total rate of photosynthesis, or energy assimilated by the plants. Plants, like all other organisms, must expend energy in the process of respiration (see Chapter 6). The rate of energy storage as organic matter after respiration is **net primary productivity.** Net primary productivity can be described by the following equation:

net primary productivity (NPP)	=	gross primary productivity (GPP)	−	respiration by autotrophs (R)

Productivity is usually expressed in units of energy per unit area per unit time: kilocalories per square meter ($kcal/m^2/yr$). However, productivity may also be expressed in units of dry organic matter: ($g/m^2/yr$). As pointed out by the ecologist Eugene Odum, in all these definitions, the term "productivity" and the phrase "rate of production" may be used inter-

changeably. Even when the word production is used, a time element is always assumed or understood, and therefore, one should always state the time interval.

Net primary productivity accumulates over time as plant biomass. The amount of accumulated organic matter found in an area at a given time is the **standing crop biomass.** Biomass is usually expressed as grams of organic matter per square meter (g/m^2) or some other appropriate unit of area. Biomass differs from productivity. Productivity is the rate at which organic matter is created by photosynthesis. Biomass is the amount present at any given time.

20.5 Energy Allocation Varies among Primary Producers

Net primary production represents the storage of organic matter in plant tissues in excess of that required for respiration. Plants budget their production for different uses, distributing it in a systematic way to leaves, twigs, stems, bark, flowers, seeds, and other tissues. How much is allocated to each component is a function of both the life form of the plant (see Focus on Ecology 13.1: Plant Life Form Classification in Chapter 13) and environmental conditions (Chapter 6). The patterns of allocation of the dominant plants within the ecosystem will largely define the physical structure of the ecosystem (see Chapter 13, Section 13.5).

Annuals begin their life cycles in the spring with the germination of overwintering seeds. In regions with distinct dry and wet seasons, germination occurs with the onset of the rainy season. With only one growing season in which to complete its life cycle, an annual has to allocate its photosynthates first to leaves. Leaves, in turn, become involved in photosynthesis, which replenishes the supply of photosynthates and increases plant biomass. At the time of flowering, the plant decreases the amount of energy allocated to leaves and diverts most of its photosynthates to reproduction. For example, in the sunflower, the biomass of leaves declines from approximately 60 percent of the total plant weight during the period of growth to 10 to 20 percent by the time the seeds are ripe. When in bloom, the sunflower allocates 90 percent of its photosynthates to the flower head and the remainder to the leaves, stem, and roots.

Perennial plants maintain a vegetative structure over several years. They begin their life cycles like annuals, but once established, they allocate their energy in a very different manner. Before perennials

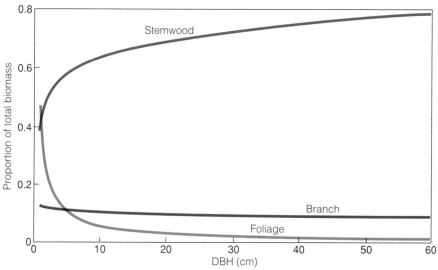

FIGURE 20.2 Changes in the proportion of biomass in various plant tissues for white oak trees (*Quercus alba*) as a function of tree size (diameter of the trunk at 1.5 meters above the ground—dbh).

expend any energy on reproduction, they divert photosynthate to the roots. This allocation to roots is in excess of that required for the development of roots for the uptake of nutrients and water from the soil. In some species, such as the skunk cabbage (*Symplocarpus foetidus)*, the roots develop into large storage organs. Energy stored in the roots makes up a reserve upon which the plants draw when they begin growth the following growing season. When they are ready to flower, perennials divert energy from storage to the production of flowers and fruit. As the flowers fade and the fruits ripen, the plant once more sends photosynthate to the roots to build up the reserves they will need for the following spring.

Trees and woody shrubs can survive for a long time, which greatly influences the manner in which they allocate energy. Early in life, leaves make up more than one-half of their biomass (dry weight); but as trees age, they accumulate more woody growth. Trunks and stems become thicker and heavier, and the ratio of leaves to woody tissue changes (Figure 20.2). Eventually, leaves account for only 1 to 5 percent of the total mass of the tree. The production system (the leaf mass) that supplies the energy is considerably less than the biomass it supports. Thus, as the woody plant grows, much of the energy goes into support and maintenance (respiration), which increases as the plant ages.

When deciduous trees leaf out in the spring, they expend up to one-third of their reserve energy on the growth and expansion of leaves. This expenditure is repaid as the leaves carry out photosynthesis during the spring and summer. After leaves, trees give preference to flowers; then transport tissues, new leaf buds, and deposits of starch in roots and bark; and finally new flower buds.

Evergreen trees have a somewhat different approach. Because their photosynthetic tissues (leaves) can function year-round when temperature and moisture conditions permit, they do not need to draw on root reserves for new growth in the spring. They can afford to wait until later in the growing season to produce new shoots. Then evergreens can draw upon energy built up earlier in the spring. For the same reason, new growth develops rapidly and matures within a few weeks.

Reproduction and vegetative growth compete for energy allotments. If photosynthesis is limited, vegetative growth gets first claim. Because the energy reproduction demand is high—up to 15 percent in pines, 20 percent in deciduous trees, and 35 percent or more in fruit trees—trees can afford an abundance of fruit only periodically, once every 2 to 3 years in deciduous trees and 2 to 6 years in conifers.

The proportionate allocation of net production to above-ground and below-ground biomass, or shoot and root tissues, tells much about different ecosystems. In Chapter 6 we explored the influence of environmental conditions (light, water, and nutrients) on the allocation of biomass to the growth of shoot and root tissues. Low light conditions favor the allocation of energy to the production of leaves and stem (shoot) at the expense of roots. A reduction in water or nutrient availability results in the reverse, a greater allocation of energy to the production of roots at the expense of leaves and stem. As a result, ecosystems that are associated with low rainfall or low soil fertility have a lower shoot-to-root ratio than those associated with high below-ground resources. For example, plants of tundra ecosystems (see Chapter 26), living in an environment with a long, cold winter and a short growing season, have shoot-to-root ratios from 1:3 to 1:10. Prairie grasslands of the midwestern region of North America have a shoot-to-root ratio of 1:3, indicative of cold winters and limited moisture supply. Plants of arid regions have a low shoot-to-root ratio.

Within the forest ecosystem, different components of the plant community exhibit different patterns of above-ground and below-ground allocation. For the Hubbard Brook Forest in New Hampshire, the shoot-to-root ratio for trees is 1:0.213; for shrubs, 1:0.5; and for herbaceous vegetation, 1:1.

20.6 Environmental Controls on Primary Production

Productivity of terrestrial ecosystems is influenced by climate. Measured estimates of net primary productivity for a variety of terrestrial ecosystems are plotted in Figure 20.3 against the mean annual precipitation (Figure 20.3a) and mean annual temperature (Figure 20.3b) for each site. As the graphs show, net primary productivity increases with increasing temperature and rainfall. The reason is that temperature and precipitation influence the rate of photosynthesis, the amount of leaf area that can be supported, and the duration of the growing season.

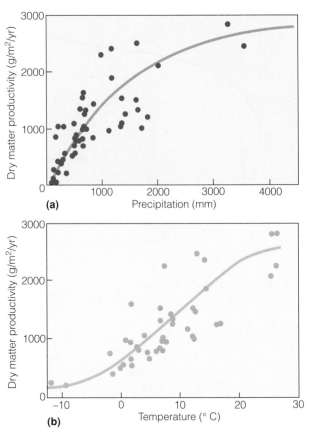

FIGURE 20.3 Net primary productivity for a variety of terrestrial ecosystems (a) as a function of mean annual precipitation and (b) as a function of mean annual temperature. (Adapted from Reichle 1970.)

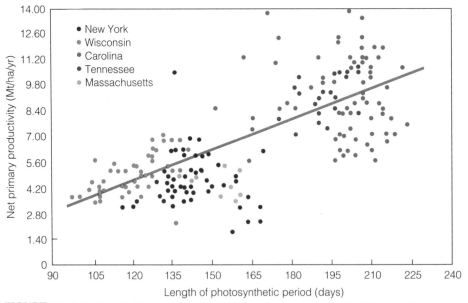

FIGURE 20.4 Relationship between net primary productivity and the length of the growing season for deciduous forest stands in eastern North America. Each point represents a single forest site. The (regression) line represents the general trend of increasing productivity with increasing length of the growing season (largely a function of latitude). (From Leith 1975.)

Primary productivity is a function of the rate of photosynthesis (energy capture) and the total surface area of leaves that are photosynthesizing. Rates of photosynthesis are limited by extremely cold and hot temperatures. Within the range of temperatures that are tolerated, rates of photosynthesis rise with temperature (see Chapter 6). In addition, in the temperate zones, the length of the growing season is defined as the period when temperatures are sufficiently warm to support photosynthesis and net primary productivity. As a result, warmer temperatures typically both support higher rates of photosynthesis and are associated with a longer growing season (Figure 20.4).

As we discussed in Chapter 6, for photosynthesis and productivity to occur, the plant must open the stomata to take in CO_2. When the stomata are open, water is lost from the leaf to the surrounding air. For the plant to keep the stomata open, roots must replace the lost water. The higher the rainfall, the more water is available for transpiration. The amount of water available to the plant will therefore limit both the rate of photosynthesis and the amount of leaves (surface area that is transpiring) that can be supported. The combination of these factors determines the rate of primary productivity.

Although the two graphs in Figure 20.3 show independent effects of temperature and precipitation on primary productivity, in reality the influence of these two factors is closely related. Warm temperatures mean high water demand. If temperatures are warm but water availability is low, productivity will also be low. Conversely, if water availability is high but temperatures are low, productivity will be low. It is the combination of warm temperatures and an adequate water supply for transpiration that gives the highest primary productivity. This pattern is reflected in Figure 20.5, which relates the net primary productivity of various ecosystems to estimates of evapotranspiration. Evapotranspiration is the combined value of surface evaporation and transpiration. It reflects both the demand and the supply of water to the ecosystem. The demand is a function of incoming radiation and temperature, whereas the supply is a function of precipitation (see Chapter 4, Section 4.7).

The influence of climate on primary productivity in terrestrial ecosystems is reflected in the global patterns presented in Figure 20.6. These patterns of productivity reflect the global patterns of temperature and precipitation presented in Chapter 3 (see Figures 3.5 and 3.18). In addition, measured estimates of net primary productivity for a variety of ecosystems are summarized in Table 20.1. The highest terrestrial production and biomass are in the rain forests of the warm, wet tropical regions (but see Focus on Ecology

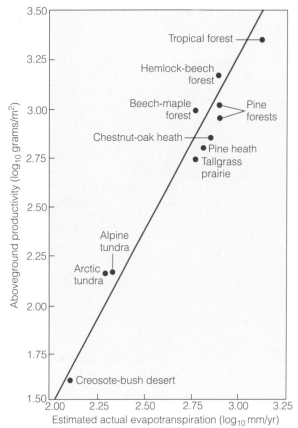

FIGURE 20.5 Relationship between above-ground net primary productivity and (actual) evapotranspiration for a range of terrestrial ecosystems. Evapotranspiration in the combination of evaporation and transpiration at a site and depends upon both precipitation and temperature (see Chapter 4). (From R. H. MacArthur and J. H. Connell, *The Biology of Populations,* p. 175, Fig. 7.3. New York: John Wiley, 1966. Used by permission.)

20.1: Productivity of a Cornfield). Deserts, where temperatures are high but rainfall is low, have very low production and biomass. So do tundra ecosystems, where precipitation is high relative to demand but temperatures are low.

In addition to climate, the availability of essential nutrients required for plant growth (see Chapter 4, Table 4.1) has a direct effect on ecosystem productivity. John Pastor of the University of Minnesota, together with colleagues, examined the role of nitrogen availability on patterns of primary productivity in different forest types on Blackhawk Island, Wisconsin. Their results clearly show the relationship between nitrogen mineralization rate (nitrogen availability) and above-ground primary productivity (Figure 20.7a). Similar increases in primary productivity with nutrient availability have been observed in lake ecosystems (Figure 20.7b).

In the oceans, phyloplankton (primary producers) are near the surface. Nutrients in the deeper waters must be transported to the surface waters,

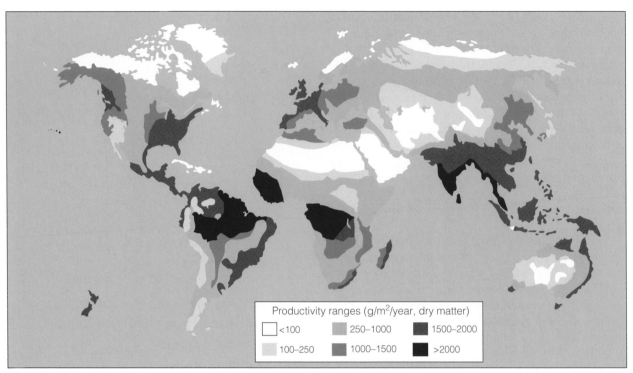

FIGURE 20.6 A global map of primary productivity for the terrestrial surface (ecosystems).

TABLE 20.1

Net Primary Production and Plant Biomass of World Ecosystems

Ecosystems (in order of productivity)	Area (10⁶ km²)	Mean Net Primary Production per Unit Area (g/m²/yr)	World Net Primary Production (10⁹ Mt/yr)	Mean Biomass per Unit Area (kg/m²)
Continental				
Tropical rain forest	17.0	2000.0	34.00	44.00
Tropical seasonal forest	7.5	1500.0	11.30	36.00
Temperate evergreen forest	5.0	1300.0	6.40	36.00
Temperate deciduous forest	7.0	1200.0	8.40	30.00
Boreal forest	12.0	800.0	9.50	20.00
Savanna	15.0	700.0	10.40	4.00
Cultivated land	14.0	644.0	9.10	1.10
Woodland and shrubland	8.0	600.0	4.90	6.80
Temperate grassland	9.0	500.0	4.40	1.60
Tundra and alpine meadow	8.0	144.0	1.10	0.67
Desert shrub	18.0	71.0	1.30	0.67
Rock, ice, sand	24.0	3.3	0.09	0.02
Swamp and marsh	2.0	2500.0	4.90	15.00
Lake and stream	2.5	500.0	1.30	0.02
Total continental	149.0	720.0	107.09	12.30
Marine				
Algal beds and reefs	0.6	2000.0	1.10	2.00
Estuaries	1.4	1800.0	2.40	1.00
Upwelling zones	0.4	500.0	0.22	0.02
Continental shelf	26.6	360.0	9.60	0.01
Open ocean	332.0	127.0	42.00	0.003
Total marine	361.0	153.0	55.32	0.01
World total	510.0	320.0	162.41	3.62

Source: (After Whittaker 1975.)

PRODUCTIVITY OF A CORNFIELD

About 30 percent of global land area has been converted to agroecosystems. These monocultural systems devoted to one crop species have replaced natural ecosystems with their diversity of species. Corn has replaced tallgrass prairie (see Chapter 25) in the midwestern United States. Corn is an annual that allocates much of its energy to seed production and a minimal amount to roots. The prairie grasses it replaced allocate much of their primary production to roots. For this reason, the cornfield has a much higher above-ground annual production.

The first effort at estimating primary production took place not in a natural ecosystem but in an Illinois cornfield in the mid-1920s. In 1926 agronomist Edgar Transeau published an estimated energy budget for a cornfield. It provided a foundation for future studies. He considered the amount of incoming solar radiation, the amount of energy fixed in gross production and lost through respiration and transpiration, and the amount of water transpired. He determined the allocation of net primary production to stalk, root, and grain, and estimated the efficiency of photosynthesis.

As a baseline, Transeau used a 0.405-hectare cornfield containing 10,000 plants that produced 100 bushels of corn with a dry weight of 2160 kg. At maturity the corn plant contains 80 percent water and 20 percent dry matter. Carbon fixed by photosynthesis makes up 45 percent of dry matter. The dry weight of the corn stalks was 600 grams, of which 216 grams was in grain, 200 grams in stalk, 140 grams in leaves, and 44 grams in roots. Thus, corn allocated 36 percent of its net primary production to grain.

Total carbon fixed by 10,000 plants was 2675 kg. The glucose equivalent of this carbon is 6687 kg. During the season, the corn plants lost through respiration 2045 kg of glucose or 3000 kg of carbon, about one-fourth of the energy absorbed in photosynthesis. Over the growing season, the corn plants manufactured 8732 kg of glucose. Transeau estimated that the amount of energy it took to produce 1 kg of glucose was 3760 kcal. The total solar energy available to the Illinois corn plant is 2043 million kcal. The total energy consumed by the corn plant in photosynthesis was 33 million kcal. Thus, the corn plants used 1.6 percent of the available energy in photosynthesis. This amount may seem low, but consider that the corn plant uses only about 20 percent of the total light spectrum in photosynthesis. That raises the photosynthetic efficiency to about 8 percent.

Not all of the energy lost by the corn plant is through respiration. Transpiration is another source. The Illinois cornfield evaporated about 276 kg of water for every kilogram of its dry weight. Through the growing season, the cornfield lost 1.5 million kg of water, or 408,000 gallons, enough to cover a hectare to a depth of about 1 meter. The energy necessary to evaporate 1 kg of water at the average temperature during the growing season is 593 kcal. As a result, 910 million kcal or 44.5 percent of available energy was expended in transpiration.

Transeau's final energy budget looked like this:

Total energy available	2043 million kcal
Used in photosynthesis	33 million kcal
Used in transpiration	910 million kcal
Total energy consumed	943 million kcal
Energy not used by plants	1100 million kcal
Energy released by respiration	8 million kcal
Efficiency of photosynthesis (assimilation efficiency)	1.6 percent
Efficiency at 20 percent of usable light spectrum	8.0 percent

Transeau's cornfield of the 1920s was exceptional in its productivity. Prior to 1950, corn allocated about 14 percent of its production to grain, and yields were more on the order of 100 bushels per hectare. Transeau's cornfield, growing on still highly fertile prairie soil, allocated 36 percent of its net primary production to grain. Since the introduction of hybrid corn in the 1950s, corn production has tripled. This increased production resulted from increased leaf area of corn plants, their ability to grow at densities of 50,000 to 70,000 plants per hectare, and heavy subsidies of fertilizers and pesticides. Although photosynthetic efficiency remains about the same, corn plants now allocate about 45 percent of their net production to grain.

Thus cornfields equal or exceed the productivity of the most highly productive natural ecosystems, the tropical rain forest and estuaries. All have one feature in common. They are heavily subsidized, the tropical forest by a complex interaction of temperature and high evaporation (see Chapter 25), estuaries by tidal flow and terrestrial inputs (see Chapter 27), the cornfield by fertilizer, protective chemicals, and soil manipulation. This high cropland productivity, however, comes at a cost of low diversity, nutrient losses (see Chapter 21), and soil erosion (see Chapter 5).

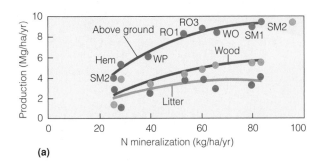

(a)

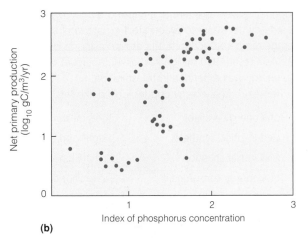

(b)

FIGURE 20.7 Relationship between net primary production and nutrient availability. (a) Above-ground productivity (total, wood, and leaf litter) increases with increasing nitrogen availability (N mineralization rate) for a variety of forest ecosystems on Blackhawk Island, Wisconsin. (From Pastor et al. 1984.) (b) Net primary productivity of phytoplankton in lake ecosystems increases as a function of phosphorus concentration. (From Schindler 1978.)

where light is available to drive photosynthesis. As a result, nutrients—particularly nitrogen, phosphorus, and iron—are a major limitation on primary productivity in the oceans (see Chapters 21 and 28). The most productive waters of the oceans are coastal environments (Figure 20.8 and Table 20.1) and coral reefs, for two reasons. First, shallow waters mean less vertical separation, greater transport of nutrients to surface waters, and changing tides. Second, coastal waters receive a large input of nutrients carried from terrestrial ecosystems by rivers and streams (see Chapters 21 and 27).

20.7 Primary Production Varies with Time

Primary production also varies within an ecosystem with time and age. Both photosynthesis and plant growth are directly influenced by seasonal variations in environmental conditions. Regions with cold winters or distinct wet and dry seasons have a period of plant dormancy when primary productivity ceases. In the wet regions of the tropics where conditions are favorable for plant growth year-round, there is little seasonal variation in primary productivity.

Year-to-year variations also occur as a result of climatic variation and disturbances such as herbivory and fire. For example, grassland productivity of a site may vary by a factor of 8 between wet and dry years. Overgrazing of grasslands by cattle and sheep or defoliation of forests by such insects as the gypsy moth can seriously reduce net primary production. Fire in grasslands may increase productivity in wet years, but reduce it in dry years. An insufficient supply of nutrients, especially nitrogen and phosphorus, can limit net productivity, as can pollutants, both atmospheric and water-borne.

Net primary productivity also varies with age, particularly in ecosystems that are dominated by woody vegetation. As the age of a forest stand increases, more and more of the living biomass is in woody tissue, while the leaf area remains constant. As the stand ages, rates of both photosynthesis and respiration decline. In addition, more of the gross production (photosynthates) goes for maintenance (respiration of woody tissues), and less remains for growth. The pattern is well illustrated by Douglas-fir in the Pacific Northwest of the United States. Seventy percent of the gross production of a 20- to 40-year-old forest stand accumulates as stored biomass (net production). In a 450-year-old stand, only 6 to 7 percent of gross photosynthesis is available for net production. The result is a pattern of increasing primary productivity during the early stages of stand development, followed by a decline as the forest ages and the standing biomass increases (Figure 20.9).

20.8 Primary Productivity Limits Secondary Production

Net primary production is the energy available to the heterotrophic component of the ecosystem. All of it is eventually consumed, by either herbivores or decomposers, but often it is not all utilized within the same ecosystem. Humans or other agents, such as wind or water currents, may disperse the net primary production of any given ecosystem to another food chain outside the ecosystem. For example, about 45 percent of the net production of a salt marsh is lost to adjacent estuarine waters (see Chapters 21 and 27).

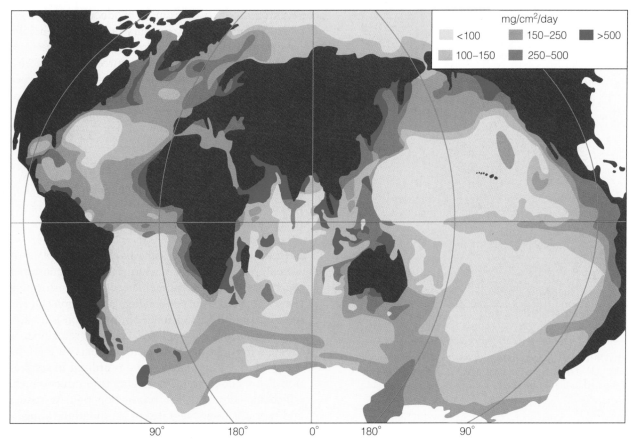

FIGURE 20.8 Geographic variation in primary productivity of the world's oceans. Note that the highest productivity is along the coastal regions, while areas of lowest productivity are in the open ocean (see Chapter 27 for discussion).

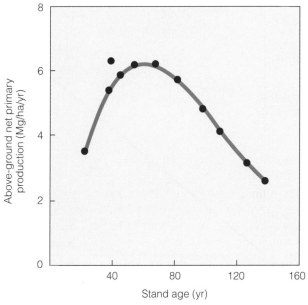

FIGURE 20.9 Changes in above-ground net primary productivity with age for a stand of white spruce trees *(Picea abies)* in the region of Karelia, Russia. (From Ryan et al. 1997; adapted from data in DeAngelis et al. 1980.)

Some energy in the form of plant material, once consumed, passes from the body as feces and urine (Figure 20.10). Of the energy assimilated, part is utilized as heat for metabolism. The remainder is available for maintenance—capturing or harvesting food, performing muscular work, and keeping up with wear and tear on the animal's body. The energy used for maintenance is eventually lost to the surrounding environment as heat.

The energy left over from maintenance and respiration goes into production, including both the growth of new tissues and the production of young. This net energy of production is called **secondary production.** Secondary production is greatest when the birthrate of the population and the growth rate of individuals are highest.

Secondary production depends on primary production for energy. Therefore, any environmental constraint on primary productivity, such as climate, will also constrain secondary productivity within the ecosystem. For example, Figure 20.11a shows the observed relationship between mean annual rainfall

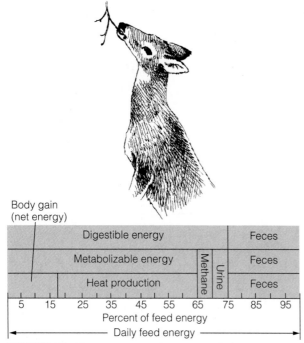

Body gain
(net energy)

Digestible energy			Feces
Metabolizable energy	Methane	Urine	Feces
Heat production			Feces

Percent of feed energy
5 15 25 35 45 55 65 75 85 95

Daily feed energy

FIGURE 20.10 End products of energy metabolism in the white-tailed deer. Note the small amount of net energy gained compared to that lost as heat, gas, urine, and feces.

and the productivity of large herbivores in African ecosystems. The increase in large herbivore production with increasing rainfall is a direct result of the corresponding increase in net primary productivity, which provides the food source for the herbivore populations. A similar relationship between phyto-

plankton production (primary productivity) and zooplankton production (secondary productivity) in lake ecosystems is shown in Figure 20.11b.

Herbivores are the energy source for carnivores. Just as when the herbivore eats a plant, the carnivore cannot use all of the energy in its food for production. Metabolic losses for respiration, heat production, and maintenance occur.

20.9 Consumers Vary in Efficiency of Production

Even if they feed on the same tissues (either plant or animal) as a source of energy, not all consumer organisms have the same efficiency of transforming energy consumed into secondary production. Of the food ingested by a consumer *(I)*, a portion is assimilated across the gut wall *(A)*, and the remainder is expelled from the body as waste products *(W)*. Of the energy that is assimilated, some is utilized in respiration *(R)*, while the remainder goes to production *(P)*, which includes both the production of new tissues and reproduction. The ratio of assimilation to ingestion *(A/I)*, the assimilation efficiency, is a measure of the efficiency with which the consumer extracts energy from food. The ratio of production to assimilation *(P/A)*, the production efficiency, is a measure of the efficiency with which the consumer incorporates assimilated energy into secondary production.

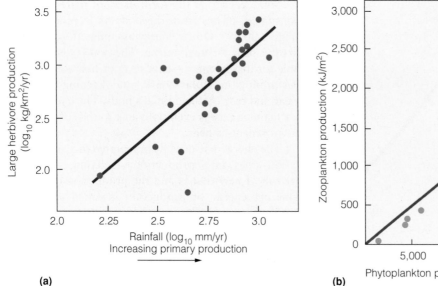

(a)

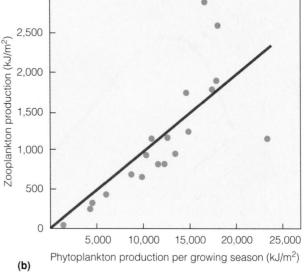

(b)

FIGURE 20.11 The relationship between primary and secondary productivity: (a) rainfall (which affects primary productivity; see Figure 20.3a) and secondary productivity of large mammalian herbivores in Africa (adapted from Coe et al. 1976); (b) phytoplankton (primary) and zooplankton (secondary) productivity in lake ecosystems. (Adapted from Brylinsky and Mann 1973.)

TABLE 20.2

Secondary Production of Selected Consumers (kcal/m²/yr)

Species	Ingestion (I)	Assimilation (A)	Respiration (R)	Production (P)	A/I	P/I
Harvester ant (h)	34.50	31.00	30.90	0.10	0.90	0.003
Plant hopper (h)	41.30	27.50	20.50	7.00	0.67	0.169
Salt marsh grasshopper (h)	3.71	1.37	0.86	0.51	0.37	0.137
Spider, small < 1 mg (c)	12.60	11.90	10.00	0.91	0.94	0.072
Spider, large > 10 mg (c)	7.40	7.00	7.30	−3.00	0.95	—
Savanna sparrow (o)	4.00	3.60	3.60	0.00	0.90	0.000
Old-field mouse (h)	7.40	6.70	6.60	0.10	0.91	0.014
Ground squirrel (h)	5.60	3.80	3.69	0.11	0.68	0.020
Meadow mouse (h)	21.29	17.50	17.00	—	0.82	—
African elephant (h)	71.60	32.00	32.00	8.00	0.45	0.112
Weasel (c)	5.80	5.50	—	—	0.95	—

Note: h = herbivore; o = omnivore; c = carnivore.

The ability of a consumer to convert the energy it ingests varies with species and the type of consumer (Table 20.2). Insects that feed on plant tissues, such as grasshoppers, are more efficient producers than insects that feed on plant juices, such as aphids. Larval stages of insects are more efficient producers than the adult stage. Endotherms have a high assimilation efficiency, but because they use about 98 percent of that energy in metabolism, they have poor production efficiency. Ectotherms use about 79 percent of their assimilation in metabolism. They convert a greater portion of their assimilated energy into biomass than do endotherms. The difference, however, is balanced by assimilation efficiency. Ectotherms have an efficiency of around 30 percent in digesting food, whereas endotherms have an efficiency of around 70 percent.

20.10 Ecosystems Have Two Major Food Chains

Energy fixed by plants is the base on which the rest of life on Earth depends. This energy stored by plants is passed along through the ecosystem in a series of steps of eating and being eaten known as a food chain (see Chapter 13, Section 13.4). Feeding relationships within a food chain are defined in terms of trophic or consumer levels (see Chapter 17, Section 17.4). From a functional rather than a species point of view, all organisms that obtain their energy in the same num-ber of steps from the autotrophs or primary producers belong to the same trophic level in the ecosystem. The first trophic level belongs to the primary producers, the second level to the herbivores (**first-level consumers**), and the higher levels to the carnivores (**second-level consumers**). Some consumers occupy a single trophic level, but many others, such as omnivores, occupy more than one trophic level (see Chapter 8, Section 8.2).

Food chains are descriptive. They represent a more abstract expression of the food webs presented in Chapters 13 and 17. Major feeding groups are defined based on a common source of energy, such as autotrophs, herbivores, and carnivores. Each feeding group is then linked to others in a manner that represents the flow of energy. A simple food chain is presented in Chapter 17, Figure 17.6. Boxes represent the three feeding groups—autotrophs, herbivores, and carnivores. The arrows linking the boxes represent the direction in which energy flows.

Within any ecosystem there are two major food chains, the **grazing food chain** and the **detrital food chain** (Figure 20.12). The distinction between these two food chains is the source of energy for the first-level consumers, the herbivores. In the grazing food chain, the source of energy is living plant biomass or net primary production. In the detrital food chain, the source of energy is dead organic matter or detritus. In turn, the herbivores in each food chain are the source of energy for the carnivores, and so on. The grazing food chain is easier to see. Cattle grazing on pastureland, deer browsing in the forest, rabbits

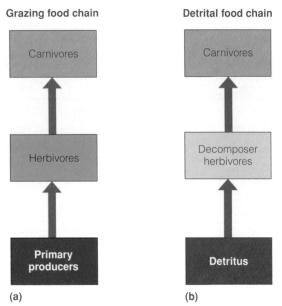

Grazing food chain

Carnivores ← Herbivores ← Primary producers

Detrital food chain

Carnivores ← Decomposer herbivores ← Detritus

(a)   **(b)**

FIGURE 20.12 Two parts of any ecosystem: (a) a grazing food chain and (b) a detrital food chain.

feeding in old fields, and insects feeding on garden crops represent first-level consumers of the grazing food chain. In spite of its conspicuousness, the grazing food chain is not the major one in terrestrial and many aquatic ecosystems. Only in some open-water aquatic ecosystems do the grazing herbivores play a dominant role in energy flow.

In terrestrial ecosystems, only a small portion of primary production goes into the grazing food chain. Over a three-year period, herbivores used only 2.6 percent of the net primary production of a temperate deciduous forest, although grazing insects made holes in the leaves, eating away 7.2 percent of the photosynthetic surface. R. D. Andrews of Colorado State University and his associates studied energy flow through a shortgrass prairie ecosystem. By controlling the amount of cattle grazing, the scientists were able to examine ungrazed, lightly grazed, and heavily grazed plots. Even on heavily grazed grassland, cattle consumed only 30 to 50 percent of above-ground net primary production. Cattle return about 40 to 50 percent of energy they consume to the detrital food chain as feces.

Although above-ground herbivores are the conspicuous grazers, below-ground herbivores can have a pronounced effect on primary production. Andrews and his associates found that the below-ground herbivores—mainly nematodes (Nematoda), scarab beetles (Scarabaeidae), and adult ground beetles (Carabidae)—accounted for 81.7 percent of total herbivore assimilation on the ungrazed plots, 49.5 percent on the lightly grazed plots, and 29.1 percent on the heavily grazed plots.

The detrital food chain is common to all ecosystems. In terrestrial and littoral (shoreline) ecosystems, it is the major pathway of energy flow, because grazing herbivores utilize so little of the net primary production. Of the total amount of energy fixed by photosynthesis in a temperate deciduous forest, approximately 50 percent of gross production goes into maintenance and respiration, 13 percent becomes new tissue, 2 percent is consumed by herbivores, and 35 percent goes directly into the decomposer food chain. Two-thirds to three-fourths of gross production in a grassland ecosystem that is ungrazed by domestic animals returns to the soil as dead plant material, and less than one-fourth is consumed by herbivores. About one-half of the quantity consumed by herbivores returns to the soil as feces. In the salt marsh ecosystem, the dominant grazing herbivore, the grasshopper, consumes just 22 percent of net primary production.

Forest litter, the habitat of an array of detritus-feeding invertebrates, is a good place to seek an example of a detrital food web. One such food web (Figure 20.13) involves five groups of litter feeders, in this case macrodecomposers: millipedes (Diplopoda), orbatid mites (Cryptostigmata), springtails (Collembola), cave crickets (Orthoptera), and pulmonate snails (Pulmonata). Of these, the mites and springtails are the most important litter feeders. These herbivores are preyed upon by small spiders (Araneidae) and predatory mites (Mesostigmata). The spiders also feed on the predatory mites. Carabid beetles feed upon springtails, snails, small spiders, and cave crickets. Medium-size spiders also eat the crickets and are eaten by the beetles. Birds and small mammals, members of the grazing food chain, consume the beetles, snails, and spiders. In this manner, predation links the detrital food chain to the grazing food chain at higher consumer levels.

20.11 Energy Flows Through Trophic Levels

When ecologists trace the flow of energy through an ecosystem, they have to track the flow of energy between trophic levels. They also have to define the links between the two food chains, grazing and detrital, and measure the losses from the ecosystem through respiration (R).

Figure 20.14 combines the food chains to produce a generalized model of trophic structure and energy flow through an ecosystem. The two food chains are linked. The initial source of energy for the detrital food chain is the input of waste materials and dead

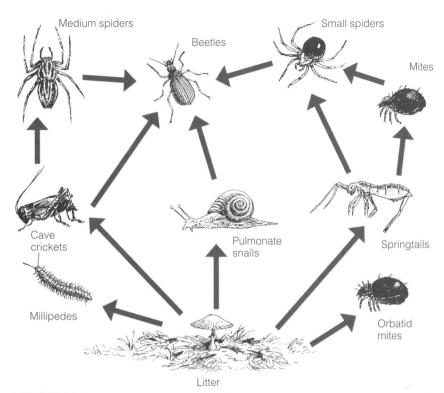

FIGURE 20.13 Detrital food chain involving forest litter-dwelling invertebrates in an Appalachian yellow-poplar forest.

organic matter from the grazing food chain. This linkage appears as a series of arrows from each of the trophic levels in the grazing food chain leading to the box designated as detritus or dead organic matter. There is one notable difference in the flow of energy

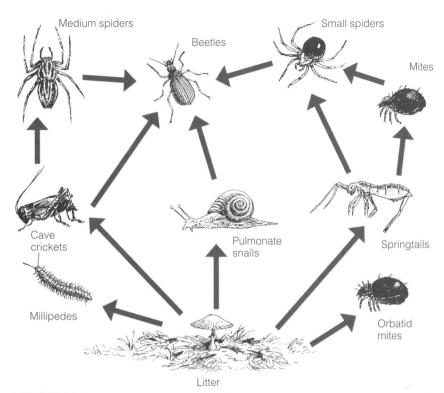

Grazing food chain   **Detrital food chain**

FIGURE 20.14 Grazing and detrital food chains from Figure 20.12 combined, showing their connections.

between trophic levels in the grazing and decomposer food chains. In the grazing food chain, the flow is unidirectional, with net primary production providing the energy source for herbivores, herbivores providing the energy for carnivores, and so on. In the decomposer food chain, the flow of energy is not unidirectional. The waste materials and dead organic matter (organisms) in each of the consumer trophic levels are "recycled," returning as an input to the dead organic matter box at the base of the detrital food chain. In addition, the higher trophic levels of the detrital food chain provide energy for higher trophic levels (through predation) of the grazing food chain.

To quantify the flux of energy through the ecosystem, we need to return to the processes involved in secondary production discussed in Section 20.9: consumption, ingestion, assimilation, respiration, and production. We will diagram a single trophic compartment (Figure 20.15a). The energy available to a given trophic level (designated as n) is the production of the next lower level ($n - 1$); for example, net primary production (P_{n-1}) is the available energy for grazing herbivores (trophic level n). Some proportion of that productivity is consumed or ingested (I); the remainder makes its way to the dead organic matter of the detrital food chain. Some portion of the energy consumed is assimilated by the organisms (A), and the remainder is lost as waste materials (W) to the

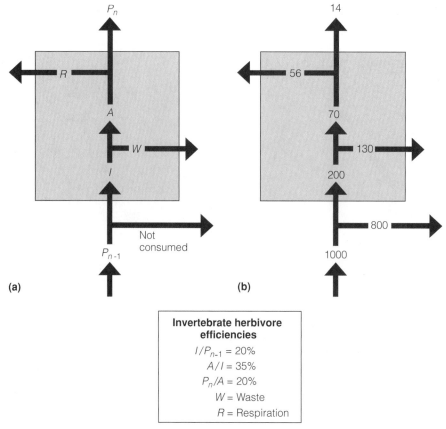

FIGURE 20.15 (a) Energy flow within a single trophic compartment. (b) A quantified example of energy flow through that compartment for an invertebrate herbivore. Values are kcal.

detritus food chain. Of the energy assimilated, some is lost to respiration, shown as the arrow labeled R that is leaving the upper left corner of the box, and the remainder goes to production (P_n).

We quantify this flow with the formulas in Quantifying Ecology 20.1: Ecological Efficiencies. Notice the consumption efficiency, the ratio of ingestion to production (I_n/P_{n-1}). The consumption efficiency defines the amount of available energy being consumed. The assimilation efficiency (A/I) defines the proportion of the energy ingested that is assimilated and the proportion that is lost as waste material. The production efficiency (P/A) defines the relative proportions of the assimilated energy that go to production and respiration. Sample values of these efficiencies for an invertebrate herbivore in the grazing food chain are provided in Figure 20.15. Using these efficiency values, we can track the fate of 1000 kcal of energy available to herbivores in the form of net primary productivity through the herbivore trophic level (Figure 20.15b).

If we apply efficiency values for each trophic level in the grazing and detrital food chains, we will know the flow of energy through the whole ecosystem. The production from each trophic level provides the input to the next higher level, while unconsumed production, waste products, and dead individuals from each trophic level provide input into the dead organic matter compartment. The entire flow of energy through the ecosystem is a function of the initial transformation of solar energy into net primary productivity. All energy entering the ecosystem as net primary productivity eventually is lost through respiration.

20.12 Energy Decreases in Each Successive Trophic Level

The quantity of energy flowing into a trophic level decreases with each increasing trophic level. This pattern occurs because not all energy is used for production. An ecological rule of thumb allows a magnitude-of-10 reduction in energy as it passes

Ecological Efficiencies

Assimilation efficiency (within a trophic level) =

$$\frac{\text{Assimilation}}{\text{Ingestion}} \quad \frac{A}{I}$$

Production efficiency =

$$\frac{\text{Production}}{\text{Assimilation}} \quad \frac{P}{A}$$

Growth efficiency =

$$\frac{\text{Production}}{\text{Ingestion}} \quad \frac{P}{I}$$

Consumption efficiency =

$$\frac{\text{Consumption at trophic level } n}{\text{Production at trophic level } n-1} \quad \frac{I_n}{P_n - 1}$$

from one trophic level to another. If herbivores eat 1000 kcal of plant energy, about 100 kcal will convert into herbivore tissue, 10 kcal into first-level carnivore production, and 1 kcal into second-level carnivore production. However, data suggest that a 90 percent loss of energy from one trophic level to another may be too high.

Certainly, a wide range in the efficiency of conversion (assimilation and production efficiencies) exists among different feeding groups (see Table 20.2). Production efficiency in photosynthetic organisms (net production/solar radiation) is low, ranging from 0.34 percent in some phytoplankton to 0.8 to 0.9 percent in grassland vegetation. Plant production consumed by herbivores is used with varying efficiency. Herbivores consuming green plants are wasteful feeders, but not nearly as wasteful as those feeding on plant sap.

Assimilation efficiencies vary widely among ectotherms and endotherms (Table 20.3). Endotherms are much more efficient than ectotherms. However, carnivorous animals, even ectothermic ones, have high assimilation efficiency. Predatory spiders feeding on invertebrates have assimilation efficiencies of over 90 percent (see Table 20.2). Because of high maintenance and respiratory costs, endotherms have low production efficiency compared to ectotherms. Only about 2 to 10 percent of the energy consumed by herbivore endotherms goes into biomass production, less than the 10 percent average suggested by the rule of thumb. However, ectotherms convert about 17 percent of their consumption to herbivore biomass. On midwestern grasslands, average herbivore production efficiency, involving mostly ectotherms, ranged from 5.3 to 16.5 percent (Table 20.4). Production efficiency on the carnivore level ranged from 13 to 24 percent.

Transfer of energy from one trophic level to another tells the real story, but such data are hard to collect. The ratio of primary (phytoplankton) to secondary (zooplankton) production in open freshwater ecosystems is about 7.1:1, and the ratio of herbivore zooplankton production to carnivore zooplankton production is 2.1:1. Efficiencies are lower in the benthic community—2.2 for herbivores and 0.3 for carnivores.

The proportion of production consumed by the next higher trophic level, the energy transfer efficiency, also varies greatly. Among invertebrate consumers on a shortgrass plain, the proportion is about 9 percent for herbivores, 38 percent for above-ground predators, and 56 percent for below-ground predators.

TABLE 20.3

Assimilation and Production Efficiencies for Endotherms and Ectotherms

Efficiency	All Endotherms	Grazing Arthropods	Sap-feeding Herbivores	Lepidoptera	All Ectotherms
Assimilation					
A/I	77.5 ± 6.4	37.7 ± 3.5	48.9 ± 4.5	46.2 ± 4.0	41.9 ± 2.3
Production					
P/I	2.0 ± 0.46	16.6 ± 1.2	13.5 ± 1.8	22.8 ± 1.4	17.7 ± 1.0
P/A	2.46 ± 0.46	45.0 ± 1.9	29.2 ± 4.8	50.0 ± 3.9	44.6 ± 2.1

TABLE 20.4

Consumption Efficiency (Secondary Production/Secondary Consumption)

Habitat	Growing Season (days)	Producers		Herbivores		Carnivores	
		Production (kcal/m²)	Efficiency (%)	Production (kcal/m²)	Efficiency (%)	Production (kcal/m²)	Efficiency (%)
Shortgrass plains	206	3767	0.8	53	11.9	6	13.2
Midgrass prairie	200	3591	0.9	127	16.5	37	23.7
Tallgrass prairie	275	5022	0.9	162	5.3	15	13.9

20.13 Patterns of Energy Flow Through Food Chains Vary among Ecosystems

Although the general model of energy flow presented in Figure 20.1 pertains to all ecosystems, the relative importance of the two major food chains and the rate at which energy flows through the various trophic levels can vary widely among different types of ecosystems. In most terrestrial and shallow-water ecosystems, with their high standing crop and relatively low harvest of primary production, the detrital food chain is dominant. In deepwater aquatic ecosystems, with their low biomass, rapid turnover of organisms, and high rate of harvest, the grazing food chain may be dominant.

A comparison of the general patterns of energy flow through four distinct ecosystem types is presented

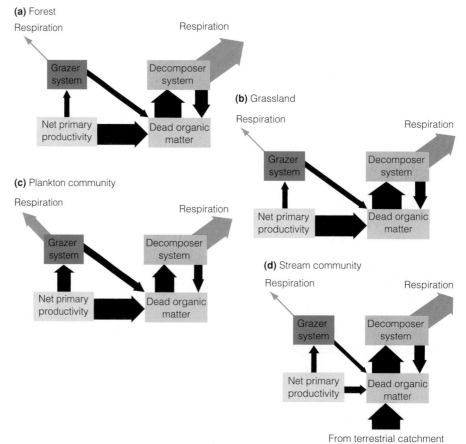

FIGURE 20.16 General patterns of energy flow through four ecosystems: (a) forest, (b) terrestrial grassland, (c) ocean (phytoplankton community), and (d) stream. Relative size of boxes and arrows are proportional to the relative magnitude of compartments and flow. (From Begon et al. 1986.)

in Figure 20.16. The relative size of each box represents the amount of energy in each trophic level of the food chain, and the arrows represent the relative flow of energy between trophic levels. The open-water ecosystem and associated phytoplankton community have the highest consumption efficiency (proportion of NPP consumed), with the grazing food chain playing a greater role in energy flow than in the other three ecosystem types. In terrestrial ecosystems, the grazing food chain is much more important in grassland than in forest ecosystems. In forest ecosystems, the vast majority of net primary productivity is not consumed as living tissues by herbivores; rather, it is stored as woody standing crop biomass that eventually makes its way to the detrital food chain as dead organic matter. Stream ecosystems have extremely low net primary productivity, and the grazing food chain is minor (see Chapter 26). The detrital food chain dominates and depends on inputs of dead organic matter from adjacent terrestrial ecosystems.

20.14 Ecological Pyramids Characterize Distribution of Energy

If we sum all of the biomass or energy contained in each trophic level, we can construct pyramids for the ecosystem (Figure 20.17). The pyramid of biomass indicates by weight, or other means of measuring living material, the total bulk of organisms or fixed energy present at any one time—the standing crop. Because some energy or material is lost at each successive trophic level, the total mass supported at each level is limited by the rate at which energy is being stored at the next lower level. In general, the biomass of producers must be greater than that of the herbivores they support, and the biomass of herbivores must be greater than that of carnivores. That circumstance results in a narrowing pyramid for most ecosystems.

This arrangement does not hold for all ecosystems. In such ecosystems as lakes and open seas, primary production is concentrated in the microscopic algae. These organisms have a short life cycle and rapid reproduction. They are heavily grazed by herbivorous zooplankton that are larger and longer-lived. As a result, despite the high productivity of algae, their biomass is low compared to that of zooplankton herbivores.

The energy pyramid indicates only the amount of energy flow at each level. The base on which it is constructed is the quantity of organisms produced per unit time. Stated differently, it is the rate at which

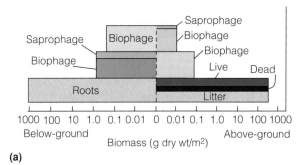

(a)

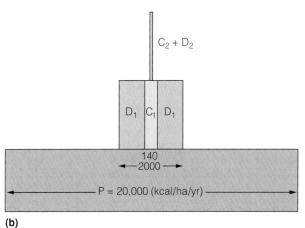

(b)

FIGURE 20.17 Examples of ecological pyramids. The detrital and grazing food chains have been collapsed into the same trophic levels. (a) A pyramid of biomass for a northern shortgrass prairie. The base of the pyramid represents biomass (g dry weight/m²) of producers; the second (middle) level, first-level consumers; and the top, second-level consumers. The dashed vertical line separates above-ground biomass (right) and below-ground biomass (left). The trophic level magnitudes are plotted on a horizontal logarithmic scale. The compartments are divided into live, standing dead, and litter biomass, and into consumers of live biomass (biophages) and dead biomass (saprophages). (b) A pyramid of energy for the Lamto Savanna, Ivory Coast. P is primary production; C_1, first-level consumers; C_2, second-level consumers; D_1, decomposers of vegetable matter; and D_2, decomposers of animal matter.

food material passes through the food chain. Some organisms have a small biomass, but the total energy they assimilate and pass on may be considerably greater than that of organisms with a much larger biomass. On a pyramid of biomass, these organisms would appear much less important in the ecosystem than they really are. Energy pyramids narrow because less energy is transferred from each level than was paid into it, in accordance with the second law of thermodynamics. In instances in which the producers have less bulk than consumers, as in open water, the energy they store and pass on must be greater than that of the next higher level. Otherwise, the biomass that producers support could not be greater than that of the producers themselves. This high energy flow is maintained by a rapid turnover of individuals rather than an increase in total mass.

Summary

Ecosystem Concept (20.1–20.2) The biological community and the abiotic environment unite to form a single interactive system, the ecosystem. This system exchanges energy and matter with the surrounding environment (20.1). All ecosystems consist of three basic components—autotrophs, consumers, and nonliving matter—that exchange energy and material. The driving force is the energy of the Sun. This energy, harnessed by producers, flows from producers to consumers and decomposers and eventually dissipates as heat (20.2).

Laws of Thermodynamics (20.3) Energy flow in ecosystems supports life. Energy is governed by the laws of thermodynamics. The first law states that although energy can be transformed, it can be neither created nor destroyed. The second law states that as energy is transferred or transformed, a portion ceases to be usable. As energy moves through an ecosystem, much of it is lost as heat of respiration. Energy is degraded from a more organized to a less organized state, or entropy. However, a continuous flux of energy from the Sun prevents ecosystems from running down.

Primary Production (20.4–20.7) The flow of energy through an ecosystem starts with the harnessing of sunlight by green plants, the rate of which is referred to as primary productivity. The total rate of photosynthesis by plants is gross primary productivity. The rate of energy storage after plants have met their respiratory need is net primary productivity, in the form of plant biomass. The rate of primary productivity is measured in units of weight per unit area per unit time (20.4). Energy fixed by a plant is allocated to different parts of the plant and to reproduction. How much is allocated to each component is a function of both the life form of the plant and environmental conditions. The patterns of allocation of the dominant plants within the ecosystem will largely define the physical structure of the ecosystem. The amount of energy allocated to aboveground stems and below-ground roots tells much about different ecosystems. A low shoot-to-root ratio suggests a stressful environment for plants (20.5).

Productivity of terrestrial ecosystems is influenced by climate, especially temperature and precipitation. Temperature influences the rate of photosynthesis, and the amount of available water limits photosynthesis and the amount of leaves that can be supported. Warm, wet conditions make the tropical rain forest the most productive terrestrial ecosystem.

Nutrient availability directly influences patterns of primary productivity. It is the most pervasive influence on the productivity of oceans. The most productive ecosystems are shallow coastal waters, coral reefs, and estuaries, where nutrients are more available (20.6).

Primary production in an ecosystem varies with time. Seasonal and year-to-year variations in moisture and temperature directly influence primary production. In ecosystems dominated by woody vegetation, net primary production declines with age. As the ratio of woody biomass to foliage increases, more of gross production goes into maintenance (20.7).

Secondary Production (20.8–20.9) Net primary production is available to consumers directly as plant tissue or indirectly through animal tissue. Once consumed and assimilated, energy is diverted to maintenance, growth, and reproduction and to feces, urine, and gas. Change in biomass, including both weight change and reproduction, is secondary production. Secondary production depends upon primary production. Any environmental constraint on primary production will constrain secondary production in the ecosystem (20.8). Efficiency of production varies. Endotherms have high assimilation efficiency but low production efficiency, because they have to expend so much energy in maintenance. Ectotherms have low assimilation efficiency but high production efficiency; they put more energy into growth (20.9).

Food Chains and Energy Flow (20.10–20.13) A basic function of the ecosystem is the flow of energy from the Sun through various consumers to its final dissipation in a series of energy transfers known as the food chain. The various members of a food web can be grouped into categories called trophic or feeding levels. Autotrophs occupy the first trophic level. Herbivores that feed on autotrophs make up the next trophic level. Carnivores that feed on herbivores make up the third and higher trophic levels.

Energy flow in ecosystems takes two routes: one through the grazing food chain, the other through the detritus food chain. The bulk of production is used by organisms that feed on dead organic matter. The two food chains are linked at higher trophic levels of each through carnivory and omnivory (20.10).

At each trophic level, efficiency is defined as consumption efficiency, the amount of available energy being consumed; assimilation efficiency, the portion of energy ingested that is assimilated and not lost as waste material; and production efficiency, the portion of assimilated energy that goes to growth and respiration (20.11). The efficiency of conversion varies among ectotherms and endotherms and between herbivores and carnivores. In general, ectotherms have low assimilation efficiency and high production efficiency, whereas endotherms have the reverse (20.12).

The loss of energy at each transfer limits the number of trophic levels or steps in the food chain to four or five. At each level, biomass usually declines. A plot of the total weight of individuals at each successive level produces a tapering pyramid. In aquatic ecosystems, however, where there is a rapid turnover of small aquatic producers, the pyramid of biomass becomes inverted (20.13).

Study Questions

1. What is an ecosystem, and what are its basic components?

2. Distinguish between potential and kinetic energy.

3. How do the first and second laws of thermodynamics relate to ecology?

4. Define primary production, primary productivity, gross primary production, net primary production, and secondary production.

5. How does climate influence primary production?

6. What world ecosystems have high and low net productivity? Why?

7. How does secondary production differ from primary production?

8. Define assimilation efficiency and production efficiency and relate these terms to the flow of energy through a trophic level.

9. What is the difference in energy allocation and production efficiency between endotherms and ectotherms?

10. What is a food chain?

11. What are the two major food chains, and how are they related?

12. Why is decomposition more than just the traditional end point in the food chain?

13. What is a trophic level? Relate the levels to ecological pyramids.

A Rocky Mountain alpine tundra carpeted with cotton grass *(Eriophorum angustifolium)* in full bloom.

Nutrient Cycling

OBJECTIVES

On completion of this chapter, you should be able to:

- Define two types of biogeochemical cycles.

- Outline a general model of nutrient cycling within an ecosystem.

- Discuss how the rates of primary productivity and decomposition influence the nutrient cycle.

- Explain how climate influences the rate of nutrient cycling in ecosystems.

- Contrast nutrient cycling in terrestrial and open-water ecosystems.

- Discuss how the continuous and directional flow of water in stream ecosystems influences nutrient cycling.

- Describe how estuarine ecosystems conserve nutrients.

The living world depends on a flow of energy and the cycling of nutrients through the ecosystem. Energy and nutrients are tightly linked in organic matter; one cannot be separated from the other. Their linkage begins in photosynthesis, in which plants use solar energy to fix CO_2 into organic carbon compounds. Carbon together with a variety of essential nutrients (see Table 4.1) makes up organic matter, the tissues of plants and animals. Because of this linkage between energy and nutrients, the general model of energy flow through an ecosystem presented in Chapter 20 provides a basic framework for examining the flow of matter through ecosystems.

21.1 All Nutrients Follow Biogeochemical Cycles

All nutrients flow from the nonliving to the living and back to the nonliving components of the ecosystem in a more or less cyclic path known as a **biogeochemical** cycle (from *bio,* "living," *geo* for the rocks and soil, and *chemical* for the processes involved). The important players in all nutrient cycles are the green plants, which organize the nutrients into biologically useful compounds; the decomposers, which return them to their simple elemental state; and the air and water, which transport nutrients between the abiotic and living components of the ecosystem. Without these components, no cyclic flow of nutrients would exist.

There are two basic types of biogeochemical cycles: gaseous and sedimentary. In gaseous cycles, the main reservoirs of nutrients are the atmosphere and the oceans. For this reason, gaseous cycles are pronouncedly global. The gases most important for life are nitrogen, oxygen, and carbon dioxide. These three gases in stable quantities of 78, 21, and 0.03 percent respectively are the dominant components of Earth's atmosphere. (See Focus on Ecology 21.1: The Gaia Hypothesis.)

In sedimentary cycles, the main reservoir is the soil, rocks, and minerals. The mineral elements that living organisms require come initially from inorganic sources. Available forms occur as salts dissolved in soil water or in lakes, streams, and seas. The mineral cycle varies from one element to another, but essentially it consists of two phases: the rock phase and the salt solution phase. Mineral salts come directly from Earth's crust through weathering (see Chapter 5, Section 5.3). The soluble salts then enter the water cycle. With water, they move through the soil to streams and lakes and eventually reach the seas, where they remain indefinitely. Other salts

return to Earth's crust through sedimentation. They become incorporated into salt beds, silts, and limestone. After weathering, they enter the cycle again.

There are many different kinds of sedimentary cycles. Cycles such as the sulfur cycle are a hybrid between the gaseous and the sedimentary because they have reservoirs not only in Earth's crust but also in the atmosphere. Other cycles, such as the phosphorus cycle, are wholly sedimentary; the element is released from rock and deposited in both the shallow and deep sediments of the sea.

Both gaseous and sedimentary cycles involve biological and nonbiological agents; both are driven by the flow of energy through the ecosystem; and both are tied to the water cycle (see Chapter 4, Section 4.7). Water is the medium by which elements and other materials move through the ecosystem. Without the cycling of water, biogeochemical cycles would cease.

Although the biogeochemical cycles of the various essential nutrients required by autotrophs and heterotrophs differ in detail, from the perspective of the ecosystem, all biogeochemical cycles have a common structure, sharing three basic components: inputs, internal cycling, and outputs (Figure 21.1). We shall look at this general model of nutrient cycling in ecosystems first, and then examine specific biogeochemical cycles in more detail in the following chapter.

21.2 Nutrients Enter the Ecosystem via Inputs

The input of nutrients to the ecosystem depends on the type of biogeochemical cycle. Nutrients with a gaseous cycle, such as carbon and nitrogen, enter the ecosystem via the atmosphere. In contrast, nutrients such as calcium and phosphorus have sedimentary cycles, with inputs dependent on the weathering of rocks and minerals (see Chapter 5, Section 5.3). The process of soil formation and the resulting soil characteristics have a major influence on processes involved in nutrient release and retention (Chapter 5). Many soil materials are deficient in nutrients on which plants depend, affecting both plants and herbivores.

Supplementing nutrients in the soil are nutrients carried by rain, snow, air currents, and animals. Precipitation brings appreciable quantities of nutrients, called **wetfall**. Some of these nutrients, such as tiny dust particles of calcium and sea salt, form the nuclei of raindrops; others wash out of the atmosphere as the rain falls. Some nutrients are brought in by airborne particles and aerosols, collectively called **dryfall**.

THE GAIA HYPOTHESIS

Earth's atmosphere is extremely different from that predicted for a nonliving Earth and from that of other planets in the solar system. Those atmospheres are dominated by carbon dioxide and possess only a trace of oxygen. There are two views of the formation of Earth's atmosphere. One is that physical forces interacted to form life-sustaining conditions, and life then evolved to adapt to those conditions. The other is that organisms evolved in partnership with the physical environment. From the beginning, these organisms helped to control geochemical cycles. For example, photosynthetic algae in the early oceans first released O_2 into the atmosphere. Photosynthetic organisms in the oceans still provide 70 percent of our atmosphere's oxygen.

The constancy of Earth's atmosphere over 3.6 billion years, with its high O_2 and low CO_2 content and moderate temperatures, suggests some feedback system. It prompted James Lovelock—physical scientist, engineer, and inventor of instruments to measure the Martian environment—and microbiologist Lynn Margulis to postulate the Gaia hypothesis (*Gaia* is the Greek word for "Earth goddess") of global biogeochemical homeostasis. The theory says that Earth's biosphere, atmosphere, oceans, and soil together make up a feedback system that maintains an optimal physical and chemical environment for life on Earth. This feedback system could not have developed nor could be maintained without the critical buffering activity of early life forms and continued coordinated activity of plants and other photosynthetic organisms. Together they damp the fluctuations of the physical environment that would occur in the absence of a well-organized living system.

No control mechanisms have been discovered, but microorganisms are the only life forms that could function like a chemostat, making Earth one large cybernetic system. For example, maintenance of 21 percent O_2 in the atmosphere—which maximizes aerobic metabolism just below the level that would make Earth's vegetation flammable—is possibly the outcome of microbial activity. Microbial production of CH_4 from the small amount of carbonaceous living matter buried each year might keep oxygen in check.

Not all ecologists and atmospheric scientists accept the Gaia hypothesis, but it does help us understand the behavior of ecosystems and the interactions of biogeochemical cycling. Evidence seems to indicate that organisms do play a dynamic role in determining the composition of many chemicals in the soil, water, and atmosphere. We need to look no farther than the tremendous impact we humans have had on the physical aspects of Earth.

Between 70 and 90 percent of rainfall striking the forest canopy reaches the forest floor. As it drips through the canopy (throughfall) and runs down the stems (stemflow), rainwater picks up and carries with it nutrients deposited as dust on leaves and stems together with nutrients leached from them. Therefore, rainfall reaching the forest floor is richer in calcium, sodium, potassium, and other nutrients than rain falling in the open at the same time.

The major sources of nutrients for aquatic life are inputs from the surrounding land in the form of drainage water, detritus, and sediment, and from precipitation. Flowing-water aquatic systems (streams and rivers) are highly dependent on a steady input of dead organic matter from the watersheds through which they flow (Chapter 27).

21.3 Nutrients Are Recycled Within the Ecosystem

Primary productivity in ecosystems depends on the uptake of essential mineral (inorganic) nutrients by plants and their incorporation into living tissues. Nutrients in organic form, stored in living tissues, represent a significant proportion of the total nutrient pool in most ecosystems. As these living tissues senesce, the nutrients are returned to the soil or sediments in the form of dead organic matter. Various microbial decomposers transform the organic nutrients into a mineral form, a process called mineralization (see Chapter 7), and the nutrients are once again available to the plants for uptake and incorporation

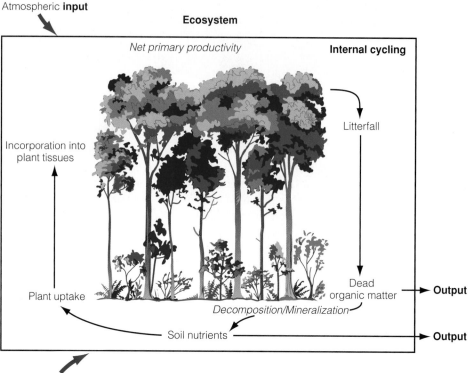

Atmospheric **input**

Ecosystem

Net primary productivity   **Internal cycling**

Litterfall

Incorporation into
plant tissues

Dead
organic matter   **Output**

Plant uptake

Decomposition/Mineralization

Soil nutrients   **Output**

Input from the weathering
of rocks and minerals

FIGURE 21.1 A generalized model of nutrient cycling in a terrestrial ecosystem. The three common components of inputs, internal cycling, and outputs are shown in bold. The key ecosystem processes of net primary productivity and decomposition are italicized.

into new tissues. This process is called **internal cycling** and is an essential feature of all ecosystems. It represents a recycling of nutrients within the ecosystem (see Focus on Ecology 21.2: Hot Ecology).

Only a small fraction of the nutrient pool is involved in short-term annual cycling of nutrients in the forest ecosystem. Nutrients taken up by trees are returned to the forest floor by litterfall, throughfall, and stemflow (see Figure 21.1). A significant portion of the nutrient uptake is stored in tree limbs, trunk, bark, and roots as accumulated biomass. This portion is effectively removed from short-term cycling. Some of the nutrients accumulate in the litter and in the living biomass of consumer organisms, including decomposers of the forest floor, from which they are recycled at various rates. Nutrients accumulated in soil organic matter have a key role in recycling because they prevent rapid losses from the ecosystem. Large quantities of nutrients are bound tightly in this organic matter structure; they are not readily available until released by activities of decomposers.

Open-water ecosystems, such as lakes and ponds, lack the long-term biological retention of nutrients typical of forested systems. Nutrient availability

depends heavily on the turnover of nutrients in phytoplankton and zooplankton. Major long-term storage takes place in deep bottom sediments, where nutrients may be unavailable for a long time. Retention of nutrients in flowing-water ecosystems is difficult, but it is aided by logs and rocks that hold detritus in place, by algal uptake of nutrients, and by aquatic invertebrates (see Section 21.8 and Chapter 27).

21.4 Key Ecosystem Processes Influence the Rate of Nutrient Cycling

You can see from Figure 21.1 that the internal cycling of nutrients through the ecosystem depends on the processes of primary production and decomposition. Primary productivity determines the rate of nutrient transfer from inorganic to organic form (nutrient uptake), and decomposition determines the rate of transformation of organic nutrients into

HOT ECOLOGY

Ecologists have developed a good understanding of the processes controlling the cycling of nutrients within ecosystems. Although time consuming, quantifying the amount of nutrients in various components of the ecosystem at any one time can be accomplished by sampling the soil, plants, and other organisms and determining their nutrient concentrations. Quantifying the rates of exchange between the various components of the ecosystem is infinitely more difficult. During the 1950s and 1960s, an interdisciplinary field of radiation ecology developed. It was the product of the "nuclear age" that emerged with the development of nuclear weapons in the Manhattan Project and their use in World War II. Radiation ecology was concerned not only with studying the effects of ionizing radiation, but also with the development and use of radioactive compounds that could be used as tracers to examine the movement of nutrients through ecosystems.

In the early 1960s, J. P. Witherspoon of Oak Ridge National Laboratory conducted a pioneering study using radioisotopes of elements to quantify the cycling of nutrients through an ecosystem. The object was to follow the pathway of a radiolabeled trace element (micronutrient) through a forest ecosystem. Cesium behaves like potassium. It is highly mobile, cycles rapidly in an ionic form, and is easily leached from plant surfaces by rainfall. Moreover, because a known quantity of the element could be traced, the amounts of the element apportioned to wood, twigs, and leaves could be determined.

Witherspoon inoculated the trunks (boles) of 12 white oak *(Quercus alba)* trees with 20 microcuries (μC) of ^{134}Cs. He followed gains, losses, and transfers of this isotope in the trees and soil. About 40 percent of the ^{134}Cs inoculated into the oaks in April moved into the leaves in early June (Figure A). Leaching of radiocesium from the leaves began when the first rains fell after inoculation. By September, this loss amounted to 13 percent of the maximum concentration in the leaves. Seventy percent of this rainwater loss reached the mineral soil; the remaining 30 percent found its way into the litter and understory. When the leaves fell in autumn, they carried with them twice as much radiocesium as had leached from the crown. Over the winter, half was leached to the mineral soil. Of the radiocesium in the soil, 92 percent still remained in the upper 10 cm nearly 2 years after inoculation. Eight percent of the cesium was confined to an area within the crown perimeter, and 19 percent was located in a small area about the trunk. In spring, cesium retained over winter in the wood and minimal transfers from the soil and litter moved back into the leaves. This quantified study provided early insights and a general model of internal cycling and retention of elements in forest trees.

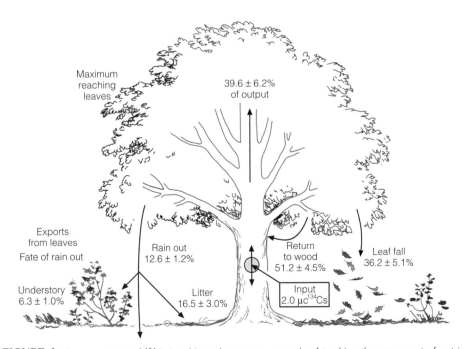

FIGURE A The movement of ^{134}Cs in white oak trees, an example of tracking the movement of nutrients through the ecosystem using radioactive isotopes. The figures represent an average of 12 trees at the end of the 1960 growing season.

inorganic form (nutrient release). Therefore, the rates at which these two processes occur directly influences the rates at which nutrients cycle through the ecosystem. But how do these two key processes interact to limit the rate of the internal cycling of nutrients through the ecosystem? The answer lies in their interdependence.

For example, consider the cycling of nitrogen, an essential nutrient for plant growth. The direct link between soil nitrogen availability, rate of nitrogen uptake by plant roots, and the resulting leaf nitrogen concentrations was presented in Chapter 6 (see Figure 6.17). The maximum rate of photosynthesis is strongly correlated with nitrogen concentrations in the leaves (see Figure 6.16) because certain compounds directly involved in photosynthesis (e.g., rubisco and chlorophyll) contain a large portion of leaf nitrogen. Thus, availability of nitrogen in the soil will directly affect rates of ecosystem primary productivity via its influence on photosynthesis and carbon uptake.

A low availability of soil nitrogen reduces not only net primary production (the total production of plant tissues), but also the nitrogen concentration of the plant tissues that are produced (again, see Figures 6.16 and 6.17). Thus, the reduced availability of soil nitrogen influences the input of dead organic matter to the decomposer food chain (see Chapter 20, Section 20.10) by reducing both the total quantity of dead organic matter produced and its nutrient concentration. The net effect is a lower input of nitrogen in the form of dead organic matter.

Both the quantity and quality of organic matter as a food source for decomposers directly relate to the rate of decomposition and nitrogen mineralization (nutrient release) (Chapter 7, Section 7.7). Lower nutrient concentrations in the dead organic matter promote immobilization of nutrients from the soil and water to meet the nutrient demands of the decomposer populations. This immobilization effectively reduces nutrient availability to the plants, adversely affecting primary productivity.

You can now appreciate the feedback system that exists in the internal cycling of nutrients within an ecosystem (Figure 21.2). Reduced nutrient availability can have the combined effect of reducing both the nutrient concentration of plant tissues (primarily leaf tissues) and net primary productivity. This reduction lowers the total amount of nutrients returned to the soil in dead organic matter. The reduced quantity and quality (nutrient concentration) of organic matter entering the decomposer food chain increases immobilization and reduces the availability of nutrients for uptake by plants. In effect, low nutrient availability begets low nutrient availability. Conversely, high nutrient availability

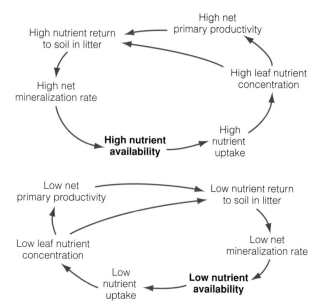

FIGURE 21.2 Feedback that occurs between nutrient availability, net primary productivity, and nutrient release in decomposition for initial conditions of low and high nutrient availability. (After Chapin 1980.)

encourages high plant tissue concentrations and high net primary productivity. In turn, the associated high quantity and quality of dead organic matter encourages high rates of net mineralization and nutrient supply in the soil.

21.5 Both Climate and Plant Characteristics Influence the Rate of Nutrient Cycling

In addition to its role in the weathering of rocks and minerals and soil formation (see Chapter 5, Section 5.3), climate directly affects the rate of nutrient cycling in ecosystems by influencing rates of primary production (see Figure 20.3 and 20.5) and decomposition (see Figure 7.10). Both increase as conditions become warmer and wetter. The net effect is a faster rate of nutrient cycling in warm, wet environments, such as a tropical rain forest, than in cooler (temperate or boreal forest) or drier (grassland) ecosystems.

Nutrient cycling in an ecosystem is also influenced by the nature of its organisms. Organisms such as phytoplankton and zooplankton in aquatic systems are short-lived, grow rapidly, absorb nutrients quickly, and just as quickly return them to water as available nutrients. Other organisms, such as forest trees, are large, grow slowly, and store large quantities of nutrients in their biomass for much longer periods. This portion is effectively removed from short-term cycling.

Each species contributes differently to the overall nutrient cycling within an ecosystem. Trees and shrubs, for example, sequester varying amounts of elements in short- and long-term nutrient pools in their structural components—wood, bark, twigs, roots, and leaves. Short- and long-term cycles are defined by the rate at which the different structural components decompose. Nutrients sequestered in slower-decomposing tissues such as bark and wood are released more slowly than those contained in leaves, twigs, and fine roots.

Species also differ in their concentrations of nutrients and rate at which they decompose (see Chapter 7, Sections 7.4 and 7.7). These differences in litter quality will directly influence the rates of decomposition, nutrient mineralization and subsequent primary productivity of the ecosystem. John Pastor of the University of Minnesota and colleagues examined above-ground production and nutrient cycling (nitrogen and phosphorus) in a series of forest stands along a gradient of soil texture on Blackhawk Island, Wisconsin. Tree species producing higher-quality litter (lower ratio of carbon to nitrogen; see Chapter 7, Section 7.7) dominated sites with a progressively finer soil texture (silt and clay content; see Chapter 5, Section 5.6 and 5.7). The higher-quality litter resulted in a higher rate of nutrient mineralization (release) (Figure 21.3a). Higher rates of nutrient availability in turn resulted in a higher rate of primary productivity (see Figure 20.7) and nutrient return in litterfall (Figure 21.3b; also see Figure 21.2). The net effect was to increase the rate at which nitrogen and phosphorus cycle through the forest stands. The changes in species composition and litter quality along the soil gradient were directly related to the influence of soil texture on plant available moisture (see Chapter 5, Section 5.7).

21.6 Outputs Represent a Loss of Nutrients from the Ecosystem

The export of nutrients from the ecosystem represents a loss that must be offset by inputs if a net decline is not to occur. Export can occur in a variety of ways depending on the nature of the specific biogeochemical cycle. Carbon is exported to the atmos-

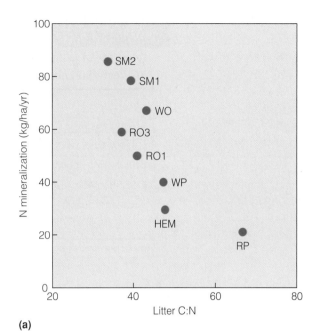

(a)

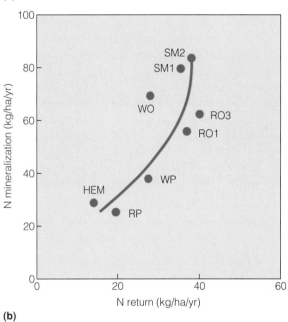

(b)

FIGURE 21.3 Relationship between (a) litter quality (C:N) and nitrogen mineralization rate (N availability), and (b) nitrogen mineralization rate and nitrogen returned in annual litterfall for a variety of forest ecosystems on Blackhawk Island, Wisconsin. The abbreviations refer to the dominant trees in each stand: Hem = hemlock; RP = red pine; RO = red oak; WO = white oak; SM = sugar maple; WP = white pine. (From Pastor et al. 1984.)

phere in the form of CO_2 via the process of respiration by all living organisms (Chapter 6, Chapter 20, Section 20.10). Likewise, a variety of microbial and plant processes result in the transformation of nutrients to a gaseous phase that can subsequently be transported from the ecosystem in the atmosphere. Examples of these processes will be provided in the

TABLE 21.1

Comparison of the Average Annual Yield and Nutrient Removal by a 16-Year-Old Loblolly Pine Plantation with That of Agricultural Crops

Crop	Yield (Tons/ha)	REMOVAL (kg/ha)			
		N	P	K	Ca
Loblolly pine, whole tree	14.5	17.5	2.4	12.6	12.8
Loblolly pine, pulpwood	8.75	6.5	0.9	5.1	6.4
Corn (grain)	11.75	130.5	29.7	37.3	—
Soybeans (beans)	3.0	145.0	14.85	46.7	—
Alfalfa (forage)	10.0	212.5	23.25	185.6	75.4

Source: Data from Jorgensen and Wells 1986.

following chapter, which examines specific biogeochemical cycles.

Transport of nutrients from the ecosystem can also occur in the form of organic matter. Organic matter from a forested watershed can be carried from the ecosystem through surface flow of water in streams and rivers. The input of organic carbon from terrestrial ecosystems constitutes the majority of energy input into stream ecosystems (see Chapter 27). Organic matter can also be transferred between ecosystems by herbivores. Moose feeding on aquatic plants can transport and deposit nutrients to adjacent terrestrial ecosystems in the form of feces. Conversely, the hippopotamus (*Hippopotamus*

amphibius) feeds at night on herbaceous vegetation adjacent to the body of water in which it resides. Large quantities of nutrients are then transported in the form of feces and other wastes to the water.

Organic matter has a key role in recycling nutrients because it prevents rapid losses from the system. Large quantities of nutrients are bound tightly in organic matter structure; they are not readily available until released by activities of decomposers. However, some nutrients are leached from the soil and carried out of the ecosystem by underground water flow to streams. These losses may be balanced by inputs to the ecosystem, such as the weathering of rocks and minerals (see Chapter 5).

Considerable quantities of nutrients are withdrawn permanently from ecosystems by harvesting (Table 21.1; also see Chapter 18, Section 18.4), especially in farming and logging, as biomass is directly removed from the ecosystem. In such ecosystems, these losses must be replaced by the application of

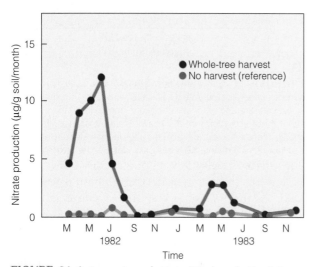

FIGURE 21.4 Comparison of nitrate (NO_3^-) production following logging for a loblolly pine (*Pinus taeda*) plantation in the southeastern United States. Data for the reference stand (no harvest) are compared with those of a whole-tree harvest clear-cut. (Adapted from Vitousek 1992.)

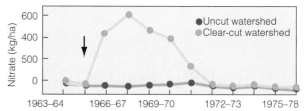

FIGURE 21.5 Temporal changes in the annual net export of nitrate in streamwater for two forested watersheds in Hubbard Brook, New Hampshire. The forest on one watershed was clearcut (note arrow), while the other forest remained undisturbed. Note the large increase in concentrations of nitrate in the stream on the clear-cut watershed. This increase in export is due to increased decomposition and nitrogen mineralization following the removal of trees. The nitrogen was then leached into the surface and groundwater. (Adapted from Likens and Borman 1995.)

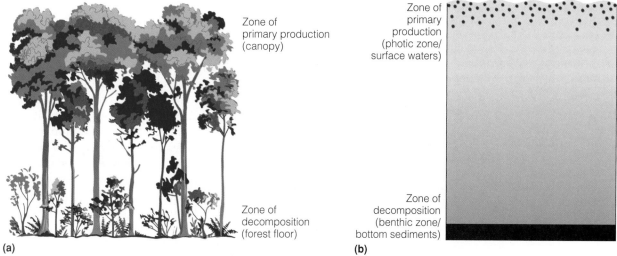

Zone of
primary production
(canopy)

Zone of
decomposition
(forest floor)

Zone of
primary
production
(photic zone/
surface waters)

Zone of
decomposition
(benthic zone/
bottom sediments)

(a)   **(b)**

FIGURE 21.6 Comparison of the vertical zones of production and decomposition in (a) a terrestrial (forest) and (b) an open-water (lake) ecosystem. Note that in the terrestrial ecosystem the two zones are linked by the vegetation (trees). However, this is not the case in the lake ecosystem.

fertilizer; otherwise, the ecosystem becomes impoverished. In addition to the nutrients removed directly through biomass removal, logging can also result in the transport of nutrients from the ecosystem by altering processes involved in internal cycling. The removal of trees in clear-cutting and other forest management practices increases the amount of radiation (including direct sunlight) reaching the soil surface. The resulting increase in soil temperatures promotes decomposition (see Chapter 7) and results in an increase in net mineralization rates (Figure 21.4). This increase in nutrient availability in the soil occurs at the same time that demand for nutrients is low because plants have been removed and net primary productivity is low. As a result, there is a dramatic increase in the leaching of nutrients from the ecosystem in surface waters (Figure 21.5). This export of nutrients from the ecosystem results from decoupling the two processes of nutrient release in decomposition and nutrient uptake in net primary productivity.

Depending on its intensity, fire kills vegetation and converts varying proportions of the biomass and soil organic matter to ash (see Chapter 19 for a discussion of fire). In addition to the loss of nutrients through volatilization and airborne particles, the addition of ash changes the chemical and biological properties of the soil. Many nutrients become readily available, and nitrogen in ash is subject to rapid mineralization. If not taken up by vegetation during recovery, nutrients may be lost from the ecosystem through leaching and erosion. Stream-water runoff is often greatest after fire because of reduced water demand for transpiration. High nutrient availability

in the soil coupled with high runoff can lead to large nutrient losses from the ecosystem.

21.7 Nutrient Cycling Differs Between Terrestrial and Aquatic Ecosystems

The process of nutrient cycling is an essential feature of all ecosystems and represents a direct (cyclic) link between net primary productivity and decomposition. However, the nature of this link varies among ecosystems, particularly between terrestrial and aquatic ecosystems.

In virtually all ecosystems, there is a vertical separation between the zones of production (photosynthesis) and decomposition (Figure 21.6). In terrestrial ecosystems, the plants themselves bridge this physical separation between the zone of decomposition at the soil surface and the zone of productivity in the plant canopy; the plants physically exist in both zones. The root systems provide access to the nutrients made available in the soil through decomposition, and the vascular system within the plant transports these nutrients to the sites of production (canopy).

In aquatic ecosystems, this is not always the case. In shallow-water environments of the shoreline, emergent vegetation such as cattails, cordgrasses *(Spartina)*, and sedges are rooted in the sediments. Here, as in terrestrial ecosystems, the zone

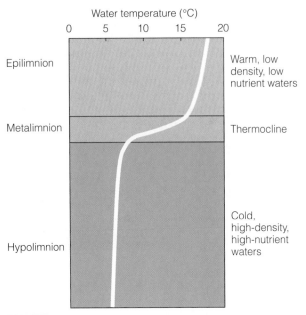

Epilimnion — Warm, low density, low nutrient waters

Metalimnion — Thermocline

Hypolimnion — Cold, high-density, high-nutrient waters

FIGURE 21.7 Vertical structure of an open-water ecosystem such as a lake or an ocean. The vertical profile can be divided into three distinct layers. The epilimnion or surface water is a layer of warm, oxygen-rich water. This hypolimnion is the deep, cold oxygen-poor layer. The transition zone between these two layers is called the metalimnion and is characterized by a dramatic shift in temperature, called the thermocline.

of decomposition and production are linked directly by the plants. Likewise, submerged vegetation, such as seagrasses, is rooted in the sediments, and the plants extend up the water column into the photic zone (see Chapter 13, Figure 13.2), the shallower waters where light levels support higher productivity. However, as water depths increase, primary production is dominated by free-floating phytoplankton within the upper waters (photic zone). Here exists a physical separation between the zones of decomposition in the bottom sediments and waters, or benthic zone, and the surface waters where temperatures and light availability support primary productivity. This physical separation between the zones where nutrients become available through decomposition and the zone of productivity where nutrients are needed to support photosynthesis and plant growth is a major factor controlling the productivity of open-water ecosystems (e.g., lakes and oceans).

To understand how nutrients are transported vertically from the deeper waters to the surface, where temperature and light conditions can support primary productivity, we must first examine the vertical structure of the physical environment in open-water ecosystems first presented in Chapter 4 (Section 4.11). As presented briefly in both Chapters 4 and

13, the vertical structure of open-water ecosystems, such as lakes or oceans, can be divided into three rather distinct zones: the epilimnion, the metalimnion, and the hypolimnion (Figure 21.7). The epilimnion, or surface water, is relatively warm as a result of the interception of solar radiation. In addition, the oxygen content is relatively high due to the diffusion of oxygen from the atmosphere into the surface waters. In contrast, the hypolimnion, or deep water, is cold and relatively low in oxygen. The metalimnion is a transition zone between the surface and deep waters and is characterized by a steep temperature gradient called the thermocline. The thermocline is the transition zone between the warmer surface waters and the deeper cold waters. In effect, the vertical structure can be represented as a warm, low-density surface layer of water on top of a denser cold layer of deep water, separated by the rather thin zone of the thermocline. This vertical structure and physical separation of the epilimnion and hypolimnion have an important influence on the distribution of nutrients and subsequent patterns of primary productivity in aquatic ecosystems. The colder deep waters are relatively nutrient-rich, but temperature and light conditions cannot support high productivity. In contrast, the surface waters are relatively nutrient-poor; however, this is the zone where temperatures and light can support high productivity.

Although winds blowing over the water surface cause turbulence that mixes the waters of the epilimnion, this mixing does not extend into the colder, deeper waters because of the thermocline. As autumn and winter approach in the temperate and polar zones, the amount of solar radiation reaching the water surface decreases and the temperature of the surface water declines. As the water temperature of the epilimnion approaches that of the hypolimnion, the thermocline breaks down, and mixing throughout the profile can take place (Figure 21.8). If surface waters become cooler than the deeper waters, they will begin to sink, displacing deep waters to the surface. This process is called turnover. With the breakdown of the thermocline and mixing of the water column, nutrients are brought up from the bottom to the surface waters. With the onset of spring, increasing temperatures and light in the epilimnion give rise to a peak in productivity with the increased availability of nutrients in the surface waters. As the spring and summer progress, the nutrients in the surface water are used, reducing the nutrient content of the water, and a subsequent decline in productivity occurs. The resulting annual cycle of productivity in these ecosystems (Figure 21.9) is a direct function of the dynamics of the thermocline and the resulting behavior of the vertical distribution of nutrients.

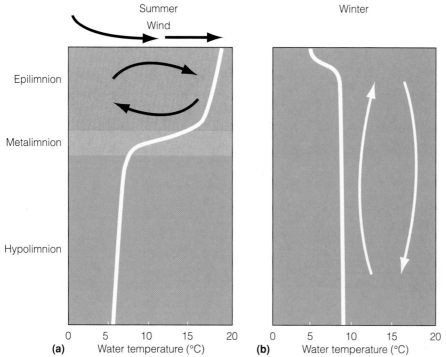

FIGURE 21.8 Seasonal dynamics in the vertical structure of an open-water aquatic ecosystem in the Temperate Zone. Winds mix the waters within the epilimnion during the summer (a), but the thermocline isolates this mixing to the surface waters. With the breakdown of the thermocline during the winter months (b), turnover occurs, allowing the entire water column to become mixed. This mixing allows nutrients in the epilimnion to be brought up to the surface waters.

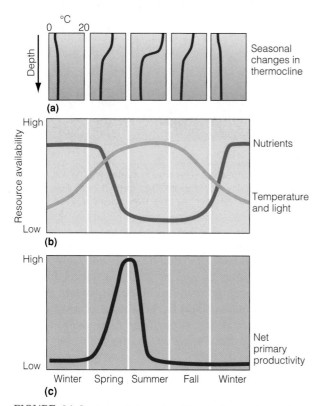

FIGURE 21.9 Seasonal dynamics of (a) the thermocline and associated changes in (b) the availability of light and nutrients, and (c) net primary productivity of the surface waters.

21.8 Water Flow Influences Nutrient Cycling in Streams and Rivers

Inputs in the form of dead organic matter from adjacent terrestrial ecosystems (leaves and woody debris), rainwater, and subsurface seepage bring nutrients into streams. Although the internal cycling of nutrients follows the same general pathway as that discussed for both terrestrial and open-water ecosystems (also see Figure 21.1), the continuous, directional movement of water alters the nature of nutrient cycling in stream ecosystems. Jack Webster of the University of Georgia was the first to note that because nutrients are continuously being transported downstream, a spiral rather than a cycle better represents the cycling of nutrients. He coined the term **nutrient spiraling** to describe this process.

Nutrients in terrestrial and open-water ecosystems are recycled more or less in place. An atom of nutrients passes from the soil or water column to plants and consumers and back to soil or water in the form of dead organic matter. Then it is recycled within the same location within the ecosystem, although losses do occur. Cycling essentially involves

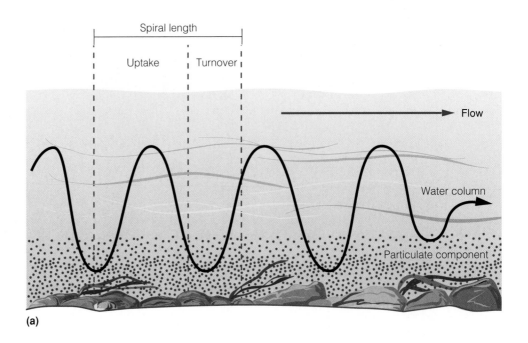

(a)

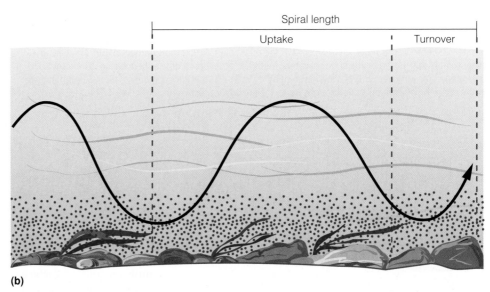

(b)

FIGURE 21.10 Nutrient spiraling in a stream ecosystem between organic matter and the water column. Uptake and turnover take place as nutrients flow downstream. The tighter the spiraling, the longer the nutrients remain in place. (a) Tight spiraling. (b) Open spiraling. (Newbold et al. 1982.)

time. Flowing water has an added element, a spatial cycle. Nutrients in the form of organic matter are constantly being carried downstream. How quickly these materials are carried downstream depends on how fast the water moves and what physical and biological means hold nutrients in place. Physical retention involves storage in wood detritus such as logs and snags, in debris caught in pools behind logs and boulders, in sediments, and within patches of aquatic vegetation. Biological retention occurs through the uptake and storage of nutrients in animal and plant tissue.

The processes of recycling, retention, and downstream transport may be pictured as a spiral lying horizontally (longitudinally) in a stream (Figure 21.10). One cycle in the spiral is the uptake of an atom of nutrient, its passage through the food chain, and its return to water, where it is available for reuse. Spiraling is measured as the distance needed for completion of one cycle. The longer the distance required, the more open the spiral; the shorter the distance, the tighter the spiral. If dead leaves and other debris are physically held in place long enough to allow organisms to process the organic matter, the spiral is tight. This type of physical retention is especially important in fast headwater streams, which can rapidly lose organic matter downstream. Organisms can function to both open and tighten the spiral. Organisms that shred and fragment the organic matter can open the spiral by facilitating the transport of organic materials downstream. Other organisms tighten the spiral by physically storing dead organic matter.

J. D. Newbold and colleagues at Oak Ridge National Laboratory in Tennessee experimentally determined how quickly phosphorus moved downstream in a small woodland brook, Walker Branch. They determined that phosphorus moved downstream at the rate of 10.4 meters a day and cycled once every 18.4 days. The average downstream distance of one cycle (spiral) was 190 meters. In other words, one atom of phosphorus on the average completed one cycle from the water compartment and back again for every 190 meters of downstream travel.

FIGURE 21.11 Coastal ecosystem in Chesapeake Bay, Maryland, showing estuary with salt marsh vegetation in the foreground.

21.9 Land and Marine Environments Influence Nutrient Cycling in Coastal Ecosystems

Coastal ecosystems are among the most productive environments. Water from most streams and rivers eventually drains into the oceans; and the place where this freshwater joins saltwater is called an estuary. Estuaries are semi-enclosed parts of the coastal ocean where seawater is diluted and partially mixed with water coming from the land. As the rivers meet the ocean, the current velocity drops, and sediments are deposited within a short distance (referred to as a sediment trap; see Figure 21.11). The buildup of sediments creates alluvial plains about the estuary, giving rise to mudflats and salt marshes that are dominated by grasses and small shrubs rooted in the mud and sediments (Figure 21.12; see Chapters 27 and 28 for detailed descriptions of these ecosystems). The cycling of nutrients in these ecosystems differs from that of terrestrial, open-water, and stream ecosystems discussed thus far. In a way it combines features of each. As with terrestrial ecosystems, the dominant plants are rooted in the sediments and therefore function to link the zones of decomposition and primary production (see Figure 21.5). Submerged plants take up nutrients from both the sediments and directly from the water column. As with streams and rivers, the directional (horizontal) flow of water functions to

transport both organic matter (energy) and nutrients both into (inputs) and out of (outputs) the ecosystem.

Nutrients are carried into the coastal marshes by precipitation, surface water (streams and rivers), and groundwater. In addition, the rise and fall of water depth with the tidal cycle serves to flush out salts and other toxins from the marshes and brings in nutrients from the coastal waters, a process referred to as the tidal subsidy. It also serves to replace oxygen-depleted waters within the surface sediments with oxygenated water.

The salt marsh is a detrital system, with only a small portion of primary production being consumed by herbivores. Almost three-quarters of the detritus (dead organic matter) produced in the salt-marsh ecosystem is broken down by bacteria and fungi. Nearly 50 percent of the total net primary productivity is lost through respiration by microbial decomposers. The low oxygen content of the sediments favors anaerobic bacteria. They can carry on their metabolic functions without oxygen, by using inorganic compounds such as sulfates rather than oxygen in the process of fermentation (Chapter 7).

A substantial portion (usually 20 to 45 percent) of the net primary production of a salt marsh is exported to adjacent estuaries. The exact nature of this exchange is a function of the geomorphology of the basin (its shape and nature of the opening to the sea), and the magnitude of tidal and freshwater flows (fluxes). Each salt marsh apparently differs in the way carbon and other nutrients move through the food web and in the route taken and amount of nutrients that are exported. Some salt marshes are dependent on tidal exchanges and import more than they export, whereas others export more than they import. A portion may be exported to the estuary as

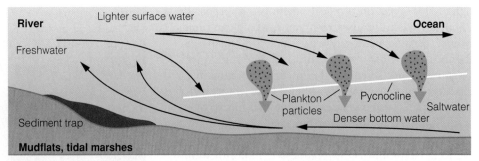

FIGURE 21.12 Circulation of freshwater and saltwater in an estuary creates a nutrient trap. A salty wedge of intruding seawater on the bottom produces a surface flow of lighter fresh water and a counterflow of heavier brackish water. These layers are physically separated by variations in water density arising from both salt concentration and temperature differences. The zone of maximum vertical difference in density is called the pycnocline, which functions much like the thermocline in lake ecosystems. Living and dead particles settle through the pycnocline into the countercurrent and are carried up the estuary along with their nutrient content, conserving the nutrients within the estuary rather than flushing them out to sea. (Correll 1978.)

mineral nutrients, or physically as detritus, as bacteria, or as fish, crabs, and intertidal organisms within the food web.

Although inflowing water from rivers and coastal marshes carries mineral nutrients into the estuary, primary production is regulated more by internal nutrient cycling than by external sources. This internal cycling involves the release of nutrients through decomposition and mineralization within the bottom sediments, as well as excretion of mineralized nutrients by herbivorous zooplankton (see Chapter 7, Section 7.8).

As with the tidal marshes, nutrients and oxygen are carried into the estuary by the tides. Typical estuaries maintain a "salt wedge" of intruding seawater on the bottom, producing a surface flow of freshwater and a counterflow of more brackish, heavier water (Figure 21.12). These layers are physically separated by variations in water density arising from both salt concentration and temperature differences.

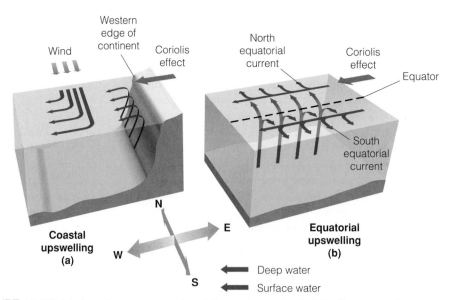

FIGURE 21.13 (a) Along the western margins of the continents, the Coriolis effect causes the surface waters to move offshore (solid arrows). The movement of surface waters offshore results in an upwelling of deeper, more nutrient-rich waters to the surface (blue arrows). Example shown is for the Northern Hemisphere. (b) Along the equator, the Coriolis effect acts to pull the westward flowing currents to the north and south (solid arrows), resulting in an upwelling of deeper, more nutrient-rich waters to the surface. (From: *Marine Biology*, J. Nybakken 1993.)

The zone of maximum vertical difference in density is called the **pycnocline** (see Figure 21.12), which functions much like the thermocline (see Section 21.7) in lake ecosystems. Living and dead particles settle through the pycnocline into the countercurrent and are carried up the estuary along with their nutrient content, conserving the nutrients within the estuary rather than being flushed out to sea.

The regular movement of freshwater and saltwater into the estuary, coupled with the shallowness and turbulence, generally allows for sufficient vertical mixing to occur. In deeper estuaries, a thermocline can form during the summer months. In this case, the seasonal pattern of vertical mixing and nutrient cycling will be similar to the pattern discussed for open-water ecosystems in Section 21.7.

21.10 Surface Ocean Currents Bring About Vertical Transport of Nutrients

The global pattern of ocean suface currents presented in Chapter 3 (Section 3.5, Figure 3.16) influences patterns of surface water temperature, productivity, and nutrient cycling. The Coriolis effect drives the patterns of suface currents. But how deep does this lateral movement of water extend vertically into the water column? In general, the lateral flow is limited to the upper 100 meters, but in certain regions the lateral movement can bring about a vertical circulation or **upwelling** of water. Along the western margins of the continents, the surface currents flow along the coastline and toward the equator (see Figure 3.16). At the same time, these surface waters are pushed offshore by the Coriolis effect. The movement of surface waters offshore results in deeper, more nutrient-rich waters being transported vertically to the surface (Figure 21.13a).

Surface currents give rise to a similar pattern of upwelling in the equatorial waters. As the two equatorial currents flow west, they are deflected to the right north of the equator, and to the left south of the equator. Where this occurs, subsurface water is transported vertically, bringing cold waters, rich in nutrients, to the surface (Figure 21.13b). These regions of nutrient-rich waters are highly productive (see Figure 20.8) and support some of the world's most important fisheries.

CHAPTER REVIEW

Summary

Biogeochemical Cycles (21.1) Nutrients flow from the living to the nonliving components of the ecosystem and back in a perpetual cycle. By means of these cycles, plants and animals obtain nutrients necessary for their survival and growth. There are two basic types of biogeochemical cycles: the gaseous, represented by oxygen, carbon, and nitrogen, whose major pools are in the atmosphere; and the sedimentary, represented by the sulfur and phosphorus cycles, whose major pools are in Earth's crust. The sedimentary cycles involve two phases, salt solution and rock. Minerals become available through the weathering of Earth's crust, enter the water cycle as a salt solution, take diverse pathways through the ecosystem, and ultimately return to Earth's crust through sedimentation.

All nutrient cycles have a common structure, sharing three basic components: inputs, internal cycling, and outputs.

Nutrient Cycling (21.2–21.6) The input of nutrients to the ecosystem depends on the type of biogeochemical cycle, gaseous or sedimentary. The availability of essential nutrients in terrestrial ecosystems depends heavily on the nature of the soil. Supplementing nutrients in the soil are nutrients carried by rain, snow, air currents, and animals. The major sources of nutrients for aquatic life are inputs from the surrounding land in the form of drainage water, detritus, and sediments **(21.2)**.

As plants take up nutrients from the soil or water, they become incorporated into living tissues, organic matter. As the tissues senesce, the dead organic matter is returned to the soil or sediment surface. Various decomposers transform the organic nutrients into mineral form, and they are once again available for uptake by plants. This process is called internal cycling **(21.3)**. The rate at which nutrients cycle through the ecosystem is directly related to the rates of primary productivity (nutrient uptake) and decomposition (nutrient release) **(21.4)**. Environmental factors that influence these two processes will affect the rate at which nutrients cycle through the ecosystem **(21.5)**.

The export of nutrients from the ecosystem represents a loss that must be offset by inputs if a net decline is not to occur. Export can occur in a variety of ways, depending on the nature of the specific biogeochemical cycle. A major means of transportation is in the form of organic matter carried by surface flow of water in streams and rivers. Leaching of dissolved nutrients from soils into surface water and groundwater also represents a significant export in some ecosystems. Harvesting of biomass in forestry and agriculture represents a permanent withdrawal from the ecosystem. Fire is also a major source of nutrient export in some terrestrial ecosystems (21.6).

Comparison of Terrestrial and Aquatic Ecosystems (21.7) There is typically a vertical separation between the zones of primary production and decomposition. In terrestrial and shallow-water ecosystems, plants function to bridge this gap. In open-water ecosystems, there is a physical separation between these zones that limits nutrient availability in the surface waters. The thermocline functions to limit the movement of nutrients from the bottom (benthic) zone (cold) to the surface (warm) waters. During the winter season, the thermocline breaks down, allowing for a mixing of the water column and the movement of nutrients into the surface waters. This seasonality of the thermocline and mixing of the water column controls seasonal patterns of productivity in these ecosystems.

Stream Ecosystems (21.8) The continuous, directional movement of water alters the nature of nutrient cycling in stream ecosystems. Because nutrients are continuously being transported downstream, a spiral rather than a cycle better represents the cycling of nutrients. One cycle in the spiral is the uptake of an atom of nutrient, its passage through the food chain, and its return to water, where it is available for reuse. The length of the cycle is related to the flow rate of the stream and the physical and biological mechanisms available for nutrient retention.

Coastal Ecosystems (21.9–21.10) Water from most streams and rivers eventually drain into the oceans, giving rise to estuary and salt-marsh ecosystems along the coastal environment. As with streams and rivers, the directional flow of water functions to transport both organic matter and nutrients both into and out of the ecosystem. The rise and fall of water depth with the tidal cycle serves to flush out salts and other toxins from the marshes and brings in nutrients from the coastal waters. The combined effect of the inward (toward the coast) movement of saltwater together with the outward flow of freshwater is to develop a countercurrent that carries both living and dead particles

and the nutrients they contain back toward the coastline. This mechanism functions to conserve nutrients within the estuary and salt-marsh ecosystems.

The global pattern of surface currents brings about the transport of deep, nutrient-rich waters to the surface in coastal regions. As surface currents move waters away from the western coastal margins, deep water moves to the surface carrying nutrients with it. A similar pattern of upwelling occurs in the equatorial regions of the oceans where surface currents move to the north and south (21.10).

Review Questions

1. What are the two types of biogeochemical cycles? What are their distinguishing characteristics?

2. What are some of the inputs and outputs in a nutrient cycle?

3. What is the relationship among net primary production, decomposition, and nutrient cycling?

4. Contrast nutrient cycling in terrestrial and open-water aquatic ecosystems. What is the outstanding difference?

5. How does the continuous, directional flow of water influence the cycling of nutrients in stream ecosystems?

6. What mechanism functions to conserve nutrients in estuary ecosystems?

7. In natural ecosystems, nutrients are recycled in place; however, few ecosystems are natural and nutrients are often removed from ecosystems by human actions such as timber harvest and food production. What role do we as individuals play in nutrient cycling? What was the source of the nutrients you consumed yesterday? Make a list of the foods you ate, their point of origin, and their nutrient content.

8. Discuss methods of garbage disposal in relation to nutrient cycling and in relation to input and output. Are we impoverishing the very ecosystems upon which we depend for food?

9. What are the major shortcomings of chemical fertilizers in relation to nutrient cycling?

Impala *(Aepyceros melampus)* standing in the shade of acacia trees. Their urine and droppings make the impala important contributors to the internal nitrogen cycle of these trees.

Biogeochemical Cycles

OBJECTIVES

On completion of this chapter, you should be able to:

- Explain the role of photosynthesis and respiration in the carbon cycle.
- Discuss how atmospheric concentrations of CO_2 change both diurnally and seasonally.
- Outline the major pools in the global carbon cycle.
- Discuss the role of bacteria in the nitrogen cycle.
- Contrast the nitrogen and phosphorus cycles.
- Discuss the role of photosynthesis and respiration in the oxygen cycle.
- Explain how the various biogeochemical cycles are linked in nutrient cycling.

In the previous chapter, we presented a general model of nutrient cycling in ecosystems. The cycling of nutrients and energy occurs within all ecosystems, and it is most often studied as a local process—that is, the internal cycling of nutrients within the ecosystem and the identification of exchanges both to (inputs) and from (outputs) the ecosystem. Through these processes of exchange, the biogeochemical cycles of differing ecosystems are linked. Often, the output from one ecosystem represents an input to another, as with the case of the export of nutrients from terrestrial to aquatic ecosystems. The processes of exchange of nutrients among ecosystems require viewing the biogeochemical cycles from a much broader spatial framework than that of a single ecosystem. This is particularly true of those nutrients that possess a gaseous cycle. Because the main pools of these nutrients are the atmosphere and the ocean, they have pronouncedly global circulation patterns. In this chapter, we will explore the cycling of carbon, nitrogen, phosphorus, sulfur, and oxygen, examining the specific processes involved in their movement through the ecosystem. We will then expand our model of biogeochemical cycling to provide a framework for understanding the global cycling of these elements, which are critical to life.

22.1 The Carbon Cycle Is Closely Tied to Energy Flow

Carbon is a basic constituent of all organic compounds and is involved in the fixation of energy by photosynthesis (see Chapter 6). Carbon is so closely tied to energy flow that the two are inseparable. In fact, we typically express ecosystem productivity in terms of grams of carbon fixed per square meter per year (see Chapter 20, Section 20.4).

The source of all carbon in both living organisms and fossil deposits is carbon dioxide (CO_2) in the atmosphere and the waters of Earth. Photosynthesis draws CO_2 from the air and water into the living component of the ecosystem. Just as energy flows through the grazing food chain, carbon passes to herbivores and then to carnivores. Primary producers and consumers release carbon back to the atmosphere in the form of CO_2 by respiration. The carbon in plant and animal tissue eventually goes to the dead organic matter reservoir. Decomposers release it to the atmosphere through respiration.

Figure 22.1 shows the cycling of carbon through a terrestrial ecosystem. The difference between the rate of carbon uptake by plants in photosynthesis and the rate of carbon release by respiration is the net primary productivity (in units of carbon). The difference between the rate of carbon uptake in photosynthesis and the rate of carbon loss due to autotrophic and heterotrophic respiration is the **net ecosystem productivity.**

The rate at which carbon cycles through the ecosystem is determined by a number of processes, particularly the rates of primary productivity and decomposition. Both processes are strongly influenced by environmental conditions such as temperature and precipitation (see Section 20.6). In warm, wet ecosystems such as a tropical rain forest, rates of productivity and decomposition are high, and carbon cycles through the ecosystem quickly. In cool, dry ecosystems, the process is slower. In ecosystems where temperatures are very low, decomposition is slow, and dead organic matter accumulates (see Chapter 25). In swamps and marshes, where dead material falls into the water, organic material does not completely decompose. Stored as raw humus or peat (see Chapter 29), carbon circulates very slowly. Over geologic time, this buildup of partially decomposed organic matter in swamps and marshes has formed fossil fuels (oil, coal, and natural gas).

Similar cycling takes place in freshwater and marine environments (see Figure 22.1). Phytoplankton uses the carbon dioxide that diffuses into the upper layers of water or is present as carbonates and converts it into plant tissue. The carbon then passes from the primary producers through the aquatic food chain. Carbon dioxide produced through respiration is either reutilized or reintroduced to the atmosphere by diffusion from the water surface to the surrounding air (see Chapter 4, Section 4.16 for discussion of the carbon dioxide-carbonate system in aquatic ecosystems).

Significant portions of carbon can be bound as carbonates in the bodies of mollusks and foraminifers and incorporated into their exoskeletons (shells, etc.). Some of these carbonates dissolve back into solution, while some become buried in the bottom mud at varying depths when the organisms die. Isolated from biotic activity, this carbon is removed from cycling. Incorporated into bottom sediments, over geologic time, it may appear in coral reefs and limestone rocks.

22.2 The Cycling of Carbon Varies Daily and Seasonally

If you were to measure the concentration of carbon dioxide in the atmosphere above and within a forest on a summer day, you would discover that it fluctuates

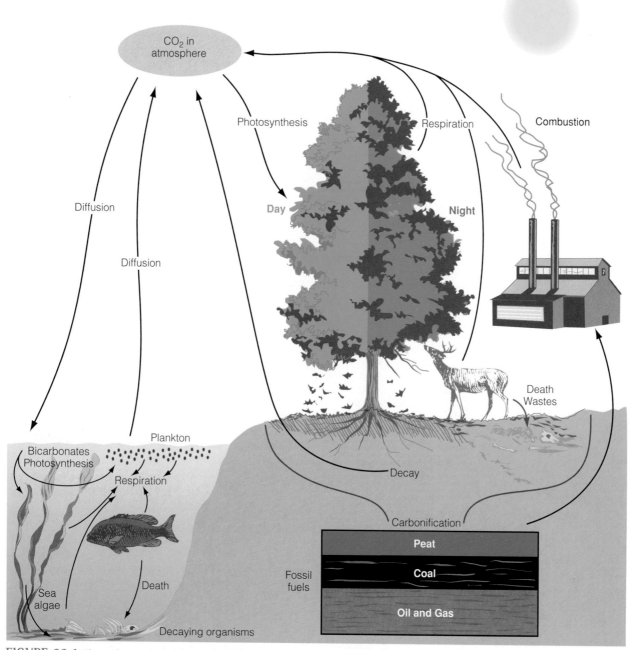

FIGURE 22.1 The carbon cycle as it occurs in both terrestrial and aquatic ecosystems.

throughout the day (Figure 22.2). At daylight when photosynthesis begins, plants start to withdraw carbon dioxide from the air, and the concentration declines sharply. By afternoon when the temperature is increasing and relative humidity is decreasing, the rate of photosynthesis declines, and the concentration of carbon dioxide in the air surrounding the canopy increases. By sunset, photosynthesis ceases (carbon dioxide is no longer is being withdrawn from the atmosphere), respiration increases, and the atmospheric concentration of carbon dioxide increases

sharply. A similar diurnal fluctuation takes place in aquatic ecosystems.

Likewise, there is a seasonal fluctuation in the production and utilization of carbon dioxide that relates both to temperature and to the timing of the growing and dormant seasons (Figure 22.3). With the onset of the growing season when the landscape is greening, the atmospheric concentration begins to drop as plants withdraw carbon dioxide through photosynthesis. As the growing season reaches its end, photosynthesis declines or ceases, respiration is

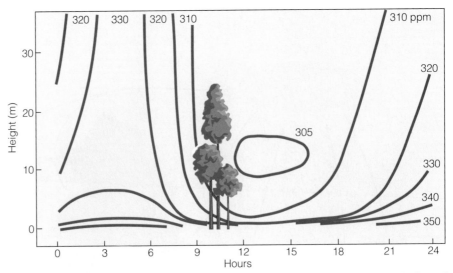

Figure 22.2 Daily flux of CO_2 in a forest. Note the consistently high level of CO_2 on the forest floor, the site of microbial respiration. Atmospheric CO_2 in the forest is lowest from midmorning to late afternoon. CO_2 levels are highest at night, when photosynthesis shuts down and respiration pumps CO_2 into the atmosphere. (From Baumgartner 1968 after Miller and Rusch 1960.)

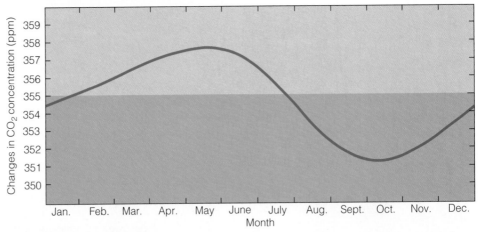

FIGURE 22.3 Variation in atmospheric concentration of CO_2 during a typical year at Barrow, Alaska. Concentrations are highest during the winter months, declining with the onset of photosynthesis during the growing season. (Pearman and Hyson 1981.)

the dominant process, and atmospheric concentrations of carbon dioxide rise. Although these patterns of seasonal rise and decline occur in both aquatic and terrestrial ecosystems, the fluctuations are much greater in terrestrial environments. As a result, these fluctuations in atmospheric concentrations of carbon dioxide are more pronounced in the Northern Hemisphere with its much larger land area.

22.3 The Global Carbon Cycle Involves Exchanges among the Atmosphere, Oceans, and Land

The carbon budget of Earth is closely linked to the atmosphere, land, and oceans and to the mass movements of air around the planet (see Chapter 3). Earth contains about 10^{23} grams of carbon, or 100 million

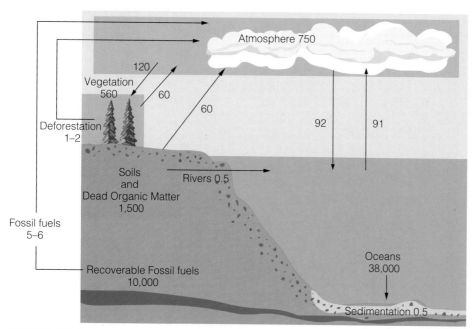

FIGURE 22.4 The global carbon cycle. Boxes show the sizes of the major pools of carbon, and arrows indicate the major exchanges among them. All values are in Gt of carbon, and exchanges are on an annual time scale. The largest pool of carbon, geologic, is not included because of the slow rates (geologic time scale) of transfer with other active pools. (Edmonds 1992.)

Gt (Gt is a gigaton, equal to 1 billion [10^9] metric tons or 10^{15} grams). All but a small fraction is buried in sedimentary rocks and is not actively involved in the global carbon cycle. The carbon pool involved in the global carbon cycle (Figure 22.4) amounts to an estimated 55,000 Gt. Fossil fuels, created by the burial of partially decomposed organic matter, account for an estimated 10,000 Gt. The oceans contain the vast majority of the active carbon pool, about 38,000 Gt, mostly as bicarbonate and carbonate ions (see Chapter 4, Section 4.16). Dead organic matter in the oceans accounts for 1650 Gt of carbon, and living matter, mostly phytoplankton (primary producers), 3 Gt. The terrestrial biosphere (all terrestrial ecosystems) contains an estimated 1500 Gt of carbon as dead organic matter and 560 Gt as living matter (biomass). The atmosphere holds about 750 Gt of carbon.

In the ocean, the surface water acts as the site of main exchange of carbon between atmosphere and ocean. The ability of the surface waters to take up CO_2 is governed largely by the reaction of CO_2 with the carbonate ion to form bicarbonates (see Chapter 4, Section 4.16). In the surface water, carbon circulates physically by means of currents and biologically through photosynthesis by phytoplankton and movement through the food chain. The net exchange of CO_2 between the oceans and atmosphere due to both physical and biological processes results in a net uptake of 1 Gt per year by the oceans, while burial in

sediments accounts for a net loss of 0.5 Gt of carbon per year.

The uptake of CO_2 from the atmosphere by terrestrial ecosystems is governed by gross production (photosynthesis). Losses are a function of autotrophic and heterotrophic respiration, the latter being dominated by microbial decomposers. Until recently, exchanges of CO_2 between the landmass and the atmosphere (uptake in photosynthesis and release by respiration/decomposition) were believed to be nearly in equilibrium (see Figure 22.4). However, more recent research suggests that the terrestrial surface is acting as a carbon sink, with a net uptake of CO_2 from the atmosphere (Chapter 30).

Of considerable importance in the terrestrial carbon cycle are the relative proportions of carbon stored in soils and in living vegetation (biomass). Carbon stored in soils includes dead organic matter on the soil surface and in the underlying mineral soil. Estimates place the amount of soil carbon at 1500 Gt, compared with 560 Gt in living biomass.

The average amount of carbon per volume of soil increases from the tropical regions poleward to the boreal forest and tundra (see Chapters 24–26). Low values for the tropical forest reflect high rates of decomposition, which compensate for high productivity and litterfall. Frozen tundra soil and waterlogged soils of swamps and marshes have the greatest accumulation of dead organic matter because moisture, low temperature, or both inhibit decay.

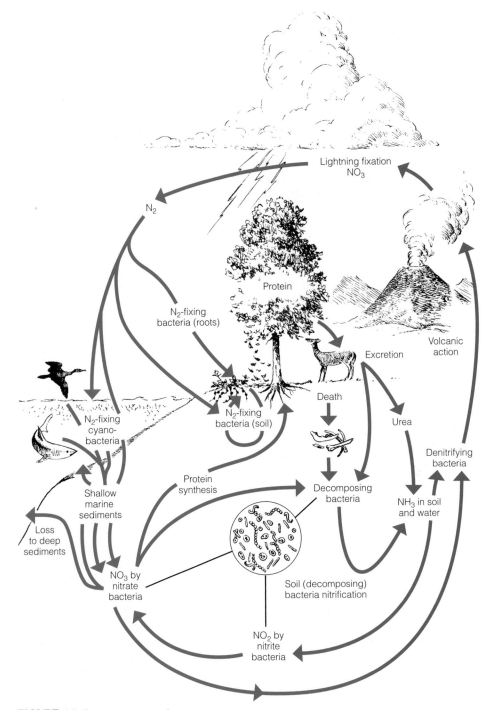

FIGURE 22.5 The nitrogen cycle.

22.4 The Nitrogen Cycle Begins with Fixing Atmospheric Nitrogen

Nitrogen is an essential constituent of protein, which is a building block of all living tissue. Nitrogen is generally available to plants in only two chemical forms: ammonium (NH_4^+) and nitrate (NO_3^-). Thus, although Earth's atmosphere is almost 80 percent nitrogen gas, it is in a form (N_2) that is not available for uptake (assimilation) by plants. Nitrogen enters the ecosystem via two pathways; the relative importance of each varies greatly among ecosystems (Figure 22.5). The first is atmospheric deposition. This can be in wetfall, such as rain, snow, or even cloud and fog droplets, and in dryfall, such as aerosols and particulates (see Chapter 21, Section 21.2). Regardless of the form of atmospheric deposition, nitrogen in this pathway is being supplied in a form that is already available for uptake by plants (also see Chapter 23).

The second pathway for nitrogen to enter ecosystems is via nitrogen fixation. This fixation comes about in two ways. One is high-energy fixation. Cosmic radiation, meteorite trails, and lightning provide the high energy needed to combine nitrogen with the oxygen and hydrogen of water. The resulting ammonia and nitrates are carried to Earth's surface in rainwater. Estimates suggest that less than 0.4 kg N/ha comes to Earth annually in this manner. About two-thirds of this amount comes as ammonia and one-third as nitric acid (HNO_3).

The second method of fixation is biological. This method produces approximately 10 kg N/yr for each hectare of Earth's land surface, or roughly 90 percent of the fixed nitrogen contributed each year. This fixation is accomplished by symbiotic bacteria living in mutualistic association with plants (see Chapter 16, Section 16.11 and Figure 16.7), by free-living aerobic bacteria, and by cyanobacteria (blue-green algae). Fixation splits molecular nitrogen (N_2) into two atoms of free N. The free N atoms then combine with hydrogen to form two molecules of ammonia (NH_3). The process of fixation requires considerable energy. To fix 1 gram of nitrogen, nitrogen-fixing bacteria associated with the root system of a plant must expend about 10 grams of glucose, a simple sugar produced by the plant in photosynthesis.

In agricultural ecosystems, leguminous plants of approximately 200 species are the preeminent nitrogen fixers. In nonagricultural systems, some 12,000 species, from cyanobacteria to nodule-bearing plants, are responsible for nitrogen fixation. Also contributing to the fixation of nitrogen are free-living soil bacteria. The most prominent of the 15 known genera are the aerobic *Azotobacter* and the anaerobic *Clostridium*. Cyanobacteria (blue-green algae) are another important group of largely nonsymbiotic nitrogen fixers. Of some 40 known species, the most common are in the genera *Nostoc* and *Calothrix*, which are found both in soil and in aquatic habitats. Certain lichens (*Collema tunaeforme* and *Peltigera rufescens*) are also implicated in nitrogen fixation. Lichens with nitrogen-fixing ability possess nitrogen-fixing cyanobacteria as their algal component.

Ammonium in the soil can be used directly by plants. In addition to atmospheric deposition, NH_4^+ occurs in the soil as a product of microbial decomposition or organic matter (see Chapter 7, Section 7.7), wherein NH_3 is released as a waste product of microbial activity. This process is called **ammonification** (Figure 22.6). Most soils have an excess of

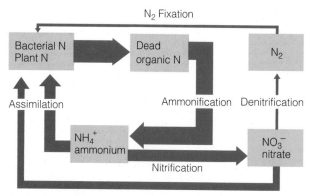

FIGURE 22.6 The bacterial processes involved in nitrogen cycling.

H^+ (slightly acidic; see Chapter 4, Section 4.16), and the NH_3 is rapidly converted to ammonium (NH_4^+). Interestingly, since NH_3 is a gas, the transfer of nitrogen back to the atmosphere (volatilization) can occur in soils with a pH close to 7 (neutral)—low concentration of H^+ ions to convert ammonia to ammonium. Volatilization can be especially pronounced in agricultural areas where both nitrogen fertilizers and lime (to decrease soil acidity) are used extensively.

In some ecosystems, plant roots must compete for NH_4^+ with two groups of aerobic bacteria, which use it as part of their metabolism (see Figure 22.6). The first group (*Nitrosomonas*) oxidizes NH_4^+ to NO_2^-, and a second group (*Nitrobacter*) oxidizes NO_2^- to NO_3^-. This process is called **nitrification**. Once nitrate is produced, several things can happen to it. First, plant roots can take it up. Second, **denitrification** can occur under anaerobic (lacking oxygen) conditions, when another group of bacteria (*Pseudomonas*) chemically reduces NO_3^- to N_2O and N_2. These gases are then returned to the atmosphere. The anaerobic conditions necessary for denitrification are generally rare in most terrestrial ecosystems (but can occur seasonally). These conditions, however, are common in wetland ecosystems and in the bottom sediments of open-water aquatic ecosystems (see Chapters 27–29).

Finally, nitrate is the most common form of nitrogen exported from terrestrial ecosystems in streamwater (see Chapter 21, Section 21.4, Figure 21.5), although in undisturbed ecosystems this is usually quite small because of the great demand for nitrogen. Indeed, the amount (magnitude) of nitrogen recycled within the ecosystem is usually much greater than the amount either entering or leaving the ecosystem through inputs and outputs (see Chapter 21, Section 21.1).

Since both nitrogen fixation and nitrification are processes mediated by bacteria, they are influenced

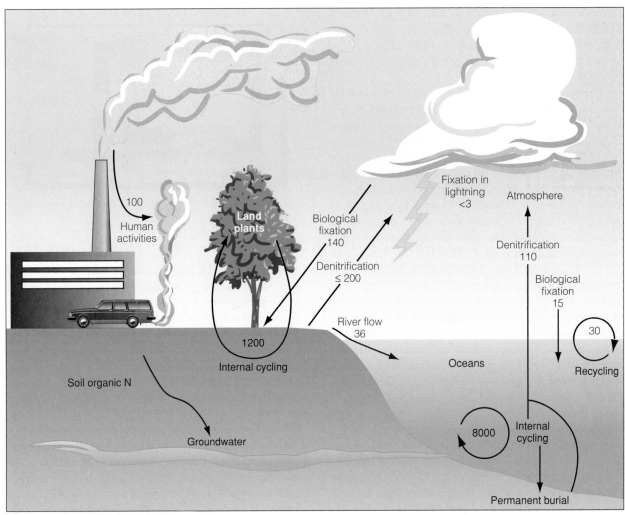

FIGURE 22.7 The global nitrogen cycle. Each flux is shown in units of 10^{12} grams N/year. (From Schlesinger 1997.)

by a variety of environmental conditions, such as temperature and moisture (see Chapter 21, Section 21.5). However, one of the more important factors is soil pH (see Chapter 4, Section 4.16). Both processes are usually limited to a great extent in extremely acidic soils due to the inhibition of bacteria under those conditions.

Although the inputs of nitrogen can vary, the internal cycling of nitrogen is fairly similar from ecosystem to ecosystem. It involves the assimilation of ammonium and nitrate by plants and the return of nitrogen to the soil, sediments, and water via the decomposition of dead organic matter.

The global nitrogen cycle follows the pathway of the local nitrogen cycle presented above, only on a grander scale (Figure 22.7). The atmosphere is the largest pool, containing 3.9×10^{21} grams. Comparatively small amounts of nitrogen are found in the bio-

mass (3.5×10^{15} grams) and soils ($95\text{--}140 \times 10^{15}$ grams) of terrestrial ecosystems. Global estimates of denitrification in terrestrial ecosystems vary widely but are on the order of 200×10^{12} grams year^{-1}, of which over half occurs in wetland ecosystems.

The major sources of nitrogen to the world's oceans are dissolved forms in the freshwater drainage from rivers (36×10^{12} grams year^{-1}) and inputs in precipitation (30×10^{12} grams year^{-1}). Biological fixation accounts for another 15×10^{12} grams year^{-1}. Denitrification accounts for an estimated flux of 110×10^{12} grams N year^{-1} from the world's oceans to the atmosphere.

There are small but steady losses from the biosphere to the deep sediments of the ocean and to sedimentary rocks. In return, there is a small addition of new nitrogen from the weathering of igneous rocks and from volcanic activity.

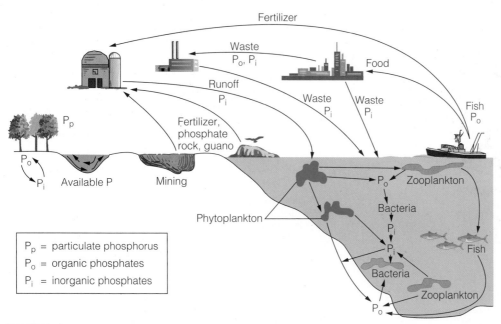

P_p = particulate phosphorus
P_o = organic phosphates
P_i = inorganic phosphates

FIGURE 22.8 The phosphorus cycle in aquatic and terrestrial ecosystems.

22.5 The Phosphorus Cycle Has No Atmospheric Reservoir

Phosphorus occurs in only very minute amounts in the atmosphere. Therefore, the phosphorus cycle can follow the water (hydrological) cycle only part of the way, from land to sea (Figure 22.8). Because phosphorus lost from the ecosystem in this fashion is not returned via the biogeochemical cycle, phosphorus is in short supply under undisturbed natural conditions. Phosphorus's natural scarcity in aquatic ecosystems is emphasized by the explosive growth of algae in water receiving heavy discharges of phosphorus-rich wastes.

The main reservoirs of phosphorus are rock and natural phosphate deposits. Phosphorus is released from these rocks and minerals by weathering, leaching, erosion, and mining for use as agricultural fertilizers. Nearly all of the phosphorus in terrestrial ecosystems comes from the weathering of calcium phosphate minerals. In most soils, only a small fraction of the total phosphorus is available to plants. The major process regulating phosphorus availability for net primary production is the internal cycling of phosphorus from organic to inorganic forms (nutrient cycling; see Chapter 21). Some of the available phosphorus in terrestrial ecosystems escapes and is exported to lakes and seas.

In marine and freshwater ecosystems, the phosphorus cycle moves through three states: particulate organic phosphorus, dissolved organic phosphates, and inorganic phosphates. Organic phosphates are taken up quickly by all forms of phytoplankton, which are eaten in turn by zooplankton and detritus-feeding organisms (see also Chapter 27). Zooplankton may excrete as much phosphorus daily as it stores in its biomass, returning it to the cycle. More than half of the phosphorus zooplankton excretes is inorganic phosphate, which is taken up by phytoplankton. The remainder of the phosphorus in aquatic ecosystems is in organic compounds that may be utilized by bacteria, which fail to regenerate much dissolved inorganic phosphate. Bacteria are consumed by the microbial grazers, which then excrete the phosphate they ingest. Part of the phosphate is deposited in shallow sediments and part in deep water. In the ocean upwelling (see Chapter 21), the movement of deep waters to the surface brings some phosphates from the dark depths to shallow waters where light is available to drive the process of photosynthesis. These phosphates are taken up by phytoplankton. Part of the phosphorus contained in the bodies of plants and animals sinks to the bottom and is deposited in the sediments. As a result, surface waters may become depleted of phosphorus and the deep waters become saturated. Much of this phosphorus becomes locked up for long periods of time in the bottom sediments, while some is returned to the surface waters by upwelling (see Chapter 21).

The global phosphorus cycle (Figure 22.9) is unique among the major biogeochemical cycles in having no significant atmospheric component,

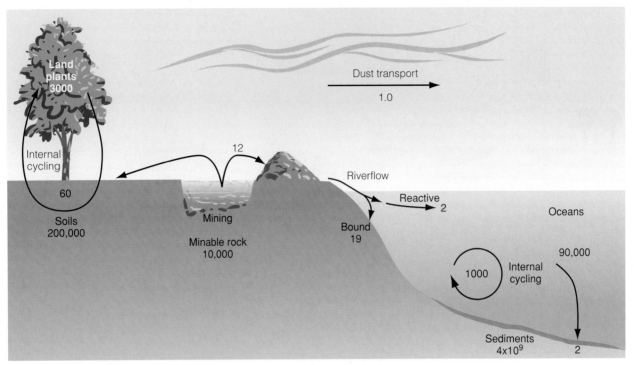

FIGURE 22.9 The global phosphorus cycle. Each flux is shown in units of 10^{12} grams P/year. (From Schlesinger 1997.)

although airborne transport of P in soil dust and sea spray is on the order of 1×10^{12} grams P per year.

Rivers transport approximately 21×10^{12} grams P per year to the oceans, but only about 10 percent of this amount is available for net primary productivity. The remainder is deposited in sediments. The concentration of phosphorus in the ocean waters is low, but the large volume of water results in a significant global pool of phosphorus. The turnover of organic phosphorus in the surface waters is on the order of days, and the vast majority of phosphorus taken up in primary production is decomposed and mineralized (internally cycled) in the surface waters. However, approximately 2×10^{12} grams year^{-1} is deposited in the ocean sediments. On a geological time scale, uplifting and subsequent weathering return this phosphorus to the active cycle.

22.6 The Sulfur Cycle Is Both Sedimentary and Gaseous

The sulfur cycle has both sedimentary and gaseous phases (Figure 22.10). In the long-term sedimentary phase, sulfur is tied up in organic and inorganic deposits, released by weathering and decomposition,

and carried to terrestrial ecosystems in salt solution. The gaseous phase of the cycle permits the circulation of sulfur on a global scale.

Sulfur enters the atmosphere from several sources: the combustion of fossil fuels, volcanic eruptions, exchange at the surface of the oceans, and gases released by decomposition. It enters the atmosphere initially as hydrogen sulfide (H_2S), which quickly interacts with oxygen to form sulfur dioxide (SO_2). Atmospheric sulfur dioxide, soluble in water, is carried back to the surface in rainwater as weak sulfuric acid (H_2SO_4). Whatever the source, sulfur in a soluble form is taken up by plants and incorporated through a series of metabolic processes, starting with photosynthesis, into sulfur-bearing amino acids. From the producers, sulfur in amino acid is transferred to consumers.

Excretion and death carry sulfur from living material back to the soil and to the bottom of ponds, lakes, and seas, where bacteria release it as hydrogen sulfite or sulfate. One group, the colorless sulfur bacteria, both reduces hydrogen sulfide to elemental sulfur and oxidizes it to sulfuric acid. Green and purple bacteria, in the presence of light, utilize hydrogen sulfide in the process of photosynthesis. Best known are the purple bacteria found in salt marshes and in the mudflats of estuaries. These organisms are able to transform hydrogen sulfide into sulfate, which is then recirculated and taken up by producers or used by bacteria

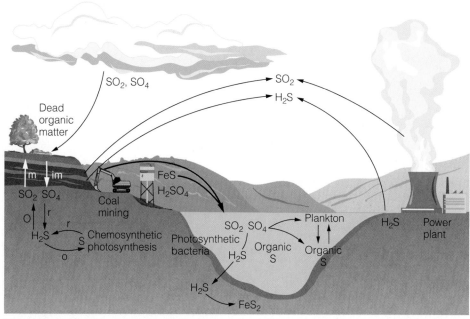

FIGURE 22.10 The sulfur cycle. Note the two components: sedimentary and gaseous. Major sources from human activity are the burning of fossil fuels and acidic drainage from coal mines.

that further transform the sulfates. Green bacteria can transform hydrogen sulfide into elemental sulfur.

Sulfur, in the presence of iron and under anaerobic conditions, will precipitate as ferrous sulfide (FeS_2). This compound is highly insoluble in neutral and low pH (acidic) conditions, and it is firmly held in mud and wet soil. Sedimentary rocks containing ferrous sulfide, called pyritic rocks, may overlie coal deposits. Exposed to air during deep and surface mining for coal, the ferrous sulfide reacts with oxygen. In the presence of water, it produces ferrous sulfate ($FeSO_4$) and sulfuric acid.

In this manner, sulfur in pyritic rocks, suddenly exposed to weathering by human activities, discharges sulfuric acid, ferrous sulfate, and other sulfur compounds into aquatic ecosystems. These compounds destroy aquatic life. They have converted hundreds of kilometers of streams in the eastern United States to lifeless, highly acidic water (see Chapter 23).

22.7 The Global Sulfur Cycle Is Poorly Understood

The global sulfur cycle is presented in Figure 22.11. Although a great deal of research now focuses on the sulfur cycle, particularly the role of human inputs, our understanding of the global sulfur cycle is primitive.

The gaseous phase of the sulfur cycle permits circulation on a global scale. The annual flux of sulfur compounds through the atmosphere is on the order of 300×10^{12} grams. The atmosphere contains not only sulfur dioxide and hydrogen sulfide but sulfate particles as well. The sulfate particles become part of dry deposition (dryfall); the gaseous forms combine with moisture and are transported in precipitation (wetfall).

The oceans are a large source of aerosols that contain sulfate (SO_4); however, most are redeposited in the oceans as precipitation and dryfall (see Figure 22.11). Dimethylsulfide ($(CH_3)_2S$) is the major gas emitted from the oceans that is generated by biological processes. Estimates of 16×10^{12} grams S year^{-1} make it the largest natural source of sulfur gases released to the atmosphere.

A variety of biological sources of sulfur emissions from terrestrial ecosystems exist, but collectively these represent a minor flux to the atmosphere. The dominant sulfur gas emitted from freshwater wetlands and anoxic (oxygen-depleted) soils is hydrogen sulfide (H_2S). Emissions from plants are poorly understood, but forest fires emit on the order of 3×10^{12} grams S annually. It is almost impossible to estimate the biological turnover of sulfur dioxide because of the complicated cycling within the biosphere. Estimates of the net annual assimilation of sulfur by marine plants is on the order of 130×10^{12} grams. Adding the anaerobic oxidation of organic matter (see Chapter 7, Section 7.1) brings the total to an estimated 200×10^{12} grams.

Volcanic activity also contributes to the global biogeochemical cycle of sulfur. Major events, such as

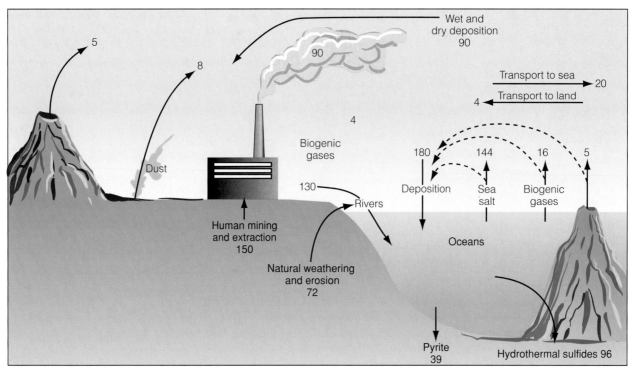

FIGURE 22.11 The global sulfur cycle. Each flux is shown in units of 10^{12} grams S/year. (From Schlesinger 1997.)

the eruption of Mt. Pinatubo in 1991, release on the order of $5–10 \times 10^{12}$ grams S. When volcanic activity is averaged over a long period, the annual global flux is on the order of 10×10^{12} grams S.

Human activity plays a dominant role in the biogeochemical cycle of sulfur. Thus, to complete the picture of the global sulfur cycle, we must examine the inputs due to industrial activity—a topic explored in the following chapter.

22.8 The Oxygen Cycle Is Largely under Biological Control

The major source of free oxygen (O_2) that supports life is the atmosphere. There are two significant sources of atmospheric oxygen. One is the breakup of water vapor through a process driven by sunlight. In this reaction, the water molecules (H_2O) are disassociated to produce hydrogen and oxygen. Most of the hydrogen escapes into space. If the hydrogen did not escape, it would recombine with the oxygen to form water vapor again.

The other source of oxygen is photosynthesis, active only since life began on Earth (Figure 22.12). Oxygen is produced by green plants (see Chapter 6, Section 6.1) and consumed by both plants and animals (Sections 6.6 and 8.4). Because photosynthesis and aerobic respiration involve the alternate release and utilization of oxygen, one would seem to balance the other, so no significant quantity of oxygen would accumulate in the atmosphere. Nevertheless, at some time in Earth's history, the amount of oxygen introduced into the atmosphere had to exceed the amount taken up in respiration (including the decay of organic matter) and geological processes, such as the oxidation of sedimentary rocks. Part of the oxygen present in the atmosphere is from the past imbalance between photosynthesis and respiration in plants. Undecomposed organic matter in the form of fossil fuels and carbon in sedimentary rocks represent a net positive flux of oxygen to the atmosphere. The amount of stored carbon suggests that 15×10^{20} grams of oxygen has been available to the atmosphere, 15 times as much as is now present (10×10^{20} grams).

The other main reservoirs of oxygen are water and carbon dioxide. All the reservoirs are linked through photosynthesis. Oxygen is also biologically exchangeable in such compounds as nitrates and sulfates, which organisms transform to ammonia and hydrogen sulfide (see Sections 22.4 and 22.6).

Because oxygen is so reactive, its cycling in the ecosystem is complex. As a constituent of carbon dioxide, it circulates throughout the ecosystem. Some carbon dioxide combines with calcium to form carbonates. Oxygen combines with nitrogen compounds

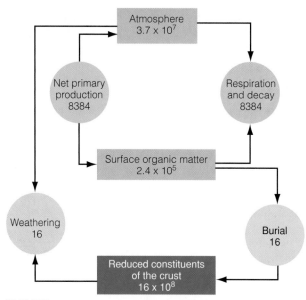

FIGURE 22.12 A simple model for the global biogeochemical cycle of O_2. Data are expressed in units of 10^{12} moles of O_2 per year or the equivalent amount of reduced compounds. Note that a small misbalance in the ratio of photosynthesis to respiration can result in a net storage of reduced organic materials in the crust and an accumulation of O_2 in the atmosphere. (From Schlesinger 1997.)

to form nitrates, with iron to form ferric oxides, and with other minerals to form various oxides. In these states, oxygen is temporarily withdrawn from circulation. In photosynthesis, the freed oxygen is split from the water molecule. The oxygen is then reconstituted into water during plant and animal respiration. Part of the atmospheric oxygen is reduced to ozone (O^3) by high-energy ultraviolet radiation, a topic discussed in the following chapter.

22.9 The Various Biogeochemical Cycles Are Linked

Although we have introduced each of the major biogeochemical cycles independently, they are all linked in various ways. In specific cases, they are linked through their common membership in compounds that form an important component of their cycles. Examples are the links between calcium and phosphorus in the mineral apatite, a phosphate of calcium, and the link between nitrogen and oxygen in nitrate. In general, cycled nutrients are all components of living organisms and constituents of organic matter. As a result, they travel together in their odyssey through internal cycling.

Autotrophs and heterotrophs require nutrients in different proportions for different processes. For example (as indicated in the equation for photosynthesis in Chapter 6), photosynthesis utilizes 6 moles of water (H_2O) and produces 6 moles of oxygen (O_2) for every 6 moles of CO_2 that is transformed into 1 mole of sugar $(CH_2O)_6$. The proportions of hydrogen, oxygen, and carbon involved in photosynthesis are fixed. Likewise, a fixed quantity of nitrogen is required to produce a mole of rubisco, the enzyme that catalyzes the fixation of CO_2 in photosynthesis (see Chapter 6, Section 6.1). Therefore, the nitrogen content of a rubisco molecule is the same in every plant, independent of species or environment. The same is true for the variety of amino acids, proteins, and other nitrogenous compounds that are essential for the synthesis of plant cells and tissues. The branch of chemistry dealing with the quantitative relationships of elements in combination is called **stoichiometry**. The stoichiometric relationships among various elements involved in processes related to carbon uptake and plant growth have an important influence on the cycling of nutrients in ecosystems.

Because of similar relationships among the variety of macronutrients and micronutrients required by plants, the limitation of one nutrient can affect the cycling of all the others. As an example, consider the link between carbon and nitrogen presented earlier in Chapter 21. Although the nitrogen content of a rubisco molecule is the same in every plant independent of species or environment, plants can differ in the concentration of rubisco found in their leaves and, therefore, in their concentration of nitrogen (grams N/gram dry weight). Plants growing under low nitrogen availability will have a lower rate of nitrogen uptake and less nitrogen for the production of rubisco and other essential nitrogen-based compounds. In turn, the lower concentrations of rubisco result in lower rates of photosynthesis and carbon gain (see Chapter 6, Section 6.10). In turn, the lower concentrations of nitrogen in the leaf litter influence the relative rates of immobilization and mineralization and subsequent nitrogen release to the soil in decomposition (see Chapter 7, Section 7.7). In this manner, nitrogen availability and uptake by plants influence the rate at which carbon and other essential plant nutrients cycle through the ecosystem.

Conversely, the variety of other essential nutrients and environmental factors that directly influence primary productivity, and thus the demand for nitrogen, influence the rate of nitrogen cycling through the ecosystem. In fact, the cycles of all essential nutrients for plant and animal growth are linked due to the stoichiometric relationships that define the mixture of chemicals that make up all living matter.

Summary

The Carbon Cycle (22.1–22.3) The carbon cycle is inseparable from energy flow. Carbon is assimilated as carbon dioxide by plants, consumed in the form of plant and animal tissue by heterotrophs, released through respiration, mineralized by decomposers, accumulated in standing biomass, and withdrawn into long-term reserves. The rate at which carbon cycles through the ecosystem depends on the rates of primary productivity and decomposition. Both processes are faster in warm, wet ecosystems. In swamps and marshes, organic material stored as raw humus or peat circulates slowly, forming oil, coal, and natural gas. Similar cycling takes place in freshwater and marine environments (22.1).

Cycling of carbon exhibits daily and seasonal fluctuations. Carbon dioxide builds up at night, when respiration increases. During the day, plants withdraw carbon dioxide from the air, and its concentration drops sharply. During the growing season, atmospheric concentration drops (22.2).

The carbon budget of Earth is closely linked to the atmosphere, land, and oceans and to the mass movements of air around the planet. In the ocean, the surface water acts as the main site of exchange of carbon between atmosphere and ocean. The ability of the surface waters to take up CO_2 is governed largely by the reaction of CO_2 with the carbonate ion to form bicarbonates. The uptake of CO_2 from the atmosphere by terrestrial ecosystems is governed by gross production (photosynthesis). Losses are a function of autotrophic and heterotrophic respiration, the latter being dominated by microbial decomposers (22.3).

The Nitrogen Cycle (22.4) The nitrogen cycle is characterized by the fixation of atmospheric nitrogen by mutualistic nitrogen-fixing bacteria associated with roots of many plants, largely legumes, and cyanobacteria. Other processes are ammonification, the breakdown of amino acids by decomposer organisms to produce ammonia; nitrification, the bacterial oxidation of ammonia to nitrate and nitrates; and denitrification, the reduction of nitrates to gaseous nitrogen.

The global nitrogen cycle follows the pathway of the local nitrogen cycle presented above, only on a grander scale. The atmosphere is the largest pool, with comparatively small amounts of nitrogen found in the biomass and soils of terrestrial ecosystems. The major sources of nitrogen to the world's oceans are dissolved forms in the freshwater drainage from rivers and inputs in precipitation.

The Phosphorus Cycle (22.5) The phosphorus cycle is wholly sedimentary. The main reservoirs of phosphorus are rock and natural phosphate deposits. The terrestrial phosphorus cycle follows the typical biogeochemical pathways. In marine and freshwater ecosystems, however, the phosphorus cycle moves through three states: particulate organic phosphorus, dissolved organic phosphates, and inorganic phosphates. Involved in the cycling are phytoplankton, zooplankton, bacteria, and microbial grazers.

The global phosphorus cycle is unique among the major biogeochemical cycles in having no significant atmospheric component, although airborne transport of P occurs in the form of soil dust and sea spray. Nearly all of the phosphorus in terrestrial ecosystems is derived from the weathering of calcium phosphate minerals. The transfer of phosphorus from terrestrial to aquatic ecosystems is low under natural conditions; however, the large-scale application of phosphate fertilizers and the disposal of sewage and wastewater to aquatic ecosystems results in a large input of phosphorus to aquatic ecosystems.

The Sulfur Cycle (22.6–22.7) Sulfur has both gaseous and sedimentary phases. Sedimentary sulfur comes from the weathering of rocks, runoff, and decomposition of organic matter. Sources of gaseous sulfur are decomposition of organic matter, evaporation of oceans, and volcanic eruptions. A significant portion of the sulfur released to the atmosphere is a by-product of the burning of fossil fuels. Sulfur enters the atmosphere mostly as hydrogen sulfide, which quickly oxidizes to sulfur dioxide, SO_2. Sulfur dioxide reacts with moisture in the atmosphere to form sulfuric acid, carried to Earth in precipitation. Plants incorporate it into sulfur-bearing amino acids. Consumption, excretion, and death carry sulfur back to soil and aquatic sediments, where bacteria release it in inorganic form (22.6).

The global sulfur cycle is a combination of gaseous and sedimentary cycles, because sulfur has reservoirs in Earth's crust and in the atmosphere. The sulfur cycle involves a long-term sedimentary phase in which sulfur is tied up in organic and inorganic deposits, is released by weathering and decomposition, and is carried to terrestrial and aquatic ecosystems in salt solution. The bulk of sulfur first appears in the gaseous phase as a volatile gas, hydrogen sulfide (H_2S), in the atmosphere, which quickly oxidizes to form sulfur dioxide. Once in soluble form, sulfur is taken up by plants and incorporated into organic compounds. Excretion and death carry sulfur in living material back to the soil and to the bottoms of ponds, lakes, and seas, where sulfate-reducing bacteria release it as hydrogen sulfide or as a sulfate (22.7).

The Oxygen Cycle (22.8) Oxygen, the by-product of photosynthesis, is very active chemically. It combines with a wide range of chemicals in Earth's crust, and it reacts spontaneously with organic compounds and reduced substances. It is involved in the oxidation of carbohydrates in the process of respiration, releasing energy, carbon dioxide, and water. The current atmospheric pool of oxygen is maintained in a dynamic equilibrium between the production of oxygen in photosynthesis and its consumption in respiration. An important constituent of the atmospheric reservoir of oxygen is ozone (O_3).

Biogeochemical Cycles are Linked (22.9) All of the major biogeochemical cycles are linked, as the nutrients that cycle are all components of living organisms, constituents of organic matter. The stoichiometric relationships among various elements involved in plant processes related to carbon uptake and plant growth have an important influence on the cycling of nutrients in ecosystems.

Review Questions

1. How are the processes of photosynthesis and decomposition involved in the carbon cycle?

2. In the temperate zone, is the atmospheric concentration of carbon dioxide higher during the day or night? Why?

3. Characterize the following processes in the nitrogen cycle: fixation, ammonification, nitrification, and denitrification.

4. What biological and nonbiological mechanisms are responsible for nitrogen fixation?

5. What is the source of sulfur in the sulfur cycle? Why does the sulfur cycle have characteristics of both sedimentary and gaseous cycles?

6. What is the source of phosphorus in the phosphorus cycle?

7. What is the major source of phosphorus input into aquatic ecosystems?

8. What is the role of photosynthesis and decomposition in the oxygen cycle?

Pollution at the Boise Cascade Paper Mill, Rumford, Maine. Paper mills are a significant local source of pollutants to both adjacent air and watersheds.

Human Intrusions into Biogeochemical Cycles

OBJECTIVES

Upon completion of this chapter, you should be able to:

- Describe the role of nitrogen and sulfur as environmental pollutants.
- Explain the formation of acid deposition.
- Discuss the effects of acid deposition on terrestrial and aquatic ecosystems.
- Explain why human intrusions into the phosphorus cycle differ so greatly from intrusions into the nitrogen and sulfur cycles.
- Explain how nitrogen pollution interacts with ozone formation.
- Discuss the sources of heavy metals and their effects on ecosystems.
- Describe how chlorinated hydrocarbons cycle through and affect ecosystems.

Throughout our history, human activities have influenced the biogeochemical cycles. Aboriginal fires, grazing, timber cutting, agricultural activities, drainage of marshes, and the building of cities have altered the biogeochemical cycles of both terrestrial and aquatic ecosystems. These intrusions, however, have historically been more regional than global. Only since the Industrial Revolution—with the developments of industry and agriculture, expansion of cities, and extensive land clearing accompanied by rapid population growth—have we overburdened natural cycles and spread our intrusion to a global scale.

23.1 Human Activities Are Altering the Global Carbon Cycle

The exchange of carbon among land, sea, and atmosphere in the global carbon cycle (see Chapter 22, Section 22.3) has been disturbed by a rapid injection of carbon dioxide into the atmosphere from the burning of fossil fuels and the clearing of forests. Clearing increases the input of CO_2 from burning trees and decomposing organic matter. Adding to the problem of increasing carbon dioxide are increases in other atmospheric greenhouse gases, especially methane (CH_4). Its major sources are ruminant animals, microbial decomposition in swamps, marshes, and tundra, and industrial gases released to the atmosphere. Atmospheric methane has approximately doubled over the past 200 years. This increase is linked to human population growth and to increased cattle ranching and rice paddy production.

The rising atmospheric concentrations of CO_2 and methane have the potential to alter the global energy balance and subsequently the global climate system. The implication of rising atmospheric concentrations of CO_2 and possible climate change are the focus of Chapter 30.

23.2 Human-Produced Emissions of Nitrogen Act as Pollutants

The saying that too much of a good thing is harmful certainly applies to nitrogen. It is one of the most essential nutrients for life, and up to a point it increases the fertility of soil and water. But we humans, through our myriad of activities such as agriculture, urban development, and the combustion of fossil fuels, have altered the global nitrogen cycle. The mobility of nitrogen is such that local inputs spread regionally and globally, with potentially serious environmental consequences.

The leaching of nitrates from terrestrial environments into streams and rivers and the natural input of nitrogen oxides into the atmosphere have always occurred. During nitrification and denitrification, microorganisms in marine, freshwater aquatic, and terrestrial ecosystems release nitrous oxide (N_2O) to the atmosphere (see Chapter 22, Section 22.4). Tropical forests and woodlands alone release three-fourths of the natural global flux of nitrous oxide. For several decades, however, nitrous oxides have been increasing at the rate of 0.2 to 0.3 percent a year. Most of this increase comes from human activity. In fact, human alteration to the nitrogen cycle has doubled the rate of nitrogen input into both aquatic and terrestrial ecosystems.

Major human sources of nitrogen input are agriculture, industry, and automobiles. The first major intrusion probably came from agriculture when people began burning forests and clearing land for crops and pasture. Conversion of natural grasslands into grain fields has caused a steady decline in the nitrogen content of their soils. Breaking up and mixing soil increases the rate of decomposition of deep organic matter, releasing nitrates and nitrous oxides. Harvesting crops and logging result in a heavy loss of nitrogen from agricultural and forest ecosystems, not only in the material removed (see Chapter 21, Table 21.1), but also in nitrate losses from the soil. Heavy application of chemical fertilizers to croplands disturbs the natural balance between nitrogen fixation and denitrification. A considerable portion of nitrogen fertilizers is lost as nitrates to groundwater and runoff that find their way into aquatic ecosystems. As they move through the soil to streams, nitrates carry calcium with them. With calcium gone, inorganic aluminum in the soil increases (see Chapter 5). Excess nitrogen not leached out as nitrates is removed by microbial denitrification, increasing atmospheric levels of nitrous oxide. In time, about 10 percent of applied nitrogen fertilizers evaporate into the atmosphere. Added to these inputs are nitrogenous inputs from animal wastes at concentrated livestock feeding yards, from municipal sewage treatment plants, and from chemical fertilizer plants. Excess nitrates in aquatic situations reduce water quality, present a human health problem, harm aquatic life, and lead to the eutrophication of lakes and estuaries.

Automobile exhaust and industrial high-temperature combustion add nitrous oxide (N_2O), nitric oxide (NO), and nitrogen dioxide (NO_2) to the

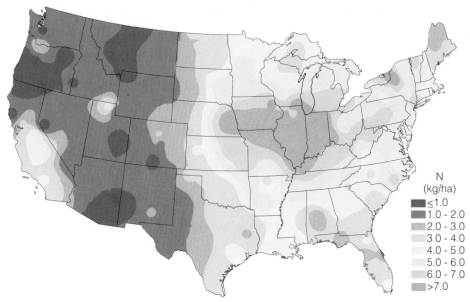

FIGURE 23.1 Estimated inorganic nitrogen deposition from nitrate and ammonium, 1998. (National Atmosphere Deposition Program.)

N
(kg/ha)
- ≤1.0
- 1.0 - 2.0
- 2.0 - 3.0
- 3.0 - 4.0
- 4.0 - 5.0
- 5.0 - 6.0
- 6.0 - 7.0
- >7.0

atmosphere. These oxides can reside in the atmosphere for 20 years, drifting slowly up to the stratosphere. There, ultraviolet light reduces nitrous oxide to nitric oxide and atomic oxygen (O). Atomic oxygen reacts with oxygen (O_2) to form ozone (O_3) (see Section 23.6).

One outcome of this massive atmospheric input is an increased deposition of nitrogen (Figure 23.1), which benefits ecosystems that are traditionally nitrogen-limited, especially northern and high-altitude forests. Because nitrogen is limiting, these forests are efficient at retaining and recycling nitrogen from precipitation and organic matter. Only a few lose a significant amount of nitrates to streams. Now many of these forests are receiving more nitrogen in the form of ammonium and nitrates than the trees and their associated microbial populations can handle and accumulate, a phenomenon known as nitrogen saturation.

The first response to increased availability of nitrogen in a nitrogen-limited ecosystem is increased growth. Evidence suggests, however, that mounting levels of nitrogen lead to the decline and dieback of coniferous forests at high elevations (Figure 23.2). If the increased growth in foliage continues into the summer, the late new growth may not have time to become frost-hardened, and it is killed during the winter. Overstimulated by nitrogen, tree growth exceeds the availability of other necessary nutrients in the soil, particularly phosphorus and calcium, and the tree begins to experience nutrient deficiencies. Experimental evidence suggests that the production of fine roots and mycorrhizae, which take up nutrients from the soil, is lower on sites rich in nutrients, especially nitrogen, than in nutrient-poor soils, and root turnover is higher. Trees on nutrient-poor soils have a longer life, a higher-density root system, and a lower turnover of root biomass, a condition that helps them to scavenge nutrients from poor soils. Thus, as nitrogen levels increase, root biomass decreases, further inhibiting the uptake of nutrients other than nitrogen and impairing the ability of trees to take up water from the soil during periods of drought.

23.3 Human-Produced Sulfur Dioxide Is a Major Air Pollutant

Of all the atmospheric gases, sulfur dioxide (SO_2) is most strongly implicated in air pollution. Major sources fall into two categories: natural and human-made. Natural sources include microbial activity, volcanoes, sea spray, and weathering (see Chapter 22, Section 22.6). These sources make up about 60 percent of sulfur emissions to the atmosphere. Human-made emissions comprise the remaining 40 percent.

Of the human-made emissions, 68 percent comes from the burning of fossil fuels, and 40 percent of that from the burning of coal. Natural emissions are widely distributed about the globe; human-made emissions are concentrated regionally. Ninety percent of those inputs come from the urban and industrialized areas of Europe, North America, India, and the

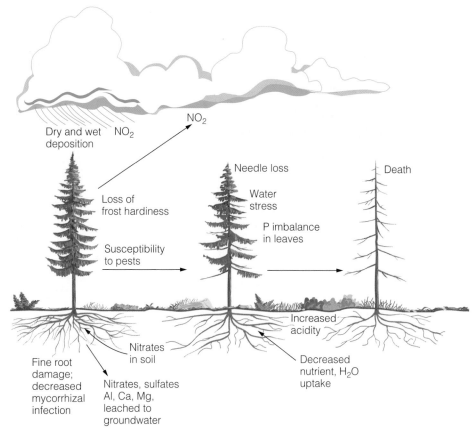

FIGURE 23.2 Effects of excess nitrogen in nutrient-limited forest ecosystems. The forest dies back, and the ecosystem changes from a nitrogen sink to a nitrogen source.

Far East. Europe and North America release between 110 and 125 million metric tons of sulfur into the atmosphere every year. In these regions, industrial input far exceeds natural inputs. Because of the imposition of emission controls, new technologies, and changes in patterns of fuel consumption, sulfur dioxide emissions have been declining. However, if coal consumption increases worldwide as predicted, sulfur dioxide levels could rise above former levels.

Sulfur dioxide produces acute toxicity and major damage to vegetation in areas surrounding the source of emission. Examples are the destruction of forest cover in the vicinity of iron smelters at Sudbury and Wana, Ontario, and copper smelters at Duckberry and Copperhill, Tennessee. Damage away from point sources is more long-term and insidious. Tall stacks at power plants and industrial complexes have greatly reduced local damage, but they eject into the atmosphere gases that are carried by upper-level winds to points far removed. On the way, some of the gases and particulate matter drop out close to the source. In the moving plume, sulfur dioxide combines with atmospheric moisture to

form sulfuric acid (see Chapter 22, Section 22.5), which falls on land and water and forms a significant part of acid rain.

Sulfur dioxide in the atmosphere affects both humans and plants. In a concentration of a few parts per million, it irritates the respiratory tract. In a fine mist or adsorbed onto small particles, sulfur dioxide moves into the lungs and attacks sensitive tissues. High concentrations (over 1000 millimicrons/m^3) have caused a number of air pollution disasters characterized by higher-than-expected death rates and increased incidence of bronchial asthma.

Sulfur dioxide injures exposed plants or kills them outright. Acidic aerosols present during periods of fog, light rain, and high relative humidity, together with moderate temperatures, do most of the injury. External surfaces of leaves absorb the aerosols. When dry, leaves and needles take up sulfur dioxide through the stomata. In the leaf, the sulfur dioxide rapidly reacts with moisture to form sulfuric acid. Symptoms of sulfur damage are a bleached look to deciduous leaves and red-brown needles on conifers, partial defoliation, and reduced growth.

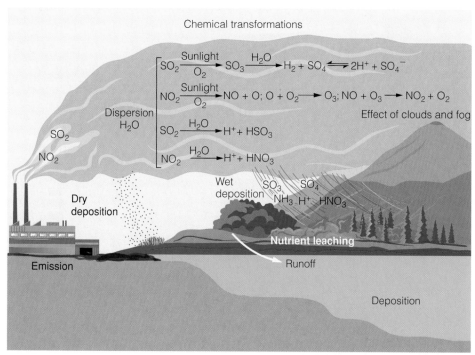

FIGURE 23.3 Formation of acid deposition. Excessive sulfur and nitrogen in several forms are poured into the atmosphere. They are then converted to sulfites (SO_3), sulfates (SO_4), sulfuric acid (H_2SO_4), and nitric acid (HNO_3) and carried to Earth.

23.4 The Sulfur and Nitrogen Cycles Produce Acid Deposition

The gaseous sulfur dioxide component of the sulfur cycle and the nitrogen oxides of the nitrogen cycle mix in the atmosphere. Some of this mixture returns to Earth as particulate matter and airborne gases, known as **dry deposition.** A major portion is transported away from the sourcing in a direction strongly influenced by the general atmospheric circulation. During their atmospheric transport, SO_2 and NO_2 and their oxidative products participate in complex reactions involving hydrogen chloride and other compounds, oxygen, and water vapor. These reactions dilute solutions of strong acids, notably nitric acid and sulfuric acid (Figure 23.3). Eventually they come to Earth in acidic rain, snow, and fog, known as **wet deposition.**

Rainwater is naturally acidic. Unpolluted rainwater, considered pure water, has a pH of 5.6, but rarely is rainwater pure. Even in regions not subject to industrial pollution, atmospheric moisture is exposed to varying amounts of acids of natural origin, so precipitation has a pH of about 5. In regions extending hundreds of kilometers around centers of human activity,

however, the pH of precipitation is much lower—3.5 to 4.5, or occasionally lower still (Figure 23.4).

Acid rain has been with us for well over a century. In 1872, when the Industrial Revolution was in full swing, Robert Angus Smith introduced the term to describe the rainfall around Manchester, England. In some areas, acid rain was probably worse then than it is now, but it was more localized. Only when we constructed large stacks in industrial and power plants, sending emissions much higher into atmospheric circulation, did acid rain become a regional problem. The industrial midwestern United States and Ohio River valley send acidic pollutants to eastern Canada and the northeastern United States; eastern Canada sends them to the northeastern United States; and the industrialized regions of central Europe and the United Kingdom send them to Scandinavia.

Little evidence exists to show that acid rain has a direct effect on most plants, but it is associated with the mortality of the red spruce at high elevations and with the decline of the sugar maple in northeastern North America. When intercepted by vegetation, acid rain leaches nutrients, particularly calcium, magnesium, and potassium, from the leaves and needles. Such leaching, a normal process in nutrient cycling, has little effect on trees' health, provided the trees can replace the lost nutrients by uptake from the soil.

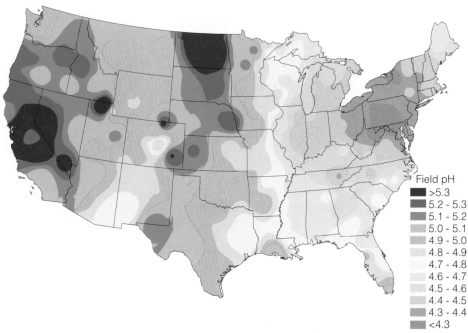

FIGURE 23.4 Hydrogen ion concentration as pH from measurements made at the field laboratories, 1998. (National Atmosphere Data Program.)

Field pH
>5.3
5.2 - 5.3
5.1 - 5.2
5.0 - 5.1
4.9 - 5.0
4.8 - 4.9
4.7 - 4.8
4.6 - 4.7
4.5 - 4.6
4.4 - 4.5
4.3 - 4.4
<4.3

Mists and fog, however, frequently envelop forests at high elevations. Cloud droplets are more acidic and hold higher concentrations of other pollutants than rainfall. When immersed in fog, needle-leaved conifers comb moisture out of the air, and their wet surfaces permit the uptake of the pollutants the moisture contains. As the water evaporates, it leaves behind high concentrations of pollutants, some of which wash off during the next rain and fall on the soil. Added to this wet deposition is a dusting of dry deposition that leaches nutrients from the leaves.

23.5 Acid Deposition Impacts Soils and Aquatic Systems

Acid precipitation has its greatest impact on soils that are low in cations and poorly buffered (see Chapter 5). Such soils, mostly podzolic (see Chapter 5) and derived largely from granitic bedrock, are characteristic of the eastern and upper north-central United States, the southeastern United States, Canada, and northern Europe. Although not yet affected, parts of Asia, Africa, and South America are also acid-sensitive. In all these regions, terrestrial ecosystems are nutrient-poor and their soils acidic. Over a period of time, acid precipitation can have

adverse effects. It increases the leaching of calcium, magnesium, and potassium from the receiving soil and replaces these cations with hydrogen ions, further increasing soil acidity. Acid precipitation can reduce the solubility and availability of phosphorus and the rate of nitrogen fixation. If the rate of leaching outstrips the replacement of these nutrients by weathering, acid precipitation upsets the nutrient balance of trees and other vegetation. Further, acid rain can inhibit the activity of fungi and bacteria in the soil. By so doing, it reduces the rates of humus production, mineralization, and fixation of nutrients. All of these interactions cause nutrient-deficient soils.

The most serious effect, however, is the mobilization of toxic elements in the soil, particularly aluminum and manganese, as they are replaced by hydrogen ions on soil particles (see Chapter 5). Aluminum affects the structure and function of fine roots and interferes with the uptake of calcium from the soil. It also suppresses growth of the cambium in trees, which in turn reduces the formation of new sapwood. As sapwood growth in conifers declines, the ratio between living sapwood and dead heartwood declines. When sapwood forms less than 25 percent of the cross-section of a tree, that tree succumbs (see Focus on Ecology 23.1: Air Pollution and Forest Decline).

Acid deposition has its most pronounced effects on aquatic ecosystems in acid-sensitive regions. Acidic inputs to aquatic ecosystems come directly from

AIR POLLUTION AND FOREST DECLINE

Over much of Europe, forests, especially coniferous ones, are declining and dying. In Germany, 80 percent of the fir, 54 percent of both spruce and pine, and 60 percent of beech and oak show symptoms of damage—defoliation, thinned crowns, yellowed and discolored foliage, premature leaf fall, and death. Over central Europe, Norway spruce, pine, beech, and oak suffer from chlorosis and defoliation. Fourteen percent of all Swiss forests, 22 percent of Austria's forests, and 29 percent of Holland's forests show symptoms of decline.

The story is the same in North America. Pine forests in southern California northeast of Los Angeles and San Diego suffer from chlorosis, decrease in radial growth, and death. In the high Appalachians from North Carolina to Vermont, red spruce has been declining in growth and dying since the 1960s. In the Great Smoky Mountains, one-half of red spruce and Fraser fir are dead. In the southern United States, loblolly pine and slash pine are experiencing a 30 to 50 percent decline in radial growth and increasing mortality. Sugar maples in the northeastern United States and eastern Canada are showing crown dieback and bark peeling on larger branches. In the province of Quebec, dieback affects 52 percent of the sugar maples. Throughout northeastern North America, white pine shows needle discoloration and decreased growth in height, diameter, and needle length, all symptoms of ozone damage.

Forest decline is not a new phenomenon. During the past two centuries, our forests have experienced several declines, with different species affected. What sets the current decline apart from all others is differences among the symptoms in the past and the similarity of symptoms among species today. Past declines could be attributed to natural stresses, such as drought and disease. What causes forest decline and dieback today is not established, but the widespread similarity of symptoms suggests a common cause—air pollution. All of the affected forests are in the path of pollutants from industrial and urban sources.

Forests close to the point of origin of pollutants experience the most direct effects of air pollution, and their decline and death can be directly attributed to it. Little evidence exists that acid precipitation alone is the cause of forest decline and death at distant points. The effects of acid deposition, however, can so weaken trees that they succumb to other stresses such as drought and insect attack. The stressed stands of Fraser fir in the Great Smoky Mountains succumb to the attacks of the introduced balsam woolly adelgid (*Adelgid picea*). The once deep, fragrant stands of Fraser fir, especially on the windward side and peaks of the Great Smoky Mountains, are now stands of skeleton trees (Figure A).

Air pollution and acid rain also are altering succession by changing the species composition of forests. Just as the chestnut blight shifted dominance in the central hardwood forest from chestnut to oaks, so is air pollution shifting dominance from pines and other conifers to deciduous trees more tolerant of air pollution.

FIGURE A Forest decline in the Great Smoky Mountains. Air pollution makes these trees vulnerable to the attacks of the balsam woolly adelgid.

rainfall and snowmelt and indirectly from the soils of the surrounding watershed. Acidic water leached from soils increases the nutrient levels of streams and lakes in a watershed, provided the soil has appreciable reserves of calcium (see Chapter 27). As acid rain percolates through the soil, it is neutralized while releasing basic ions and carrying them to streams and lakes. Snowmelt and spring rainwater flowing over the surface of frozen ground and following old root channels and animal burrows into receiving waters often cancel such enrichment. Such inflow discharges much of the winter precipitation in a slug of acidic waters. The receiving water then can become acidic in spite of the buffering effects of the soil.

In the water, sulfate and nitrogen ions replace bicarbonate ions, pH declines, and the concentration of metallic ions, especially aluminum, increases. When the pH of the groundwater and surface water in the surrounding watershed is 5 or lower, high concentrations of aluminum ions are carried to lakes and streams. Aluminum then tends to precipitate the dark humics, increasing the transparency of the water. Increased light penetration into the water can stimulate phytoplankton production and the growth of benthic algae and bryophytes, but species richness and abundance of zooplankton and macroinvertebrates decrease.

Although adult fish and some other aquatic organisms can tolerate high acidity, a combination of high acidity and a high level of aluminum, a typical situation during snowmelt, can kill them. At the level of 0.1 to 0.3 mg/l, aluminum retards growth and gonadal development of fish and increases their mortality.

Filling depressions and temporary ponds with snowmelt and surface runoff, even in regions without acidified lakes and streams, inhibits the reproduction of frogs and salamanders, whose eggs and larvae are sensitive to acidic water. This effect may account in part for the rapid decline of amphibians. Acidic waters are toxic to invertebrates also, either killing them directly or interfering with calcium metabolism, causing crustaceans to lose the ability to recalcify their shells after molting. As recruitment fails and food declines, fish life disappears from the affected waters.

Acidification already has damaged many lakes and streams in the northeastern United States, Canada, Norway, Sweden, and the United Kingdom. Of 1469 lakes surveyed in the Adirondack Mountains of New York, 346 (24 percent) had no fish, which in turn affects the presence of fish-eating birds such as herons. Unassisted biological recovery probably will never happen. Restoration will depend upon restocking components of the original lake community. Even with that help, restoration of the original food chains and community structure will be difficult, if not impossible.

Added to the damage to forests, crops, and aquatic ecosystems is the enormous $75–80 billion annual cost attributed to the corrosion of bridges, highways, buildings, and monuments exposed to the atmosphere. Iron and its alloys are rusted and weakened by chlorides and sulfides, aluminum by chlorides, copper and its alloys by sulfides, and masonry and marble by sulfides and other atmospheric aerosols, whose corrosive effects appear to be enhanced by ozone and solar radiation.

23.6 Pollutants Influence the Dynamics of Ozone

Ozone (O_3) is an ambivalent atmospheric gas. In the stratosphere, 10 to 40 km above Earth, it shields the planet from biologically harmful ultraviolet radiation. Close to the ground, ozone is a damaging pollutant, cutting visibility, irritating eyes and respiratory systems, and injuring or killing plant life. In the stratosphere, ozone is diminished by its reaction with human-caused pollutants. In the troposphere, ozone is born from the union of nitrogen oxides with oxygen in the presence of sunlight.

A cycling reaction requiring sunlight maintains ozone in the stratosphere. Solar radiation breaks the O—O bond in O_2. Freed oxygen atoms rapidly combine with O_2 to form O_3. At the same time, a reverse reaction consumes ozone to form O and O_2. Under natural conditions in the stratosphere, a balance exists between the rates of ozone formation and destruction. In recent times, however, a number of human-caused and some biologically derived catalysts injected into the stratosphere have been reactive enough to reduce stratospheric ozone. Among them are chlorofluorocarbons (CFCs); methane (CH_4), both natural and human-caused; and nitrous oxide from denitrification and synthetic nitrogen fertilizer. Of particular concern is chlorine monoxide (ClO) derived from chlorofluorocarbons in aerosol spray propellants (banned in the United States), refrigerants, solvents, and other sources. This form of chlorine can break down ozone.

In 1985 atmospheric scientists discovered a pronounced springtime thinning in the ozone layer over Antarctica. They questioned whether the thinning is due to pollution or is a natural effect of upper-level winds. Detection of chlorine monoxide in the hole points to pollution. Reduction of the ozone layer has adverse ecological effects on Earth. It alters DNA and increases skin cancer. In addition, changes in the ozone layer could increase the temperature of the lower atmosphere, change air circulation patterns, and contribute to the greenhouse effect (see Chapter 30).

Down in the troposphere, nitrogen oxides and volatile hydrocarbons react with O_2 in the presence of sunlight to form ozone and a number of secondary pollutants, such as formaldehydes, aldehydes, and peroxacetylnitrates, known as PAN. All of these substances collectively form photochemical smog.

Ozone, relatively insoluble in water, readily diffuses through stomatal cavities, where it reacts rapidly. The degree of reaction varies among crop plants and forest trees. Highly sensitive tobacco becomes flecked with white lesions; bean leaves show stippling and bleached areas; leaves of woody plants have red-dish-brown lesions. Plants especially sensitive to ozone include white pine and ponderosa pine, red spruce, alfalfa, oats, spinach, and tomato. Foliar damage reduces the photosynthetic capacity of plants, annual radial growth in trees, and nutrient retention in foliage; predisposes trees to insect and fungal infections; and is associated with forest decline.

Other pollutants, especially PAN, are extremely toxic. By destroying some of the lower epidermal and internal cells of leaves, they interfere with plants' metabolic processes and reduce growth. Damage to crop plants in the United States alone is an estimated $2 billion a year.

23.7 Human Intrusion in the Phosphorus Cycle Involves Its Redistribution

Phosphorus has no atmospheric reservoir. Cycling by the way of the hydrological cycle is limited, and a significant fraction is stored in the soil. Yet in spite of its rather tight local and regional cycling, humans have managed to intrude the phosphorus cycle with significant environmental impacts.

Human intrusion starts with the mining of phosphate rock, transporting it in fertilizer and animal feeds, and using it in food production and manufacture of detergents and other products. Some of the phosphorus applied as phosphate fertilizer becomes incorporated into plant and animal tissue, is transported far from the point of origin, and is released in waste when food is processed or consumed. The rest of the phosphorus reacts with calcium, iron, and aluminum in the soil, becomes immobilized as insoluble salts, and accumulates. Erosion of upland soils carries phosphorus on soil particles into aquatic systems. Added to this is phosphorus concentrated in the wastes of food-processing plants and animal feedlots. Great quantities are released by urban sewage systems. Primary sewage treatment is only 30 percent effective in removing phosphorus, so 70 percent remains in the effluent and is added to waterways. In freshwater and estuarine ecosystems, vegetation and phytoplankton take up phosphorus rapidly, causing a sudden increase in algal biomass. Eventually this phosphorus settles in bottom sediments of ponds, lakes, and oceans.

Local and regional inputs of nitrogen and sulfur from human activity join the global cycles via atmospheric circulation and the water cycle. In sharp contrast, human inputs to the phosphorus cycle involve the mining of deposits at a limited number of local sites. Phosphorus in its various altered forms for human use is then commercially transported globally, where it enters terrestrial and aquatic ecosystems. The amount is considerable. Elena Bennett, Steve Carpenter, and Nina Caraco of the University of Wisconsin-Madison have estimated the human input into the global phosphorus budget. In preindustrial time, before the mining of phosphate, the input of phosphorus, mainly by weathering, ranged between 10 and 15 Tg/year (1 teragram = 1 million metric tons). Added to this is the current 18.5 Tg/year input of mined phosphorus to surface soils, resulting in an almost threefold increase in the discharge of phosphorus from the world's rivers into the oceans. This increased transport of phosphorus from terrestrial sources to aquatic systems leads to decline in water quality, oxygen depletion in affected lakes and estuaries (see Chapter 27 on eutrophication), extirpation of native plants, and loss of biological diversity.

23.8 Heavy Metals Also Cycle Through Ecosystems

Heavy metals such as mercury, cadmium, chromium, and lead, which are toxic to life in varying amounts, have their own biogeochemical cycles. Metals such as mercury and cadmium are associated with industrial pollution, runoff from agricultural fields, toxic dumps, and landfills. Discharged into rivers or lakes or seeping into groundwater, these elements contaminate water supplies and build up in food chains. The serious problems they create, such as the contamination of fish and adverse effects on reproduction of some birds, are local; but we feel their effects over a much wider area.

Other heavy metals, notably lead, join sulfur and nitrogen in atmospheric circulation, moving great distances from the point of origin. Automobiles burning leaded gasoline poured most of the lead into the air until the use of unleaded fuel was mandated in the United States, Japan, Brazil, and the European Economic Community. Mining, smelting, and refining of lead, lead-consuming industries, coal combustion, burning of refuse and sewage sludge, and burning and decay of lead-painted surfaces add additional quantities to the atmosphere.

Emitted into the air as very small particles, less than 0.5 μm, lead is widely distributed to all parts of Earth. Areas near point sources of pollution,

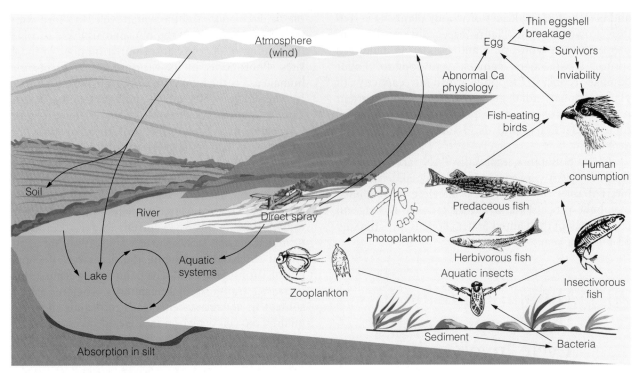

FIGURE 23.5 The movement of chlorinated hydrocarbons in terrestrial and aquatic systems. The initial input comes from spraying on vegetation. A large portion fails to reach the ground and is carried on water droplets and particulate matter through the atmosphere.

particularly roadsides, receive the heaviest deposition. Urban areas close to industry and heavy automobile traffic may have a flux rate greater than 3000 g/ha/year, whereas remote areas may have a flux rate of less than 20 g/ha/year.

Lead particles settle on the surface of the soil and on vegetation. Forest canopies are particularly efficient at collecting lead from the atmosphere as dry deposition. Lead accumulates in the canopies during the summer and is carried to the ground by rain as throughfall and stemflow and by leaf fall in autumn. In the forest soil, lead becomes bound to organic matter in the litter layer and reacts with sulfate, phosphate, and carbonate anions in the soil. In such an insoluble form, lead moves slowly, if at all, into the lower horizons. It persists in the upper soil layer for about 5000 years.

Once in the soil and on plants, lead enters the food chain. Plant roots take up lead from the soil, and leaves pick it up from contaminated air or from particulate matter on the leaf. Lead is then taken up by herbivorous insects and grazing mammals, who pass it on to higher consumers. This uptake through the food chain is most pronounced along roadsides. Microbial systems also pick up lead and immobilize substantial quantities of it.

The long-term increase in concentrations of atmospheric lead in the industrialized areas of Earth has resulted in significant increases of lead in humans.

The average concentration of lead in bodies of adults and children in the United States is 100 times greater than the natural burden, and existing rates of lead absorption are 30 times the level in preindustrial society. An intake of lead can cause mental retardation, palsy, partial paralysis, loss of hearing, and death.

23.9 Chlorinated Hydrocarbons Establish Cycles

Biogeochemical cycles are nearly as old as Earth. Now substances that humans manufacture and release to the environment have established their own biogeochemical cycles that mimic natural ones. These substances include chlorinated hydrocarbons—pesticides (see Section 18.13), dioxin-containing herbicides, and the industrial chemical PCB.

The substance that first received attention was DDT, the dangers of which were widely exposed by Rachel Carson in her book *Silent Spring* (1962). The detection of DDT in the tissues of animals in the Antarctic, far removed from any applied source of the insecticide, emphasized the fact that chlorinated hydrocarbons do indeed establish their own global biogeochemical cycles that disperse them around the globe. DDT has been banned in the United States and most industrialized nations, but it still is used extensively in other parts of the world, notably Central and South America and Asia. We can use it to illustrate the cycling of chlorinated hydrocarbons.

Chlorinated hydrocarbons have certain characteristics that make global circulation possible (Figure 23.5). They are highly soluble in fats or lipids and poorly soluble in water. Therefore, they tend to accumulate in plants and animals. They are persistent and stable, but undergo some degradation. DDT, for example, degrades to DDE and has a half-life of approximately 20 years. It has a vapor pressure high enough to ensure direct losses from plants. It can become adsorbed to particles or remain as a vapor, and in either state it can be transported by atmospheric circulation. It can return to land or sea in rainwater.

Insecticides are applied on a large scale by aerial spraying. One-half or more of the toxicant applied in this manner is dispersed to the atmosphere and never reaches the ground. If the vegetative cover is dense, only about 20 percent reaches the ground. On the ground, some of the pesticide is lost through volatilization, chemical degradation, bacterial decomposition, and runoff.

DDT sprayed on forests and croplands enters streams and lakes, where it is subject to further distribution and dilution as it moves downstream. Insecticides released in oil solutions penetrate to the bottom and kill fish and aquatic invertebrates. Trapped in the bottom rubble and mud, the insecticide may continue to circulate and kill for some days. DDT in oil solutions floats on the surface and is moved about by the wind. Eventually it reaches the ocean, where it may concentrate in surface slicks. These slicks, which attract plankton, are carried across the seas by ocean currents. In the ocean, part of the DDT residue may circulate in the mixed layer. Some of it may be transferred to the deep waters, and more may be lost through the sedimentation of organic matter.

The major movement of pesticide residues takes place in the atmosphere. Not only does the atmosphere receive the bulk of the pesticide spray, but it also picks up the fractions volatized from soil, water, and vegetation. The capacity of the atmosphere to hold DDT is increased greatly by the adsorption of residues on fine particulate matter. Thus, the atmosphere becomes a large circulating reservoir of DDT and other chlorinated hydrocarbons. Residues are removed from the atmosphere by chemical degradation, diffusion across the air-sea interface, and mostly by rainfall and dry deposition.

The high solubility of DDT in lipids leads to its concentration through the food chain. Most of the DDT contained in ingested food is retained in the fatty tissue of consumers. Because it breaks down slowly, DDT accumulates to high and even toxic levels. DDT so concentrated is passed on to the next trophic level. The carnivores on the top level of the food chain receive massive amounts of pesticides. With high concentrations of DDT in their tissues, consumers die or experience impaired reproduction and genetic damage. For example, a residue level of 5 ppm in the fatty tissues of the ovaries of freshwater trout causes 100 percent dieoff of the fry. They pick up lethal doses as they use up the yolk sac.

DDT and its degraded products interfere with calcium metabolism in birds. Chlorinated hydrocarbons block ion transport by inhibiting the enzyme ATP synthase, which provides the energy needed to transport ionic calcium across membranes. DDT also inhibits the enzyme carbonic anhydrase. This enzyme is essential for the deposition of calcium carbonate in the eggshell and the maintenance of a pH gradient across the membrane of the shell gland.

The ecological problem of pesticides has not lessened. Chlorinated hydrocarbons, including DDT, are used extensively in Central and South America and Asia. There the problems associated with these pesticides persist. They affect not only the year-round fauna but also migratory birds from the Northern Hemisphere that come in contact with pesticides on their wintering grounds. Quantities of pesticides are sent back north to the United States, Canada, and Europe on fruits and vegetables grown for the winter market.

The use of biocides (pesticides and herbicides) other than DDT continues unabated in the United States. Herbicides make up 60 percent of this usage, insecticides 24 percent, and fungicides 16 percent. Of these biocides, 341 million kg are used on agricultural crops and pastures, 55 million by government and industry, 4 million on forests, and a surprising 55 million in and around urban and suburban homes. The most concentrated use is around homes, not in agricultural fields. The dosage of pesticides around homes is 14 kg/ha, compared to 3 kg/ha on agricultural crops. As much as one-third of these household pesticides is never used and is thrown into the trash and ultimately into the environment. These excess pesticides, together with losses from croplands and roadsides, expose pests to widespread selective pressures, increasing their resistance to pesticides.

Finally, we must consider the effects on human health. Humans are widely exposed to pesticides when they eat fresh fruits and vegetables, when they spray homes and gardens, when they swim in contaminated waters, and in other ways. Those who work directly with pesticides are particularly at risk. Annually in the United States 45,000 humans are poisoned to some degree by pesticides. Pesticides that mimic the effects of estrogen may increase human infertility and birth defects.

Summary

Altering Global Carbon Cycle (23.1) The concentration of carbon dioxide in the atmosphere is increasing as a result of the burning of fossil fuels and from the clearing of forests.

Nitrogen Emissions (23.2) Humans pour nitrogen oxides into the atmosphere and nitrates into aquatic ecosystems. The major sources of nitrogen oxides are automobiles and the burning of fossil fuels. Large quantities of nitrates are added to aquatic systems by fertilizers, animal wastes, and sewage effluents. Increased deposition of nitrogen on high-elevation coniferous forests overstimulates growth and produces nutrient deficiencies.

Sulfur Emissions (23.3) Sulfur enters the atmosphere mostly as hydrogen sulfide, which quickly oxidizes to sulfur dioxide, SO_2. Sulfur dioxide reacts with moisture in the atmosphere to form sulfuric acid, carried to Earth in precipitation. Plants incorporate it into sulfur-bearing amino acids. Consumption, excretion, and death carry sulfur back to soil and aquatic sediments, where bacteria release it in inorganic form. Sulfur is an essential nutrient, but sulfur dioxide is a major atmospheric pollutant, damaging and killing plants, causing respiratory afflictions, and adding to acid deposition.

Acid Deposition (23.4–23.5) Acid deposition develops when sulfur dioxide and nitrogen oxides combine with water and hydrogen in the atmosphere to produce sulfuric and nitric acids. These acids reach Earth as wet deposition in the form of acidic rain, snow, and fog and as dry deposition in the form of particulate matter and gases (**23.4**). The acidification of lakes and streams kills fish, crustaceans, and insects. Acid precipitation promotes forest decline by increasing the acidity of poorly buffered soils, nutrient depletion, and aluminum toxicity in the soil, and by inhibiting the activity of soil fungi and bacteria. Acid deposition can so weaken trees that they succumb to other stresses. Acid pollution also causes crop losses and ruins roads, bridges, buildings, and monuments (**23.5**).

Ozone Production (23.6) Ozone (O_3), related to the oxygen cycle, is produced by photochemical reactions in the atmosphere. In the stratosphere, ozone is essential to reduce the influx of harmful ultraviolet radiation to Earth. However, human-produced chlorofluorocarbons and other pollutants rise to the stratosphere and destroy this ozone. In the troposphere, ozone is created from nitrogen oxides by photochemical reactions. With related pollutants, it forms smog toxic to vegetation.

Redistribution of Phosphorus (23.7) Human intrusion into the phosphorus cycle begins with the mining of phosphate rock. Phosphorus is used as fertilizer, as well as for food production and the manufacture of detergents and other products. Humans overload the phosphorus cycle by applying phosphate fertilizers and disposing of wastes, which eventually make their way to aquatic ecosystems. This movement from terrestrial sources to aquatic systems leads to eutrophication, decline in water quality, oxygen depletion in affected lakes and estuaries, extirpation of native plants, and loss of diversity.

Cycling of Heavy Metals (23.8) Heavy metals toxic to life follow their own biogeochemical cycles. Some cycles are local; others, such as that of lead, can become regional and even global. Lead is one of the most pervasive heavy metals. Mining, smelting, lead-consuming industries, combustion of coal, burning of leaded fuel, and decay of lead-painted surfaces release small particles of lead into the atmosphere. These particles settle on soil and vegetation and pass through the food chain. Lead may remain in upper levels of soil for millennia.

Chlorinated Hydrocarbons (23.9) Of serious consequence globally is the insecticidal use of chlorinated hydrocarbons. Following biogeochemical cycles, these pesticides have contaminated global ecosystems. Because they become concentrated at higher trophic levels, chlorinated hydrocarbons affect predaceous animals most, interfering with their reproduction. Such insecticides kill more than the targeted pest species, reducing species diversity. They endanger human health.

Review Questions

1. Why is excess nitrogen a pollutant?

2. What are the major sources of nitrogen pollution? Discuss the relationship between excessive nitrogen and forest decline.

3. What is acid deposition? What effects does it have on soils and aquatic ecosystems?

4. What is the main source of phosphorus input into aquatic ecosystems?

5. Explain the paradox of ozone: its beneficial role in the stratosphere and its harmful effects in the lower atmosphere.

6. Give two examples of heavy metals that are released as industrial pollutants.

7. What is the impact of chlorinated hydrocarbons on ecosystems?

Time to Rethink the Lawn

It's any weekend day from April to October. America heads for the garage or garden shed to start the lawn mower. The tasks are well defined: mow, weed, fertilize, lime, aerate, and irrigate. The enemy is identified: crabgrass, dandelions, grubs, moles, ants, and Japanese beetles are among their ranks. The attack begins with herbicides and insecticides. It is an ongoing crusade, and the Holy Grail is the perfect lawn—the envy of the neighborhood (Figure A).

The reason for America's obsession with the lawn is not clear, but theories abound. Some say it's rooted in early 19th-century British ideas of natural beauty. Evolutionary psychologists suggest that the love of lawns is genetically encoded— a molecular memory of the time when progenitors of the human species left their arboreal dwellings for Africa's savannas. Whatever the reason, the lawns of 58 million American households combine to cover 20 million acres of the United States—an area roughly the size of Pennsylvania. Nor is this obsession inexpensive. Americans spend an estimated $30 billion per year on lawn care (enough to provide 1.5 million families with an income above poverty level). Despite the enormous economic expenditures, the real cost of creating and maintaining our lawns is perhaps best measured not in dollars but by the cumulative effects on environmental and human health.

Lawns are human-made ecosystems. To create a lawn, a bare piece of land is required. In the eastern regions of the United States, this means clearing forests. Most of the nitrogen and other essential nutrients in a forest are in the living vegetation and dead organic matter on the soil surface. When these are removed (by bull-dozing or burning), chemical fertilizers must be applied in order for the topsoil to support a lawn. The nutrients in synthetic fertilizers are in an inorganic form that is readily available for uptake by grass plants. However, because these nutrients are not easily stored in the soil, they leach into ground or surface water. In contrast, nutrients in the organic matter that was previously on the surface of forest soil became available slowly through decomposition and nutrient mineralization (the transformation of nutrients from organic to inorganic form). Very little was lost from the ecosystem through leaching.

The problem of nutrients leaching from denuded forest soil is compounded when homeowners remove mowed grass clippings for esthetic reasons. Removal of clippings may result in a loss of up to 100 pounds of nitrogen per acre of lawn per year. Without the natural recycling of nutrients via decomposition of grass clippings, additional fertilizer needs to be added to maintain the lawn. Furthermore, to ensure maximum productivity, lawn owners often add more fertilizer than plants are capable of assimilating. This practice often leads to beautiful lawns, but it can create serious environmental problems.

Excess nutrients have a variety of negative effects on a lawn. Too much nitrogen can increase a grass plant's vulnerability to disease, reduce its ability to withstand extreme temperatures and drought, and discourage the activity of beneficial soil microorganisms. In addition, some synthetic fertilizers acidify the soil, reducing the uptake of magnesium, calcium, and potassium (elements important to biological and chemical processes) in plant and soil organisms.

The environmental impact of overfertilization is not limited to the lawns; it also affects ecosystems that communicate with lawns through the exchange of energy and matter. Nitrous oxide, a product of the breakdown of ammonia (a form of nitrogen commonly used as lawn fertilizer), is implicated in contributing to global warming. When leached from the soil into waterways, the nitrogen and phosphorus in fertilizers can lead to the eutrophication of aquatic ecosystems. This process begins with excessive growth of water plants and ends in a smelly body of

FIGURE A A well-manicured lawn.

water deprived of oxygen and the loss of many life forms. Chemical contamination of wells is a notable problem for humans, with nitrate from lawn fertilizers being the most common pollutant. Recent EPA surveys indicate that 1.2 percent of community water systems and 2.4 percent of rural domestic wells nationwide contain concentrations of nitrate that exceed public health standards. High concentrations of nitrate in drinking water may cause birth defects, cancer, nervous system impairments, and "blue baby syndrome," a condition in which the oxygen content in the infant's blood falls to dangerously low levels.

Further environmental damage occurs from lawn owners' adversarial relationships with nature. In a lawn owner's eyes, three groups of organisms threaten the lawn: animals (such as moles and insects), weeds, and fungi. They may bring disease or simply change the appearance of the lawn, disrupting the smooth, even carpet of green. Since the 1950s, many homeowners have waged war on these enemies with pesticides, specifically with rodenticides, insecticides, herbicides, and fungicides. In contrast, naturally occurring ecosystems protect themselves against disease and insect outbreaks in many different ways. Some plants, such as milkweed, produce chemicals in their leaves that make them unpalatable. Many insect populations are held in check by predators or diseases that are absent—due to pesticide use—in lawn ecosystems. Although pesticides may hold down populations of unwanted lawn pests for a time, they also result in the death of beneficial organisms—such as birds, earthworms, some insects, bacteria, and fungi—important to the health of the lawn. In addition, pesticides may persist in the envi-

ronment for a long time with their lethal capabilities intact, even as they travel by wind, surface runoff, or seepage through the soil to wells and reservoirs used for public water supplies, wetlands, streams, rivers, and lakes, and even to marine environments.

Pesticide contamination of groundwater is less documented than fertilizer pollution but is of growing concern. Detectable levels of pesticides or their chemical breakdown products have been found in 10 percent of the wells in community water systems. Of the various types of pesticides, the most is known about insecticides. Many insecticides work by blocking communication between cells of the nervous system, but the newest ones, such as *Bt* (for *Bacillus thuringiensis*), prevent insects from maturing. The latter are considered to be nontoxic to vertebrates because they are naturally occurring insecticides that function by affecting the insect's growth hormones. However, we do not know their long-term effects when used liberally, and we ought to keep in mind that many invertebrate hormones are similar or identical in molecular structure to vertebrate hormones. Furthermore, history has shown that when we use nature's products for our own purposes, rarely do we understand the full implications of our interference with the environment.

Left unattended, our lawns would quickly be invaded by a variety of native plant species, eventually giving way to the ecosystem that formerly occupied the site. To

stave off the inexorable processes of succession, herbicides are applied to lawns. Only recently has it become apparent that herbicides, like fertilizers and insecticides, are also cycling throughout the environment. Because the data are both scarce and new, the public health implications of these findings are unclear. Nevertheless, herbicides are among the suspects for the suddenly high incidence of deformities observed in amphibians.

Only 5 percent of total yearly herbicide use in the United States is attributable to homeowners, but this 5 percent amounts to an astounding 33 million pounds. Herbicides come in a multitude of formulations. Some are selective, killing either broad-leaved plants or grasses, but not both; others are nonselective, killing all vegetation. The most toxic herbicides cause death if ingested by damaging cellular components; others prevent the germination of seeds. Herbicides work by a variety of

When we think of managed ecosystems, we typically envision agricultural fields or forest plantations, yet the American lawn is the most expensive and management-intensive of all ecosystems.

means including inhibiting cell division; electron transport systems; or the synthesis of lipids, chlorophyll, or key enzymes. Because traces of herbicide residues are being detected in unexpected and unwanted places, including our drinking water, we might be wise to exercise caution in their application until enough time has passed for us to understand their cycling in the environment and their potential toxicity.

Another problem with lawns is that they need watering. The United States is a nation where population increases have combined with increased per capita consumption of water to generate a water crisis. Since 1950, the rate of public water use (which excludes agricultural and self-supplied industrial use) grew at more than twice the rate of our population increase. Water tables are falling, and stream flow is decreasing in many river basins. For instance, water tables in the Dallas–Fort Worth area have fallen more than 400 feet in the last 30 years. Some states in the West are now battling over water rights among themselves and with Mexico, and water shortages are now also frequent on the typically much wetter East Coast. With lawn watering accounting for up to 60 percent of urban water use in the West and 30 percent in the East, it's not surprising that an increasing number of communities are restricting the use of water for lawns.

Grass grows from below ground in what is probably an evolutionary adaptation to grazing animals. We respond by using mechanized grazers—lawnmowers—to keep the grass at a uniformly short height. This, among other reasons, is why lawns are solidly linked to environmental issues that surround fossil fuel consumption, including smog, acid rain, oil spills, destruction of the ozone layer, and global warming. The most obvious use of fossil fuels for lawn management is in running the fleet of mechanized equipment, including mowers, aerators, leaf blowers, weed whackers, and edgers. A staggering 580 million gallons of fuel is consumed annually just by gasoline-powered mowers alone.

Fossil fuels are also employed to manufacture and transport inorganic fertilizers and pesticides. The principal nutrients in lawn fertilizers are nitrogen, phosphorus, and potassium. Natural gas is a reagent in the production of ammonia, the most common source of nitrogen in inorganic fertilizers. Furthermore, industrial nitrogen fixation requires a great deal of heat and pressure, which are supplied by energy from fossil fuels. Fossil fuels are also used to mine and refine potassium and phosphorus. The bag of fertilizer you might buy in Virginia may have originated, in part, in Peru, Utah, and Saudi Arabia. Still more fossil fuel is required to power the ships, trains, trucks, and cars to move these products from their source to the lawn. The same is true for pesticides. Thus, calculation of the environmental costs of these products involves not only their direct chemical effects but also their hidden costs, such as consumption of fossil fuels.

When we think of managed ecosystems, we typically envision agricultural fields or forest plantations, yet the American lawn is the most expensive and management-intensive of all ecosystems. The endless nature of our lawn maintenance cycle is clearly evident: water and nutrients promote growth; herbicides lessen competition from weeds; rapid growth requires frequent mowing; pesticides inhibit not only grass predators but also organisms that decompose clippings; to achieve the desired appearance, clippings are removed; the removal of clippings requires fertilizers to replace nutrients, which promotes growth, etc. Changes are underway, with landscapers attempting to incorporate more native species of grasses and other herbaceous vegetation adapted to the local environmental conditions and requiring less maintenance. It is now becoming fashionable to create meadows and mixed-plant gardens that will attract wildlife in place of unbroken expanses of lawn (Figure B). Research is ongoing to find nontoxic pesticides that do not easily cycle through the environment. Growing concerns about environmental pollution, energy conservation, and scarcity of water are leading to restrictions on water use, engine emissions, and pesticide applications. Although the weekend ritual of firing up the lawn mower to once again do battle with the forces of nature is likely to continue for some generations to come, the pressure is on to reduce, if not altogether abandon, the crusade for the perfect lawn.

FIGURE B An alternative to the lawn.

PART VII

Biogeography
and Biodiversity

A diversity of plant life can be seen looking down through the canopy of a tropical rain forest in Queensland, Australia.

Biogeography and Biodiversity

OBJECTIVES

On completion of this chapter, you should be able to:

- Discuss the link between climate and ecosystem classification.
- Contrast the different bases for classifying terrestrial and aquatic ecosystems.
- Describe the dominant plant forms in the major terrestrial biomes.
- Explain how the different aquatic ecosystems are linked by the water cycle.
- Discuss how global patterns of species diversity vary latitudinally.
- Discuss some of the possible mechanisms for regional patterns of diversity.

From the window of a transcontinental flight from Boston to San Francisco, the ecology-minded passenger finds the view quite revealing. Far below, the pattern of vegetation changes from the mixed coniferous-hardwood forests of the Northeast to the oak forests of the central Appalachians, dotted with dark patches of spruce at higher elevations. As the flight continues westward, forest cover becomes fragmented, merging with midwestern cropland of corn, soybean, and wheat, land that was once the domain of tallgrass prairie. Wheat fields yield to high-elevation shortgrass plains, and the plains give way to the coniferous forests of the Rocky Mountains, capped by tundra and snowfields. Beyond the mountains to the southwest lie tan-colored deserts. On a comparable trip of less than 8 hours over any continent, the airborne ecologist can observe a similar range of vegetation that would have taken early naturalists and explorers, such as Friedrich Humboldt, Charles Darwin, Alfred Wallace, and Joseph Hooker, months or even years to discover (see Chapter 1).

These early naturalist-explorers made two very important, general observations. First, they noted that the plant communities occupying different regions of the world were often very different in appearance: rain forests, deciduous forests, coniferous forests, shrublands, grasslands, and desert. Second, those regions that do support similar types of plant communities—rain forests, for example—are characterized by similar climates. Consider the photos of three desert regions in Australia, Africa, and south-central Asia shown in Figure 24.1. All three regions have a similar climate with low, seasonal precipitation. Although the species that occupy these three shrub-desert ecosystems are quite distinct, the physical structures of the plant communities are similar—that is, they are similar in appearance.

This observed similarity in the structure of plant communities led 19th-century plant geographers, notably J. F. Schouw, A. de Candolle, and A. F. W. Schimper, to correlate the distribution of vegetation to climate. They noted that the world could be divided into zones that represent broad categories of vegetation, categories that are defined on the basis of having a similar physical appearance (physiognomy)—deserts; grassland; and coniferous, temperate, and tropical forests. They called these categories **formations**. Secondly, they noted that regions of the globe occupied by a given type of vegetation formation were characterized by a similar climate. Their observations led to the development of a general understanding of the factors controlling the distribution of vegetation at a global scale.

These early studies relating the distribution of plant life—and later animal life (see Focus on Ecol-

(a)

(b)

(c)

FIGURE 24.1 Shrub desert ecosystems: (a) Karoo Desert in Southern Africa, (b) Kara-Kum Desert in Turkmenistan, Central Asia, and (c) western New South Wales, Australia. All three sites are characterized by low, seasonal precipitation.

ogy 24.1: Animal Distribution: The Biogeographical Realms)—with climate developed into the field of biogeography. **Biogeography** is the study of the spatial distribution of organisms, both past and present. Its goal is to describe and understand the many patterns in the distribution of species and larger taxonomic groups (genus, family, etc.). Historical biogeography is concerned with the reconstruction of the origin, dispersal, and extinction of various taxonomic groups. Ecological biogeography is concerned with studying the distribution of contemporary organisms. In this chapter, we will focus on the regional- to continental-scale patterns of biological diversity.

24.1 The Classifications of Ecosystems and Climate Have a Linked History

The early explorer-naturalists and plant geographers produced maps showing the geographic distribution of various plant formations. Their maps suggested that the world could be neatly divided into regions or zones defined by these general classes of vegetation: formations such as rain forest, grassland, or desert. These maps raise the question of why such regular patterns should exist. A working hypothesis emerged that stated that the existence of consistent plant formations and the positions of their boundaries were the result of climate—namely, temperature and available water. In 1874, the French botanist and taxonomist Alphonse de Candolle put forward one of the earliest systems for classifying formations. De Candolle sought to categorize the plants indigenous to each formation by their requirements for heat and moisture (Table 24.1). Further progress in an explanation for plant formations was hindered by the lack of adequate data on regional climate patterns. During the 19th century, the distribution of weather stations was extremely limited, with virtually no systematic collection of climate data for many, if not most, regions of the world. However, once plant geographers had developed vegetation maps for most of the world's regions, scientists studying climate proposed using their maps as a base for climate maps. More than half a century after the publication of de Candolle's work on plant distribution, Vladimir Köppen (1844, 1900, 1918) used de Candolle's classification of vegetation to found the modern system of classifying climate (Figure 24.2). The link between climate and the distribution of vegetation (and ecosystems) had become permanently established. Every system of ecosystem classification that followed would have climate as its central component.

Patterns of vegetation zonation along elevation gradients in mountainous regions provided early plant geographers with additional insight into the relationship between climate and ecosystem distribution. In 1884, G. Hart Merriam developed his system of life zones after observing the sharp zonation of vegetation

TABLE 24.1

Classification of Plant Types and Associated Vegetation Formations (Ecosystems) Developed by the Plant Geographer Alphonse de Candolle (1855)

De Candolle's Plant Group	Postulated Plant Requirements	Formation	Köppen Climatic Division
Megatherms (most-heat)	Continuous high temperature and abundant moisture	Tropical rain forest	A (rainy with no winter)
Xerophiles (dry-loving)	Tolerate drought, need minimum hot season	Hot desert such as Sonoran	B (dry)
Mesotherms (middle-heat)	Moderate temperature and moderate moisture	Temperate deciduous forest	C (rainy with mild winters)
Microtherms (little-heat)	Less heat, less moisture Tolerate long cold winters	Boreal forest	D (rainy climates with severe winters)
Hekistotherms (least-heat)	Tolerate polar regions "beyond tree-line"	Tundra	E (polar climates with no warm season)

Associated Köppen climate divisions are shown for comparison.

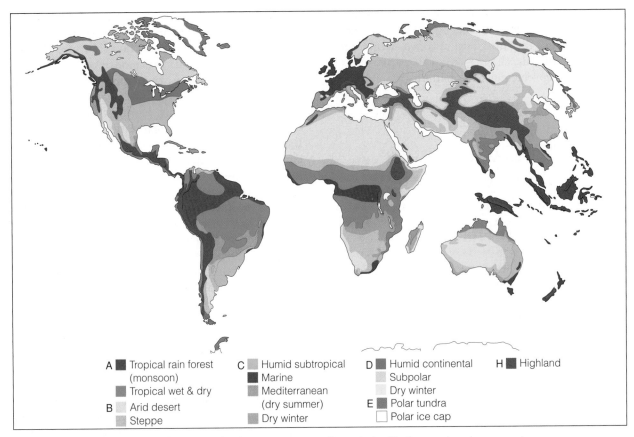

FIGURE 24.2 Climates of Earth mapped under the Köppen system. Köppen's classification was based on the assumption that the boundaries between different types of vegetation and ecosystems were set by climate. His climate maps were essentially vegetation maps using the classification system of plant geographer de Candolle (see Table 24.1).

Legend:

A ■ Tropical rain forest (monsoon)
   ■ Tropical wet & dry
B ■ Arid desert
   ■ Steppe
C ■ Humid subtropical
   ■ Marine
   ■ Mediterranean (dry summer)
   ■ Dry winter
D ■ Humid continental
   ■ Subpolar
   ■ Dry winter
E ■ Polar tundra
   □ Polar ice cap
H ■ Highland

on San Francisco Mountain in Arizona. Impressed with the importance of changes in temperature with elevation as an influence on plant and animal distribution, he formulated two "rules" regarding temperature control. The first states that the northward distribution of terrestrial plants and animals is governed by the integration of temperature over the entire season of growth and reproduction: the duration and warmth of the growing season. The second rule states that temperatures during the hottest part of the year govern the southward distribution: the species' tolerance to heat. Using these rules, Merriam divided the North American continent into three primary transcontinental regions: the Boreal, the Austral, and the Tropical. He then subdivided these regions into zones. Although not used today, his life zones are reflected in the temperature hardiness zones mapped and widely used in horticultural and planting guides. Merriam's approach of relating plant distribution to an integrated index of temperature over the growing season is also widely used in agriculture and forestry (see Focus on Ecology 6.2).

In 1939, the ecologists F. E. Clements and V. E. Shelford introduced an approach for combining both plant and animal distribution into a single classification system. They began with the broad concept of plant formations developed by their predecessors and combined the animal life associated with these formations into a single biotic unit (as with the concept of community). Clements and Shelford called these biotic units **biomes.** Each biome is characterized by a uniform type of vegetation, such as grassland or coniferous forest. The approach worked well because of the dependency of animal life on vegetation to provide suitable habitat and basic resources.

Depending on how finely you want to classify biomes, there are at least nine major terrestrial biome types: tundra, taiga, temperate forest, temperate rain forest, tropical rain forest, savanna, temperate grasslands, chaparral, and desert (Figure 24.3). When the plant ecologist Robert Whittaker plotted these biome types on gradients of mean annual temperature and mean annual precipitation, he found they formed a distinctive climatic pattern, as graphed in Figure

24.4. As the graph indicates, boundaries between biomes are broad and often indistinct as one vegetation type blends into another. In addition to climate, topography, soils, and exposure to disturbances such as fire can influence which one of the several biome types occupies an area.

Although scientists have devised many systems for classifying vegetation formations and ecosystems, the life zone system developed by L. R. Holdridge in 1947 was the first to quantify the relationship between climate and ecosystem distribution. The Holdridge life zone system involves three levels of classification: (1) *life zones* defined by climate; (2) *associations* or subdivisions of life zones based on local environmental conditions; and (3) local subdivisions based on actual land cover or land use. Holdridge's term *association* refers to a unique ecosystem, a distinctive habitat or physical environment and its associated community of plants and animals. Holdridge based his classification on three assumptions: (1) interaction between temperature and rainfall determine vegetation patterns; (2) geographic boundaries of vegetation correspond closely to boundaries between climatic zones; and (3) mature, stable vegetation types represent discrete units that are recognizable throughout the world.

The Holdridge classification divides the world into life zones represented by a series of hexagons arranged in a triangular coordinate system (Figure 24.5). Three climate variables, or indices, determine the classification: average daily biotemperature, annual precipitation, and potential evapotranspiration ratio (PET ratio). Biotemperature is calculated by summing the average daily temperature over the year, but only for those days where the value is above above 0°C (freezing), then dividing the sum by the number of days in the year (365). Biotemperature provides an index similar to growing degree-days discussed in Chapter 6. The third variable in the classification, PET ratio, is the ratio of annual potential evapotranspiration (see Chapter 4, Section 4.7) and annual precipitation. Holdridge assumed that potential evapotranspiration is a simple function of biotemperature (PET = 58.93 × biotemperature), so the two primary variables for the classification are biotemperature and annual precipitation.

Identical axes for annual precipitation for two sides of the equilateral triangle are shown in Figure 24.5. The PET ratio forms the third side, and an axis of biotemperature is oriented perpendicular to its base. By striking equal intervals on these logarithmic axes, hexagons are formed that designate the various life zones. One additional division in the classification relates to the occurrence of frost (freezing temperatures): the line that divides hexagons between biotemperatures of 12°C and 24°C into warm temperate (frost) and subtropical (no frost) zones.

Provided with climate data to calculate the two primary variables—biotemperature and annual precipitation—for a given location, the model will predict the corresponding life zone. In this manner, the model provides an explicit hypothesis that plant geographers can test against observations of the distribution of vegetation or ecosystems within a region.

Since the publication of Holdridge's life zone model in 1947, scientists have proposed various systems for classifying the regional and global distribution of ecosystems and vegetation formations (see Focus on Ecology 24.2: Ecoregions: Classifying and Mapping Ecosystems for Resource Management). While these models differ in the way they categorize ecosystems and correlate with climate, they all involve some mathematically explicit method of mapping vegetation and/or ecosystems to features of the physical environment, such as climate and geology, that vary geographically. The motive for developing most of these classification models has been to explore the consequences of past and future climate change on the distribution of vegetation and ecosystems on both a continental and global scale (see Chapter 30). Plant geographers apply these modern approaches to mapping vegetation using geographically distributed climate data, and climatologists, particularly paleoclimatologists, use them to map climate using geographic data on plant distribution (see Focus on Ecology 24.3: Predicting Climate from Pollen).

24.2 Terrestrial Ecosystems Reflect Adaptations of the Dominant Plant Life Forms

The broad categories of biomes shown in Figure 24.4 are based on the dominant plant life forms of the associated plant communities (see Chapter 13, Sections 13.5 and 13.6). The relative contribution of three general plant life forms—trees, shrubs, and grasses—form the basis for the broad categories of terrestrial ecosystems represented by these biome types: forest, woodland/savanna, shrubland, and grassland (Figure 24.6). A closed canopy of trees characterizes forest ecosystems. Woodland and savanna ecosystems are characterized by the codominance of grasses and trees (or shrubs). As the names

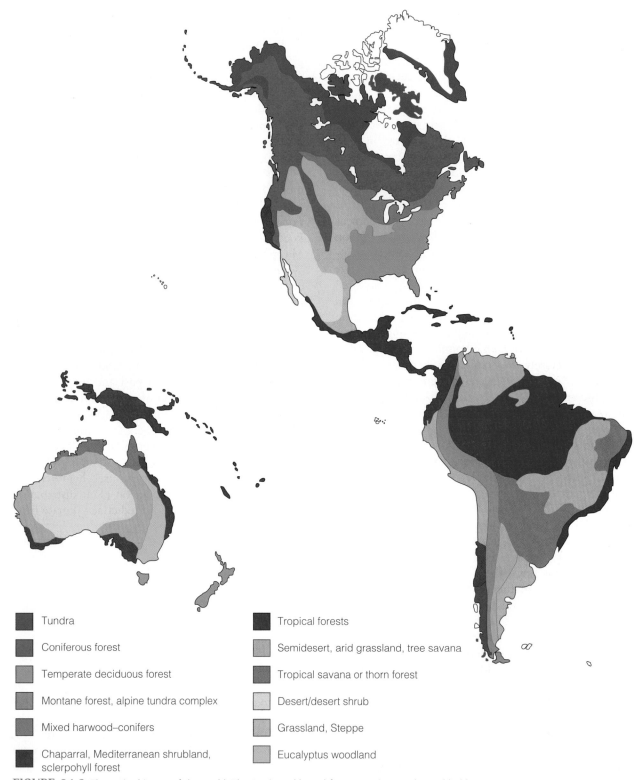

Tundra

Coniferous forest

Temperate deciduous forest

Montane forest, alpine tundra complex

Mixed harwood–conifers

Chaparral, Mediterranean shrubland, sclerpohyll forest

Tropical forests

Semidesert, arid grassland, tree savana

Tropical savana or thorn forest

Desert/desert shrub

Grassland, Steppe

Eucalyptus woodland

FIGURE 24.3 The major biomes of the world. The Arctic and boreal forest are circumpolar and hold many taxonomically related and functionally similar species. Other biomes possess different genera and species that function as ecological equivalents.

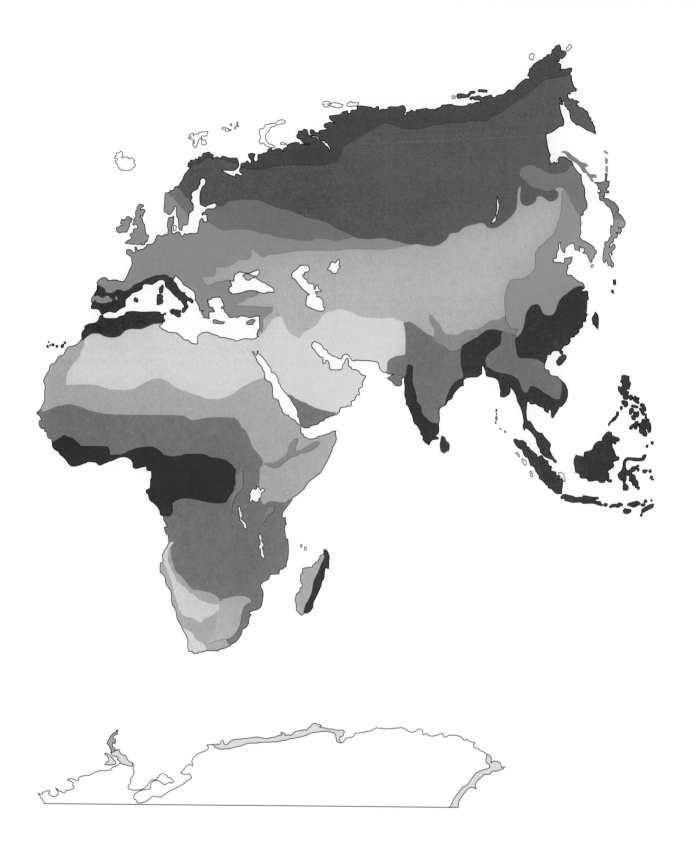

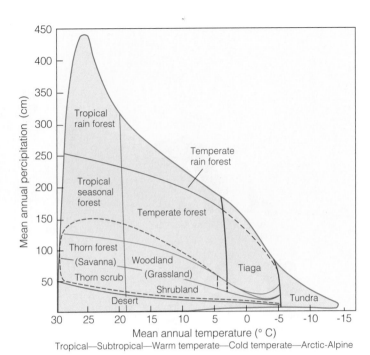

FIGURE 24.4 The pattern of world plant biomes in relation to temperature and moisture. Where the climate varies, soil can shift the balance between types. The dashed line encloses environments in which either grassland or one of the types dominated by woody plants may prevail (Whittaker 1970).

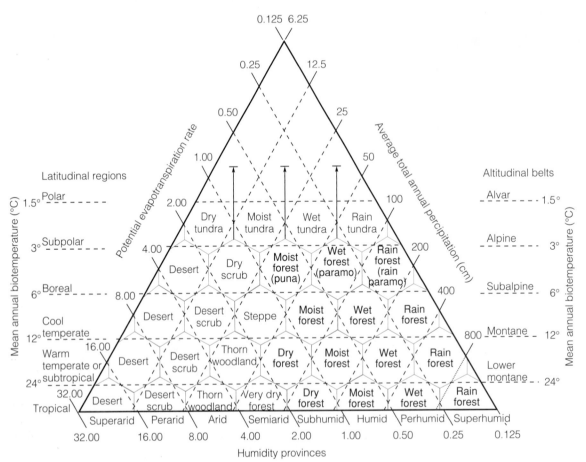

FIGURE 24.5 The Holdridge life zone system for classifying plant formations. The life zones are determined by a gradient of biotemperatures with latitude and altitude, annual precipitation, and the PET ratio: the ratio of potential evapotranspiration to annual precipitation. See text for detailed description. (After Holdridge 1947, 1967.)

FIGURE 24.6 Terrestrial ecosystems are classified based on vegetation structure, more specifically, the dominant life form(s) of plants. (a) Forests are characterized by a closed/continuous canopy of trees; (b) woodlands and savanna ecosystems are characterized by the coexistence of grasses and woody vegetation, the latter typically being an open canopy of trees (or shrubs); (c) shrubs dominate the semiarid shrubland ecosystems; and (d) grasses (either annual or perennial) are the dominant plant life form in grassland ecosystems.

imply, shrubs are the dominant plant form in shrublands and grasses in grasslands. Desert is a general category used to describe the relative absence of plant cover.

Given the general categories of ecosystems based on the relative dominance of trees, shrubs, and grasses, the question becomes: Why are there consistent patterns in the distribution and abundance of these three dominant plant life forms that relate to climate and the physical environment? The answer lies in the adaptations that these three very different plant life

forms represent, and the advantages and constraints that arise from these adaptations under different environmental conditions. Although the broad categories of grasses, shrubs, and trees each represent a diverse range of species and characteristics, they represent fundamentally different patterns of carbon allocation and morphology. Grasses allocate less carbon to the production of supportive tissues (stems) than do woody plants (shrubs and trees), enabling them to maintain a higher proportion of their biomass in photosynthetic tissues (leaves). For woody

ANIMAL DISTRIBUTION: THE BIOGEOGRAPHICAL REALMS

Zoogeographers lagged behind plant geographers in their study of distribution. Complicating their studies of animal distribution were the great number of animal species and the lack of a clear relationship between distribution and climate. Ultimately, zoogeographers did accumulate basic information on the global distribution of animals. Alfred Wallace, also known for having developed the same general theory of evolution as Darwin, provided the major synthesis of animal distribution. Wallace's realms, with some modification, still stand today.

There are six biogeographical realms, each more or less embracing a major continental land mass and separated by oceans, mountain ranges, or desert (see Figure A). They are the Palearctic, the Nearctic, the Neotropical, the Ethiopian, the Oriental, and the Australian. Because some zoogeographers consider the Neotropical and the Australian realms to be so different from the rest of the world, these two are often considered as realms equal to the other four combined. Then there are just three realms: Neogea (Neotropical), Notogea (Australian), and Metagea (everything else). Each region possesses a certain distinction and uniformity in the taxonomic units it contains, although each shares some of the families of animals with other regions. Each has at some time in Earth's history had some land connection with another, across which animals and plants could pass.

Two realms, the Palearctic and Nearctic, are quite closely related. In fact, the two are often considered as one, the Holarctic. Both are much alike in their faunal composition, and together they share, particularly in the north, such animals as the wolf, hare, moose (called elk in Europe), stag (called elk in North America), caribou, wolverine, and bison.

Below the coniferous forest belt, the two regions become more distinct. The Palearctic is not rich in vertebrate fauna, of which few are endemic. Palearctic reptiles are few and are usually related to those of the African and Oriental tropics. The Nearctic, in contrast, is the home of many reptiles and has more endemic families of vertebrates. The Nearctic fauna is a complex of New World tropical and Old World temperate families. The Palearctic is a complex of Old World tropical and New World temperate families.

Isolated until 15 million years ago, the fauna of the Neotropical is most distinctive and varied. In fact, about half of the South American mammals, such as the tapir and llama, are descendants of North American invaders, whereas the only South American mammals to survive in North America are the opossum and the porcupine. Lacking in the Neotropical is a well-developed ungulate fauna of the plains, characteristic of North America and Africa. However, the Neotropical is rich in endemic families of vertebrates. Of the 32 families of mammals, excluding bats, 16 are restricted to the Neotropical. In addition, 5 families of bats, including the vampire, are endemic.

The Ethiopian realm embraces tropical forests in central Africa, and savanna, grasslands, and desert in the mountains of east Africa. During the Miocene and Pliocene epochs, Africa, Arabia, and India shared a moist climate and a continuous land bridge, which allowed animals to move freely among them. That connection accounts for some similarity in the fauna between the Ethiopian and Oriental regions. Of all the regions, the Ethiopian contains the most varied vertebrate fauna, and in endemic families it is second only to the Neotropical.

Of the tropical realms, the Oriental, once covered with lush forests, possesses the fewest endemic species and lacks a variety of widespread families. It is rich in primate species, including two families confined to the region, the tree shrews and tarsiers.

Perhaps the most interesting and the strangest region, and certainly the most impoverished in vertebrate species, is the Australian. Partly tropical and partly south temperate, this region is noted for its lack of a land connection with other regions; the few freshwater fish, amphibians, and reptiles; native placental mammals restricted to bats, mice, and rats; and the dominance of marsupials. Included are the monotremes, with two egg-laying families, the duck-billed platypus and the spiny anteaters. The marsupials have become diverse and have evolved ways of life similar to those of the placental mammals of other regions.

---

plants, shrubs allocate relatively less to stems than do trees. The production of woody tissue provides the advantage of height and access to light, but it also has the associated cost of maintenance and respiration. If this cost cannot be offset by carbon gain through photosynthesis, the plant will not be able to maintain a positive carbon balance and will die (see Chapter 6, Figure 6.5). As a result, as environmental

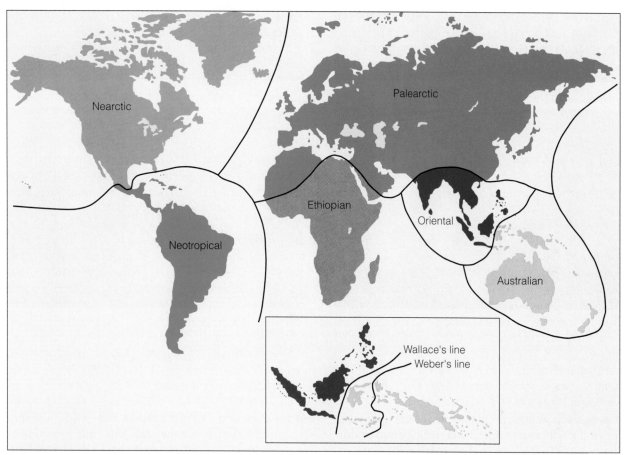

FIGURE A The biogeographical realms of the world. Inset: No definite boundary exists between the Oriental and Australian regions, where the islands of the Malay Archipelago stretch toward Australia. Two lines have been proposed to separate the two regions. One, Wallace's line, runs from the Philippines through Borneo and the Celebes. The other, Weber's line, lies east of Wallace's line. It separates the islands with the majority of Oriental animals from those with a majority of Australian animals. Because the islands between these two lines are a transitional zone, some zoogeographers call the area Wallacea.

conditions become adverse for photosynthesis (dry, low nutrient, or short growing season and cold temperatures), trees will decline in both stature and density until they are no longer able to persist as a component of the plant community.

Within the broad classes of forest and woodland ecosystems in which trees are dominant or codominant, leaf form is another plant characteristic that ecologists use to classify ecosystems. Leaves can be classified into two broad categories based on their longevity. Leaves that live for only a single year or growing season are classified as deciduous, while those living beyond a year are called evergreen. The deciduous leaf is characteristic of environments with a distinct growing season. Leaves are typically shed at the end of the growing season, and then regrown at the beginning of the next. Deciduous leaf type is further divided into two categories based on the period

FIGURE 24.7 Examples of winter and drought-deciduous trees. Temperate deciduous forest in central Virginia during the (a) summer and (b) winter seasons. Semi-arid savanna/woodland in Zimbabwe, Africa, during the (c) rainy and (d) dry seasons.

of dormancy. Winter-deciduous leaves are characteristic of the temperate regions where the period of dormancy corresponds to low (below freezing) temperatures (Figure 24.7a, b). Drought-deciduous leaves are characteristic of environments with seasonal rainfall in the subtropical and tropical regions, where leaves are shed during the dry period (Figure 24.7c, d). The advantage of the deciduous habit is that the plant does not have the additional cost of maintenance and respiration during the period of the year when environmental conditions restrict photosynthesis.

Evergreen leaves can likewise be classified into two broad categories. The broadleaf-evergreen leaf type (Figure 24.8a) is characteristic of environments with no distinct growing season where photosynthesis and growth continue year-round. The needle-leaf evergreen form (Figure 24.8b) is characteristic of environments where the growing season is very short (northern latitudes) or nutrient availability severely constrains photosynthesis and plant growth. A simple economic model has been proposed to explain the adaptation of this leaf form (see Chapter 6). The production of a leaf has a "cost" to the plant that can be

defined in terms of the carbon and other nutrients required to construct the leaf. The time required to pay back the cost of production (carbon) will be a function of the rate of net photosynthesis (carbon gain). If environmental conditions result in low rates of net photosynthesis, the period of time required to pay back the cost of production will be longer. If the rate of photosynthesis is low enough, it may not be possible to pay back the cost over the period of a single growing season. A plant adapted to such environmental conditions cannot "afford" a deciduous leaf form, which requires producing new leaves every year. The leaves of evergreens, however, may survive for a number of years. So under this model, we can view the needle-leaf evergreen as a plant adapted for survival in an environment with a distinct growing season, where conditions limit the ability of the plant to produce enough carbon through photosynthesis during the growing season to pay for the cost of producing the leaves.

Combining this simple classification of plant life forms and leaf type with the constraints imposed by the physical environment, we can understand the

(a)   (b)

FIGURE 24.8 Examples of evergreen trees. (a) Broadleaf evergreen trees dominate the canopy of this tropical rain forest in Queensland, Australia. (b) Needle-leaf evergreen trees (foxtail pine) inhabit the high-altitude zone of the Sierra Nevada Mountains in western North America.

distribution of biome types relative to the axes of temperature and precipitation shown in Figure 24.4. Ecosystems characteristic of warm, wet climates with no distinct seasonality are dominated by broadleaf evergreen trees and are called tropical (and subtropical) rain forest. As conditions become drier, with a distinct dry season, the broadleaf evergreen habit gives way to drought-deciduous trees that characterize the seasonal tropical forests. As precipitation declines further, the stature and density of these trees declines, giving rise to the woodlands and savannas that are characterized by the coexistence of trees (shrubs) and grasses. As precipitation further declines, trees can no longer be supported, giving rise to the arid shrublands (thorn scrub).

The temperature axis represents the latitudinal gradient from the equator to the poles. Moving from the broadleaf evergreen forests of the wet tropics into the cooler, seasonal environments of the temperate regions, the dominant trees are winter-deciduous. These are the regions of temperate deciduous forest. In areas of the temperate region where precipitation is insufficient to support trees, grasses dominate, giving rise to the prairies of North America and the steppes of Eurasia. Moving poleward, the temperate deciduous forests give way to the needle-leaf dominated forests of the boreal region (conifer forest or tiaga). As temperatures become more extreme and the growing season shorter, trees can no longer be supported and the short-stature shrubs and grasses characteristic of the tundra dominate the landscape. The worldwide distribution of terrestrial biomes shown in Figure 24.3 map the relationships outlined in Figure 24.4 onto the global patterns of temperature and precipitation described in Chapter 3.

24.3 Aquatic Ecosystems Are Classified on the Basis of Physical Features

Whereas scientists classify terrestrial ecosystems according to their dominant plant life forms, classification of aquatic ecosystems is largely based on features of the physical environment. One of the major features that influence the adaptations of organisms to their aquatic environment is water salinity (see Chapter 4, Section 4.17). For this reason, scientists divide aquatic ecosystems into two major categories—freshwater and saltwater (or marine). These categories are further divided into a number of ecosystem types based on substrate, depth and flow of water, and the type of dominant organisms (typically plants). All aquatic ecosystems, both freshwater and marine, are linked either directly or indirectly as components of the hydrologic cycle, as illustrated in Figure 24.9 (also see Chapter 4, Section 4.7).

Freshwater ecosystems are classified on the basis of depth and flow of water, physical features that in turn influence the vertical profiles of temperature, light, oxygen, and nutrients. Water evaporated from both the oceans and terrestrial environments falls as precipitation. That portion of precipitation that remains on the land surface—that is, that does not infiltrate into the soil or evaporate—follows a path determined by gravity and topography—more specifically, geomorphology. Flowing water ecosystems begin as streams. These streams, in turn, coalesce into rivers as they follow the topography of the landscape, or they collect in basins and floodplains to

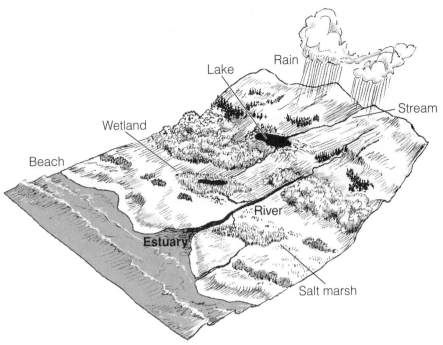

FIGURE 24.9 Idealized landscape/seascape showing the linkages among the various types of aquatic ecosystems via the water cycle.

form nonflowing water ecosystems, such as ponds, lakes, and inland wetlands. Rivers eventually flow to the coast, forming estuaries, the transition from freshwater to marine. Generally defined as a partially enclosed coastal embayment where freshwater and saltwater mix, this broad category includes bay, lagoon, and fjord ecosystems.

Ecologists subdivide marine ecosystems into two broad categories of coastal and open-water systems. Coastal environments, the transition zone between land and sea, give rise to a diverse array of ecosystems. Two transitional areas exist, one between the terrestrial and marine environment, and the other between the marine and freshwater. The transition between terrestrial and marine environments is the intertidal zone, that shore area lying between the extremes of high and low tide. In this zone, different substrates—rock, sand, and mud—support very distinct communities. Where rivers flow into the coastal waters, the deposition of sediments creates shallow-water ecosystems that undergo varying degrees of inundation during the tidal cycle. Salt-marsh ecosystems, dominated by grasses and small shrubs rooted in the mud and sediments, occupy these habitats in the temperate regions, giving way to mangrove forests in the tropics.

Ecologists subdivide the waters of the open ocean into vertical and horizontal zones. The entire area of the open ocean is referred to as the pelagic zone. The waters that overlie the continental shelves are referred to as the neritic zone and all other open waters as the oceanic zone. The general term for the sea bottom is the benthic zone. We discuss further vertical subdivisions within the neritic and oceanic zones in Chapter 28.

24.4 Regional and Global Patterns of Species Diversity Vary Geographically

Earth's ecosystems support an amazing diversity of species. In fact, scientists have identified and named approximately 1.4 million species (Figure 24.10) (see the Ecological Application after Ch. 19: Asteroids, Bulldozers, and Biodiversity). Because new species are continuously being discovered, quantifying the actual number of species inhabiting Earth is an ongoing exercise. Some scientists, such as Harvard biologist E. O. Wilson, believe that the actual number of species may be closer to 10 million.

The 1.4 million species that have been identified are not distributed equally across Earth's surface. There are distinct geographic patterns of species

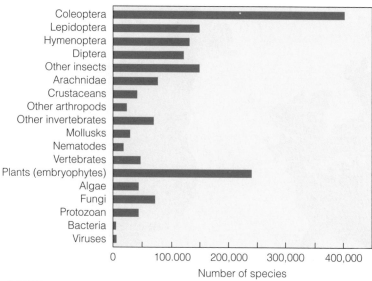

FIGURE 24.10 Number of living species of all kinds of organisms currently known. Species are classified by major taxonomic groups. Insects and plants dominate the diversity of living organisms.

richness (number of species). In general, the number of terrestrial species decreases as one moves away from the equator toward the poles. The three maps in Figure 24.11 illustrate distinct geographic patterns of species richness for trees, mammals, and birds in North America. Note the overall decrease in species richness as one moves from south to north. This pattern is more obvious if we plot species richness as a function of latitude (Figure 24.12).

The decline in species diversity as you move northward in the North American continent is part of a global pattern of declining diversity (both terrestrial and marine) from the equator north and southward toward the poles. Although scientists do not know the exact mechanisms underlying the geographic pattern of species diversity, they have put forward a variety of hypotheses related to a range of factors, including the age of the community, spatial heterogeneity of the environment, stability of the climate over time, and ecosystem productivity. Of the various hypotheses proposed to account for global patterns of species diversity, the most easily interpreted are those explicitly relating to environmental features such as climate and availability of essential resources, which are known to directly influence basic plant and animal processes.

24.5 Species Richness in Terrestrial Ecosystems Correlates with Climate and Productivity

D. J. Currie and V. Paquin of the University of Ottawa, Canada, examined the relationship between patterns of species richness in North America tree species and several variables describing regional differences in climate. They found that although variation in species richness correlates to climatic factors such as integrated measures of annual temperature, solar radiation, and precipitation, it most strongly correlates with estimates of actual evapotranspiration (AET) (Figure 24.13).

Recall from earlier discussions in Chapters 4 and 20 that actual AET is the flux of water from the terrestrial surface to the atmosphere through both evaporation and transpiration. It is a function of both the atmospheric demand for water brought about by the input of solar energy to the surface and the supply of water from precipitation. The pattern of increasing species richness with increasing AET parallels the positive correlation between AET and net primary productivity presented earlier (Figure 20.5), suggesting a relationship between plant diversity and primary productivity. In other words, environmental conditions favorable for photosynthesis and plant growth may well give rise to increased plant diversity over evolutionary time.

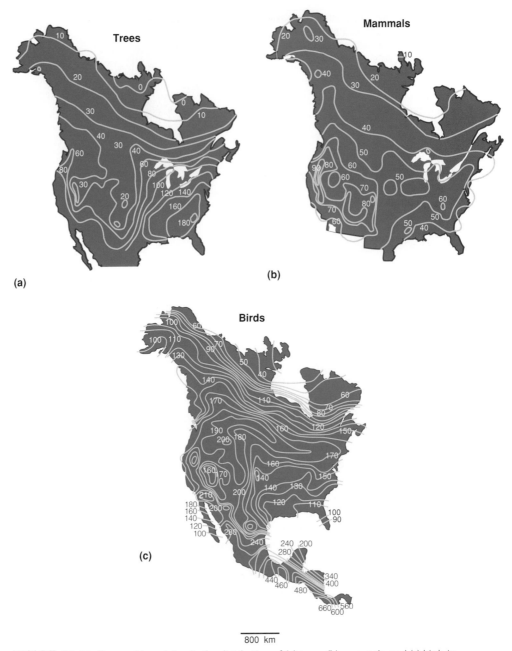

Trees

Mammals

(a)

(b)

Birds

(c)

800 km

FIGURE 24.11 Geographic variation in the distribution of (a) trees, (b) mammals, and (c) birds in North America. Contour lines connect points with about the same number of species. (After Currie 1991.)

Currie also demonstrated a relationship between climate and species richness of vertebrates across North America. He found a positive correlation between potential evapotranspiration (PET), an index of integrated energy availability, and regional patterns of species richness for both mammals and birds across North America (Figure 24.14). Although the correlation between PET and vertebrate species richness offers hope of a mechanistic interpretation other than latitude, PET is correlated with temperature, solar radiation, precipitation, humidity, and a host of other abiotic factors that vary with latitude. Like actual evapotranspiration, PET also correlates with plant productivity. In fact, Currie reported a positive correlation between vertebrate diversity and plant species diversity. Given the correlation between plant diversity and net primary productivity, the correlation between plant and animal diversity may relate to the positive relationship between primary and secondary productivity discussed in Chapter 20 (Section 20.8). This suggests a relationship between secondary productivity and animal diversity similar

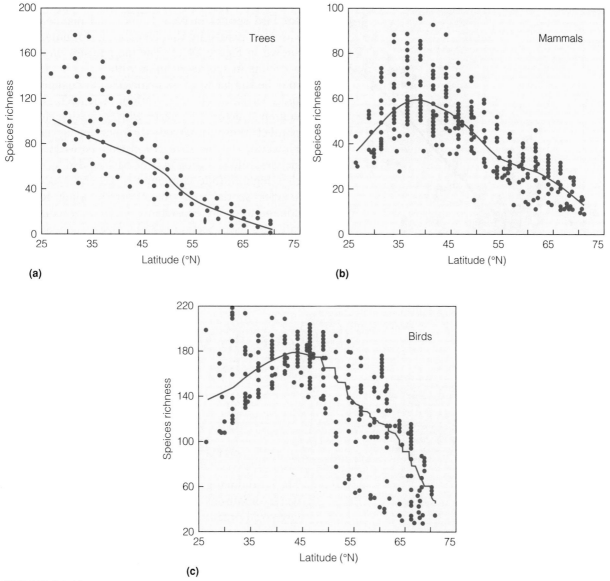

FIGURE 24.12 North American latitudinal gradients in species richness for (a) trees, (b) mammals, and (c) birds based on cells of 2.5 x 2.5° latitude/longitude. Species richness per cell is based on range maps for individual species. (After Currie 1991.)

to that observed between primary productivity and plant diversity.

Animal diversity is linked to plant diversity because variety in plant species provides both a variety of potential food sources as well as suitable habitat for animals. Increased structural diversity within plant communities, as measured by foliage height diversity (see Figure 17.14), provides a wider range of microhabitats and associated resources, and consequently supports a greater variety of animal species. This relationship between habitat heterogeneity and species diversity is not limited to animals. Environmental heterogeneity also gives rise to increased plant species diversity. For example, the diverse topography of mountainous regions generally supports more species than the consistent terrain of flatlands. From east to west in North America, the number of species of trees (Figure 24.11a), as well as breeding land birds (Figure 24.11c) and mammals (Figure 24.11b), increases. This increased diversity on an east-west gradient relates to an increased diversity of the environment both horizontally and altitudinally.

Although mountainous regions may support more species than flatlands, within mountainous regions scientists have observed a pattern of decreasing species richness with increasing elevation.

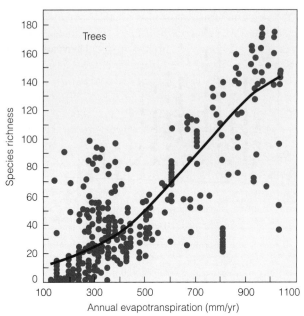

FIGURE 24.13 Relationship between annual measure of actual evapotranspiration (AET) and tree species richness for North America. (From Currie and Paquin 1987.)

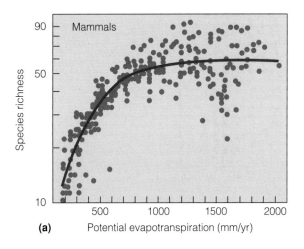

(a)

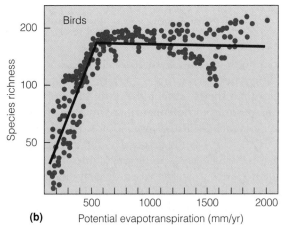

(b)

FIGURE 24.14 Relationship between annual estimate of potential evapotranspiration and species richness of (a) mammals and (b) birds in North America. (From Currie 1991.)

Patterns of species richness with increasing elevation for bird species in New Guinea and mammals and vascular plants in the Himalayan Mountains are shown in Figure 24.15. The mechanisms underlying a decline in species richness with a rise in elevation may be similar to those involved with changing latitude. Variations in temperature, PET, AET, and vegetation structure that occur with increasing elevation parallel those observed with increasing latitude. However, the negative correlation between species richness and elevation may be confounded by the fact that high-altitude communities generally occupy a smaller spatial area than corresponding lowland communities in ecosystems located at equivalent latitudes (see Section 19.3). These high-elevation communities are also likely to be isolated from similar communities, suggesting the importance of immigration in enabling populations to persist over time (see Section 19.4).

24.6 In Marine Environments, There Is an Inverse Relationship Between Productivity and Diversity

The latitudinal gradients of species richness for saltwater organisms are similar to those observed for terrestrial organisms (Figure 24.16). The correlation between patterns of species richness and productivity are not as straightforward as those observed in terrestrial environments. In fact, the general latitudinal gradient of productivity in the oceans is the reverse of that observed on land (see Figures 20.6 and 20.8). With the exception of localized areas of upwelling (see Section 21.10), the primary productivity of the oceans increases from the equator to the poles. This relationship suggests an inverse relationship between productivity and diversity—the opposite of that observed for terrestrial environments.

Circumstantial evidence points to the importance of seasonality, rather than total annual productivity, as a factor influencing local patterns of diversity for both pelagic and benthic species. Observations show that as the influence of seasonal fluctuations in temperature on primary productivity increase (see Figure 28.2), species richness declines and species dominance increases. Recall from Chapter 21 that primary productivity in the ocean waters is influenced by the seasonal dynamics of the thermocline and vertical transport of nutrients from the deep to surface waters (see Figure 21.8). In northern latitudes,

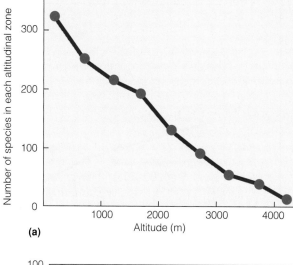

(a)

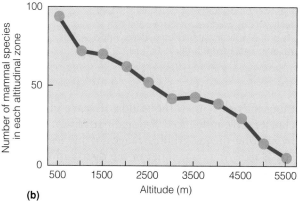

(b)

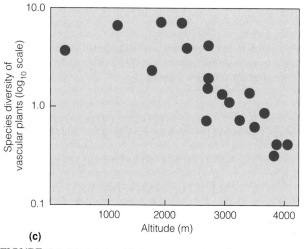

(c)

FIGURE 24.15 Relationship between species richness and altitude for (a) bird species in New Guinea (from Kikkawa and Williams 1971), (b) mammal species in the Himalayas (from Hunter and Yonzon 1992), and (c) vascular plants in the Himalayas. (From Whittaker 1977.)

the seasonal formation and breakdown of the thermocline results in the productivity of surface waters varying between very high (spring and summer) and very low (winter) (see Chapters 21 and 28). By contrast, the permanent presence of a thermocline in the tropical ocean waters results in a low but continuous pattern of primary productivity throughout the year. Seasonal variation in the temperature of the surface waters functions to increase primary productivity, whereas the lack of seasonal variation functions to support a higher diversity of life.

Scientists have hypothesized that geologic history has been an important factor influencing latitudinal patterns of species richness within the oceans. Some suggest that glaciation that occurred during the Quaternary Period may be responsible for lower diversity observed at the higher latitudes (see Section 13.14). During this period, sea ice covered the Norwegian Sea and the northern reaches of the Atlantic. As a result, the poleward decline in regional diversity may to some extent represent a recovery from the effects of glaciation.

24.7 Species Diversity Is a Function of Processes Operating at Many Scales

The discussion of species diversity, even at the broad geographic scale that has been the focus of this chapter, is complicated by a variety of factors that relate directly to topics presented in earlier chapters. For example, in Chapter 13 we discussed species diversity of individual communities. Ecologists define species diversity at this spatial scale as **local** (alpha) **diversity.** Quantifying local patterns of diversity is hindered by the often-difficult task of defining community boundaries. In addition, the relationship between species diversity and area discussed in Chapter 19 makes it difficult to compare patterns of species diversity among communities and ecosystems that differ in size, for example, lake ecosystems of varying size. Local patterns of plant and animal diversity also change over time during succession (see Figures 17.13 and 17.15), further compounding the difficulty in drawing comparisons among communities.

Total species diversity (or species richness) across all the communities within a geographic area is called **regional** (gamma) **diversity.** Diversity at this scale corresponds to the patterns depicted in Figures 24.11 and 24.16. The comparison of even these broad-scale patterns of diversity at a continental or global scale

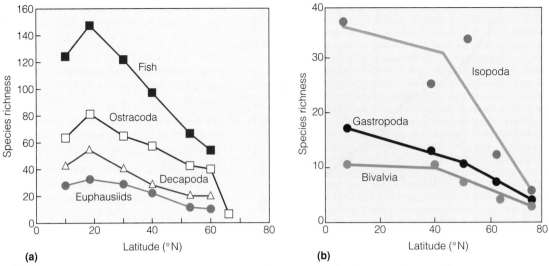

FIGURE 24.16 Latitudinal gradient of species richness: (a) Four groups of pelagic organisms caught at six stations along 20° W (longitude) in the northeast Atlantic Ocean. Samples were collected over a period of 14 days (day and night) from the top 2000 meters of the vertical water column. (From Angel 1991.) (b) Four groups of benthic organisms in the North Atlantic. (Adapted from Rex et al. 1993.)

can be confounded by time. Latitudinal patterns of diversity for pelagic species illustrate this point. Data presented in Figure 24.16a comes from the Institute of Oceanographic Sciences Deacon Laboratory in Surrey, England. At the time of sampling, the major front (boundary) between the South Atlantic Central Water and the North Atlantic Central Water occurred at 18°N. Because these two major regions of the Atlantic Ocean differ in physical characteristics, each supports a different fauna, which come together within this boundary zone. Thus, the peak in species diversity observed for the three groups of organisms at 18°N and graphed in Figure 24.16a represents the rich mix of species characteristic of the edge effect described in Chapter 19. This boundary zone of maximum diversity shifts its geographic position seasonally by several degrees, and thus its location changes with the seasons.

Although not apparent in the map shown in Figure 24.11, regional estimates of bird species diversity in eastern North America are likewise seasonally dependent. Over 50 percent of the bird species that nest and breed in this region during the spring and summer months are migratory, residing farther south in North, Central, and South America during the fall and winter months. Patterns of species migration alter seasonal patterns of regional diversity for a wide range of taxonomic groups.

Changes in regional diversity also occur over geologic timescales. Over timescales of tens to hundreds of millions of years, evolution drives changes in patterns of diversity through the emergence and extinction of species (see Figure 10.12). On a timescale of

thousands to tens of thousands of years, changes in climate have influenced regional patterns of diversity by shifting the geographic ranges of species. In eastern North America, shifts in the distribution of tree species following the last glacial maximum around 20,000 years BP represent one such example (see Figures 13.16 and 13.17). The geographic ranges of many species in North America continue to shift, influencing both local and regional patterns of diversity. Possible change in the geographic distributions of both plant and animal species in response to future changes in Earth's climate is a key area of research on global change (see Chapter 30).

24.8 Charles Darwin Was a Student of Geography: Adaptation Revisited

It is no coincidence that Charles Darwin, the man responsible for the theory of natural selection, was a student of geography. As the naturalist assigned to *H.M.S. Beagle,* he was charged with cataloging and describing the plants and animals encountered during the voyage around the world. Darwin's duties as naturalist gave him the opportunity to observe the many forms of life, living and fossilized, indigenous to the different islands and continents to which he traveled. He was struck by both the diversity of life encountered and the similarity in form and function of organisms inhabiting geographically distinct, but

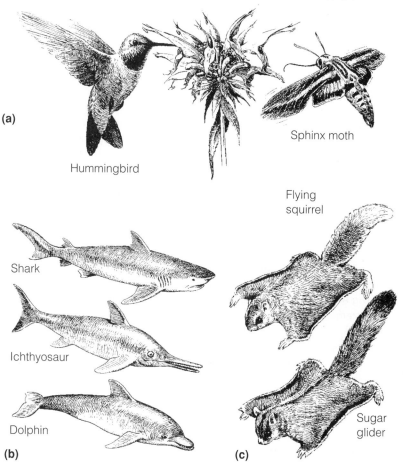

(a)

Hummingbird

Sphinx moth

Flying squirrel

Shark

Ichthyosaur

Dolphin

Sugar glider

(b) **(c)**

FIGURE 24.17 Convergent evolution of dissimilar species showing relationships between form and function. (a) The hummingbird and the sphinx moth both feed on nectar, the hummingbird by day and the sphinx moth by night. Both have bills and mouthparts adapted for probing flowers; both have rapid, hovering modes of flight. (b) A prehistoric marine reptile, the ichthyosaur; the modern shark; and a marine mammal, the dolphin. All have the same streamlined shape for fast movement through water. (c) A North American rodent, the flying squirrel, and an Australian marsupial, the sugar glider, both have flat, bushy tails and extensions of the skin between the foreleg and the hindleg that enable them to glide down from one tree limb to another.

environmentally similar, regions. Although the primary focus of Darwin's theory of natural selection is to explain the mechanism by which diversity arises from common ancestry, it offers a second, less-discussed explanation—convergent evolution.

Evolutionary biologists use the term **convergent evolution** to describe the occurrence of a similar characteristic in two different species that has not been derived from a recent common ancestor (Figure 24.17). Found in both plants and animals, convergent evolution explains the similar characteristics of vegetation in geographically distinct regions having similar climates and geology. Recall that natural selection is essentially a two-step process (see Chapter 2): the production of variation among individuals within the population, and the elimination from the population of those individuals that are

"inferior." It is the random processes of genetic mutation and recombination that give rise to variation, but it is the necessity of adaptation to the prevailing (and often changing) environmental conditions that guides the invisible hand of design. As so clearly stated by Ernst Mayr, one of the most prominent and influential evolutionary biologists of our time: "By adopting natural selection, Darwin settled the several-thousand-year-old argument among philosophers over chance or necessity. Change on earth is a result of both, the first step being dominated by randomness, the second by necessity."

The physical processes discussed in Chapters 3 and 4 give rise to the broad-scale patterns of climate and abiotic environment on Earth. These processes transcend oceans, continents, mountain ranges, and the geographic barriers that function as the mechanisms

ECOREGIONS: CLASSIFYING AND MAPPING ECOSYSTEMS FOR RESOURCE MANAGEMENT

A recent approach to the classification and mapping of the biotic world is the concept of the ecoregion, developed largely by Robert G. Bailey of the U.S. Forest Service. He based it in part on the ecoregion concept of a Canadian geographer, J. M. Crowley. **Ecoregions** are major ecosystems that result from predictable patterns of climate as influenced by latitude, global position, and altitude (see Chapter 3). Each ecoregion is a continuous geographical area across which the interactions of climate, soil, and topography are sufficiently uniform to permit the development of similar types or forms of vegetation. Thus, the ecoregion approach provides a means of mapping any large portion of Earth's surface over which the ecosystems have certain characteristics in common. On this basis, ecoregions occur in predictable locations in different parts of the world.

The classification scheme involves a synthesis of climate, vegetation, and soil types into a single, geographic, hierarchical classification. The largest category is the **domain,** a subcontinental area of broad climatic similarity. There are 4 domains: polar, humid temperate, humid tropical, and dry. The first 3 are temperature defined; dry is defined by moisture. Domains are broken down into 14 **divisions** based on the seasonality of precipitation or degree of coldness and dryness. Divisions are subdivided into **provinces** that correspond to broad vegetation regions having a uniform regional climate and the same distinct vegetation and soils (Figure A). Provinces are further subdivided into smaller and smaller categories that are useful for land classification and mapping at a local scale.

Whereas most ecosystem classification systems concentrate on terrestrial ecosystems, the ecoregion concept

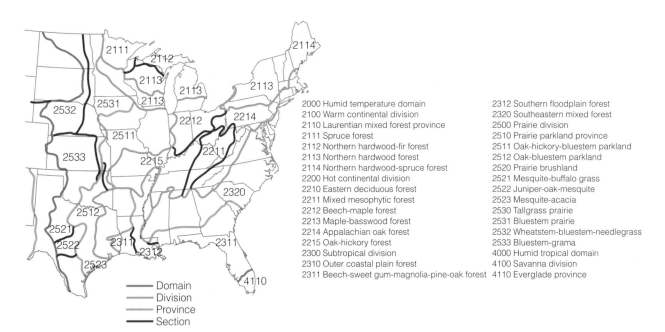

2000 Humid temperature domain	2312 Southern floodplain forest
2100 Warm continental division	2320 Southeastern mixed forest
2110 Laurentian mixed forest province	2500 Prairie division
2111 Spruce forest	2510 Prairie parkland province
2112 Northern hardwood-fir forest	2511 Oak-hickory-bluestem parkland
2113 Northern hardwood forest	2512 Oak-bluestem parkland
2114 Northern hardwood-spruce forest	2520 Prairie brushland
2200 Hot continental division	2521 Mesquite-buffalo grass
2210 Eastern deciduous forest	2522 Juniper-oak-mesquite
2211 Mixed mesophytic forest	2523 Mesquite-acacia
2212 Beech-maple forest	2530 Tallgrass prairie
2213 Maple-basswood forest	2531 Bluestem prairie
2214 Appalachian oak forest	2532 Wheatstem-bluestem-needlegrass
2215 Oak-hickory forest	2533 Bluestem-grama
2300 Subtropical division	4000 Humid tropical domain
2310 Outer coastal plain forest	4100 Savanna division
2311 Beech-sweet gum-magnolia-pine-oak forest	4110 Everglade province

——— Domain
——— Division
——— Province
——— Section

FIGURE A The subdivision of ecoregions in the eastern and central United States to the edge of the humid temperate domain. (From Bailey 1978.)

of isolation in the process of evolution. The similarity between tropical rain forests in the Amazon Basin of South America and Africa's Congo Basin reflects the similar form and function of the plant species that have adapted to these distant ecosystems—species that

evolved independently but under similar constraints imposed by the physical environment.

Parallel form and function in different species is the product of chance, necessity, and the essential ingredient of time. This is the phenomenon that embraces both the terrestrial and the oceanic. Ocean ecoregions result from the interaction of large-scale features of the climate and ocean currents. These interactions determine the major units within the polar, temperate, and tropical domains (Figure B). The polar domain (shown in blues) is characterized by water that is low in temperature and salinity, and rich in plankton. The temperate domain (shown in greens) embraces the middle latitudes between the poleward limits of the tropics and the equatorward limits of pack ice in winter. It is characterized by mixed waters whose currents correspond to wind direction around the subtropical atmospheric high-pressure cells (see Chapter 3, Figure 3.10). The tropical domain (shown in browns) is characterized by ocean water that is high in temperature and salinity, and low in productivity. These 3 domains consist of 14 divisions dif-

ferentiated by combinations of circulation, temperature, salinity, and the presence of upwellings. Like terrestrial divisions, each oceanic division occurs in several parts of the world that are broadly similar in physical and biological characteristics.

The ecoregion approach was initially developed to provide a system of ecosystem classification for resource management. The approach provides a means of studying the problems of resource management on a regional basis, and it provides a framework for the organization and retrieval of resource inventory data. The National Science Foundation Long Term Ecological Research (LTER) program takes place within an ecoregion framework, as do a variety of other governmental and nongovernmental programs that examine aspects of conservation and resource management.

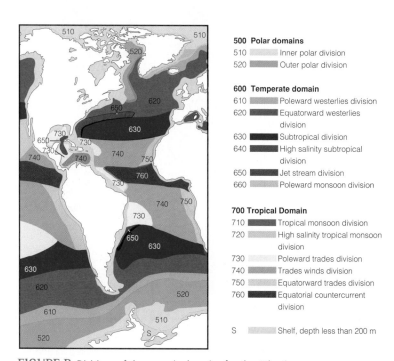

500 Polar domains
510 Inner polar division
520 Outer polar division

600 Temperate domain
610 Poleward westerlies division
620 Equatorward westerlies division
630 Subtropical division
640 High salinity subtropical division
650 Jet stream division
660 Poleward monsoon division

700 Tropical Domain
710 Tropical monsoon division
720 High salinity tropical monsoon division
730 Poleward trades division
740 Trades winds division
750 Equatorward trades division
760 Equatorial countercurrent division

S Shelf, depth less than 200 m

FIGURE B Divisions of the oceanic domains for the Atlantic.

formed the basis for classifying terrestrial ecosystems and biomes at the start of this chapter. So now we have come full circle—from adaptation at the level of the individual to the geography of ecosystems. A

common thread provided by Darwin's theory of natural selection links the topics in between. Ecology is, indeed, a broad scientific discipline by necessity, not simply by design.

PREDICTING CLIMATE FROM POLLEN

The development and use of classification systems that relate plant distribution to climate have a long history in ecological research, a history that continues today. Predictive models that relate plant distribution to climate are the main tools being used by ecologists to explore the implications of future climate change on the distribution of terrestrial ecosystems (see Chapter 30 for discussion and examples). If these models can be used to predict plant distribution for a given climate, couldn't you reverse the process and predict climate for an area if you know the distribution and abundance of plants? At first this may seem like an odd question. If you can collect data to describe the vegetation at a site, why don't you just measure the climate? That is true if you are interested in the present patterns of climate at a location, but what if you are interested in the past climate—say 10,000 years ago?

Although the plants that occupy a given location reflect the local climate within the recent past (the period over which they have occupied the location), like the trail of bread crumbs left by Hansel and Gretel in the woods, plants leave clues about their past presence at a location. As noted in Chapter 13 (Section 13.14), paleoecologists use the pollen deposited by plants in the sediments of lakes and ponds as clues to the past presence and

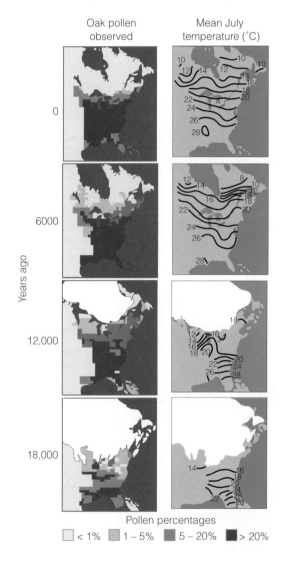

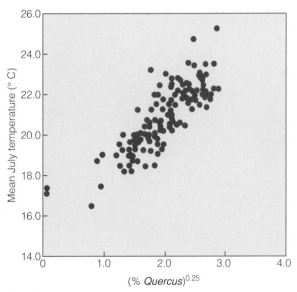

FIGURE A Relationship between the current deposition of oak (*Quercus* species) pollen and the July mean temperature for collection sites (lakes and ponds) in eastern North America. Pollen is expressed as a percentage raised to the 0.25 power. Each point represents a sample site. (Prentice et al. 1991.)

FIGURE B Maps (isopleths) of observed oak pollen data and predicted July mean temperatures for eastern North America from 18,000 BP to present. (Bartlein et al. 1984.)

abundance of plant species within the area. By examining the presence and quantity of pollen deposited in sediment layers and dating the sediments with radiocarbon, scientists can reconstruct the distribution and abundance of various plant groups within the region of the lake or pond (see Chapter 13, Figure 13.16). Colin Prentice, P. Bartlein, and Tom Webb reconstructed the distribution of six different plant groups (spruce, birch, northern pines, southern pines, oak, and prairie forbs) in eastern North America over the past 18,000 years using this approach.

By relating the current production and deposition of pollen by these plant groups with current climate data collected at the sites (Figure A), Prentice and colleagues developed a regression model relating pollen deposition and climate. The scientists then applied the model to the patterns of pollen distribution and abundance for the six plant types through time (past 18,000 years), enabling them to reconstruct the associated features of the climate (Figure B). This approach is one of the numerous techniques used by scientists to reconstruct past climates.

CHAPTER REVIEW

Summary

Ecosystem Classification (24.1–24.3) Starting with the early plant geographers of the mid-19th century, many approaches have been developed to classify ecosystems into broad categories that relate to climate. One such classification is the biome system, which groups plants and animals of the world into integral units characterized by distinctive life. Boundaries of biomes coincide with the boundaries of major plant formations of the world. By including both plants and animals as a total unit that evolved together, the biome permits recognition of the close relationship among all living things (**24.1**).

Terrestrial ecosystems are classified based on vegetation structure, which is typically defined on the basis of the dominant plant life forms (grasses, shrubs, and trees). Constraints imposed on the adaptations of these major plant life forms to features of the physical environment (climate and soils) determine their patterns of dominance along gradients of temperature and moisture. The patterns of plant life form dominance along these two gradients determine the corresponding distribution of the ecosystems after which they are named (**24.2**).

Aquatic ecosystems are classified largely on features of the physical environment. One of the major features influencing the adaptations of organisms that inhabit aquatic ecosystems is water salinity. For this reason, aquatic ecosystems are divided into two major categories—saltwater

(or marine) and freshwater. These two major categories are further divided into a number of ecosystem types based on the depth and flow of water, substrate, and the type of organisms (typically plants) that dominate. The variety of aquatic ecosystems, freshwater and marine, are physically linked as components of the water cycle (**24.3**).

Patterns of Species Richness (24.4–24.6) The approximately 1.4 million species that scientists have identified are not distributed evenly over Earth's surface. In general, species richness decreases from the equator toward the poles for both aquatic and terrestrial organisms. A variety of hypotheses have been put forward to explain these patterns, including the role of climate (**24.4**).

Regionally, plant species richness is correlated with actual evapotranspiration, suggesting a positive relationship between species richness and net primary productivity. Species richness of terrestrial vertebrates is correlated with potential evapotranspiration, an integrated measure of energy input to the ecosystem (**24.5**).

Patterns of species diversity for both pelagic and benthic organisms appears to be influenced by the seasonality of primary productivity rather than by total productivity per se (**24.6**).

Local and Regional Diversity (24.7) Ecologists define diversity within a community or ecosystem as local or alpha diversity. Quantifying local patterns of diversity is hindered by difficulties in defining community boundaries, the species diversity–area relationship, and changes in diversity during succession. The total diversity (or species richness) across all communities within a geographic area is called regional or gamma diversity.

Convergent Evolution (24.8) Within the broad-scale patterns of climate and abiotic environments across Earth exist many geographically distinct but environmentally similar regions. Within these geographically distinct regions, one often observes species that have independently evolved a similarity in form and function in response to similar selection pressures. This evolved resemblance that is not the consequence of a recent common ancestry is known as convergent evolution.

Review Questions

1. What is the basis for classifying terrestrial ecosystems?

2. How does climate influence the distribution of plant life forms (grasses, shrubs, and trees)?

3. What is the basis for classifying aquatic ecosystems?

4. How does species richness vary with latitude? With elevation?

5. How does plant species richness vary with climate?

6. How is climate correlated with regional patterns of vertebrate diversity?

7. How does environmental heterogeneity influence patterns of species diversity?

8. How is primary productivity related to species richness in the oceans?

9. Contrast local (alpha) and regional (gamma) diversity.

10. What are some factors that make it difficult to quantify patterns of species diversity/richness?

Spectacular fall color is a hallmark of the eastern deciduous mixed hardwood forest, Harvey Pond, Madrid, Maine.

Terrestrial Ecosystems 1: Forests, Woodlands, and Savannas

OBJECTIVES

Upon completion of this chapter, you should be able to:

- Describe the major latitudinal forest zones.
- Discuss the major environmental constraints on the geographic distribution of forest ecosystems.
- Describe the types and general features of tropical, broadleaf deciduous, and coniferous forest ecosystems.
- Contrast patterns of net productivity and decomposition in these forest ecosystems.
- Compare and contrast the structure of coniferous and deciduous forest ecosystems.
- Explain the role of permafrost on boreal forest ecosystems.
- Contrast the structure of forest, woodland, and savanna ecosystems.

Ecosystems in which trees are either the dominant or codominant plant life form include a wide range of habitats. Where soil moisture is sufficient, trees form a closed canopy with a complex vertical structure, largely due to the uneven size and age structure of the tree populations. These are the forest ecosystems, the most widespread and diverse of all vegetation types of the world. Forests grow in distinct latitudinal bands around the Northern Hemisphere from the polar circle to the equator. Progressing southward from the tundra are consecutive belts of coniferous, temperate deciduous, and, around the equator, tropical forests. The latter are most extensive in the Southern Hemisphere. Within this broad geographical range, forest distribution is limited by the constraint of adequate soil moisture to support a closed canopy of trees. As a result of this global distribution and the influence of temperature and seasonality on tree form (see Chapter 24), plant geographers conveniently classify forests into three broad geographic (latitudinal) groups: tropical, temperate (often subdivided into warm and cool), and boreal (Figure 25.1).

25.1 The Structure and Function of Forest Ecosystems Vary with Climate

Temperature and moisture are the primary factors limiting the geographic distribution of the various forest ecosystems (see Chapter 24). Within the warm, aseasonal environment of the wet tropics, tropical rain forests of broadleaf evergreen trees are found, giving way to seasonal and dry tropical forests in regions of seasonal precipitation. The distinct warm and cold seasons of the temperate regions result in a transition to deciduous and then coniferous forests as the flux of solar radiation and length of the growing season decline moving poleward.

The gradient of dominant tree form and forest type with changing climate is accompanied by changes in the structure and function of these ecosystems (Figure 25.2). Both maximum tree height and stand density decline with temperature and precipitation (more specifically, AET), resulting in a decline in standing biomass (B). A combination of reduced leaf area, photosynthesis, and length of growing season function to reduce both gross and net primary productivity (P) with declining temperature and moisture (see Chapter 20, Section 20.6, Figures 20.3 to 20.5). Rates of decomposition are similarly affected by temperature and soil moisture, declining with net primary productivity. The combined influence of climate on net primary productivity and decomposition gives rise to a distinct pattern of carbon and nutrient cycling across these different forest ecosystems.

Both net primary productivity and the rate of decomposition increase as conditions become warmer and wetter. This relationship is reflected in the positive correlation between actual evapotranspiration and both net primary productivity and rate of decomposition (presented in Chapter 21). The net effect is a faster rate of nutrient cycling in warm, wet environments, such as a tropical rain forest, than in cooler (temperate or boreal forest) ecosystems.

Although both net primary productivity and rate of decomposition increase with increasing actual evapotranspiration, an examination of rate of input and standing mass of dead organic matter on the soil surface in a variety of forest ecosystems shows that these two processes exhibit slightly different responses to climate. This difference has a major influence on the cycling of nutrients in these ecosystems, as well as the relative distribution of carbon and nutrients stored in living biomass and dead organic matter (litter and soils). The annual input of dead organic matter (plant litter) and the standing mass of dead organic matter on the soil surface for a number of forest ecosystems are plotted in Figure 25.3. Note that there is an inverse relationship between the production of dead organic matter (litter) and the amount of dead organic matter on the forest floor. The warmer, wetter conditions of the tropical rain forest result in high rates of net primary productivity and subsequently higher annual rates of litter input to the forest floor. However, these ecosystems are characterized by a very low mass of litter on the forest floor. This difference between rate of input and standing mass is a direct result of high rates of decomposition. To put it another way, decomposers are consuming the dead organic matter at about the same rate at which it is falling to the forest floor. In contrast, boreal forests have very low rates of net primary productivity and, subsequently, low rates of litter input. However, the extremely low rate of decomposition, a result of cold temperatures, allows a buildup of dead organic matter on the forest floor.

At first, the inverse relationship between rate of input and standing mass of dead organic material shown in Figure 25.3 might seem odd in that both rate of net primary production (and associated litter input to the soil surface) and decomposition decrease moving in latitude from the tropical rain forest to the temperate and boreal forest regions. However, this inverse relationship results from a greater slowdown of decomposition from declining temperatures relative to the slowing of net primary productivity (and

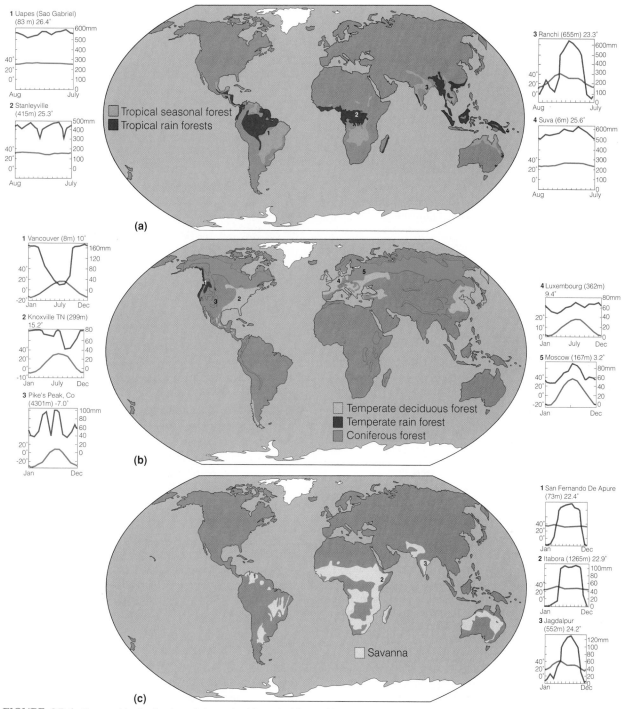

FIGURE 25.1 Geographic distribution of the major (a) tropical forest (b) temperate and coniferous forests and (c) savanna ecosystems of the world, and associated climate diagrams showing the long-term patterns of monthly temperature and precipitation for selected locations.

litter input). As a consequence of this difference between the rate of litter input and the rate of decomposition, more nutrients are tied up in dead organic matter in the cooler, more northern forest ecosystems, resulting in a slower rate of internal nutrient cycling. This broad-scale influence of climate on the internal cycling of nutrients within forest ecosystems can be compounded by differences in species characteristics that directly affect rates of decomposition and nutrient release from plant litter, particularly leaf litter (see Chapter 21, Section 21.5, Figure 21.2 for discussion and illustration).

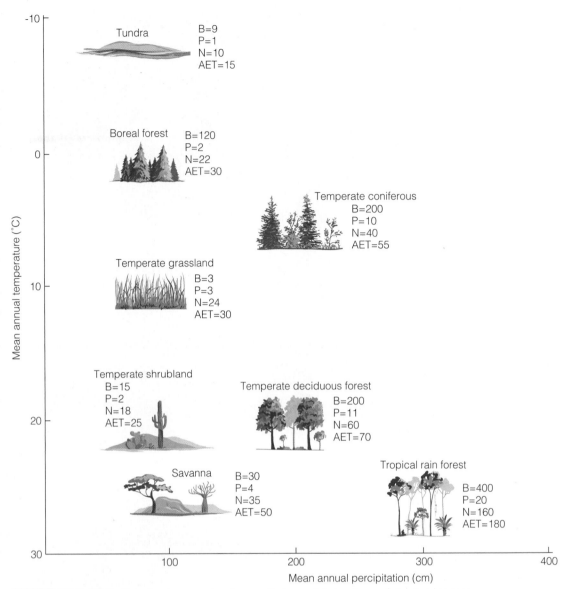

FIGURE 25.2 Distribution of major ecosystem types as their associated patterns of standing biomass (B: Mg/ha), primary production (P: Mg/ha/yr), actual evapotranspiration (AET: cm/yr) and nitrogen uptake by plants (N: kg/ha/yr) relative to mean annual temperature (°C) and precipitation (cm/yr). (After Etherington 1982; Aber and Melillo 1991.)

25.2 The Seasonality of Rainfall Influences the Nature of Tropical Forest Ecosystems

Plant geographers divide tropical forests into a number of types and subtypes. One type grades into another, with no sharp boundary. The prototype of tropical forests is the tropical rain forest, restricted mostly to the equatorial climatic zone between latitudes 10°N and 10°S. The rain forest occupies those regions of the world where the temperatures are warm through the year and rainfall, measured in meters, occurs almost daily. Although there is little annual variation in temperature and precipitation, daily variations in heat, rainfall, and humidity are great.

Tropical forests do not form a continuous belt around the equator. They are discontinuous, broken up by differences in precipitation (governed by the direction of the winds) and in the landmasses. Most tropical forests, for example, grow below the altitude of 1000 meters, and they are absent from the eastern part of equatorial Africa, the northwest part of South America, and the southern tip of India, all places that experience long seasonal periods of drought.

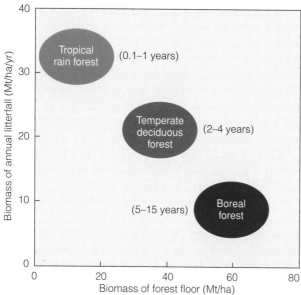

FIGURE 25.3 Relationship between the standing mass of dead organic matter (biomass) on the forest floor and the annual rate of litterfall (dead organic matter input) for three types of forest ecosystems. Tropical rain forests, found in wet, warm environments, have a high rate of litterfall but low biomass on the forest floor, the result of high rates of decomposition. In contrast, the boreal forests found in cool, wet environments have a low annual input of litterfall but a high forest floor biomass, because of the very low rates of decomposition and the accumulation of dead organic matter on the forest floor. Values in parentheses are the average turnover rates of litter (time for litter to fully decompose).

FIGURE 25.4 An interior view of an Amazonian rain forest. Note the vigorous undergrowth and vines.

FIGURE 25.5 A dry tropical forest in Costa Rica. Most of the unique dry tropical forests in Central America have disappeared as a result of land-clearing schemes.

Tropical rain forests fall into three main groups. The largest and most continuous is in the Amazon Basin of South America. The second is in the Indo-Malaysian area from the west coast of India and southeast China through Malaysia and Java to New Guinea. This forest has the greatest diversity of plant species. The third forest is in West Africa around the Gulf of Guinea, into the Congo basin. Smaller rain forests occur on the eastern coast of Australia, the windward side of the Hawaiian Islands, the South Sea Islands, and the east coast of Madagascar. There the trees are mainly evergreen.

Within the rain forest are many subtypes. The most luxuriant is the multilayered lowland rain forest (Figure 25.4; also see pg. 467). At higher elevations, the lowland forest grades into mountain forest, with abundant undergrowth, tree ferns, and small palms. The mountain forest gives way to cloud forest, with understory thickets and trees burdened with epiphytes. Although it receives less rainfall, the cloud forest is continually wrapped in clouds and mists. Fingers of rain forest, called gallery forests, follow river courses into savannas. Swamp forests occupy perennially wet soils, and peat forests grow on nutrient-poor ones.

The rain forest transitions into tropical and subtropical seasonal forests, also called semi-evergreen

and semi-deciduous forests. Although they retain many characteristics of the rain forest, these forests are subject to seasonal droughts lasting from 2 to 4 months. Some 30 percent of the trees of the upper canopy lose their leaves during the dry period, but the lower canopy trees and understory retain their leaves through the year. In the Indo-Malaysian forest about a month after the coming of the monsoon rains, rising temperature triggers the emergence of new leaves. In other regions, leaves emerge about a month before the rainy season begins, often coinciding with flowering. Fruits develop at the beginning of the dry season. Like the rain forest, the seasonal forest has lowland and mountain types.

Largely overlooked in our concern over tropical forests are the dry tropical forests (Figure 25.5). The largest proportion of dry tropical forest is in Africa and on tropical islands, where it comprises about 80 percent of the forested area. About 22 percent of South American and 55 percent of Central American forested areas are dry tropical forest. Much of the original forest is gone, especially in Central America and India. It has been converted to agricultural and grazing land, or it has regressed through disturbance to thorn woodland, savanna, and grassland.

Dry tropical forests undergo a seasonal dry period, the length of which is based on latitude. The more distant the forest is from the equator, the longer the dry season, up to 8 months (see Chapter 3). During the dry period, the trees and shrubs drop their leaves. Before the start of the rainy season, which may be much wetter than the wettest time in the rain forest, the trees begin to leaf. During the rainy season, the landscape becomes uniformly green.

25.3 Tropical Rain Forests Have a Complex Biological and Physical Structure

Tropical rain forests are noted for their diversity of plant and animal life. Tree species number in the thousands. A 10-square-kilometer area of tropical rain forest may contain 1500 species of flowering plants and up to 750 species of trees. The richest area is the lowland tropical forest of peninsular Malaysia, which contains some 7900 species. There one of the major groups, the Dipterocarpaceae, contains 9 genera and 155 species, of which 27 are endemic. (The Asian dipterocarps have 12 genera and 470 species.) Tropical rain forests also account for an estimated several million species of flora and fauna, one-half of all known plant and animal species, and 20 to 25 percent of all known arthropods.

The tropical rain forest divides into five general layers (Figure 25.6), which are most apparent in the undisturbed forest. Stratification, however, is often poorly defined because the growth plan of many tree species is the same, differing only in size. The uppermost layer consists of emergent trees over 40 to 80 meters high, whose deep crowns billow above the rest of the forest to form a discontinuous canopy. The second layer, consisting of mop-crowned trees, forms another, lower, discontinuous canopy. Not clearly separate from one another, these two layers form an almost complete canopy. The third layer, the lowest tree stratum, is made up of trees with conical crowns. It is continuous, often the deepest layer, and well defined. The fourth layer, usually poorly developed in deep shade, consists of shrubs, young trees, tall herbs, and ferns. Many of these plants have elongated, downward-curving leaf blades, called drip tips. These blades apparently enable the leaves to rid themselves of excess water in their permanently wet environment, increase transpiration, and reduce nutrient leaching. The fifth stratum is the ground layer of tree seedlings and low herbaceous plants and ferns.

FIGURE 25.6 Vertical stratification of a tropical rain forest.

FIGURE 25.7 Planklike buttresses help support tall rain forest trees.

Reaching upward as much as 5 meters from the wide-spreading roots of many species of large trees are thin, strong, planklike outgrowths called buttresses (Figure 25.7). These buttresses function as prop roots, providing support for trees rooted in soil that offers poor anchorage.

25.4 The Mature Tropical Forest Is a Mosaic of Continually Changing Vegetation

The mature tropical forest, like the mature temperate forest, is a mosaic of continually changing vegetation. Death of tall trees, brought about by senescence, lightning, windstorms, hurricanes, defoliation by caterpillars, and other causes, creates gaps (see Chapter 19), which shade-intolerant pioneer species quickly fill. These trees are replaced eventually by shade-tolerant late successional species, perhaps over a period of 100 years. Continuous random disturbances across the forest, however, ensure persistence of the species in the mature forest. A high frequency of tree fall may account for the low density of large trees (1 meter dbh and larger) in mature rain forests. Enhancing this diversity are local changes in soil, topography, and drainage that support varying arrays of species.

Most tropical rain forest trees reach full height when they have achieved only about one-third to one-half of their maximum bole diameter. Layering comes about when a group of species of similar mature height dominates a stand. Layering is also influenced by crown shape, which in turn correlates with tree growth. Young trees still growing in height have a single stem and a tall narrow crown; they are monopodial. As many species mature, large limbs diverge from the upper stem or trunk; they become sympodial. This change happens when the bud of the main stem axis ceases to grow and the lateral buds take over their role. The process repeats itself, adding to crown growth and producing a pattern that suggests the spokes of an umbrella. Looking up into the canopy, the observer gains the impression that crowns of trees fit together like a jigsaw puzzle, with the pieces about a meter apart. This growth pattern is called crown shyness (Figure 25.8). Among the dipterocarps of the Far East, this reiteration of the growth pattern within the crown to produce many dense subcrowns gives a cauliflowerlike appearance to the canopy. Within the crown, there is minimal overlap among the leaves; each is positioned to receive the maximum amount of light available.

A conspicuous part of the rain forest is plant life dependent on trees for support. Such plants include epiphytes, climbers, and stranglers. Climbers, the lianas, are vines with fine stringlike to massive cable-like stems that reach the tops of trees and expand into the form and size of a tree crown. They may loop to the ground and ascend again. Climbers grow prolifically in openings, giving rise to the image of the impenetrable jungle. This popular image applies more to secondary forest, second growth that develops where primary forest has been disturbed.

Stranglers and epiphytes share some characteristics. Stranglers start life as epiphytes. As they grow, they send roots to the ground and increase in number and girth until they eventually encompass the host tree and claim the crown and limbs as support for their own leafy growth. Epiphytes inhabit niches on the trunks, limbs, branches, and even leaves of trees, shrubs, and climbers. Their roots are aerial. Epiphytes come in various types. One group, the microepiphytes, consists of mosses, lichens, and algae. Macroepiphytes, such as orchids, bromeliads, and members of the Ericaceae, are vascular plants. Their roots never reach the ground. They attach themselves to a tree and take up nutrients from the air, rainwater, and organic debris on their supports. Some of the epiphytes are important in recycling minerals leached from the canopy.

The floor of the tropical rain forest is thickly laced with roots, both large and small, forming a dense mat on the ground. Except for a few that reach down to weathered parent rock, rain forest roots are shallowly concentrated in the upper 0.3 meter of soil, where inorganic nutrients are available. Associated with the fine roots are mycorrhizae (see Chapter 16) that aid in the uptake of nutrients from the decaying organic matter.

FIGURE 25.8 A view through the canopy of a secondary forest reveals crown shyness. The crowns are separated from one another, giving the impression of a jigsaw puzzle.

Layering of vegetation influences the internal microclimate of the forest. Crowns of emergent trees experience conditions similar to open land. The level of CO_2 and amount of humidity increase going down through the canopy, and temperature and evaporation decrease. From the ground up to about 1 meter, the levels of CO_2 are high, and humidity at over 90 percent is oppressive. Temperature on the average is 6°C cooler inside than outside the forest, with a strong nocturnal inversion. Although light penetrates the upper canopy, the lianas, epiphytes, and lower tree layers block out most of it. The amount of light that reaches the floor of a Malaysian rain forest is about 2 to 3 percent of incident radiation, and half of that comes from sunflecks; about 6 percent comes from breaks in the canopy; and 44 percent comes from reflected and transmitted light.

25.5 Stratification of Vegetation Supports a Diversity of Animal Life

Stratification of animal life in the tropical rain forest is pronounced. J. L. Harrison in 1962 described six distinct feeding communities of birds and mammals in a lowland tropical forest of the Far East. (1) A group feeding above the canopy is made up mostly of insectivorous and some carnivorous birds and bats. (2) A canopy group consists of a large variety of birds, fruit bats, and other species of mammals that eat leaves, fruit, and nectar. A few are insectivorous and mixed feeders. (3) Below the canopy, in a zone of

tree trunks, is a world of flying mammals, birds, and insectivorous bats. (4) Also in the middle canopy are scansorial mammals, such as squirrels that range up and down the trunks, entering the canopy, and the ground zone to feed on the fruits of epiphytes, on insects, and on other animals. (5) The forest floor is occupied by large herbivores, such as the gaur, tapir, and elephant, which feed on ground vegetation and low-hanging leaves, and their attendant carnivores, such as leopards and tigers, all of which range over a large area. (6) The final feeding stratum includes the small ground and undergrowth animals, birds and small mammals capable of some climbing, that search the ground litter and lower parts of tree trunks for food. This stratum includes insectivorous, herbivorous, carnivorous, and omnivorous feeders.

The enormous diversity, but low species population density, of animal life in the tropical rain forest mirrors the great diversity of microhabitats and resources. Insect fauna number in the millions, with many species still to be discovered. Invertebrate life is distributed not only vertically, but horizontally as well. Many species are restricted to certain forests, and within them inhabit only certain plants. Numerous species live in the epiphytes or in small pools of water caught in epiphytic plants, where they may be joined by small canopy-dwelling frogs.

Nearly 90 percent of all nonhuman primates live in the tropical rain forests of the world. Sixty-four species of New World primates, small with prehensile tails, live in the trees. The Indo-Malaysian forests are inhabited by a number of primates, many of which are restricted to certain regions. The orangutan, an arboreal ape, is confined to the island of Borneo. Peninsular Malaysia has 7 species of nonhuman primates, including 3 gibbons, 2 langurs, and 2 macaques. The long-tailed macaque is common to disturbed or secondary forests, and the pig-tailed macaque is a terrestrial species, adaptable to human settlements. The tropical forest of Africa is home to mountain gorillas and chimpanzees. The diminished rain forest of Madagascar holds 39 species of lemurs.

25.6 Deciduous Forests Characterize the Wetter Environments of the Warm Temperate Regions

The deciduous forest once covered large areas of Europe and China, parts of North and South America, and the highlands of Central America.

FIGURE 25.9 An Appalachian hardwood forest in spring, dominated by oaks and yellow-poplar, with an understory of red-bud in bloom.

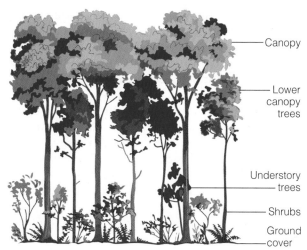

FIGURE 25.10 Stratification in an eastern deciduous forest.

The deciduous forests of Europe and Asia have large-ly disappeared, cleared over the centuries for agri-culture. What remains is seminatural, except for pockets in the more mountainous regions of central Europe. The two major forest types there are beech-oak-hornbeam and oak-hornbeam. Beech *(Fagus syl-vatica)* is the most uniformly distributed tree throughout central Europe. Beech forests, which grow from lowlands into the mountains, are charac-terized by dense canopy and poorly developed understory. Occupying damper, more acid soils are oak-hornbeam forests dominated by pedunculate oak *(Quercus robur)* and sessile oak *(Q. petraea)*. The Atlantic deciduous forest, originally dominated by beech, oaks, ash *(Fraxinus* spp.), and birch *(Betula* spp.), exists only in a seminatural state. Because of glacial history, the species diversity of the European deciduous forests does not compare with that of North America or China. The Asiatic broadleaf forest, found in eastern China, Japan, Taiwan, and Korea, is similar to the North American deciduous forest and contains a number of plant species of the same genera as those found in North America and Western Europe.

In North America, the temperate deciduous forest reaches its greatest development in the mixed meso-phytic forest of the central Appalachians, where the number of temperate tree species is unsurpassed by any other area in the world (Figure 25.9). In eastern North America, the deciduous forest consists of a number of associations, including the mixed meso-phytic forest of the unglaciated Appalachian plateau; the beech-maple and northern hardwood forests (with pine and hemlock) in northern regions that eventually grade into the boreal forest; the maple-basswood forests of the lake states; the oak-chestnut (now oak, since the die-off of the American chestnut) or central hardwood forests, which cover most of the Appalachian Mountains; the magnolia-oak forests of the Gulf Coast states; and the oak-hickory forests of the Ozarks.

Fingers of forest extend along the rivers of the prairies, plains, and semiarid regions of the south-western United States and Mexico. These forests, growing on the floodplains and banks of rivers and streams, are known as riparian woodlands. Growing on rich, moist, alluvial soil, riparian ecosystems are highly productive, add diversity to the landscape, and provide habitat for a number of wildlife species disproportionate to their area.

25.7 Four Distinct Layers Characterize the Vertical Structure of a Temperate Deciduous Forest

Highly developed, uneven-aged deciduous forests usually have four strata (Figure 25.10). The upper canopy consists of dominant and codominant trees, below which is the lower tree canopy and then the shrub layer. The ground layer has herbs, ferns, and mosses.

Even-aged stands, the results of fire, clear-cut logging, and other large-scale disturbances (see Chapters 18 and 19), often have poorly developed strata beneath the canopy because of dense shade. The low tree and shrub strata are thin, and the ground layer is poorly developed, except in small, open areas.

The physical stratification of the forest influences the microclimate within the forest. The highest temperatures are in the upper canopy because this stratum intercepts solar radiation. Temperatures tend to decrease through the lower strata. The most rapid decline takes place from the leaf litter down through the soil.

Humidity in the forest interior is high in summer because of plant transpiration and poor air circulation. During the day, when the air warms and its water-holding capacity increases, relative humidity is lowest. At night, when temperature and moisture-holding capacities are low, relative humidity rises. The lowest humidity in the forest is a few feet above the canopy, where air circulation is best. The highest humidity is near the forest floor, kept that way by the evaporation of moisture from the ground and settling of cold air from the strata above.

Variation of humidity within the forest is influenced in part by the degree to which the lower strata are developed. Leaves add moisture to the immediate surrounding air; well-developed strata with more leaves have higher humidity. Thus, layers of increasing and decreasing humidity may exist from the floor to the canopy.

Bathed in full sunlight, the uppermost layer of the canopy is the brightest part of the forest. Down through the forest strata, light intensity dims. In an oak forest, only about 6 percent of the total midday sunlight reaches the forest floor; the forest floor is about 0.4 percent as bright as the upper canopy.

Light intensity within the forest varies seasonally (see Chapter 4). The forest floor receives its maximum illumination during early spring before the leaves appear; a second, lower peak of maximum illumination during the growing season occurs in the fall. The darkest period is midsummer. Light intensity during summer is highly variable from point to point and time to time as the Sun shines through gaps in the canopy. Sun flecks can influence the distribution of herbaceous vegetation on the forest floor.

25.8 Vertical Structure in the Temperate Forest Influences the Diversity and Distribution of Life

In general, the diversity of animal life is associated with stratification and the growth forms of plants (see Chapter 17). Some animals, particularly forest arthropods, are associated with or spend the major part of their lives in a single stratum; others range over two or more strata. The greatest concentration and diversity of life in the forest occurs on and just below the ground layer. Many animals, the soil and litter invertebrates in particular, remain in the subterranean stratum. Others, such as mice, shrews, ground squirrels, and forest salamanders, burrow into the soil or litter for shelter and food. Larger mammals live on the ground layer and feed on herbs, shrubs, and low trees. Birds move rather freely among several strata, but favor one layer over another. Some occupy the ground layer but move into the upper strata to feed, roost, or advertise territory.

Other species occupy the upper strata of the shrub, low tree, and canopy layers. The red-eyed vireo *(Vireo olivaceus),* the most abundant bird of the eastern deciduous forests of North America, inhabits the lower tree stratum, and the wood pewee *(Contopus virens)* the lower canopy. The black-throated green warbler *(Dendroica virens)* and scarlet tanager *(Piranga olivacea)* live in the upper canopy. Squirrels are mammalian inhabitants of the canopy, and woodpeckers, nuthatches, and creepers live amid the tree trunks between shrubs and the canopy.

25.9 Subtropical Areas Support Temperate Evergreen Forests

Extensive mixed forests of both broadleaf evergreen and coniferous trees occur in several subtropical areas of the world. Such forests include the eucalyptus forests of Australia (Figure 25.11), paramo forests and anacardia gallery forests of South America and New Caledonia, and false beech (*Nothofagus* spp.) forests of Patagonia and New Zealand. Representatives of temperate evergreen forests also occur in the Caribbean and on the North American continent along the Gulf Coast, in the

FIGURE 25.11 A eucalyptus forest in the Grampian Mountains of Australia.

hummocks of the Florida Everglades, and in the Florida Keys. Depending on location, these forests are characterized by oaks, magnolias, gumbo-limbo *(Bursera simaruba),* and royal and cabbage palms.

25.10 Temperate Coniferous Forests Inhabit a Variety of Different Environments

Although all coniferous forests share the common feature of being dominated by trees having the needle-leaf evergreen growth form, this general category of forests includes a diverse array of forest types and associated habitats. Many temperate coniferous forests are in the mountains and are referred to as montane forests. In Central Europe, extensive coniferous forests dominated by Norway spruce *(Picea abies)* cover the slopes up to the subalpine zone in the Carpathian Mountains and the Alps (Figure 25.12a). In North America, several coniferous forest associations blanket the Rocky, Wasatch, Sierra Nevada, and Cascade Mountains. In the southwestern United States, such forests occur at between 2500 and 4200 meter elevation, and in the northern United States and Canada at between 1700 and 3500 meter elevation. At high elevations in the Rocky Mountains, where winters are long and snowfall is heavy, grows a subalpine forest dominated by Engelmann spruce *(Picea engelmannii)* and subalpine fir *(Abies lasiocarpa)* (Figure 25.12b). Mid-elevations have stands of Douglas-fir, and lower elevations are dominated

by open stands of ponderosa pine *(Pinus ponderosa)* (Figure 25.12c) and dense stands of the early successional conifer, lodgepole pine *(P. contorta).*

Similar forests grow in the Sierra and Cascades. There high-elevation forests consist largely of mountain hemlock *(Tsuga mertensiana),* red fir *(Abies magnifica),* and lodgepole pine. We also find sugar pine *(Pinus lambertiana),* incense cedar *(Libocedrus decurrens),* and the largest tree of all, the giant sequoia *(Sequoiadendron giganteum),* which grows only in scattered groves on the west slopes of the California Sierra.

An early successional deciduous tree species common both to the montane and the boreal forest (Section 25.13) is quaking aspen *(Populus tremuloides).* It is the most widespread tree in North America (Figure 25.12d).

Coniferous forests are also found in lower-elevation environments within the cool temperate region. Pines form extensive stands in both Eurasia and North America. A major component of the Eurasian boreal forest (see Section 25.13), Scots pine *(Pinus sylvestris)* is also widespread through central Europe, where it grows from the lowlands to the tree line in the mountains. It occurs largely as planted or seminatural stands in southern England and western France.

Coniferous forests dominated by *Pinus* are also found in the warmer temperate coastal regions of the southeast United States. Unlike the Scots pine forests of Eurasia, the pine forests of the coastal plains of the South Atlantic and Gulf states are considered a seral (successional) stage of the temperate deciduous forest, because without disturbance from fire and logging they give way to it. These pines maintain their presence by possessing a competitive advantage over hardwoods on nutrient-poor, dry, sandy soil and by their adaptation to a fire regime. At the northern end of the coastal pine forest in New Jersey, pitch pine *(Pinus rigida)* is the dominant species. Farther south, loblolly *(P. taeda),* longleaf *(P. australis)* (Figure 25.13), and slash *(P. elliottii)* pine are most abundant.

South of Alaska, the coniferous forest of the coastal regions differs from the northern boreal forest (see Section 25.13) both floristically and ecologically. The reasons for the change are both climatic and topographic. Moisture-laden winds move in from the Pacific, meet the barrier of the Coast Range, and rise abruptly. Suddenly cooled, the moisture in the air is released as rain and snow in amounts up to 635 cm/year. During the summer, when winds shift to the northwest, the air is cooled over chilly northern seas. Although rainfall is low, cool air brings in heavy fog, which collects on the forest foliage and drips to the ground to add 127 cm or more of moisture.

FIGURE 25.12 Some coniferous forest types. (a) A Norway spruce forest in the Carpathian Mountains of central Europe. (b) Rocky Mountain subalpine forest dominated by subalpine fir *(Abies lasiocarpa)*. This tree grows with Engelmann spruce and mountain hemlock. (c) A montane coniferous forest in the Rocky Mountains. The dry lower slopes support ponderosa pine; the upper slopes are cloaked with Douglas-fir. (d) Quaking aspen *(Populus tremuloides)* is the dominant deciduous tree in the western montane forest of North America.

This land of superabundant moisture, high humidity, and warm temperatures supports the temperate rain forest, a community of luxuriant vegetation dominated by conifers adapted to wet, mild winters; dry, warm summers; and nutrient-poor soils. The forests are dominated by western hemlock *(T. heterophylla)*, mountain hemlock, Pacific silver fir *(Abies amabilis)*, and Douglas-fir, all trees with high foliage and stem biomass (Figure 25.14). Farther south, where precipitation is lower, grows the redwood *(Sequoia sempervirens)* forest, occupying a strip of land about 724 km wide.

25.11 Coniferous Forests Have a Simple Vertical Structure That Varies with the Dominant Species

Coniferous forests fall into three broad classes of growth form and growth behavior: (1) pines with straight, cylindrical trunks, whorled spreading branches, and a crown density that varies with the species from the dense crowns of red and white pine

FIGURE 25.13 A longleaf pine *(Pinus palustris)* forest in Florida. Regular burning, which simulates natural ground fires, eliminates the buildup of unburned fuel that could cause a devastating fire if ignited.

FIGURE 25.14 Old-growth forest in the Pacific Northwest. Typical is the Douglas-fir stand with an abundant western hemlock understory. Such forests are the home of the spotted owl and murrelet.

to the open, thin crowns of Virginia, jack, Scots, and lodgepole pine; (2) spire-shaped evergreens, including spruce, fir, Douglas-fir, and (with some exceptions) the cedars, with more-or-less tall pyramidal crowns, gradually tapering trunks, and whorled, horizontal branches; and (3) deciduous conifers such as larch *(Larix* spp.) and bald cypress, with pyramidal, open crowns that shed their needles annually. Growth form and behavior influence animal life and other aspects of coniferous ecosystems.

Vertical stratification in coniferous forests is not well developed. Because of a high crown density, the lower strata are poorly developed in spruce and fir forests, and the ground layer consists largely of ferns and mosses with few herbs. The maximum canopy development in spire-shaped conifers is about one-third down from the open crown, a profile different from that of pines. Pine forests with a well-developed high canopy lack lower strata. The litter layer in coniferous forests is usually deep and poorly decomposed, resting on top of, instead of mixing in with, the mineral soil.

This poor vertical stratification influences the environmental stratification within the stand. When you walk into a spruce or fir forest, you are struck by the sharp diminution in light. Light intensity is progressively reduced through the canopy to only a fraction of full sunlight. The upper crown of spruce and firs, a zone of widely spaced narrow spires, is open and well lighted, whereas the lower crown is dense and intercepts most of the solar radiation. Most pines form a dense upper canopy that excludes so much sunlight that lower strata cannot develop. Open-crowned pines allow more light to reach the

forest floor, stimulating a grassy or shrubby understory. Because conifers retain their foliage through the year, the degree of light interceptions in coniferous forests is about the same throughout the year. Illumination is greater during midsummer, when the Sun's rays are most direct, and lower in winter, when the intensity of incident sunlight is the lowest.

The temperature profile of a coniferous forest also varies with the growth form. For example, in forests of sprucelike trees, temperatures tend to be coolest in the upper canopy, perhaps because of greater air circulation, and hottest in the lower canopy.

25.12 Animal Life in Coniferous Forests Is Varied

Animal life in the coniferous forest varies widely, depending on the nature of the stand. Mites dominate invertebrate fauna of the soil litter. Earthworm species are few and their numbers low. Insect populations, although not diverse, are high in numbers and, encouraged by the homogeneity of the stands, are often destructive. Sawflies *(Neodiprion),* for example, attack a wide variety of pines, including pitch, Virginia, shortleaf, and loblolly, and the southern pine beetle *(Dendroctonus frontalis)* can reach outbreak proportions in southern pinelands.

A number of bird species are closely associated with coniferous forests. In North America, they

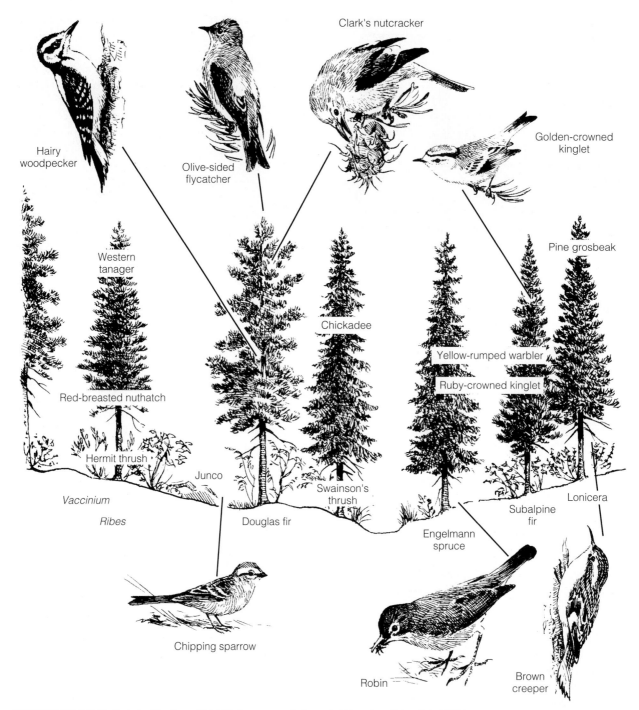

FIGURE 25.15 Vertical stratification of bird life in a spruce-fir forest in Wyoming. (After Salt 1967.)

include chickadees, kinglets, pine siskin, crossbills, purple finch, and hermit thrush (Figure 25.15). Related species, the tits and grosbeaks, are common to European coniferous forests.

Except for strictly boreal species such as the pine marten and lynx, mammals have much less affinity for coniferous forests. Most are associated with both coniferous and deciduous forests; the white-tailed deer, moose, black bear, and mountain lion are examples. Their north-south distribution seems to be limited more by climate, especially temperature, than by vegetation. The red squirrel, commonly associated with coniferous forests, is quite common in deciduous woodlands in the southern part of its range.

25.13 Boreal Forests Are the Dominant Forests of the Northern Latitudes

The largest vegetation formation on Earth is the boreal forest or taiga. This belt of coniferous forest encompassing the high latitudes of the Northern Hemisphere covers about 11 percent of Earth's terrestrial surface. Its northern limit is roughly along the July 13 isotherm, the southern extent of the Arctic front in summer, which also marks the beginning of the northward-stretching tundra. Its southern limit, much less abrupt, is more or less marked by the winter position of the Arctic front, roughly just north of 58°N latitude. In North America, the boreal forest covers much of Alaska and Canada and spills into northern New England with a finger extending down the high Appalachians. In Eurasia, the boreal forest begins in Scotland and Scandinavia and extends across the continent, covering much of Siberia, to northern Japan.

Four major vegetation zones make up the taiga: the forest-tundra ecotone with open stands of stunted spruce, lichens, and moss; the open boreal woodland with stands of lichens and black spruce; the main boreal forest with continuous stands of spruce and pine broken by poplar and birch on disturbed areas; and the boreal-mixed forest ecotone where the boreal forest grades into the mixed forest of southern Canada and the northern United States. Occupying, for the most part, glaciated land, the taiga is also a region of cold lakes, bogs, rivers, and alder thickets.

In Europe, the forest is dominated by Norway spruce *(Picea abies)*, Scots pine (Figure 25.16), and downy birch *(Betula pubescens)*; in Siberia by Siberian spruce *(Picea obovata)*, Siberian stone pine *(Pinus sibirica)*, and larch *(Larix sibirica)*; and in the Far East by Yeddo spruce *(Picea jezoensis)*. The North American taiga, richer in species, has four genera of conifers *(Picea, Abies, Pinus,* and *Larix)* and two genera of deciduous trees *(Populus* and *Betula)*. Dominant tree species include black spruce *(Picea mariana)* (Figure 25.17) and jack pine *(Pinus banksiana)*.

25.14 The Seasonal Freezing and Thawing of Soil Influences the Structure and Dynamics of Boreal Ecosystems

A cold continental climate with strong seasonal variation dominates the taiga, or boreal forest. The summers are short, cool, and moist, and the winters are prolonged, harsh, and dry, with a long-lasting snowfall. The driest winters and the most extreme seasonal fluctuations are in interior Alaska and central Siberia, which experience as much as 100°C difference in seasonal temperature extremes.

Much of the taiga is under the controlling influence of permafrost, which impedes infiltration and maintains high soil moisture. **Permafrost** is the perennially frozen subsurface that may be hundreds of meters deep. It develops where the ground temperatures remain below 0°C for many years. Its upper layers thaw in summer and refreeze in winter. Because

FIGURE 25.16 Scots pine is dominant in the Scandinavian taiga.

FIGURE 25.17 Black spruce is a dominant conifer in the North American taiga.

the permafrost is impervious to water, it forces all water to remain and move above it. Thus the ground stays soggy even though precipitation is low, enabling plants to exist in the driest parts of the Arctic.

Vegetation and its accumulated organic matter protect the permafrost by shading and insulation, which reduce the warming and retard thawing of the soil in summer. Any natural or human disturbance, however slight, can cause the permafrost to melt. Thus, vegetation and its organic debris impede the thawing of the permafrost and act to conserve it.

In turn, permafrost chills the soil, retarding the general growth of both aboveground and belowground parts of plants, limiting the activity of soil microorganisms, and diminishing the aeration and nutrient content of the soil. The effect becomes more pronounced the closer the permafrost is to the surface of the soil, where it contributes to the formation of shallow root systems.

25.15 The Boreal Forest Has a Relatively Simple Structure

The taiga appears as an endless sweep of sameness—a blanket of spire-shaped evergreens over the landscape. The appearance of the land can be deceptive, because variations in slope, aspect, topography, drainage, and permafrost add variety to the vegetation. In the North American taiga, black spruce with its low nutrient requirements and ability to tolerate wet soils occupies cold, wet, north-facing slopes and bottomlands. White spruce (Picea glauca) and birch grow on permafrost-free south-facing slopes, and jack pine grows on the high, drier, warmer sites. Areas swept by fires come back in early successional hardwoods—quaking aspen, balsam poplar (Populus balsamifera), paper birch (Betula papyrifera), and jack pine.

The boreal forest conifers fall into three growth forms: (1) the spire-shaped spruces and fir, with an open, narrow upper canopy and a dense lower canopy that casts a deep shade on the forest floor; (2) the open, thin, light-penetrating upper canopy of pines; and (3) the deciduous larch. Only a thick carpet of mosses grows in the dense shade of spruce, whereas under pine, light-tolerant lichens replace the shade-loving mosses.

The conifers are well suited to the cold taiga environment. The narrow, needlelike leaves with their thickened cuticles and sunken stomata reduce transpiration and assist in moisture conservation during periods of summer drought and winter freeze. Because they retain several years' growth of foliage at any one time, conifers can start photosynthesis quickly when environmental conditions are favorable.

Permafrost imposes its own authority on patterns of vegetation, which paradoxically encourages its formation. Permafrost impedes soil drainage, chills the soil, reduces its depth, slows decomposition, and reduces the availability of nutrients. Trees grow best where permafrost lies deep beneath the soil or is absent altogether; but the taiga trees worsen the situation for themselves by encouraging and maintaining the permafrost. Stands of spruce shade the ground and encourage a heavy growth of moss. Moss and an accumulation of the fine litter of undecomposed needles insulate the soil, immobilize nutrients, and increase soil moisture. The colder the soil becomes, the closer to the surface the permafrost moves, and the more shallow the soil becomes. With little space in which to anchor their roots, the trees are subjected to frost heaving. During early periods of warm weather, the roots encased in frozen soil are unable to replace the moisture lost through the crowns. The result is winter kill.

In stands of pine, larch, and scattered spruce, conditions do not improve. Heavy mats of lichens replace mosses on the dry, nutrient-poor, highly acidic soil. Lichens retain soil moisture through the growing season, encouraging growth of trees on sites that otherwise would be too dry; but the thick lichen mat also insulates the soil, chilling it and inhibiting the decomposition of organic matter. In spite of these effects, lichens do appear to improve tree growth.

Fires are recurring events in the taiga. During periods of drought, fires can sweep over hundreds of thousands of hectares. All of the boreal species, both broadleaf trees and conifers, are well adapted to fire. Unless too severe, fire provides a seedbed for regeneration of trees. Light surface burns favor early successional hardwood species. More severe fires eliminate hardwood competition and favor spruce and jack pine regeneration.

25.16 The Boreal Forest Has a Unique Animal Community

Caribou are the major herbivores of the boreal forest zone. Inhabiting open spruce-lichen woodlands, caribou are wide-ranging and feed on grasses, sedges, and especially lichens. Joining the caribou is the

(a) **(b)**

FIGURE 25.18 Animal life in the boreal forest. (a) The bull moose is a dominant herbivore. (b) The lynx preys on snowshoe hare and other small mammals.

moose, the largest of all deer (Figure 25.18a). Called elk in Eurasia, the somewhat solitary moose is a lowland mammal feeding on aquatic and emergent vegetation as well as alder and willow. Competing with moose for browse is the snowshoe hare. The arboreal red squirrel inhabits the conifers and feeds on young pollen-bearing cones and seeds of spruce and fir; and the quill-bearing porcupine *(Erethizon dorsatum)* feeds on leaves, twigs, and the inner bark of trees. Major mammalian predators are the wolf, which feeds on caribou and moose; the lynx (Figure 25.18b), which preys on snowshoe hare and other small mammals; the pine marten *(Martes americana),* the major predator on red squirrels; and other species of weasels. Ruffed grouse and spruce grouse *(Dendragapus canadensis)* are conspicuous birds of the North American boreal forest, and capercaillie *(Tetrao* spp.) and hazel grouse *(Tetrastes bonasia)* in the Eurasian taiga. Crossbills *(Loxia* spp.) and siskins *(Carduelis* spp.) extract conifer seeds from cones and occasionally move south in winter when the food supply fails. Importantly, the taiga is the nesting ground of many species of both neotropical and tropical migrant warblers. Major avian predators include numerous species of owls, such as the great gray owl *(Strix nebulosa),* and goshawks *(Accipiter* spp.).

Of great ecological and economic importance are major herbivorous insects, the larch sawfly *(Pristiphora erichsonii),* pine sawfly *(Neodiprion sertifer),* and spruce budworm *(Choristoneura fumiferana).* Although major food items for the insectivorous summer birds, these insects experience periodic outbreaks and defoliate and kill large expanses of forest.

25.17 At High Elevations, the Forest Is Reduced to Stunted Trees

At high altitudes where the winds are too steady and strong for all but low ground-hugging plants, the forest is reduced to pockets of stunted, wind-shaped trees. This area where forest gives way to tundra is the **Krummholz,** or "crooked wood" (Figure 25.19). The Krummholz in the North American alpine region is best developed in the Appalachian and Adirondack mountains. On the high ridges, trees begin to show signs of stunting far below the timber line. As you climb upward, stunting increases until spruces and birches, deformed and semiprostrate, form carpets 0.6 to 1 meter high, impossible to walk through but often dense enough to walk upon. Where strong winds come in from a constant direction, the trees are sheared until the tops resemble close-cropped heads, although the trees on the lee of the clumps grow taller than those on the windward side.

In the Rocky Mountains, the Krummholz is much less marked, for there the timber line ends abruptly with little lessening of height. Most of the trees are flagged; that is, the branches remain only on the lee side. In the alpine regions of Europe, the Krummholz is characterized by dwarf mountain pine *(Pinus mugo),* which forms dense thickets 1 to 2 meters high on calcareous soils. On acid soils, it is replaced by dwarf juniper *(Juniperus communis* subsp. *nana).* In the Australian Brindabella Range, the Krummholz is

(a)

(b)

FIGURE 25.19 The Krummholz. (a) In the Rocky Mountains, the tree line is sharply defined. Note the narrow pockets of stunted trees. (b) The tree line in the Australian Alps in the Brindabella Range is marked by low, twisted snow gum *(Eucalyptus pauciflora)* in protected pockets that give way to low-growing plants.

marked by pockets of low, twisted snow gum *(Eucalyptus pauciflora)*.

Though wind, cold, and winter desiccation are regarded as the cause of the dwarfed and misshapen condition of the trees, the ability of some tree species to show a Krummholz effect is genetically determined, with species such as mountain *(mugo)* pine adapted to high alpine conditions. Eventually, conditions become too severe even for the prostrate forms, and trees drop out entirely, except for those that have taken root behind the protection of high rocks. Tundra vegetation then takes over completely (see Chapter 26).

25.18 Reduced Soil Moisture Gives Rise to Woodlands

In areas of the temperate zone where soil moisture is not able to support a closed canopy of trees, a form of dry forest or woodland replaces forest ecosystems. Woodlands are defined by their distinct physical structure. Tree diversity is low, with the canopy often composed of only one or two species. Low soil moisture and the shade-intolerant habit (see Chapter 6) of these tree species restrict regeneration and the development of a multiple layer canopy. The shrub layer and ground vegetation are taxonomically distinct from the canopy, the latter typically being dominated by grasses. The tree and shrub species can be characterized by either drought-deciduous leaves or small,

hard xeromorphic (evergreen) leaves that can survive the dry season.

In western parts of North America where the climate is too dry for montane coniferous forest, we find the woodlands characterized by open-growth small trees with a well-developed understory of grass or shrubs. They contain needle-leaved trees, deciduous broadleaf trees, sclerophylls, or any combination. An outstanding example is the piñon-juniper woodland (Figure 25.20a). This ecosystem occurs in the southwestern United States, primarily Utah, Arizona, New Mexico, and Colorado. In southern Arizona, New Mexico, and northern Mexico grow oak-juniper and oak woodlands (Figure 25.20b), and in the Rocky Mountains, in particular, there are oak-sagebrush woodlands. In the Central Valley of California, characterized by low, winter precipitation, grows still another type—blue oak woodlands with grassy undergrowth (Figure 25.20c).

In eastern North America, bur oak woodlands form a transition from the oak-hickory forests to the east and the tall-grass prairie (grasslands) to the west, a result of the westward gradient of declining precipitation (see Chapter 3, Figure 3.18). Open stands of quaking aspen form a similar transition zone to the north in Canada; and in the plains of Eastern Europe, open woodlands of English oak *(Quercus robur)* form the transition zone into the steppes (grasslands) of central Eurasia. The open nature of the tree canopy gives these woodlands a parklike appearance, although agricultural lands have displaced most of these woodland ecosystems in both North America and Eurasia.

(a)

(b)

(c)

FIGURE 25.20 (a) A piñon pine *(Pinus edulis)* and juniper *(Juniperus osteosperma)* woodland in Utah, typical of temperate woodland in the southwestern United States. Seeds of piñon and fruits of juniper were staples in the diet of southwestern Native Americans. (b) Oak-juniper woodlands, Patagonia-Sonoita Ck, Arizona, can be found in various areas of the southwestern United States and northern Mexico. (c) Blue oak woodlands are found in California's Central Valley.

25.19 Savannas Form the Transition from Forest to Grassland in the Tropics

The one ecosystem that defies any general description is the tropical savanna. The problem is an old one, involving even its name. The word in its several origins, largely Spanish, referred to grasslands or plains; but over time the word was applied to an array of vegetation types representing a continuum of increasing cover of woody vegetation, from open grassland to widely spaced shrubs or trees to closed woodland (Figure 25.21). Moisture appears to control the density of woody vegetation, a function of both rainfall (amount and distribution) and the texture, structure, and water-holding capacity of soil (Figure 25.22).

Savannas cover much of central and southern Africa, western India, northern Australia, large areas of northwestern Brazil where they are known as *cerrados,* Colombia and Venezuela where they are called *llanos,* and to a more limited extent Malaysia. Some savannas are natural. Others are seminatural, brought about and maintained by centuries of human interference. In the African savannas, in particular, it is difficult to separate the effects of humans from the effects of climate. The savannas of central India, however, are the result of human degradation of original forested land.

Savannas, in spite of their vegetational differences, exhibit a certain set of characteristics. Savannas occur on land surfaces of little relief, often old alluvial plains. The soils are low in nutrients, due in part to infertile parent material and a long period of weathering. Savanna regions are associated with a warm continental climate with precipitation ranging between 500 mm and 2000 mm (see Figure 25.1). Precipitation exhibits extreme seasonal fluctuations;

(a)

(b)

(c)

(d)

FIGURE 25.21 Savanna ecosystems in Africa. (a) Open grass savanna with giraffe on the Maasai Mara reserve in Kenya. (b) Shrub savanna in Namibia, (southwestern) Africa. (c) Tree savanna with a well-developed growth of *Acacia* in the background and a bull elephant at its edge. (d) Savanna thorn woodland cloaks a hillside in South Africa.

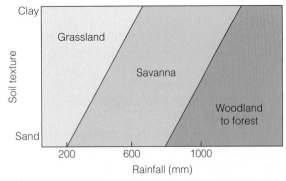

FIGURE 25.22 Diagram showing the interaction between annual precipitation and soil texture in defining the transition from woodland to savanna and grassland (desert scrub/shrub savanna) in southern Africa. Access by plants to soil moisture is more limited on the heavy textured soils (clays) than on the coarser sands, requiring greater annual precipitation to support the productivity of woody plants.

in South American savannas, in particular, the soil-water regime may fluctuate from excessively wet to extremely dry, often below the permanent wilting point. Savannas are subject to recurrent fires, and the dominant vegetation is fire-adapted. Grass cover with or without woody vegetation is always present. When present, the woody component is short-lived, with individuals seldom surviving for more than several decades (except for the African baobab trees).

25.20 Savannas Have a Distinct Physical Structure

The major and most essential stratum of the savanna ecosystem is grass, mostly bunch or tussock, with no vertical structure; its biomass decreases with height. A woody component adds one or two more vertical layers, ranging from about 50 to 80 cm when small woody shrubs are present to about 8 meters in the tree savannas. Highly developed root systems make up the larger part of the living herbaceous biomass. The root system is concentrated in the upper 10 cm but can extend down to about 30 cm. Savanna trees have extensive horizontal roots that go below the layer of grass roots. Competition may exist between grass and woody vegetation for soil moisture, but more intense competition takes place among trees, accounting for the spacing patterns of woody vegetation.

In contrast to the poorly developed vertical structure is a well-developed, although often unapparent, horizontal structure. The tussock grasses form an array of clumps set in a matrix of open ground, creating patches of low vegetation with frequent changes in microclimatic conditions. The addition of woody growth, the widely spaced shrubs and trees, increases horizontal structure extending to the soil. Trees add some organic matter and nutrients to the soil beneath them; reduce evapotranspiration, resulting in increased herbaceous and woody shrub growth; and provide patches of shade. On the African savanna in particular, large grazing herbivores rest in the shade during the heat of the day and concentrate nutrients from dung and urine beneath them.

Breaking up the monotony of the savannas are numerous marshy depressions that support wetland wildlife. Large ribbons of riverine or gallery forests weave through the savannas. The gallery forests support a diversity of wildlife and provide forage in the dry season for buffalo, waterbuck, and other large ungulates.

25.21 Savannas Support a Diversity of Herbivores

Savannas are capable of supporting a large and varied assemblage of herbivores, invertebrate and vertebrate, grazing and browsing. Dominant herbivores are the invertebrates, including acrid mites, acridid grasshoppers, seed-eating ants, and detrital-feeding dung beetles and termites. Savanna vegetation supports an incredible number of insects: flies, grasshoppers, locusts, crickets, carabid beetles, and especially termites and ants, which dominate insect life. Insect abundance is seasonal and is strongly affected by burning, which can reduce populations by more than 60 percent.

In South American savannas, there is a strong element of grazing ungulates represented by pampas deer and the capybara *(Hydrochaeris hydrochaeris)*. Granivorous, insectivorous, and frugivorous birds are also an important component of the consumer community.

The African savanna, visually at least, is dominated by a large and diverse ungulate fauna of at least 60 species that partition the vegetative resource among them. Some, such as the wildebeest and zebra, are migratory during the dry season. Others, such as the impala, partially disperse during the dry season. Still others, such as the giraffe and Grant's gazelle, have little or no seasonal dispersal. Among the ungulates, zebras and wildebeest are generalist grazers. Zebras, especially during the migratory period, feed on upper grass leaves, which are low in protein. Wildebeest feed on the more nutritious grasses, and the small gazelles, being more refined feeders, live on the lower grasses left behind, especially the new short growth at the beginning of the rainy season. Other ungulates, such as giraffe (Figure 25.23), Thompson's gazelle, kudu, and black rhino, are woody browsers. A close interaction exists among the grazing herbivores, and intensive grazing pressure by one species can affect the populations of others. In spite of their visual dominance, large ungulates consume only about 10 percent of primary production.

FIGURE 25.23 Two major woody browsers of the African savanna. The giraffe feeds on the tall woody growth; the endangered black rhino feeds on low shrubs.

Putting the level of consumption aside, herbivores have short-term and long-term impacts on the savanna. Over the short term, the grazing ungulates affect vegetation structure. Elephants can convert woodland to grassland, and large concentrations of grazers can turn grassland to eroded, bare ground. The species composition and structure of the African savanna vegetation would be different if it were not subject to heavy grazing, which alters competitive interactions among plants. Heavy grazing that reduces grass cover can cause an increase in the abundance of woody growth.

Living on the ungulate fauna is an array of carnivores, including the lion, leopard, cheetah, hyena, and wild dog. Subsisting on leftover prey are scavengers, including vultures and jackals.

25.22 The Seasonality of Rainfall Drives Productivity and Nutrient Cycling in Savanna Ecosystems

Because of the wide diversity of savanna types and limited studies, it is difficult to make any strong generalizations about primary production. Probably, a wide range of production exists between grass savanna on one end of the gradient and tree savanna and woodlands on the other.

At the beginning of the wet season, moisture facilitates decomposition and the release of nutrients (net mineralization; see Chapter 7) from dead organic matter accumulated in the dry season and stimulates nutrient translocation from the roots. This action is followed by a quick flush of growth into grass and woody plants. Nutrient movement between soil and vegetation is generally higher under the trees than in the open because of greater organic matter accumulation and reduced evaporation, which keeps the soil moist.

Savanna trees, especially the African acacias, exhibit tight internal cycling. Nitrogen concentration in the leaves, for example, decreases as the dry season approaches, with maximum withdrawal (retranslocation; see Chapter 6) before leaf fall. The trees transfer some of the nitrogen into new woody growth, but much of it goes to the root reserve, where it is available to stimulate the flush of new season growth. A similar tight circulation exists in neotropical savannas. Most of the nitrogen in the dry above-ground biomass is lost to the atmosphere by volatilization if fire sweeps the savanna; otherwise, a fraction will be transferred to the soil through leaching from rainwater.

The influence of large herbivores on nutrient dynamics over the long term is debatable, but the impact of ants and especially termites cannot be questioned. Ants and termites consume and break down plant litter and modify the soil. Mound-building termites excavate and move tons of soil, mixing mineral soil with organic matter. Some species construct extensive subterranean galleries, and others accumulate organic matter. Comprising over 50 percent of soil biomass, termites have a considerable impact on the physical and chemical properties of savanna soil.

CHAPTER REVIEW

Summary

Climate and Forest Ecosystems (25.1) Interaction between moisture and temperature is the primary factor limiting the nature and geographic distribution of forest ecosystems. The gradient of dominant tree forms and forest types with changing climate is accompanied by changes in structure and function of these ecosystems. The combined influence of climate on net primary productivity and decomposition give rise to a distinct pattern of carbon and nutrient cycling across the different forest ecosystems. At one extreme, tropical forests possess a large standing-crop biomass that ties up great quantities of nutrients. Much of the mineral cycling takes place between rapid decomposition of litter and the rapid uptake of the nutrients it contains. In the boreal forest, standing biomass is relatively small, and pro-ductivity is low because of low nutrient availability, slow decomposition of organic matter, and cold temperatures.

Tropical and Subtropical Seasonal Forests and Dry Tropical Forests (25.2–25.5) Seasonality of rainfall determines the types of tropical forests. Rain forests are associated with high aseasonal rainfall; subtropical seasonal forests experience droughts of two to four months. Dry tropical forests undergo varying lengths of dry season that end with a rainy season (25.2). Tropical rain forests noted for their enormous diversity of life divide into five general layers: emergent trees, high upper canopy, low tree stratum, shrub understory, and a ground layer of herbs and ferns. Conspicuous parts of the rain forest are the lianas or climbing vines, epiphytes growing up in the trees, and stranglers that grow downward from the canopy to the ground. Many large trees develop buttresses for support (25.3). Horizontally, the rain forest is a mosaic of continually changing vegetation that adds to its diversity brought about by disturbance and death of trees. Vertical layering of strata results when groups of species of similar mature height dominate a stand (25.4). Reflecting this stratification is the stratification of animal life into six pronounced feeding groups from the canopy down to the ground layer (25.5).

Broadleaf Deciduous Forests (25.6–25.9) Broadleaf deciduous forests are found in the wetter environments of the warm temperate region. They once covered large areas of Europe and China; the remnants are largely seminatural. In North America, deciduous forests are still widespread. They include a number of types such as beech-maple and oak-hickory; the greatest development is in the mixed mesophytic forest of the unglaciated Appalachians (25.6). Well-developed deciduous forests have four strata: upper canopy, lower canopy, shrub layer, and ground layer. This physical stratification affects the microclimate within the stand, including temperature, humidity, and light intensity (25.7). Vertical structure influences the diversity and distribution of life in the in the forest. Certain species are associated with each stratum (25.8). Temperate broadleaf evergreen forests of southern North America and the eucalyptus forests of northern Australia occupy subtropical regions (25.9).

Temperate Coniferous Forests (25.10–25.12) The coniferous forests include broadly the montane pine forests and lower-elevation pine forests of Eurasia and North America, and temperate rain forests of the Pacific Northwest (25.10). Vertical structure of coniferous forests is not well developed and varies with the dominant species. The ground layer consists largely of ferns and mosses, and the litter layer is deep and poorly decomposed (25.11). The growth form and characteristics of various coniferous species that make up the forest create different environments that dictate the animals that live within it. There is an array of forest types (25.12).

Boreal Forests (25.13–25.16) North of the temperate coniferous and deciduous forest is the circumpolar taiga or boreal forest, the largest vegetation formation on Earth. Characterized by a cold continental climate, the taiga consists of four major zones: the forest ecotone, open boreal woodland, main boreal forest, and boreal-mixed forest ecotone (25.13). Permafrost, the maintenance of which is influenced by tree and ground cover, has a strong influence on the pattern of vegetation, as do recurring fires (25.14). The boreal forest is dominated by spruces and pines, with successional communities of birch and poplar. Ground cover below spruce is mostly moss; in open spruce stands and pine, the cover is mostly lichens (25.15). Major herbivores include caribou, moose (called elk in Europe), and snowshoe hare. Preying on them is a colorful assortment of predators, including the wolf, lynx, and pine martin. The taiga is also the nesting ground of neotropical and tropical birds and the habitat of northern seed-eating birds such as crossbills (25.16).

Krummholz (25.17) At high elevations in mountainous regions, forest gives way to stunted trees and low ground-hugging plants. Known as the Krummholz, it is a place of high winds and cold temperatures. Where conditions at the highest elevations become too severe, the stunted trees are replaced by tundra vegetation.

Woodlands (25.18) In regions in the temperate zone where soil moisture is insufficient to support a closed canopy forest, dry forests or woodlands replace forest ecosystems. The canopy is open, and dominated by only one or two species such as oak, piñon pine, or juniper. Woodlands possess a well-developed understory of grass or shrubs and an open, parklike appearance.

Savanna (25.19–25.22) Savannas are grasslands with woody vegetation. They are characteristic of regions with alternating wet and dry seasons. Difficult to characterize precisely, savannas range from grass with occasional trees to shrubs to communities where trees form an almost continuous canopy. The latter grades into woodland with an understory of grass. Savannas of Africa and Australia may have evolved under human influences of fire and the impact of grazing animals. All savannas in one way or another have been altered by humans, primarily through overgrazing (25.19). The physical structure of savannas is distinct. The vertical structure is poorly developed, whereas the horizontal structure—consisting of clumps of tussock grass, widely spaced shrubs, and trees—is well developed (25.20). Savannas support a large and varied assemblage of herbivores, invertebrate and vertebrate. Visually, the African savanna is dominated by large and diverse ungulate fauna and associated carnivores. The large herbivores can influence vegetation structure (25.21).

Seasonality of rainfall influences productivity and nutrient cycling in savannas. Most production and decomposition takes place through the wet season. Much of the nutrient pool is tied up in plant and animal biomass, but nutrient turnover is high, with little accumulation of organic matter (25.22).

Review Questions

1. Name and characterize the types of tropical forests.

2. What are the major strata in the tropical forest?

3. What are emergents, lianas, and epiphytes? What are their positions in the rain forest?

4. What are buttresses?

5. What is crown shyness?

6. If tropical forest soils are so nutrient-poor, how can they support such a high plant biomass and diversity?

7. Discuss the types of deciduous and coniferous forests.

8. How does structural stratification of a forest affect its diversity of animal life?

9. What are the major life forms among conifers, and what effect do they have on the structure of coniferous forests?

10. Where and under what climatic conditions would you expect to find woodland ecosystems?

11. What are the distinguishing features of the boreal forest?

12. Speculate on why the boreal forest holds such a large number of important fur-bearing mammals.

13. What characterizes the vegetation of the Krummholz?

14. What distinguishes savannas from grasslands in structure and function?

15. Where are the world's savannas located? Under what climatic conditions have they developed?

16. What is the relationship of humans to savannas?

17. Speculate on why the African savannas have such a great diversity of animal life compared to New World savannas.

The midgrass (often called mixed-grass) prairie stands between the tallgrass prairie and the shortgrass plains.

Terrestrial Ecosystems II: Grassland, Shrubland, Desert, and Tundra

OBJECTIVES

Upon completion of this chapter, you should be able to:

- Describe the major characteristics of grasslands, shrublands, desert, and tundra.
- Discuss the distribution of these ecosystems as they relate to climate.
- Describe the types and major features of grassland.
- Contrast above-ground and below-ground production in grassland ecosystems.
- Characterize the major types of shrubland.
- Discuss how spatial heterogeneity of vegetation structure influences nutrient cycling in shrub ecosystems.
- Describe the characteristics of the tundra ecosystem.
- Contrast the Arctic and alpine tundra ecosystems.

As precipitation and soil moisture decline, the woodlands of the temperate zone give way to grassland (Figure 26.1), aided by the restrictive effect of fire on the establishment of woody plants (see Chapter 19, Section 19.7. In areas where topography allows, trees and other woody plants follow the drainage lines where runoff accumulates, forming ribbons of trees and shrubs across the landscape. The characteristics of these grasslands vary with climate, declining in stature and productivity as temperature, precipitation, and soils interact to reduce the availability of soil moisture.

Moving geographically from the cool temperate into the warm temperate and subtropical regions, reduced precipitation is compounded by increasing temperatures, further restricting the availability of soil moisture. These conditions give rise to the semiarid and arid shrublands. In the tropical regions, semiarid shrublands form part of the woodland-savanna-grassland continuum (see Chapter 25, Section 25.19, Figure 25.22). With decreasing soil moisture, trees give way to shrubs as the dominant woody plants in savannas, with the terms *shrub-savanna* and *semiarid shrubland* being synonymous.

In the high latitudes and high-elevation environments, the combined effects of low temperature and related aridity result in a transition from boreal forest to a unique form of shrubland—tundra. As extreme temperatures, either hot or cold, and increasing aridity further restrict plant productivity, desert ecosystems, with their unique array of plant and animal life, occupy all seven continents.

26.1 Increasing Aridity Marks the Transition from Forest to Grassland in the Temperate Zone

At one time, grasslands covered about 42 percent of the land surface of Earth. In the Northern Hemisphere, great expanses of grassland covered the mid-continent of North America and extended across the central part of Eurasia. In the Southern Hemisphere, grasses covered much of the southern tip of South America and the high plateau of southern Africa. Today, grasslands probably occupy less than 12 percent of the land surface, with most of them plowed under for cropland and degraded by overgrazing.

All grasslands have in common a climate characterized by rainfall between 250 and 800 mm (too light to support a heavy forest and too great to result in a desert), a high rate of evaporation, and periodic severe droughts. They share a rolling to flat terrain. Grazing and burrowing species are the dominant animals. Most grasslands require periodic fires for maintenance, renewal, and elimination of woody growth.

Grasses have a mode of growth that adapts them to grazing and fire. The grass plant consists of leafy shoots called tillers. Each shoot has a leaflike blade or lamina, the base of which has a tubelike sheath. These tillers grow from short underground stems, which grow upward only when the plant begins flowering. Tillers that group closely about a central stem and buds make up bunch or tussock grasses. Species that spread lateral buds on underground stems, producing a sod, are sod or turf grasses (Figure 26.2). Associated with grasses are a variety of legumes and composite plants.

26.2 The Character of Grassland Ecosystems Varies with Climate and Geography

The grasslands most familiar to a majority of us are hayfields and pasturelands created and maintained by human efforts and referred to as **domestic grasslands.** Mostly, they occupy forested land cleared for settlements and agriculture. Some domestic grasslands, especially in Britain, Switzerland, and Scandinavia, have existed for centuries, becoming a climax community supporting distinctive vegetation. In other areas, such as eastern North America, abandoned agricultural grasslands revert to forest.

Domestic grasslands are permanent, rotational (plowed every few years for other crops), or rough. The last are marginal, unimproved, semiwild lands used principally for grazing. Many successional grasslands fit this category (see Chapter 13, Section 13.12).

In North America, native grasslands once covered much of the interior between the Rocky Mountains and the eastern deciduous forest. There were three main types, distinguished by the height of the dominant species, influenced by climate and rainfall.

Tallgrass prairie occupied a narrow belt running north and south next to the deciduous forest of eastern North America (Figure 26.3a). It was well developed in a region that could support forests. Oak-hickory forests did extend into the grassland along streams and rivers, on well-drained soils, in sandy

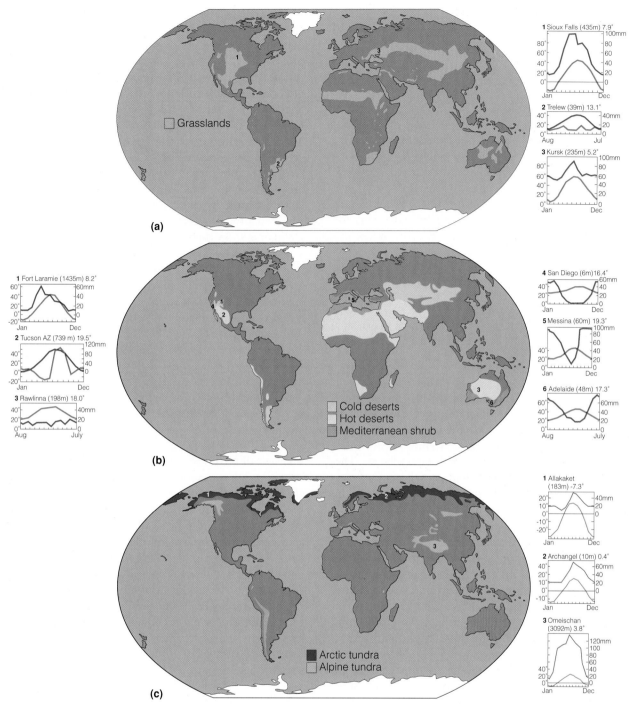

FIGURE 26.1 Geographic distribution of the major (a) grassland, (b) desert and mediterranean shrub, and (c) tundra ecosystems of the world, and associated climate diagrams showing the long-term patterns of monthly temperature and precipitation for selected locations.

areas, and on hills. Fires, often set by Native Americans in the fall, stimulated a vigorous growth of grass and eliminated the encroaching forest. Little tallgrass prairie remains.

Big bluestem *(Andropogon gerardi)*, growing 1 meter tall with flowering stalks 1 to 3.5 meters tall,

was the dominant grass of moist soils and occupied the valleys of rivers and streams and the lower slopes of hills. The drier uplands were dominated by bunch-forming needlegrass *(Stipa)*, side-oats grama *(Bouteloua curtipendula)*, and dropseed *(Sporobolus* spp.).

FIGURE 26.2 Growth forms and root penetration (maximum depth about 2.5 m) of a sod grass (right) and a bunchgrass (left).

West of the tallgrass prairie was the **mixed-grass prairie** (Figure 26.3b). Typical of the Great Plains, it embraced largely the needlegrass-grama grass *(Bouteloua stipa)* community. Extremes in precipitation changed its makeup from year to year. The grasses were largely bunch and cool-season species that began their growth in early April, flowered in June, and matured in late July and August. The drier uplands grew such a diversity of composites that they were nicknamed "daisy land."

South and west of the mixed prairie and grading into the desert are the **shortgrass plains,** one grassland that has remained somewhat intact (Figure 26.3c). The shortgrass plains reflect a climate in which rainfall is infrequent and light (up to 400 mm in the west and 500 mm in the east), humidity low, winds high, and evaporation rapid. The shallow-rooted grasses utilize moisture in the upper soil layer, beneath which the roots do not penetrate. Sod-forming blue grama *(Bouteloua gracilis)* and buffalo grass *(Buchloe dactyloides)* dominate the shortgrass plains. Because of the dense sod, few forbs grow on the plains, but prominent among them are lupines *(Lupinus* spp.).

From southeastern Texas to southern Arizona and south into Mexico lies the **desert grassland,** similar in many respects to the shortgrass plains, except that three-awn grass *(Aristida* spp.) replaces buffalo grass. Composed largely of bunchgrasses, desert grasslands are widely interspersed with other vegetation types, such as oak savanna and mesquite. The climate is hot

and dry. Rain falls only during two seasons, summer (July and August) and winter (December to February), in amounts that vary from 300 mm to 400 mm in the west and 500 mm in the east; but evaporation is rapid, up to 2000 mm per year. Vegetation puts on most of its annual growth in August.

Confined largely to the Central Valley of California is **annual grassland.** It is associated with Mediterranean-type climate (see Section 26.9), characterized by rainy winters and hot, dry summers. Growth occurs during early spring, and most plants are dormant in summer, turning the hills a dry tan color accented by the deep green foliage of scattered California oaks (see the discussion of woodlands in Chapter 25, Section 25.18). The original vegetation was perennial grasses dominated by purple needlegrass *(Stipa pulchra),* but since settlement, native grasses have been replaced by vigorous annual species well-adapted to a Mediterranean-type climate. Dominant species are wild oats *(Avena fatua)* and slender oats *(Avena barbata).*

At one time, the great grasslands of the Eurasian continent extended from eastern Europe to western Siberia south to Kazakhstan. These **steppes,** treeless except for ribbons and patches of forest, are divided into four belts of latitude, from the mesic meadow steppes in the north to semiarid grasslands in the south. The meadow steppes occupy a region in which the rainfall is 500 to 600 mm, extending south from the taiga. Dominated by bunch-forming fescues *(Festuca)* and feather grass *(Stipa)* along with many species of daisy (Compositae), the meadow steppes were once outstandingly beautiful in spring and early summer. Little remains of meadow steppes, which have been turned under the plow for cereal grains. Farther south where rainfall is 400 to 500 mm, tussock-forming species of *Stipa* dominate, and flowering herbs are fewer. In the central Asian steppes with their cold, dry spring, no ephemeral plants exist, and grasses give way to woody and herbaceous species of drought-resistant *Artemisia.* About the Black Sea and in Kazakhstan, where the humidity is higher, steppe vegetation is dominated by large feather grasses and sheep's fescue *(Festuca ovina)* and by ephemeral spring plants such as tulips *(Tulipa).*

In the Southern Hemisphere, the major grasslands exist in southern Africa and southern South America. Known as **pampas,** the South American grasslands extend westward in a large semicircle from Buenos Aires to cover about 15 percent of Argentina. In the eastern part of the pampas, rainfall exceeds 900 mm, well distributed throughout the year. In this humid east, tallgrasses dominate pampas. South and west, where rainfall is about 450 mm, semidesert vegetation becomes prominent. South

(a)

(b)

(c)

FIGURE 26.3 North American grasslands: (a) a remnant tallgrass prairie in Iowa; (b) the mixed-grass prairie, which has been called "daisyland" because of the diversity of its wild flowers; (c) shortgrass steppe in western Wyoming.

into Patagonia, where the rainfall averages about 250 mm, the pampas change to open steppe grasses dominated by *Stipa* and *Festuca* and xerophytic cushion plants. These pampas have been modified by the introduction of European forage grasses and alfalfa *(Medicago sativa),* and the eastern tallgrass pampas have been converted to wheat and corn.

The pampas of Argentina occupy the lowlands; by contrast, the **velds** of southern Africa (not to be confused with the savanna) occupy the eastern part of a high plateau 1500 to 2000 meters above sea level in the Transvaal and the Orange Free State. Most of the rainfall comes in the summer, brought in by moist air masses from the Indian Ocean. The heaviest rainfall is in the east, the lowest in the west

where the grasslands grade into the semiarid shrubland known as the karoo (see Section 26.13).

Australia has four types of grasslands: arid tussock grassland in the northern part of the continent, where the rainfall averages between 200 and 500 mm, mostly in the summer; arid hummock grasslands dominated by *Triodia* and *Plectrachne* in areas with less than 200 mm rainfall; coastal grasslands dominated by *Sporobolus* in the tropical summer rainfall region; and subhumid grasslands dominated by such grasses as *Poa* and kangaroo grass *(Themeda)* along coastal areas, where rainfall is between 500 and 1000 mm. Fertilization, introduced grasses and legumes, and sheep grazing have changed most of these grasslands.

26.3 The Vertical Structure of a Grassland Changes with the Season

The most visible feature of grassland is the tall, green, ephemeral herbaceous growth that develops in spring and dies back in autumn. One of the three strata in the grassland, it arises from the crowns, nodes, and rosettes of plants hugging the soil. The ground layer and the below-ground root layer are the other two major strata of grasslands.

The herbaceous layer, consisting of both grasses and forbs, has three or more sublayers that are more or less variable in height according to the grassland type (Figure 26.4). Low-growing and ground-hugging plants such as dandelion, strawberry, and mosses make up the first layer. As the growing season progresses, these plants become hidden beneath the middle and upper layers. The middle layer consists of shorter grasses and such forbs as wild mustard and daisy. The upper layer consists of leaves and flowering stems of tallgrasses and the leafy stalks and flowers of forbs.

The ground layer is easy to see in late winter and early spring. Exposed to high light, the plants respond to warmth and moisture. As the grasses and forbs grow taller and shade the ground, light intensity reaching the ground layer decreases. Temperature declines, relative humidity increases, and wind flow decreases, creating a region of calm near the ground. Conditions on grazed lands are much different. Because the grass cover is closely cropped, the ground layer continues to receive much higher solar radiation, higher temperatures, and greater wind velocity.

Grasslands that are unmowed, unburned, and ungrazed accumulate a thick layer of mulch (on a lawn, it is called thatch). The oldest bottom layer, humic mulch, consists of decayed and fragmented remains of fresh mulch; the top layer consists of fresh herbage, leafy and largely undecayed, that is deposited throughout the growing season. As the mat increases in depth, it retains more moisture, creating favorable conditions for microbial activity. Three or four years must pass before natural grassland mulch decomposes completely.

Grazing reduces the mulch layer, as do fire and mowing. Light grazing tends to increase the weight of decayed humic mulch at the expense of fresh mulch; moderate grazing increases compaction, which favors microbial activity and a subsequent reduction in both fresh and humic mulch. Heavy grazing greatly reduces mulch accumulation. Burning reduces both fresh and humic mulch, but the mulch structure returns after a fire on lightly grazed and ungrazed lands. Mowing greatly reduces both fresh mulch and humic mulch. Haylands have a minimal amount of mulch on the ground.

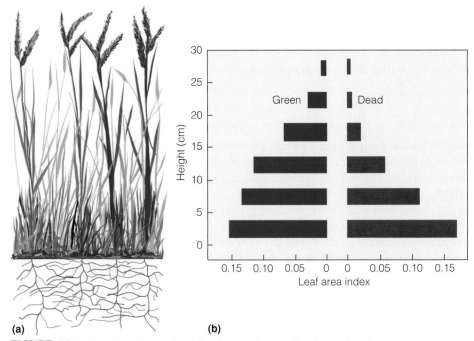

FIGURE 26.4 (a) Profile of a grassland showing physical stratification during the summer, energy flow, structure, and stratification of the physical environment. (b) Leaf area indexes, as indicated by bar graphs, at different levels in a grassland for both green and dead plant structures.

The amount of mulch is ecologically important. Mulch increases soil moisture through its effects on infiltration and evaporation; it decreases runoff and erosion, stabilizes soil temperatures, and improves conditions for seed germination. How much mulch is necessary is the question. Where mulch can accumulate to the proper degree, grassland maintains itself. Heavy mulches can suppress growth of grasses and allow the invasion of forbs and woody vegetation. In areas of no accumulation, grassland regresses to weedy plants. Deep litter provides habitat for meadow mice and certain ground-nesting birds such as the bobolink, but inhibits the presence of others.

The root layer is more highly developed in grasslands than in any other major community. Half or more of the plant is hidden beneath the soil; in winter, roots represent almost the total grass plant. Most roots are fibrous and occupy rather uniformly the upper 15 cm or so of the soil profile; they decrease in abundance with depth. The depth to which roots extend is considerable. Little bluestem, for example, reaches 1 to 2 meters and forms a dense mat as deep as 0.8 meter. In addition, many grasses possess underground stems or rhizomes that serve both to propagate the plant and to store food. Constantly dying, roots add finely divided organic matter to the mineral soil.

Roots develop in three or more zones. Some plants are shallow-rooted and seldom extend much below 0.5 meter. Others go well below the shallow-rooted species but seldom more than 1.5 meters. Deep-rooted plants extend even farther into the soil and absorb little moisture from the surface soil. Thus, plant roots absorb nutrients from different depths in the soil at different times, depending on moisture.

26.4 Grasslands Support a Diversity of Animal Life

Natural or domestic, grasslands support similar forms of life, vertebrate and invertebrate. The invertebrate life includes an incredible number and variety of species and occupies all strata during some time of the year. During winter in temperate grasslands, insect life is confined largely to soil, litter, and grass crowns where it exists as eggs or pupae. In spring, soil occupants are chiefly earthworms and ants, the latter being the most prevalent, if not the most conspicuous. The ground and mulch layers harbor scavenger carabid beetles and predaceous spiders, of which the majority are hunters rather than web builders. Life in the herbaceous layer varies as the strata become more pronounced from spring to fall. Here, invertebrate life is most abundant and varied. Homoptera, Coleoptera, Diptera, Hymenoptera, and Hemiptera

are all represented. Insect life reaches two highs during the year, a major peak in summer and a less-defined one in the fall.

Large grazing ungulates and burrowing mammals are the most conspicuous vertebrates. All of the world's native grasslands support similar forms. The North American grasslands once were dominated by huge migratory herds of bison, numbering in the millions (Figure 26.5), and the forb-consuming pronghorned antelope *(Antilocarpa americana)*. The most common burrowing rodent was the prairie dog *(Cynomys* spp.), which along with gophers *(Thomomys* and *Geomys)* and the mound-building harvester ants appeared to be instrumental in the development and maintenance of the ecological structure of the shortgrass prairie.

The Eurasian steppes lack herds of large ungulates. The western steppes are home to the small migratory goat antelope, the saiga *(Saiga tartarica)*, characterized by a large proboscislike nose, which increases in size in the male during rut. Nearly extinct in 1917, it now numbers over 1 million animals. Farther east lives the Mongolian gazelle *(Procarpa gutterosa)* and several species of rare wild horses. The dominant burrowing animals are the bobak marmot *(Marmota bobak)*, which looks like an oversized prairie dog; the sousliks or ground squirrels *(Citellus* spp.); and the common hamster *(Cricetus cricetus)*.

The Argentine pampas also lack a large ungulate fauna. The two major large herbivores are the pampas deer *(Ozotoceras bezoarticus)* and farther south the guanaco *(Lama guanaco)*, small relatives of the camel, greatly reduced in number compared to historical times. Major burrowing rodents are the viscacha *(Lagostomus maximus)* and the Patagonian

FIGURE 26.5 Bison, which once roamed the shortgrass plains in countless numbers, epitomize the North American grasslands.

hare or mara *(Dolichotis patagonium),* a monogamous, cavylike rodent with the long ears of a hare and the body and long legs of an antelope.

The African grassveld once supported great migratory herds of antelope and zebra along with their associated carnivores—the lion, leopard, and hyena. Burrowing rodents include the kangaroolike springhare *(Pedetes capensis)* and the gerbil *(Tatera brantsii),* and a most interesting carnivore, the meerkat *(Cynictis penicillata),* whose burrowing habits suggest those of the prairie dog. The rodents remain, but the great ungulate herds have been destroyed and replaced with sheep, cattle, and horses.

The Australian marsupial mammals evolved many forms that are the ecological equivalents of placental grassland mammals. The dominant grazing animals are a number of species of kangaroos, especially the red kangaroo *(Macropus rufus)* and the gray kangaroo *(M. giganteus).* The wombats *(Vombatus)* occupy the ecological niche of the viscachas of the pampas and the gophers of the prairies.

Three of the world's grasslands evolved unique unrelated birds with a poor ability to fly, large size, and high running speed. Australia has the emu *(Dromiceius casuarius),* the pampas the rhea *(Rhea americana),* and Africa the ostrich *(Struthio).* New Zealand, whose grasslands (not discussed) lacked herbivorous mammals, had flocks of the now-extinct grass-consuming moa *(Dinorus).* Although the grasslands of the Northern Hemisphere lack such large birds, the European steppes do have the large great bustard *(Otis tarda),* weighing up to 16 kg. Its numbers have been reduced by loss of habitat to agriculture.

Vertebrate life in seral and tame grasslands is strongly affected by human management. Mowing hayfields destroys habitat at a critical nesting time. Losses to birds, rabbits, and mice from mechanical injury and predation on exposed nests are often heavy, but most species will remain on the area to complete or reattempt nesting. Early mowing at the very start of the nesting season eliminates nesting cover and forces the animals elsewhere. Early mowing is one of the reasons for the sharp decline in grassland birds. Pasturelands more often than not are so badly overgrazed that they support little vertebrate life. The two most common inhabitants in eastern North America are the killdeer *(Charadrius vociferus)* and the horned lark *(Eremophila alpestris).*

26.5 Productivity and Nutrient Cycling in Grasslands Is Governed by Moisture

Grasslands are adapted to periods of drought and survive under low rainfall, but grasses grow best under optimal moisture and temperature. Grasslands do the poorest where precipitation is the lowest and temperatures are high; they do best where the mean annual precipitation is greater than 800 mm and mean annual temperature is above 15°C. Production, however, is most directly related to precipitation (Figure 26.6). The greater the mean annual precipitation, the greater the above-ground production. This increased production comes about because increased moisture reduces water stress and enhances the uptake of nutrients.

We associate grassland production with the above-ground growth of grass, but much of the net production is below-ground (Figure 26.7). Grasslands send most of their production into the roots. Except for a short period of maximum above-ground biomass during the growing season, below-ground biomass is two to three times that above-ground. Seventy-five to 85 percent of grassland photosynthate is translocated to the roots for storage.

26.6 Grasslands Evolved under Grazing Pressure

Grasslands have evolved under grazing pressures of ungulates since the Cenozoic. Their structure and growth habits reflect this selective pressure. Critical growth tissues are at or below ground surface, protected from grazing and fire. As the grazers clip and

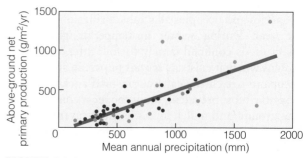

FIGURE 26.6 Relationship between above-ground primary production and mean annual precipitation for 52 grassland sites around the world. North American grasslands are indicated by dark green dots.

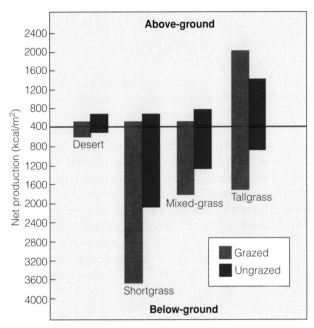

FIGURE 26.7 Above-ground and below-ground net primary production for grazed and ungrazed North American grassland types.

eat the leaves, grasses respond by increasing the photosynthetic rate in remaining tissue, stimulating new growth, and reallocating nutrients and photosynthates from one part of the plant to another, especially from roots to stems. In addition, grazers recycle nutrients in grass through dung and urine.

Grassland ecosystems respond to grazing in still another way by changing species composition. Some grasses and forbs tend to disappear, while other species increase. On desert grasslands of North America, black grama is replaced by weedy species. On shortgrass plains, which are the most stable under grazing pressure, blue grama and prickly pear increase. On mixed-grass prairies, midgrasses decrease and shortgrasses and sedges increase. On tallgrass sites, tallgrasses disappear and little bluestem and tall dropseed increase; if grazing pressure is heavy, the site may be invaded by the weedy Japanese chess. In domestic pasturelands, heavy grazing increases the amount of unpalatable forbs, such as thistles and ironweed.

We equate grazing pressure with above-ground herbivores such as cattle and rabbits and invertebrate grazers such as grasshoppers because they are so conspicuous. In reality, much more intense grazing takes place underground. The above-ground biomass of invertebrates—including plant consumers, saprovores, and predators—ranges from 1 to 50 g/m², whereas grazing mammals amount to about 2 to 5 g/m². Below-ground invertebrates,

most of them nematodes, exceed 135 g/m². They account for 90, 95, and 93 percent of all below-ground herbivory, carnivory, and saprophagous activity, respectively.

Not only is a large proportion of primary production eaten below ground, but a greater proportion is also used at each trophic level there. Some invertebrate above-ground consumers, particularly grasshoppers, are wasteful. The amount of above-ground vegetation they detach or otherwise kill about equals that consumed by vertebrate grazing herbivores.

Invertebrate consumers are also highly inefficient in assimilating ingested material and deposit much of their intake as highly soluble feces, or frass. The nutrients feces contain return rapidly to the system. Large grazing herbivores return a portion of their intake as dung, which is fed upon by a well-developed coprophagous fauna that speeds the decay of manure and accelerates the activity of bacteria in feces.

Most of the primary production, however, goes to the decomposers, dominated by fungi whose biomass is 2 to 7 times that of bacteria. Overall, the decomposer biomass exceeds that of invertebrates.

Central to the cycling of nutrients in grasslands is mulch or detritus, a large standing crop of which can have detrimental effects on nitrogen cycling, particularly in tallgrass ecosystems. Detritus intercepts rainfall, from which microbes can assimilate inorganic nitrogen directly before it reaches plant roots, while the mulch itself inhibits nitrogen fixation by free-living nitrogen-fixing microbes. By insulating the soil surface from solar radiation, mulch reduces production of new roots and inhibits the activity of soil microbes and invertebrates. Periodic grassland fires clear away the mulch layer and release nutrients in detritus to the soil, but nitrogen equal to about 2 years of nitrogen inputs to the system through rainfall is lost to the atmosphere. Fires, however, stimulate the growth of nitrogen-fixing leguminous forbs and improve conditions for earthworms.

26.7 Shrubland Ecosystems Occupy the Arid and Semiarid Regions of the Tropics

In the forested regions of the world, plant communities in which shrubs and small-stature trees are the dominant woody vegetation are characteristic of the early stages (or seres) of secondary succession

FIGURE 26.8 In eastern North America and in northern and western Europe, shrublands are usually successional communities in forested regions. This old field in southern New York State is in early stages of shrub invasion.

26.8 Shrubs Are Difficult to Characterize

Shrubland—plant communities where the shrub growth form is either dominant or codominant—is a difficult type of ecosystem to categorize, in large part because of the difficulty in characterizing the term *shrub* itself. Shrubs have, as W. G. McGinnes points out, a "problem in establishing their identity." They constitute neither a taxonomic nor an evolutionary category. A rough definition is that a shrub is a plant with multiple woody, persistent stems but no central trunk and a height from 4.5 to 8 meters. However, under severe environmental conditions, many trees will not exceed that size. Some trees, particularly coppice stands, are multistemmed, and some shrubs have large, single stems. Shrubs may have evolved either from trees or from herbs.

The success of shrubs depends on their ability to compete for nutrients, energy, and space. In certain environments, shrubs have many advantages. They invest less energy and nutrients in above-ground parts than trees. Their structural modifications improve light interception, heat dissipation, and evaporation. The more arid the site, the more common is drought-deciduousness and the less common is evergreenness. The multistemmed forms influence interception of moisture and stemflow, increasing or decreasing infiltration into the soil. Because most shrubs can get their roots down quickly and form extensive root systems, they use moisture deep in the soil. This feature gives them a competitive advantage over trees and grasses in regions where soil recharges between growing seasons. Because they do not have a high root-shoot ratio, shrubs draw fewer nutrients into above-ground biomass and more into roots. Their perennial nature allows immobilization of limiting nutrients and slows nutrient recycling, favoring further shrub invasion of grasslands.

following disturbance (Figure 26.8). These seral shrub communities are ephemeral, lasting only about 15 to 20 years (see Chapter 13, Section 13.12).

In the tropics, ecosystems in which shrubs are a major component are part of the woodland-savanna-grassland-desert gradient. As precipitation declines across the landscape, shrubs replace trees as the dominant woody vegetation, a pattern that continues into the scrub or shrub deserts of the subtropical and tropical regions (see Section 26.13).

In the temperate region, shrubby vegetation covers large portions of the arid and semiarid landscape. In addition, shrubland covers parts of temperate regions because historical disturbances of landscapes have seriously affected their potential to support forest. Among such shrub-dominated human-induced communities are the moors of Scotland and the macchia of South America.

26.9 Winter Rainfall Characterizes the Mediterranean Shrublands of the World

In five regions of the world, lying for the most part between 32° and 40° north and south of the equator, are areas with a mediterranean climate: the semiarid

(a)

(c)

(b)

FIGURE 26.9 Mediterranean-type shrublands. (a) Mattoral community in southern Spain dominated by Holm oak (Quercusilex) and French lavender. (b) Southern California mixed chaparral is highly prone to fire. (c) The shrub mallee in Victoria, Australia, is an example of a mediterranean-type shrubland dominated by Eucalyptus. Note the canopy structure and the open understory of grass at the beginning of the spring rains. This type of vegetation supports a rich diversity of bird life.

regions of western North America, the regions bordering the Mediterranean Sea, central Chile, the Cape region of South Africa, and southwestern and southern Australia. The mediterranean climate has hot, dry summers with at least one month of protracted drought and cool, moist winters. About 65 percent of the annual precipitation falls during the winter months, and for at least one month the temperature remains below 15°C.

All five areas support similar-looking communities of xeric broadleaf evergreen shrubs and dwarf trees known as sclerophyll (*scleros,* "hard"; *phyll,* "leaf") vegetation with an herbaceous understory (Figure 26.9 a–c). Sclerophyllous vegetation possesses small leaves, thickened cuticles, glandular hairs, and sunken stomata. In the Northern Hemisphere, this vegetation evolved from tropical floras in dry summer climates that began during the Pleistocene. Vegetation in each of the mediterranean systems also

shares adaptations to fire and to low nutrient levels in the soil.

Despite convergence, each area has distinct flora and fauna. In the Mediterranean region, shrub vegetation falls into three major types. The **garrigue,** resulting from degradation of pine forests, includes several types of dwarf-shrub communities less than 0.5 meter high, dominated by aromatic evergreen shrublets on well-drained to dry, calcareous soil. The **maquis,** replacing cork oak forests, is a dense evergreen sclerophyllous shrub community where the climate is moist. The **mattoral** appears to be equivalent to the North American chaparral (Figure 26.9a). The Chilean mediterranean system, also called mattoral, varies from the coast to the foothills of the Andean cordillera.

In North America, the sclerophyllous shrub community is known as **chaparral,** a word of Spanish origin meaning a thicket of shrubby evergreen oaks

(Figure 26.9b). California chaparral is dominated by scrub oak (*Quercus berberidifolia*) and chamise (*Adenostoma fasciculatum*). Another shrub type, also designated as chaparral, is associated with the Rocky Mountain foothills. It differs from California chaparral in two ways. It is dominated by Gambel oak (*Q. gambelii*) and other species and lacks chamise; it is summer-active and winter-deciduous, whereas California chapparal is evergreen, winter-active, and summer-dormant.

In their original presettlement state, both the Mediterranean and California mediterranean plant communities were dominated by oaks, both shrub and tree. Natural oak forests remain now only in scattered patches. In Spain, cork oak is a plantation tree, and many oak forests throughout the region have been converted to olive plantations. In California, four major oak forest communities included two endemics, the evergreen blue oak (*Q. douglassii*) and valley oak (*Q. lobata*). The ranges of both have been greatly reduced by settlement.

Much of the South African mediterranean shrubland is heathland, discussed later. The rest, dominated by a broad-sclerophyll woody shrub, goes by the names of strandveld, coastal renosterveld, and inland renosterveld.

In southwest Australia, the mediterranean shrub country, known as **mallee,** is dominated by low-growing *Eucalyptus*, 5 to 8 meters high with broad sclerophyllous leaves. There are six types of mallee ecosystems, which intergrade. Three of them fall into mediterranean-type ecosystems, with a grassy and herbaceous understory (Figure 26.9c). The other three types occur on nutrient-poor soils and fall under the category of heathland shrubs. Razed by fire at irregular intervals, the mallee retains a summer growth rhythm evolved in the subtropical Tertiary, which is out of phase with the mediterranean climate of the area. Growth takes place during the summer, the driest part of the year, using extensive root systems to draw on water conserved in the soil during the wet winter and spring.

For the most part, mediterranean-type shrublands lack an understory and ground litter, are highly inflammable, and are heavy seeders. Many species require the heat and scarring action of fire to induce germination. Others sprout vigorously after a fire.

For centuries, periodic fires have roared through mediterranean-type vegetation, clearing away the old growth, making way for the new, and recycling nutrients through the ecosystem. When humans intruded on this type of vegetation, they changed the fire regime, either by attempting to exclude fire completely or by overburning. In the absence of fire, chaparral grows tall and dense and yearly adds more leaves and twigs to those already on the ground. During the dry season, the shrubs, even though alive, nearly explode when ignited. Once they are set on fire by lightning or humans, an inferno follows.

After fire, the land returns either to lush green sprouts coming up from buried root crowns or to grass, if a seed source is nearby. New grass and vigorous young sprouts are excellent food for deer, sheep, and cattle. As the sprout growth matures, chaparral becomes dense, the canopy closes, litter accumulates, and the stage is set for another fire.

Precipitation, temperature, soil moisture, and nutrients are major influences on the function of mediterranean-type systems. Precipitation falls mostly in the cool winter months, and most of the plant growth and flowering is concentrated in spring, much of it at the end of the rainy season. Then the plants have to respond to the dry season, which imposes a great deal of environmental stress. How they respond is reflected in plant growth forms. Aridity, cold temperatures, a short growing season, long periods of drought, and a low nutrient supply favor shrubs. Where drought periods are shorter and neither temperature nor nutrients strongly limit growth, trees grow. If the physiological costs of growing new leaves each year are less than the costs of maintaining the same leaves, the plants possess a deciduous habit. If the costs are greater, the plants are evergreen.

Soils of mediterranean-type ecosystems are low in nutrients and are especially deficient in nitrogen and phosphorus. During the dry period, nitrogen and other nutrients accumulate beneath the woody plants and remain fixed as the topsoil dries out. Wetting of the soil during the winter stimulates a flush of microbial activity involving decomposition of humus and mineralization of nitrogen and carbon. The concentration of nutrients stimulates a flush of growth. If heavy rains suddenly enter dry topsoil, quantities of nutrients may be lost by leaching and erosion.

Some plants of the mediterranean systems conserve nutrients. *Ceanothus,* an early successional species of California chaparral, is a nonleguminous nitrogen-fixer (see Sections 6.23 and 16.11). In the Australian mallee, *Atriplex vesticana*, a dominant plant, lowers the nitrogen content of the surrounding soil during the growing season and concentrates nitrogen directly beneath it through litterfall. It transfers nitrogen and phosphorus from its leaves to its stems before leaf fall.

26.10 Heathlands Are a Taxonomically Distinct Form of Shrubland

Typically, heathlands have been associated with cool to cold temperate climatic regions of northwestern Europe. It was probably coincidental that the original name came to refer to land dominated by *Ericaceae* (the heath family). The word *heath* comes from the German *heide,* meaning "an uncultivated stretch of land," regardless of the vegetation.

Heathlands are found in all parts of the world, from the tropics to polar regions and from lowland to alpine altitudes. Heathland flora probably evolved in the Mesozoic in the eastern to central portion of Gondwanaland and retained most of its characteristics. It expanded from Africa into western Europe and the northern part of Eurasia and North America, from India into southeastern Asia, and from Australasia into the Malay Archipelago.

Heathland vegetation is an assemblage of dense to mid-dense growth of ancient or primitive genera adapted to fire (Figure 26.10). Heathland shrubs have leaves with thick cuticles, sunken stomata, thick-walled cells, and hard and waxy upper surfaces. Many species have leaves with small surface area—less than 25 mm²—and others roll their edges in toward the midrib. Although mostly associated with this distinct plant comunity, many heathland shrubs are usually present as shrubby understory in other ecosystems, such as the deciduous forest.

Heathlands invariably occur on nutrient-poor soils especially deficient in phosphorus and nitrogen. Although heathlands are most extensive in the Arctic regions, they are also prominent in the mediterranean-type regions of South Africa, where they are known as *fynbos* and in southeastern and western Australia. In subtropical to tropical climates, true heathlands are confined to alpine areas and to lowland, poor soils subject to seasonal waterlogging. Some heathlands, such as the heather-dominated moors of Scotland, are human-induced and maintained only by periodic fires.

There are two distinct heathland ecosystems: dry heathlands and wet heathlands. Dry heathlands are on well-drained soils subject to seasonal drought, and wet heathlands are subject to seasonal waterlogging. In wet heathlands, grasses and sedges may become codominant with heathland shrubs; and in extreme wet heathlands the grass component is suppressed by *Sphagnum* moss. Because both foliage cover and height vary considerably with the habitat, heathlands are divided according to height of the uppermost

FIGURE 26.10 Heathlands dominated by ericaceous shrubs have a similar appearance around the world. Typical is this heathland along the southern coast of Australia.

stratum: shrubs taller than 2 m, scrub; shrubs 1–2 m, tall heathland; shrubs 25–100 cm, heathland; and shrubs less than 25 cm, dwarf heathland.

26.11 Heterogeneity Is a Characteristic of Shrubland Structure

Shrub ecosystems are characterized by woody structure, increased stratification over grasslands, dense branching on a fine scale, and low height, up to 8 meters. Typically, there are three layers—a broken upper canopy, an irregular low shrub canopy, and a grass/herbaceous layer—but the presence of these layers varies. Dense shrubland may have only a canopy layer, and stratification often decreases as the shrubs reach maximum height. This condition is particularly true in seral shrublands. Horizontal patterns vary with the vegetation type. Heathlands may exhibit little patchiness across the landscape. Mediterranean-type shrublands, notably the mattoral and the mallee, may be very patchy, with woody growth well interspersed with open areas of grass. Greatest patchiness probably occurs in seral shrublands, with scattered clumps of invading shrubs and trees.

The patchy nature of shrublands has a direct influence on soil characteristics and nutrient cycling. The lateral root system of shrubs typically extends well beyond the perimeter defined by their canopy. This is especially true in the semiarid and arid shrublands. Nutrients taken up from the surrounding soil and incorporated into leaf (and fine woody) tissues are deposited in the area below the canopy upon senescence, where the dead organic matter is broken down and nutrients released to the soil in the process of decomposition. Nutrients are in effect mined from the root zone and redeposited in the zone of the canopy, resulting in a higher concentration of soil organic matter and net mineralization in the zone defined by the canopy and litterfall. This heterogeneity in soil properties in turn influences the distribution and productivity of plants in the herbaceous (or grass) layer.

26.12 Shrub Ecosystems Support a Unique Animal Community

Because of this structure, shrub communities have distinctive animal life. Seral shrub communities support not only species common to shrubby edges of forest and shrubby borders of fields but also a number of species dependent on them, such as bobwhite quail, cottontail rabbit, prairie warbler *(Dendroica discolor)*, and yellow-breasted chat *(Ictera virens)*. In Great Britain, some shrub communities, especially hedgerows, have been stable for centuries, and many forms of animal life, invertebrate and vertebrate, have become adapted to or dependent on them. Among these species are the whitethroat *(Sylvia communis)*, linnet *(Acanthis cannabina)*, blackbird *(Turdus merula)*, and yellowhammer *(Emberiza citrinella)*.

Shrub communities have a complex of animal life that varies with the region. Within the mediterranean-type shrublands and heathlands, similarity in habitat structure and in the nature and number of niches has resulted in pronounced parallel and convergent evolution among bird species and some lizard species, especially between the Chilean mattoral and the California chaparral. North American chaparral and sagebrush communities support mule deer *(Odocoileus hemionus)*, coyotes *(Canis latrans)*, a variety of rodents, jackrabbits *(Lepus* spp.), and sage grouse (Figure 26.11) *(Centrocercus urophasianus)*. The Australian mallee is rich in birds, including the endemic mallee-fowl *(Leipoa ocellata)*, which incubates its eggs in a large mound. Among the mammalian life are the gray kangaroo (Figure 26.12) *(Macropus giganteus)* and various species of wallaby. Rash clearance of mallee vegetation is endangering

FIGURE 26.11 Male sage grouse in courtship display.

FIGURE 26.12 Gray kangaroos are the dominant herbivore in the Australian shrublands.

mallee wildlife as well as affecting millions of migrant, nectar-feeding birds supported by the mallee in spring.

26.13 Deserts Represent a Diverse Group of Ecosystems

Geographers define deserts as land where evaporation exceeds rainfall. No specific amount of rainfall serves as a criterion; deserts range from extremely arid regions to those with sufficient moisture to support a variety of life. Deserts have been classified according to rainfall into semideserts, ones that have precipitation between 150 and 300 to 400 mm per year; true deserts, regions with rainfall below 150 mm per year; and extreme deserts, areas with rainfall below 70 mm per year (Figure 26.13).

(a)

FIGURE 26.13 Three examples of hot deserts. (a) The Chihuahuan Desert in Nuevo Leon, Mexico. The substrate of this desert is sand-sized particles of gypsum. (b) The edge of the Great Victorian Desert in Australia. This desert and the Chihuahuan Desert are dominated by woody, brittle-stemmed shrubs. (c) Dunes in the Saudi Arabian desert near Riyadh. Note the extreme sparseness of vegetation.

(b)

(c)

Deserts, which occupy about 26 percent of the continental area, occur in two distinct belts between 15° and 35° latitude in both the Northern and Southern Hemispheres—the Tropic of Cancer and the Tropic of Capricorn.

Deserts are the result of several forces. One force that leads to the formation of deserts is the movement of air masses (see Chapter 3, Section 3.7). High-pressure areas alter the course of rain. The high-pressure cell off the coast of California and Mexico deflects rainstorms moving south from Alaska to the east and prevents moisture from reaching the Southwest. In winter, high-pressure areas move southward, allowing winter rains to reach southern California and parts of the North American desert. Winds blowing over cold waters become cold also. They carry little moisture and produce little rain. Thus the west coast of California and Baja California, the Namib Desert on coastal southwest Africa, and the coastal edge of the Atacama in Chile may be shrouded in mist, yet remain extremely dry.

Mountain ranges also play a role in desert formation by causing a rain shadow on their lee side. The high Sierras and Cascade Mountains intercept rain from the Pacific and help maintain the arid conditions of the North American desert. The low eastern highlands of Australia block the southeast trade winds from the interior. Other deserts, such as the Gobi and the interior of the Sahara, are so remote from the ocean that all of the water has been wrung from the winds by the time they reach those regions.

All deserts have in common low rainfall, high evaporation (from 7 to 50 times as much as precipitation), and a wide daily range in temperature from hot by day to cool by night. Low humidity allows up to 90 percent of solar radiation to penetrate the atmosphere and heat the ground. At night, the desert yields the accumulated heat of the day back to the atmosphere. Rain, when it falls, is often heavy and, unable to soak into the dry earth, rushes off in torrents to basins below.

Deserts are not the same everywhere. Differences in moisture, temperature, soil drainage, topography, alkalinity, and salinity create variations in vegetation cover, dominant plants, and groups of associated species. There are hot deserts and cool deserts, extreme deserts and semideserts, ones with sufficient moisture to verge on being grasslands or shrublands, and gradations between those extremes within continental deserts.

There is a certain degree of similarity among hot deserts and cold deserts of the world. The cold deserts—including the Great Basin of North America and the Gobi, Takla Makan, and Turkestan deserts of Asia—and high elevations of hot deserts are dominated by *Artemisia* and chenopod shrubs, and may be considered shrub steppes or desert scrub (Figure 26.14). In the Great Basin of North America, the northern, cool, arid region lying west of the Rocky Mountains, is the northern desert scrub. The climate is continental, with warm summers and prolonged cold winters. Its physiognomy differs greatly from the southern hot desert, and the dominant vegetation is shrub. The vegetation falls into two main associations: one is sagebrush, dominated by *Artemisia tridentata,* which often forms pure stands; the other is shadscale, *Atriplex conifertifolia,* a C_4 species, and other chenopods, halophytes tolerant of saline soils.

(a)

(b)

FIGURE 26.14 Two examples of desert scrub. (a) The northern desert shrubland in Wyoming is dominated by sagebrush. Although this desert is classified as cold desert, sagebrush forms one of the most important shrub types in North America. (b) Saltbrush shrubland in Victoria, Australia, is dominated by *Atriplex*. It is an ecological equivalent of the shrublands of the Great Basin in North America.

A similar type of desert scrub exists in the semi-arid inland of southwestern Australia. Numerous chenopod species, particularly the saltbushes of the genera *Atriplex* and *Maireana,* form extensive low shrublands on low riverine plains.

The hot deserts range from no or scattered vegetation to ones with some combination of chenopods, dwarf shrubs, and succulents. The deserts of southwestern North America—the Mojave, the Sonoran, and the Chihuahuan—are dominated by creosote bush *(Larrea divaricata)* and bur sage *(Franseria* spp.). Areas of favorable moisture support tall growths of *Acacia* spp., saguaro *(Cereus giganteus),* palo verde *(Cercidium* spp.), and ocotillo *(Fouquieria* spp.).

FIGURE 26.15 Organpipe cactus and saguaro (in background) dominate this part of the Sonoran Desert in the southwestern United States.

26.14 Desert Ecosystems Have a Simple Physical Structure

The topography of the desert, unobscured by vegetation, is stark and, paradoxically, partially shaped by water. The unprotected soil erodes easily during violent storms and is further cut away by the wind. Alluvial fans stretch away from eroded, angular peaks of more resistant rocks. They join to form deep expanses of debris, the **bajadas.** Eventually, the slopes level off to low basins, or **playas,** which receive waters that rush down from the hills and water-cut canyons, or **arroyos.** These basins hold temporary lakes after the rains, but water soon evaporates and leaves behind a dry bed of glistening salt.

Woody-stemmed and soft brittle-stemmed shrubs are characteristic desert plants. In a matrix of shrubs grows a wide assortment of other plants: yucca, cacti, small trees, and ephemerals. In the Sonoran, Peru-Chilean, South African Karoo, and southern Namib deserts, large succulents rise above the shrub level and change the appearance of the desert far out of proportion to their numbers. The giant saguaro, the most massive of all cacti, grows on the bajadas of the Sonoran desert (Figure 26.15). Ironwood, smoketree, and palo verde grow best along the banks of intermittent streams, not so much because they require the moisture but because their hard-coated seeds must be scraped and bruised by the grinding action of sand and gravel during flash floods before they can germinate.

Both plants and animals are adapted to the scarcity of water either by drought evasion or by drought resistance (see Chapter 6). Plant drought-evaders flower only in the presence of moisture. They persist as seeds during drought periods, ready to sprout, flower, and produce seeds when moisture and temperature are favorable. There are two periods of flowering in the North American deserts: after winter rains come in from the Pacific Northwest, and after summer rains move up from the south out of the Gulf of Mexico. Some species flower only after winter rains, others only after summer rains, but a few bloom during both seasons. If no rains come, these ephemeral species do not grow. Drought-evading animals, like their plant counterparts, adopt an annual lifestyle or go into estivation or some other stage of dormancy during the dry season. For example, the spadefoot toad *(Scaphiopus)* remains underground in a gelatinous-lined cell, making brief reproductive appearances during periods of winter and summer rains. If extreme drought develops during the breeding season, birds fail to nest and lizards do not reproduce.

Below-ground biomass in the desert can be as patchy as the above-ground biomass. Desert plants may be deep-rooted woody shrubs, such as mesquite *(Prosopis* spp.) and *Tamarix,* whose taproots reach the water table, rendering them independent of water supplied by rainfall. Some, such as *Larrea* and *Atriplex,* are deep-rooted perennials with superficial laterals that extend as far as 15 to 30 meters from the stems. Other perennials, such as the various species of cactus, have shallow roots, often extending no more than a few centimeters below the surface. Ephemerals have shallow and poorly branched roots reaching a depth of about 30 cm, where they pick up moisture quickly from light rains.

The desert floor is stark, a raw mineral substrate of various types devoid of a continuous litter layer. Dead leaves, bud scales, and dead twigs, mostly associated with drought-resistant species that shed them to reduce transpiring surfaces, accumulate in wind-protected areas beneath the plants and in depressions in the soil.

26.15 Moisture Limits Productivity and Nutrient Cycling in Desert Ecosystems

Primary production in the desert depends on the proportion of available water used and the efficiency of its use. Data from various deserts in the world suggest that annual primary production of above-ground vegetation varies from 30 to 200 g/m^2. Below-ground production is also low but greater than above-ground production. It ranges from 100 to 400 g/m^2 in arid regions and from 250 to 1000 g/m^2 in semiarid regions.

The amount of biomass that accumulates and the ratio of annual production to biomass depend on the dominant type of vegetation. In those deserts in which trees, shrubs, and cactuslike plants dominate, annual production is about 10 to 20 percent of the total above-ground standing crop biomass. Annual or ephemeral communities have a 100 percent turnover of both roots and above-ground foliage, and their annual production is the same as peak biomass. In general, desert plants do not have a high root biomass relative to above-ground shoot biomass.

Adding to primary production in the desert are lichens, green algae, and cyanobacteria, abundant as soil crusts. Cyanobacteria have unusually high rates of nitrogen fixation, but less than one-half of their total nitrogen input becomes part of higher plants. Approximately 70 percent of the nitrogen is short-circuited back to the atmosphere as volatilized ammonia and as N$_2$ from denitrification, speeded by dry alkaline soils.

Nutrient cycling in arid ecosystems is tight. Two major nutrients, phosphorus and nitrogen, are in short supply; much of them is tied up in plant biomass, living and dead. Desert plants tend to retain certain elements in stems and roots, particularly nitrogen, phosophorus, and with some, potassium, before shedding any parts. The nutrients remaining in the shed parts collect and decompose beneath the plants, where microclimate conditions created by the plants favor biological activity. The soil is further enriched by animals attracted to the shade. The plants, in effect, create islands of fertility beneath themselves.

26.16 Adapted to Aridity, Desert Animal Life Is Diverse

In spite of their aridity, desert ecosystems support a surprising diversity of animal life, including a wide assortment of beetles, ants, locusts, lizards, snakes, birds, and mammals. The mammals are notably herbivorous species (Figure 26.16). Grazing herbivores of the desert tend to be generalists and opportunists in their mode of feeding. They consume a wide range of species, plant types, and parts. Desert sheep feed on succulents and ephemerals when available and then switch to woody browse during the dry period. As a last resort, herbivores consume dead litter and lichens. Small herbivores—the desert rodents, particularly the

(a)

(b)

FIGURE 26.16 Typical herbivores of the desert. (a) A five-toed jerboa (*Alactaga* spp.), a Middle Eastern rodent whose long ears radiate heat. (b) The desert oryx *(Oryx gazella)* copes with the desert environment by reducing unnecessary energy expenditure.

family Heteromyidae, and ants—tend to be granivores, feeding largely on seeds, and are important in the dynamics of desert ecosystems.

Herbivores can have a pronounced impact on desert vegetation, especially if they are more abundant than the range's capacity to support them. Once grazers have eaten the annual production, they eat plant reserves, especially during long dry periods. Overbrowsing can so weaken the plant that the vegetation is destroyed or irreparably damaged. Areas protected from grazing, especially grazing by goats and sheep, have a higher biomass and a greater percentage of palatable species than grazed areas.

Native plant-eating herbivores in a shrubby desert, under most conditions, consume only a small part of the above-ground primary production, but seed-eating herbivores can eat most of the seed production, close to 90 percent of it. This consumption can have a pronounced effect on plant composition and plant populations.

Desert carnivores, like the herbivores, are opportunistic feeders, with few specialists. Most desert carnivores, such as foxes and coyotes, have mixed diets that include leaves and fruits; even insectivorous birds and rodents eat some plant material. Omnivory, rather than carnivory and complex food webs, seems to be the rule in desert ecosystems.

The detrital food chain seems to be less important in the desert than in other ecosystems. Fungi and actinomycetes are the most prominent decomposers. Microbial decomposition, like the blooming of ephemerals, is limited to short periods when moisture is available. For this reason, dry litter tends to accumulate until the detrital biomass may be greater than the above-ground living biomass. Most of the ephemeral biomass disappears through grazing, weathering, and erosion. Decomposition proceeds mostly through detritus-feeding arthropods such as termites that ingest and break down woody tissue in their guts. In some deserts, considerable amounts of nutrients are locked up in termite structures, to be released when the structure is destroyed. Other important detritivores are acarids and isopods.

26.17 Low Precipitation and Cold Temperatures Characterize the Tundra

Encircling the top of the Northern Hemisphere is a frozen plain, clothed in sedges, heaths, and willows. Called the **tundra**, its name comes from the Finnish

FIGURE 26.17 The wide expanse of the Arctic tundra in the Northwest Territories of Canada.

tunturi, meaning "a treeless plain." At lower latitudes, similar landscapes, the alpine tundra, occur in the mountains of the world. Arctic or alpine, the tundra is characterized by low temperatures, a short growing season, and low precipitation (cold air carries little water vapor).

The Arctic tundra is a land dotted with lakes and crossed by streams (Figure 26.17). Where the ground is low and moist, extensive bogs exist. On high drier areas and places exposed to the wind, vegetation is scant and scattered, and the ground is bare and rock-covered. These regions are the fell-fields, an anglicization of the Danish *fjoeldmark,* or rock deserts. Lichen-covered, the fell-fields are most characteristic of highly exposed alpine tundra. The Arctic tundra falls into two broad types: *tundra* with 100 percent cover and wet to moist soil, and *polar desert* with less than 5 percent cover and dry soil.

Conditions unique to the Arctic tundra are a product of at least three interacting forces: permafrost (permanent frozen layer of soil; see Chapter 25, Section 25.14), vegetation, and the transfer of heat. Vegetation and its accumulated organic matter protect the permafrost by shading and insulation, which reduce the warming and retard thawing of the soil in summer. In turn, permafrost chills the soil, retarding the general growth of both above-ground and below-ground parts of plants, limiting the activity of soil microorganisms, and diminishing the aeration and nutrient content of the soil. Alternate freezing and thawing of the upper layer of soil create the unique, symmetrically patterned landforms typical of the tundra. This action of frost pushes stones and other material upward and outward from the mass to form a patterned surface. Frost hummocks, frost boils, and earth stripes, all typical nonsorted or

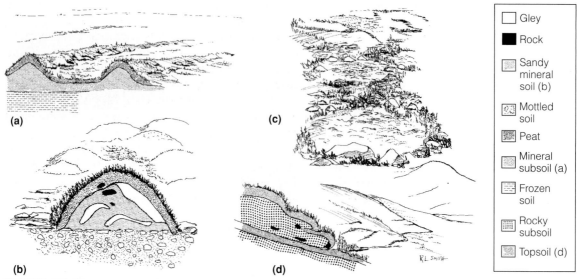

	Gley
	Rock
	Sandy mineral soil (b)
	Mottled soil
	Peat
	Mineral subsoil (a)
	Frozen soil
	Rocky subsoil
	Topsoil (d)

FIGURE 26.18 Patterned landforms typical of the tundra region: (a) unsorted earth stripes; (b) frost hummocks; (c) sorted stone nets and polygons; and (d) a solifluction terrace.

irregular patterns, are associated with seasonally high water tables (Figure 26.18). Sorted or regular patterns appear on better-drained sites. Best known are stone polygons, whose size is related to frost intensity and size of the material. On sloping ground, creep, frost thrusting, and downward flow of soil changes polygons into sorted stripes running downhill. Mass movement of supersaturated soil over the permafrost forms solifluction terraces, or "flowing soil." This gradual downward creep of soils and rocks eventually rounds off ridges and other irregularities in topography. The molding of the landscape by frost action,

called cryoplanation, is far more important than erosion in wearing down the Arctic landscape.

Alpine tundra (Figure 26.19) has little permafrost, confined mostly to very high elevations; but frost-induced processes, such as small solifluction terraces and stone polygons, are present nevertheless. Lacking permafrost, soils are drier; only in alpine wet meadows and bogs do soil moisture conditions compare with those of the Arctic. Precipitation, especially snowfall and humidity, is higher in the alpine regions than in the Arctic tundra, but steep topography induces a rapid runoff of water.

(a)

(b)

FIGURE 26.19 Alpine tundra (a) in the Rocky Mountains and (b) in the Australian Alps. Although the species of vegetation are different, the growth forms are convergent and the physiognomy is similar.

26.18 Simple Structure and Low Diversity Characterize the Vegetation of the Tundra

In spite of its distinctive climate and many endemic species, the tundra does not possess a vegetation type unique to itself. Structurally, the vegetation of the tundra is simple. The number of species tends to be few; the growth is slow; and most of the biomass and functional activity are confined to a few groups. In the Arctic, only those species able to withstand constant disturbance of the soil, buffeting by the wind, and abrasion from wind-carried particles of soil and ice can survive. In the alpine tundra, the environment is even more severe for plants. It is a land of strong winds, snow, cold, and widely fluctuating temperatures. During the summer, the temperature on the surface of the soil ranges from 40°C to 0°C. The atmosphere is thin, so light intensity, especially ultraviolet, is high on clear days.

Although it appears homogeneous, the pattern of vegetation is patchy. A combination of microrelief, snowmelt, frost heaving, and aspect, among other conditions, produces an endless change in plant associations from spot to spot. In the Arctic tundra, low ground is covered with a cotton grass, sedge, dwarf heath, sphagnum moss complex. Well-drained sites support heath shrubs, dwarf willows and birches, dryland sedges and rushes, herbs, mosses, and lichens. The driest and most exposed sites—the flat-topped domes, rolling hills, and low-lying terraces, all usually covered with coarse, rocky material and subject to extreme action by frost—support only sparse vegetation, often confined to small depressions (Figure 26.20). Plant cover consists of scattered heaths and mats of dryads, as well as crustose and foliose lichens growing on the rocks.

The alpine tundra is a land of rock-strewn slopes, bogs, alpine meadows, and shrubby thickets. Cushion and mat-forming plants, rare in the Arctic, are important in the alpine tundra. Low and ground-hugging, they are able to withstand the buffeting of the wind, and their cushionlike blanket traps heat. The interior of the cushion may be 20°C warmer than the surrounding air, a microclimate that insects use.

In spite of similar conditions, only about 20 percent of the plant species of the Arctic and Rocky Mountain alpine tundra are the same, and they are of different ecotypes. Lacking in the Rocky Mountain alpine tundra are heaths and heavy growth of lichens and mosses between other plants; lichens are confined mostly to rocks, and the ground is bare between plants. The alpine tundra of the Appalachian Mountains, however, is dominated by heaths and sedge meadows, as is the alpine tundra of Europe, and mosses are common. The Australian alpine region supports a growth of heaths on rocky sites, and wet areas are covered with sphagnum bogs, cushion heaths, and sod tussock grasslands.

Alpine plant communities are not restricted to the northern and southern temperate regions of Earth.

FIGURE 26.20 Alpine sorrel (Oxyria digyna) growing on the Arctic tundra.

FIGURE 26.21 Tropical alpine tundra on Mt. Kenya, East Africa, with giant rosettes of *Lobelia*. Plant height tends to increase with elevation.

They also exist above the tree line of the high mountains in tropical regions: Central America, South America, Africa, Borneo, New Guinea, Java, Sumatra, and Hawaii. Tropical alpine vegetation and its environment contrast with those of temperate alpine regions. Tropical alpine regions undergo great seasonal variation in rainfall and cloud cover but little seasonal variation in mean daily temperature. Instead, there is a strong daily fluctuation in temperature, from below-freezing conditions at night to hot, summerlike temperatures during the day. Such a diurnal freeze-thaw cycle is unique to tropical alpine regions.

Tropical alpine tundras support tussock grasses, small-leaved shrubs, and heaths; but the one feature that sets tropical alpine vegetation apart from the temperate alpine is the presence of unbranched or little-branched, giant, treelike rosette plants (Figure 26.21), many of which belong to the family Compositae. Although the genera and species differ among the tropical alpine areas (for example, *Senecio* species in Africa and *Espeletia* species in the Andes), their growth forms and physiology are strikingly similar, suggesting convergent evolution. These species are the antithesis of the usual low-growing tundra species. The higher the elevation, the taller these plants grow, reaching up to 6 meters. Many of these giant rosettes have well-developed water-storing pith in the xylem; retain dead rosette leaves about the stem, apparently as insulation against the cold; secrete a mucilaginous fluid about the bases of leaves that seems to function as a heat-storage device (water has a high heat-storing capacity); and possess dense pubescent hairs that reduce convective loss of heat.

Arctic plants propagate themselves almost entirely by vegetative means, although viable seeds many hundreds of years old exist in the soil. Alpine plants, including those of the tropics, propagate mostly by seeds. The short-lived adventitious roots of Arctic plants are short and parallel to the rhizomes. Adventitious roots of alpine plants are long-lived, long, and unimpeded by permafrost, able to penetrate to considerable depths.

In both Arctic and alpine tundras, topographic location and snow cover delimit a number of plant communities. On the Arctic tundra, steep, south-facing slopes and river bottoms support the most luxuriant and tallest shrubs, grasses, and legumes, whereas cotton grass dominates the gentle north-facing and south-facing slopes, reflecting higher air and soil temperatures and greater snow depth. Pockets of heavy snow create two types of plant habitat in both Arctic and alpine tundra, the snow patch and the snow bed. Snow-patch communities occur where wind-driven snow collects in shallow depressions and protects the plants beneath. Snow beds are found where large masses of snow accumulate because of topographic peculiarities. Not only does the deep snow protect the plants beneath, but the meltwater from the slowly retreating snowbank provides a continuous supply of water throughout the growing season. Snow-bed plants have a short growing season, but they break into leaf and flower quickly because of the advanced stage of growth beneath the snow.

Microbiota, the bacteria and fungi, live near the surface and tolerate cold temperatures. Bacteria are active at −7.5°C, and fungal growth continues at 0°C. Fungi, however, can break down plant structural carbohydrates at temperatures below 0°C, whereas aerobic bacteria at 0°C are restricted to nonstructural carbohydrates and products of fungal decomposition. Fungi and aerobic bacteria live mostly in the upper 7 to 10 cm of soil. Below that depth, only anaerobic bacteria exist.

26.19 Low Temperatures and Extreme Seasonality Control Tundra Productivity and Nutrient Cycling

Primary production on the tundra is low, the consequence of low temperature, and a short growing season ranging from 50 to 60 days in the high Arctic to

160 days in the low-latitude alpine tundra. These physical constraints result in a low availability of nutrients that further function to reduce productivity.

Plants are photosynthetically active on the Arctic tundra about 3 months out of the year. As quickly as snow cover disappears, plants start photosynthetic activity, but it is limited initially because plant leaves are poorly developed. Alpine species, however, undergo a rapid burst of growth following snowmelt, at the expense of below-ground root and rhizome carbohydrate reserves.

Arctic plants make maximum use of the growing season and light by carrying on photosynthesis during the 24-hour daylight period, even at midnight when light is one-tenth that of noon. They rarely become light-saturated, and they possess a high leaf area index (0.5 to 1.0). The nearly erect leaves of some Arctic plants permit the almost complete interception of the slanting rays of Arctic sun.

Much of the photosynthate goes into the production of new growth, but about one month before the growing season ends, plants cease to allocate photosynthate to above-ground biomass. They withdraw nutrients from the leaves and move them to roots and below-ground biomass, sequestering 10 times the amount stored by temperate grasslands. In general, alpine tundras are more productive than Arctic tundras.

Structurally, most of the tundra vegetation is underground. Root-to-shoot ratios of vascular plants range from 3:1 to 10:1. Roots are concentrated in the upper soil that thaws during the summer, and above-ground parts seldom grow taller than 30 cm. It is not surprising, then, that the net annual production of above-ground vegetation ranges from 40 to 110 g/m^2, whereas net annual below-ground production ranges from 130 to 360 g/m^2.

Arctic and alpine tundras are short on nutrients because the short growing season, cold temperatures, and low precipitation slow weathering and restrict decomposition. Nutrient cycling has to be conservative and tight, with minimal loss outside the system. Of all the terrestrial ecosystems, the tundra has the smallest proportion of its nutrient capital in live biomass. Dead organic matter functions as the nutrient pool, but most of it is not directly available to plants. To conserve nutrients, vascular plants retain and reincorporate nutrients, especially nitrogen, phosphorus, potassium, and calcium, in their tissues rather than release them to decomposers (retranslocation: see Chapter 6, Section 6.10).

Because the tundra soil does not store available nutrients in any great quantity, plants depend on the release of nutrients from decomposition, the uptake of which is often aided by mycorrhizae. Decomposition is stimulated by rising temperature in spring.

Because 60 percent of the active roots are in the upper 5 cm of soil, the root mass, once thawed, takes up most of the nutrients.

Leaching or removal of nutrients is minimal, occurring mostly at the beginning of the growing season. Melting snow releases nutrients frozen over winter in the litter, excreta of animals, and microbes. Spring rains leach nutrients from dead plant material remaining from the previous growing season. In summer, vascular plants leak nutrients, particularly phosphorus and potassium, from the cuticle of the leaves to the surface, where they are washed off by summer rains. Mosses often capture these nutrients before they reach the soil and release them slowly, thus functioning as a temporary nutrient sink.

A rapid upward movement of nutrients early in the season at the expense of below-ground biomass supports fast shoot growth. Although more nutrients become available as the depth of thaw increases, they usually are in short supply. Tundra plants respond to the shortage, especially of nitrogen and phosphorus, by producing only a small biomass of leaves and stems that is well supplied with nutrients. Six weeks into the growing season, plants start to send nutrients below-ground. As the cold approaches, the above-ground tissues die, and their dead parts add to the accumulation of organic matter. Nutrients leached from the dead leaves are accumulated by mosses or are frozen into place until the following summer's snowmelt.

The two nutrients most limiting are nitrogen and phosphorus. The major sources of nitrogen are precipitation and biological fixation. Nitrogen-fixers are anaerobic and free-living aerobic bacteria; cyanobacteria in soil, water, and the foliage of mosses where they live epiphytically; and lichens. Precipitation adds nearly as much nitrogen. Phosphorus comes from decomposition of organic matter and animal feces.

Animals contribute to the release of nutrients by either stimulating or short-circuiting decomposition. Soil invertebrates consume nearly all of the microbial populations near the surface of the soil. Grazing herbivores, especially lemmings (Figure 26.22), eat large amounts of above-ground production and distribute nutrients across the tundra through their droppings. During a cyclic high, lemmings may consume over 25 percent of the above-ground primary production, or 10 percent of total plant production. They return about 70 percent as feces. Muskoxen, being selective grazers, restrict their feeding activity to certain areas, from which they may remove up to 85 percent of the herbage available. Overall, this removal amounts to only about 15 percent of production. Decomposition of their dung is slow, taking 5 to 12 years. Grazing herbivores also fell standing litter and live plant biomass, improving conditions for decomposers.

FIGURE 26.22 The brown lemming *(Lemmus sibiricus)* peers out of its tunnel beneath tundra vegetation.

FIGURE 26.23 The yellow-bellied marmot *(Marmota flaviventris)* lives in rocky dens in the alpine tundra.

26.20 Low in Diversity, Tundra Animals Are Well Adapted to the Cold

The tundra world holds fascinating animal life even though the diversity of species is low. Invertebrate fauna are concentrated near the surface, where there are abundant populations of segmented whiteworms (Enchytraeidae), collembolas, and flies (Diptera), chiefly craneflies. Summer in the Arctic tundra brings hordes of blackflies, deerflies, and mosquitoes. In alpine regions, flies and mosquitoes are scarce, but collembolas, beetles, grasshoppers, and butterflies are common. Because of ever-present winds, butterflies keep close to the ground; other insects have short wings or no wings at all. Insect development is slow; some butterflies may take two years to mature, and grasshoppers three.

Dominant vertebrates on the Arctic tundra are herbivores, including the lemming, Arctic hare, caribou, and musk-ox. Although caribou have the greatest herbivore biomass, lemmings, which breed through the year, undergo three- to four-year cycles. At their peak, they may reach densities as great as 125 to 250 per hectare, consuming three to six times as much forage as caribou. Arctic hares that feed on willows disperse over the range in winter and congregate in more restricted areas in summer. Caribou are extensive grazers, spreading out over the tundra in summer to feed on sedges. Musk-oxen are more intensive grazers, restricted to more localized areas where they feed on sedges, grasses, and dwarf willow. Herbivorous birds are few, dominated by ptarmigan and migratory geese.

The major Arctic carnivore is the wolf, which preys on the muskox, caribou, and, when they are abundant, lemmings. Medium-sized to small predators include the Arctic fox *(Alopex lagopus)* that preys on the Arctic hare and several species of weasel that prey on lemmings. Also feeding on lemmings are snowy owls *(Nyctea scandiaca)* and the hawklike jaegers *(Stercorarius* spp.). Sandpipers, plovers, longspurs, and waterfowl, which nest on the wide expanse of ponds and boggy ground, feed heavily on insects.

The alpine tundra, which extends upward like islands in mountain ranges, is small in area and contains few characteristic species. The alpine regions of western North America are inhabited by the hay-cutting pika, marmots (Figure 26.23), mountain woodchucks that hibernate over winter, mountain goats (not goats at all but related to the alpine-dwelling chamois of South America), mountain sheep, elk, voles, and pocket gophers. Eurasian alpine mammals include marmots and wild goats. The African alpine tundra is the home of the rock hyrax *(Procavia capensis)*.

Summary

Nature and Structure of Grassland (26.1–26.4) Natural grasslands occupy regions where rainfall is between 250 mm and 800 mm a year, but they are not exclusively climatic. Many exist through the intervention of fire and human activity (**26.1**). Once covering extensive areas of the globe, natural grasslands have shrunk to a fraction of their original size because of conversion to cropland and grazing lands. Conversions of forests into agricultural lands and the planting of hay and pasturelands extended grasslands into once-forested regions. Grasslands consist of sod formers, bunch grass, or both; and both types have a mode of growth that adapts them to grazing and fire. Grasslands vary with climate and geography. Domestic or managed grasslands occupy land once cleared of forests. Native grasslands of North America, influenced by declining precipitation from east to west, consist of tallgrass prairie, mixed-grass prairie, shortgrass plains, and desert grasslands. Eurasia has steppes; South America the pampas and the veldt of southern South Africa (**26.2**). Grassland consists of an ephemeral herbaceous layer that arises from crowns, nodes, and rosettes of plants hugging the ground above and has three or more sublayers, the ground layer, and below-ground root layer. Depending upon their fire history and degree of grazing and mowing, grasslands accumulate a layer of mulch that retains moisture, influences the character of and composition of plant life, and provides shelter and nesting sites for some animals. A highly developed root layer that makes up more than half of the plant extends to a considerable depth into the soil (**26.3**). Grasslands support a diversity of animal life dominated by herbivorous species, both invertebrate and vertebrate. Grasslands once supported herds of large grazing ungulates such as bison in North America and migratory herds of antelope and zebra (**26.4**).

Productivity and Evolution of Grasslands (26.5–26.6) Productivity varies considerably, influenced greatly by precipitation. It ranges from 400 kcal/m^2/yr in semiarid grasslands to 30 times as much in subhumid, domestic, and cultivated grasslands. The bulk of production goes underground to roots (**26.5**). Grasslands evolved under the selective pressure of grazing. Thus, up to a point, grazing stimulates primary production. Although the most conspicuous grazers are large herbivores, the major consumers are invertebrates. The heaviest consumption takes place below-ground, where the dominant herbivores are nematodes. Most of the primary production goes to the decomposers. Essential to cycling of nutrients is mulch. Nutrients are cycled rapidly. A significant quantity goes to roots to be moved above-ground to next year's growth (**26.6**).

Shrubland (26.7–26.12) Shrubby vegetation covers much of the arid and semiarid landscape as well as shrub-dominated, human-induced communities in parts of the temperate regions. Successional shrublands occupy land in transition from grassland to forest (**26.7**). Shrublands characteristically have a densely branched woody structure and low height. The success of shrubs depends upon their ability to compete for nutrients, energy, and space. In semiarid situations, shrubs have numerous competitive advantages, including structural modifications that affect light interception, heat loss, and evaporation (**26.8**). Shrublands, which go by different names in various parts of the world, dominate regions with a mediterranean-type climate in which winters are mild and wet and summers long, hot, and dry. Growth in mediterranean-type shrublands is concentrated at the end of the wet season, when nutrients in solution and a relative abundance of moisture produce a flush of vegetation. Nutrient cycling, especially of nitrogen and phosphorus, is tight. Many plants retranslocate nutrients from leaves to stem and roots before leaf fall. Others concentrate nutrients in litterfall, which plants take up quickly in the wet season (**26.9**). Heathlands, dominated by ericaceous shrubs, are associated with cool to cold temperate climates of northern regions of the Northern Hemisphere and Arctic, subalpine, and alpine regions of the world. Heathlands invariably occur on nutrient-poor soils especially deficient in phosphorus and nitrogen (**26.10**). Shrublands, especially mediterranean-type and successional shrublands, exhibit considerable patchiness. This patchy nature has a direct influence on soil characteristics and nutrient cycling and thus on the distribution and productivity of plants in the herbaceous layer beneath the shrub canopy (**26.11**). Shrublands support a variety of animal life adapted to and dependent on the shrub community (**26.12**).

Deserts (26.13–26.16) Deserts occupy about one-seventh of Earth's land surface and are largely confined to two worldwide belts around the Tropic of Capricorn and the Tropic of Cancer. Deserts result from the climatic patterns of Earth, rain-blocking mountain ranges, and remoteness from oceanic moisture. Two broad types of deserts exist: cool deserts exemplified by the Great Basin of North America, and hot deserts, like the Sahara (**26.13**). Structurally, deserts are very simple, with scattered shrubs, ephemeral plants, and open, stark topography. In this harsh environment, plants and animals have evolved ways of circumventing aridity and high temperatures by becoming either drought-evaders or drought-resistors (**26.14**). Moisture limits primary productivity and rates of decomposition. Important contributors to production are lichens, green algae, and crusts of cyanobacteria. Nutrient cycling is tight, and nitrogen and phosphorus are in short supply (**26.15**). In spite of their aridity, deserts support a diversity of animal life, notably opportunistic herbivorous species and carnivores. The detrital food chain is less important than in other ecosystems. (**26.16**)

Tundra (26.17–26.20) The Arctic tundra that extends beyond the tree line of the far north of the Northern Hemisphere and the alpine tundra of high mountain ranges in the lower latitudes are at once similar and dissimilar.

Both are characterized by low temperature, low precipitation, and a short growing season. Both possess a frost-molded landscape. The Arctic tundra has a perpetually frozen subsurface, the permafrost, which is rare in the alpine tundra (26.17). Both Arctic and alpine tundra possess plant species whose growth forms are low and whose growth rates are slow. Arctic plants require longer periods of daylight than alpine plants. Over much of the Arctic, the dominant vegetation is cotton grass, sedge, and dwarf heaths. In the alpine tundra, cushion and mat-forming plants able to withstand buffeting by the wind dominate exposed sites (26.18). Net primary production is low, and most plant growth occurs underground. In spite of an assemblage of grazing ungulates and rodents, most of the production goes to decomposers. Decomposition, however, is slow, resulting in an accumulation of peat, which locks up nutrient supplies. Nutrient cycling, especially of N and P, is necessarily conservative and tight, operating on small pools of soluble and exchangeable nutrients. Most of the cycled nutrients are concentrated in roots, translocated to growing shoots above-ground early in the season, and replaced by production and subsequent return of nutrients to the roots later in the growing season. Animals aid in the cycling of nutrients by consuming some of the primary production and distributing dung and droppings across the tundra (26.19). The animal community is low in diversity, but unique. Summer in the Arctic brings hordes of insects, proving a rich food source for sandpipers, plovers, and waterfowl. Dominant vertebrates are the lemming, Arctic hare, caribou, and musk-ox. Major carnivores are the wolf, Arctic fox, and snowy owl (26.20).

Review Questions

1. What is the difference between a bunchgrass and a sod grass?

2. What characteristics do all grasslands have in common?

3. Why does the root system assume such importance in the grassland ecosystem?

4. What is the role of mulch in grassland ecosystems?

5. How have grasses adapted to grazing?

6. What characterizes grassland animal life?

7. Contrast the production and function of the above-ground and below-ground components of grassland ecosystems. Why does so much production and nutrient cycling take place below-ground?

8. What characteristics do seral and climax shrublands share?

9. Worldwide, what are the major types of shrublands, and where are they located?

10. Describe mediterranean-type shrublands.

11. What climatic forces lead to deserts? Human influences?

12. In what general ways are plants and animals of the desert adapted to aridity?

13. What is unique about nutrient cycling in the desert ecosystem?

14. What physical and biological features characterize the tundra?

15. How does the alpine tundra differ from the Arctic tundra?

16. What are the major contrasts between alpine tundras of temperate and tropical regions?

17. What is the relationship among permafrost, plant life, and nutrient cycling in the tundra?

18. Where does the bulk of net primary production of the Arctic tundra accumulate, and why?

19. What is the role of herbivores in the tundra ecosystem?

Pond pine *(Pinus serotina)* and a diversity of floating and emergent aquatic plants dominate this area of Cypress swamp in Spring, Charleston, South Carolina.

Freshwater and Estuarine Ecosystems

OBJECTIVES

On completion of this chapter, you should be able to:

- Discuss horizontal and vertical zonation of life in lakes and ponds.

- Describe the physical characteristics of flowing-water ecosystems.

- Compare fast streams with slow streams and rivers.

- Explain the role of detritus in flowing-water ecosystems.

- Discuss the problems that channelization causes.

- Explain the meaning of regulated rivers, and discuss the impact of dams on flowing-water rivers.

- Define an estuary and describe its characteristics.

- Relate freshwater and tidal inputs to the functioning of estuaries.

Of the precipitation that reaches the land surface, that which does not infiltrate into the soil or return to the atmosphere through evaporation and transpiration is called surface water (Chapter 4). Surface water moves along the landscape, following a path determined by gravity and topography. Streams coalesce into rivers as they follow the topography of the landscape. In basins and floodplains, ponds, lakes, and wetlands form. Rivers eventually flow to the coast, forming the transition from freshwater to marine. The area of land that a stream or river drains is its **watershed.** Each watershed is different, characterized by vegetative cover, geology, soils, topography, and land use. Streams and rivers provide the drainage pathways. Ponds, lakes, and wetlands act as the catch basins, or **catchments.**

Freshwater ecosystems are classified on the basis of the depth and flow of water, which in turn influence the vertical profiles of temperature, light, oxygen, and nutrients. Flowing-water, or **lotic,** ecosystems include streams and rivers. Nonflowing water, or **lentic,** ecosystems include ponds, lakes, and inland wetlands. The study of freshwater ecosystems is known as limnology.

27.1 Lakes Have Many Origins

Lakes and ponds are inland depressions containing standing water (Figure 27.1). They vary in depth from 1 meter to over 2000 meters. They range in size from small ponds of less than a hectare to large lakes covering thousands of square kilometers. Ponds are small bodies of water so shallow that rooted plants can grow over much of the bottom. Some lakes are so large that they mimic marine environments. Most ponds and lakes have outlet streams, and both may be more-or-less temporary features on the landscape, geologically speaking.

Some lakes formed by glacial erosion and deposition. Abrading slopes in high mountain valleys, glaciers carved basins that filled with water from rain and melting snow to form tarns. Retreating valley glaciers left behind crescent-shaped ridges of rock debris that dammed up water behind them. Numerous shallow kettle lakes and potholes were left behind by the glaciers that covered much of northern North America and northern Eurasia.

Lakes also form when silt, driftwood, and other debris deposited in beds of slow-moving streams dam up water behind them. Loops of streams that meander over flat valleys and floodplains often become cut off, forming crescent-shaped oxbow lakes.

Shifts in Earth's crust, uplifting mountains or displacing rock strata, sometimes develop water-filled depressions. Craters of some extinct volcanoes have also become lakes. Landslides block streams and valleys to form new lakes and ponds. In any given area, all natural ponds and lakes have the same geological origin and similar characteristics; but because of varying depths at time of origin, they may represent several stages of development.

Many lakes and ponds form through nongeological activity. Beavers dam streams to make shallow, but often extensive, ponds. Humans create huge lakes by damming rivers and streams for power, irrigation, or water storage and construct smaller ponds and marshes for recreation, fishing, and wildlife. Quarries and strip mines form other ponds.

27.2 Lakes Have Well-Defined Physical Characteristics

Unlike most terrestrial ecosystems, lakes and ponds have well-defined boundaries—the shoreline, the sides of the basin, the surface of the water, and the bottom sediments. Within these boundaries, environmental conditions vary from one pond or lake to another. However, all still-water ecosystems share certain characteristics. Life in still-water ecosystems depends on light. The amount of light penetrating the water is influenced not only by natural attenuation (see Chapter 4, Section 4.2), but also by silt and other material carried into the lake and by the growth of phytoplankton. Temperatures vary seasonally and with depth. Oxygen can be limiting, especially in summer, because only a small proportion of the water is in direct contact with air, and decomposition on the bottom consumes it. These variations in oxygen, temperature, and light strongly influence the distribution and adaptations of life in lakes and ponds (see Chapter 4, Sections 4.5 and 4.6).

Ponds and lakes may be divided into both vertical and horizontal strata based on penetration of light and photosynthetic activity (Figure 27.2). The horizontal zones are obvious to the eye; the vertical ones, influenced by depth of light penetration, are not. Surrounding most lakes and ponds and engulfing some ponds completely is the **littoral zone** or shallow-water zone, in which light reaches the bottom, stimulating the growth of rooted plants. Beyond the littoral is open water, the **limnetic zone,** which extends to the depth of light penetration. Inhabiting it are plant and animal plankton and

(a)

(b)

(c)

(d)

(e)

FIGURE 27.1 Lakes and ponds fill basins or depressions in the land. (a) A rock basin glacial lake or tarn in the Rocky Mountains. (b) A swampy tundra in Siberia is dotted with numerous ponds and lakes. (c) An oxbow lake formed when a bend in the river was cut off from the main channel. (d) A beaver dam forms a pond in this Colorado Rocky Mountain meadow. (e) A human-constructed old New England millpond. Note the floating vegetation.

nekton, free-swimming organisms such as fish that can move about freely. Beyond the depth of effective light penetration is the **profundal zone.** Its beginning is marked by the **compensation level** of light, the point at which respiration balances photosynthesis.

The profundal zone depends on a rain of organic material from the limnetic zone for energy. Common to both the littoral and profundal zones is the third vertical stratum, the **benthic zone** or bottom region, which is the place of decomposition. Although these

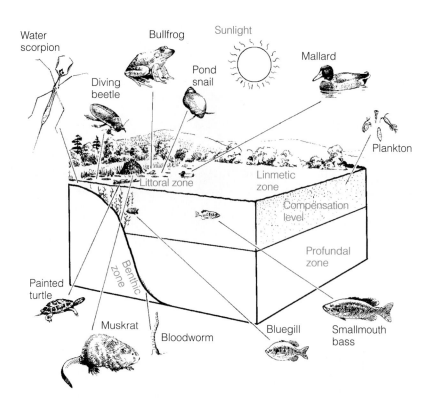

FIGURE 27.2 The major zones of a lake in midsummer: littoral, limnetic, profundal, and benthic. The compensation level is the depth at which light levels are such that gross production in photosynthesis is equal to respiration, so net production (primary) is zero. The organisms shown are typical of the various zones in a lake community.

zones are named and often described separately, all are closely dependent on one another in the dynamics of lake ecosystems.

27.3 The Littoral Zone Supports the Richest Diversity of Life

Aquatic life is richest and most abundant in the shallow water about the edges and in other places within lakes and ponds where sediments have accumulated on the bottom, decreasing water depth. Dominating these areas is emergent vegetation, plants whose roots are anchored in the bottom mud, whose lower stems are immersed in water, and whose upper stems and leaves stand above water (Figure 27.3). The distribution and variety of plants vary with water depth and fluctuation of water levels. Very shallow depths support spike rushes and small sedges; deeper water is occupied by plants with narrow tubular or linear leaves, such as bulrushes, reeds, and cattails. With them are associated such broadleaf emergents as pickerelweed (*Pontederia* spp.) and arrowhead (*Sagittaria* spp.). Beyond the emergents and occupying even deeper water is a zone of floating plants such as pondweed *(Potamogeton)* and pond lily *(Nuphar* spp.). Many of these floating plants have poorly developed root systems but highly developed aerating

systems. In depths too great for floating plants live submerged plants, such as certain species of pondweed. Lacking cuticles, these plants absorb nutrients and gases directly from the water through thin and finely dissected or ribbonlike leaves.

Associated with the emergents and floating plants is a rich community of organisms, among them hydras, snails, protozoans, and sponges. Insects include dragonflies and diving insects such as water boatmen and diving beetles that carry a bubble of air with them when they go underwater in search of prey. Fish such as pickerel and sunfish find shelter, food, and protection among the emergent and floating plants. Fish of lakes and ponds lack strong lateral muscles characteristic of fish living in swift water, and some, such as sunfish, have compressed bodies that permit them to move with ease through the masses of aquatic plants. The littoral zone contributes heavily to the large input of organic matter into the system.

27.4 Planktonic Organisms Dominate Life in the Limnetic Zone

People usually associate the open water of a lake with fish, but the main forms of life in the limnetic zone are minute protistan and animal organisms—phytoplankton and zooplankton (Figure 27.4).

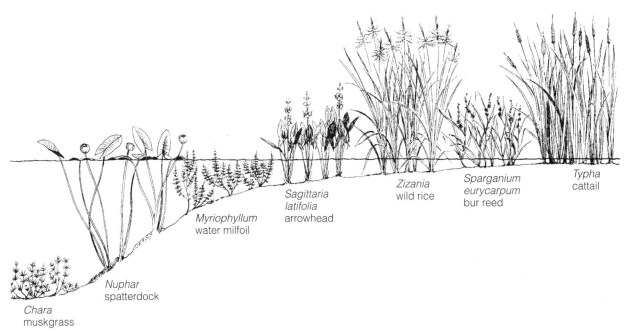

FIGURE 27.3 Zonation of emergent, floating, and submerged vegetation at the edge of a lake or pond. Such zonation does not necessarily reflect successional stages, but is rather a response to water depth (see Chapter 14, Figure 14.13).

Because the tiny protists that make up the phytoplankton—including desmids, diatoms, and filamentous algae—carry on photosynthesis in open water, they are the base on which the rest of life in open water depends. Their presence is noticeable during bloom, the season when the populations of these protists are most abundant. Suspended with the phytoplankton are small animals, mostly tiny crustaceans that graze on minute protists. These animals form an important link in energy flow in the limnetic zone.

Light sets the lower limit at which phytoplankton can survive, so populations of these protists concentrate in the epilimnion. Because the zooplankton feeds on these minute protists, it too is concentrated in the limnetic zone. By its own growth, phytoplankton limits light penetration into the water. As summer progresses, growth lessens the depth at which phytoplankton can live. As the zone becomes shallower, phytoplankton can absorb more light, increasing organic production.

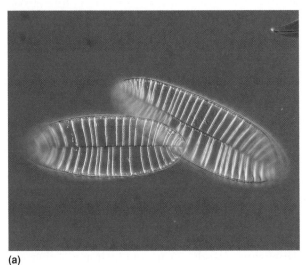

(a)

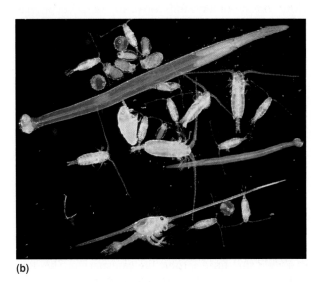

(b)

FIGURE 27.4 (a) Phytoplankton. (b) Zooplankton.

Within the limits of light penetration, the depth at which various species of phytoplankton can live depends on the optimum conditions for their development. Some phytoplankton species live just below the surface; others are more abundant a few feet beneath; and those requiring colder temperatures live deeper still. Cold-water plankton, in fact, is restricted to lakes in which phytoplankton growth is scarce in the epilimnion and in which the oxygen content of the deepwater is not depleted by decomposition of organic matter.

Animal plankton may be seasonally stratified because it is capable of independent movement. In winter, some animal plankton species distribute themselves evenly to considerable depths; in summer, they concentrate in layers most favorable to them and to their stages of development. At that season, zooplankton species undertake a vertical migration during some part of the 24-hour period. Depending on the species, they rest in the deepwater or on the bottom and move up to the surface during the alternate period to feed on phytoplankton.

During the spring and fall overturns (see Chapter 21, Section 21.7), plankton is carried downward, but at the same time nutrients released by decomposition on the bottom are carried upward to the impoverished surface layers. In spring, when surface waters warm and stratification develops, phytoplankton has access to both nutrients and light. A spring bloom develops, followed by a rapid depletion of nutrients and a reduction in planktonic populations, especially in shallow water.

Fish make up most of the nekton in the limnetic zone. Their distribution is influenced mostly by food supply, oxygen, and temperature. During the summer, largemouth bass, pike, and muskellunge inhabit the warmer epilimnion waters where food is abundant. In winter, they retreat to deeper water. Lake trout, on the other hand, move to greater depths as summer advances. During the spring and fall overturn, when oxygen and temperature are fairly uniform throughout, both warm-water and cold-water species occupy all levels.

27.5 Life in the Profundal Zone Is Restricted

Life in the profundal zone depends not only on the supply of energy and nutrients from the limnetic zone above, but also on the temperature and availability of oxygen. In highly productive waters, oxygen may be limiting because the decomposer organisms so deplete it that little aerobic life can survive. The profundal zone of a deep lake is much larger in proportion to total volume, so production of the epilimnion is relatively low, and decomposition does not deplete the oxygen. In these lakes, the profundal zone supports some life, particularly fish, some plankton, and such organisms as certain cladocerans that live in the bottom ooze. Some zooplankton may occupy this zone during some part of the day, but migrate up to the surface to feed. Only during spring and fall overturns, when organisms from the upper layers enter this zone, is life abundant in profundal waters.

Easily decomposed substances drifting down through the profundal zone are partly mineralized while sinking. The remaining organic debris—dead bodies of plants and animals of the open water, and decomposing plant matter from shallow-water areas—settles on the bottom. Together with quantities of material washed in, they make up the bottom sediments, the habitat of benthic organisms.

27.6 The Benthic Zone Is the Site of Decomposition

The bottom ooze is a region of great biological activity, so great, in fact, that the oxygen curves for lakes and ponds show a sharp drop in the profundal water just above the bottom (see Chapter 4, Figure 4.14). Because the organic muck is so low in oxygen, the dominant organisms there are anaerobic bacteria. Under anaerobic conditions, however, decomposition cannot proceed to inorganic end products. When the amounts of organic matter reaching the bottom are greater than can be utilized by bottom fauna, they form an odoriferous muck rich in hydrogen sulfide and methane. Thus, lakes and ponds with highly productive limnetic and littoral zones have an impoverished fauna on the profundal bottom. Life in the bottom ooze is most abundant in lakes with a deep hypolimnion in which some oxygen is still available.

As the water becomes shallower, the benthos changes. The action of water, plant growth, drift materials, and recent organic deposits modifies the bottom material—stones, rubble, gravel, marl, and clay. Increased oxygen, light, and food encourage a richness of life not found on the profundal bottom.

Closely associated with the benthic community are organisms collectively called **periphyton** or **aufwuchs.** They are attached to or move on a submerged substrate, but do not penetrate it. Small aufwuchs communities colonize the leaves of submerged aquatic plants, sticks, rocks, and debris. Periphyton, mostly algae and diatoms, living on plants are fast-growing and lightly attached. Because

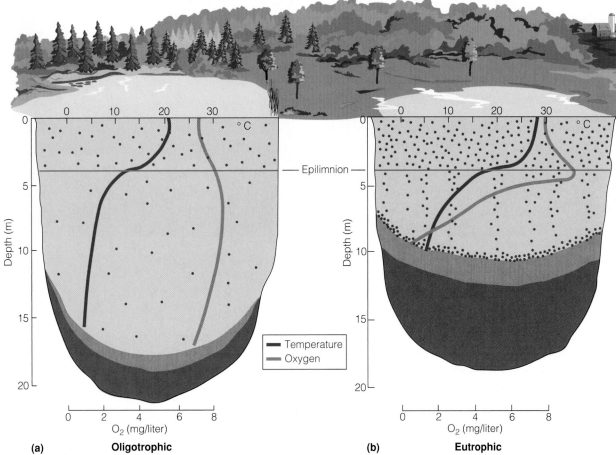

FIGURE 27.5 A comparison of eutrophic and oligotrophic lakes. (a) The oligotrophic lake is deep and has relatively cool water in the epilimnion. The hypolimnion is well supplied with oxygen. Organic matter that drifts to the bottom falls through a relatively large volume of water. The watershed surrounding the lake is largely oligotrophic, dominated by coniferous forests on thin and acidic soils. (b) The eutrophic lake is shallow and warm with a heavy growth of phytoplankton (suggested by blue dots), and oxygen is nearly depleted in the deeper water. The amount of organic detritus is large in relation to the volume of water. The watershed surrounding the lake is eutrophic, consisting of a nutrient-rich deciduous forest and farmland.

the substrate is so short-lived, the associated periphyton rarely lives for more than one summer. Aufwuchs on stones, wood, and debris form a more crustlike growth of cyanobacteria, diatoms, water moss, and sponges.

27.7 The Character of a Lake Reflects Its Surrounding Landscape

Because a close relationship exists between land and water ecosystems, lakes reflect the character of the landscapes in which they reside. Primarily through the hydrological cycle, one feeds on the other. The water that falls on land runs from the surface or moves through the soil to enter streams, springs, and eventually lakes. The water carries with it silt and nutrients in solution. Human activities, including road construction, logging, mining, construction, and agriculture, add another heavy load of silt and nutrients, especially nitrogen, phosphorus, and organic matter. These inputs enrich aquatic systems, a process called **eutrophication.** The term **eutrophy** (from the Greek *eutrophos*, "well nourished") means a condition of being nutrient-rich.

A typical eutrophic lake (Figure 27.5a) has a high surface-to-volume ratio—that is, the surface area is large relative to depth. Nutrient-rich deciduous forest and farmland surround it. An abundance of nutrients, especially nitrogen and phosphorus, flowing into the lake stimulates a heavy growth of algae and

(a)

(b)

FIGURE 27.6 (a) A eutrophic lake. Note the floating algal mats on the water surface. (b) An oligotrophic lake in Maine.

other aquatic plants. Increased photosynthetic production leads to an increased regeneration of nutrients and organic compounds, stimulating even further growth.

Phytoplankton concentrates in the warm upper layer of the water, giving it a murky green cast (Figure 27.6a). The turbidity reduces light penetration and restricts biological productivity to a narrow layer of surface water. Algae, inflowing organic debris and sediment, and remains of rooted plants drift to the bottom, adding to the highly organic sediments. Bacteria partially convert this dead organic matter into inorganic substances. The activities of these decomposers deplete the oxygen supply of the bottom sediments and deepwater to the point at which this region of the lake cannot support aerobic life. The number of bottom species declines, although the biomass and number of organisms remain high.

In contrast to eutrophic lakes and ponds are oligotrophic ones. **Oligotrophy** is the condition of being nutrient-poor. Oligotrophic lakes have a low surface-to-volume ratio (Figure 27.5b). The water is clear and appears blue to blue-green in the sunlight (Figure 27.6b). The epilimnion is cool; the hypolimnion is relatively high in oxygen; and the bottom sediments are largely inorganic. The nutrient content of the water, however, is low; and although nitrogen may be abundant, phosphorus is highly limited. A low input of nutrients from surrounding terrestrial ecosystems and other external sources is mostly responsible for this condition. Typically, coniferous forests on thin, acid soil dominate the watershed.

Low availability of nutrients causes low production of organic matter, particularly phytoplankton.

Low organic matter production leaves little for decomposers, so oxygen concentration remains high in the hypolimnion. These oxidizing conditions are responsible for the low release of nutrients from the sediment. The lack of decomposable organic matter means low bacterial populations and slow rates of microbial metabolism.

Although the numbers of organisms in oligotrophic lakes and ponds may be low, species diversity is often high. Members of the salmon family often predominate within the fish community.

Lakes that receive large amounts of organic matter from surrounding land, particularly in the form of humic materials that stain the water brown, are called **dystrophic** (from *dystrophos*, "ill-nourished"). Although the productivity of dystrophic lakes is considered low, they are low only in planktonic production. Dystrophic lakes generally have highly productive littoral zones, particularly those that develop bog flora. This littoral vegetation dominates the metabolism of the lake, providing a source of both dissolved and particulate organic matter.

Many lakes and ponds receive an inflow of nutrients, especially nitrogen and phosphorus from surrounding housing and industrial developments, runoff from agricultural and urban areas, and other sources. This input can turn an oligotrophic lake with low algal growth into a eutrophic one with considerable growth of algae. It can change an autrophic lake into a **hypertrophic** body of water overloaded with nutrients. This accelerated enrichment causes chemical and environmental changes and major shifts in plant and animal life. We call this process **cultural eutrophication.**

27.8 Velocity Influences the Characteristics of Flowing–Water Communities

Even the largest rivers begin somewhere back in the hinterlands as springs or seepage areas, becoming headwater brooks and streams, or they arise as outlets of ponds or lakes. A very few emerge full-blown from glaciers. As a brook drains away from its source, it flows in a direction and manner dictated by the lay of the land and underlying rock formations. Its course may be determined by the original slope; or water, seeking the least resistant route to lower land, may follow joints and fissures in bedrock near the surface and shallow depressions in the ground. Whatever its direction, water concentrates in rills that erode small furrows, which soon grow into gullies. Moving downstream, especially where the gradient is steep, the moving water carries a load of debris collected from its surroundings that cuts the channel wider and deeper. Sooner or later, the stream deposits this material on its bed or along its banks. In mountainous areas, erosion continues to eat away at the head of the gully, cutting backward into the slope and increasing the drainage area. Joining the new stream are other small streams, spring seeps, and surface water.

Just below its source, the stream may be small, straight, and swift, with waterfalls and rapids. Farther downstream, where the gradient is less, velocity decreases, meanders become common, and the stream deposits its load of sediment as silt, sand, or mud. At flood time, a stream drops its load of sediment on surrounding level land, over which floodwaters spread to form floodplain deposits. These floodplains are a part of a stream or river channel used at the time of high water—a fact few people recognize.

Where a stream flows into a lake or a river into the sea, the velocity of water is suddenly checked. The river then is forced to deposit its load of sediment in a fan-shaped area around its mouth to form a delta. Here its course is carved into a number of channels, which are blocked or opened with subsequent deposits. As a result, the delta becomes an area of small lakes, swamps, and marshy islands. Material the river fails to deposit in the delta is carried out to open water and deposited on the bottom.

Because streams become larger on their course to rivers and are joined along the way by many others, we can classify them according to order (Figure 27.7). A small headwater stream without any tributaries is a first-order stream. When two streams of the same

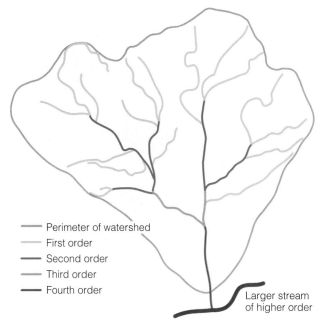

— Perimeter of watershed
— First order
— Second order
— Third order
— Fourth order

Larger stream of higher order

FIGURE 27.7 Stream orders within a watershed.

order join, the stream becomes one of higher order. If two first-order streams unite, the resulting stream becomes a second-order one; and when two second-order streams unite, the stream becomes a third-order one. The order of a stream can increase only when a stream of the same order joins it. It cannot increase with the entry of a lower-order stream. In general, headwater streams are orders 1 to 3; medium-sized streams, 4 to 6; and rivers, greater than 6.

27.9 Flowing–Water Ecosystems Vary in Structure and Types of Habitats

The velocity of a current molds the character and the structure of a stream. The shape and steepness of the stream channel, its width and depth and the roughness of the bottom, and the intensity of rainfall and rapidity of snowmelt affect velocity. Fast streams (Figure 27.8) are those whose velocity of flow is 50 cm per second or higher. At this velocity, the current will remove all particles less than 5 mm in diameter and will leave behind a stony bottom. High water increases the velocity; it moves bottom stones and rubble, scours the streambed, and cuts new banks and channels. As the gradient decreases and the width, depth, and volume of water increase, silt and decaying organic matter accumulate on the

FIGURE 27.8 A fast mountain stream. The gradient is steep, and the bottom is largely bedrock.

FIGURE 27.9 A slow stream is deeper and has a lower slope gradient.

bottom. The character of the stream changes from fast water to slow (Figure 27.9), with an associated change in species composition.

Flowing-water ecosystems often alternate two different but related habitats, the turbulent riffle and the quiet pool (Figure 27.10). Processes occurring in the rapids above influence the waters of the pool, and the waters of the rapids are likewise influenced by events in the pool.

Riffles are the sites of primary production in the stream. Here the periphyton or aufwuchs assume dominance. Periphyton, which occupies a position of the same importance as phytoplankton in lakes and ponds, consists chiefly of diatoms, cyanobacteria, and water moss.

Above and below the riffles are the pools. Here the environment differs in chemistry, intensity of cur-

rent, and depth. Just as the riffles are the sites of organic production, so the pools are the sites of decomposition. They are catch basins of organic materials, for here the velocity of the current is reduced enough to allow part of the load to settle. Pools are the major sites of carbon dioxide production during the summer and fall. That work is necessary for the maintenance of a constant supply of bicarbonate in solution. Without pools, photosynthesis in the riffles would deplete the bicarbonates and result in smaller and smaller quantities of available carbon dioxide downstream.

27.10 Life Is Highly Adapted to Flowing Water

Living in moving water, inhabitants of streams and rivers have a major problem remaining in place and not being swept downstream. They have evolved unique adaptations for dealing with life in the current (Figure 27.11a). A streamlined form, which offers less resistance to water current, is typical of many animals of fast water, such as the dace and the brook trout. The larval forms of many species of insects cling to the undersurfaces of stones, where the current is weak. They possess extremely flattened bodies and broad, flat limbs that allow the current to flow over them. Typical are many species of mayflies and stoneflies. Other forms, such as the blackfly *(Simuliidae)* larvae, attach themselves in one way or another to the substrate, and they obtain food by straining particles carried to them by the current. The larvae of certain species of caddisflies construct protective cases of sand or small pebbles and cement

FIGURE 27.10 Two different but related habitats in a stream: a riffle (background) and a pool (foreground).

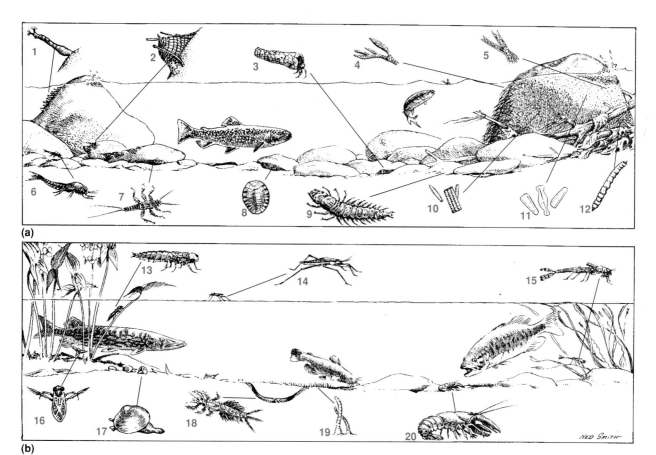

FIGURE 27.11 Life in a fast stream (a) and a slow stream (b). Fast stream: (1) blackfly larva (Simuliidae); (2) net-spinning caddisfly (*Hydropsyche* spp.); (3) stone case of caddisfly; (4) water moss *(Fontinalis)*; (5) algae *(Ulothrix)*; (6) mayfly nymph *(Isonychia)*; (7) stonefly nymph (*Perla* spp.); (8) water penny *(Psephenus)*; (9) hellgrammite (dobsonfly larva, *Corydalis cornuta*); (10) diatoms *(Diatoma)*; (11) diatoms *(Gomphonema)*; (12) cranefly larva (Tipulidae). Slow stream: (13) dragonfly nymph (Odonata, Anisoptera); (14) water strider *(Gerris)*; (15) damselfly larva (Odonata, Zygoptera); (16) water boatman (Corixidae); (17) fingernail clam *(Sphaerium)*; (18) burrowing mayfly nymph *(Hexegenia)*; (19) bloodworm (Oligochaeta, *Tubifex* spp.); (20) crayfish (*Cambarus* spp.). The fish in the fast stream is a brook trout *(Salvelinas fontinalis)*. The fish in the slow stream are, from left to right: northern pike *(Esox lucius)*, bullhead *(Ameiurus melas)*, and smallmouth bass *(Micropterus dolommieu)*.

them to the bottoms of stones. Larvae of net-spinning caddisflies *(Hydropsyche)* firmly attach to stones funnel-shaped food-collecting nets whose open ends face upstream. Sticky undersurfaces help snails and planarians cling tightly and move about on stones and rubble in the current.

Among the plants, water moss *(Fontinalis)* and heavily branched filamentous algae cling to rocks by strong holdfasts. Other algae grow in cushionlike colonies or closely appressed sheets that are covered with a slippery, gelatinous coating and follow the contours of stones and rocks.

All animal inhabitants of fast-water streams require high, near-saturation concentrations of oxygen and moving water to keep their absorbing and respiratory surfaces in continuous contact with oxygenated water. Otherwise, a closely adhering film of liquid impoverished of oxygen forms a cloak about their bodies.

In slow-flowing streams where current is at a minimum, streamlined forms of fish give way to species such as smallmouth bass (Figure 27.11b), shiners, and darters. They trade strong lateral muscles needed in fast current for compressed bodies that enable them to move through beds of aquatic vegetation. Pulmonate snails and burrowing mayflies replace rubble-dwelling insect larvae. Bottom-feeding fish, such as catfish, feed on life in the silty bottom, and back swimmers and water striders inhabit sluggish stretches and still backwaters.

Invertebrate inhabitants fall into four major groups by their feeding habits (Figure 27.12). **Shredders** make up one large group of insect larvae. Among these shredders are craneflies *(Tipulidae)*,

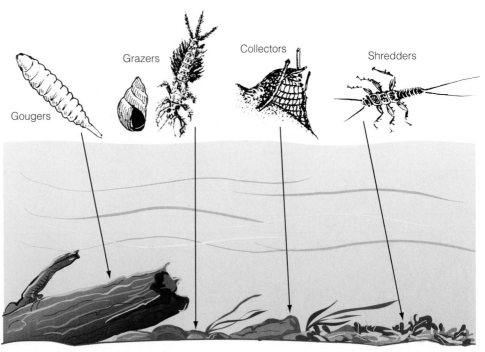

FIGURE 27.12 The four major feeding groups within the stream community: shredders, collectors, grazers, and gougers. Processing the leaves and other particulate matter are bacteria and fungi.

caddisflies *(Trichoptera)*, and stoneflies *(Plecoptera)*. They feed on coarse particulate organic matter (CPOM), mostly leaves that fall into the stream and are softened by water and colonized by bacteria and fungi. The shredders break down the CPOM, feeding on the material not so much for the energy it contains but for the bacteria and fungi growing on it. Shredders assimilate about 40 percent of the material they ingest and pass off 60 percent as feces.

Broken up by the shredders and partially decomposed by microbes, the leaves, along with invertebrate feces, become part of the fine particulate organic matter, shortened to FPOM. Drifting downstream and settling on the stream bottom, FPOM is picked up by another feeding group of stream invertebrates, the **filtering** and **gathering collectors.** The filtering collectors include, among others, the larvae of blackflies *(Simuliidae)*, with filtering fans, and net-spinning caddisflies. Gathering collectors, such as the larvae of midges, pick up particles from stream bottom sediments. Collectors obtain much of their nutrition from bacteria associated with the fine detrital particles.

While shredders and collectors feed on detrital material, another group feeds on the algal coating of stones and rubble. These are the **grazers,** which include the beetle larvae, water penny *(Psephenus* spp.), and a number of mobile caddisfly larvae. Much of the material they scrape loose enters the drift as FPOM. Another group, associated with woody

debris, are the **gougers,** invertebrates that burrow into water-logged limbs and trunks of fallen trees.

Feeding on the detrital feeders and grazers are predaceous insect larvae such as the powerful dobsonfly larvae *(Corydalus cornutus)* and fish such as the sculpin *(Cottus)* and trout. Even these predators do not depend solely on aquatic insects; they also feed heavily on terrestrial invertebrates that fall or wash into the stream.

Because of current, quantities of CPOM, FPOM, and invertebrates tend to drift downstream to form a traveling benthos. This drift is a normal process in streams, even in the absence of high water and abnormal currents. Drift is so characteristic of streams that a mean rate of drift can serve as an index of the production rate of a stream.

27.11 The Flowing-Water Ecosystem Is a Continuum of Changing Environments

From its headwaters to its mouth, the flowing-water ecosystem is a continuum of changing environmental conditions (Figure 27.13). Headwater streams

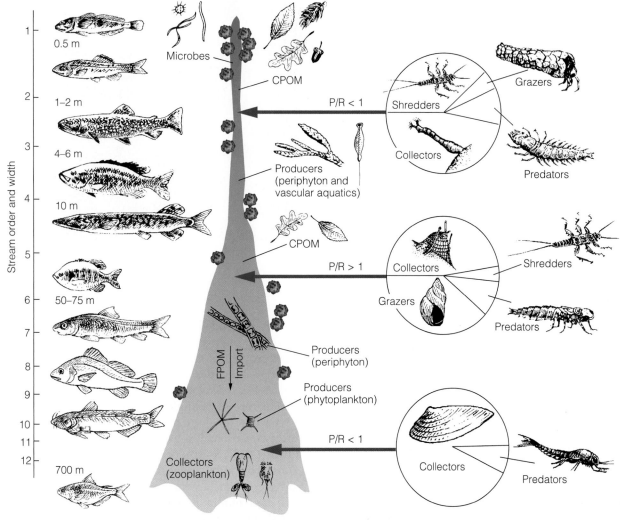

FIGURE 27.13 Changes in consumer groups along the river continuum. Stream order and width (m) are shown on the axis to the left of the figure. The headwater stream is strongly heterotrophic, dependent on terrestrial input of detritus. The dominant consumers are shredders and collectors. As stream size increases, the input of organic matter shifts to primary production by algae and rooted vascular plants. The major consumers are now collectors and grazers. As the stream grows into a river, the lotic system shifts back to heterotrophy. A phytoplankton population may develop. The consumers are mostly bottom-dwelling collectors. The fish community likewise changes as you move downstream (from top to bottom as shown: sculpin, darter, brook trout, smallmouth bass, pickerel, sunfish, sucker, freshwater drum, catfish, and shad).

(orders 1 to 3) are usually swift, cold, and in shaded forested regions. They are strongly heterotrophic, heavily dependent on the input of detritus from terrestrial streamside vegetation, which contributes more than 90 percent of the organic input. Even when headwater streams are exposed to sunlight and autotrophic production exceeds heterotrophic inputs, organic matter produced invariably enters detrital food chains. Dominant organisms are shredders, processing large-sized litter and feeding on CPOM, and collectors, processors of FPOM. Populations of grazers are minimal, reflecting the small amount of autotrophic production, and predators are mostly small fish—sculpins, darters, and

trout. Headwater streams, then, are accumulators, processors, and transporters of particulate organic matter of terrestrial origin. As a result, the ratio of gross primary production to community respiration is less than 1.

As streams increase in width to medium-sized creeks and rivers (orders 4 to 6), the importance of riparian vegetation and its detrital input decreases. Exposed to the Sun, water temperature increases; and as the gradient declines, the current slows. These changes bring about a shift from dependence on terrestrial input of particulate organic matter to primary production by algae and rooted aquatic plants. Gross primary production now exceeds community respiration. Because of the lack of CPOM, shredders disappear; collectors, feeding on FPOM transported downstream, and grazers, feeding on autotrophic production, become the dominant consumers. Predators show little increase in biomass but shift from cold-water species to warm-water species, including bottom-feeding fish such as suckers and catfish.

As the stream order increases from 6 through 10 and higher, riverine conditions develop. The channel is wider and deeper. The volume of flow increases, and the current becomes slower. Sediments accumulate on the bottom. Both riparian and autotrophic production decrease with a gradual shift back to heterotrophy. A basic energy source is FPOM utilized by bottom-dwelling collectors, now the dominant consumers. However, slow, deep water and dissolved organic matter (DOM) support a minimal phytoplankton and associated zooplankton population.

Throughout the downstream continuum, the community capitalizes on upstream feeding inefficiency. Downstream adjustments in production and the physical environment are reflected in changes in consumer groups. Through the continuum, the ecosystem achieves some balance between the forces of stability, such as natural obstructions in flow, and the forces of instability, such as flooding, drought, and temperature fluctuations.

27.12 By Regulating Flows, Dams Interfere with the River Continuum

Dams constructed across rivers and steams interrupt and regulate the natural flows of rivers and streams (Figure 27.14), which profoundly affects the river's hydrology, ecology, and biology. Dams change the environment in which lotic organisms live, more often than not to their detriment.

Under normal conditions free-flowing streams and rivers experience seasonal fluctuation in flow. Snowmelt and early spring rains bring scouring high water; summer brings low water levels that expose some of the streambed and speed decomposition of organic matter along the edges. Life has adapted to these seasonal changes. Damming a river or stream interrupts both nutrient spiraling and the river continuum.

Downstream flow is greatly reduced as a pool of water fills behind the dam, developing characteristics similar to those of a natural lake, yet retaining some features of the lotic system, such as a constant inflow of water. Heavily fertilized by decaying material on

FIGURE 27.14 Dammed river.

the newly flooded land, the lake develops a heavy bloom of phytoplankton and, in tropical regions, dense growths of floating plants. Species of fish, often introduced exotics adapted to lakelike conditions, replace fish of flowing water.

The type of pool allowed to develop depends on the purpose of the dam and has a strong effect on downstream conditions. Single-purpose dams serve only for flood control or water storage; multipurpose dams provide hydroelectric power, irrigation water, and recreation, among other uses. A flood-control dam has a minimum pool; the dam fills only during a flood, at which time inflow exceeds outflow. Engineers release the water slowly to minimize downstream flooding. In time, the water behind the dam recedes to the original pool depth. During and after floods, the river below carries a strong flow for some time, scouring the riverbed. During normal times, flow below the dam is stabilized. If the dam is for water storage, the reservoir holds its maximum pool; but during periods of water shortage and drought, drawdown of the pool can be considerable, exposing large expanses of shoreline for a long time and stressing or killing littoral life. Only a minimal quantity of water is released downstream, usually an amount required by law, if such exists. Hydroelectric and multipurpose dams hold a variable amount of water, determined by consumer needs. During periods of power production, pulsed releases are strong enough to wipe out or dislodge benthic life downstream, which under the best of conditions has a difficult time becoming established.

Reservoirs with large pools of water become stratified, with well-developed epilimnions, metalimnions, and hypolimnions. If water is discharged from the upper layer of the reservoir, the effect of the flow downstream is similar to that of a natural lake. Warm, nutrient-rich, well-oxygenated water creates highly favorable conditions for some species of fish below the spillway and on downstream. If the discharge is from the cold hypolimnion, downstream receives cold, oxygen-poor water carrying an accumulation of iron and other minerals and a concentration of soluble organic materials. Such conditions inimical to stream life may persist for hundreds of kilometers downstream before the river reaches anything near normal conditions. Gated selective withdrawal structures or induced artificial circulation to increase oxygen concentration reduce such problems at some dams.

The impact of dams is compounded when a series of multipurpose dams is built on a river. The amount of water released and moving downstream becomes less with each dam, until eventually all available water is consumed and the river simply dries up. That is the situation on the Colorado River, the most regulated river in the world. The river is nearly dry by the time it reaches Mexico.

Associated with dam-building is stream and river channelization, the dredging and straightening of streams and rivers for flood control, navigation, and agricultural development. Thousands of kilometers of meandering, productive, fish-filled streams and rivers have been channeled into sterile, unattractive drainage ditches. Such channelization has eliminated streamside vegetation and important wildlife habitat and has destroyed associated wetlands (see Chapter 29). Newly cut channels support little bottom fauna, and fish lack food, shelter, and breeding sites. Paradoxically, channelization, designed to speed water from the uplands to the rivers and sea, actually intensifies downstream flooding because it increases the volume and rapidity of flow.

27.13 Rivers Flow into the Sea, Forming Estuaries

Waters of most streams and rivers eventually drain into the sea; and the place where this freshwater joins saltwater is called an **estuary**. Estuaries are semi-enclosed parts of the coastal ocean where seawater is diluted and partially mixed with water coming from the land (Figure 27.15, also Figure 21.12). Estuaries differ in size, shape, and volume of water flow, all influenced by the geology of the region in which they occur.

As the river reaches the encroaching sea, sediments drop in the quiet water. They accumulate to form deltas in the upper reaches of the mouth and shorten the estuary. When silt and mud accumula-

FIGURE 27.15 Estuary on the east coast of Australia.

tions become high enough to be exposed at low tide, tidal flats develop. These flats divide and braid the original channel of the estuary. At the same time, ocean currents and tides erode the coastline and deposit material on the seaward side of the estuary, also shortening the mouth. If more material is deposited than is carried away, barrier beaches, islands, and brackish lagoons appear (Figure 27.16).

The one-way flow of streams and rivers into the estuary meets the inflowing and outflowing tides. This meeting sets up a complex of currents that varies with the season, amount of rainfall, tidal oscillations, and winds. The interaction of freshwater and saltwater influences the salinity of the estuarine environment (see Chapter 4, Section 4.17).

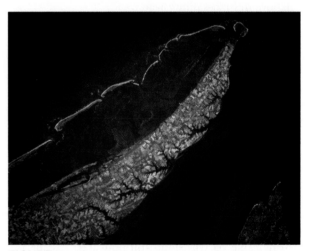

FIGURE 27.16 Color infrared, southeast-looking image (October 1995) of Cape Charles and the southern portion of the Delmarva Peninsula, Virginia. The western shore of the narrow peninsula facing Chesapeake Bay is irregular and marshy. The eastern shore facing the Atlantic Ocean is fairly straight with sandy beaches and offshore sandy bars and small islands. Waves and winds from Atlantic Ocean storms are constantly changing the shape of these barrier islands.

Temperatures in estuaries fluctuate considerably, both daily and seasonally. Waters are heated by the Sun, as well as by inflowing and tidal currents. High tide on the mudflats may heat or cool the water, depending on the season. The upper layer of estuarine water may be cooler in winter and warmer in summer than the bottom, a condition that, as in a lake, will cause spring and autumn overturns.

Mixing waters of different salinities and temperatures creates a countercurrent flow that works as a nutrient trap (see Chapter 21, Section 21.9; also Figure 21.12). Inflowing river waters more often than not impoverish rather than fertilize the estuary, except for phosphorous. Instead, nutrients and oxygen are carried into the estuary by the tides. If vertical mixing takes place, these nutrients are not swept back out to sea but circulate up and down among organisms, water, and bottom sediments.

27.14 Estuarine Organisms Encounter Both Flowing Water and Salinity

Organisms inhabiting the estuary face two problems—maintenance of position and adjustment to changing salinity. Most estuarine organisms are benthic. They attach to the bottom, bury themselves in the mud, or occupy crevices and crannies about sessile organisms. Mobile inhabitants are mostly crustaceans and fish, largely young of species that spawn offshore in high-salinity water.

Planktonic organisms are wholly at the mercy of the currents. Because the seaward movements of stream flow and ebb tide transport plankton out to sea, the rate of circulation or flushing time determines the size of the plankton population. If the circulation is too vigorous, the plankton population may be small. Phytoplankton in summer are mostly near the surface and in low-salinity water. In winter, phytoplankton are more uniformly distributed. For plankton to become endemic in an estuary, reproduction and recruitment must balance loss from physical dispersal.

Salinity dictates the distribution of life in the estuary. Essentially, the organisms of the estuary are marine, able to withstand full seawater. Except for anadromous fish, no freshwater organisms live there. Some estuarine inhabitants cannot withstand lowered salinities, and these species decline along a salinity gradient. Sessile and slightly motile organisms have an optimum salinity range within which they

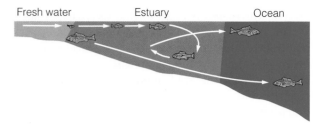

FIGURE 27.17 The relationship of a semi-anadromous fish, the striped bass, to the estuary. Adults live in the marine environment, but young fish grow up in the estuary.

grow best. When salinities vary on either side of this range, populations decline.

Size influences the range of motile species within estuarine waters, particularly fish. Some species, such as the striped bass *(Morone saxatilis),* spawn near the interface of fresh and low-salinity water (Figure 27.17). The larvae and young fish move downstream to more saline waters as they mature. Thus, for the striped bass the estuary serves as both a nursery and as a feeding ground for the young. Anadromous species such as the shad *(Alosa)* spawn in freshwater, but the young fish spend their first summer in the estuary, then move out to the open sea. Species such as the croaker (Sciaenidae) spawn at the mouth of the estuary, but the larvae are transported upstream to feed in plankton-rich, low-salinity areas. Still others, such as the bluefish *(Pomatomus saltatrix),* move into the estuary to feed. In general, marine species drop out toward freshwater and are not replaced by freshwater forms. In fact, the mean number of species decreases progressively from the mouth of the estuary to upstream.

Salinity changes often affect larval forms more severely than adults. Larval veligers of the oyster drill *Thais* succumb to low salinity more easily than the adults. A sudden influx of freshwater, especially after hurricanes or a heavy rainfall, sharply lowers the salinity and causes a high mortality of oysters and their associates.

The oyster bed and oyster reef are the outstanding communities of the estuary. The oyster is the dominant organism about which life revolves. Oysters may be attached to every hard object in the intertidal zone, or they may form reefs, areas where clusters of living organisms grow cemented to the almost buried shells of past generations. Oyster reefs usually lie at right angles to tidal currents, which bring planktonic food, carry away wastes, and sweep the oysters clean of sediment and debris. Closely associated with oysters are encrusting organisms such as sponges, barnacles, and bryozoans, which attach themselves to oyster shells and depend on the oyster or algae for food.

In shallow estuarine waters, rooted aquatics such as widgeon grass and eelgrass *(Zostera marina)* assume major importance. These aquatic plants are complex systems supporting a large number of epiphytic organisms. Such communities are important to certain vertebrate grazers, such as Brant and Canada geese, the black swan in Australia, and sea turtles, and they provide a nursery ground for shrimp and bay scallops.

CHAPTER REVIEW

Summary

Lakes Defined (27.1–27.2) Lakes and pond ecosystems are bodies of water that fill a depression in the landscape. They are formed by many processes ranging from glacial and geological to human activities. Geologically speaking, lakes and ponds are ephemeral features. In time, most of them fill, get smaller, and may finally be replaced by a terrestrial ecosystem (27.1). A nearly self-contained ecosystem, a lake exhibits both vertical and horizontal gradients. Seasonal stratifications in light, temperature, and dissolved gases influence the distribution of life in a lake (27.2).

Zonation of Life in Lakes (27.3–27.6) The area where light penetrates to the bottom of the lake, a zone called the littoral, is occupied by rooted plants (27.3). Beyond this is the open-water or limnetic zone inhabited by plant and animal plankton and fish (27.4). Below the depth of effective light penetration is the profundal region, where the diversity of life varies with temperature and oxygen supply (27.5). The bottom or benthic zone is a place of intense biological activity, for here decomposition of organic matter takes place. Anaerobic bacteria are dominant on the bottom beneath the profundal water, whereas the benthic zone of the littoral is rich in decomposer organisms and detritus feeders (27.6).

Nutrient Input into Lakes (27.7) Lakes may be classified as eutrophic or nutrient-rich, oligotrophic or nutrient-poor, and dystrophic, acidic and rich in humic material. The type is strongly influenced by the surrounding landscape in which the lake is situated. Most lakes are subject to cultural eutrophication, which is the rapid addition of nutrients, especially nitrogen and phosphorus from sewage, industrial wastes, and agricultural runoff.

Flowing-Water Habitat (27.8–27.10) Currents and their dependence on detrital material from surrounding terrestrial ecosystems set flowing-water ecosystems apart from other aquatic systems. Current shapes the life in streams and rivers and carries nutrients and other materials downstream (27.8). Flowing-water ecosystems change longitudinally in flow and size from headwater streams to rivers. They may be fast or slow, with stream habitats characterized by a series of riffles and pools (27.9). Organisms well adapted to living in the current inhabit fast-water streams. They may be streamlined in shape, flattened to conceal themselves in crevices and underneath rocks or attach to rocks and other substrates. In slow-flowing streams where current is minimal, streamlined forms of fish are replaced by fish with compressed bodies that can move through aquatic vegetation and by burrowing invertebrates that inhabit the silty bottom. Stream invertebrates fall into four major groups that feed on detrital material: shredders, collectors, grazers, and gougers (27.10).

River Continuum (27.11–27.12) Life in flowing water reflects a continuum of changing environmental conditions from headwater streams to the river mouth. Headwater streams depend on inputs of detrital material. As stream size increases, algae and rooted plants become important energy sources, reflected in changing species composition of fish and invertebrate life. Large rivers depend on fine particulate matter and dissolved organic matter as sources of energy and nutrients. Life there is dominated by filter feeders and bottom-feeding fish (27.11). Interrupting the downstream continuum of rivers are dams that impound water and regulate the flow, completely changing the character of affected streams and rivers. Dams hold back sediments from floodplains and riparian habitats and allow buildup of sediments that normally would have been deposited on the floodplains and riparian habitats, increase and decrease the intensity and volume of flow, and affect fish and invertebrate populations (27.12).

The Estuary (27.13–27.14) In estuaries, where freshwater meets the sea, the nature and distribution of life are determined by salinity (27.13). As salinity declines from the estuary up through the river, so do estuarine fauna, mainly marine species. The estuary serves as a nursery for many marine organisms, particularly a number of commercially important finfish and shellfish, for here the young can develop protected from predators and competing species unable to tolerate lower salinity. The value of estuaries is being compromised by pollution and development (27.14).

Study Questions

1. What is the origin of lakes and ponds?

2. What characterizes the epilimnion, hypolimnion, and metalimnion of lakes?

3. What distinguishes the littoral zone from the limnetic, and the limnetic from the profundal?

4. What conditions distinguish the benthic zone from the other strata, and what is its role in the lake ecosystem?

5. Distinguish among oligotrophy, eutrophy, and dystrophy.

6. What is cultural eutrophication?

7. What physical characteristics are unique to flowing-water ecosystems?

8. In what way are organisms adapted to flowing water? How do adaptations change as fast streams become slow?

9. Characterize the major feeding groups of stream invertebrates and their roles in the food web.

10. How do downstream systems relate to upstream systems?

11. What is meant by a river continuum?

12. What are regulated rivers? How do they affect the river continuum?

13. What types of dams exist in your area? What have been their ecological and economic benefits and costs? Describe the downstream conditions.

14. Select one major dam—for example, the dams on the Columbia River, the Glen Canyon Dam, the Aswan High Dam in Egypt, the Ord River Dam of tropical Australia, the Kariba Dam on the Zambezi River, or the large dams on the River Volga. How is it harmful? How is it helpful?

15. What is an estuary?

16. How does tidal overmixing work?

17. How is the estuary a nutrient trap?

18. Why is the estuary so vulnerable to pollution?

CHAPTER 28

Marine life at the Great Barrier Reef.

Marine Ecosystems

OBJECTIVES

On completion of this chapter, you should be able to:

- Describe the major zones in the sea and their relationship to temperature stratification.

- Discuss how the seasonal dynamics of the thermocline influences geographic differences in the productivity of the oceans.

- Describe the major features and zonation of rocky, sandy, and muddy shores and coral reefs.

- Discuss the adaptations of life to an intertidal environment.

- Explain the role of disturbance in intertidal and coral reef ecosystems.

- Describe the formation of coral reefs.

- Explain the symbiotic relationship between coral anthozoans and algae and its ecological significance.

Freshwater rivers eventually empty into the oceans, and terrestrial ecosystems end at the edge of the sea. For some distance, there is a region of transition. Rivers enter the saline waters of the ocean, creating a gradient of salinity. That gradient provides a habitat for organisms uniquely adapted to the half-world between saltwater and freshwater (Chapter 27, Sections 27.13 and 27.14). The coastal regions exposed to the open sea are inhabited by other organisms able to live in the often-severe environments dominated by tides. Beyond lies the open ocean—the shallow seas overlying continental shelves and the deep oceans.

The marine environment is marked by a number of differences from the freshwater world. It is large, occupying 70 percent of Earth's surface, and it is deep, in places nearly 10 km. The surface area lighted by the Sun is small compared to the total volume of water. This small volume of sunlit water and the dilute solution of nutrients limit primary production. All of the seas are interconnected by currents, dominated by waves, influenced by tides, and characterized by salinity (see Chapter 4, Sections 4.7 and 4.9).

28.1 Oceans Exhibit Zonation and Stratification

Just as lakes exhibit stratification and zonation, so do the seas. The ocean itself has two main divisions: the **pelagic,** or whole body of water, and the **benthic zone,** or bottom region (Figure 28.1). The pelagic is further divided into two provinces: the **neritic province,** water that overlies the continental shelf,

and the **oceanic province.** Because conditions change with depth, the pelagic is divided into three vertical layers or zones. From the surface to about 200 meters is the **photic zone,** in which there are sharp gradients in illumination, temperature, and salinity. From 200 to 1000 meters is the **mesopelagic zone,** where little light penetrates and the temperature gradient is more even and gradual, without much seasonal variation. It contains an oxygen-minimum layer and often the maximum concentration of nitrate and phosphate. Below the mesopelagic is the **bathypelagic zone,** where darkness is virtually complete, except for bioluminescence; temperature is low, and the pressure is great.

The upper layers of ocean water are thermally stratified (see Chapter 4, Section 4.11, and Chapter 21, Figures 21.7–21.9). Depths below 200 meters are usually thermally stable. In high and low latitudes, temperatures remain fairly constant throughout the year. Polar seas, covered with ice most of the year, exhibit no thermocline in winter, spring, and fall (Figure 28.2). The waters are well mixed, and nutrients are not limiting. Slight stratification takes place in the polar summer (July and August). At that time, the ice melts, the water warms enough, and light is sufficient to support a bloom of phytoplankton.

In tropical seas, the upper waters are well lighted, and the continuous input of solar energy maintains a high temperature throughout the year. Light and temperature are optimal for phytoplankton production, but the waters are permanently stratified. That prevents mixing and upward circulation of nutrients. The result is low productivity (see Figure 28.2).

In temperate seas, thermal structure changes seasonally, reflecting the amount of light and solar thermal energy entering the water. Water in the summer is thermally stratified, with no mixing. In spring and

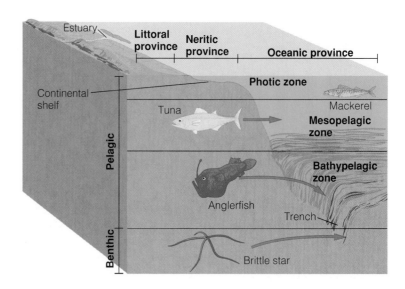

FIGURE 28.1 Major regions of the ocean.

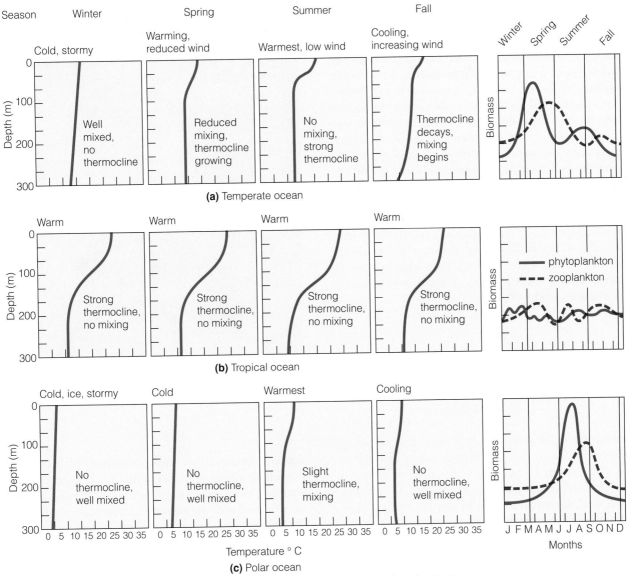

FIGURE 28.2 Thermal profiles, extent of vertical mixing, and associated patterns of productivity in (a) temperate, (b) tropical, and (c) polar oceans during the four seasons of the year. (Nybakken 1988.)

fall, when the surface water warms and cools, respectively, thermal stratification decreases, and the waters mix to varying degrees, recharging nutrients in the surface waters (see Figure 28.2).

28.2 Pelagic Communities Are Hidden Beneath the Surface

Viewed from the deck of a ship or from an airplane, the open sea appears to be monotonously the same. Nowhere can you detect any strong pattern of life or well-defined communities, as you can over land. The reason is that pelagic ecosystems lack the supporting structures and framework of large dominant plant life, and their major herbivores are not large, conspicuous mammals like elephants and deer, but tiny zooplankton.

There is a reason for the smallness of sea plants. Surrounded by a chemical medium that contains in varying quantities the nutrients necessary for life, they absorb their food directly from the water. The smaller the organism, the greater the surface-to-volume ratio. More surface area is exposed for the absorption of nutrients and solar energy. Seawater is so dense that there is little need for supporting structures.

Nevertheless, differences based on physical characteristics and life forms do allow a division of the sea into ecological regions. The Arctic Ocean lies north of the landmass in the Northern Hemisphere and is open to the Atlantic and Pacific oceans via the Bering Strait. It holds novel forms of life, as does the Southern or Antarctic Ocean, which lies around the continent of Antarctica and is open to three oceans, the Atlantic, Pacific, and Indian. Warm oceanic waters making up the Atlantic and Pacific have their own distinctive communities. The deep-sea benthic ecosystems are quite different from the lighted waters. Other marine ecosystems include the coral reefs and upwelling systems off the coasts of California, Peru, northwest Africa, southwest Africa, India, and Pakistan. Important and distinctive are the shelf-sea ecosystems, such as the Georges and Grand Banks of the North Atlantic. Shallow, productive, and nutrient-rich, they support a diversity of fish and invertebrate life. All these regions vary, from oligotrophic to eutrophic and from polar to tropical, while sharing functioning food webs.

28.3 Phytoplankton Dominates Oceanic Surface Waters

Requiring light, phytoplankton is restricted to the upper surface waters. Light penetration varies from tens to hundreds of meters. Because of seasonal, annual, and geographic variations in light, temperature, and nutrients, as well as grazing by zooplankton, the distribution and species composition of phytoplankton vary from ocean to ocean and place to place within them.

Each ocean or region within an ocean appears to have its own dominant forms (Figure 28.3). Littoral and neritic waters and regions of upwelling are richer in plankton than are mid-oceans. In regions of downwelling, the dinoflagellates, a large, diverse group characterized by two whiplike flagellae, concentrate near the surface in areas of low turbulence. They attain their greatest abundance in warmer waters. In summer, they may concentrate in the surface waters in such numbers that they color it red or brown. Often toxic to vertebrates, such concentrations of dinoflagellates are responsible for red tides. In regions of upwelling, the dominant forms of phytoplankton are diatoms. Enclosed in a silica case, diatoms are particularly abundant in Arctic waters.

Smaller than diatoms, the **nanoplankton** make up the largest biomass in temperate and tropical waters. Most abundant are the tiny prochlorophytes. Distributed in all waters except the polar seas, the coccolithophores are a major source of primary pro-

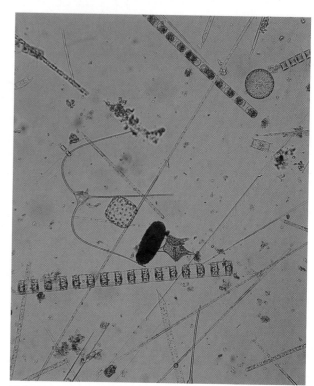

FIGURE 28.3 An array of marine phytoplankton, including diatoms and dinoflagellates.

duction. Because they have calcareous plates and a threadlike appendage, coccolithophores can swim. Droplets of oil aid in buoyancy and serve as a means of storing food. In equatorial currents in shallow seas, the concentration of phytoplankton is variable. Where both lateral and vertical circulation of water is rapid, the composition reflects in part the ability of the species to grow, reproduce, and survive under local conditions.

28.4 Dominant Herbivores Are Zooplankton

Carbohydrate production by photosynthetic nanoplankton is the base on which the life of the seas rests. Conversion of primary production into animal tissue is the work of herbivorous zooplankton, the most important of which are the copepods. To feed on the minute phytoplankton, most of the grazing herbivores must also be small, between 0.5 and 5.0 mm. In the oceans, most of the grazing herbivores are ciliate protozoans and members of the genera *Calanus*, *Acartia*, *Temora*, and *Metridia*, probably the most abundant animals in the world. The single most abundant copepods are *Calanus finmarchicus* and its close

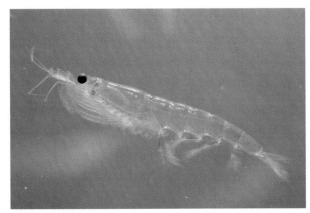

FIGURE 28.4 Small euphausiid shrimps called krill are eaten by baleen whales and are essential to the marine food chain.

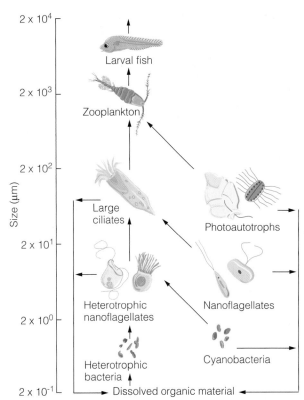

FIGURE 28.5 Diagrammatic representation of the microbial loop and its relationship to the plankton food web. Autotrophs are on the right side of the diagram and heterotrophs on the left (After T. Fenchel 1988).

relative *C. helgolandicus.* In the Antarctic, shrimplike euphausiids, or krill, fed on by baleen whales and penguins, are the dominant herbivores (Figure 28.4). Feeding on the herbivorous zooplankton is the carnivorous zooplankton, which includes such organisms as the larval forms of comb jellies *(Ctenophora)* and arrowworms *(Chaetognatha).* The herbivorous copepods, then, are the link in the food web between the phytoplankton and the second-level consumers.

However, part of the food web begins not with the phytoplankton, but with organisms even smaller. Bacteria and protists, both heterotrophic and photosynthetic, make up one-half of the biomass of the sea and are responsible for the largest part of energy flow in pelagic systems. Photosynthetic nanoflagellates (2–20 μm) and cyanobacteria (1–2 μm) are responsible for a large part of photosynthesis in the sea. These cells excrete a substantial fraction of their photosynthate in the form of dissolved organic material that heterotrophic bacteria use. Populations of such bacteria are dense, around 1 million cells per milliliter of seawater. These bacteria account for about 20 percent of primary production. Bacterioplankton have a high assimilation efficiency, up to 60 percent, as compared to the efficiency rate of 10 to 20 percent for larger organisms. Combined with their large population numbers, this means that bacteria are converting a large amount of dissolved organic carbon (DOC) into particulate organic carbon (POC) in the form of bacterial cells. The bacteria, in turn, are consumed by different heterotrophic nanoplankton organisms and are therefore channeling a significant amount of energy into the phytoplankton food webs. This has been termed the **microbial loop** (Figure 28.5).

The composition of zooplankton varies. Zooplankton fall into two main groups based on size. These are the larger net zooplankton (so called because they cannot pass through plankton nets) and the smaller microzooplankton. Zooplankton of the continental shelf contain a large portion of the larvae of fish and benthic organisms. They include a greater diversity of species, reflecting a greater diversity of environmental and chemical conditions. The open ocean, more homogeneous and nutrient-poor, supports a less diverse zooplankton. Zooplanktonic species of polar and temperate waters, having spent the winter in a dormant state in the deepwater, rise to the surface during short periods of diatom blooms to reproduce. In temperate regions, distribution and abundance depend on the temperature of the water. In tropical regions where temperature is nearly uniform, zooplankton are not so restricted, and reproduction occurs throughout the year.

Like phytoplankton, zooplankton live mainly at the mercy of the currents; but possessing sufficient swimming power, many forms of zooplankton exercise some control. Some species migrate vertically each day to arrive at a preferred level of light intensity. As darkness falls, zooplankton rapidly rise to the surface to feed on phytoplankton. At dawn, they move back down to preferred depths.

28.5 Nekton Comprise Swimming Organisms

Feeding on zooplankton and passing energy along to higher trophic levels are the nekton, swimming organisms that can move at will in the water column. They range in size from small fish to large predatory sharks and whales, seals, and marine birds such as penguins. Some of the predatory fish, such as tuna, are more or less restricted to the photic zone. Others are found in the deeper mesopelagic and bathypelagic zones or move between them, as the sperm whale does. Although the size of prey typically increases with the predator size (see Chapter 15), some of the largest nekton organisms in the sea, the baleen whales, feed on disproportionately small prey, euphausiids or krill (see Figure 28.4). By contrast, the sperm whale attacks very large prey, the giant squid.

Living in a world that lacks refuges against predation or sites for ambush, inhabitants of the pelagic zone have evolved various means of defense and of securing prey. Among them are the stinging cells of the jellyfish, streamlined shapes that allow speed both for escape and for pursuit, unusual coloration, advanced sonar, a highly developed sense of smell, and social organization involving schools or packs. Some animals, such as baleen whales, have specialized structures that permit them to strain krill and other plankton from the water. Others, such as the sperm whale and certain seals, dive to great depths to secure food. Some phytoplankton has the ability to light up darkened seas, and fish take advantage of that bioluminescence to detect their prey.

Residents of the deep also have special adaptations for securing food. Some, like the zooplankton, swim to the surface to feed by night; others remain in the dimly lit or dark waters. Darkly pigmented and weak-bodied, many of the deep-sea fish depend on luminescent lures, mimicry of prey, extensible jaws, and expandable abdomens (which enable them to consume large items of food). Although most of the fish are small (usually 15 cm or less in length), the region is inhabited by rarely seen large species such as the giant squid. In the mesopelagic region, bioluminescence reaches its greatest development—two-thirds of the species produce light. Fish have rows of luminous organs along their sides and lighted lures that enable them to bait prey and recognize other individuals of the same species. Bioluminescence is not restricted to fish. Squid and euphausiids possess searchlightlike structures complete with lens and iris; and squid and shrimp discharge luminous clouds to escape predators.

28.6 Benthos Is a World of Its Own

The term *benthic* refers to the floor of the sea, and **benthos** refers to plants and animals that live there. There is a gradual transition of life from the benthos on the rocky and sandy shores to that in the ocean's depths. From the tide line to the abyss, organisms that colonize the bottom are influenced by the substrate. Where the bottom is rocky or hard, the populations consist largely of organisms that live on the surface of the substrate, the **epifauna** and the **epiflora**. Where the bottom is largely covered with sediment, most of the inhabitants, chiefly animals, live within the deposits and are known collectively as the **infauna**. The kind of organism that burrows into the substrate is influenced by particle size. The mode of burrowing is often specialized for a certain type of substrate.

The substrate varies with the depth of the ocean and with the relationship of the benthic region to land areas and continental shelves. Near the coast, bottom sediments derive from the weathering and erosion of land areas along with organic matter from marine life. The sediments of deepwater are fine-textured material that varies with depth and with the types of organisms in overlying waters. Although we call these sediments organic, they contain little decomposable carbon, consisting largely of skeletal fragments of planktonic organisms. In general, with regional variations, organic deposits down to 4000 meters are rich in calcareous matter. Below 4000 meters, hydrostatic pressure causes some forms of calcium carbonate to dissolve. At 6000 meters and lower, sediments contain even less organic matter and consist largely of red clays rich in aluminum oxides and silica.

Within the sediments are layers that relate to oxidation-reduction reactions. The surface, or oxidized layer, yellowish in color, is relatively rich in oxygen, ferric oxides, nitrates, and nitrites. It supports the bulk of benthic animals, such as polychaete worms, bivalves, and copepods, and a rich growth of aerobic bacteria. Below this surface is a grayish transition zone to the black layer, characterized by a lack of oxygen, iron in the ferrous state, nitrogen in the form of ammonia, and hydrogen sulfide. Anaerobic bacteria, chiefly reducers of sulfates, inhabit this layer.

In a world of darkness, no photosynthesis takes place, so the bottom community is strictly heterotrophic (except in vent areas), depending entirely on the organic matter that finally reaches the bottom as a source of energy. In spite of the darkness and depth, the benthic communities support a high diversity of species. In the shallow benthic regions, the recorded number of polychaete worms is over 250 species and of pericarid crustaceans (shrimplike mysidaceans, cumaceans, the small tanaidaceans, and isopods) well over 100. But the deep-sea benthos supports a surprisingly higher diversity. The number in over 500 samples taken from only 50 m^2 was 707 species of polychaetes and 426 species of pericarid crustaceans. Most of the species are small, but among the deep-sea amphipods, isopods, and copepods, a few are large by crustacean standards, reaching 15 to 42 cm in length.

There are several hypotheses about this deep-sea diversity. One attributes it to the lack of widespread disturbance or environmental extremes. The temperature is nearly constant, and the bottom is not stirred by storms, as the shallow bottoms are. Small local disturbances are created as the crustaceans and other bottom dwellers move, creating mounds and pits on the surface, interrupting the smoothness, and adding diversity. Increasing the variety of living conditions is the patchy distribution of food. The benthos depends on the rain of organic matter drifting to the bottom, which is small and scattered. Patches of dead phytoplankton and the bodies of dead whales, seals, birds, fish, and invertebrates all provide a diversity of food for different feeding groups and species. The low input and patchy distribution of food probably mean less interference among species.

Bottom organisms have four feeding strategies. They may filter suspended material from the water, as the stalked cnidarian do; they may collect food particles that settle on the surface of the sediment, as sea cucumbers do; they may be selective or unselective deposit feeders, as the polychaete worms are; or they may be predatory, like the brittle stars and the spiderlike pycnogonids.

Important in the benthic food chain are the bacteria of the sediments. Common where large quantities of organic matter are present, bacteria may reach several tenths of a gram per square meter in the topmost layer of silt. Bacteria synthesize protein from dissolved nutrients, and in turn become a source of protein, fat, and oils for deposit feeders.

ids that heat the surrounding water to 8°C to 16°C, considerably higher than the 2°C ambient water. Since then, oceanographers have discovered similar vents on other volcanic ridges along fast-spreading centers of ocean floor, particularly in the Atlantic and eastern Pacific.

Vents form when cold seawater flows down through fissures and cracks in the basaltic lava floor deep into the underlying crust. The waters react chemically with the hot basalt, giving up some minerals but becoming enriched with others, such as copper, iron, sulfur, and zinc. Heated to a high temperature, the water reemerges through mineralized chimneys rising up to 13 meters above the sea floor. Among the chimneys are white smokers and black smokers (Figure 28.6). White-smoker chimneys rich in zinc sulfides issue a milky fluid under 300°C. Black smokers, narrower chimneys rich in copper sulfides, issue jets of clear water from 300°C to over 450°C that are soon blackened by precipitation of fine-grained sulfur-mineral particles.

Associated with these vents is a rich diversity of unique deep-sea life confined within a few meters of the vent system. They include giant clams, mussels, polychaete worms that encrust the white smokers, crabs, and vestimentiferan worms lacking digestive systems.

The primary producers are chemosynthetic bacteria, which oxidize reduced sulfur compounds such as H_2S to release energy, which they use to form organic matter from carbon dioxide. Primary consumers, the clams, mussels, and worms, filter bacteria from water and graze on bacterial film on rocks. The giant clam *Calyptogena magnifica* and the large vestimentiferan worm *Riftia pachyptila* possess symbiotic chemosynthetic bacteria. These bacteria need reduced sulfide, which the blood of these animals carries. *Riftia* has in its blood a sulfide-binding protein that concentrates sulfide from the environment and transports it to the bacteria. Such concentrations would poison other animals, but the sulfide-bearing protein of the worm and apparently the clam has a high affinity for free sulfides, preventing them from accumulating in the blood and entering the cells.

28.7 Hydrothermal Vents Are Unique Benthic Ecosystems

In 1977 oceanographers first discovered high-temperature deep-sea springs along volcanic ridges in the ocean floor of the Pacific near the Galápagos Islands. These springs vent jets of hydrothermal flu-

28.8 Life on the Seashore Is Complex

Where the land meets the sea, we find the fascinating and complex world of the seashore. Rocky, sandy, or muddy, protected or pounded by incoming swells, all shores have one feature in common: they are alternately exposed and submerged by the tides (see Chapter 4, Section 4.9). Roughly, the region of the

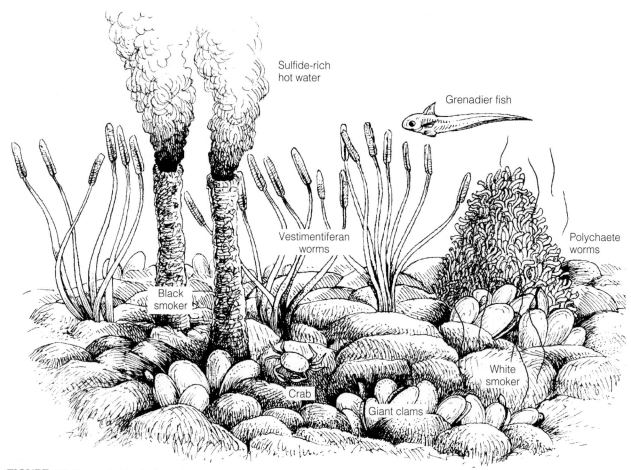

FIGURE 28.6 A typical hydrothermal vent mound, resting on flows of black basaltic lava.

seashore is bounded on one side by the height of extreme high tide and on the other by the height of extreme low tide. Within these confines, conditions change from hour to hour with the ebb and flow of the tides. At flood tide the seashore is a water world; at ebb tide it belongs to the terrestrial environment, with its extremes in temperature, moisture, and solar radiation. In spite of all this change, seashore inhabitants are essentially marine organisms, adapted to withstand some degree of exposure to the air for varying periods of time.

28.9 Tides and Waves Influence Life on Rocky Shores

As the sea recedes at ebb tide, rocks, glistening and dripping with water, begin to appear. Life hidden by tidal water emerges into the open air, layer by layer. The uppermost layers of life are exposed to air, wide temperature fluctuations, intense solar radiation, and desiccation for a considerable period, while the lowest fringes on the intertidal shore may be exposed only briefly before the flood tide submerges them again. These varying conditions result in one of the most striking features of the rocky shore, the zonation of life (Figure 28.7). This zonation differs from place to place as a result of local variations in aspect, substrate, wave action, light intensity, shore profile, exposure to prevailing winds, climatic differences, and the like, but the same general features are always present. All rocky shores have three basic zones, each characterized by dominant organisms (Figure 28.8).

Where the land ends and the seashore begins is hard to determine. The approach to a rocky shore from the landward side is marked by a gradual transition from lichens and other land plants to marine life dependent at least partly on the tidal waters (Figure 28.9). The first major change from land shows up on the **supralittoral fringe,** where saltwater comes only once every fortnight on the spring tides. It is marked by the black zone, a patchlike or beltlike encrustation of *Verrucaria* lichens,

FIGURE 28.7 The broad zones of life exposed at low tide on the rocky shore of the Bay of Fundy in Canada. Note the heavy growth of knotted wrack on the lower portion and the white zone of barnacles above.

the cyanobacteria *Calothrix*, and the green alga *Entophysalis*. Living under conditions that few other plants could survive, these organisms, enclosed in slimy, gelatinous sheaths, and their associated lichens represent an essentially nonmarine community. Common to this black zone are periwinkles, basically marine animals that graze on wet algae covering the rocks. On European shores lives a similarly adapted species, the rock periwinkle, the most highly resistant to desiccation of all the shore animals.

Below the black zone lies the **littoral zone,** covered and uncovered daily by the tides. The littoral tends to be divided into subzones. In the upper

reaches, barnacles are most abundant. The oyster, the blue mussel, and the limpets appear in the middle and lower portions of the littoral, as does the common periwinkle.

Occupying the lower half of the littoral zone (midlittoral) of colder climates and in places overlying the barnacles is an ancient group of plants, the brown algae, commonly known as rockweeds (*Fucus* spp.) and wrack *(Ascophyllum nodosum).* Rockweeds attain their finest growth on protected shores, where they may grow 2 meters long; on wave-whipped shores, they are considerably shorter.

The lower reaches of the littoral zone may be occupied by blue mussels instead of rockweeds, particularly on shores where hard surfaces have been covered in part by sand and mud. No other shore animals grow in such abundance; the blue-black shells packed closely together may blanket the area.

Near the lower reaches of the littoral zone, mussels may grow in association with red algae, *Gigartina,* a low-growing, carpetlike plant. Algae and mussels together often form a tight mat over the rocks. Here, well protected from waves in the dense growth, live infant starfish, sea urchins, brittle stars, and bryozoan sea mats or sea lace *(Membranipora).*

The lowest part of the littoral zone, uncovered only at the spring tides and not even then if wave action is strong, is the **infralittoral fringe.** This zone, exposed for short periods of time, consists of forests of the large brown alga, *Laminaria,* one of the kelps, with a rich undergrowth of smaller plants and animals among the holdfasts.

Beyond the infralittoral fringe is the permanently submerged **infralittoral zone.** This zone is principally

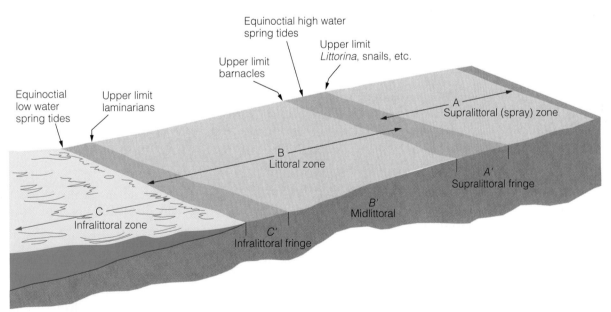

FIGURE 28.8 Basic zonation of Atlantic rocky shores. Refer to this diagram while studying Figure 28.9.

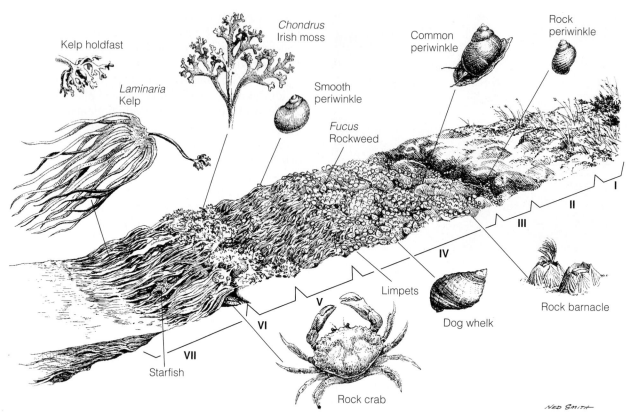

FIGURE 28.9 Zonation on a rocky shore along the North Atlantic. Compare with Figure 28.8. I, land: lichens, herbs, grasses; II, bare rock; III, black algae and rock periwinkle (*Littorina*) zone; IV, barnacle (*Balanus*) zone: barnacles, dog whelks, common periwinkles, mussels, limpets; V, fucoid zone: rockweed (*Fucus*) and smooth periwinkles; VI, Irish moss (*Chondrus*) zone; VII, kelp (*Laminaria*) zone.

neritic and benthic. It contains a wide variety of fauna depending on the substrate, the presence of protruding rocks, gradients in turbulence, oxygen tensions, light, and temperature.

Grazing, predation, competition, larval settlement, and wave action heavily influence the pattern of life on rocky shores. Waves bring inshore a steady supply of nutrients and carry away organic material. They keep the fronds of seaweeds in constant motion, moving them in and out of shadow and sunlight, allowing more even distribution of incident light and thus more efficient photosynthesis. By dislodging organisms, both plants and invertebrates, from the rocky substrate, waves open up space for colonization by algae and invertebrates and reduce strong interspecific competition (see Chapter 13, Section 13.11). Heavy wave action reduces the activity of such predators as starfish and sea urchins that feed on sessile intertidal invertebrates. In effect, disturbance influences community structure (see Chapter 13).

Where heavy waves beat on New England's intertidal rocky shores, periwinkles are rare, permitting a more vigorous growth of algae. Lack of grazing favors ephemeral algal species such as *Ulva* and *Enteromorpha*, whereas grazing allows the perennial *Fucus* to become established. On the New England coast, the mussel *Mytilus edulis* outcompetes barnacles and algae; but predation by the starfish *Asterias* and the snail *Nucella lapillus* prevents dominance by mussels except on the most wave-beaten areas. A similar situation exists on the Washington State coast. Barnacles of several species tend to outcompete and displace algal species, but in turn the dominant mussel *M. californianus* destroys the barnacles by overgrowing them. However, where present, the predatory starfish *Pisaster ochraceus* prevents the mussel from completely overgrowing barnacles. The end result of such interactions of the physical and the biotic is a patchy distribution of life across the rocky intertidal shore.

The ebbing tide leaves behind pools of water in rock crevices, in rocky basins, and in depressions (Figure 28.10). They represent distinct habitats, which differ considerably from exposed rock and the open sea, and even differ among themselves. At low

FIGURE 28.10 Tidal pools fill depressions along this length of rocky coastline in Maine.

tide, all pools are subject to wide and sudden fluctuations in temperature and salinity. Changes are most marked in shallow pools. Under the summer sun, the temperature may rise above the maximum many organisms can tolerate. As water evaporates, especially in the shallower pools, salt crystals may appear around the edges. When rain or land drainage brings freshwater to the pools, salinity may decrease. In deep pools, such freshwater tends to form a layer on top, developing a strong salinity stratification in which the bottom layer and its inhabitants are little affected. If algal growth is considerable, oxygen will be high during the daylight hours but low at night, a situation that rarely occurs at sea. The rise of CO_2 at night lowers the pH.

Pools near low tide are influenced most briefly by the rise and fall of tides; those that lie near and above the high tide line are exposed the longest and undergo the widest fluctuations. Some may be recharged with seawater only by the splash of breaking waves or occasional high spring tides. Regardless of their position on the shore, most pools suddenly return to sea conditions on the rising tide and experience drastic and instantaneous changes in temperature, salinity, and pH. Life in the tidal pools must be able to withstand those fluctuations.

28.10 Sandy and Muddy Shores Are Harsh Environments

Sandy and muddy shores appear devoid of life at low tide, in sharp contrast to the life-studded rocky shore; but the sand and black mud are not as barren as they seem, for beneath them life lurks, waiting for the next high tide.

The sandy shore is a harsh environment; indeed, the matrix of this seaside environment is a product of the harsh and relentless weathering of rock, both inland and along the shore. Through eons, rivers and waves carry the products of rock weathering and deposit them as sand along the edge of the sea. The size of the sand particles deposited influences the nature of the sandy beach, water retention during low tide, and the ability of animals to burrow through it. Beaches with steep slopes are usually made up of larger sand grains and are subject to more wave action. Beaches exposed to high waves are generally flattened, for much of the material is carried away from the beach to deeper water, and fine sand is left behind (Figure 28.11). Sand grains

FIGURE 28.11 A long stretch of sandy beach washed by waves on the southern Australian coast. Although the beach appears barren, life is abundant beneath the sand.

of all sizes, especially the finer particles in which capillary action is the greatest, are more or less cushioned by a film of water, reducing further wearing action. Sand's retention of water at low tide is one of the outstanding environmental features of the sandy shore.

In sheltered areas of the coast, the slope of the beach may be so gradual that the surface appears to be flat. Because of the flatness, the outgoing tidal currents are slow, leaving behind a residue of organic material settled from the water. In these situations, mudflats develop.

28.11 Life on Sandy Shores Is Largely Hidden

Life on the sand is almost impossible. Sand provides no surface for attachment of seaweeds and their associated fauna; and the crabs, worms, and snails characteristic of rocky crevices find no protection there. Life, then, is forced to live beneath the sand.

Life on sandy and muddy beaches consists of epifauna and infauna. Most infauna either occupy permanent or semipermanent tubes within the sand or mud or are able to burrow rapidly into the substrate. Multicellular infauna obtain oxygen either by gaseous exchange with the water through their outer covering or by breathing through gills and elaborate respiratory siphons.

Within the sand and mud live vast numbers of **meiofauna** with a size range between 0.5 mm and 62 μm, including copepods, ostracods, nematodes, and gastrotrichs. Interstitial fauna are generally elon-

gated forms with setae, spines, or tubercles greatly reduced. The great majority of these organisms do not have pelagic larval stages. These animals feed mostly on algae, bacteria, and detritus. Interstitial life, best developed on the more sheltered beaches, shows seasonal variations, reaching maximum development in summer months.

Sandy beaches also exhibit zonation related to the tides (Figure 28.12), but you must discover it by digging. Sandy and muddy shores divide roughly into supralittoral, littoral, and infralittoral zones, based on animal organisms, although a universal pattern similar to that of the rocky shore is lacking. Pale, sand-colored ghost crabs and beach hoppers occupy the upper beach, the supralittoral. The intertidal beach, the littoral, is a zone where true marine life appears. Although sandy shores lack the variety found on rocky shores, the populations of individual species of largely burrowing animals often are enormous. An array of animals, among them starfish and the related sand dollar, can be found above the low-tide line and in the infralittoral.

Organisms living within the sand and mud do not experience the same violent fluctuations in temperature as do those on rocky shores. Although the surface temperature of the sand at midday may be 10°C or more higher than the returning seawater, the temperature a few centimeters below remains almost constant throughout the year. Nor is there a great fluctuation in salinity, even when freshwater runs over the surface of the sand. Below 25 cm, salinity is little affected.

Associated with these essentially herbivorous animals are predators, whether the tide is in or out. Near and below the low-tide line live predatory gastropods, which prey on bivalves beneath the sand. In the same area lurk predatory portunid crabs such as the blue crab and green crab, which feed on mole crabs, clams, and other organisms. They move back and forth with the tides. The incoming tides also bring other predators, such as killifish and silversides. As the tide recedes, gulls and shorebirds scurry across the sand and mudflats to hunt for food.

28.12 Organic Matter Is the Basis of Life on Sandy Shores

For life on the sandy shore to exist, it needs some food base. That food base is an accumulation of organic matter. Most sandy beaches contain a certain amount of detritus from seaweeds, dead animals, feces, and material blown from inland. This organic

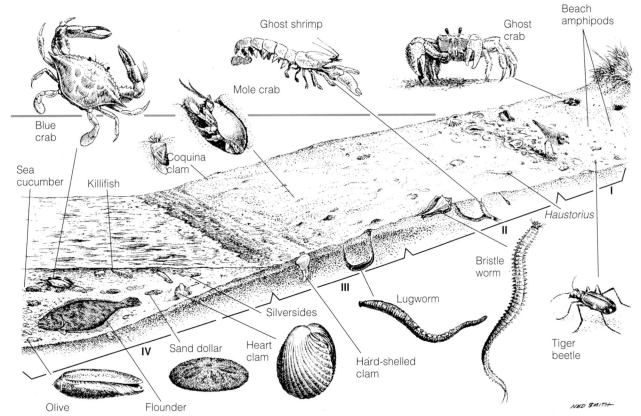

Blue crab

Sea cucumber

Killifish

Coquina clam

Mole crab

Ghost shrimp

Ghost crab

Beach amphipods

Haustorius

Bristle worm

Lugworm

Tiger beetle

Silversides

Hard-shelled clam

Sand dollar

Heart clam

Olive

Flounder

NED SMITH

FIGURE 28.12 Life on a sandy beach along the mid-Atlantic coast. Although strong zonation is absent, organisms still change on a gradient from land to sea. I, supratidal zone: ghost crabs and sand fleas; II, flat beach zone: ghost shrimp, bristle worms, clams; III, intratidal zone: clams, lugworms, mole crabs; IV, subtidal zone: sand dollar, blue crab. The blue line indicates high tide level.

matter accumulates within the sand, especially in sheltered areas. In fact, an inverse relationship exists between the turbulence of the water and the amount of organic matter on the beach, with accumulation reaching its maximum on the mudflats.

Organic matter clogs the spaces between the grains of sand and binds them together. As water moves down through the sand, it loses oxygen from both the respiration of bacteria and the oxidation of chemical substances, especially ferrous compounds. The point within the mud or sand at which water loses all its oxygen is a region of stagnation and oxygen deficiency. A layer of dark iron sulfides of variable depth marks it. On mudflats, such conditions exist almost to the surface.

Thus, the energy base for sandy beach and mudflat fauna is organic matter. Much of it becomes available by bacterial decomposition, which goes on at the greatest rate at low tide. Bacteria concentrate around organic matter in the sand, where they escape the diluting effects of water. Each high tide dissolves and washes away into the sea the products of decomposition and brings in more organic matter to be con-

sumed by bacteria that in turn become food for higher-level consumers.

A number of deposit-feeding organisms ingest organic matter largely as a means of obtaining bacteria. Prominent among them are numerous nematodes and copepods *(Harpacticoida),* the polychaete worm *Nereis,* and the gastropod mollusks. Deposit feeders on sandy beaches obtain their food by burrowing through the sand and ingesting the substrate to feed on the organic matter it contains. Most common among them is the lugworm *Arenicola,* which is responsible for the conspicuous coiled and cone-shaped casts on the beach.

Other sandy beach animals are filter feeders, obtaining their food by sorting particles of organic matter from tidal water. Two of these "surf fishers," advancing and retreating with the tide, are the mole crab *Emerita* and the coquina clam *Donax.*

Because of their dependence on imported organic matter and their heterotrophic nature, perhaps we should not consider sandy beaches and mudflats as separate ecosystems. In effect, they are part of a larger coastal ecosystem involving the salt marsh, the

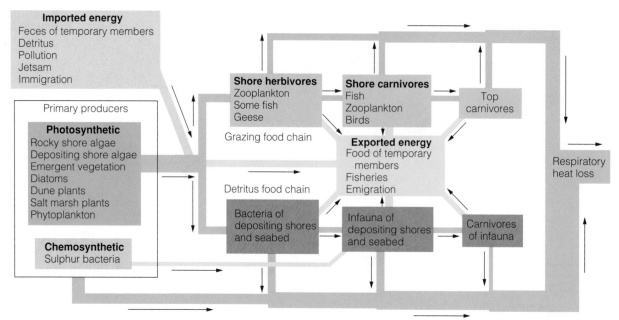

FIGURE 28.13 Diagram of the coastal ecosystem, a supraecosystem consisting of the shore, the fringing terrestrial regions, and the sublittoral zones. It connects two food webs: that of the rocky shore with its algae, herbivores, and zooplankton, and the detrital food webs involving the bacteria of the depositing shore and sublittoral muds and the dependent detritivores and carnivores. Coastal ecosystems are extremely productive; because energy imports exceed exports, the system is continuously gaining energy.

estuary, and coastal waters (Figure 28.13). Sandy shores and mudflats act as sinks for energy and nutrients because the energy they utilize comes not from primary production within the ecosystem, but from organic matter that originates outside the area.

28.13 Coral Reefs Are Oases in Tropical Seas

Lying in the warm, shallow waters about tropical islands and continental landmasses are colorful rich oases in nutrient-poor seas, the coral reefs (Figure 28.14). They are a unique accumulation of dead skeletal material built up by carbonate-secreting organisms, living coral (Cnidaria, Anthozoa), coralline red algae (Rhodophyta, Corallinaceae), green calcerous algae *(Halimeda)*, foraminifera, and mollusks. Built only underwater at shallow depths, coral reefs need a stable foundation upon which to grow, provided by shallow continental shelves and submerged volcanoes.

Coral reefs are of three types, with many gradations among them. (1) Fringing reefs grow seaward from the rocky shores of islands and continents. (2) Barrier reefs parallel shorelines of continents and

FIGURE 28.14 A rich diversity of coral species, algae, and colorful fish occupy this reef in Fiji in the south Pacific Ocean.

islands and are separated from land by shallow lagoons. (3) Atolls are horseshoe-shaped rings of coral reefs and islands surrounding lagoons, formed when volcanic mountains subsided. Such lagoons are about 40 meters deep, usually connect to the open sea by breaks in the reef, and may hold small islands of patch reefs. Reefs build up to sea level.

28.14 Built by Colonies of Coral Animals, Coral Reefs Are Complex Ecosystems

Coral reefs are complex ecosystems that begin with the complexity of the corals themselves. Corals are modular animals, anemone-like cylindrical polyps with prey-capturing tentacles surrounding the opening or mouth. Most corals form sessile colonies supported on the tops of dead colonies and cease growth when they reach the surface of the water. There are close relationships between coral animals and algae (see Chapter 16, Focus on Ecology 16.2, Figure B). In the tissues of the gastrodermal layer live **zooxanthellae,** symbiotic, photosynthetically active, endozoic dinoflagellate algae upon which coral depend for most efficient growth. On the calcareous skeletons live still other kinds of algae, both the encrusting red and green coralline species and filamentous species, including turf algae. Also associated with coral growth are mollusks, such as giant clams *(Tridacna, Hippopus)*, echinoderms, crustaceans, polychaete worms, sponges, and a diverse array of fishes, both herbivorous and predatory.

Coral are partially photosynthetic and partially heterotrophic. During the day, zooxanthellae carry on photosynthesis and directly transfer organic material to coral tissue. At night, coral polyps feed on zooplankton, securing phosphates and nitrates and other nutrients for the anthozoans and their symbiotic algae. Thus, nutrients are recycled in place between the anthozoans and the algae. In addition, carbon dioxide concentrations in animal tissue enable the coral to extract calcium carbonate to build exoskeletons.

Adding to the productivity of the coral are crustose coralline algae, turf algae, macroalgae, sea grass, sponges, phytoplankton, and a large bacterial population. Coral reefs are among the most highly productive ecosystems on Earth. Net productivity ranges from 1500 to 5000 g $C/m^2/year$, compared to 15 to 50 g $C/m^2/year$ for the surrounding ocean. Because the coralline community acts as a nutrient trap, offshore coral reefs are oases of productivity within a nutrient-poor sea.

This productivity and the varied habitats within the reef support a high diversity of life—thousands of kinds of invertebrates (some of which, such as sea urchins, feed on coral animals and algae), many kinds of herbivorous fish that graze on algae, and hundreds of predatory species. Some of these predators, such as the puffers and filefish, are corallivores, feeding on coral polyps. Others lie in ambush for prey in coralline caverns. In addition, there is a wide array of symbionts, such as cleaning fish and crustaceans that pick parasites and detritus from larger fish and invertebrates.

28.15 Zonation of Coral Reefs Results from Biotic and Abiotic Interactions

The zonation and diversity of coral species depend on an interaction of depth, light, predation, competition, and disturbance. Light sets the depth at which the zooxanthellae can survive. Diversity is lowest at the crest near the surface, where only species such as massive pillar-shaped corals tolerant of intense or frequent disturbance by waves can survive. Diversity increases with depth to a maximum of about 20 meters, a region occupied by brain, crustose, and delicate branched and fan corals; then it decreases as light attenuates, eliminating shade-intolerant species. Imposed upon this gradient are other abiotic and biotic disturbances that vary in intensity and decrease with depth. Growth rates of coral-containing photosynthetic zooxanthellae are highest in shallow depths; and a few species, especially the branching corals, can dominate the reef by overgrowing and shading the crustose corals and algae. Disturbances by wave action, storms, and grazing reduce the rate of competitive displacement among corals. Heavy grazing of overgrowing algae by sea urchins and fish, such as parrotfish, increases encrusting coralline algae. Light grazing allows rapidly growing filamentous and foliose algal species to eliminate crustose algae.

Intense disturbance or loss of species can have pronounced, even disastrous, effects on coral reefs. In one population irruption, a major coral predator, the crown-of-thorns starfish *Acanthaster planci*, ate nearly all of the corals on certain reefs in the western Pacific, destroying the reefs. Another disaster involved the most important predator of coral and algae in the Caribbean, the black sea urchin *Diadema antillarium*. Although it preyed mostly on newly settled young coral, the urchin did create algae-free settling areas for the larvae. An epizootic disease destroyed 95 percent of the *Diadema* population in 1983–1984, shortly after Hurricane Allen had scattered two dominant elkhorn coral species. The absence of black sea urchin allowed dense growths of microalgae to smother newly developing colonies of the corals. The result was the population collapse of these once dominant species along the north Jamaican coast.

28.16 Pollution Endangers Open Sea and Coastal Ecosystems

The vastness and depth of the ocean, far removed from direct human contact, make it a seemingly ideal place for disposing of sewage; sewage sludge; industrial wastes; garbage; toxic substances such as mercury, cadmium, and radioactive wastes; construction materials; and junk from cars to military shells and old ships. Discharge of sewage from southern California cities out to the mainland shelf has contaminated a 3640 km^2 area of ocean bottom. This pollution has degraded benthic invertebrates, killed beds of kelp, and caused disease in fish. The bottom of a 105 km^2 area of the New York Bight is covered with black toxic sludge. Although some of the metallic junk has provided sheltering reefs for fish, other material entangles or poisons marine birds and mammals.

Oil has become a major pollutant of the seas. Leakage from rusting oceanic pipelines, errors in handling oil cargo, accidents involving tankers and barges, sinking of ships during wars, illegal washing of tanker bilges, and spills and leakage from offshore drilling rigs all contribute to a growing oil-pollution problem that affects both the open sea and intertidal ecosystems. We know little about the effect oil has on the benthos, where it comes in direct contact with life in the deep. We do know that oil reaching the bottom of bays and harbors kills bottom life, and that certain fractions of oil soluble in water are highly toxic to many forms of marine life.

Oil is probably the most dramatic intertidal pollutant. Intertidal life suffers most when oil spills off the coast reach the shore. The damage to intertidal life depends on the extent and intensity of the spill and the type of oil: heavy crude oil (involved in the Persian Gulf and Prince William Sound spills), diesel oil, or split oils.

Controlling oil pollution on shores is difficult because incoming tides wash more oil onto the shore. Millions of sea birds such as cormorants and diving ducks and thousands of marine mammals have been victims of oil pollution. A heavy coating of oil mats feathers, impairing the ability of birds to fly and swim; reduces the feathers' water repellency; and destroys the insulating effects of down. During cold weather, a spot of oil as small as a button over a vital organ can induce a bird's death from hypothermia. As they preen to clean their feathers, birds ingest fatal amounts of oil. Marine mammals, especially seals and sea otters, fare no better. Oil destroys the insulating effects of fur, clogs ears and nostrils, and

irritates eyes. Many such mammals succumb to hypothermia and dehydration.

Less visible to humans is the damage done to invertebrate and plant life of the shores. Particularly vulnerable to smothering oil pollution on sandy shores are interstitial life and sand crabs. Even after the sand appears to be scrubbed by tides, a layer of oil may still reside several meters below the surface. Some intertidal invertebrates, particularly mollusks, appear to be resistant to oil pollution, but their flesh becomes tainted, and the oil is passed along the food chain. On rocky shores, barnacles resist oil pollution, but the grazing periwinkles, limpets, and whelks are particularly vulnerable. Oil works its way beneath their shells, causing these gastropods to lose their hold, and they are carried away by the tides. As oil eliminates these grazing invertebrates, algae and seaweeds, particularly the red algae, colonize the barren patches. They, too, become encrusted with oil and are torn away by the waves. Aside from damage to particular species, oil pollution in the intertidal zone means a great reduction in species diversity, a simplification of the food web, and an increase in the populations of resistant species.

In addition to pollution, human intrusion onto the seashore also has serious effects on intertidal wildlife, especially that of sandy shores. Beach-nesting birds such as the piping plover (*Charadrius melodus*) and the least tern (*Sterna antillarum*) are so disturbed by bathers and dune buggies that both species are in danger of extinction. Other terns and shore birds are subjected to competition for nest sites and to egg predation by rapidly growing populations of large gulls that are highly tolerant of humans and thrive on human garbage. Sea turtles and the horseshoe crab (*Limulus polyphemus*), dependent on sandy beaches for nesting sites, find themselves evicted and are declining rapidly for that reason. Furthermore, each incoming tide brings onto the beaches feces-contaminated water that makes them unhealthy for humans and wildlife alike. Tides also carry in old fishing lines, plastic debris, blobs of oil, and other wastes hazardous to wildlife.

Coral reefs have not escaped human disturbance. Dredging, sewage pollution, and overfishing have degraded coral reefs. The most devastating human activity in reefs is overfishing, especially in the Philippines and the Caribbean. Overfishing removes both algal-feeding herbivorous fish and predatory fish that feed on the competitive algal-feeding sea urchins (*Diadema* spp.). Their removal upsets the complex food webs and interspecific relations on the reef and allows algae to overgrow the coral structures. The heavy growth of algae precludes the settlement of coral on the reef. Dynamiting reefs to capture fish for the marine aquarium trade destroys both the reefs and their fish populations.

Recovery of coral from disturbances is slow, often requiring 25 to 100 years. The rate of recovery depends upon the extent of disturbance, the nearness of a source of larvae, and favorable currents to sweep the larvae to substrates free of algal growth.

Summary

Open Seas (28.1–28.2) The marine environment is characterized by salinity, waves, tides, depth, and vastness. Like lakes, oceans experiences both zonation of life and stratification of temperature. The open sea can be divided into three vertical zones. The bathypelagic is the deepest, void of sunlight and inhabited by darkly pigmented, weak-bodied animals possessing bioluminescence. Above it lies the dimly lit mesopelagic zone, inhabited by characteristic species such as certain sharks and squid. Both the bathypelagic and mesopelagic depend upon a rain of detrital material from the upper lighted zone, the photic, for their energy. The sea bottom, or benthic region, is inhabited by its own unique fauna adapted to life in total darkness and high pressure (**28.1**). Because they lack large biomass, pelagic communities dominated by phytoplankton are hidden beneath the surface. These communities vary latitudinally, and patterns of productivity are influenced by the seasonal dynamics of the thermocline (**28.2**).

Ocean Life (28.3–28.6) Phytoplankton dominate the surface waters. The littoral and neritic zones are richer in plankton than the mid-ocean. Nanoplankton, which make up the largest biomass in temperate and tropical waters, are the major source of primary production (**28.3**). Feeding on phytoplankton are herbivorous zooplankton, the most important of which are copepods. They are preyed upon by carnivorous zooplankton. The greatest diversity of zooplankton, including larval forms of fish, occur in the water over coastal shelves and upwellings; the least diversity occurs in the open ocean (**28.4**). Making up the larger life forms are the free-swimming nekton, ranging from small fish to sharks and whales. Nekton organisms have various means of defense and securing prey (**28.5**). Benthic organisms, those that live on the ocean floor, vary with depth and substrate. They are strictly heterotrophic and depend on organic matter that drifts to the bottom. They include filter feeders, collectors, deposit feeders, and predators. Benthic organisms possess the highest diversity because of the more stable environment, with local disturbance creating a diversity of habitat patches and a patchy distribution of food (**28.6**).

Hydrothermal Vents (28.7) Along volcanic ridges are hydrothermal vents inhabited by novel and newly discovered forms of life, including crabs, clams, and worms. Chemosynthetic bacteria that use sulfates as an energy source account for primary production for these hydrothermal vent communities.

Intertidal Life (28.8–28.12) Sandy shores and rocky coasts are places where sea meets the land. The drift line marks the farthest advance of the tide on sandy shores. On rocky shores, a zone of black algal growth marks the tide line (**28.8**). The most striking feature of the rocky shore, zonation of life, results from alternate exposure and submergence by the tides. The black zone marks the supralittoral, the upper part of which is flooded once only every two weeks by spring tides. Submerged daily by the tides is the littoral, characterized by barnacles, periwinkles, mussels, and fucoid seaweeds. Uncovered only at spring tides is the infralittoral, which is dominated by large brown laminarian seaweeds, Irish moss, and starfish. Distribution and diversity of life across the rocky shore are also influenced by wave action, competition herbivory, and predation. Left behind by outgoing tides are tidal pools. These are distinct habitats subject over a 24-hour period to wide fluctuations in temperature and salinity, and inhabited by varying numbers of organisms, dependent upon the amount of emergence and exposure (**28.9**).

Sandy beaches are a product of weathering of rock. Exposed to wave action, they are subject to deposition and wearing away of the sandy substrate (**28.10**). Sandy and muddy shores appear barren of life at low tide; but beneath the sand and the mud, conditions are more amenable to life than on the rocky shore. Zonation of life is hidden beneath the surface (**28.11**). The energy base for sandy and muddy shores is organic matter carried in by the tides and made available by bacterial decomposition. The basic consumers, then, are bacteria, which in turn are a major source of food for both deposit-feeding and filter-feeding organisms (**28.12**).

Coral Reefs (28.13–28.15) Nutrient-rich oases in nutrient-poor tropical seas are coral reefs (**28.13**). They are complex ecosystems based on anthozoan coral and their symbiotic endozoan dinoflagellate algae and coralline algae. Their productive and varied habitats support a high diversity of colorful invertebrate and vertebrate life (**28.14**). Zonation of coral reefs results from an interaction of light, depth, competition, grazing, and disturbance. Diversity increases with depth up to a point. Greatest diversity occurs at shallow depths (**28.15**).

Human Impacts (28.16) The oceans have become dumping grounds. Major chronic pollutants are oil released from tanker accidents, seepage from offshore drilling and other sources, toxic materials such as pesticides, and heavy metals from industrial, urban, and agricultural sources. The long-term effects of these pollutants are unknown. Intertidal communities are especially vulnerable to human activity and pollution. Chronic oil pollution and, especially, major oil spills not only cause massive economic damage; they also decimate vertebrate and invertebrate life of the intertidal communities, affect natural relationships between predator and prey, reduce species diversity, and affect community structure. Sewage and solid waste carried in by the tides periodically pollute the shores, endangering human health and wildlife. Human occupancy and recreational use of the shores have destroyed important wildlife habitats and are bringing some dependent species close to extinction.

Study Questions

1. Characterize the major regions of the ocean, both vertical and horizontal.

2. How does temperature stratification in the tropical seas differ from that in the polar seas?

3. What sea develops the most pronounced thermocline, and why?

4. What might account for the high diversity of the deep-sea benthos?

5. What are hydrothermal vents, and what makes life around them unique?

6. What is the role of phytoplankton, zooplankton, and nekton in the marine food web?

7. What is the significance of the microbial loop in the marine food web?

8. Describe the three major zones of the rocky shore.

9. What adaptations enable inhabitants of rocky shores to avoid desiccation, maintain position, and survive flooding?

10. What roles do predation and competition play in the rocky shore community?

11. C. M. Yonge, the British marine ecologist, called tide pools "microcosms of the sea" except for two significant environmental factors. What are they?

12. What are the major zones of life on sandy shores? How do they differ from those of rocky shores?

13. How does life on a sandy shore survive in the harsh environment?

14. Contrast the energy source of a sandy or muddy shore with that of the rocky shore. What are the basic consumers on sandy shores?

15. What are coral reefs, and how do they form?

16. Explain how the symbiotic relationship between algae and anthozoans influences the distribution and functioning of corals.

17. In what ways does human activity harm sandy and rocky shores?

18. Make a list of endangered and threatened shore birds. How does their status relate to human use of the seashore?

19. Explain how oil affects life on sandy shores and rocky shores. What is the difference?

20. Report on the ecological and economic effects of some major oil spills, such as the *Amoco Cadiz* off the Brittany coast of France, the *Exxon Valdez* in Prince William Sound, Alaska, and the release of oil into the Persian Gulf during the Gulf War.

CHAPTER 29

Morning mists rise up from a northern wetland in Michigan.

Wetlands

OBJECTIVES

On completion of this chapter, you should be able to:

- Define a wetland.
- Describe the various types of wetlands.
- Explain the role of hydrology and hydroperiod in wetlands.
- Explain the ecological and economic value of wetlands.
- Describe the major features of a salt marsh relative to tides and salinity.
- Discuss the unique features of a mangrove forest.
- Explain the importance of the mangrove forest to the tropical coastal ecosystem.
- Discuss effects of human activities on wetlands.

Thus far we have examined the two major classes of ecosystems, terrestrial and aquatic. The distribution of terrestrial ecosystems, defined by the dominant plant life, is related to climate. Aquatic ecosystems are classified based on the salinity, depth, and flow of water. Wetlands fall into a different category. These ecosystems function as ecotones between terrestrial and aquatic ecosystems, sharing characteristics with both. They play a major role in the landscape by providing unique habitats for a wide variety of flora and fauna. Although they cover 6 percent of Earth's surface, they are not climatically based. They are found in every climatic zone, but are local in occurrence. Only a few, such as the Everglades in Florida, the Panatal in Brazil, the Okavango in southern Africa, and the fens of England cover extensive areas of the landscape. Not until the 1960s did wetlands become a focus of study for ecologists, and experience a change in perception from wasteland to a valued component of the landscape.

29.1 What Defines a Wetland?

What is a wetland? This question seems to require only a simple answer: an area covered with water and supporting aquatic plants. Although this answer is not wrong, it is not correct either, not in this day when the life or death of a wetland rests on a precise definition.

Some wetlands are easy to distinguish. A water area supporting submerged plants such as pondweed, floating plants such as pond lily, and emergents such as cattails and sedges is unquestionably a wetland. But what about a piece of ground where the soil is more-or-less permanently wet and supports some ferns and such trees as maple that also grow on the uplands? Where do you draw the line between wetlands and uplands on this gradient of soil wetness?

Vegetation alone does not define a wetland. First, we must consider the hydrological conditions; then we may use vegetation as an indicator.

Wetlands range along a gradient from permanently flooded to periodically saturated soil (Figure 29.1) and support hydrophytic (water-loving) vegetation at some time during the growing season. Hydrophytic plants are adapted to grow in water or on soil that is periodically anaerobic (deficient in oxygen) because of excess water (see Chapter 7). Hydrophytic plants include several groups: (1) obligate wetland plants, such as the submerged pondweeds, floating pond lily, emergent cattails, and bulrushes, and trees such as baldcypress *(Taxodium distichum);* (2) facultative wetland or amphibious plants that can grow in standing water or saturated soil and rarely grow elsewhere, such as certain sedges

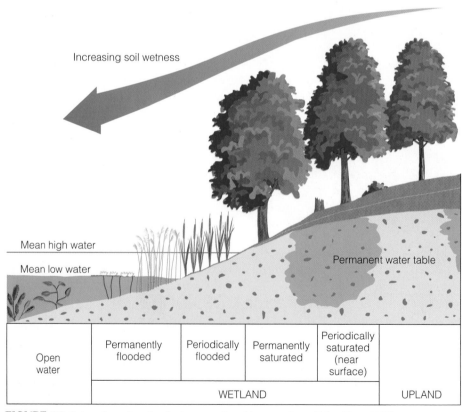

Increasing soil wetness

Mean high water

Mean low water

Permanent water table

Open water	Permanently flooded	Periodically flooded	Permanently saturated	Periodically saturated (near surface)	
	WETLAND				UPLAND

FIGURE 29.1 Location of wetlands along a soil moisture gradient. (After Tiner 1991).

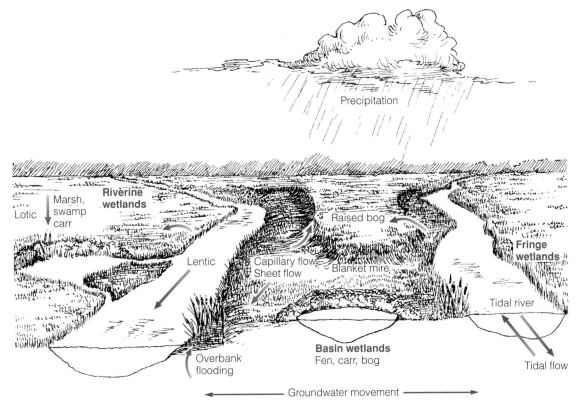

FIGURE 29.2 Water flow in various types of freshwater wetlands. (After Grosselink and Turner 1978).

and alders; (3) facultative species such as red maple *(Acer rubrum)*, which have about a 50–50 probability of growing in either wetland or nonwetland situations; and (4) facultative upland species such as beech *(Fagus grandifolia)*, which have a 1 to 30 percent probability of growing in a wetland. It is the last group of plants that is critical in determining the upper limit of a wetland on the soil moisture gradient, when species designation alone is insufficient.

For example, some species of trees usually associated with uplands adapt and grow quite well in wetland environments. An example from eastern North America is the red maple. It thrives in the drier uplands, but is also a conspicuous species in forested wetlands. In the uplands, red maple has a deep taproot; in wetland situations, the tree has a shallow root system that enables it to avoid anaerobic stress. The red maple is a facultative species that has evolved ecotypes adapted to different soil moisture conditions. Black gum *(Nyssa sylvatica)*, too, grows in both upland and wetland situations. Pitch pine *(Pinus rigida)*, associated with the drier ridgetops of the southern Appalachians, has ecotypes that grow in poorly drained soils and the muck of swamps. In fact, pitch pine is a dominant species in the wetlands of the extensive New Jersey pine barrens. Hemlock

(Tsuga canadensis), a shallow-rooted species, is at home in wetland situations.

The point is that species of vegetation alone do not define a wetland. They are important indicators, especially in the wettest situations, but ecotypes of upland species confuse the situation. It is essential to consider hydrological conditions and soil properties along with the vegetation.

29.2 Wetlands Come in Many Types

A wide variety of wetlands exist, and classifying them for management and conservation has presented problems. An old, short, but still useful classification appears in Table 29.1. A much more comprehensive classification is *Classification of Wetlands and Deepwater Habitats of the United States*.

Wetlands most commonly occur under three topographic conditions (Figure 29.2). Many develop in shallow basins ranging from upland depressions to filled-in lakes and ponds. They are basin wetlands. Other wetlands develop along shallow and periodically flooded banks of rivers and streams. They are

TABLE 29.1

Types of Wetlands

Type	Site Characteristics	Plant and Animal Populations
Inland Fresh Areas		
Seasonally flooded basins or flats	Soil covered with water or waterlogged during variable periods, but well drained during much of the growing season; in upland depressions and bottomlands	Bottomland hardwoods to herbaceous growth
Fresh meadows	Without standing water during growing season; waterlogged to within a few inches of surface	Grasses, sedges, rushes, broadleaf plants
Shallow fresh marshes	Soil waterlogged during growing season; often covered with 15 cm or more of water	Grasses, bulrushes, spike rushes, cattails, arrowhead, smartweed, pickerelweed; major waterfowl production areas
Deep fresh marshes	Soil covered with 15 cm to 1 m of water	Cattails, reeds, bulrushes, spike rushes, wild rice; principal duck breeding areas
Open freshwater	Water less than 3 m deep	Bordered by emergent vegetation such as pondweed, naiads, wild celery, water lily; brooding, feeding, nesting areas for ducks
Shrub swamps	Soil waterlogged; often covered with 15 cm or more of water	Alder, willow, buttonbush, dogwoods; nesting and feeding areas for ducks to limited extent
Wooded swamps	Soil waterlogged; often covered with 0.3 m of water; along sluggish streams, flat uplands, shallow lake basins	North: tamarack, arborvitae, spruce, red maple, silver maple; South: water oak, overcup oak, tupelo swamp, black gum, cypress
Bogs	Soil waterlogged; spongy covering of mosses	Heath shrubs, *Sphagnum*, sedges
Coastal Fresh Areas		
Shallow fresh marshes	Soil waterlogged during growing season; at high tide as much as 15 cm of water; on landward side, deep marshes along tidal rivers, sounds, deltas	Grasses and sedges; important waterfowl areas
Deep fresh marshes	At high tide covered with 15 cm to 1 m of water; along tidal rivers and bays	Cattails, wild rice, giant cutgrass
Open freshwater	Shallow portions of open water along fresh tidal rivers and sounds	Vegetation scarce or absent; important waterfowl areas

riverine wetlands. A third type, fringe wetlands, occurs along the coastal areas of large lakes and oceans. They include tidal and estuarine salt marshes and mangrove swamps.

What separates the three types in part is the direction of water flow (see Figure 29.2). Water flow in the basin wetlands is vertical, involving precipitation and capillary flow. In riverine wetlands, water flow is unidirectional. In fringe wetlands, flow is in two directions because it involves rising lake levels or tidal action. The flows bring in and carry away nutrients and sediments.

Wetlands dominated by emergent herbaceous vegetation are called **marshes** (Figure 29.3). Domi-nated by reeds, sedges, grasses, and cattails, marshes are essentially wet prairies. Forested wetlands are commonly called **swamps** (Figure 29.4). They may be deepwater swamps dominated by cypress, tupelo, and swamp oaks; or they may be shrub swamps dominated by alder and willows. Along many large river systems are extensive tracts of **bottomland** or **riparian woodlands** (Figure 29.5), which are occa-sionally or seasonally flooded by river waters, but are dry for most of the growing season.

Wetlands in which considerable amounts of water are retained by an accumulation of partially decayed organic matter are **peatlands** or **mires**. Mires fed by water moving through mineral soil, from which

Type	Site Characteristics	Plant and Animal Populations
Inland Saline Areas		
Saline flats	Flooded after periods of heavy precipitation; waterlogged within few inches of surface during the growing season	Sea blite, salt grass, saltbush; fall waterfowl feeding areas
Saline marshes	Soil waterlogged during growing season; often covered with 0.61 to 1 m of water; shallow lake basins	Alkali hard-stemmed bulrush, widgeon grass, sago pondweed; valuable waterfowl areas
Open saline water	Permanent areas of shallow saline water; depth variable	Sago pondweed, muskgrasses; important waterfowl feeding areas
Coastal Saline Areas		
Salt flats	Soil waterlogged during growing season; sites occasionally to fairly regularly covered by high tide; landward sides or islands within salt meadows and marshes	Salt grass, sea blite, saltwort
Salt meadows	Soil waterlogged during growing season; rarely covered with tide water; landward side of salt marshes	Cordgrass, salt grass, black rush; waterfowl feeding areas
Irregularly flooded salt marshes	Covered by wind tides at irregular intervals during the growing season; along shores of nearly enclosed bays, sounds, etc.	Needlerush; waterfowl cover areas
Regularly flooded salt marshes	Covered at average high tide with 15 cm or more of water; along open ocean and along sounds	Atlantic: salt marsh cordgrass; Pacific: alkali bulrush, glassworts; feeding area for ducks and geese
Sounds and bays	Portions of saltwater sounds and bays shallow enough to be diked and filled; all water landward from average low-tide line	Wintering areas for waterfowl
Mangrove swamps	Soil covered at average high tide with 15 cm to 1 m of water; along coast of southern Florida	Red and black mangroves

FIGURE 29.3 The Horicon marsh in Wisconsin is an outstanding example of a northern marsh with well-developed emergent vegetation and patches of open water—an ideal environment for wildlife.

they obtain most of their nutrients, and dominated by sedges are known as **fens**. Mires dependent largely on precipitation for their water supply and nutrients and dominated by *Sphagnum* moss are called **bogs** (Figure 29.6). Mires that develop on upland situations where decomposed, compressed peat forms a barrier to the downward movement of water, resulting in a perched water table above mineral soil, are **blanket mires** and **raised bogs** (Figure 29.7). Raised bogs are popularly known as **moors**. Because bogs depend on precipitation for nutrient inputs, they are highly deficient in mineral salts and low in pH. Bogs also develop when a lake basin fills with sediments and organic matter carried by inflowing water. These sediments divert water around the lake basin and raise the surface of the mire above the influence of groundwater. Other bogs form when a lake basin fills in from above rather than from below, creating a floating mat of peat

(a)

(b)

FIGURE 29.4 (a) A cypress deepwater swamp in the southern United States. (b) An alder *(Alnus)* shrub swamp with an herbaceous understory of skunk cabbage *(Symplocarpus foetidus)*.

FIGURE 29.5 A riparian forest in Alabama.

FIGURE 29.6 An upland black spruce-tamarack bog in the Adirondack Mountains of New York.

over open water. Such bogs are often termed **quaking** (Figure 29.8).

On the alluvial plains around the estuary and in the shelter of spits and offshore bars and islands along the coast are **salt marshes** (Figure 29.9). Although salt marshes look like acres of waving grass, they are a complex of distinctive and clearly demarked plant associations. These include, among others, low marsh dominated by salt marsh cord grass, high marsh dominated by various associations of salt marsh hay grass, spike grass, and the short form of salt marsh cord grass. The reason for this complexity is the spatial heterogeneity resulting from tides and salinity.

Replacing salt marshes on tidal flats in tropical regions are **mangrove forests** or **mangals** (Figure 29.10), which cover 60 to 75 percent of the coastline of the tropical regions. Mangals develop where wave action is absent, sediments accumulate, and the muds are anoxic. They extend landward to the highest vertical tidal range, where they may be only periodically flooded. The dominant plants are mangroves, which include 8 families and 12 genera dominated by *Rhizophora, Avicennia, Bruguiera,* and *Sonneratia.* Growing with them are other salt-tolerant plants, mostly shrubs. Mangals reach their greatest development and have the most species in the Indo-Malaysian region.

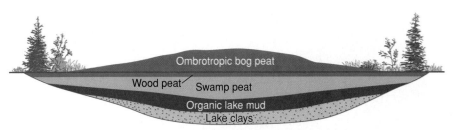

FIGURE 29.7 A raised bog develops when an accumulation of peat rises above the surrounding landscape.

In growth form, mangroves range from short, prostrate forms to timber-size trees 30 meters high. All mangroves have shallow, widely spreading roots, many with prop roots coming from trunk and limbs (Figure 29.11). Many species have root extensions called pneumatophores that take in oxygen for the roots. Their fleshy leaves, although tough, are often succulent and may have salt glands. Many mangrove species have a unique method of reproduction, vivipary. Following fertilization, the embryos undergo uninterrupted development with no true resting seed. They grow into a seedling on the tree, drop to the water, and float upright until they reach water shallow enough for their roots to penetrate the mud.

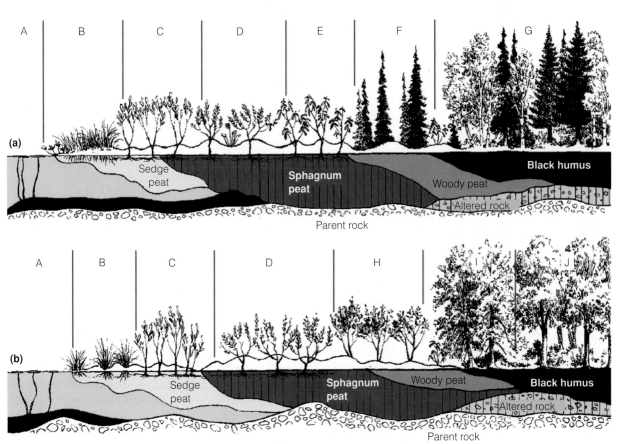

FIGURE 29.8 (a) Transect through a quaking bog, showing zones of vegetation, sphagnum mounds, peat deposits, and floating mats. A, pond lily in open water; B, buckbean *(Menyanthes trifoliata)* and sedge; C, sweetgale *(Myrica gale);* D, leatherleaf *(Chamaedaphne calyculata);* E, Labrador tea *(Ledum groenlandicum);* F, black spruce; G, birch-black spruce-balsam fir forest. (b) An alternative vegetational sequence. H, alder; I, aspen, red maple; J, mixed deciduous forest.

FIGURE 29.9 A salt marsh on the Bay of Fundy, New Brunswick, Canada. The marsh, not well developed, is dominated by *Puccinellia* rather than *Spartina*. Compare with the salt marsh on the Virginia coast (Figures 29.14 and 29.15).

29.3 Hydrology Defines the Structure of Freshwater Wetlands

The structure of a wetland is influenced by the phenomenon that creates it—its hydrology. Hydrology has two components. One is the physical aspect of water and its movement: precipitation, surface and subsurface flow, the direction and kinetic energy of water, and the chemistry of the water. The other is **hydroperiod,** which includes the duration, frequency, depth, and season of flooding. The length of the hydroperiod varies among types of wetlands. Basin wetlands have a long hydroperiod. They usually flood during periods of high rainfall and draw down during dry periods. Both phenomena appear to be essential to the long-term existence of wetlands. Riverine wetlands have a short period of flooding associated with peak stream flow. The hydroperiod of fringe wetlands, influenced by wind and lake waves, may be short and regular, and does not undergo the seasonal fluctuation characteristic of many basin marshes.

Hydroperiod influences plant composition, for it affects germination, survival, and mortality at various stages of the plants' life cycles. The effect of

hydroperiod is most pronounced in basin wetlands, especially those of the prairie regions of North America. In basins deep enough to have standing water throughout periods of drought (called potholes in the prairie region), the dominant plants will be submergents. If the wetland goes dry annually or during a period of drought, tall or midheight emergent species such as cattails will dominate the marsh. If the pothole is shallow and flooded only briefly in the spring, grasses, sedges, and forbs will make up a wet-meadow community.

If the basin is sufficiently deep toward its center and large enough, zones of vegetation may develop, ranging from submerged plants to deepwater emergents such as cattails and bulrushes, shallow-water emergents, and wet-ground species such as spike rush. Zonation reflects the response of plants to hydroperiod. Those areas of wetland subjected to a long hydroperiod will support submerged and deepwater emergents; those with a short hydroperiod and shallow water are occupied by shallow-water emergents and wet-ground plants.

Periods of drought and wetness can induce vegetation cycles associated with changes in water levels. Periods of above-normal precipitation can raise the water level and drown the emergents to create a lake marsh dominated by submergents. During a drought, the marsh bottom is exposed by receding water,

FIGURE 29.10 Mangroves replace tidal marshes in tropical regions.

FIGURE 29.11 Shallow-rooted mangroves have prop roots descending from the trunk and branches.

stimulating the germination of seeds of emergents and mudflat annuals. When water levels rise again, the mudflat species drown, and the emergents survive and spread vegetatively.

Peatlands differ from other freshwater wetlands in that their rate of organic production exceeds the rate of decomposition, and much of the production accumulates as peat. In northern regions, acid-forming, water-holding sphagnum mosses add new growth on top of the accumulating remains of past

moss generations; and their spongelike ability to hold water increases water retention on the site. As the peat blanket thickens, the water-saturated mat of moss and associated vegetation is raised above and insulated from mineral soil. The peat mat then becomes its own reservoir of water, creating a perched water table.

Peat bogs and mires generally form under oligotrophic and dystrophic conditions (see Chapter 27, Section 27.7). Although usually associated with and most abundant in boreal regions of the Northern Hemisphere, peatlands also exist in tropical and subtropical regions. They develop in mountainous regions or in lowland or estuarine regions where hydrological situations encourage an accumulation of partly decayed organic matter. Examples are the Everglades in Florida and the pocosins of the southeastern U.S. coastal plains.

29.4 Tides and Salinity Dictate the Structure of Salt Marshes

In contrast to the freshwater wetland, salt marsh structure is dictated by tides and salinity. Together they create complex, distinctive, and clearly demarked plant associations. Two times a day, the salt-marsh plants on the outermost fringes are submerged in salty water and then exposed to full sun. Their roots extend into poorly drained, poorly aerated soil in which the soil solution contains varying concentrations of salt. Only plant species with a wide range of salt tolerances can survive such conditions.

From the edge of the sea to the high land, zones of vegetation distinctive in form and color develop, reflecting a microtopography that lifts the plants to various heights within and above high tide (Figure 29.12). Most conspicuous on the seaward edge of East Coast marshes and along tidal creeks are the deep green growths of salt marsh cordgrass, *Spartina alterniflora,* that dominate the low marsh. Stiff, leafy, up to 3 meters tall, and submerged in saltwater at each high tide, salt marsh cordgrass forms a marginal strip between the open mud to the front and the high marsh behind. No litter accumulates beneath the stand. Strong tidal currents sweep the floor of the *Spartina* clean, leaving only thick, black mud.

Spartina alterniflora is well adapted to grow on the intertidal flats, of which it has sole possession. It has a high tolerance for saltwater and is able to live in a semisubmerged state. It can live in a saline

environment by selectively concentrating sodium chloride in its cells at a level higher than the surrounding saltwater, thus maintaining its osmotic integrity. To rid itself of excessive salts, salt marsh cordgrass has special salt-secreting cells in the leaves. Water excreted with the salt evaporates, leaving sparkling crystals on the surface of leaves to be washed off by tidal water. To get air to its roots, buried in anaerobic mud, cordgrass has hollow tubes leading from the leaf to the root, through which oxygen diffuses.

Above and behind the low marsh is the high marsh, standing at the level of mean high water. At this level, the tall salt marsh cordgrass gives way rather sharply to a short form, yellowish, almost chlorotic in appearance, contrasting with the tall, dark green form. This short form seemingly represents a phenotypic plastic response to environmental conditions of the high marsh. The high marsh has a higher salinity, a decreased input of nutrients, and an accumulation of toxic wastes, the result of lower tidal exchange rates. The shorter, more open canopy of the high marsh results in higher leaf and soil temperatures and a higher rate of evaporation than in the low marsh. These conditions, along with the

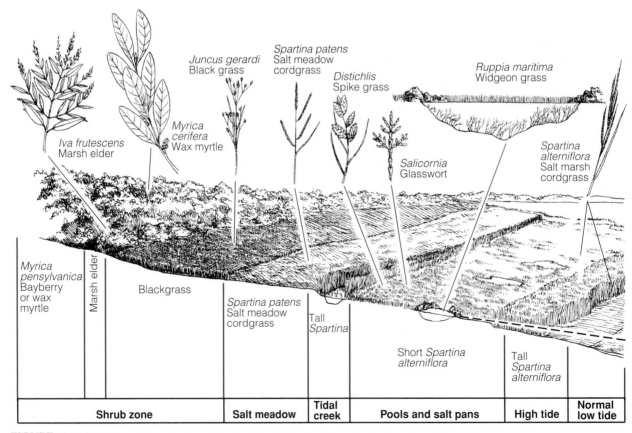

FIGURE 29.12 Patterns of zonation in an idealized New England salt marsh, showing the relationship of plant distribution to microtopography and submergence.

FIGURE 29.13 Glasswort dominates highly saline areas on the salt marsh. The plant, which turns red in fall, is a major food of overwintering geese.

FIGURE 29.14 The distinctive cowlick sweep of salt meadow cordgrass next to a stand of saltgrass. Bayberry in the background marks the shrub zone.

decreased dominance of *Spartina*, allow other marsh plants to grow. Here are the fleshy, translucent glassworts *(Salicornia* spp.) (Figure 29.13) that turn bright red in fall, sea lavender *(Limonium carolinianum)*, spearscale *(Atriplex patula)*, and sea blite *(Suaeda maritima)*.

Where the microelevation is about 5 cm above mean high water, short *Spartina alterniflora* and its associates are replaced by salt meadow cordgrass *(Spartina patens)* and an associate spikegrass or saltgrass, *Distichlis spicata*. Salt meadow cordgrass is a fine grass that grows so densely and forms such a tight mat that few other plants can grow with it (Figure 29.14).

As the microelevation rises several more centimeters above mean high tide and if there is some intrusion of freshwater, *Spartina* and *Distichlis* may be replaced by two species of black needlerush or black grass *(Juncus roemerianus* and *Juncus gerardi)*, so-called because their dark green color becomes almost black in the fall. Beyond the black grass and often replacing it is a shrubby growth of marsh elder *(Iva frutescens)* and groundsel *(Baccharis halimifolia)*. These shrubs tend to invade the high marsh where a slight rise in microelevation exists, but such invasions are often short-lived, as storm tides sweep in and kill the plants. On the upland fringe grow bayberry *(Myrica pensylvanica)* and the pink-flowering sea holly *(Hibiscus palustris)*.

Two conspicuous features of a salt marsh are the meandering creeks and the salt pans. The creeks form an intricate system of drainage channels that carry tidal waters back out to sea (Figure 29.15). The salt pans are circular to elliptical depressions. At high

FIGURE 29.15 A tidal creek at high tide on the high marsh. Tall *Spartina* grows along the banks.

tide, they are flooded. At low tide, they remain filled with saltwater. If they are shallow enough, the water may evaporate completely, leaving an accumulating concentration of salt on the mud.

Salt pans support distinctive vegetation, which varies with the depth of the water and the salt concentration. Pools with a firm bottom and sufficient depth to retain tidal water support dense growths of widgeon grass *(Ruppia maritima)*, whose small, black seeds are relished by waterfowl.

Shallow depressions in which water evaporates are covered with a heavy algal crust and crystallized salt. The edges of these salt flats may be invaded by

FIGURE 29.16 A salt pan or pool in the high marsh.

glasswort, spikegrass, or even short *Spartina alterni-flora* (Figure 29.16).

The exposed banks of tidal creeks that braid through the salt marshes support a dense population of mud algae, the diatoms and dinoflagellates, photosynthetically active all year. Photosynthesis is highest in summer during high tides and in winter during low tides, when the Sun warms the mud. Some of the algae are washed out at ebb tide and become part of the estuarine plankton available to such filter feeders as oysters.

The salt marsh described is typical of the North American Atlantic coast. Many variations exist locally and latitudinally around the world. North America has several distinctive types. Arctic salt marshes have few species. East and Gulf Coast marshes dominated by *Spartina* and *Juncus* reach their best development on the heavy silt deposits from North Carolina south. Pacific coast marshes, poorly developed, are dominated up to mean high-water level by *Spartina foliosa*, and above that by *Salicornia*. Salt marshes on the western coast of Europe and in Great Britain are dominated by *Salicornia* and *Puccinellia*. Those of northwestern Europe are dominated by *Spartina angelica*, a hybrid polyploid species that spontaneously formed from the native *Spartina townsendii*, and *Spartina alterniflora*, introduced from North America. Because of its vigorous growth, *S. angelica* is widely planted to stabilize and bind coastal mudflats.

29.5 Range and Duration of Tidal Flooding Dictate the Formation of Mangrove Swamps

The range and duration of tidal flooding strongly influence the formation and appearance of mangrove forests. Mangals become established in areas where the lack of wave action allows fine sediments or mud to accumulate. The tangle of prop roots and pneumatophores further slows the movement of tidal waters, allowing more sediments to settle out. Land begins to march seaward, followed by colonizing mangroves.

A feature of this response is a change in vegetation from seaward edge to true terrestrial environment. Although often used as an example, the mangals of the Americas, especially Florida, have the least pronounced changes, largely because of the few species involved. The pioneering red mangrove (*Rhizophora mangle*) occupies the seaward edge and receives the deepest tidal flooding. Backing the red mangroves are black mangroves (*Avicennia germinanas*), shallowly flooded by high tides. White mangroves (*Laguncularia racemosa*), along with buttonwood (*Conocarpus erectus*), a nonmangrove species that acts as a transition to terrestrial vegetation, dominate the landward edge.

29.6 Freshwater Wetlands Support a Rich Diversity of Life

Biologically, freshwater wetlands are among the richest and most interesting ecosystems. They support a diverse community of benthic, limnetic, and littoral invertebrates, especially crustaceans and insects. These invertebrates, along with small fishes, provide a food base for waterfowl, herons, gulls, and other birds, and supply the fat-rich nutrients that ducks need for egg production and the growth of young. Amphibians and reptiles—notably frogs, toads, and turtles—inhabit the emergent growth, soft mud, and open water of the marsh and swamp.

Herbivores make up a conspicuous component of animal life. Microcrustaceans filter algae from the water column. Snails eat algae growing on the leaves and litter; geese graze on new emergent growth; coots, mallards, and other surface-feeding ducks feed on algal mats. The dominant herbivore in the prairie marshes is the muskrat (Ondatra zibethicus). During population highs, muskrats can eliminate emergent vegetation, creating "eat-outs" and transforming an emergent-dominated marsh into an open-water one. Introduced into Eurasia, the muskrat has become the major herbivore in many marshes on that continent. Muskrats are the major prey for mink, the dominant carnivore on the marshes. Other predators include raccoon, fox, weasel, and skunk, which can seriously reduce the reproductive success of waterfowl on small marshes surrounded by agricultural land.

29.7 Animal Life in the Salt Marsh Is Adapted to Tidal Rhythms

Animal life of the salt marsh, not noted for its diversity, is outstanding for its interesting organisms. Some of the inhabitants are permanent residents in sand and mud; others are seasonal visitors; and most are transients that come to feed at high and low tide.

Three dominant animals of the low marsh are the ribbed mussel (Modiolus demissus), fiddler crab (Uca pugilator and U. pugnax), and marsh periwinkle (Littorina spp.). The marsh periwinkle, related to the periwinkles of the rocky shore, moves up and down the stems of Spartina and onto the mud to feed on algae and detritus. Buried halfway in the mud is the ribbed mussel (see Figure 7.1a). At low tide, the

mussel is closed; at high tide, the mussel opens to filter particles from the water, accepting some and rejecting others in a mucous ribbon known as pseudofeces.

Running across the marsh at low tide like a vast herd of tiny cattle are the fiddler crabs (see Figure 7.1b). Among the marsh animals, they are the most adaptable. They have both lungs and gills. They can endure high tides and cold winters without oxygen. They have a saltwater control system that enables them to move from diluted seawater to briny pools. Omnivorous feeders, they eat detrital material, algae, and small animals. Fiddler crabs live in burrows, marked by mounds of freshly dug, marble-sized pellets. Like earthworms, burrowing crabs bring up nutrients to the surface.

Prominent about the base of Spartina and under debris are the sandhoppers (Orchestia). These detritus-feeding amphipods may be abundant and are important in the diet of some marsh birds. Three conspicuous vertebrate residents of the low marsh of eastern North America are the diamond-backed terrapin (Malaclemys terrapin), the clapper rail (Rallus longirostris), and the seaside sparrow (Ammospiza maritima). The terrapin feeds on fiddler crabs, small mollusks, marine worms, and dead fish. The rail eats fiddler crabs and sandhoppers. The sparrow eats sandhoppers and other small invertebrates.

On the high marsh, animal life changes as suddenly as the vegetation. The small, coffee-bean-colored pulmonate snail Melampus, found by the thousands under the low grass, replaces the marsh periwinkle. Meadow voles (Microtus), have a maze of runways within the matted growth. Replacing the clapper rail and seaside sparrow on the high marsh are the willet (Catoptrophorus semipalmatus) and the seaside sharp-tailed sparrow (Ammospiza caudacuta).

Among the shrubby fringes of the marsh, dense growth of marsh elder, groundsel, and wax myrtle gives nesting cover for blackbirds and provides sites for heron rookeries. Remote stands of these shrubs support the nests of smaller herons and egrets, whereas tall, dead pines and human-made structures support the nests of fish-eating osprey.

Low tide brings a host of predaceous animals onto the marsh to feed. Herons, egrets, gulls, terns, willets, ibis, raccoons, and others spread over the exposed marsh floor and muddy banks of tidal creeks to feed. At high tide, the food web changes as the tide waters flood the marsh. Such fish as silversides (Menidia menidia), killifish (Fundulus heteroclitus), and four-spined stickleback (Apeltes quadracus), restricted to channel waters at low tide, spread over the marsh at high tide, as does the blue crab.

29.8 Mangroves Are Rich in Animal Life

Mangals are faunally rich, with a unique mix of terrestrial and marine life. Living and nesting in the upper branches are numerous species of birds, particularly herons and bitterns. As in the salt marsh, *Littorina* snails live on the prop roots and trunks of mangrove trees. Attached to the stems and prop roots are barnacles, and on the mud at the base of the roots are detritus-feeding snails (Figure 29.17). Fiddler crabs and tropical land crabs burrow into the mud during low tide and live on prop roots and high ground during high tide. In the Indo-Malaysian mangrove forests live mud skippers, fish of the genus *Periophthalmus*, with modified eyes set high on the head. They live in burrows in the mud and crawl about on the top of it. In many ways, they act more like amphibians than fish. The sheltered waters about the roots provide a nursery and haven for the larvae and young of crabs, shrimp, and fish.

29.9 Wetlands Have Long Been Considered Wastelands

For centuries, we have looked at wetlands as forbidding, mysterious places, sources of pestilence, home to dangerous and pestiferous insects, the abode of slimy, sinister creatures that rise out of swamp waters. They have been looked upon as places that should be drained for more productive uses by human standards: agricultural land, solid waste

FIGURE 29.17 The root system of the red mangrove is used as nursing areas by many marine animals. Detritus-feeding snails, barnacles, and oysters attach themselves to the mangrove prop roots.

dumps, housing, industrial developments, and roads. The Romans drained the great marshes around the Tiber to make room for the city of Rome. William Byrd described the Great Dismal Swamp on the Virginia–North Carolina border as a "horrible desert, the foul damps ascend without ceasing." In spite of the enormous amount of vacant dry land available in 1763, a corporation called the Dismal Swamp Land Company, owned in part by George Washington, failed in an attempt to drain the western end of the swamp for farmland. Although severely affected over the past 200 years, much of the swamp still remains as a wildlife refuge.

The rationales for drainage are many. The most persuasive relates to agriculture. Drainage of wetlands opens many hectares of rich organic soil for crop production. In the prairie country, agriculturalists viewed the innumerable potholes as an impediment to efficient farming. Draining them tidies up fields and allows unhindered use of large agricultural machinery. There are other reasons, too. Wetlands are viewed as an economic liability by landowners and local governments. They produce no economic return, and they provide little tax revenue. Many regard the wildlife that wetlands support as threats to grain crops. Elsewhere, wetlands are considered valueless lands, at best filled in and used for development. Some major wetlands have been in the way of dams. For example, the large Pymatuning Lake in the states of Pennsylvania and Ohio covers a 4200 hectare sphagnum-tamarack bog. Peat bogs in the northern United States, Canada, Ireland, and northern Europe are excavated for fuel, horticultural peat, and organic soil. In some areas, such exploitation threatens to wipe out peatland ecosystems.

Many remaining wetlands, especially in the north-central and southwestern United States, are contaminated and degraded by pesticides and heavy metals carried into them by surface and subsurface drainage and sediments from surrounding croplands. Although inputs of nitrogen and phosphorus increase the productivity of wetlands, a concentration of herbicides, pesticides, and heavy metals poisons the water, destroys invertebrate life, and has debilitating effects on wildlife, including deformities, lowered reproduction, and death. Waterfowl in wetlands scattered throughout agricultural lands are also more subject to predation, and without access to natural upland vegetation they breed less successfully.

Fifty-one percent of the population of the contiguous United States and 70 percent of all humanity live within 80 km of the coastlines. With so much humanity clustered near the coasts, it is obvious why coastal ecosystems are threatened and are disappearing rapidly. In spite of some efforts at regulation and acquisition at state and federal levels to slow the loss, coastal wetlands in the United States are still disappearing at a rate of 8000 hectares a year. Commonly regarded as economic wastelands, salt marshes have been and are being ditched, drained, and filled for real estate development (everyone likes to live at the water's edge), industrial development, and agriculture. Reclamation of marshes for agriculture is most extensive in Europe, where the high marsh is enclosed within a sea wall and drained. Most of the marshland and tideland in Holland has been reclaimed in this fashion. Many coastal cities such as Boston, Amsterdam, and much of London have been built on filled-in marshes. Salt marshes close to urban and industrial developments become polluted with spillages of oil, which becomes easily trapped within the vegetation.

The most rapid loss of wetlands in the United States is taking place on the coasts of Texas, Mississippi, and Louisiana. Two-thirds of the loss has been to deepwater, the remaining third to urban development. At the present rate of destruction, about 125 km^2/year, most of the Louisiana coastal marshes will be gone in two decades. Canals laced through the marshes to accommodate oil-drilling rigs allow the intrusion of saltwater and disrupt natural hydrology. Flood-control structures along the Mississippi interfere with the buildup of sediments in the delta areas necessary to maintain salt marshes along a sinking coastline. This sinking is further aggravated by extracting oil, gas, and groundwater, which allows the surface to subside.

Losses of coastal wetlands have a pronounced effect on the salt marsh and associated estuarine ecosystems. They are the nursery ground for commercial and recreational fisheries. There is, for example, a correlation between the expanse of coastal marsh and shrimp production in Gulf coastal waters. Oysters and blue crabs are marsh-dependent, and the decline of these important species relates to loss of salt marshes. Coastal marshes are major wintering grounds for waterfowl. One-half of the migratory waterfowl of the Mississippi Flyway depend on Gulf Coast wetlands, and the bulk of the snow goose population winters on the coastal marshes from the Chesapeake Bay to North Carolina. These geese may remove by grazing or uprooting nearly 60 percent of the below-ground production of marsh vegetation. Forced concentration of these wintering migratory birds into shrinking salt-marsh habitats could jeopardize marsh vegetation and the future.

In tropical coastal regions, humans since antiquity have exploited mangrove, mostly for firewood. Such exploitation probably destroyed the mangrove forests of the Red Sea and Persian Gulf. Large mangroves have been cut for fence posts, poles, and timber. In the Indo-Malaysian region, where mangroves are an important timber resource, silvicultural management systems have been developed for them. Pulp of certain species is used in the manufacture of rayon, lacquers, and cellulose acetate, and tannin is extracted from the bark. In the past, at least, bark and fruit have been used as a source of medicinals for treatment of rheumatism, boils, and eye infections. In addition, mangroves stabilize and protect the coast against erosion and shelter and support important commercial fisheries.

In spite of their value, mangrove forests have been and are being destroyed by filling and dredging for commercial development, marinas, and condominiums. They even are used as solid-waste dumps. In parts of the Pacific, especially the Philippines, mangals have been cleared for rice lands and for mariculture—raising fish, crabs, and prawns in brackish pools. The most massive destruction of mangals took place in Vietnam, where herbicidal spraying during the Vietnam War destroyed an estimated 100,000 hectares. The mangals have never recovered.

Our disregard for wetlands and their values is underscored by the destruction we have imposed on them. Wetlands, both forested and nonforested, once made up about 3 percent of Earth's surface, but much of that area, especially in the Northern Hemisphere, has been converted to other land uses. In colonial times, the area now embraced by the 50 United States contained some 392 million acres of wetlands. Of these, 221 million acres were in the lower 48 states, 170 million acres in Alaska, and 59,000 acres in Hawaii. Now, over 200 years later, Alaska has lost a fraction under 1 percent, Hawaii 12 percent, and the lower 48 states well over 50 percent of their wetlands. Among the states, California has lost 91 percent of its wetlands. Wetlands, which once made up 5 percent of that state's total land area, have shrunk to one-half of 1 percent. Over the continental United States, the 392 million acres of wetland have decreased to 274 million acres (Figure 29.18), and many of these remnants are degraded.

The loss of wetlands has reached a point where both environmental and socioeconomic values—including waterfowl habitat, groundwater supply and quality, floodwater storage, and sediment trapping—are in jeopardy. Although the United States has made some progress toward preserving the remaining wetlands through legislative action and land purchase, the future of freshwater wetlands is not secure. Apathy, hostility toward wetland preservation, political maneuvering, court decisions, and arguments over what constitutes a wetland allow the continued destruction of wetlands at a rate of over 200,000 hectares per year.

(a) 1780s

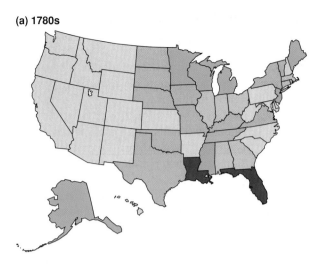

(b) 1980s

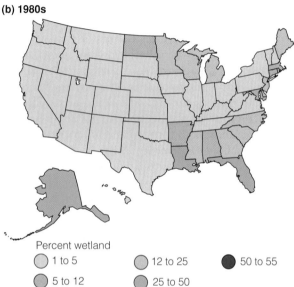

Percent wetland
- 1 to 5
- 5 to 12
- 12 to 25
- 25 to 50
- 50 to 55

FIGURE 29.18 The loss of wetland in the United States over 200 years (From U.S. Department of Agriculture).

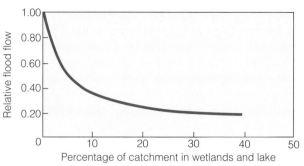

FIGURE 29.19 The influence of different percentages of wetlands in a watershed on relative flood flows in Wisconsin.

29.10 Wetlands Are Important Ecologically and Economically

Wetlands assume an ecological and economic importance out of proportion to their size. Their major contribution is to the hydrology of a region. Basin wetlands, in particular, are groundwater recharge points. They hold rainwater, snowmelt, and surface runoff in their basins and discharge the water slowly into the aquifers. These same basins also function as natural flood-control reservoirs. As little as 5 percent of the watershed or catchment area in wetlands can reduce flood flows by as much as 50 percent (Figure 29.19).

Wetlands act as water-filtration systems. Wetland vegetation takes up excessive nitrogen, phosphorus, sulfates, copper, iron, and other heavy metals brought by surface runoff and inflow, incorporates them into plant biomass, and deposits much of them in anaerobic bottom muds. Because of this ability to filter out heavy metals and reduce pH, we are beginning to treat urban wastewater and drainage from surface mines by diverting these flows into natural or specially created wetlands.

Wetlands contribute to the human economy in other ways. They provide places of recreation, sources of horticultural peat and timber—notably baldcypress and bottomland hardwoods in the southern United States—and sites for growing cranberries in the northeastern United States. These uses, however, tend to interfere with the natural function and integrity of the wetland ecosystems.

Wetlands are vitally important as wildlife nesting and wintering habitat. Many species of wildlife, some of them endangered, are dependent on wetlands. Worldwide, wetlands are home to numerous species of amphibians and reptiles, including alligators and crocodiles. Many species of fish of the Amazon and other tropical rivers depend on seasonal flooding of riverine swamps and floodplains to forage for terrestrial foods and to spawn. Waterfowl, wading birds, gulls and terns, herons, and storks depend on marshes and wooded swamps for nesting and foraging. In fact, the prairie pothole region of the north-central United States and Canada is used by two-thirds of the continent's 10 to 12 million waterfowl as a nesting area. Waterfowl in concentrated numbers use southern marshes and swamps as wintering habitat. Moose, hippopotamus, waterbuck, otter, and muskrat are mammalian inhabitants of wetlands.

Summary

Wetland Characteristics and Types (29.1–29.2) Wetlands can be defined as a community of hydrophytic plants occupying a gradient of soil wetness from permanently flooded to periodically saturated during the growing season. Hydrophytic plants are those adapted to grow in water or on soil that is periodically deficient in oxygen (29.1). Wetlands dominated by grasses and herbaceous hydrophytes are marshes. Those dominated by wooded vegetation are forested wetlands or shrub swamps. Wetlands characterized by an accumulation of peat are mires. Mires fed by water moving through the mineral soil and dominated by sedges are fens; those dominated by sphagnum moss and dependent largely on precipitation for moisture and nutrients are bogs. Bogs are characterized by blocked drainage, an accumulation of peat, and low productivity. Salt marshes, associated with estuaries and coastal areas, are dominated by salt-tolerant grasses and flooded by daily tides from the low marsh to the high marsh (29.2).

Hydrology Structures Wetlands (29.3 –29.5) The structure and function of wetlands are strongly influenced by their hydrology—both the physical movement of water and hydroperiod. Hydroperiod is the depth, frequency, and duration of flooding. Hydroperiod influence on vegetation is most evident in basin wetlands that exhibit zonation from deepwater submerged vegetation to wet-ground emergents (29.3). The interaction of salinity, tidal flow, and height produces a distinctive zonation of vegetation in the salt marsh. Salt marsh cordgrass dominates the marsh flooded by daily tides. Higher microelevations, shallowly flooded only by spring tides and subject to higher salinity, support salt meadow cordgrass and spike grass (29.4). In tropical regions, mangrove forests or mangals replace salt marshes and cover up to 70 percent of coastlines. Uniquely adapted to a tidal environment, many mangrove species have supporting prop roots that carry oxygen to the roots, and seeds that grow into seedlings on the tree and drop into the water to take root in the mud (29.5).

Wetland Wildlife (29.6–29.8) Wetlands support a diversity of wildlife. Freshwater wetlands provide essential habitats for frogs, toads, turtles, and a diversity of invertebrate life. Nesting, migrant, and wintering waterfowl depend upon them (29.6). Salt-marsh animals are adapted to tidal rhythms. Detrital feeders such as fiddler crabs and their predators are active at low tide; filter-feeding ribbed mussels are active at high tide (29.7). Mangroves support a unique mix of terrestrial and marine life. The sheltered water around prop roots provides a nursery for the larvae and young of crabs, shrimp, and fish (29.8).

Wetland Values (29.9) Wetlands have an ecologic and economic value out of proportion to their extent. They function as recharge points for groundwater aquifers, water storage basins that reduce the intensity of flooding, water-filtration systems for pollutants, sources of wood products, and habitat for a rich diversity of wetland wildlife. Mangroves are economically important as sources of wood, tannin, and food, and as protection against coastal erosion.

Threatened Wetlands (29.10) In spite of their importance, over 50 percent of the original wetlands of the United States (excluding Alaska and Hawaii) have been drained, mostly for agriculture and development. Pesticides, herbicides, heavy metals, and excessive inputs of nitrogen and phosphorus from surrounding watersheds contaminate many remaining wetlands. We are making an effort to save wetlands, but progress is slow. Because of their accessible location, salt marshes are rapidly disappearing under the pressure of commercial and residential development. Because of the extreme importance of wetlands to estuarine fisheries, the loss of coastal wetlands can have serious ecological and economic implications. Like salt marshes, mangrove swamps are being destroyed by development and drainage for agriculture.

Study Questions

1. What is a wetland? A hydrophyte?
2. How does the definition of a wetland relate to the gradient of soil wetness?
3. What are the three major types of wetlands in terms of hydrology?
4. Characterize the types of wetlands by their vegetation.
5. How do wetlands relate to the landscapes in which they occur?
6. What is the hydroperiod, and how does it relate to the structure of wetlands?
7. What influences the major structural features of a salt marsh?
8. What are the unique characteristics and adaptations of mangroves?
9. What are the typical fauna of the salt marsh?
10. Why are wetlands worth saving?
11. What is the paradox of draining wetlands, then building large flood-control dams?
12. What has been the impact of draining prairie potholes and bottomland swamps?
13. What is the ecological and economic importance of mangals?
14. What is the fate of many mangrove forests?
15. What has been the fate of wetlands in your region? How much has been lost? To what use has the drained land been put?
16. Trace the history of the attitudes toward wetlands from Roman times to the present. Why are we fearful of them?

17. How could you argue for preservation of wetlands on an economic basis to developers and local governments? Can we put monetary values on wetlands?

18. A battle often forms between those who would preserve coastal wetlands and those who would destroy them. Build a set of arguments for both sides. What side in the long term has the strongest case?

19. If you live or vacation along the coast, investigate the conditions of and attitudes toward salt marshes in that area. Of what importance are those salt marshes to wintering waterfowl? To the local economy?

20. Investigate the importance of mangrove forests to a selected area—West Africa, Australia, Indo-Malaysia, the Philippines.

CHAPTER 30

Earth as seen from Apollo 17 on its way to the moon.

Global Environmental Change

OBJECTIVES

Upon completion of this chapter, you should be able to:

- Identify the major causes of rising atmospheric CO_2 concentration.

- Describe how increasing atmospheric concentrations of CO_2 influence the oceans.

- Tell how terrestrial plants respond to elevated CO_2.

- Describe how CO_2 and other greenhouse gases influence global climate patterns.

- Discuss how changing global climate will influence the distribution and abundance of ecosystems.

- Explain how global warming will influence sea level and coastal ecosystems.

- Understand how elevated atmospheric CO_2 and changes in global climate patterns could affect agricultural production.

- Discuss the potential influence of a changing global climate on human health.

The term *environmental change* is redundant. Change is inherent in Earth's environment. Paleoecology has recorded the response of populations, communities, and ecosystems to changes in climate during periods of glacial expansion and retreat over the past 100,000 years. On an even longer time scale, the geological and fossil records recount environmental and evolutionary change since the origin of the planet.

Humans, like other species, are part of Earth's environment. Anthropologists tell us that the first humans were hunters and gatherers with low populations (see the Ecological Application after Ch. 12: Cheating Nature). At that time, humans functioned as natural predators and herbivores. With the adoption of agriculture, believed to have begun with the cultivation of grasses in the region of the Middle East some 10,000 years ago, human populations and associated conversion of land to agricultural production began to increase exponentially. That trend has continued into modern times (Figure 30.1). In the mid-1800s, the Industrial Revolution once again changed the nature of human interactions with the global environment. The demand for energy to fuel industrialization and the concentration of populations, or urbanization, brought environmental problems of unprecedented magnitude.

In this chapter, we examine a small subset of the environmental changes that are brought about by human activities and are global in scale. The nitrogen and sulfur compounds released in industrial emissions (see Chapter 23), although extensive in their distribution, are largely a regional concern. In contrast, other gaseous emissions such as carbon dioxide, methane, and chlorofluorocarbons (CFCs) mix readily and circulate globally in the atmosphere. To understand and predict the consequences of these emissions on local and regional environments, scientists must study them at a global scale. The changes in atmospheric chemistry brought about by these emissions have the potential to change Earth's climate, shifting the distribution and abundance of ecosystems and directly affecting the health and well-being of the human population.

30.1 Atmospheric Concentration of Carbon Dioxide Is Rising

The atmospheric concentration of carbon dioxide has increased by more than 25 percent over the past 100 years. The evidence for this rise comes primarily from continuous observations of atmospheric CO_2 started in 1958 at Mauna Loa, Hawaii, by Charles Keeling (Figure 30.2) and from parallel records around the world. Evidence before the direct observations of 1958 comes from a variety of sources, including the analysis of air bubbles trapped in the ice of glaciers in Greenland and Antarctica.

Reconstructing atmospheric CO_2 concentrations over the past 300 years, we see values that fluctuate between 280 and 290 ppm until the mid-1800s (Figure 30.3). After the onset of the Industrial Revolution the value increased steadily, rising exponentially by the mid-19th century onward. The change reflects the combustion of fossil fuels (coal,

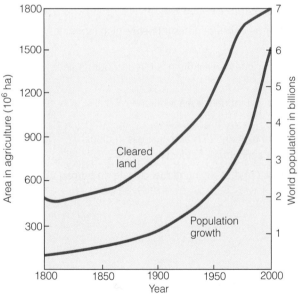

FIGURE 30.1 Changes in global population and land area cleared for agriculture over the past 200 years.

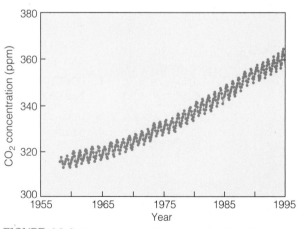

FIGURE 30.2 Concentration of atmospheric CO_2 at Mauna Loa Observatory, Hawaii. The dots depict monthly averages. (Keeling and Wohrf 1994.)

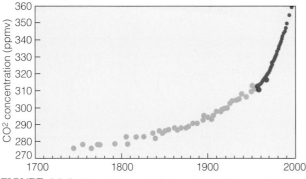

FIGURE 30.3 Historical record of atmospheric CO_2 over the past 300 years. Data prior to direct observation (1958 to present) are estimated from various techniques, including analysis of air trapped in Antarctic ice sheets. (Adapted from Watson et al. 1996.)

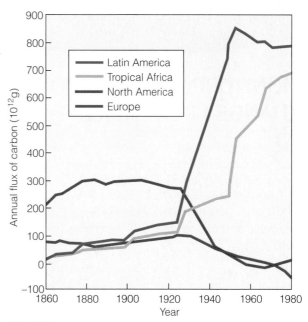

FIGURE 30.5 Historical record of annual input of CO_2 to the atmosphere from the clearing and burning of forest in North America, South America, Central America, Europe, and Africa. (From Houghton 1997.)

oil, and gas) as an energy source for industrialized nations (Figure 30.4).

The burning of fossil fuels is not the only cause of rising atmospheric CO_2 concentration. Deforestation is also a major cause (Figure 30.5). Forested lands are typically cleared and burned for farming. Although the trees may be harvested for timber or wood pulp, a large part of the biomass, litter layer, and soil organic matter is burned, releasing the carbon to the atmosphere as CO_2.

Calculations of the contribution of land clearing to atmospheric CO_2 are complex. Following timber harvest on lands managed for forest production, or

on lands that have been cultivated and then abandoned, vegetation and soil organic matter become reestablished. We calculate the net contribution to the atmosphere as the difference between CO_2 released during clearing and burning and CO_2 taken up by

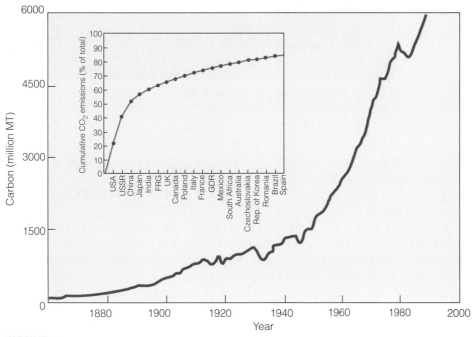

FIGURE 30.4 Historical record of annual input of CO_2 to the atmosphere from the burning of fossil fuels since 1860. The insert graph shows cumulative amounts of CO_2 emissions from the 20 largest CO_2-emitting nations in 1989 plotted as a proportion of total emissions. (Adapted from Marland and Boden 1993.)

MONITORING VEGETATION CHANGE FROM SPACE

Before the mid-1980s, scientists had to estimate vegetation cover and land use from aerial photographs and ground surveys. In the absence of these types of data, they relied on regional statistics for population or agricultural production. Now satellite remote sensing provides images of land cover on a regional and global scale.

Satellite systems sample the energy returning to space from Earth's surface in different ways. Two satellite remote sensing systems widely used for terrestrial research are LANDSAT and AVHRR (Advanced Very High Resolution Radiometer). These two systems measure different regions of the energy spectrum reflected by Earth's surface (see Section 4.1, Figure 4.2), but both use the fact that plants

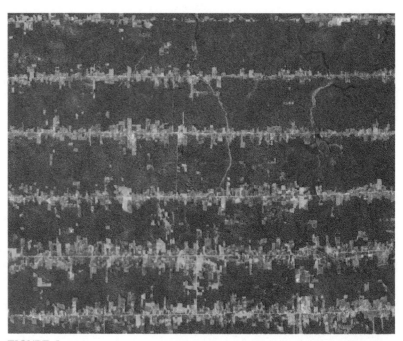

FIGURE A AVHRR image of Rondonia, Brazil.

photosynthesis and the accumulation of biomass during reestablishment. At one time, scientists used regional estimates of population growth and land use (forestry and agriculture) together with simple models of vegetation and soil succession to estimate the contribution due to changing land use. More recent estimates use satellite images to quantify these changes (see Focus on Ecology 30.1: Monitoring Vegetation Change from Space).

30.2 Tracking the Fate of CO$_2$ Emissions

Scientists estimate that the amount of carbon released to the atmosphere between 1958 and 1980 from the combustion of fossil fuel is approximately 85.5 gigatons (a gigaton is 10^9 metric tons). To put this number into perspective, if the average weight of a human is 70 kilograms (approximately 150 pounds), a gigaton would be the weight of over 14 billion people, or more than two times the world's population. This number also puts the per capita use of fossil fuels into perspective.

absorb a specific range of wavelengths (PAR, photosynthetically active radiation—see Section 6.1). The systems differ in their spatial resolution and in how often they return to a given region to record an image. LANDSAT has a spatial resolution of 30 m × 30 m and returns to a given point on Earth once every 16 days. AVHRR has a resolution of 1000 m and provides daily global coverage.

Satellite images of forest clearing in the Brazilian province of Rondonia are shown in Figures A and B. To stimulate settlement in the Amazon basin, the government constructs roads in remote areas of the rain forest. These two images show both the pattern of clearing (light-colored areas) and the difference in resolution of information between the two systems.

FIGURE B LANDSAT image of Rondonia, Brazil.

Direct measurements of atmospheric CO_2 over this same period show only a 49.9 gigaton increase in the amount of carbon in the atmosphere. The difference, 35.6 gigatons, must have flowed from the atmosphere into the other main pools in the global carbon cycle (see Figure 22.4), the oceans and terrestrial ecosystems. Determining the fate of CO_2 put into the atmosphere from the burning of fossil fuel requires input from a variety of scientific disciplines, as well as a large dose of detective work.

The process of diffusion controls uptake of carbon dioxide from the atmosphere into the oceans. Because physical processes largely control this transfer, scientists are able to make accurate estimates of the movement of CO_2 from the atmosphere into the oceans. In general, there is a net uptake of carbon from the atmosphere to the oceans (see Figure 22.4). In contrast, although the processes controlling the exchange of carbon between terrestrial ecosystems and the atmosphere are generally well understood, quantifying these processes at a regional to global scale is extremely difficult. As a result, a simple process of elimination is used to estimate global uptake of carbon by terrestrial ecosystems. Carbon that has been emitted over a specified period of time but cannot be accounted for by measurements of

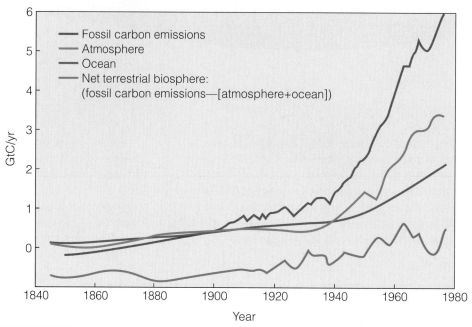

FIGURE 30.6 Emission of carbon from the burning of fossil fuels and changes in the major global reservoirs. Atmospheric values are based on direct observations and estimates from the analysis of ice cores. Ocean uptake is estimated using the GFDL ocean carbon model, and the net uptake by terrestrial ecosystems is calculated as the remaining difference. (From Houghton 1994.)

atmospheric carbon concentration or estimates of oceanic absorption is relegated to the terrestrial ecosystems (Figure 30.6).

Using this approach, Earth's terrestrial ecosystems are a net sink of carbon (in other words, uptake in photosynthesis exceeds losses due to respiration; see the global carbon cycle in Figure 22.4). Prior to this type of analysis, it was generally believed that there is no net uptake of carbon by terrestrial ecosystems at a global scale. Some studies suggest that any possible net uptake of carbon by terrestrial ecosystems may result from reforestation in the temperate regions of the Northern Hemisphere. Reforestation follows the large-scale abandonment of lands cleared for agriculture during the latter part of the 19th and early part of the 20th centuries. Although reforestation has not been proven to be the solution to the problem of balancing the global carbon cycle, it is most certainly a key component. Determining the fate of carbon input to the atmosphere through the burning of fossil fuels requires an understanding of the processes controlling the exchange of carbon among the major components of the global carbon cycle, and how the transfers might be influenced by rising atmospheric concentrations of CO_2.

30.3 Atmospheric CO_2 Concentrations Affect CO_2 Uptake by Oceans

Carbon dioxide diffuses from the atmosphere into the surface waters of the ocean, where it dissolves and undergoes a number of chemical reactions, including the transformation to carbonates and bicarbonates (Chapter 4, Section 4.16). If the concentration of CO_2 in the water were greater than that in the air, the process would be reversed, and CO_2 would diffuse from the surface waters of the ocean to the atmosphere.

As the concentration of CO_2 in the atmosphere rises, the diffusion of CO_2 into surface waters increases. Should the concentrations in the atmosphere and the surface waters become equal (if equilibrium is achieved), the net exchange would stop. However, there are two main reasons why equilibrium does not occur. First, the concentration of CO_2 in the atmosphere continues to rise as a result of fossil fuel combustion. Second, CO_2 in the surface waters transforms into bicarbonate and other carbon compounds, and plants take up both CO_2 and bicarbonates in photosynthesis; both of these processes lower CO_2 concentrations in the surface waters, helping to maintain the diffusion gradient. The ocean then acts

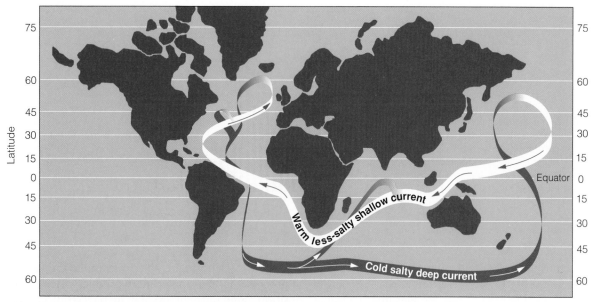

FIGURE 30.7 The major pattern of global ocean circulation referred to as the ocean conveyer system. Atlantic surface waters, flowing northward from the tropics, cool and sink when they reach sub-Arctic latitudes. After sinking, these waters become part of the huge, deep, southward countercurrent reaching all the way to the Antarctic. From there, cold, dense flows of deepwater extend far northward into the Pacific Ocean, balanced by a shallow return forming a long loop extending from the northern Pacific through the Indian Ocean and back to the Atlantic. Recent research shows that the surface route (warm water) south of Africa is weak and intermittent, effectively splitting the global circulation into two large-scale cells.

as a sink, drawing CO_2 from the atmosphere as its concentration there rises.

Given their volume, the oceans have the potential to absorb most of the carbon that is being transferred to the atmosphere by fossil fuel combustion and land clearing. This potential is not realized because the oceans do not act as a homogeneous sponge, absorbing CO_2 equally into the entire volume of water.

The oceans consist of two layers, the surface waters and deepwaters (see Figures 21.7 and 21.8). The average depth of the oceans is 2000 meters. Intercepted solar radiation warms the surface waters. Depending on the amount of radiation reaching the surface, the zone of warm water can range from 75 to 200 meters in depth. The average temperature of this surface layer is 18°C. The remainder of the vertical profile (200 to 2000 meter depth) is deepwaters, whose average temperature is 3°C. The transition between these two zones is abrupt; we call it the thermocline (see Chapter 21, Section 21.7). In effect, the oceans can be viewed as a thin layer of warm water floating on a much deeper layer of cold water. The temperature difference between these two layers leads to the separation of many processes. Turbulence caused by winds mixes the surface waters, transferring CO_2 absorbed at the surface into the waters below. Due to the thermocline, however, this mixing does not extend into the deepwaters. Mixing between surface waters and deepwaters depends on

deep ocean currents caused by the sinking of surface waters as they move toward the poles (Figure 30.7). This process occurs over hundreds of years, limiting the short-term uptake of CO_2 by the deepwaters. The result is that the amount of CO_2 that can be absorbed by the oceans over the short term is limited, despite the large volume of water.

30.4 Plants Respond to Increased Atmospheric CO_2

Carbon dioxide flows from the atmosphere into terrestrial ecosystems via photosynthesis (see Chapter 6, Section 6.1). To understand how rising atmospheric CO_2 concentrations influence the productivity of terrestrial ecosystems, we must understand how photosynthesis responds in an enriched CO_2 environment.

Recall that CO_2 diffuses from the air into the leaf through the stomatal openings (see Section 6.3). The higher the CO_2 concentration in the outside air, the greater the rate of diffusion into the leaf. A higher rate increases the availability of CO_2 for photosynthesis in the mesophyll cells of the leaf, so it generally results in a higher rate of photosynthesis. The higher rates of diffusion and photosynthesis under

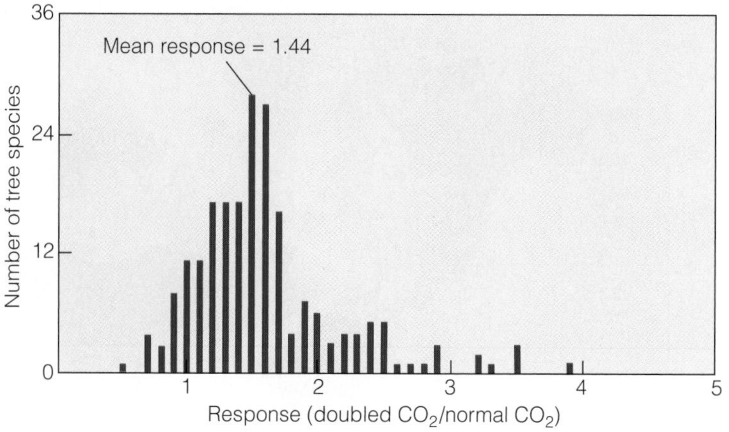

FIGURE 30.8 The change in photosynthesis for tree species grown under a doubled CO_2 concentration. Response is expressed as the ratio of observed photosynthesis under a doubled CO_2 concentration to that under normal CO_2 level. A value of 1 represents no detected increase in photosynthetic rates under a doubled CO_2 concentration. (From Amthor 1995.)

elevated atmospheric concentrations of CO_2 have been termed the "CO_2 fertilization effect." Plant species vary in their response to elevated CO_2, although most show an increase in photosynthetic rates. A review of over 200 experiments examining the photosynthetic response of tree species to a doubling of CO_2 concentration showed an average increase of 44 percent (Figure 30.8).

In addition to increased rates of photosynthesis, plants exposed to a doubled CO_2 concentration exhibit a partial closure of the stomata, reducing water loss. Thus, under elevated CO_2 levels, plants increase their water-use efficiency (carbon uptake/water loss).

The effects of long-term exposure to elevated CO_2 on plant growth and development, however, may be more complicated. In some studies, the enhanced effects of elevated CO_2 levels on plant photosynthesis have been short-lived. Some plants produce less of the photosynthetic enzyme rubisco under elevated CO_2, reducing photosynthesis to rates comparable to those measured at lower CO_2 concentrations, a phenomenon referred to as down-regulation. Other studies reveal that plants grown under increased CO_2 levels allocate less carbon to the production of leaves and more to the production of roots. In addition, plants grown under elevated CO_2 appear to produce fewer stomata on the leaf surface. The smaller leaf area and lower stomatal density reduce water loss, but also reduce growth rates.

It is uncertain whether the results observed for leaves or single plants translate into changes in the net primary productivity of terrestrial ecosystems. The availability of water or nutrients in many ecosystems may limit potential increases in plant productivity under elevated CO_2 concentrations. A number of large-scale experiments currently seek to examine the effects of elevated CO_2 on whole ecosystems (see Focus on Ecology 30.2: Raising CO_2 in a Forest). By exposing whole areas of forest and grassland to elevated CO_2, scientists are able to examine the variety of processes influencing primary productivity, decomposition, and nutrient cycling in terrestrial ecosystems. A comparison of field studies in grassland and agricultural ecosystems reveals an average increase in biomass production of 14 percent under elevated CO_2 (double ambient concentrations). However, estimates at individual sites ranged from an increase of 85 percent to a decline of almost 20 percent. These results stress the importance of the interactions of elevated CO_2 with other environmental factors, particularly temperature, moisture, and nutrient availability.

Ecosystems characteristic of low-temperature environments tend to show an initial enhancement of productivity following elevated CO_2, followed by down-regulation. A study conducted by Walter Oechel and colleagues examined the response of Arctic tundra in Alaska to elevated CO_2. They observed an initial increase in productivity, but primary productivity returned to original levels following 3 years of continuous exposure to a doubled CO_2 environment.

The largest and most persistent responses to elevated CO_2 have been observed in seasonally dry environments, where primary productivity is enhanced during years of below-average rainfall. In a study of a tallgrass prairie ecosystem in Kansas, C.E. Owensby and colleagues found no significant increase in above-ground net primary productivity (NPP) during wet years (greater-than-average rainfall) for experimental plots exposed to a doubled CO_2 concentration when compared with control plots receiving ambient CO_2. In contrast, they observed a 40 percent increase in above-ground NPP during years with average rainfall, and an 80 percent increase during years with below-average precipitation. Even though these relative increases in NPP are large, they occur in years of low NPP, so that absolute changes may be quite low.

The enhancement of primary productivity by elevated CO_2 in dry environments arises largely from

RAISING CO$_2$ IN A FOREST: WHAT HAPPENS?

Most of our understanding of how terrestrial ecosystems respond to elevated CO$_2$ comes from experiments on single plants or small groups of plants grown either in the greenhouse or under controlled conditions in the field. Although important, these experiments provide limited insight into ecosystem dynamics.

A new experimental approach modifies CO$_2$ concentrations in whole ecosystems. The new experimental approach is called FACE—**F**ree **A**ir **C**arbon Dioxide **E**nrichment. Figure A shows the FACE facility at the Duke University Forest in North Carolina. The tower reaching above the forest canopy at the center of the ring is 35 m tall. It contains instruments to measure features of the climate (including temperature, water vapor, solar radiation, and wind speed and direction) and the exchange of CO$_2$ between the atmosphere and the forest (net ecosystem productivity). The vertical pipes that form a ring around the site have a series of nozzles along their length that release CO$_2$ into the air. A computer-controlled system continuously monitors CO$_2$ concentrations in the forest (within the circle of pipes); it releases CO$_2$ from the nozzles as a function of CO$_2$ concentration, wind direction, and speed. This system provides a fairly uniform concentration of doubled CO$_2$ in the forest.

Scientists are conducting a variety of experiments at the site. They measure photosynthesis, transpiration, productivity, decomposition, and nutrient cycling and compare the experimental site with the surrounding forest.

FIGURE A The Free Air Carbon Dioxide Enrichment facility at Duke University.

the small reduction in transpiration resulting from partial closure of the stomata. These small reductions have resulted in measurable changes (increases) in soil moisture in grassland ecosystems, particularly during prolonged dry periods. Increased soil moisture both extends the growing season and increases soil microbial activity, decomposition, and nitrogen mineralization (see Chapter 7).

30.5 Greenhouse Gases Are Changing the Global Climate

Carbon dioxide in the atmosphere does not readily absorb shortwave radiation from the Sun, but it does absorb long-wave or thermal radiation (see Figure 4.2). Earth's surface emits absorbed solar radiation to the atmosphere as long-wave or thermal energy (see Chapter 3, Figure 3.1). Atmospheric CO$_2$ and water vapor trap this thermal energy, warming the atmosphere. Because it traps heat, we call carbon dioxide a greenhouse gas. Were it not for the warming effect of carbon dioxide, Earth would be several degrees cooler (see Chapter 3, Section 3.1).

As human activities increase the atmospheric concentration of CO$_2$, will they influence the global climate? Scientists estimate that at current rates of emission, the preindustrial level of 280 ppm of CO$_2$ in the atmosphere will double by the year 2020. Moreover, CO$_2$ is not the only greenhouse gas increasing as a result of human activities. Others include methane (CH$_4$), chlorofluorocarbons (CFCs), hydrogenated chlorofluorocarbons (HCFCs), nitrous oxide (N$_2$O), ozone (O$_3$), and sulfur dioxide (SO$_2$) (see Chapter 23). Although much lower in concentration, some of these gases are much more effective at trapping heat than CO$_2$. They are significant components of the total greenhouse effect.

Although the role of greenhouse gases in warming Earth's surface is well established, the specific influence that doubling the CO$_2$ concentration of the atmosphere will exert on the global climate system is much more uncertain. Atmospheric scientists

MODELING EARTH'S CLIMATE

General circulation models or GCMs are computer-based models that predict future climate patterns. The models are based on physical laws relating to the conservation of mass, energy, and momentum that are represented by mathematical equations. They use the same equations to describe the atmosphere as the weather prediction models used by meteorologists to make short-term forecasts over periods of 1 to 5 days. GCMs, however, are run for much longer periods of time—years to decades—to provide a statistical (mean and variability) picture of climate. By altering features of the atmosphere, such as the concentration of carbon dioxide, the models are able to examine the long-term consequences on the global climate system.

Modern GCMs are actually several submodels that are coupled: submodels that describe the atmosphere, land surface (and biosphere), and oceans. The ocean model simulates surface and deep ocean circulations, and is coupled to the atmosphere by the sea surface, where the transfers of water (evaporation/precipitation) and momentum occur. The land-surface model simulates transfers

between the biosphere and the atmosphere (photosynthesis, respiration, evaporation, and transpiration), and the storage and runoff of water to the oceans.

GCMs are three-dimensional models: latitude, longitude, and elevation. The horizontal resolution varies among the different GCMs, but typically ranges from 125 to 250 km. The vertical dimension that represents the atmosphere is approximately 1 km, while the typical ocean model has a vertical resolution of about 200 to 400 meters. Predictions of mean annual temperature change under a doubled atmospheric concentration of CO_2 are presented in Figure A for the CGM developed at the CSIRO research laboratory in Australia. The horizontal resolution of this model is 625 km by 350 km (latitude, longitude), as can be seen by the coarse representation of the land surface (continents).

Climate models have continued to develop over the past few decades as computing power has increased. During that time, models of the main components—atmosphere, land, and ocean—have been developed

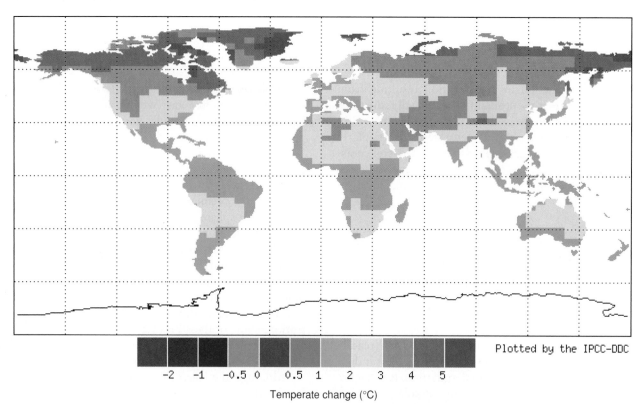

Plotted by the IPCC-DDC

Temperate change (°C)

FIGURE A Predicted change in mean annual temperature relative to the observed mean for the period 1961 to 1990. Predictions are from the CSIRO-Mk2 general circulation model using a doubled atmospheric concentration of CO_2. Values of temperature were average for the decade beginning in 2050.

separately and then gradually integrated. This coupling of the various components is a difficult process. Work is underway to incorporate a more realistic representation of the biosphere so that changes in the productivity of terrestrial ecosystems in response to elevated CO_2 and climate change can interact with the atmosphere. Most recently, components have been incorporated to represent the emissions of sulfur and how they are oxidized to form aerosol particles. The ultimate aim is to model, as much as possible, the whole of Earth's climate system. Only then can all of the components interact, allowing predictions to continuously take into account the effect of feedbacks among components that are critical to understanding and predicting the global climate system.

have developed complex computer models of Earth's climate system—called general circulation models, or GCMs for short—to help determine how increasing concentrations of greenhouse gases may influence large-scale patterns of global climate (see Focus on Ecology 30.3: Modeling Earth's Climate). Although all use the same basic physical descriptions of climate processes, GCMs at different research institutions differ in their spatial resolution and in how they describe certain features of Earth's surface and atmosphere. As a result, the models differ in their predictions (Figure 30.9).

Despite these differences, certain patterns consistently emerge. All of the models predict an increase in the average global temperature as well as a corresponding increase in global precipitation. Findings published in 2001 by the Intergovernmental Panel on Climate Change (IPCC) suggest an increase in globally averaged surface temperature in the range of 1.4°C to 5.8°C by the year 2100. These changes would not be evenly distributed over Earth's surface. Warming is expected to be greatest during the winter months and in the northern latitudes. Figure 30.10 shows the spatial variation in changes in

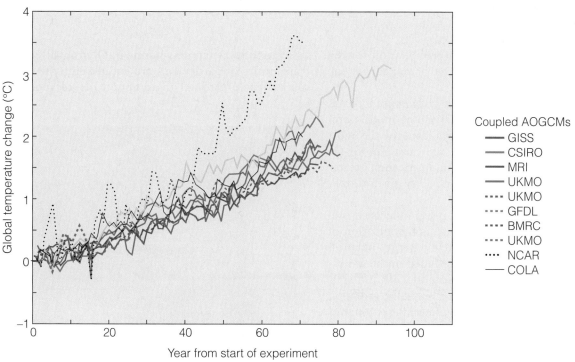

FIGURE 30.9 Comparison of predicted patterns of average global temperature change among several general circulation models. (Adapted from Kattenberg et al. 1996.)

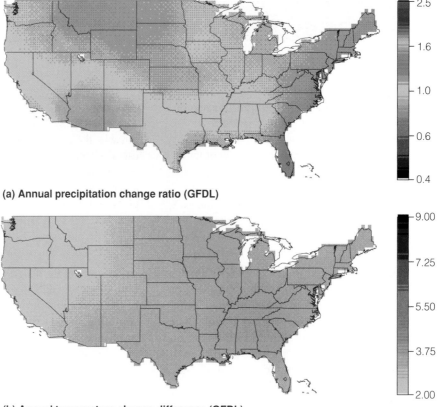

(a) Annual precipitation change ratio (GFDL)

(b) Annual temperature change difference (GFDL)

FIGURE 30.10 Changes in (a) annual temperature and (b) precipitation for a doubled CO_2 concentration estimated by the general circulation model developed by the Geophysical Fluid Dynamics Laboratory at Princeton University. Changes in temperature are expressed as absolute increases in °C. Changes in precipitation are expressed as the ratio of current to predicted annual precipitation. A value of 1.0 would imply no change, while a value of 1.5 would imply a 50 percent increase, and a value of 0.8 a 20 percent decrease. (From Vemap 1995.)

mean annual temperature and precipitation for the 48 contiguous United States as predicted by one such GCM.

Although in popular speech *greenhouse effect* is synonymous with global warming, the models predict more than just hotter days. One of the most notable predictions is an increased variability of climate, including more storms and hurricanes, greater snowfall, and increased variability in rainfall, depending on the region.

One recent development that has influenced predicted patterns of climate change is the inclusion of aerosols in calculating Earth's energy balance. Aerosols, or small particles suspended in the atmosphere, both absorb radiation from the Sun and scatter it back to space. By scattering solar radiation back to space, they function to reduce the amount of radiation reaching Earth's surface. Aerosols come from a variety of sources. In desert regions, they originate from winds blowing dust airborne. Over the oceans, aerosols come from sea spray. They also result from the burning of forests and grasslands

(referred to as biomass burning). Occasionally, large quantities of particulates are injected into the upper atmosphere through the eruption of volcanoes, as

FIGURE 30.11 A church covered in ash from the Mt. Pinatubo eruption.

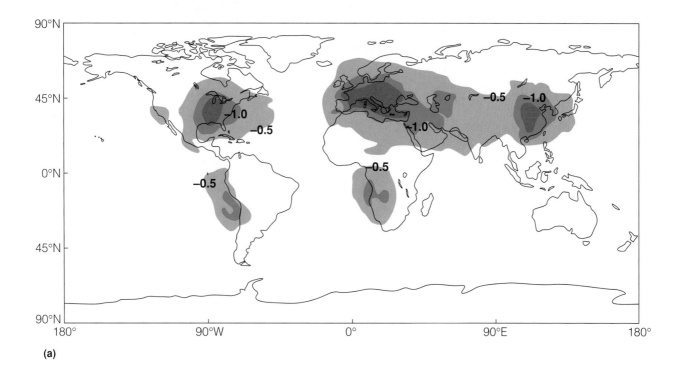

(a)

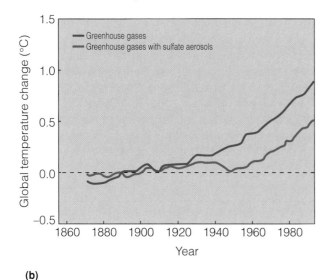

(b)

FIGURE 30.12 (a) Estimates of reduction in radiation (watts per square meter) resulting from anthropogenic sulfate aerosols in the atmosphere. The reduction is largest over regions close to the source of the emissions. (b) Predicted changes in mean global temperature for the UKMO GCM with and without the inclusion of sulfate aerosols in the simulation. (Mitchell et al. 1995.)

was the case when Mt. Pinatubo erupted in 1991 (Figure 30.11).

A major source of aerosols resulting from human activities is sulfates and soot from the burning of fossil fuels. Sulfate particles are of particular importance. They are formed from sulfur dioxide, a gas produced in large quantities by power stations that burn coal (see Chapter 23, Section 23.2). These particles remain in the atmosphere for a very short peri-

od (on average, 5 days), so their distribution is concentrated in the regions near their source (Figure 30.12a). In regions of the Northern Hemisphere, their concentration is significant and functions to offset the effects of greenhouse gases, reducing estimates of global warming (Figure 30.12b).

As models of global climate improve, there will no doubt be further changes in the patterns and the severity of changes they predict. However, the

WHO TURNED UP THE HEAT?

Is Earth's climate changing? According to the IPCC (Intergovernmental Panel on Climate Change), the answer is unequivocally "Yes." This conclusion is a result of a series of observations that allow scientists to track changes in the global climate over the past century (Figure A). Widespread direct measurements of surface temperature began around the middle of the 19th century. These direct measures from instruments such as thermometers are referred to as the "instrumental record." Observations of other surface weather variables, such as precipitation and winds, have been made for about a hundred years.

In addition to measurements made at the land surface, there are also long records of observations of sea-surface temperatures made from ships since the mid-19th century. Since the late 1970s, a network of instrumented buoys has supplemented these observations. Measurements of the upper atmosphere have been made systematically only since the late 1940s, but since the late 1970s Earth-observing satellites have been providing a continuous record of global observations for a wide variety of climate variables.

So what do these climate records reveal? The global average surface temperature has increased by 0.6°C (±0.2°C) since the late 19th century. It is very likely that the 1990s was the warmest decade and 1998 the warmest year in the instrumental record since 1861. New

analyses of daily maximum and minimum land-surface temperatures for 1950 to 1993 show that the diurnal temperature range is decreasing. On average, minimum temperatures are increasing at about twice the rate of maximum temperatures (0.2°C versus 0.1°C per decade). In other words, nighttime temperatures (minimum) have increased more than daytime temperatures (maximum) over this period.

New analyses also indicate that global ocean heat content has increased significantly since the late 1950s. More than half of the increase in heat content has occurred in the upper 300 meters of the ocean, equivalent to a rate of temperature increase in this layer of about 0.04°C per decade.

Although the consensus among scientists appears to be that the climate has changed significantly over the past century, there is continued debate over the answer to the more difficult question: "Why is it changing?" The debate centers on two points. The first relates to the nature of the instrumental data that measure the trend in land-surface temperatures. Most weather stations are located in urban areas, which are typically warmer than the surrounding rural areas (see Focus on Ecology 3.1: Urban Microclimates). Recent studies, however, have worked to remove this potential bias from the data. Current findings reported by the IPCC strongly suggests that the warming

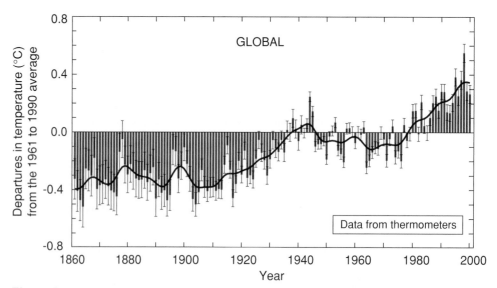

Figure A Combined annual land-surface air and sea surface temperature anomalies (°C) from 1861 to 2000. The anomalies are the differences between annual temperature for any given year and the average annual temperature for the period 1961 to 1990. Two standard errors are shown as bars for each yearly value, providing an estimate of uncertainty. (From IPCC 2001 Technical Report, Figure 2.)

trend over the past century is independent of the effects of urbanization.

The second point of debate relates to the difficulty in determining a meaningful long-term trend from an instrumental record that covers less than 2 centuries. Climate varies on a variety of time scales, and Earth has gone through periods of warming and cooling in the past. For example, the Northern Hemisphere is still recovering from the last glacial maximum some 18,000–20,000 years before the present (see Chapter 13, Section 13.14), a period when surface temperatures were much colder. Climate reconstructions of the more recent past (1000 years to present), however, suggest that the warming trend observed over the past century is consistent with that expected from the rising atmospheric concentrations of greenhouse gases. The debate will no doubt continue for years to come. The real question is: "Will the warming continue?"

physics of greenhouse gases and the consistent qualitative predictions of the GCMs lead scientists to believe that rising concentrations of atmospheric CO_2 will have a significant impact on global climate (see Focus on Ecology 30.4: Who Turned Up the Heat?).

30.6 Changes in Climate Will Affect Ecosystems at Many Levels

Climate influences almost every aspect of the ecosystem: the physiological and behavioral response of organisms (Chapters 6–8); the birth, death, and growth rates of populations (Chapters 9–11); the relative competitive abilities of species (Chapter 14); community structure (Chapters 13 and 17); productivity (Chapter 20); and cycling of nutrients (Chapters 21 and 22). Current research on greenhouse warming focuses on the response of organisms at all levels of organization: individuals, populations, communities, and ecosystems. Changes in temperature and water availability will have a direct effect on the distribution and abundance of individual species. For example, the relative abundance of 3 widely distributed European tree species is plotted as a function of mean annual temperature and rainfall in Figure 30.13. These differing environmental responses determine their distribution and abundance over the European landscape. The distribution and abundance of these 3 important tree species will change as regional patterns of temperature and precipitation change.

The potential impact of regional climate change on plant species distribution can be more clearly seen from the work of Anantha Prasad and Louis Iverson of the Northeast Research Station, U.S. Forest Service. Using data from the Inventory and Analysis Program of the U.S. Forest Service, Prasad and Iverson developed statistical models to predict the distribution of 80 different tree species inhabiting the eastern United States. Individual tree species distributions are predicted as a function of variables describing climate, soils, and topography for any location. This framework allows the investigators to predict shifts in the distribution of these tree species based on changes in temperature and precipitation for the region from a variety of GCM predictions under doubled CO_2. The predicted distributions for 3 major tree species in the eastern United States under current and doubled CO_2 climate using the Princeton GFDL GCM (see Figure 30.9) are presented in Figure 30.14. The predicted changes in temperature and precipitation will have a dramatic impact on the distribution and abundance of tree species that dominate the forest ecosystems of the eastern United States.

The distribution and abundance of animals are also directly related to features of the climate. For example, the northern limit of the winter range of the Eastern phoebe is associated with average minimum January temperatures of –4°C. The phoebe is not found in areas where temperatures drop below this value. Two lines, or isotherms, defining the region of eastern North America where average minimum January temperatures of –4°C occur are plotted in Figure 30.15. Minimum temperatures drop below –4°C in areas to the north and west of the lines, whereas temperatures are above –4°C to the south

Arolla Pine

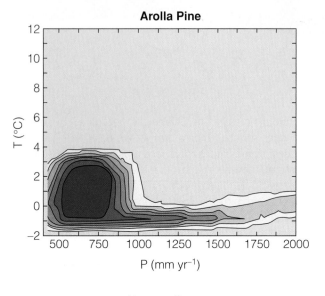

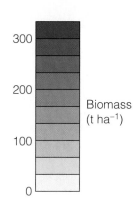

Biomass
(t ha^{-1})

Norway Spruce

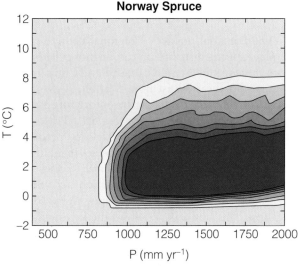

Common Beech

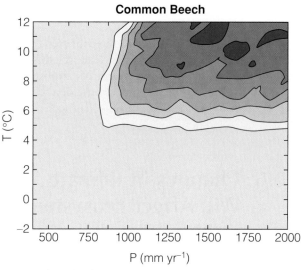

FIGURE 30.13 Abundance (biomass t/ha) of three common European tree species as it relates to mean annual temperature and precipitation. (Adapted from Miko et al. 1996.)

and east. The two isotherms show the current –4°C average minimum January temperature isotherm and the –4°C isotherm predicted by the Geophysical Fluid Dynamics Laboratory (GFDL) general circulation model for a doubled atmospheric concentration of CO_2. A change in the isotherm would be expected to result in a northern expansion of the Eastern phoebe's winter range.

Changes in climate will have a direct influence on the growth and reproductive rates of species. These changes will influence their relative competitive abilities, altering patterns of zonation and succession (see Chapters 13 and 17). Given the difficulty of experimentally changing climate conditions in the field, few studies have examined these effects. However, one such experiment was conducted in a meadow community in the Rocky Mountains of Colorado. Using

electric heaters suspended 2.6 meters above five experimental plots (Figure 30.16), scientists were able to raise soil temperature and influence soil moisture and the timing of snowmelt. In heated plots, the density of shrubs increased at the expense of grass and forb species. Results suggest that the increased warming expected under an atmosphere with a doubled concentration of CO_2 would shift the dominant vegetation of the widespread mountain meadow habitat. Shrubs would compete better in the altered environment. Such shifts have a major impact not only on plant communities, but on associated animal species as well.

In a similar approach, the International Tundra Experiment (ITEX) was established in late 1990 as a coordinated group of field experiments aimed at understanding the potential impact of warming

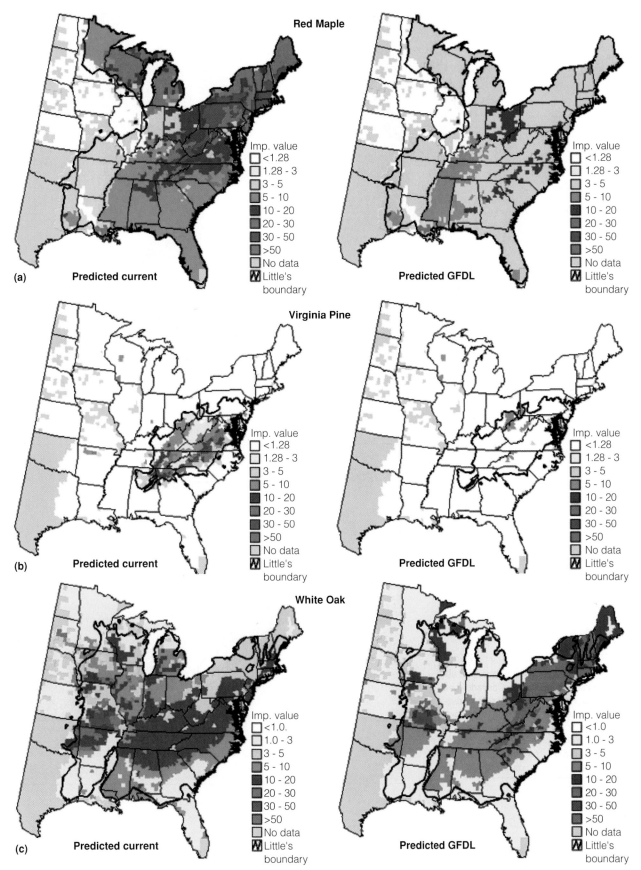

FIGURE 30.14 The distributions of (a) red maple, (b) Virginia pine, and (c) white oak under both current climate and a doubled CO_2 climate as predicted by the GFDL general circulation model. Species abundance is expressed in terms of importance value (sum of relative density, basal area, and frequency). Little's boundary refers to the observed distribution of the species as reported by Little (1977). See the text for description of the model used for predicting species distributions based on climate and site factors. (From Iverson et al. 1999.)

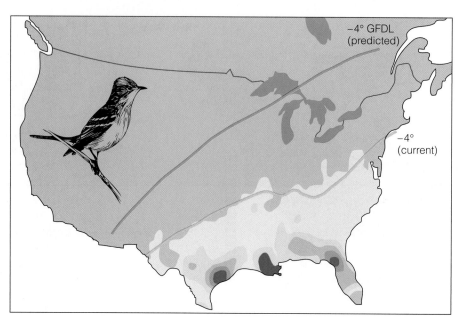

FIGURE 30.15 Map showing the existing distribution of the Eastern phoebe along the current 24°C average minimum January temperature isotherm, as well as the predicted isotherm under a changed climate. The predicted isotherm is based on changes in temperature due to a doubling of atmospheric CO_2 concentration as predicted by the Geophysical Fluid Dynamics Laboratory general circulation model (shown in Figure 30.10a). (Adapted from Root 1988.)

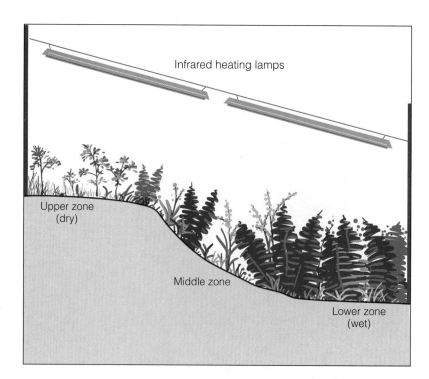

FIGURE 30.16 Experimental design used to elevate temperatures in a Colorado meadow community to examine possible effects of a global climate change. Changes in vegetation composition were monitored through time to track how the community responded to elevated temperatures and drier conditions. See the text for a discussion of results. (Adapted from Harte and Shaw 1995.)

at high latitudes on tundra ecosystems (http://www.systbot.gu.se/research/itex/itex.html). Investigators from 13 countries are applying a range of standard field techniques including passive warming of tundra vegetation using open-top chambers and manipulating snow depth to alter growing season length. Studies will examine species-, community-, and ecosystem-level responses to warming in the Arctic region.

Changes in climate also affect vegetation indirectly, through decomposition and nutrient cycling. In terrestrial ecosystems, these processes depend on temperature and available moisture (see Chapters 7 and 21). Decomposition proceeds faster under warmer, wetter conditions. An ongoing experiment at Harvard Forest in Massachusetts is examining the effect of elevated soil temperatures on rates of decomposition and nutrient cycling in a forested ecosystem. Buried heating cables raise the soil temperature by 5°C. Initial results show a 60 percent increase in rates of soil respiration (CO_2 emissions), a direct result of increased microbial and root respiration. The results are consistent with patterns of soil respiration observed in other forests in warmer regions around the world. They indicate that greenhouse warming will increase rates of decomposition and microbial respiration, leading to a significant rise in emissions of CO_2 from soils to the atmosphere.

30.7 Changing Climate Will Shift the Global Distribution of Ecosystems

Ecologists have learned a great deal about the responses of terrestrial ecosystems to changing climate conditions from the study of past climate change. Pollen samples from sediment cores taken in lake beds have allowed paleobotanists to reconstruct the vegetation of many regions during the last 20,000 years. The work of Margaret Davis in reconstructing the distribution of tree species in eastern North America since the last glacial maximum (see Section 13.14 and Figure 13.16) is a good example. Tree genera migrated northward at different rates following the retreat of the glaciers. The migration rates depended on how well the physiology and dispersal ability of a species and its competitive interactions with other tree species let it respond to changes in climate. Such studies show that the existing forest communities in eastern North America are a recent result of different responses of tree species to changing climate. As Earth's climate has changed in the past, the distribution and abundance of organisms and the communities and ecosystems they compose have changed (see Figure 13.17).

It is virtually impossible to develop experiments in the field to examine the long-term response of terrestrial ecosystems to a future climate change. This limitation means that scientists must base predictions on computer models. Perhaps the simplest but most telling of these are the biogeographical models that relate the distribution of ecosystems to climate. From the days of the early naturalists, plant ecologists have recognized the link between climate and plant distribution (see Chapter 24). For example, tropical rain forests are found in the wet tropical regions of Central and South America, Africa, Asia, and Australia. According to the biogeographical model developed by L. R. Holdridge, within these regions the distribution of tropical rain forests is limited to areas where mean annual temperatures are at or above 24°C and annual precipitation is above 2000 mm (see Section 24.1 and Figure 24.5). The regions of the tropics that meet these climate restrictions are shown on the map in Figure 30.17a. Under the changed temperature and rainfall patterns predicted by the United Kingdom Meteorological Office (UKMO) GCM for a doubled atmospheric CO_2 concentration, this distribution changes dramatically (Figure 30.17b). The region that can support tropical wet forest under this scenario shrinks by 25 percent. This decline is a direct result of drying due to higher temperatures. In some areas, the drying is a result of increased temperatures accompanied by decreased precipitation. In other areas, precipitation increases, but the increase is not sufficient to meet the increased demand for water (evaporation and transpiration) resulting from the increased temperatures. Together with the demands of agriculture and forestry (see Section 30.1), this scenario would devastate both the tropical rain forest ecosystems and the diversity of life they support. Although tropical rain forests cover only 7 percent of the total land area, they are home to over 50 percent of all terrestrial plant and animal species. Currently, deforestation in the tropics is the single major cause of species extinction, with annual rates of extinction ranging in the thousands of species. The loss of tropical rain forest predicted by the UKMO climate model would result in far more extinction.

Changes in global patterns of temperature would also affect the distribution of aquatic ecosystems. For instance, the global distribution of coral reefs is limited to the tropical waters where mean surface temperatures are at or above 20°C. Reef development is not possible where the mean minimum temperature is below 18°C. Optimal reef development occurs in waters where the mean annual temperatures are 23°C to 25°C, and some corals can tolerate temperatures up to 36°C to 40°C. A warming of the world's oceans would alter the potential range of waters in which reef development is possible, allowing for reef formation farther up the eastern coast of North America.

Ecologists are far from providing a complete analysis of the potential impacts of a global climate

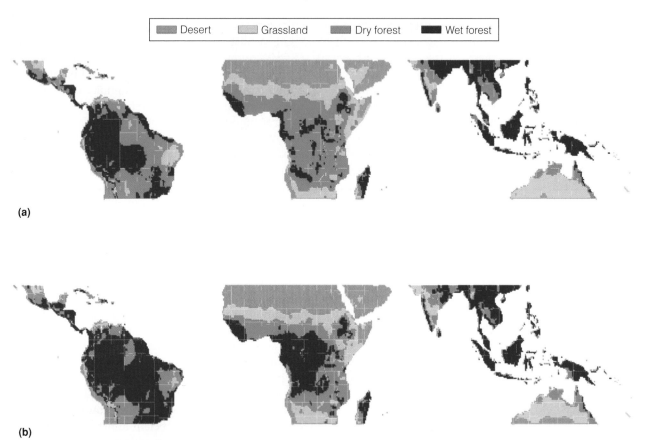

(a)

(b)

FIGURE 30.17 Maps of the areas in the tropical zone that could possibly support rain forest ecosystems as predicted by the Holdridge biogeographical model of ecosystem distribution. Map (a) is the area of tropical rain forest under current climate conditions, and (b) is the predicted area under changed climate conditions predicted by the United Kingdom Meteorological Office general circulation model for a doubled atmospheric CO_2 concentration. (Smith et al. 1992.)

change. There is little question, however, that changes in patterns of temperature and precipitation of the magnitude predicted by climate models will have a significant influence on the distribution and functioning of both terrestrial and aquatic ecosystems.

30.8 Global Warming Would Raise Sea Level and Affect Coastal Environments

During the last glacial maximum some 18,000 years ago, sea level was 100 meters lower than today. The highly productive shallow coastal waters, such as the continental shelf of eastern North America, were above sea level and were covered by terrestrial ecosystems (see Figure 13.17). As the climate warmed and the glaciers melted, sea levels rose. Over the last century, sea level has risen at a rate of 1.8 mm

per year (Figure 30.18). This is a result of the general pattern of global warming over this period and the associated thermal expansion of ocean waters and melting of glaciers. The 2001 report of the IPCC estimates that global mean sea level will rise by 0.09 to 0.88 meter between the years 1990 and 2100, but with considerable regional variation. A rise of this magnitude will have serious effects on coastal environments from the perspectives of both natural ecosystems and human populations.

A large portion of the human population lives in coastal areas; in fact, 13 of the world's 20 largest cities are located on coasts. Areas that are particularly vulnerable are delta regions, low-lying countries such as the Netherlands, Surinam, and Nigeria, and the smaller low-lying islands of the Pacific and other oceans. Bangladesh, an Asian country of about 120 million inhabitants, is located in the delta region of the Ganges, Brahmaputra, and Meghna Rivers (Figure 30.19). Approximately 25 percent of the country's population lives in areas below 3 meters above sea level, with about 7 percent of the country's habitable land and 6 million people residing in areas less than

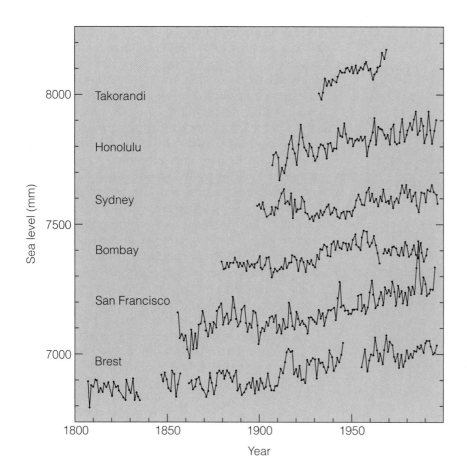

FIGURE 30.18 Long-term sea-level records from six coastal regions of the world: Takorandi (Africa), Sydney (Australia), Bombay (Asia), San Francisco (North America), and Brest (Europe). (Adapted from Houghton et al. 1996.)

1 meter above sea level. Estimates of sea-level rise in this region due to a combination of land subsidence (a result of land collapsing in response to removal of groundwater) and global warming are 1 meter by the year 2050 and 2 meters by the year 2100. Although there is great uncertainty in these estimates, the effect on Bangladesh would be devastating.

Other coastal regions of Southeast Asia and Africa would be equally affected by the predicted rise in sea level. In Egypt, about 12 percent of the arable land, with a population over 7 million, would be affected by a rise in sea level of 1 meter. In the coastal areas of eastern China, a sea-level rise of just half a meter would inundate an area of approximately 40,000 square kilometers where over 30 million people currently live.

Particularly vulnerable to a sea-level rise are small islands. Over half a million people live in the archipelagos of small islands and coral atolls. Two examples are the Maldives in the Indian Ocean and the Marshall Islands of the Pacific. These island chains lie almost entirely below 3 meters above sea level. A half-meter or more rise in sea level would not only dramatically reduce their land area, but would also

have a devastating impact on groundwater (freshwater) supply.

A sea-level rise will also have major effects on coastal ecosystems. Among these are direct inundation of low-lying wetlands and dryland areas, erosion of shorelines through loss of sediments, increased salinity of estuaries and aquifers, rising coastal water tables, and increased flooding and storm surges. Estuarine and mangrove ecosystems (see Chapters 27 to 29) would be highly susceptible to a sea-level rise of the magnitude predicted. The coastal salt marshes are dependent on the twice-daily tidal inundation of saltwater mixing with the freshwater provided by streams and rivers. The patterns of water depth, temperature, salinity, and turbidity are critical to maintaining these ecosystems. The invasion of saltwater farther into the estuary as a result of a rise in sea level would be disastrous, and could also cause salination of land adjacent to the estuary margins. Estuarine and mangrove environments are critical to coastal fisheries. Over two-thirds of the fish caught for human consumption, as well as many birds and animals, depend on the coastal marshes and mangroves for part of their life cycles.

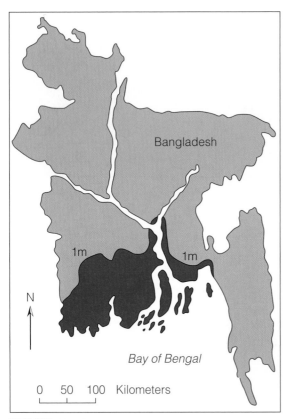

FIGURE 30.19 Land area in Bangladesh that would be submerged (dark green area of map) if sea level rose by 1 meter. (From Nicholls and Leatherman 1995.)

30.9 Climate Change Will Affect Agricultural Production

Despite technological advances in improved crop varieties and methods of irrigation, climate and weather remain the key factors determining agricultural production. Changes in the global climate patterns will exacerbate an already increasing problem of feeding the world population, which is predicted to double in size over the next half century.

The major cereal crops that feed people—wheat, corn (maize), and rice—are domesticated species. Like native species, these crops exhibit environmental tolerances to temperature and moisture that control survival, growth, and reproduction. Changes in regional climate conditions will directly influence their suitability and productivity, and therefore current patterns of agricultural production. However, these changes will be complex, with economic and social factors interacting to influence patterns of global food production and distribution.

In examining potential effects of greenhouse warming on agricultural production, both the increasing concentration of CO_2 and the changes in climate must be considered. Unlike the uncertainties observed in tree species (see Figure 30.8) and terrestrial ecosystems (Section 30.4), the results of numerous studies suggest that most crop species (and varieties) will benefit from a rise in CO_2 concentration. For example, in an experiment in Arizona, cotton and spring wheat were grown in field conditions under elevated CO_2 and irrigation. Cotton yield increased by 60 percent and wheat by over 10 percent compared with crops grown under identical field conditions and ambient concentrations of CO_2.

One of the simplest ways to evaluate the potential implications of a climate change on agriculture is to examine shifts in the geographic range of certain crop species as they relate directly to climate. For example, an average daily temperature increase of 1°C during the growing season would shift the "corn belt" (region of highest corn production) of the United States significantly to the north (Figure 30.20a). A similar analysis for the shift in suitable regions for irrigated rice production in northern Japan is shown in Figure 30.20b. In both examples, the shifts in agricultural zones imply significant changes in regional land-use patterns, with associated economic and social costs. Although analyses of this type can provide insight into changing patterns of regional agricultural production, to evaluate the actual effect on global food production and markets requires a more detailed interdisciplinary approach.

The Environmental Change Unit at Oxford University has carried out a collaborative study with agricultural scientists from 18 countries to examine both the regional and global impacts of climate change on world agricultural production. Various assumptions were made about the ability of farmers to adapt to changing environmental conditions through shifts in the species or varieties of crops grown, or changes in agricultural practices such as irrigation. The analysis also assumes a continuation of current economic growth rates, certain changes in current trade restrictions, and projected estimates of population growth.

One of the major findings is that the negative effects of climate change are to some extent compensated for by increased productivity resulting from elevated atmospheric concentrations of CO_2. The net effect of a climate change as predicted by the general circulation models, including a doubling of atmospheric CO_2 concentrations, is to reduce the global production of cereal crops by up to 5 percent. An important point to note is that this reduction is not evenly distributed across the globe or even within a given region or country (Figure 30.21).

The predicted changes would increase the current disparity in cereal crop production between devel-

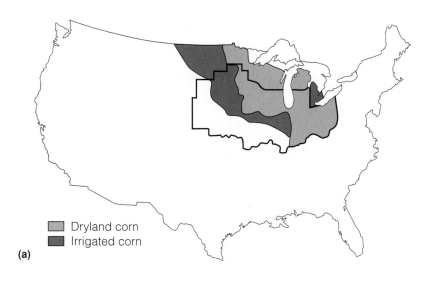

(a)

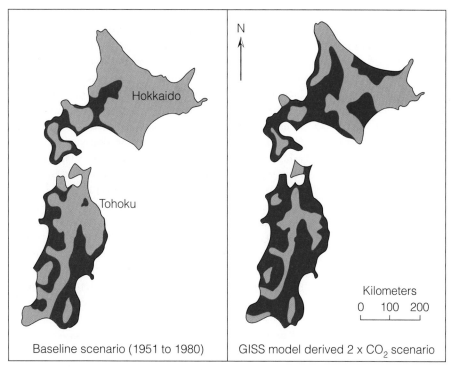

(b)

FIGURE 30.20 Regional shifts in areas suitable for crop production under a changed climate as predicted by the Goddard Institute for Space Studies GCM: (a) shift in the region suitable for corn production in the United States, (Blasing and Solomon 1983.) (b) shifts in areas suitable for irrigated rice production in northern Japan. The areas in dark green are suitable for irrigated rice production. (Yoshino et al. 1988.)

oped and developing countries. Results of the study tend to show an increase in production in developed countries, particularly in the middle latitudes (temperate regions). In contrast, production in developing nations, as a group, would decline by as much as 10 percent, with an associated increase in the population at risk of hunger. In many of these regions, climatic variability and marginal climatic conditions for agriculture are worsened under the predicted patterns of global climate change.

30.10 Climate Change Will Both Directly and Indirectly Affect Human Health

Climatic change will have a variety of both direct and indirect effects on human health. Direct effects would include increased heat stress, asthma, and a variety of

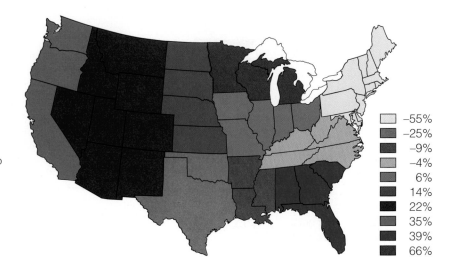

FIGURE 30.21 Changes in regional crop production by year 2060 for the United States under a climate change as predicted by the Goddard Institue for Space Studies GCM (assuming an average 3°C increase in temperature, 7 percent increase in precipitation, and 530 ppm CO_2). (Adapted from Adams et al. 1995.)

Legend:
- −55%
- −25%
- −9%
- −4%
- 6%
- 14%
- 22%
- 35%
- 39%
- 66%

other cardiovascular and respiratory ailments. Indirect health effects are likely to include increased incidence of communicable diseases, increased mortality and injury due to increased natural disasters (floods, hurricanes, etc.), and changes in diet and nutrition due to changed agricultural production.

A number of studies have examined the direct relationship between maximum summer temperatures and mortality rates (Figure 30.22). Climate change is expected to change the frequency of very hot days. For example, if average July temperature in Chicago, Illinois, were to rise by 3°C, the probability that the heat index will exceed 35°C (120° F) during the month increases from 1 in 20 to 1 in 4. In the United States, warm humid conditions during summer nights lead to the highest mortality. The greatest death toll in the United States occurred during the summer of 1936, when 4700 excess deaths were recorded due to heat-related causes. In recent decades, 1200 excess deaths occurred in Dallas during the summer of 1980, and 566 in Chicago during July 1995. Analyses of climate change scenarios show a significant rise in heat-related mortality in all regions of the United States over the next several decades (Figure 30.23). During heat waves, cardiovascular and respiratory illnesses are the major causes of mortality. The elderly and children are typically in the greatest danger during these periods.

In addition to direct heat-related mortality, the distribution and rates of transmission for a variety of infectious diseases will be influenced by changes in regional climate patterns. Disease consists of agents, such as viruses, bacteria, or protozoa, and host organisms (humans). Some diseases are transmitted to humans by intermediate organisms, or vectors (see Chapter 26). Insects are a primary vector of human disease. Although acting as carriers, the insects themselves are not affected by the disease agent. Insect-

borne viruses (referred to as arbovirus, for arthropod-borne virus) cover a wide variety of diseases. The most common insects involved in the transmission of arboviruses are mosquitoes, ticks, and blood flukes (schistosomes). Approximately 102 arboviruses can produce disease in humans. Of this number, approximately 50 percent have been isolated from mosquitoes. The insects that carry these disease agents are adapted to specific ecosystems for survival and reproduction, and exhibit specific tolerances to features of the climate, such as temperature and humidity. Changes in the climate will affect their distribution and abundance, just as is true of the Eastern phoebe (Section 30.6, Figure 30.15).

One insect-borne disease is malaria, a recurring infection produced in humans by protozoan parasites transmitted by the bite of an infected female mosquito of the genus *Anopheles*. The optimal temperature

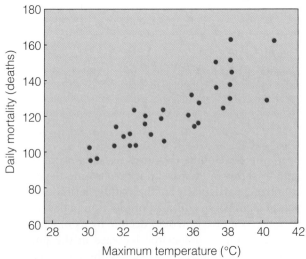

FIGURE 30.22 Relationship between maximum daily temperature and human mortality rate in Cairo, Egypt, during 1982. (Adapted from Kalkstein and Tau 1995.)

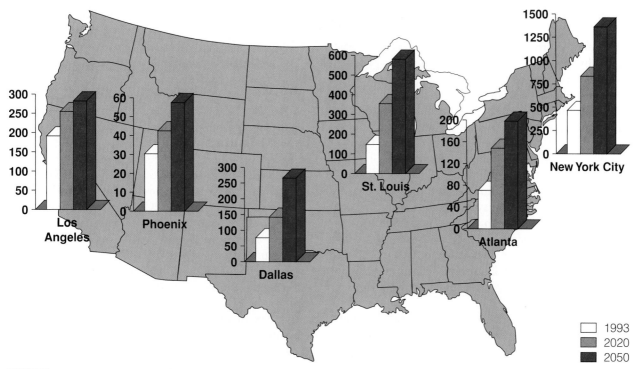

FIGURE 30.23 Average annual excess weather-related mortality for the years 1993, 2020, and 2050 in various cities of the United States. Future projections of weather-related mortality are based on changes in climate predicted by the Geophysical Fluid Dynamics Laboratory GCM. (Adapted from Kalkstein and Green 1997.)

for breeding by *Anopheles* is 20°C to 30°C, with relative humidity over 60 percent. Mosquitoes die at temperatures above 35°C and relative humidity less than 25 percent. At present, 40 percent of the world's population is at risk, and over 2 million people are killed each year by malaria. The current distribution of malaria will be extensively modified under a climate change. The expansion of the geographic range of the *Anopheles* mosquito into currently more-temperate climates is expected to increase the proportion of the world's population at risk of this infectious disease to over 60 percent by the latter part of the 21st century.

Dengue and yellow fever are related viral diseases that also are transmitted by mosquitoes. In the case of these viruses, the vector is the mosquito *Aedes aegypti*, which is adapted to the urban environment. Colonization by this mosquito is limited to areas with an average daily temperature of 10°C or greater. The virus that causes yellow fever lives in mosquitoes only when temperatures exceed 24°C under high relative humidity. Epidemics occur when mean annual temperatures exceed 20°C, making this a disease of the tropical forested regions. Yellow fever is currently prevalent across Africa and Latin America, but has been detected as far north as the midlatitude ports of Bristol, Philadelphia, and Halifax, where the mosquitoes have survived in the

water tanks of ships that have traveled from tropical regions. A climate change would have a direct influence on the distribution of both the virus and its vector, the mosquito.

30.11 Understanding Global Change Requires the Study of Ecology at a Global Scale

The increasing atmospheric concentrations of CO_2 and other greenhouse gases, and the potential changes in global climate patterns that may result, present a new class of ecological problems. To understand the effect of rising CO_2 emissions from fossil fuel burning and land clearing, we have to examine the carbon cycle on a global scale (see Figure 22.4), linking the atmosphere, hydrosphere, biosphere, and lithosphere. Although the discussion in the previous sections focuses on rising CO_2 concentrations and changes in climate on populations, communities, and ecosystems, the possible effects are not one-directional. As presented in Section 30.8, ecosystems also influence atmospheric CO_2 and regional climate patterns. For example, if climate does change as shown in Figure 30.17, the global distribution and abundance of tropical rain forests will decline dramatically. Tropical rain forests are the most productive terrestrial ecosystems on the planet. A significant decline in these ecosystems will reduce global primary productivity, the uptake of CO_2 from the atmosphere, and CO_2 storage as organic carbon in biomass. In fact, as tropical rain forests shrink, atmospheric CO_2 will increase. The drying of these regions will kill trees, increase fires, and transfer carbon stored in the living biomass to the atmosphere as CO_2 in much the same way as forest clearing does in these regions (see Section 30.1). The rise in atmospheric CO_2 will increase the greenhouse effect, further exacerbating the problem. In this case, the changes in the terrestrial surface act as a positive feedback loop to rising atmospheric concentrations of CO_2.

But if rising atmospheric CO_2 level and changing climate increase the productivity of the world's ecosystems, they will take up more CO_2 from the atmosphere. Increased productivity will function as negative feedback, drawing down atmospheric CO_2 concentrations.

In addition to changing climate indirectly through influencing atmospheric concentrations of CO_2, changes in the distribution of rain forests can influence climate directly by altering regional patterns of precipitation. In some regions, such as extensive areas of tropical rain forest, a significant portion of the precipitation is water that has been transpired from the vegetation in the area. In effect, water is being recycled locally through the hydrologic cycle (see Figure 4.11). Removal of the forest (either through deforestation or shifting distribution of ecosystems as in Figure 30.17) reduces transpiration and increases runoff to rivers that transport water from the area. Experiments using GCMs and regional climate models have examined the potential effects of large-scale deforestation in the Amazon Basin. Findings suggest that the loss of forest cover would result in a significant reduction in annual precipitation by reducing the internal cycling of water within the forest. This would effectively change the climate of the region, making it unlikely that the rain forest would be able to reestablish.

Gordon Bonan and colleagues have presented another example of a direct influence of terrestrial ecosystems on regional climate. The largest degree of warming under elevated CO_2 is in the northern latitudes (see Section 30.5). The predicted warming would significantly reduce snow cover in this region and shift the distribution of boreal forests to the north. A major factor influencing the relative absorption and reflection of shortwave radiation (solar radiation) by Earth's surface is its albedo. Albedo is an index of the ability of a surface to reflect solar radiation back to space. Snow has a high albedo, or reflectance, whereas vegetation (with its darker color) has a low albedo. Both reduced snow cover and the northern movement of the boreal forest would reduce the regional albedo, therefore increasing the amount of solar radiation absorbed by Earth's surface. This increase in the absorption of radiation would further increase regional temperatures, functioning as a positive feedback loop.

These are not simple connections. To understand the interactions among the atmosphere, oceans, and terrestrial ecosystems requires ecologists to study Earth as a single, integrated system. It is only through the development of a global ecology that ecologists, working with oceanographers and atmospheric scientists, will come to understand the potential consequences of doubling the concentration of CO_2 in the atmosphere over the next century.

CHAPTER REVIEW

Summary

Rising Atmospheric Concentrations of CO_2 (30.1–30.4) Direct observations beginning in 1958 reveal an exponential increase in the atmospheric concentration of CO_2. The rise is a direct result of fossil fuel combustion and the clearing of land for agriculture (**30.1**).

Of the CO_2 released from fossil fuel combustion and land clearing, only about 60 percent remains in the atmosphere. The remainder is taken up by the oceans and terrestrial ecosystems. Calculations of diffusion of CO_2 into the surface waters provide an estimate of uptake by the oceans. Carbon uptake by terrestrial ecosystems is calculated as the difference between inputs to the atmosphere, atmospheric concentrations, and the uptake by oceans (**30.2**).

The oceans have two layers, surface waters (0–200 meters deep) and deepwaters (>200 meters). Over 85 percent of the ocean volume is deepwater. The thermocline prevents mixing. Carbon dioxide diffuses from the atmosphere into the surface waters. The rise in atmospheric concentrations is resulting in an increased uptake of CO_2 into the surface waters. Transfer of dissolved CO_2 from the surface waters into the deepwaters takes hundreds of years. This limits the short-term uptake of CO_2 by the oceans (**30.3**).

In general, plants respond to increased atmospheric CO_2 with higher rates of photosynthesis and partial closure of stomata. These responses increase water-use efficiency. Responses to long-term exposure vary, including increased allocation of carbon to the production of roots, reduced allocation to the production of leaves, and reduction in stomatal density. Scientists are studying long-term effects on net primary productivity (30.4).

Global Climate Change (30.5–30.11) Carbon dioxide is a greenhouse gas. It traps long-wave radiation emitted from Earth's surface, warming the atmosphere. Rising atmospheric concentrations of CO_2 and other greenhouse gases could raise the global mean temperature by 1.4°C to 5.8°C by the year 2100. Warming will not be uniform over Earth. The greatest warming is predicted to come during the winter months and at northern latitudes. Increased variability in climate is predicted, including changes in precipitation and the frequency of storms. The input of sulfates and other aerosols from human sources acts to reduce the input of solar radiation to Earth's surface, thereby reducing warming (30.5).

The distribution and abundance of species will shift as temperature and precipitation change. Changes in climate will influence the competitive ability of species and thus change patterns of community zonation and succession. Ecosystem processes such as decomposition and nutrient cycling are sensitive to temperature and moisture, and changing climate will affect them (30.6). Changes in climate also will shift the distribution and abundance of both terrestrial and aquatic ecosystems. These changes in ecosystem distribution influence global patterns of plant and animal diversity (30.7).

Sea level is currently rising globally at an average rate of 1.8 mm per year. It is estimated that global warming will cause sea level to rise by 0.09 to 0.88 meter by 2100, as the polar ice caps melt and warmer ocean waters expand. A sea-level rise of this magnitude will have major effects on people living in coastal areas. In addition, rising sea level will affect coastal ecosystems such as beaches, estuaries, and mangroves (30.8).

Climate change will affect global agricultural production. Decreases in crop production from drier conditions will in part be offset by increased rates of photosynthesis under elevated atmospheric CO_2 levels; however, current models project a 5 percent decline in global production of cereal crops. This decline is not distributed evenly. Developed countries in the middle latitudes will realize a slight increase, while production in developing countries in the tropics will decline. The result will be increased hunger (30.9).

Climate change will have both direct and indirect effects on human health. Mortality rates are expected to rise as a result of heat-related deaths associated with respiratory and cardiovascular ailments. Indirect health effects include increased mortality and injury from climate-related natural disasters, as well as changes in diet and nutrition resulting from changes in agricultural production. The distribution and transmission rates of a variety of insect-borne infectious diseases that are directly related to climate, such as malaria, will also be affected (30.10).

To understand the effect of rising atmospheric concentrations of greenhouse gases and global climate change, we have to study the whole Earth as a single, complex system (30.11).

Study Questions

1. What are the major sources of greenhouse gases, especially CO_2?

2. Not all of the CO_2 released to the atmosphere remains there. What happens to the rest?

3. How does elevated CO_2 influence rates of photosynthesis and transpiration?

4. What limits the transfer of CO_2 from the surface waters of the ocean to the deepwaters?

5. Why is CO_2 called a greenhouse gas?

6. How do greenhouse gases contribute to global warming?

7. How do forest burning and land clearing affect the global climate?

8. How might changes in climate (temperature and precipitation) influence the distribution of plant and animal species?

9. How might changes in climate influence the distribution and abundance of terrestrial ecosystems?

10. How is sea level currently changing?

11. How will global warming change sea levels?

12. How might rising sea levels influence human populations? Coastal environments?

13. How might changes in climate influence agricultural production? How will rising CO_2 levels influence crop production?

14. What are some direct and indirect influences of climate change on human health?

How a Lack of Mushrooms Helped Power the Industrial Revolution

With names like Earth Star, Satyr's Beard, Big Laughing Gym, Witch's Hat, Velvety Earth Tongue, Green-headed Jelly Club, and Angel Wings, mushrooms are a diverse and mysterious lot (Figure A). Reminiscent of plants, but actually fungi, mushrooms are—depending on species—a chef's delight, a poisoner's potion, or a recycler's dream. Some, such as truffles, chanterelles, and morels, are gustatory treasures. Others, such as the *amanitas* Death Cap and (the ironically named but strikingly beautiful) Destroying Angel, have served as instruments of murder. Some, such as Corpse Finder, have unveiled evil deeds by their habit of growing atop soil harboring *corpora delecti*. Some, such as White Rot Fungus, have changed the course of Earth's environment—and human history—with their extraordinary ability to decompose the toughest parts of plants.

Although the legacy of mushrooms is fascinating, perhaps their major impact on our lives stems not from their presence but from their absence, specifically during the first few hundred million years after some plants abandoned water for a new life on land. Indeed, a lack of mushrooms during a portion of Earth's history is one major reason that we now have fuel for our cars, oil for our furnaces, and electricity for our lights.

Our modern way of life harks back to the beginning of the Industrial Revolution in the mid-18th century. It was a time that saw the development of the steam engine and a shift in labor from small workshops to factories, from humans to machines. The Industrial Revolution required huge amounts of energy, and so the source of energy shifted from the scarce resource of wood to the more abundant coal.

Coal, petroleum oil, and natural gas are referred to as fossil fuels

> "Indeed, a lack of mushrooms during a portion of Earth's history is one major reason that we now have fuel for our cars, oil for our furnaces, and electricity for our lights."

because they are derived from the fossilized remains of plants. Fossil fuel formation is an exceedingly slow process that requires special conditions. Occasionally, when plants died, their partially decomposed tissues came to rest in rock pores, fissures between rock layers, or sand beds. Later, impervious

FIGURE A Two members of the family Basidiomycetes (fungi).

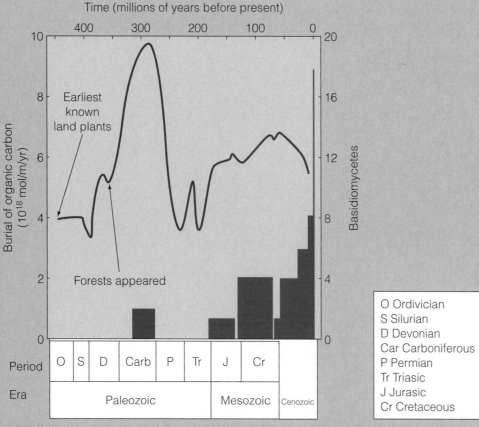

Time (millions of years before present)

FIGURE B Changes in the burial rate of organic carbon and the evolution of Basidiomycetes' diversity over geological time.

materials covered the organic matter, preventing its dispersal. Over millions of years, geological processes buried the remains of these plants deep in Earth's crust. The weight of the overlying rock subjected the plant remains to intense pressures and temperatures until, eventually, they became fossilized, or carbonized, into coal, crude oil, or natural gas.

Most of the organic carbon that gave rise to the fossil fuel reserves was deposited during the Carboniferous period some 360 to 285 million years ago (Figure B), a period termed the first Coal Age. This was a time when Earth's climate was warmer and wetter than today's and primitive land plants flourished around the margins of shallow inland seas. These primitive

forests dominated the North American and Eurasian landscapes until Earth's climate turned colder and drier and the vast forests began to disappear. A second Coal Age extended from 200 to 1.6 million years ago, but this period saw far less deposition of organic carbon than the Carboniferous.

A key to fossil fuel formation lies in the evolution of the primitive forest plants from their aquatic ancestors. The earliest land plant was most likely a multicellular green algae that became established on land over 425 million years ago. The move from an aquatic to a terrestrial environment posed a new problem for plants: how to avoid desiccation. Water that evaporates to the air from a plant's surface must be replaced with water from

the surrounding environment. The bryophytes (mosses)—some of the earliest land plants—absorbed water directly through aboveground tissues. The development of roots and an internal vascular system was still to come.

The earliest known vascular plants appear in the fossil record during the Silurian period, some 400 million years ago. By the early Carboniferous period, forests appeared. The trees that inhabited these forests depended on barklike tissues for support rather than on the woody tissues that support modern trees. This distinction is important to fossil fuel formation because bark differs from wood in a very important way: it is much higher in carbon compounds called lignins.

Although approximately 50 percent of the dry weight of plant tissues is composed of the element carbon, these carbon molecules exist in a wide variety of compounds. The major carbon compounds of plant tissues are lipids, proteins, and polysaccharides (such as cellulose and hemicellulose), and other large and complex structural molecules such as lignins. Decomposer organisms use the carbon in dead plant tissues as a source of energy. However, not all carbon compounds are of equal value as an energy resource. Lipids and proteins are the easiest to break down and yield the greatest amount of energy to decomposers. Therefore, these carbon compounds are the first to disappear during decomposition. Cellulose and hemicellulose likewise have relatively weak chemical bonds and open molecular structures that permit fairly rapid degradation, making them intermediate in energy quality to decomposer organisms. Lignins are complex carbon molecules that are extremely difficult to break down and yield little energy. Herbivores cannot digest lignins, termites are ineffective in breaking them down, and even bacteria do not significantly digest them. The decomposition of lignin compounds is carried out by the mushrooms and related fungi known collectively as the Basidiomycetes. The breakdown of lignins involves a bleaching process, known as "white rot," which degrades lignins and gives mushrooms access to nutrients bound up in these compounds. The process actually yields little or no net gain of metabolic energy to the mushrooms.

The difference in the quality of these various carbon compounds as an energy source for decomposers results in a distinctive pattern of decomposition (Figure 7.3): the higher-quality carbon compounds disappear first, until the remaining carbon in the organic matter is largely composed of lignin compounds. In fact, the dark brown organic matter called humus found on the surface of forest soils is largely made up of lignins. These lignin compounds also constitute the majority of the organic carbon that forms coal deposits.

The chemical fossils of the earliest known vascular plants contain lignin-like compounds. This places the evolution of lignins back at least to the Silurian, 400 million years ago. Because lignins are energetically expensive to synthesize, as land plants evolved they switched to using less costly carbon-based compounds, such as cellulose (the major constituent of wood), to form their supportive tissues. The majority of lignin compounds in modern plants are in the vascular tissues (veins of the leaves and the xylem and phloem of the woody tissues). This evolutionary progression toward the use of chemically simpler compounds required the passage of enormous amounts of time. Not until the Mesozoic, which began some 240 million years ago, did trees evolve the use of woody tissues for support, reducing their dependence on the more lignin-rich barklike tissues.

The dependence of Carboniferous trees on barklike tissues rather than wood for support means that the proportion of total plant carbon bound up in lignin compounds of those trees may well have been twice that of modern trees. The much higher lignin content of these early forests suggests a much slower rate of decomposition than that of today's forests. This rate was further depressed by the lack of organisms capable of degrading lignins. Although the first positive identification of Basidiomycetes is in fossil material from the Pennsylvanian epoch in the mid-Carboniferous period some 300 million years ago, it is believed that these early fungi were not decomposer organisms. The examination of thousands of coal samples has led scientists to conclude that there is no convincing evidence of the presence of fungal remains in coal deposits before the mid-Cretaceous period (130 million years ago). The presence of mushrooms and other fungi in the fossil record increases from this period onward (Figure B). This suggests that effective lignin degradation did not appear until some 200 million years after lignins evolved, or about 100 million years after the peak of organic carbon deposition. Because of the time lag between the period when plant organic matter was highest in lignin content and the evolution of organisms (mushrooms) capable of degrading lignins, a vast reserve of organic matter built up during the Carboniferous period. This reserve was eventually converted into fossil fuels and forms the lion's share of today's coal, crude petroleum, and natural gas resources.

So, next time you turn on a lamp or open the refrigerator, think for a second about how the early forests of the Carboniferous period and a 100-million-year absence of mushrooms made it all possible.

References

CHAPTER 1 The Nature of Ecology

Bates, Marston. 1956. *The nature of natural history.* New York: Random House.

———. 1960. *The forest and the sea.* New York: Random House. Both of these classic books, available in many editions, provide a delightful introduction to ecology.

Bronowski, J. 1956. *Science and human values.* Harper & Row.

Brower, J. E., and J. H. Zar. 1984. *Field and laboratory methods for general ecology.* 2d ed. Dubuque, IA: Wm. Brown.

Carson, Rachel, 1963. *Silent spring.* Boston. Houghton Mifflin. An important and pivotal book in the history of ecology.

Cloud, P. E. 1988. Gaia modified. *Science* 240:1716.

Cox, G. W. 1985. *Laboratory manual for general ecology.* 5th ed. Dubuque, IA: Wm. Brown.

Egerton, F. N., ed. 1977. *History of American ecology.* New York: Arno Press.

Golley, F. B. 1993. *A history of the ecosystem concept in ecology: More than the sum of its parts.* New Haven: Yale University Press.

Hagen, J. B. 1992. *An entangled bank: The origins of ecosystem ecology.* New Brunswick, NJ: Rutgers University Press.

Kingsland, S. 1995. *Modeling nature.* Chicago: University of Chicago Press. A history of the development of population ecology.

Langenheim, J. H. 1966. Early history and progress of women ecologists: Emphasis on research contributions. *Annual Review of Ecology and Systematics* 27:1–53. Women were important in the development of ecology.

Lovelock, J. E. 1979. *Gaia: A new look at life on earth.* New York: Oxford University Press.

McIntosh, R. P. 1985. *The background of ecology: Concept and theory.* New York: Cambridge University Press. Excellent analysis of ecology in North America.

Sheall, J. (ed.). 1988. *Seventy-five years of ecology: The British Ecological Society.* Oxford: Blackwell Scientific Publishers. History of British ecology.

Worster, D. 1977. *Nature's economy.* San Francisco: Sierra Club Books. An engaging history of ecology with a different view from that of McIntosh.

CHAPTER 2 Adaptation and Evolution

Darwin, C. 1859. *The origin of species.* Philadelphia: McKay. [Reprinted from 6th London edition.]

Desmond, A., and J. Moore. 1991. *Darwin.* New York: Warner Books.

Elton, C. S. 1927. *Animal ecology.* London: Sidgwich & Jackson.

Futuyma, D. J. 1984. *Evolutionary biology.* Sunderland, MA: Sinauer Associates. A superior treatment of the subject.

Grant, P. 1999. *Ecology and evolution of Darwin's finches.* Princeton, NJ: Princeton University Press. Updates the story of Darwin's finches. A modern field study in evolution. Should be read with Lack 1974.

Grant, V. 1971. *Plant speciation.* New York: Columbia University Press. A classic on all aspects of plant speciation.

———. 1985. *The evolutionary process: A critical review of evolutionary theory.* New York: Columbia University Press. An accessible discussion of natural selection and speciation.

Grinnell. J. 1904. The origin and distribution of the Chestnut-backed Chickadee. *Auk* 21:364–382.

Hanski, I. A., and M. E. Gilpin (eds.). 1996. *Metapopulation biology: Ecology, genetics, and evolution.* San Diego: Academic Press. Deals with various aspects of small populations.

Hartl, D. 1988. *A primer of population genetics.* Sunderland, MA: Sinauer Associates. An excellent introduction.

Highton, R. 1995. Speciation in eastern North American salamanders of the Genus *Plethodon. Annual Review of Ecology and Systematics* 26:579–600.

Hutchinson, G. E. 1959. Homage to Santa Rosalia, or why are there so many kinds of animals? *American Naturalist* 93:145–159.

———. 1978. *An introduction to population ecology.* New Haven: Yale University Press.

Lack, D. 1974. *Darwin's finches: An essay on the general biological theory of evolution.* London: Cambridge University Press. A classic in biology.

Lynch, M., J. Conery, and R. Burger. 1995. Mutation accumulation and the extinction of small populations. *American Naturalist* 146:489–518.

Mayr, E. 1963. *Animal species and evolution.* Cambridge, MA: Harvard University Press. A classic major work on animal speciation.

———. 1991. *One Long Argument.* Cambridge, MA: Harvard University Press.

Otte, D., and J. Endler (eds.). 1989. *Speciation and its consequences.* Sunderland, MA: Sinauer Associates. A series of papers that provide a major reference source on the subject.

Root, R. B. 1967. The niche exploitation of the blue-gray gnatcatcher. *Ecological Monographs* 37:317–350.

Rosensweig, M. 1995. *Species diversity in space and time.* New York: Cambridge University Press. See Chapters 5 and 6 for a clear discussion of speciation and extinction.

Weiner, J. 1994. *The beak and the finch: A story of evolution in our time.* New York: Alfred A. Knopf. A superb book on species evolution, based on Grant's studies of Darwin's finches.

White, M. J. D. 1978. *Models of speciation.* San Francisco: Freeman. A major introduction to the speciation process.

ECOLOGICAL APPLICATION: Taking the Uncertainty out of Sex

Darwin, C. 1859. *The origin of species.* Philadelphia: McKay. [Reprinted from 6th London edition.]

———. 1875. *The variation of animals and plants under domestication.* London: Murray.

Lewontin, R. 2000. *It ain't necessarily so.* New York: New York Review of Books, Inc.

Mayr, E. 1991. *One long argument.* Cambridge, MA: Harvard University Press.

Wilmut, I., K. Campbell, and C. Tudge. 2000. *The second creation: Dolly and the age of biological control.* New York: Farrar, Straus and Giroux.

CHAPTER 3 Climate

Ahrens, C. D. 2000. *Meteorology today,* 6th ed. Pacific Grove, CA: Brooks/Cole.

Berry, R. G., and R. J. Corley. 1992. *Atmosphere, weather, and climate.* 5th ed. New York: Routledge.

Gates, D. M. 1962. *Energy exchange in the biosphere.* New York: Harper & Row.

———. 1972. *Man and his environment: Climate.* New York: Harper & Row. An excellent introduction.

Geiger, R. 1965. *Climate near the ground.* Cambridge, MA: Harvard University Press. A classic book on microclimate.

Graedel, T. E., and P. J. Crutzen. 1997. *Atmosphere, climate and change.* New York: Scientific American Library.

Landsberg, H. E. 1970. Man-made climatic changes. *Science* 170:1265–1274.

Lee, R. 1978. *Forest microclimatology*. New York: Columbia University Press. Looks at climate within the forest.

Schaefer, V. J., and J. A. Day. 1981. *A field guide to the atmosphere*. Boston: Houghton Mifflin. A guide to all aspects of the ocean of air.

Schneider, S. H. 1989. *Global warming*. San Francisco: Sierra Club Books. A sound discussion of the topic.

Schroeder, M. S., and C. C. Buck. 1970. *Fire weather*. Agricultural Handbook No. 360. Washington, DC: U.S. Department of Agriculture.

Strahler, A. 1971. *The earth sciences*. New York: Harper & Row. Detailed presentation of climate, among other topics.

CHAPTER 4 Abiotic Environment

Bainbridge, R., G. C. Evans, and O. Rackham (eds.). 1966. *Light as an ecological factor*. Oxford, England: Blackwell Scientific Publications. Dated, but still an outstanding reference.

Cernusca, A. 1976. Energy exchange within individual layers of a meadow. *Oecologica* 23:148.

Clark, G. A. 1939. Utilization of solar energy by aquatic organisms. *Problems in Lake Biology* 10:27–38.

Clarkson, D. T., and J. B. Hanson. 1980. The mineral nutrition of higher plants. *Annual Review of Plant Physiology* 31:239–298.

Garrison, T. 1995. *Essentials of oceanography*. Belmont, CA: Wadsworth Publishing Company.

Gates, D. M. 1968. Energy exchange between organism and environment. In W. P. Lowry, ed. *Biometeorology*, Proc. 28th Annual Bio. Colloq. Corvallis: Oregon State University Press.

Hutchinson, B. A., and D. R. Matt. 1977. The distribution of solar radiation within a deciduous forest. *Ecological Monographs* 47:185–207.

Hynes, H. B. N. 1970. *The ecology of running water*. Toronto: University of Toronto Press.

Likens, G. E. (ed.). 1985. *An ecosystem approach to aquatic ecology: Mirror Lake and environment*. New York: Springer-Verlag.

Nybakken, J. W. 1997. *Marine biology: An ecological approach*. 4th ed. San Francisco: Addison Wesley. Chapters 1 and 6.

Reifsnyder, W. E., and H. W. Lull. 1965. *Radiant energy in relation to forests*. Technical Bulletin No. 1344. Washington, DC: U.S. Department of Agriculture.

CHAPTER 5 Soil

Boul, S. W., F. D. Hole, R. J. McCracken, and R. J. Southward. 1997. *Soil genesis and classification*. Ames, IA: Iowa State University Press. Accessible description of soil genesis, soil groups, and classification.

Brady, N. C., and R. W. Weil. 1996. *The nature and properties of soils*, 11th ed. Upper Saddle River, NJ: Prentice Hall. Excellent basic text on soils.

Brown, J. H., and W. McDonald. 1995. Livestock grazing and conservation on southwestern rangelands. *Conservation Biology* 9:1644–1647.

Brown, L. R. 1984. Conserving soil. Pages 53–73 in L. Strake, ed. *State of the world 1984*. New York: W. W. Norton.

Brown, L. R., and E. C. Wolf. 1984. *Soil erosion: Quiet crisis in the world economy*. Washington, DC: Worldwatch Institute.

Farb, P. 1959. *The living earth*. New York: Harper and Row. A small, classic, popular discussion of soil.

Fleischner, T. 1994. Ecological costs of livestock grazing in western North America. *Conservation Biology* 8:629–644.

Furley, P. A., and W. N. Newey. 1983. *Geography of the biosphere*. London: Butterworths. See Chapters 4 and 9 on soils.

Soil Survey Division Staff 1993. *Soil survey manual*. Handbook 18. Washington D.C: U.S. Government Printing Office.

Jenny, H. 1980. *The soil resource*. New York: Springer-Verlag. A definitive reference source.

Killham, K. 1994. *Soil ecology*. New York: Cambridge University Press. Excellent overview of soil ecology with emphasis on soil microbes.

Lutz, H. J., and R. F. Chandler. 1946. *Forest soils*. New York: Wiley. Dated, but still an excellent reference on native soils under forest growth.

Patton, T. R. 1996. *Soils: A new global view*. New Haven, CT: Yale University Press.

ECOLOGICAL APPLICATION: Blowin' in the Wind

Pimentel, D., C. Harvey, P. Resosudarmo, et al. 1995. Environmental and economic costs of soil erosion and conservation benefits. *Science* 267:1117–1123.

Sears, P. B. 1947. *Deserts on the march*. Norman, OK: Oklahoma University Press. A frequently reprinted classic written after the Dust Bowl of the 1930s. Much of the book still holds true.

Worster, D. 1979. *The Dust Bowl*. New York: Oxford. An in-depth look at the Dust Bowl, past and present, and how it changed the plains.

———. 1993. *The wealth of nature*. New York: Oxford University Press.

CHAPTER 6 Plant Adaptations to the Environment

Augspurger, C. K. 1982. Light requirements of neotropical tree seedlings: A comparative study of growth and survival. *Journal of Ecology* 72:777–795.

Bjorkman, O. 1973. Comparative studies on photosynthesis in higher plants. Page 53 in A. C. Geise, ed. *Photophysiology*. New York: Academic Press.

Blom, C. W. P. M., and L. A. C. J. Voesenek. 1996. Flooding: The survival strategies of plants. *Trends in Ecology and Evolution* 11:290–295.

Bradshaw, A. D., M. J. Chadwick, D. Jowett, and R. W. Snaydon. 1964. Experimental investigations into the mineral nutrition of several grass species. IV Nitrogen level. *Journal of Ecology* 52:665–676.

Caldwell, M. M., A. H. Teramura, and M. Tevini. 1989. The changing solar ultraviolet climate and the ecological consequences for higher plants. *Trends in Ecology and Evolution* 4:363–367.

Chapin, F. S., III. 1980. The mineral nutrition of wild plants. *Annual Review of Ecology and Systematics* 11:233–260.

Chazdon, R. L. 1988. Sunflecks and their importance to forest understory plants. *Advances in Ecological Research* 18:1–63.

Chazdon, R. L., and R. W. Pearcy. 1991. The importance of sunflecks for forest understory plants. *Bioscience* 41:760–765.

Dale, J. E. 1992. How do leaves grow? *Bioscience* 42:423–432.

Etherington, J. R. 1982. *Environment and plant ecology*. 2d ed. New York: Wiley.

Feldman, L. J. 1988. The habits of roots. *Bioscience* 38:612–618.

Field, C. B., and H. Mooney. 1986. The photosynthetic-nitrogen relationship in wild plants. Pages 25–55 in T. J. Givnish, ed. *On the economy of plant form and function*. Cambridge: Cambridge University Press.

Grime, J. 1971. *Plant strategies and vegetative processes*. New York: Wiley. Excellent discussion of shade tolerance.

Kirk, J. T. O. 1983. *Light and photosynthesis in aquatic systems*. New York: Cambridge University Press.

Kozlowski, T. T. 1982. Plant responses to flooding of soil. *Bioscience* 34:162–167.

———. 1984. *Flooding and plant growth*. Orlando, FL: Academic Press.

Lambers, H., F. S. Chapin III, and T. L. Pons. 1998. *Plant physiological ecology*. New York: Springer.

Larcher, W. 1996. *Physiological plant ecology*, 3rd ed. New York: Springer-Verlag. Excellent reference on plant ecophysiology.

Long, S. P., and F. I. Woodward (eds.). 1988. *Plants and temperature*. Symposia of the Society for Experimental Biology, Number 42. Cambridge, England: The Company of Biologists Limited. Advanced and comprehensive.

Mooney, H. A. 1986. Photosynthesis. Chapter 11 in M. J. Crawley, ed. *Plant ecology*. New York: Blackwell Scientific Publications. An excellent synthesis.

Mooney, H. A., O. Bjorkman, J. Ehleringer, and J. Berry. 1976. Photosynthetic capacity of in situ Death Valley plants. *Carnegie Institute Yearbook* 75: 310–413.

Mooney, H. A., and S. L. Gulmon. 1982. Constraints on leaf structure and function in reference to herbivory. *Bioscience* 32:198–206.

Parsons, T., J. R. Parsons, M.Takahashi, and B. Hargrave. 1984. *Biological oceanographic processes*. 3rd ed. New York: Pergamon Press.

Schulze, E. D., R. H. Robichaux, J. Grace, P. W. Rundel, and J. R. Ehleringer. 1987. Plant water balance. *Bioscience* 37:30–37.

Strain, B. R., and W. B. Billings (eds.). 1975. *Vegetation and the environment*. Vol. 6, *Handbook of vegetation science*. The Hague: W. Junk.

Turner, N. C., and P. J. Kramer (eds.). 1980. *Adaptation of plants to water and high temperature stress*. New York: Wiley.

Woodward, I., and T. Smith. 1993. Predictions and measurements of the maximum photosynthetic rate, A_{max}, at a global scale. Pages 491–508 in E. D. Schultz and M. M. Caldwell, eds. *Ecophysiology of photosynthesis*. Vol. 100. Berlin: Springer.

CHAPTER 7 Decomposition

Aber, J., and J. Melillo. 1982. Nitrogen immobilization in decaying hardwood leaf litter as a function of initial nitrogen and lignin content. *Canadian Journal of Botany* 60:2263–2269.

———. 1991.*Terrestrial ecosystems*. Philadelphia: Saunders College Publishing. See Chapters 12 and 13.

Anderson, J. M., and A. MacFadyen (eds.). 1976. *The role of terrestrial and aquatic organisms in the decomposition process*. Seventeenth Symposium, British Ecological Society, Oxford, England: Blackwell Scientific Publications.

Austin, A. T., and P. M. Vitousek. 2000. Precipitation, decomposition and litter decomposability of *Metrosideros polymorpha* in native forests in Hawaii. *Journal of Ecology* 88:129–138.

Fletcher, M., G. R. Gray, and J. G. Jones. 1987. *Ecology of microbial communities*. New York: Cambridge University Press. A good overview of ecology at the decomposer level.

Klap, V., P. Louchouarn, J. J. Boon, M. A. Hemminga, and J. van Soelen. 1999. Decomposition dynamics of six salt marsh halophytes as determined by cupric oxide oxidation and direct temperature-resolved mass spectrometry. *Limnology and Oceanography* 44:1458–1476.

Meentemeyer, V. 1978. Macroclimate and lignin control of decomposition. *Ecology* 59:465–472.

Melillo, J., J. Aber, and J. Muratore. 1982. Nitrogen and lignin control of hardwood leaf litter decomposition dynamics. *Ecology* 63:621–626.

Mudrick, D., M. Hoosein, R. Hicks, and E. Townsend. 1994. Decomposition of leaf litter in an Appalachian forest: Effects of leaf species, aspect, slope position and time. *Forest Ecology and Management* 68:231–250.

Odum, W. E., and M. A. Haywood. 1978. Decomposition of intertidal freshwater marsh plants. Pages 89–97 in R. E. Good, D. F. Whigham, and R. L. Simpson, eds. *Freshwater wetlands*. New York: Academic Press.

Smith, T. M. 2002. Spatial variation in leaf-litter production and decomposition within a temperate forest: The influence of species composition and diversity. *Journal of Ecology*. In review.

Staff, H., and B. Berg. 1982. Accumulation and release of plant nutrients in decomposing Scots pine needle litter. Long-term decomposition in a Scots pine forest. II. *Canadian Journal of Botany* 60:1516–1568.

Swift, M. J., O. W. Heal, and J. M. Anderson. 1979. *Decomposition in terrestrial ecosystems*. Berkeley: University of California Press.

Valiela, I. 1984. *Marine ecological processes*. New York: Springer Verlag.

Whitkamp, M., and M. L. Frank. 1969. Evolution of carbon dioxide from litter, humus and subsoil of a pine stand. *Pedobiologia* 9:358–365.

CHAPTER 8 Animal Adaptations to the Environment

Adkisson, P. L. 1966. Internal clocks and insect diapause. *Science* 154:234–241.

Aschoff, J., S. Daan, and G. A. Gross (eds.). 1982. *Vertebrate circadian rhythms*. New York: Springer-Verlag.

Beck, S. D. 1980. *Insect photoperiodism*, 2d ed. New York: Academic Press. A good review.

Belovsky, G. E., and P. F. Jordan. 1981. Sodium dynamics and adaptations of a moose population. *Journal of Mammalogy* 63:613–621.

Bioscience. 1983. Biological clocks. Special section in *Bioscience* 33:424–457.

Brady, J. 1982. *Biological timekeeping*. New York: Cambridge University Press.

Bünning, E. 1973. *The physiological clock*, 3rd ed. New York: Academic Press. A book by one of the pioneers in the field.

Cheatum, E. L., and C. W. Severinghaus. 1950. Variations in fertility of white-tailed deer related to range conditions. *Transactions of the North American Wildlife Conference* 15:170–189.

DeCoursey, P. J. 1960a. Daily light sensitivity rhythm in a rodent. *Science* 131:33–35.

———. 1960b. Phase control of activity in a rodent. *Cold Spring Harbor Symposium on Quantitative Biology* 25:49–54.

———. (ed.). 1976. *Biological rhythms in the marine environment*. Columbia: University of South Carolina Press.

Dodd, R. H. 1973. Insect nutrition: Current developments and metabolic implications. *Annual Review of Entomology* 18:381–420.

Edmunds, L. N. 1988. *Cellular and molecular bases of biological clocks*. New York: Springer-Verlag.

Farner, D. S. 1985. Annual rhythms. *Annual Review of Physiology* 47:65–82.

French, A. R. 1988. The patterns of mammalian hibernation. *American Scientist* 76:569–575.

———. 1992. Mammalian dormancy. Pages 105–121 in T. E. Tomasi and T. H. Horton, eds. *Mammalian energetics*. Ithaca, NY: Comstock.

Gilles, R. (ed.). 1979. *The mechanisms of osmoregulation in animals*. New York: Wiley.

Heinrich, B. 1979. *Bumblebee economics*. Cambridge, MA: Harvard University Press. Excellent on energetics of bumblebees.

———. (ed.). 1981. *Insect thermoregulation*. New York: Wiley. Temperature regulation by insects, including heterothermy.

———. 1996. *The thermal warriors*. Cambridge: Harvard University Press. Semipopular account of myriad strategies insects use to heat their bodies.

Hill, E. P., III. 1972. Litter size in Alabama cottontails as influenced by soil fertility. *Journal of Wildlife Management* 36:1199–1209.

Hill, R. W. 1992. The altricial/precocial contrast in the thermal relations and energetics of small mammals. Pages 122–159 in T. E. Tomasi and T. H. Horton, eds. *Mammalian energetics*. Ithaca, NY: Comstock.

Hill, R. W., and G. A. Wyse. 1989. *Animal physiology*. New York: Harper & Row. A comprehensive text strong on ecophysiology.

Johnson, C. H., and J. W. Hastings. 1986. The elusive mechanisms of the circadian clock. *American Scientist* 74:29–36.

Jones, R. L., and H. C. Hanson. 1985. *Biogeochemistry, mineral licks, and geophagy of North American ungulates*. Ames: Iowa State University Press.

Jones, R. L., and H. P. Weeks. 1985. Ca, Mg, and P in the annual diet of deer in south-central Indiana. *Journal of Wildlife Management* 49:129–133.

Lee, R. E., Jr. 1989. Insect cold-hardiness: To freeze or not to freeze. *Bioscience* 39:308–313. A good discussion of supercooling and cold hardening in insects.

Leith, H. (ed.). 1974. *Phenology and seasonality modeling*. New York: Springer-Verlag.

Lyman, C. P., A. Malan, J. S. Willis, and L. C. H. Wang. 1982. *Hibernation and torpor in mammals and birds*. New York: Academic Press.

Maloly, C. M. O. (ed.). 1979. *Comparative physiology of osmoregulation in animals*. New York: Academic Press.

Moore-Ede, M. C., F. M. Sulzman, and C. A. Fuller. 1982. *The clocks that time us.* Cambridge, MA: Harvard University Press. Circadian rhythms in humans.

Naylor, E. 1985. Tidal rhythmic behaviour of marine animals. *Symposium of Society of Experimental Biology* 39:63–93.

Palmer, J. D. 1976. *An introduction to biological rhythms.* San Diego: Academic Press.

———. 1990. The rhythmic lives of crabs. *Bioscience* 40:352–358.

———. 1995. *The biological rhythms and clocks of intertidal animals.* New York: Oxford University Press.

———. 1996. Time, tide, and living clocks of marine organisms. *American Scientist* 84:570–578.

Saunders, D. S. 1982. *Insect clocks,* 2d ed. Elmsford, NY: Pergamon Press.

Schmidt-Neilsen, K. 1997. *Animal physiology: Adaptation and environment,* 5th ed. New York: Cambridge University Press. An excellent comprehensive text strong on ecophysiology.

Storey, K. B., and J. M. Storey. 1996. Natural freezing survival in animals. *Annual Review of Ecology and Systematics* 27:365–386.

Takahashi, J. S., and M. Hoffman. 1995. Molecular biological clocks. *American Scientist* 83:158–165.

Taylor, C. R. 1972. The desert gazelle: A parody resolved. In G. M. O. Malory, ed. *Comparative physiology of desert animals.* Symposium of the Zoological Society of London No. 31. New York: Academic Press.

Weeks, H. P., Jr., and C. M. Kirkpatrick. 1978. Salt preferences and sodium drive phenology in fox squirrels and woodchuck. *Journal of Mammalogy* 59:531–542.

Weir, J. S. 1972. Spatial distribution of elephants in an African national park in relation to environmental sodium. *Oikos* 23:113.

Winfree, A. T. 1986. *The timing of biological clocks.* Scientific American Library. New York: W. H. Freeman. A popular but challenging discussion of biological clocks.

ECOLOGICAL APPLICATIONS: Saint Patrick and the Absence of Snakes in Ireland

Schall, J. J., and E. R. Pianka. 1978. Geographic trends in numbers of species. *Science* 201:679–686.

CHAPTER 9 Properties of Populations

Alexander, M. M. 1958. The place of aging in wildlife management. *American Scientist* 6:123–131.

Begon, M., J. L. Harper, and C. R. Townsend. 1996. *Ecology: Individuals, populations, and communities.* New York: Blackwell Scientific Publishers. See Chapter 4 on unitary and modular populations.

Blower, J. G., L. M. Cook, and J. A. Bishop. 1981. *Estimating the size of animal populations.* London: George Allen & Unwin.

Brower, J. E., and J. H. Zar. 1982. *Field and laboratory methods for general ecology,* 3rd ed. Dubuque, IA: William C. Brown.

Caughley, G. 1977. *Analysis of vertebrate populations.* New York: Wiley.

Engle, L. G. 1960. Yellow-poplar seedfall pattern. *Central States Forest Experiment Station Note* 143.

Hett, J., and O. L. Loucks. 1976. Age structure models of balsam fir and eastern hemlock. *Journal of Ecology* 64:1029–1044.

Krebs, C. J. 1989. *Ecological methods.* New York: Harper & Row.

MacArthur, R. H., and J. H. Connell. 1966. *The biology of populations.* New York: John Wiley.

Mueller-Dombois, D., and H. Ellenberg. 1974. *Aims and methods of vegetation ecology.* New York: Wiley.

Root, T. 1988. Energy constraints on avian distributions and abundances. *Ecology* 69:330–339.

Southwood, T. R. E. 1978. *Ecological methods.* London: Chapman and Hall.

Wiens, J. 1973. Pattern and process in grassland bird communities. *Ecological Monographs* 43:247–270.

CHAPTER 10 Population Growth

Begon, M., and M. Mortimer. 1995. *Population ecology: A unified study of plants and animals,* 2d ed. Cambridge, MA: Blackwell Scientific Publications.

Burton, J. A., and B. Pearson. 1987. *Collins guide to rare mammals of the world.* London: Collins.

Campbell, R.W. 1969. Studies on gypsy moth population dynamics. Pages 29–34 in *Forest insect population dynamics.* USDA Research Paper, NE-125.

Caughley, G. 1977. *Analysis of vertebrate populations.* New York: Wiley.

Fitzgerald, S. 1989. *International wildlife trade: Whose business is it?* Washington, DC: World Wildlife Fund. Documents why some species go extinct.

Harper, J. L. 1977. *The population biology of plants.* London: Academic Press. The major reference on plant population biology.

Kaufman, L., and Mallory, K. (eds.). 1986. *The last extinction.* Cambridge, MA: MIT Press. Excellent overview.

Krebs, C. 2000. *Ecology: The experimental analysis of distribution and abundance,* 5th edition. San Francisco: Benjamin Cummings.

Lowe, V. P. W. 1969. Population dynamics of red deer *(Cervus elapus L.)* on the Isle of Rhum. *Journal of Animal Ecology* 38:425–457.

MacLulich, D. A. 1937. *Fluctuations in the numbers of varying hare (Lepus americanus).* University of Toronto Biological Series No. 43.

Mountford, G. 1988. *Rare birds of the world.* London: Collins.

Pitelka, F. A. 1957. Some characteristics of microtine cycles in the Arctic. *Proceedings of 18th Biology Colloquium.* Salem: Oregon University Press.

Scharitz, R. R., and J. R. McCormick. 1973. Population dynamics of two competing plant species. *Ecology* 54:723–740.

Scheffer, V. C. 1951. Rise and fall of a reindeer herd. *Scientific Monthly* 73:356–362.

Shaw, J. 1985. *Introduction to wildlife management.* New York: McGraw-Hill. Discusses several concepts of carrying capacity.

Terbrough, J. 1988. *Where have all the birds gone?* Princeton, NJ: Princeton University Press. Effects of habitat fragmentation on neotropical birds.

CHAPTER 11 Intraspecific Population Regulation

Chepko-Sade, B. D., and Z. T. Halpin. 1987. *Mammalian dispersal patterns.* Chicago: University of Chicago Press.

Clatworthy, J. N. 1960. *Studies on the nature of competition between closely related species.* D.Phil. thesis, University of Oxford.

Dash, M. C., and A. K. Hota. 1980. Density effects on the survival, growth rate, and metamorphosis of *Rana tigrina* tadpoles. *Ecology* 61:1025–1028.

Fery, R. L., and J. Janick. 1971. Response of corn *(Zea mays L.)* to population pressure. *Crop Science* 11:220–224.

Fowler, C. W. 1981. Density dependence as related to life history strategy. *Ecology* 62:602–610.

Gaines, M. S., and L. R. McClenaghan, Jr. 1980. Dispersal in small mammals. *Annual Review of Ecology and Systematics* 11:163–169.

Greenwood, P. J., and P. H. Harvey. 1982. The natal and breeding dispersal in birds. *Annual Review of Ecology and Systematics* 13:1–21.

Grime, J. P. 1979. *Plant strategies and vegetative processes.* New York: Wiley. Role of stress in plant populations.

Harestad, A. S., and F. L. Bunnell. 1979. Home range and body weight—A reevaluation. *Ecology* 60:389–402.

Harper, J. L. 1977. *Population biology of plants.* New York: Academic Press.

Houston, D. B. 1982. *The northern Yellowstone elk: Ecology and management.* New York: Macmillan. Excellent population study discusses the role of density-dependent mortality.

Jones, W. T. 1989. Dispersal distance and the range of nightly movements of Merriam's kangaroo rats. *Journal of Mammalogy* 20:29–34.

King, A. A., and W. C. Schaffer. 2001. The geometry of a population cycles: A mechanistic model of snowshoe demography. *Ecology* 82:814–830.

Krebs, J., and N. B. Davies (eds.). 1991. *Behavioral ecology: An evolutionary approach,* 3rd ed. Oxford, England: Blackwell Scientific Publications. Chapters on aspects of territoriality.

Larsen, K. W., and S. Boutin. 1994. Movement, survival, and settlement of red squirrel *(Tamiasciurus hudsonicus)* offspring. *Ecology* 75:214–223.

Lett, P. F., R. K. Mohn, and D. F. Gray. 1981. Density-dependent processes and management strategy for the northwest Atlantic harp seal populations. Pages 135–158 in C. W. Fowler and T. D. Smith, eds. *Dynamics of large mammal populations.* New York: Wiley.

McCullough, D. R. 1981. Population dynamics of the Yellowstone grizzly. Pages 173–196 in C. W. Fowler and T. D. Smith, eds. *Dynamics of large mammal populations.* New York: Wiley.

Mech, L. D. 1970. *The wolf: The ecology and behavior of an endangered species.* Garden City, NY: Doubleday.

Mech, L. D., R. E. McRoberts, R. O. Peterson, and R. E. Page. 1978. Relationship of deer and moose populations to previous winter's snow. *Journal of Animal Ecology* 56:615–627.

Mloszewski, M. 1983. *The behavior and ecology of the African buffalo.* Cambridge: Cambridge University Press. Male and female social hierarchies in the buffalo.

Murdoch, W. W. 1994. Population regulation in theory and practice. *Ecology* 75: 271–287.

Myers, K., C. S. Hale, R. Mykytowycz, and R. L. Hughes. 1971. The effects of varying density and space on sociality and health in animals. Pages 148–187 in A. E. Esser, ed. *Behavior and environment: The use of space by animals and men.* New York: Plenum.

Nice, M. M. 1937. *Studies in the life history of the song sparrow.* Transactions of the Linnaean Society of New York IV. Reprint edition. New York: Dover Publications. A classic study in territoriality.

Packard, J. M., and L. D. Mech. 1983. Population regulation in wolves. Pages 151–174 in F. L. Bunnell, D. S. Eastman, and J. M. Peak, eds. *Symposium on natural regulation of wildlife populations.* Moscow, Idaho: University of Idaho.

Searcy, W. A. 1982. The evolutionary effects of mate selection. *Annual Review of Ecology and Systematics* 13:57–85.

Sinclair, A. R. E. 1977. *The African buffalo: A study of resource limitations of populations.* Chicago: University of Chicago Press.

Smith, R. L. 1963. Some ecological notes on the grasshopper sparrow. *Wilson Bulletin* 75:159–165.

Smith, S. M. 1978. The underworld in a territorial adaptive strategy for floaters. *American Naturalist* 112:570–582.

Tamarin, R. H. (ed.). 1978. *Population regulation.* Benchmark Papers. Stroudsburg, PA: Dowden, Hutchinson, and Ross. Collection of important papers.

Turchin, P. 1999. Population regulation: A synthetic view. *Oikos* 84:160–163.

Watkinson, A. R., and A. J. Davy. 1985. Population biology of salt marsh and dune annuals. *Vegetatio* 62:487–497.

Whittaker, R. H., S. A. Levi, and R. B. Root. 1973. Niche, habitat, and ecotope. *American Naturalist* 107:321–338.

Wolff, J. O. 1997. Population regulation in mammals: An evolutionary prospective. *Journal of Animal Ecology* 66:1–13.

Yoda, K., T. Kira, H. Ogawa, and K. Hozumi. 1963. Self thinning in overcrowded pure stands under cultivated and natural conditions. *Journal Biology* 14:107–129.

Zimen, E. 1981. *The wolf: A species in danger.* New York: Delacourt Press. Excellent behavioral study of the European wolf.

CHAPTER 12 Life History Patterns

Alcock, J. 1995. *Animal behavior: An evolutionary approach,* 4th ed. Sunderland, MA: Sinauer Associates. Excellent treatment of topics covered in this chapter.

Andersson, M., and Y. Iwasa. 1996. Sexual selection. *Trends in Ecology and Evolution* 11:53–58.

Ashmole, N. P. 1963. The regulation of numbers of tropical oceanic birds. *Ibis* 103b:458–473.

Bajema, C. J. (ed.). 1984. *Evolution by sexual selection theory.* Benchmark Papers in Systemic and Evolutionary Biology. New York: Scientific and Academic Editions.

Bierzychudek, P. 1982. The demography of Jack-in-the-pulpit, a forest perennial that changes sex. *Ecological Monographs* 52:335–351.

Boyce, M. S. 1984. Restitution of *r* and *K* selection as modes of density-dependent natural selection. *Annual Review of Ecology and Systematics* 15:427–447.

Buss, D. M. 1994. The strategies of human mating. *American Scientist* 82:238–249. Application of sexual selection theory to humans.

Catchpole, C. K. 1987. Bird song, sexual selection, and female choice. *Trends in Ecology and Evolution* 2:94–97.

Clutton-Brock, T. H., F. E. Guinness, and S. D. Albion. 1982. *Red deer: Behavior and ecology of two sexes.* Chicago: University of Chicago Press. An outstanding study of reproductive behavior and success based on a long-term study of known individuals.

Cody. M. L. 1966. A general theory of clutch size. *Evolution* 20:174–184.

Darwin, C. 1871. *The descent of man and selection in relation to sex.* London: Murray.

Ellstrand, N. C. 1984. Multiple paternity within the fruits of wild radish *Raphanus sativus. American Naturalist* 123:819–828.

Freeman, C. L., K. T. Harper, and E. L. Charnov. 1980. Sex change in plants: Old and new observations and new hypotheses. *Oecologica* 47:222–232.

Grime, J. P. 1977. Evidence for the existence of three primary strategies in plants and its relevance to ecological and evolutionary theory. *American Naturalist* 111:1169–1194.

———. 1979. *Plant strategies and vegetative processes.* New York: Wiley.

Gubernich, D. J., and P. H. Klopher (eds.). 1981. *Parental care in mammals.* New York: Plenum.

James, F. 1971. Ordination of habitat relationships among birds. *Wilson Bulletin* 83:215–236.

Johnsgard, P. A. 1994. *Arena birds: Sexual selection and behavior.* Washington, DC: Smithsonian Institution Press. A survey of lek behavior.

Krebs, J. R., and N. D. Davies. 1981. *An introduction to behavioral ecology.* Oxford: Blackwell Scientific Publications. Excellent discussion of topics covered in this chapter.

Lack, D. 1954. *The natural regulation of animal numbers.* Oxford: Oxford University Press (Clarendon Press).

MacArthur, R. H., and E. O. Wilson. 1967. *The theory of island biogeography.* Princeton, NJ: Princeton University Press.

McGregor, P. K., J. R. Krebs, and C. M. Perrins. 1981. Song repertoires and lifetime reproductive success in the great tit *(Parus major). American Naturalist* 118:49–59. A study of reproductive success of known male and female birds.

McNaughton, S. J. 1975. r and K selection in *Typha. American Naturalist* 109:215–261.

Mennill, D., L. M. Ratcliffe, and P. T. Boag. 2002. Female eavesdropping on male song contests in songbirds. *Science* 296:873.

Petrie, M. 1994. Improved growth and survival of offspring of peacocks with more elaborate trains. *Nature* 371:598–599.

Pianka, E. 1972. r and K selections or *b* and *d* selection? *American Naturalist* 100:65–75.

Policansky, D. 1982. Sex change in plants and animals. *Annual Review of Ecology and Systematics* 13:471–495.

Primack, R. B. 1979. Reproductive effort in annual and perennial species of *Plantago* (Plantaginaceae). *American Naturalist* 114:51–62.

Ricklefs, R. 1980. Geographical variation in clutch size in passerine birds: Ashmole's hypothesis. *Auk* 97:38–49.

Shapiro, D. Y. 1980. Serial sex changes after simultaneous removal of males from social groups of coral reef fish. *Science* 209:1136–1137.

Small, M. F. 1992. Female choice in mating. *American Scientist* 80:142–151.

Thornhill, N. W. (ed.). 1993. *The natural history of inbreeding and outbreeding: Theoretical and empirical perspectives.* Chicago: University of Chicago Press.

Thornhill, R., and J. Alcock. 1983. *The evolution of insect mating systems.* Cambridge, MA: Harvard University Press.

Warner, R. R. 1988. Sex change and size-advantage model. *Trends in Ecology and Evolution* 3:133–136.

Wasser, S. K. (ed.). 1983. *Social behavior of female vertebrates.* New York: Academic Press.

Wauters, L., and A. A. Dohondt. 1989. Body weight, longevity, and reproductive success in red squirrels (*Sciurus vulgaris*). *Journal of Animal Ecology* 58:637–651.

Werner, P. A., and W. J. Platt. 1976. Ecological relationships of co-occurring goldenrods (Solidago:Compositae). *American Naturalist* 110:959–971.

Whitham, T. G. 1980. The theory of habitat selection: Examined and extended using *Pemphigus* aphids. *American Naturalist* 115:449–466.

Willson, M. F. 1983. *Plant reproductive ecology*. New York: Wiley.

Wilson, E. O. 1980. *Sociobiology: The abridged edition*. Cambridge, MA: Harvard University Press.

CHAPTER 13 Community Structure

Anderson, S., and H. H. Shugart. 1974. *Avian community analyses of Walker Branch Watershed*. ORNL Publication No. 623. Oak Ridge, TN: Environmental Sciences Division, Oak Ridge National Laboratory.

Bond, W. J., and B. van Wilgen. 1996. *Fire and plants*. New York: Chapman and Hall. Plant population biology in relation to fire.

Bormann, F., and G. E. Likens. 1979. *Patterns and process in forested ecosystems*. New York: Springer-Verlag. A basic study of forest succession at Hubbard Brook, New Hampshire.

Braun, E. L. 1950. *Deciduous forests of eastern North America*. New York: McGraw-Hill.

Buell, M. F., and R. E. Wilbur. 1948. Life form spectra of the hardwood forests of Lake Itaska Park region, Minnesota. *Ecology* 29:352–359.

Clements, F. E. 1916. *Plant succession: An analysis of the development of vegetation*. Carnegie Inst. Wash. Publ. 242.

Davis, M. B. 1983. Holocene vegetational history of the eastern United States. Pages 166–188 in H. E. Wright, Jr., ed. *Late quarternary environments of the United States*. Vol. II, The Holocene. Minneapolis: University of Minnesota Press.

Delcourt, P. A., and H. R. Delcourt. 1981. Vegetation maps for eastern North America, 40,000 yr BP to present. Pages 123–166 in R. Romans, ed. *Geobotany*. New York: Plenum Press.

Gleason, H. A. 1926. The individualistic concept of the plant association. *Bulletin Torrey Botanical Club* 53:1–20.

Golley, F. (ed.). 1978. *Ecological succession*. Benchmark Papers. Stroudsburg, PA: Dowden, Hutchinson, and Ross. Succession as viewed over time.

Hobbie, E. A. 1994. *Nitrogen cycling during succession in Glacier Bay Alaska*. Masters Thesis. Unversity of Virginia.

Huston, M. A. 1994. *Biological diversity: The coexistence of species on changing landscapes*. New York: Cambridge University Press. Chapter 9 is a good review of succession and landscape patterns.

Maser, C., and J. M. Trappe (eds.). 1984. *The seen and unseen world of the fallen tree*. USDA Forest Service General Technical Report PNW-164.

Olson, J. S. 1958. Rates of succession and soil changes in southern Lake Michigan sand dunes. *Botanical Gazette* 119:125–170.

Oosting, H. J. 1942. An ecological analysis of the plant communities of Piedmont, North Carolina. *American Midland Naturalist* 28:1–126.

Raunkiaer, C. 1934. *The life forms of plants and statistical plant geography*. Oxford: Clarendon Press.

Ricklefs, R. E., and D. Schluter (eds.). 1993. *Ecological communities: Historical and geographical perspectives*. Chicago: University of Chicago Press.

Shannon, C. E., and W. Wiener. 1963. *The mathematical theory of communications*. Urbana: University of Illinois Press.

Simpson, E. H. 1949. Measurement of diversity. *Nature* 163:688.

Smith, R. L. 1959. Conifer plantations as wildlife habitat. *New York Fish and Game Journal* 5:101–132.

Sousa, W. P. 1979. Disturbance in marine intertidal boulder fields: The nonequilibrium maintenance of species diversity. *Ecology* 60:1225–1239.

———. 1984. Intertidal mosaics: Patch size, propagule availability, and spatially variable patterns of succession. *Ecology* 65:1918–1935.

Sprugle, D. G. 1976. Dynamic structure of wave-generated *Abies balsamea* forests in northeastern United States. *Journal of Ecology* 64:889–911.

Stern, W. L., and M. F. Buell. 1951. Life-form spectra in a New Jersey pine barren forest and Minnesota jack pine forest. *Bulletin Torrey Botanical Club* 78:61–65.

Watts, M. T. 1976. *Reading the American landscape*. New York: Macmillan. A delightful introduction to landscape ecology of the United States as influenced by humans.

West, D. C., H. H. Shugart, and D. B. Botkin (eds.). 1981. *Forest succession: Concepts and application*. New York: Springer-Verlag. Excellent reference on patterns and processes from temperate to tropical forests.

Whittaker, R. H. 1956. Vegetation of the Great Smoky Mountains. *Ecological Monographs* 26:1–80.

Wilson, E. O. 1992. *The diversity of life*. Cambridge, MA: Harvard University Press. Excellent, accessible synthesis of biodiversity.

CHAPTER 14 Interspecific Competition

American Naturalist. 1983. Interspecific competition papers. Volume 122, No. 5 (November). A group of controversial articles on interspecific competition.

Austin, M. P. 1999. A silent clash of paradigms: Some inconsistencies in community ecology. *Oikos* 86:170–178.

Austin, M. P., R. H. Groves, L. M. F. Fresco, and P. E. Laye. 1986. Relative growth of six thistle species along a nutrient gradient with multispecies competition. *Journal of Ecology* 73:667–684.

Bazzaz, F.A. 1996. *Plants in changing environments*. New York: Cambridge University Press.

Bright, C. 1998. *Life out of bounds: Bioinvasion in a borderless world*. New York: W.W. Norton.

Brown, J. C., O. J. Reichman, and D. W. Davidson. 1979. Granivory in desert ecosystems. *Annual Review of Ecology and Systematics* 10:210–227.

Brown, J. H. 1995. *Macroecology*. Chicago: University of Chicago Press. See Chapter 3 for a discussion of species, niches, and communities.

Connell, J. H. 1983. On the prevalence and relative importance of interspecific competition: Evidence from field experiments. *American Naturalist* 122:661–696.

———. 1975. Some mechanisms producing structure in natural communities: A model and evidence from field experiments. Pages 460–490 in M. L. Cody and J. Diamond, eds. *Ecology and evolution of communities*. Cambridge, MA: Harvard University Press.

Cox, G.W. 1999. *Alien species in North America and Hawaii: Impacts on natural ecosystems*. Washington, DC: Island Press.

Darwin, C. 1859. *The origin of species*. Philadelphia: McKay. [Reprinted from 6th London edition.]

Dayan, T., D. Simberloff, E. Tchernov, and Y. Yom-Tov. 1990. Feline canines: Community-wide character displacement among the small cats of Israel. *American Naturalist* 136:39–60.

Dye, P. J., and P. T. Spear. 1982. The effects of bush clearing and rainfall variability on grass yield and composition in southwest Zimbabwe. *Zimbabwe Journal of Agricultural Research* 20:103–118.

Emery, N. C., P. J. Ewanchuk, and M. D. Bertness. 2001. Competition and salt marsh plant zonation: Stress tolerators may be dominant competitors. *Ecology* 82:2471–2485.

Gause, G. F. 1934. *The struggle for existence*. Baltimore: Williams and Wilkins.

Grace, J. B., and R. G. Wetzel. 1981. Habitat partitioning and competitive displacement in cattails (*Typha*): Experimental field studies. *American Naturalist* 118:463–474.

Grant, P. 1999. *Ecology and evolution of Darwin's finches*. Princeton, NJ: Princeton University Press.

Groves, R. H., and J. D. Williams. 1975. Growth of skeleton weed (*Chondrilla juncea*) as affected by growth of subterranean clover (*Trifolium subterraneum* L.) and infection by *Puccinia chondrilla* Bubak and Syd. *Australian Journal of Agricultural Research* 26:975–983.

Gurevitch, J. L. 1986. Competition and the local distribution of the grass *Stipa neomexicana*. *Ecology* 67:46–57.

Gurevitch, J. L., L. Morrow, A. Wallace, and J. J. Walch. 1992. Meta-analysis of competition in field experiments. *American Naturalist* 140:539–572.

Heller, H. C., and D. Gates. 1971. Altitudinal zonation of chipmunks *(Eutamias)*: Energy budgets. *Ecology* 52: 424–443.

Hutchinson, G. E. 1978. *An introduction to population ecology.* New Haven: Yale University Press. Builds a strong case for interspecific competition.

Lotka, A. J. 1925. *Elements of physical biology.* Baltimore: Williams and Wilkins.

Park, T. 1954. Experimental studies of interspecies competition: 2. Temperature, humidity and competition in two species of *Trilobium. Physiological Zoology* 27:177–238.

Pianka, E. R. 1994. *Evolutionary ecology,* 5th ed. New York: HarperCollins. Good discussion on competition.

Pickett, S. T. A., and F. A. Bazzaz. 1978. Organization of an assemblage of early successional species on a soil moisture gradient. *Ecology* 59:1248–1255.

Rice, E. L. 1984. *Allelopathy.* 2d ed. Orlando, FL: Academic Press.

Schoener, T. W. 1983. Field experiments on interspecific competition. *American Naturalist* 122: 240–285.

Tilman, D. M., M. Mattson, and S. Langer. 1981. Competition and nutrient kinetics along a temperature gradient: An experimental test of of a mechanistic approach to niche theory. *Limnology and Oceanography* 26:1020–1033.

Volterra, V. 1926. Variation and fluctuations of the numbers of individuals in animal species living together. Reprinted on pages 409–448 in R. M. Chapman (1931), *Animal ecology.* New York: McGraw-Hill.

Werner, E. E., and J. D. Hall. 1976. Niche shifts in sunfishes: Experimental evidence and significance. *Science* 191:404–406.

———. 1979. Foraging efficiency and habitat switching in competing sunfish. *Ecology* 60:256–264.

Wiens, J. A. 1977. On competition and variable environments. *American Scientist* 65:590–597.

CHAPTER 15 Predation

Cooper, S. M., and N. Owen-Smith. 1986. Effects of plant spinescence on large mammalian herbivores. *Oecologica* 68:446–455.

Curio, E. 1976. *The ethology of predation.* New York: Springer-Verlag. An interesting treatment of the behavioral side of predation.

Davies, N. B. 1977. Prey selection and social behavior in wagtails *(Aves monticillidae). Journal of Animal Ecology* 46:37–57.

Errington, P. L. 1946. Predation and vertebrate populations. *Quarterly Review of Biology* 22:144–177, 221–245. A classic study of predation.

———. 1963. *Muskrat populations.* Ames, IA: Iowa State University Press. A study of a prey species and its predators.

Fleming, T. H. 1988. *The short-tailed fruit bat: A study in plant-animal interactions.* Chicago: University of Chicago Press.

Fox, L. R. 1975. Cannibalism in natural populations. *Annual Review of Ecology and Systematics* 6:87–106.

Godfray, H. C. J. 1994. *Parasitoids: Behavioral and evolutionary ecology.* Princeton: Princeton University Press.

Hassell, M. P., and R. M. May. 1974. Aggregation in predator sand insect parasites and its effect on stability. *Journal of Animal Ecology* 43:567–594.

Holling, C. S. 1959. The components of predation as revealed by a study of small mammal predation of the European sawfly. *Canadian Entomologist* 91:293–320. Classic study of functional response.

———. 1966. The functional response of invertebrate predators to prey density. *Memoirs Entomological Society of Canada* 48:1–86. Further study of functional response.

Jackson, R. R. 1992. Eight-legged tricksters. *Bioscience* 42:590–596. Cannibalism in spiders.

Jedrzejewski, W., B. Jedrzejewski, and L. Szymura. 1995. Weasel population response, home range, and predation on rodents in a deciduous forest in Poland. *Ecology* 76:179–195.

Keith, L. B., J. R. Cary, O. J. Rongstad, and M. C. Brittingham. 1984. Demography and ecology of a declining snowshoe hare population. *Journal of Wildlife Management* 90:1–43.

Korpimaki, E., and K. Norrdahl. 1991. Numerical and functional responses of kestrels, short-eared owls, and long-eared owls to vole densities. *Ecology* 72:814–826.

Krebs, C. J., R. Boonstra, S. Boutin, and A. R. E. Sinclair. 2001. What drives the 10-year cycle of snowshoe hares? *Bioscience* 51:25–35.

Krebs, J. R., and N. B. Davies (eds.). 1984. *Behavioral ecology: An evolutionary approach,* 2d ed. Oxford, England: Blackwell Scientific Publications. A theoretical approach.

Leonard, G. H., M. D. Bertness, and P. O. Yund. 1999. Crab predation, water-borne cues, and inducible defenses in the blue mussel *Mytilus edulis. Ecology* 80:1–14.

Mazncourt, de C., M. Loreau, and U. Dieckmann. 2001. Can the evolution of plant defense lead to plant-herbivore mutualism? *American Naturalist* 158: 109–123.

Mooney, H. A., and Gulmon, S. L. 1982. Constraints on leaf structure and function in reference to herbivory. *BioScience* 32:198–206.

O'Donoghue, M., S. Boutin, C. J. Krebs, G. Zuleta, D. L. Murray, and E. J. Hofer. 1998. Functional response of coyotes and lynx to the snowshoe hare cycle. *Ecology* 79:1193–1208.

Polis, G. 1981. The evolution and dynamics of intraspecific predation. *Annual Review of Ecology and Systematics* 12:225–251. Study of cannibalism.

Polis, G., and R. D. Holt. 1992. Intraguild predation: The dynamics of complex trophic interactions. *Trends in Ecology and Evolution* 7:151–154.

Proulx, M., and A. Mazumder. 1998. Reversal of grazing impact on plant species richness in nutrient-poor vs. nutrient-rich ecosystems. *Ecology* 79:2581–2592.

Relyea, R. A. 2001a. Morphological and behavioral plasticity of larval anurans in response to different predators. *Ecology* 82:523–540.

———. 2001b. The relationship between predation risk and anti-predator responses in larval anurans. *Ecology* 82:541–554.

Schnell, D. E. 1976. *Carnivorous plants of the United States and Canada.* Winston-Salem, NC: John F. Blair, Publisher.

Sinclair, A. R. E., C. J. Krebs, J. M. N. Smith, and S. Boutin. 1988. Population biology of snowshoe hares. III: Nutrition, plant secondary compounds, and food limitation. *Journal of Animal Ecology* 57:787–806.

Solomon, M. E. 1949. The natural control of animal populations. *Journal of Animal Ecology* 18:1–32.

Spenser, C. N., B. McClelland, and J. A. Stanford. 1991. Shrimp stocking, salmon collapse, and eagle displacement. *Bioscience* 41:14–21.

Stephens, D. W., and J. W. Krebs. 1987. *Foraging theory.* Princeton: Princeton University Press. Theoretical look at optimal foraging.

Taylor, R. J. 1984. *Predation.* New York: Chapman and Hall. An overview of predator-prey dynamics.

Wagner, J. D., and D. H. Wise. 1996. Cannibalism regulates densities of young wolf spiders: Evidence from field and laboratory experiments. *Ecology* 77:639–652.

Williams, K., K. G. Smith, and F. M. Stevens. 1993. Emergence of 13-year periodical cicada (Cicadidae:*Magicicada*): Phenology, mortality, and predator satiation. *Ecology* 74:1143–1152.

CHAPTER 16 Parasitism and Mutualism

Addicott, J. F. 1986. Variations in the costs and benefits of mutualism: The interaction between yucca and yucca moths. *Oeocologica* 70:486–494.

Anderson, R. C. 1963. The incidence, development, and experimental transmission of *Pneumostrongylus tenuis* Dougherty (Metastrongyloidae: Protostrongyliidae) of the menges of the white-tailed deer *(Odocoilus virginianus borealis)* in Ontario. *Canadian Journal of Zoology* 41:775–792.

Anderson, R. C., and A. K. Prestwood. 1981. Lungworms. In W. R. Davidson, ed. *Diseases and parasites of white-tailed deer.* Miscellaneous Publication No. 7. Tallahassee, FL: Tall Timbers Research Station.

Bacon, P. J. 1985. *Population dynamics of rabies in wildlife.* New York: Academic Press.

Ball, D. M., J. F. Pedersen, and G. D. Lacefield. 1993. The tall fesque endophyte. *American Scientist* 81:370–379.

Barbour, A. G., and D. Fish. 1993. The biological and social phenomenon of Lyme disease. *Science* 260:1610–1616.

Barrett, J. A. 1983. Plant-fungus symbioses. In D. Futuyma and M. Slakin, eds. *Coevolution*. Sunderland, MA: Sinauer Associates.

Barth, F. G. 1991. *Insects and flowers: The biology of a partnership*. Princeton, NJ: Princeton University Press.

Beattie, A. J., and D. C. Culver. 1981. The guild of myrmecohores in the herbaceous flora and West Virginia forests. *Ecology* 62:107–115.

Bentley, B. L. 1977. Extrafloral nectarines and protection by pugnacious bodyguards. *Annual Review of Ecology and Systematics* 8:407–427.

Boucher, D. H. (ed.). 1985. *The biology of mutualism*. Oxford, England: Oxford University Press. An excellent reference.

Boucher, D. H., S. James, and H. D. Keller. 1982. The ecology of mutualism. *Annual Review of Ecology and Systematics* 13:315–347.

Burdon, J. J. 1987. *Diseases and plant population biology*. Cambridge, England: Cambridge University Press. Effect of diseases on individual plants and populations.

Buskirk, J. V., and R. S. Ostfeld. 1995. Controlling Lyme disease by modifying the density and species composition of tick hosts. *Ecological Applications* 5:1133–1140.

Clay, K. 1988. Fungal endophytes of grasses: A defensive mutualism between plants and fungi. *Ecology* 60:10–16.

———. 1990. Fungal endophytes of grasses. *Annual Review of Ecology and Systematics* 21:275–297.

Croll, N. A. 1966. *Ecology of parasites*. Cambridge, MA: Harvard University Press.

Dobson, A. P., and E. R. Carper. 1996. Infectious diseases and human population history. *Bioscience* 46:115–125.

Doebler, S. A. 2000. The rise and fall of the honeybee. *Bioscience* 50:738–732.

Edwards, M. A., and U. McDonald. 1982. *Animal diseases in relation to animal conservation*. New York: Academic Press. Effects of disease on wild animal populations and human interactions.

Feinsinger, P. 1983. Coevolution and pollination. Pages 282–310 in D. Futuyma and M. Slatkin, eds. *Coevolution*. Sunderland, MA: Sinauer Associates.

Fenner, F., and F. N. Ratcliffe. 1965. *Myxamatosis*. Cambridge, England: Cambridge University Press.

Futuyma, D. J., and M. Slatkin (eds.). 1983. *Coevolution*. Sunderland, MA: Sinauer Associates. Papers on processes and consequences of coevolution.

Garnett, G. P., and E. C. Holmes. 1996. The ecology of emergent infectious disease. *Bioscience* 46:127–135.

Garrett, L. 1994. *The coming plague: Newly emerging diseases in a world out of balance*. New York: Farrar, Straus, and Giroux.

Grenfell, B. T. 1988. Gastrointestinal nematode parasites and the stability and productivity of intensive ruminal grazing systems. *Philosophical Transactions Royal Society, London. British Biological Science* 321:541–563.

Hamilton, W. D., and M. Zuk. 1982. Heritable true fitness and bright birds: A role for parasites? *Science* 218:384–387.

Hanzawa, F. M., A. J. Beattie, and D. C. Culvar. 1988. Directed dispersal: Demographic analysis of an ant-seed mutualism. *American Naturalist* 131:1–13.

Harley, J. L., and S. E. Smith. *The biology of mycorrhizae*. London: Academic Press.

Heithaus, E. R. 1981. Seed predation by rodents on three ant-dispersed plants. *Ecology* 63:136–145.

Hocking, B. 1975. Ant-plant mutualism: Evolution and energy. Pages 78–90 in L. E. Gilbert and P. H. Raven, eds. *Coevolution of animals and plants*. Austin: University of Texas Press.

Holmes, J. 1983. Evolutionary relationships between parasitioc helminths and their hosts. In D. J. Futuyma and M. Slakin, eds. *Coevolution*. Sunderland, MA: Sinauer Associates.

Hougen-Eitzman, D., and M. D. Rausher. 1994. Interactions between herbivorous insects and plant-insect coevolution. *American Naturalist* 143:677–697.

Howe, H. F. 1986. Seed dispersal by fruit-eating birds and mammals. Pages 123–190 in D. R. Murra, ed. *Seed dispersal*. Sydney: Academic Press.

Howe, H. F., and L. C. Westley. 1988. *Ecological relationships of plants and animals*. New York: Oxford University Press. An outstanding presentation of herbivory, pollination, and other relationships.

Hutchins, H. E. 1994. Role of various animals in the dispersal and establishment of white-barked pine in the Rocky Mountains, USA. Pages 163–171 in W. C. Schmidt and F-K. Holtmeier comps. 1994. *Proceedings–International workshop on subalpine stone pines and their environment: the status of our knowledge*; 1992 September 5–11; St. Moritz, Switzerland. General Technical Report INT-GTR-309, Ogden, UT: USDA Forest Service, Intermountain Research Station.

Iowa, K., and M. D. Rausher. 1997. Evolution of plant resistance to multiple herbivores: Quantifying diffuse evolution. *American Naturalist* 149:316–335.

Janzen, D. H. 1966. Coevolution of mutualism between ants and acacias in Central America. *Evolution* 20:249–275.

Korstian, C. F., and P. W. Stickel. 1927. *The natural replacement of blight-killed chestnut*. USDA Miscellaneous Circular 100. Washington, DC: U.S. Government Printing Office.

Lafferty, K. D., and A. K. Morris. 1996. Altered behavior of parasitized killifish increases susceptibility to predation by bird final hosts. *Ecology* 77:1390–1397.

Lindstrom, E. R., et al. 1994. Disease reveals the predator: Sarcoptic mange, red fox predation, and prey populations. *Ecology* 74:1041–1049.

Margulis, L. 1982. *Early life*. Boston: Science Books International.

———. 1998. *Symbiotic planet*. New York: Basic Books.

Maser, C., J. M. Trappe, and R. A. Nussbaum. 1978. Fungal-small mammal interrelationship with emphasis on Oregon coniferous forests. *Ecology* 59:799–809.

Mattes, H. 1994. Coevolutional aspects of stone pines and nutcrackers. Pages 31–35 in W. C. Schmidt and F-K. Holtmeier comps. 1994. *Proceedings–International workshop on subalpine stone pines and their environment: the status of our knowledge*; 1992 September 5–11; St. Moritz, Switzerland. General Technical Report INT-GTR-309, Ogden, UT: USDA Forest Service, Intermountain Research Station.

May, R. M. 1983. Parasitic infections as regulators of animal populations. *American Scientist* 71:36–45.

May, R. M., and R. M. Anderson. 1979. Population biology of infective diseases: Part II. *Nature* 280:455–461.

McNeill, W. H. 1976. *Plagues and peoples*. New York: Doubleday. The role of infectious diseases in the course of history.

Moore, J. 1995. The behavior of parasitized animals. *Bioscience* 45:89–96.

Muscatine, L., and J. W. Porter. 1977. Reef corals: Mutualistic symbioses adapted to nutrient-poor environments. *Bioscience* 27:454–460.

Ostfeld, R. S. 1997. The ecology of Lyme disease risk. *American Scientist* 85:338–346.

Ostfeld, R. S., C. G. Jones, and J. O. Wolff. 1996. Of mice and mast. *Bioscience* 46:323–330.

Randolph, S. E. 1975. Patterns of the distribution of the tick *Ioxodes trianguliceps birula* on its host. *Journal of Animal Ecology* 44:451–474.

Real, L. A. (ed.). 1983. *Pollination biology*. Orlando, FL: Academic Press.

———. 1996. Sustainability and the ecology of infectious disease. *Bioscience* 46:88–97.

Schall, J. J. 1983. Lizard malaria: Cost to vertebrate host's reproductive success. *Parasitology* 87:1–6.

Sheldon, B. C., and S. Verhulst. 1996. Ecological immunity: Costly parasite defenses and trade-offs in evolutionary ecology. *Trends in Ecology and Evolution* 11:317–321.

Sinclair, A. R. E. 1977. *The African buffalo: A study of resource limitation of populations*. Chicago: University of Chicago Press.

Spielman, A., M. I. Wilson, J. F. Levine, and J. Piesman. 1985. Ecology of *Ixodes domini*-borne human babesiosis and Lyme disease. *Annual Review of Entomology* 30:439–460.

Tomback, D. F. 1994. Ecological relationship between Clark's nutcracker and four wingless-seed *Strobus* pines of western

North America. Pages 221–224 in W. C. Schmidt and F-K. Holtmeier, comps. 1994. *Proceedings–International workshop on subalpine stone pines and their environment: the status of our knowledge; 1992 September 5–11; St. Moritz, Switzerland. General Technical Report INT-GTR-309, Ogden, UT: USDA Forest Service, Intermountain Research Station.

Tomback, D. F. 1982. Dispersal of white-barked pine seeds by Clark's nutcracker: a mutualism hypothesis. *Journal of Animal Ecology* 51:451–457.

Tomback, D. F. and Y. B. Linhart. 1990. The evolution of bird-dispersed pines. *Evolutionary Ecology* 4:185–219.

Wakelin, D. 1997. Parasites and the immune system. *Bioscience* 47:32–40.

Warner, R. E. 1968.The role of introduced diseases in the extinction of endemic Hawaiian avifauna. *Condor* 70:101–120.

Wilson, E. O. 1975. *Sociobiology: The new synthesis.* Cambridge, MA: Harvard University Press. See Chapter 17, Social Symbiosis, for a detailed discussion of social parasitism.

Woodard, T. N., R. J. Gutierrez, and W. H. Rutherford. 1974. Bighorn lamb production, survival and mortality in south-central Colorado. *Journal of Wildlife Management* 28:381–391.

Zuk, M. 1991. Parasites and bright birds: New data and new predictions. Pages 317–327 in J. E. Loge and M. Zuk, eds. *Bird-parasite interactions.* Oxford: Oxford University Press.

CHAPTER 17 Processes Shaping Communities

Austin, M. P., and T. M. Smith. 1989. A new model of the continuum concept. *Vegetatio* 83:35–47.

Bazzaz, F. A. 1979. The physiological ecology of plant succession. *Annual Review of Ecology and Systematics* 10:351–371.

Berger, J., P. B. Stacey, L. Bellis, and M. P. Johnson. A mammalian predator-prey imbalance: Grizzly bear and wolf extinction affect avian neotropical migrants. *Ecological Applications* 11:947–960.

Billings, W. D. 1938. The structure and development of oldfield shortleaf pine stands and certain associated physical properties of the soil. *Ecological Monographs* 8:437–499.

Briand, F. 1983. Environmental control of food web structure. *Ecology* 64:253–263.

Clements, F. E. 1916. *Plant succession: An analysis of the development of vegetation.* Carnegie Inst. Wash. Publ. 242.

———. 1936. Nature and structure of the climax. *Journal of Ecology* 24:252–284.

Cohen, J. E. 1989. Food webs and community structure. Pages 181–202 in J. Roughgarden, R. May, and S. Levin eds. *Perspectives in ecological theory.* Princeton, NJ: Princeton University Press.

Connell, J. H. 1978. Diversity in tropical rain forests and coral reefs. *Science* 199:1302–1310.

Connell, J. H., and R. O. Slatyer. 1977. Mechanisms of succession in natural communities and their role in community stability and organization. *American Naturalist* 111:1119–1144.

Cowles, H. C. 1899. The ecological relations of the vegetation on the sand dunes of Lake Michigan. *Botanical Gazette* 27:95–117, 167–202, 281–308, 361–391.

Crocker, R. L., and J. Major. 1955. Soil development in relation to vegetation and surface age at Glacier Bay, Alaska. *Journal of Ecology* 43:427–488.

Dodson, S. I. 1970. Complementary feeding niches sustained by size-selective predation. *Limnology and Oceanography* 15:131–137.

Egler, F. 1954. Vegetation science concepts. I. Initial floristic composition, a factor in old field vegetation development. *Vegetatio* 14:412–417.

Ehrlich, P. R., and A. Ehrlich. 1981. *Extinction.* New York: Random House.

Elton, C. 1927. *Animal ecology.* London: Sidgwick & Jackson. A classic, with original descriptions of food chains and ecological pyramids.

Emery, N. C., P. J. Ewanchuk, and M. D. Bertness. 2001. Competition and salt marsh plant zonation: Stress tolerators may be dominant competitors. *Ecology* 82:2471–2485.

Fowler, N. L. 1981. Competition and coexistence in a North Carolina grassland II: The effects of the experimental removal of species. *Journal of Ecology* 69:843–854.

Groves, R. H., and J. D. Williams. 1975. Growth of skeleton weed (*Chondrilla juncea* L.) as affected by growth of subterranean clover (*Trifolium subterraneum* L.) and infection by *Puccinia chondrilla* Bubak and Syd. *Australian Journal of Agricultural Research* 26:975–983.

Hairston, N. G., F. E. Smith, and L. B. Slobodkin. 1960. Community structure, population control, and competition. *American Naturalist* 94:421–425.

Huston, M. 1979. A general hypothesis of species diversity. *American Naturalist* 113:81–101.

———. 1994. *Biological diversity: The coexistence of species on changing landscapes.* New York: Cambridge University Press.

Huston, M., and T. M. Smith. 1987. Plant succession: Life history and competition. *American Naturalist* 130:168–198.

Krebs, C. 2001. *Ecology,* 5th ed. San Francisco: Benjamin Cummings.

Krebs, C. J., S. Boutin, R. Boonstra, A. R. E. Sinclair, J. N. M. Smith, M. R. T. Dale, K. Martin, and R. Turkington. 1995. Impact of food and predation on the snowshoe hare cycle. *Science* 269:1112–1115.

Krebs, C. J., S. Boutin, and R. Boonstra (eds.). 2001. *Vertebrate community dynamics in the Kluane Boreal Forest.* Oxford: Oxford University Press.

MacArthur, R. H., and J. W. MacArthur. 1961. On bird species diversity. *Ecology* 42:594–598.

Oosting, H. J. 1942. An ecological analysis of the plant communities of Piedmont, North Carolina. *American Midland Naturalist* 28:1–26.

Paine, R. T. 1966. Food web complexity and species diversity. *American Naturalist* 100:65–75.

———. 1969. The *Pisaster-Tegula* interaction: Prey patches, predator food preference and intertidal community structure. *Ecology* 50:950–961.

Pastor, J., R. J. Naiman, B. Dewey, and P. McInnes. 1988. Moose, microbes, and boreal forest. *Bioscience* 38:770–777.

Pimm, S. L. 1982. *Food webs.* London: Chapman and Hall. A detailed discussion of aspects of food web theory.

———. 1991. *The balance of nature?* Chicago: University of Chicago Press.

Power, M. E., M. J. Matthews, and A. J. Stewart. 1985. Grazing minnows, piscivorous bass, and stream algae: Dynamics of a strong interaction. *Ecology* 66:1448–1456.

Smith, T. M., and M. Huston. 1987. A theory of spatial and emporal dynamics of plant communities. *Vegetatio* 3:49–69.

Smith, T. M., F. I. Woodward, and H. H. Shugart, (eds.). 1996. *Plant functional types: Their relevance to ecosystem properties and global change.* Cambridge, England: Cambridge University Press.

Walker, B. H. 1992. Biodiversity and ecological redundancy. *Conservation Biology* 6:18–23.

Warming, J. E. B. 1909. *Oecology of plants.* Oxford: Oxford University Press.

Whittaker, R. H. 1960. Vegetation of the Siskiyou Mountains, Oregon and California. *Ecological Monographs* 23:41–78.

———. 1975. *Communities and ecosystems.* New York: Macmillan.

Whittaker, R. H., and G. M. Woodwell. 1968. Dimension and production relations of trees and shrubs in the Brookhaven forest, New York. *Ecology* 56:1–25.

———. 1969. Structure and function, production, and diversity of the oak-pine forest at Brookhaven, New York. *Journal of Ecology* 57:155–174.

Yodzis, P. 1989. *Introduction to theoretical ecology.* New York: Harper & Row.

CHAPTER 18 Human Interactions with Communities

Baden, J. A., and D. Leal. 1990. *The Yellowstone primer.* San Francisco: Pacific Research Institute for Public Policy. An excellent review of the problems of Yellowstone Park, a microcosm of problems facing all parks and reserves.

Bricklemeyer, E. C., Jr., S. Ludicello, and H. J. Hartmann. 1989. Discarded catch in U. S. commercial marine fisheries. Pages 259–295 in *Audubon wildlife report 1989/1990*. San Diego: Academic Press. Excellent review of current fishery regulations and the discard problem.

Carson, R. 1962. *Silent spring*. Boston: Houghton Mifflin.

Caughley, G. 1976. Wildlife management and the dynamics of ungulate populations. *Applied Biology* 1:183–246. Good introduction to sustained yield in wildlife management.

Cottam, G. 1990. Community dynamics on an artificial prairie. Pages 257–270 in W. R. Jordan III, M. E. Gilpin, and J. D. Aber, eds. *Restoration ecology: A synthetic approach to ecological research*. Cambridge: Cambridge University Press.

Debach, P. 1974. *Biological control by natural enemies*. London: Cambridge University Press. Somewhat dated, but an excellent introduction. Chapters on classic biological control successes.

DiSilvestro, R. L. 1989. *The endangered kingdom: The struggle to save America's wildlife*. New York: Wiley. Case histories of restoration and endangerment of selected wildlife species, as well as an assessment of wildlife management.

Dover, M. J., and B. A. Croft. 1986. Pesticide resistance and public policy. *Bioscience* 36:78–85. Discusses pesticide resistance and effective pest control.

Ellis, R. 1991. *Men and whales*. New York: Alfred Knopf. Excellent, well-illustrated history of whaling and the human-whale relationship.

Ellstrand, N. C., and C. A. Hoffman. 1990. Hybridization as an avenue of escape for engineered genes. *Bioscience* 40:438–442.

Fitzgerald, S. 1989. *International wildlife trade: Whose business is it?* Washington, DC: World Wildlife Fund. Why endangered species need protection; a review of illegal trade in wildlife species.

Gambell, R. 1976. Population biology and the management of whales. *Applied Biology* 1:237–343. A short history of whaling, population dynamics of species, and the failure of attempts to apply sustained yield management.

Harris, L. D. 1984. *The fragmented forest*. Chicago: University of Chicago Press. Integrates island biogeography and forest management in the preservation of biotic diversity.

Hoffman, C. A. 1990. Ecological risks of genetic engineering of crop plants. *Bioscience* 40:434–436.

Horn, D. J. 1988. *Ecological approach to pest management*. New York: Guilford. Good review; many examples.

Huffaker, C. B., and R. L. Rabb (eds.). 1984. *Ecological entomology*. New York: Wiley. Excellent chapters on evolutionary process in insects, natural control of insects, and application of ecology to insect population management.

Jordan, W. R., III, M. E. Gilpin, and J. D. Aber (eds.). 1990. *Restoration ecology: A synthetic approach to ecological research*. Cambridge: Cambridge University Press.

Kleiman, D. G. 1989. Reintroduction of captive mammals for conservation. *Bioscience* 39:152–161. Authoritative discussion of guidelines for introducing endangered species into the wild.

Kot, M. 2001. *Elements of mathematical ecology*. Cambridge: Cambridge University Press.

Meffe, G. K., and C. R. Carroll. 1997. *Principles of conservation biology*, 2d ed. Sunderland, MA: Sinauer Associates.

Pimentel, D., H. Acquay, M. Biltonen, et al. 1992. Environmental and economic costs of pesticide use. *Bioscience* 42:750–760.

Prescott-Allen, C., and R. Prescott-Allen. 1986. *The first resource: Wild species in North American economy*. New Haven: Yale University Press.

Regier, H. A., and G. L. Baskerville. 1986. Sustainable development of regional ecosystems degraded by exploitive development. Pages 74–103 in W. E. Clark and R. E. Munn, eds. *Sustainable development of the biosphere*. Cambridge, England: Cambridge University Press. Focuses on recovery of degraded forestry and fishery resource systems.

Regier, H. A., and W. L. Hartman. 1973. Lake Erie's fish community: 150 years of cultural stress. *Science* 180:1248–1255. History of Lake Erie's fisheries.

Russell, E. 2001. *War and nature*. Cambridge: Cambridge University Press.

Small, G. 1976. *The blue whale*. New York: Columbia University Press. The authoritative and depressing history of the blue whale.

Soule, M. E. (ed.). 1986. *Conservation biology: The science of scarcity and diversity*. Sunderland, MA: Sinauer Associates. Considers various aspects of conserving endangered populations.

Soule, M. E., and B. A. Wilcox (eds.). 1980. *Conservation biology: An evolutionary ecological perspective*. Sunderland, MA: Sinauer Associates. Excellent overview of the problems of conserving small populations, including captive propagation.

Thatcher, R. C., J. L. Searcy, J. E. Coster, and G. D. Hertel (eds.). 1986. *The southern pine beetle*. USDA Forest Service Science and Education Tech. Bull. 1631. Washington, DC: U.S. Department of Agriculture. A colorful, accessible explanation of all aspects of IPM in action against the southern pine beetle.

Trefethen, J. B. 1975. *An American crusade for wildlife*. New York: Winchester Press. An excellent history of restoration of wildlife and the politics involved.

Walters, C. 1986. *Adaptive management of renewable resources*. New York: Macmillan. Advanced but challenging approaches to managing renewable resources.

Western, D., and M. C. Pearl. 1988. *Conservation for the twenty-first century*. New York: Oxford University Press. Focuses on problems in and approaches to wildlife conservation. Excellent reference.

Whitcomb, R. F., J. F. Lynch, P. A. Opler, and C. S. Robbins. 1976. Island biogeography and conservation: Strategy and limitations. *Science* 193:1030–1032.

Zaret, T. M., and R. T. Paine. 1973. Species introduction in a tropical lake. *Science* 182:449–455.

CHAPTER 19 Landscape Ecology

Aber, J. D., and J. M. Melillo. 1991. *Terrestrial ecosystems*. Philadelphia: Saunders College Publishing.

Bormann, F. H., and G. E. Likens. 1979. *Pattern and process in a forested ecosystem*. New York: Springer-Verlag. A good discussion of nutrient cycling in the forest.

Brown, J. H., and A. Kodrich-Brown. 1977. Turnover rates in insular biogeography: Effect of immigration on extinction. *Ecology* 58:445–449.

Chepko-Sade, B. D., and Z. T. Halpin (eds.). 1987. *Mammalian dispersal patterns: The effect of social structure on population genetics*. Chicago: University of Chicago Press.

Christensen, N. L. 1977. Fire and soil-plant nutrient relations in a pine-wiregrass savanna on the coastal plain of North Carolina. *Oecologia* 31:27–44.

Curtis, J. T. 1959. *The vegetation of Wisconsin*. Madison, WI: University of Wisconsin.

Delcourt, H. R. 1987. The impact of prehistoric agriculture and land occupation on natural vegetation. *Trends in Ecology and Evolution* 2:39–44.

Fahrig, L., and G. Merriam. 1994. Conservation of fragmented populations. *Conservation Biology* 8:50–59.

Forman, R. T. T. 1995. *Land mosaics: The ecology of landscapes and regions*. New York: Cambridge University Press. Ecology of heterogenous land areas; interaction between humans and natural processes on the land.

Forman, R. T. T., and M. Gordon. 1981. Patches and structural components for a landscape ecology. *Bioscience* 31:733–740.

Frankel, O. H., and M. E. Soule. 1981. *Conservation and evolution*. Cambridge, England: Cambridge University Press.

Freemark, K. E., and H. G. Merriam. 1986. Importance of area and habitat heterogenity to bird assemblages in temperate forest fragments. *Biological Conservation* 36:115–141.

Galli, A. E., C. F. Leck, and R. T. T. Forman. 1976. Avian distribution patterns in forest islands of different sizes in central New Jersey. *Auk* 93:356–364.

Gray, J. T., and W. H. Schlesinger. 1981. Nutrient cycling in mediterranean-type ecosystems. In P. C. Miller, ed. *Resource use by chaparral and matorral*. New York: Springer Verlag.

Guthery, F. S., and R. L. Bingham. 1992. On Leopold's principle of edge. *Wildlife Society Bulletin* 20:340–344.

Hanski, I. A., and M. E. Gilpin (eds.). 1997. *Metapopulation biology: Ecology, genetics, and evolution.* San Diego: Academic Press. Major reference on theory and application.

Harris, L. D. 1984. *The fragmented forest.* Chicago: University of Chicago Press. An important reference.

———. 1988. Edge effects and the conservation of biological diversity. *Conservation Biology* 2:212–215.

Harrison, S., D. D. Murphy, and P. R. Ehrlich. 1988. Distribution of the Bay checkerspot butterfly, *Euphydryas editha bayensis:* Evidence for a metapopulation model. *American Naturalist* 132:360–382.

Heinselman, M. L. 1973. Fire in the virgin forest of the Boundary Waters Canoe Area, Minnesota. *Quarternary Research* 3:329–382.

Heinselman, M. L., and T. M. Casey. 1981a. Fire and succession in the conifer forests of northern North America. Pages 374–405 in D. E. West, H. H. Shugart, and D. B. Botkin, eds. *Forest succession: Concepts and applications.* New York: Springer-Verlag.

Johnson, H. B. 1976. *Order upon the land.* Oxford: Oxford University Press.

Kerbes, R. H., P. M. Kotanen, and R. L. Jeffries. 1990. Destruction of wetland habitats by lesser snow geese: A keystone species on the west coast of Hudson Bay. *Journal of Applied Ecology* 27: 242–258.

Leopold, A. 1933. *Game management.* New York: Scribner.

Levenson, J. B. 1981. Woodlots as biogeographic islands in southeastern Wisconsin. Pages 13–39 in R. L. Burgess and D. M. Sharpe, eds. *Forest island dynamics in man-dominated landscapes.* New York: Springer-Verlag.

MacArthur, R. H., and E. O. Wilson. 1967. *The theory of island biography.* Princeton: Princeton University Press.

Marquis, D. A. 1974. The impact of deer browsing on Allegheny hardwood regeneration. USDA Forest Service Research Paper NE-308.

———. 1981. Effect of deer browsing on timber production in Allegheny hardwood forests of northwestern Pennsylvania. USDA Forest Service Research Paper NE-475.

McCollough, D. R. (ed.). 1996. *Metapopulations and wildlife conservation.* Washington, DC: Island Press. Application of metapopulation theory to wildlife conservation.

Mooney, H. A., et al. (eds.). 1981. *Proceedings: Symposia on fire regimes and ecosystem properties.* General Technical Report WO-26. Washington, DC: USDA Forest Service.

Naiman, R. J., J. M. Melillo, and J. E. Hobbie. 1986. Ecosystem alteration of boreal forest streams by beaver *(Castor canadensis). Ecology* 67:1254–1269.

Ranney, J. W. 1977. Forest island edges—Their structure, development, and importance to regional forest ecosysem dynamics. EDFB/IBP Cont. No. 77/1, Oak Ridge National Laboratory, Oak Ridge, TN.

Ranney, J. W., M. C. Brunner, and J. B. Levenson. 1981. The importance of edge in the structure and dynamics of forest lands. Pages 67–91 in R. L. Burgess and D. M. Sharpe, eds. *Forest island dynamics in man-dominated landscapes* (Ecological Studies No. 41). New York: Springer-Verlag.

Robbins, C. S., D. K. Dawson, and B. A. Dowell. 1989. Habitat area requirements of breeding forest birds of the Middle Atlantic States. *Wildlife Monographs* 103.

Romme, W. H., and D. H. Knight. 1982. Landscape diversity: The concept applied to Yellowstone Park. *Bioscience* 32:664–670.

Rosenberg, D. K., B. R. Noon, and E. C. Meslow. 1997. Biological corridors: Forms, function and efficiency. *Bioscience* 47:677–687.

Runkle, J. R. 1985. Disturbance regimes in temperate forests. Pages 17–34 in S. T. A. Pickett and P. S. White, eds. *Ecology of natural disturbance and patch dynamics.* Orlando: Academic Press.

Schonewald-Cox, C. M., S. M. Chambers, B. MacBryde, and W. L. Thomas (eds.). 1983. *Genetics and conservation.* Menlo Park, CA: Benjamin Cummings. A reference manual for managing wild animal and plant populations.

Schwartz, M. W. 1997. *Conservation in highly fragmented landscapes.* New York: Chapman & Hall.

Simberloff, D. S. 1974. Equilibrium theory of island biogeography and ecology. *Annual Review of Ecology and Systematics* 5:161–182.

Simberloff, D. S., and E. O Wilson. 1969. Experimental zoogeography of islands. The colonization of empty islands. *Ecology* 50:278–296.

Soule, M. (ed.). 1986a. *Conservation biology: The science of scarcity and diversity.* Sunderland, MA: Sinauer Associates. A sourcebook on the ecology of small populations and extinction.

———. 1986b. *Viable populations for conservation.* Cambridge, England: Cambridge University Press. Discussion of minimum viable population concepts and management of small populations.

Temple, S. A. 1986. Predicting impacts of habitat fragmentation on forest birds: A comparison of two models. Pages 301–304 in J. Verner, M. L. Morrison, and C. T. Ralph, eds. *Wildlife 2000: Modeling habitat relations of terrestrial vertebrates.* Madison, WI: University of Wisconsin Press.

Thomas, J. W., R. G. Anderson, C. Master, and E. L. Bull. 1979. Snags. Pages 60–77 in J. W. Thomas, ed. *Wildlife habitats in managed forests* (The Blue Mountains of Oregon and Washington), USDA Forest Service Agricultural Handbook No. 553.

Thornhill, N. W. (ed.). 1993. *The natural history of inbreeding and outbreeding.* Chicago: University of Chicago Press. A collection of highly interesting papers covering the topic as well as some aspects of small populations and speciation.

Turner, M. 1989. Landscape ecology: the effects of pattern and process. *Annual Review of Ecology and Systematics* 20:171–197.

———. 1998. Landscape ecology. In S. I. Dodson, T. F. H. Allen, S. R. Carpenter, A. R. Ives, R. L. Jeanne, J. F. Kitchell, N. E. Langston, and M. G. Turner. *Ecology.* Oxford: Oxford University Press.

Urban, D., R. V. O'Neill, and H. H. Shugart. 1987. Landscape ecology. *Bioscience* 37:119–127.

Walker, B. 1989. Diversity and stability in ecosystem conservation. Pages 125–130 in D. Western and M. C. Pearl, eds. *Conservation for the twenty-first century.* New York: Oxford.

Watt, A. S. 1947. Pattern and process in the plant community. *Journal of Ecology* 35:1–22.

Western, D., and M. C. Pearl (eds.). 1989. *Conservation for the twenty-first century.* New York: Oxford University Press. Excellent overview of problems and possible solutions to the conservation of species and their habitat. A major reference.

Whitcomb, R. F., C. S. Robbins, J. F. Lynch, B. L. Whitcomb, M. K. Klimkiewicz, and D. Bystrak. 1981. Effects of forest fragmentation on avifauna of the eastern deciduous forest. Pages 3–13 in R. L. Burgess and D. M. Sharpe, eds. *Forest island dynamics in man-dominated landscapes.* Orlando, FL: Academic Press.

Wiens, J. A. 1973. Pattern and process in grassland bird communities. *Ecological Monographs* 43:237–270.

Wiens, J. A., and J. T. Rotenberry. 1981. Habitat associations and community structure of birds in shrub-steppe environments. *Ecological Monographs* 51:21–41.

Wilcove, D. S. 1986. Nest predation in forest tracts and the decline of migratory songbirds. *Ecology* 66:1211–1214.

Williamson, M. 1981. *Island populations.* Oxford, England: Oxford University Press.

Wing, L. D., and I. D. Buss. 1970. Elephants and forests. *Wildlife Monographs* 19.

Zedler, P. H., C. R. Gautier, and G. S. McMaster. 1983.Vegetation change in response to extreme events: The effect of a short interval between fires in California chaparrel and coastal shrub. *Ecology* 64:809–818.

ECOLOGICAL APPLICATIONS: Asteroids, Bulldozers and Biodiversity

Culotta, E. 1995. Many suspects to blame in Madagascar extinctions. *Science* 268:156–159.

Green, G. N., and R. W. Sussman. 1990. Deforestation history of the eastern rain forests of Madagascar from satellite images. *Science* 248:212–215.

Myers, N. 1991. Tropical deforestation: The latest situation. *Bioscience* 41:282.

———. 1986. Tropical deforestation and a mega-extinction spasm. Pages 349–409 in M. E. Soule, ed. *Conservation biology: The science of scarcity and diversity.* Sunderland, MA: Sinauer Associates.

Tattersal, I. 1993. Madagascar's lemurs. *Scientific American* 268:110–117.

Wilson, E. O. 1992. *The diversity of life.* Cambridge, MA: The Belknap Press of Harvard University Press.

Wilson, E. O., and F. M. Peter (eds.). 1988. *Biodiversity.* Washington DC: National Academy Press.

CHAPTER 20 Ecosystem Productivity

Aber, J. D., and J. M. Melillo. 1991. *Terrestrial ecosystems.* Philadelphia: Saunders College Publishing.

Anderson, J. M., and A. MacFadyen (eds.). 1976. *The role of terrestrial and aquatic organisms in the decomposition process.* Seventeenth Symposium, British Ecological Society. Oxford, England: Blackwell Scientific Publications.

Andrews, R. D., D. C. Coleman, J. E. Ellis, and J. S. Singh. 1975. Energy flow relationships in a shortgrass prairie ecosystem. Pages 22–28 in *Proceedings 1st International Congress of Ecology.* The Hague: W. Junk Publishers.

Begon, M., J. L. Harper, and C. R. Townsend. 1996. *Ecology: Individuals, populations, and communities.* New York: Blackwell Scientific Publishers.

Brylinsky, M. 1980. Estimating the productivity of lakes and reservoirs. Pages 411–418 in E. D. Le Cren and R. H. Lowe-McConnell, eds. *The functioning of freshwater ecosystems.* International Biological Programme No. 22. Cambridge, England: Cambridge University Press.

Coe, M. J., D. H. Cummings, and J. Phillipson. 1976. Biomass and production of large African herbivores in relation to rainfall and primary production. *Oecologica* 22:341–354.

Cooper, J. P. 1975. *Photosynthesis and productivity in different environments.* New York: Cambridge University Press.

Cowan, R. L. 1962. Physiology of nutrition of deer. Pages 1–8 in *Proceedings 1st National White-tailed Deer Disease Symposium.*

DeAngelis, D. L., R. H. Gardner, and H. H. Shugart. 1980. Productivity of forest ecosystems studies during the IBP: The woodlands data set. In D. E. Reichle, ed. *Dynamic properties of forest ecosystems.* International Biological Programme 23. Cambridge, England: Cambridge University Press.

Elton, C. 1927. *Animal ecology.* London: Sidgwick & Jackson.

Gates, D. M. 1985. *Energy and ecology.* Sunderland, MA: Sinauer Associates.

Golley, F. B., and H. Leith. 1972. Basis of organic production in the tropics. Pages 1–26 in P. M. Golley and F. H. Golley, eds. *Tropical ecology with an emphasis on organic production.* Athens: University of Georgia Press.

Leith, H. 1973. Primary production: Terrestrial ecosystems. *Human Ecology* 1:303–332.

———. 1975. Primary productivity in ecosystems: Comparative analysis of global patterns. Pages 67–88 in W. H. van Dobben and R. H. Lowe-McConnell, eds. *Unifying concepts in ecology.* The Hague: Junk.

Leith, H., and R. H. Whittaker (eds.). 1975. *Primary productivity in the biosphere.* Ecological Studies Vol. 14. New York: Springer-Verlag.

MacArthur, R. H., and J. H. Connell. 1966. *The biology of populations.* New York: Wiley.

National Academy of Science. 1975. *Productivity of world ecosystems.* Washington, DC: National Academy of Science. Good summary of world primary production.

Odum, E. P. 1983. *Basic Ecology.* Philadelphia: Saunders.

Pastor, J., J. D. Aber, C. A. McClaugherty, and J. M. Melillo. 1984. Aboveground production and N and P cycling along a nitrogen mineralization gradient on Blackhawk Island, Wisconsin. *Ecology* 65:256–268.

Petrusewicz, K. (ed.). 1967. *Secondary productivity of terrestrial ecosystems.* Warsaw, Poland: Pantsworve Wydawnictwo Naukowe.

Phillipson, J. J. 1966. *Ecological energetics.* New York: St. Martin's Press. Although dated, this is still an excellent introduction to energy in ecosystems.

Reichle, D. E. *Analysis of temperate forest ecosystems.* New York: Springer Verlag.

———. (ed.). 1981. *Dynamic properties of forest ecosystems.* Cambridge, England: Cambridge University Press.

Ryan, M. G., D. Binkley, and J. H. Fownes. 1997. Age-related decline in forest productivity: pattern and process. *Advances in Ecological Research* 27:213–262.

Schindler, D. W. 1977. Evolution of phosphorus limitation in lakes. *Science* 195:260–262.

———. 1978. Factors regulating phytoplankton production and standing crop in the world's freshwaters. *Limnology and Oceanography* 23:478–486.

Transeau, E. N. 1926. The accumulation of energy by plants. *Ohio Journal of Science* 26:1–10.

Whittaker, R. H., and G. E. Likens. 1973. Carbon in the biota. In G. M. Woodwell and E. V. Pecan, eds. *Carbon and the biosphere conference 72501.* Springfield, VA: National Technical Information Service.

Wiegert, R. C. (ed.). 1976. *Ecological energetics.* Benchmark Papers. Stroudsburg, PA: Dowden, Hutchinson & Ross. Collection of papers surveying the development of the concept.

CHAPTER 21 Nutrient Cycling

Aber, J. D., and J. M. Melillo. 1991. *Terrestrial ecosystems.* Philadelphia: Saunders College Publishing.

Chapin, F. S. 1990. The mineral nutrition of wild plants. *Annual Review of Ecology and Systematics* 11:233–260.

Correll, D. L. 1978. Estuarine productivity. *Bioscience* 28:646–650.

Fenchel, T. 1988. Marine plankton food chains. *Annual Review of Ecology and Systematics* 19:19–38.

Fenchel, T., and T. H. Blackburn. 1979. *Bacteria and mineral cycling.* London: Academic Press.

Fenchel, T., and P. Harrison. 1976. The significance of bacterial grazing and mineral cycling for the decomposition of particulate detritus. Pages 285–321 in J. M. Anderson and A. MacFadyen, eds. *The role of terrestrial and aquatic organisms in the decomposition process.* New York: Blackwell.

Fletcher, M., G. R. Gray, and J. G. Jones. 1987. *Ecology of microbial communities.* New York: Cambridge University Press.

Likens, G. E., and F. H. Bormann. 1974. Linkages between terrestrial and aquatic ecosystems. *Bioscience* 24(8):447–456.

Lovelock, J. 1979. *GAIA: A new look at life on Earth.* Oxford: Oxford University Press.

Newbold, J. D., R. V. O'Neill, J. W. Elwood, and W. Van Winkle. 1982. Nutrient spiraling in streams: Implications for nutrient and invertebrate activity. *American Naturalist* 20:628–652.

Nybakken, J. W. 1988. *Marine biology: An ecological approach,* 2d ed. New York: Harper and Row.

Pastor, J., J. D. Aber, C. A. McClaugherty, and J. M. Melillo. 1984. Aboveground production and N and P cycling along a nitrogen mineralization gradient on Blackhawk Island, Wisconsin. *Ecology* 65:256–268.

Swift, M. J., O. W. Heal, and J. M. Anderson. 1978. *Decomposition in terrestrial ecosystems.* Oxford: Blackwell Scientific Publications.

Vitousek, P. M., S. W. Andariese, P. A. Matson, L. Morris, and R. L. Sanford. 1992. Effects of harvest intensity, site preparation, and herbicide use on soil nitrogen transformations in a young loblolly pine plantation. *Forest Ecology and Management* 49:277–292.

Webster, J. R. 1975. Analysis of potassium and calcium dynamics in stream ecosystems on three Appalachian watersheds of contrasting vegetation. Ph.D. thesis, University of Georgia, Athens, GA.

Webster, J. R., D. J. D'angelo, and G. T. Peters. 1991. Nitrate and phosphate uptake in streams at Coweeta Hydrological Laboratory. *Vehn. Internationale Verein Limnologie* 24:1681–1686.

Witherspoon, J. P. 1964. Cycling of cesium-134 in white oak trees. *Ecological Monographs* 34:403–420.

CHAPTER 22 Biogeochemical Cycles

Alexander, M. (ed.). 1980. *Biological nitrogen fixation*. New York: Plenum. Ecology and physiology of nitrogen-fixing organisms.

Bolin, B., E. T. Degens, S. Kempe, and P. Ketner (eds.). 1979. *The global carbon cycle*. SCOPE 13. New York: John Wiley. A comprehensive study of the carbon cycle.

Bormann, F. H., and G. E. Likens. 1979. *Pattern and process in a forested ecosystem*. New York: Springer-Verlag. A good discussion of nutrient cycling in the forest.

Pomeroy, L. R. 1974. *Cycles of essential elements*. Benchmark Papers in Ecology. Stroudsburg: PA: Dowden, Hutchinson, and Ross. Collection of important papers on mineral cycling.

Post, W. M., T. H. Peng, W. R. Emanuel, A. W. King, V. H. Dale, and D. L. DeAngelis. 1990. The global carbon cycle. *American Scientist* 78:310–326. An excellent updated review of the global carbon cycle—what we know and do not know about it.

Schlesinger, W. H. 1997. *Biogeochemistry: An analysis of global change*, 2d ed. London: Academic Press.

Sprent, J. I. 1988. *The ecology of the nitrogen cycle*. New York: Cambridge University Press. Various processes and magnitudes of the nitrogen cycle.

CHAPTER 23 Human Intrusions into Biogeochemical Cycles

Aber, J. D., K. N. Nadelhoffer, P. Steudler, and J. M. Melillo. 1989. Nitrogen saturation in northern forest ecosystems. *Bioscience* 39:378–386.

Binkley, D., C. T. Driscoll, H. L. Allen, P. Schoeneberger, and D. McAvoy. 1989. *Acidic deposition and forest soils: Context and case studies in southeastern United States*. New York: Springer-Verlag.

Dover, M. J., and B. A. Croft. 1986. Pesticide resistance and public policy. *Bioscience* 36:78–91.

MacKenzie, J. J., and M. T. El-Ashry. 1989. *Air pollution's toll on forests and crops*. New Haven: Yale University Press.

Pimentel, D., and C. A. Edwards. 1982. Pesticides and ecosystems. *Bioscience* 32:595–600. A summary.

Pimentel, D., and L. Levitan. 1986. Pesticides: Amounts applied and amounts reaching pests. *Bioscience* 36:86–91.

Schulze, E. D., O. L. Lange, and O. Oren. 1989. *Forest decline and air pollution: A study of spruce* (Picea abies) *on acid soils*. New York: Springer-Verlag.

Smith, W. H. 1990. *Air pollution and forests: Interaction between air contaminants and forest ecosystems*, 2d ed. New York: Springer-Verlag. A thorough and indispensable reference source.

Wellburn, A. 1988. *Air pollution and acid rain: The biological impact*. Essex, England: Longman Scientific and Technical and New York: Wiley. A basic introduction to the topic.

ECOLOGICAL APPLICATIONS: Time to Rethink the Lawn

Bormann, F. H., D. Balmori, and G. T. Geballe. 1993. *Redesigning the American lawn*. New Haven, CT: Yale University Press.

CHAPTER 24 Biogeography and Biodiversity

Angel, M. V. 1991. Variations in time and space: Is biogeography relevant to studies of long-time scale change? *Journal of the Marine Biological Association of the United Kingdom* 71:191–206.

Bailey, R. G. 1996. *Ecosystem geography*. New York: Springer.

Bailey, R. W. 1978. *Description of the ecoregions of the United States*. Ogden, UT: USDA Forest Service Intermountain Region.

Bartlein, P. J., T. Webb, and E. C. Fleri. 1984. Climate response surfaces from pollen data for some eastern North American taxa. *Journal of Biogeography* 13:35–57.

Clements, F. E., and V. E. Shelford. 1939. *Bio-ecology*. New York: McGraw-Hill.

Currie, D. J. 1991. Energy and large-scale biogeographical patterns of animal and plant species richness. *American Naturalist* 137:27–49.

Currie, D. J., and V. Paquin. 1987. Large-scale biogeographical patterns of species richness in trees. *Nature* (London) 329:326–327.

DeCandolle, A. P. A. 1874. Constitution dans le Regne Vegetal de Groupes Physiologiques Applicables a l Geographie Ancienne et Moderne. *Archives des Sciences Physiques et Naturelles.* May, Geneva.

Holdridge, L. R. 1947. Determination of wild plant formations from simple climatic data. *Science* 105:367–368.

———. 1967. Determination of world plant formation from simple climatic data. *Science* 130:572.

Hunter, M. L., and P. Yonzon. 1992. Altitudinal distributions of birds, mammals, people, forests and parks in Nepal. *Conservation Biology* 7:420–423.

Huston, M. 1994. *Biological diversity: The coexistence of species on changing landscapes*. New York: Cambridge University Press.

Kikkawa, J., and W. T. Williams. 1971. Altitudinal distribution of land birds in New Guinea. *Search* 2:64–69.

Koppen, W. 1918. Klassifikation der Klimate nach Temperatur, Niederschlag, und Jahres lauf. *Petermann's Mitteilungen* 64:193–203, 243–248.

Merriam, C. H. 1890. Results of a biological survey of the San Francisco mountain region and the desert of the Little Colorado, Arizona. *North American Fauna* 1–136.

———. 1899. Zone temperatures. *Science* 9:116.

Prentice, I. C., P. J. Bartlein, and T. Webb. 1991. Vegetation and climate change in eastern North America since the last glacial maximum: A response to continuous climate forcing. *Ecology* 72:2038–2056.

Rex, M. A., C. T. Stuart, R. R. Hessler, J. A. Allen, H. A. Sanders, and G. D. F. Wilson. 1993. Global-scale latitudinal patterns of species diversity in the deep-sea benthos. *Nature* 365:636–639.

Schimper, A. F. W. 1903. *Plant-geography upon a physiological basis*. Oxford: Clarendon.

Whittaker, R. H. 1975. *Communities and ecosystems*. London: Collier-Macmillan.

———. 1977. Evolution of species diversity in land communities. *Evolutionary Biology* 10:1–67.

Wilson, E. O. 1999. *The diversity of life*. New York: Norton.

CHAPTER 25 Terrestrial Ecosystems 1: Forests, Woodlands, and Savannas

Tropical Forests

Ashton, P. S. 1988. Dipterocarp biology as a window to understanding of tropical forest structure. *Annual Review of Ecology and Systematics* 19: 347–370.

Bawa, K. S. 1990. Plant-pollinator interactions in tropical rain forests. *Annual Review of Ecology and Systematics* 21:399–422.

Buschbacker, R. J. 1986. Tropical deforestation and pasture development. *Bioscience* 36:22–28. Good review of the various aspects of the problem.

Colchester, M. 1989. *Pirates, squatters and poachers: The political ecology of the dispossession of the native peoples of Sarawak*. London: Survival International. Excellent review of the complex problems surrounding forestry, economic development, and welfare and rights of native peoples.

Collins, M. (ed.). 1990. *The last rain forests: A world conservation atlas*. New York: Oxford University Press. Authoritative, comprehensive world guide to all aspects of the tropical rain forest.

Denslow, J. S. 1987. Tropical rainforest gaps and tree species diversity. *Annual Review of Ecology and Systematics* 18:431–451.

Erwin, T. L. 1988. The tropical forest canopy: The heart of biotic diversity. In E. O. Wilson and F. M. Peters, eds. *Biodiversity*. Washington, DC: National Academy Press.

Fleming, T. H., R. Breitwisch, and G. H. Whitesides. 1987. Patterns of tropical vertebrate frugivore diversity. *Annual Review of Ecology and Systematics* 18:71–90. Concise review of tropical frugivory.

Furtado, J. I., and K. Ruddle. 1986. The future of tropical forests. Pages 145–171 in N. Polunin, ed. *Ecosystem theory and application.* New York: Wiley.

Golley, F. B. (ed.). 1983. *Tropical forest ecosystems: Structure and function.* Ecosystems of the World No. 14A. Amsterdam: Elsevier Scientific Publishing Company. An important reference.

Jansen, D. J. 1986. The future of tropical ecology. *Annual Review of Ecology and Systematics* 17:305–324. Discusses approaches needed to maintain the integrity of tropical ecosystems.

Jordan, C. F., and J. R. Kline. 1972. Mineral cycling: Some basic concepts and their application in a tropical rain forest. *Annual Review of Ecology and Systematics* 3:33–50.

Lathwell, D. J., and T. L. Grove. 1986. Soil-plant relations in the tropics. *Annual Review of Ecology and Systematics* 17:1–16.

Leigh, E. G., Jr. 1975. Structure and climate in tropical rain forest. *Annual Review of Ecology and Systematics* 6:67–86. Concise review of effects of climate on lowland and montane rain forests.

Martin, C. 1991. *The rainforests of West Africa.* New York: Birkhauser. A comprehensive treatment of the ecology and conservation of African rain forests.

Murphy, P. G., and A. E. Lugo. 1986. Ecology of tropical dry forests. *Annual Review of Ecology and Systematics* 17:67–88.

Myers, N. 1983. *A wealth of wild species.* Boulder, CO: Westview Press. Excellent reference on the value of tropical plants to human welfare.

Richards, P. W. 1996. *The tropical rain forest: An ecological study,* 2d ed. New York: Cambridge University Press. A thoroughly revised edition of a classic.

Simpson, B. B., and J. Haffer. 1978. Spatial patterns in the Amazonian rain forest. *Annual Review of Ecology and Systematics* 9:497–518.

Sutton, S. L., T. C. Whitmore, and A. C. Chadwick (eds.). 1983. *Tropical rain forests: Ecology and management.* Oxford, England: Blackwell.

Tomlinson, P. B. 1987. Architecture of tropical plants. *Annual Review of Ecology and Systematics* 18:1–21.

Walter, H. 1971. *Ecology of tropical and subtropical vegetation.* Edinburgh: Oliver & Boyd.

Whitmore, T. C. 1984. *Tropical rainforests of the Far East.* Oxford: Clarendon Press. The authoritative reference on Southeast Asian rain forests.

———. 1990. *An introduction to tropical rain forests.* New York: Oxford University Press. An excellent reference on the structure and function of the rain forest.

Temperate Forests

Bormann, F. H., and G. E. Likens. 1979. *Pattern and process in a forest ecosystem.* New York: Springer-Verlag. Excellent discussion of forest ecosystem functions, with special reference to Hubbard Brook.

Brinson, M. M., B. L. Swift, R. C. Plantico, and J. S. Barclay. 1981. *Riparian ecosystems: Their ecology and status.* FWS/OBS-81/17. Kearneysville, WV: Eastern Energy and Land Use Team, U.S. Fish and Wildlife Service. General information on riparian ecosystems is difficult to find. This is an excellent introduction.

Curtis, J. T. 1959. *The vegetation of Wisconsin.* Madison: University of Wisconsin Press. A classic study of forest vegetation of the lake states.

Davis, M. B. (ed.). 1996. *Eastern old-growth forests: Prospects for rediscovery and recovery.* Covelo, CA: Island Press. First book on old-growth forests in eastern North America.

Duvigneaud, P. (ed.). 1971. *Productivity of forest ecosystems.* Paris: UNESCO. Somewhat dated, but an excellent reference on the subject.

Edmonds, R. L. (ed.). 1981. *Analysis of coniferous forest ecosystems in the western United States.* Stroudsburg, PA: Dowden, Hutchinson & Ross. Strong on function.

Franklin, J. F., and C. T. Dyrness. 1973. *Natural vegetation of Oregon and Washington.* USDA Forest Service Gen. Tech. Rept. PNW 8. Corvallis, OR: USDA Forest Service. Excellent description of old-growth forests and other vegetation.

Irland, L. C. 1982. *Wildland and woodlots: The story of New England's forests.* Hanover, NH: University Press of New England. Reviews the changing character and problems of New England's forests since colonial days.

Lang, G. E., W. A. Reiners, and L. H. Pike. 1980. Structure and biomass dynamics of epiphytic lichen communities of balsam fir forests in New Hampshire. *Ecology* 63:541–550.

Lansky, M. 1992. *Beyond the beauty strip: Saving what's left of our forests.* Gardiner, ME: Tilbury House. A detailed and critical examination of forestry practices in the boreal forests of the northeastern United States, applicable elsewhere.

Mather, G. A. S. 1990. *Global forest resources.* Portland, OR: Timber Press. Worldwide overview of forests, forest use, and management or lack thereof.

Norse, E. A. 1990. *Ancient forests of the Pacific Northwest.* Washington, DC: Island Press. Excellent description of the forest and its problems.

Packham, J. R., and D. J. L. Harding. 1982. *Ecology of woodland processes.* London: Edward Arnold. A concise, pleasant introduction to woodland ecology. Although it considers North American examples, it is Europe-oriented, which makes it an excellent introduction to European forest ecology.

Perlan, J. 1991. *A forest journey: The role of wood in the development of civilization.* Cambridge, MA: Harvard University Press. Destruction of the temperate and mediterranean forest and its impact on western civilization.

Peterken, G. F. 1996. *Natural woodlands: Ecology and conservation in north temperate regions.* New York: Cambridge University Press. How woods grow, die, and regenerate in the absence of human interference. An excellent book on old-growth forests.

Polunin, O., and M. Walters. 1985. *A guide to the vegetation of Britain and Europe.* New York: Oxford University Press. Outstanding, well-illustrated guide to seminatural and natural vegetation, including detailed description of forests.

Reichle, D. E. (ed.). 1981. *Dynamic properties of forest ecosystems.* Cambridge, England: Cambridge University Press. The major reference source on functions of forest ecosystems throughout the world.

USDA Forest Service. 1980. *Environmental consequences of timber harvesting in Rocky Mountain coniferous forests.* USDA Forest Service Gen. Tech. Rept. INT-90. Ogden, UT: Intermountain Forest and Range Experiment Station. Technical, detailed assessment of timber harvesting in the mountains.

Boreal Forest

Bonan, G. B., and H. H. Shugart. 1989. Environmental factors and ecological processes in boreal forests. *Annual Review of Ecology and Systematics* 20:1–18. An informative review paper.

Knystautas, A. 1987. *The natural history of the USSR.* New York: McGraw-Hill. Contains an interesting, well-illustrated description of the Siberian taiga.

Larsen, J. A. 1980. *The boreal ecosystem.* New York: Academic Press. A summary of what we know about the ecology of the boreal ecosystem.

———. 1989. *The northern forest border in Canada and Alaska.* Ecological Studies 70. New York: Springer-Verlag. An interesting description of the biotic communities and ecological relationships of the forest-tundra ecotone.

Polunin, O., and M. Walters. 1985. *A guide to the vegetation of Britain and Europe.* Oxford, England: Oxford University Press. Excellent.

Woodland

Plumb, T. R. 1979. *Proceedings of the symposium on the ecology, management, and utilization of California oaks.* General Technical Report PSW-44. Berkeley, CA: Pacific Southwest Forest and Range Experiment Station, Forest Service, U.S. Dept. of Agriculture. Thorough coverage of this important and degraded mediterranean ecosystem.

Savannas

Bourliere, F. (ed.). 1983. *Tropical savannas: Ecosystems of the world* 13. Amsterdam: Elsevier Scientific Publishers. A major reference work on savannas.

Sarimiento, G. 1984. *Ecology of neotropical savannas.* Cambridge, MA: Harvard University Press. A major study of South American savannas.

Sinclair, A. R. E., and M. Norton-Griffiths (eds.). 1979. *Serengeti: Dynamics of an ecosystem.* A study of the African savanna ecosystem. Excellent.

Sinclair, A. R. E., and P. Arcese (eds.). 1995. *Serengeti II: Dynamics, management, and conservation of an ecosystem.* Chicago: University of Chicago Press. A masterful study of an ecosystem.

Tothill, J. C., and J. J. Mott (eds.). 1985. *Ecology and management of the world's ecosystems.* Canberra: Australian Academy of Science. Group of papers providing in-depth review of savannas and their management. An important reference.

CHAPTER 26 Terrestrial Ecosystems II: Grasslands, Shrublands, Desert, and Tundra

Grasslands

Breymeyer, A., and G. Van Dyne (eds.). 1980. *Grasslands, systems analysis and man.* Cambridge University Press. Synthesis of International Biological Programme studies of grasslands; comprehensive.

Callenback, E. 1996. *Bring back the buffalo: A sustainable future for America's Great Plains.* Covelo, CA: Island Press. An ecological approach to the management of the shortgrass plains.

Coupland, R. T. (ed). 1979. *Grassland ecosystems of the world: Analysis of grasslands and their uses.* Cambridge, England: Cambridge University Press. Good reference.

Duffy, E. 1974. *Grassland ecology and wildlife management.* London: Chapman and Hall. Examines ecology of domestic grasslands.

Falt, J. H. 1976. Energetics of a suburban lawn ecosystem. *Ecology* 57:141–150. Ecological concepts applied to the lawn. An interesting study.

French, N. (ed.). 1979. *Perspectives on grassland ecology.* New York: Springer-Verlag. A good summary of grassland ecology.

Hodgson, J., and A. W. Illius (eds.). 1966. *Ecology and management of grazing systems.* New York: CAB International. Synthesis of grassland ecology and application of principles to management of domestic grassland and rangeland.

Levin, S. A. (ed.). 1993. Forum: Grazing theory and rangeland management. *Ecological Applications* 3:1–38.

Manning, D. 1995. *Grassland: History, biology, politics, and promise of the American prairie.* New York: Penguin Books.

Reichman, O. J. 1987. *Konza prairie: Tallgrass natural history.* Lawrence, KS: University of Kansas Press. A good description of tallgrass prairie based on studies of one of the few remaining tracts.

Risser, P. G., et al. 1981. *The true prairie ecosystem.* Stroudsburg, PA: Dowden, Hutchinson, & Ross. Ecology of midgrass and tallgrass prairies.

Weaver, J. E. 1954. *North American prairie.* Lincoln, NE: Johnson. An old but important study of prairie vegetation by an outstanding botanist familiar with the original condition of the prairies.

Shrublands

Castri, F. Di, D. W. Goodall, and R. L. Specht (eds.). 1981. *Mediterranean-type shrublands.* Ecosystems of the World No. 11. Amsterdam: Elsevier Scientific. A more comprehensive treatment of mediterranean ecosystems. A major reference.

Castri, F. Di., and H. A. Mooney (eds.). 1973. *Mediterranean-type ecosystems.* Ecological Studies No. 7. New York: Springer-Verlag. An excellent basic introduction to all aspects of mediterranean ecosystems.

Groves, R. H. 1981. *Australian vegetation.* Cambridge, England: Cambridge University Press. Concise descriptions of the diverse plant communities of Australia.

McKell, C. M. (ed.). 1983. *The biology and utilization of shrubs.* Orlando, FL: Academic Press. A major reference on shrubs worldwide.

Polunin, O., and M. Walters. 1985. *A guide to the vegetation of Britain and Europe.* Oxford, England: Oxford University Press. An outstanding survey of the plant communities of Europe as far as the Russian border; well-illustrated.

Specht, R. L. (ed.). 1979, 1981. *Heathlands and related shrublands.* Ecosystems of the World 9A and 9B. Amsterdam: Elsevier Scientific. A major reference on heathland communities worldwide.

Desert

Brown, G. W., Jr. (ed.). 1976–1977. *Desert biology,* 2 vol. New York: Academic Press. Basic information on biology and physical features of world deserts.

Evenardi, M., I. Noy-Meir, and D. Goodall (eds.). 1985–1986. *Hot deserts and arid shrublands of the world.* Ecosystems of the World 12A and 12B. Amsterdam: Elsevier Scientific. A major reference work on world deserts.

Wagner, F. H. 1980. *Wildlife of the deserts.* New York: Harry W. Abrams. An excellent, well-illustrated introduction to desert animals' adaptation to arid environments.

Tundra

Bliss, L. C., O. H. Heal, and J. J. Moore (eds.). 1981. *Tundra ecosystems: A comparative analysis.* New York: Cambridge University Press. A major reference.

Brown, J., P. C. Miller, L. L. Tieszen, and F. L. Bunnell. 1980. *An arctic ecosystem: The coastal tundra at Barrow, Alaska.* Stroudsburg, PA: Dowden, Hutchinson & Ross.

Furley, P. A., and W. A. Newey. 1983. *Geography of the biosphere.* London: Butterworths. Chapter 10 gives a detailed discussion of the tundra biome.

Rosswall, T., and O. W. Heal (eds.). 1975. *Structure and function of tundra ecosystems.* Ecological Bulletins 20. Stockholm: Swedish Natural Sciences Research Council.

Smith, A. P., and T. P. Young. 1987. Tropical alpine plant ecology. *Annual Review of Ecology and Systematics* 18:137–158. An informative review paper.

Sonesson, M. (ed.). 1980. *Ecology of a subarctic mire.* Ecological Bulletins 30. Stockholm: Swedish Natural Sciences Research Council. A detailed study of structure and function.

Wielgolaski, F. E. (ed.). 1975. *Fennoscandian tundra ecosystems.* Part I, *Plants and microorganisms;* Part II, *Animals and systems analysis.* New York: Springer-Verlag.

Zwinger, A. H., and B. E. Willard. 1972. *Land above the trees.* New York: Harper & Row.

CHAPTER 27 Freshwater and Estuarine Ecosystems

Barnes, R. S. K., and K. H. Mann. 1980. *Fundamentals of aquatic ecosystems.* New York: Blackwell Scientific Publishers. A solid reference.

Brock, T. D. 1985. *A eutrophic lake: Lake Mendota, Wisconsin.* New York: Springer-Verlag. A study of the process of eutrophication.

Carpenter, S. A. 1980. Enrichment of Lake Wingra, Wisconsin, by submerged macrophyte decay. *Ecology* 61:1145–1155.

Carpenter, S. A., and J. F. Kitchell. 1984. Plankton community structure and limnetic primary production. *American Naturalist* 124:159–172.

Carpenter, S. A., J. F. Kitchell, and J. Hodgson. 1985. Cascading trophic interactions and lake productivity. *Bioscience* 35:634–639. Carpenter's three papers provide an excellent introduction to the functioning of lake ecosystems.

Cummins, K. W. 1974. Structure and function of stream ecosystems. *Bioscience* 24:631–641.

———. 1979. Feeding ecology of stream invertebrates. *Annual Review of Ecology and Systematics* 10:147–172. A good review of functional aspects of stream organisms.

Gore, J. A., and G. E. Petts. 1989. *Alternatives in regulated river management.* Boca Raton, FL: CRC Press. Focuses attention on ways to reduce ecological effects of river regulation.

Hutchinson, G. E. 1957–1967. *A treatise on limnology.* Vol. 1, *Geography, physics, and chemistry.* Vol. 2, *Introduction to lake biology and limnoplankton.* New York: Wiley. A classic reference.

Hynes, H. B. N. 1970. *The ecology of running water.* Toronto: University of Toronto Press. Dated, but a classic and valuable work; a major reference.

Ketchum, B. H. (ed.). 1983. *Estuaries and enclosed seas.* Ecosystems of the World 26. Amsterdam: Elsevier. Major reference on world estuarine structure and function.

Likens, G. E. (ed.). 1985. *An ecosystem approach to aquatic ecology: Mirror Lake and environment.* New York: Springer-Verlag. An outstanding detailed study of a lake ecosystem and its interactions with the surrounding terrestrial communities.

Macan, T. T. 1970. *Biological studies of English lakes.* New York: Elsevier.

———. 1973. *Ponds and lakes.* New York: Crane, Russak. Macan's books are classic studies of English lakes.

———. 1974. *Freshwater ecology,* 2d ed. New York: John Wiley. & Sons. Strong on the physical aspects of freshwater ecosystems.

Maitland, P. S. 1978. *Biology of fresh waters.* New York: John Wiley & Sons. Although dated, still a good basic introduction.

McLusky, D. S. 1971. *The ecology of estuaries.* London: Heinemann Educational.

———. 1989. *The estuarine ecosystem,* 2d ed. New York: Chapman & Hall. Clearly describes the structure and function of estuarine ecosystems.

Meyer, J. L. 1990. A blackwater perspective on riverine ecosystems. *Bioscience* 40:643–651. A detailed look at stream ecosystem function, especially food webs.

Petts, G. E. 1984. *Impounded rivers: Perspectives for ecological management.* New York: Wiley. Detailed analysis of the effects of dams on the world's rivers, especially downstream.

Rich, P. H., and R. G. Wetzel. 1978. Detritus in the lake ecosystem. *American Naturalist* 112:57–71. Details the role of detritus in the economy of a lake ecosystem.

Stanford, J. A., and A. P. Covich (eds.). 1988. Community structure and function in temperate and tropical streams. *Journal of North American Benthological Society* 7:261–529. A valuable special issue of the journal.

Valiela, I. 1984. *Marine biology processes.* New York: Springer-Verlag. An advanced synthesis.

Vannote, R. L., G. W. Minshall, K. W. Cummins, J. R. Sedell, and C. E. Cushing. 1980. The river continuum concept. *Canadian Journal of Fisheries and Aquatic Science* 37:130–137.

Wiley, M. 1976. *Estuarine processes.* New York: Academic Press.

CHAPTER 28 Marine Ecosystems

Oceans

Grassle, J. F. 1985. Hydrothermal vent animals: Distribution and biology. *Science* 229:713–717.

———. 1989. Species diversity in deep-sea communities. *Trends in Ecology and Evolution* 4:12–15.

———. 1991. Deep-sea benthic diversity. *Bioscience* 41:464–469.

Gross, M. G. 1982. *Oceanography: A view of Earth,* 3rd ed. Englewood Cliffs, NJ: Prentice-Hall. An excellent reference on the physical aspects of the sea.

Hardy, A. 1971. *The open sea: Its natural history.* Boston: Houghton Mifflin. A classic introduction.

Hayman, R. M., and R. C. McDonald. 1985. The ecology of deep sea hot springs. *American Scientist* 73:441–449.

Kinne, O. (ed.). 1978. *Marine ecology,* 5 vol. A major and often technical reference source.

Marshall, N. B. 1980. *Deep-sea biology: Developments and perspectives.* New York: Garland STMP Press.

National Academy of Sciences. 1985. *Oil in the sea: Inputs, fates, and effects.* Washington, DC: National Academy Press.

Nelson-Smith, A. 1972. *Oil pollution and marine ecology.* London: Elek Science. Good, concise basic reference.

Nybakken, J. W. 1997. *Marine biology: An ecological approach,* 4th ed. Menlo Park, CA: Benjamin/Cummings. A solid reference on marine life and ecosystems.

Powell, M. A., and C. N. Somero. 1983. Blood components prevent sulfide poisoning of respiration of the hydrothermal vent tubeworm *Riftia pachyptila. Science* 219:297–299.

Rex, M. A. 1981. Community structure in the deep-sea benthos. *Annual Review of Ecology and Systematics* 12:331–353. A good review of deep-sea species diversity.

Steele, J. 1974. *The structure of a marine ecosystem.* Cambridge, MA: Harvard University Press.

Intertidal Zone and Coral Reef

Bertness, M. D. 1999. *The ecology of Atlantic shorelines.* Sunderland MA: Sinauer Associates.

Carson, R. 1955. *The edge of the sea.* Boston: Houghton Mifflin. A classic introduction.

Dayton, P. 1971. Competition, disturbance, and community organization: The provision and subsequent utilization of space in a rocky intertidal community. *Ecological Monographs* 45:137–159.

Eltringham, S. K. 1971. *Life in mud and sand.* New York: Crane, Russak. A concise, informative introduction.

Hiatt, R. W., and D. W. Strasburg. 1960. Ecological relationships of the fish fauna on coral reefs of the Marshall Islands. *Ecological Monographs* 30:66–120. An important, well-illustrated reference on coral reef fish.

Huston, M. 1985. Patterns of species diversity on coral reefs. *Annual Review of Ecology and Systematics.* 16:149–177. An important reference on the roles of light and disturbance.

Jackson, J. B. C. 1991. Adaptation and diversity of reef corals. *Bioscience* 41:475–482. Relates patterns of species distribution to life history and disturbance.

Jones, O. A., and R. Endean (eds.). 1973, 1976. *Biology and geology of coral reefs.* Vols. II, III. New York: Academic Press.

Leigh, E. G., Jr. 1987. Wave energy and intertidal productivity. *Proceedings of National Academy of Science* 84:1314.

Lessios, H. A. 1988. Mass mortality of *Diadema antillarum* in the Caribbean: What have we learned? *Annual Review of Ecology and Systematics* 19:371–393. Detailed review of the ecological impact of the die-off.

Lubchenco, J. 1978. Algal zonation in the New England rocky intertidal community: An experimental analysis. *Ecology* 61:333–344. An outstanding study and informative reference.

Mathieson, A. C., and P. H. Nienhuis (eds.). 1991. *Intertidal and littoral ecosystems.* Ecosystems of the world 24. Amsterdam: Elsevier. A detailed, broad survey of the intertidal and littoral zones of the world.

Moore, P. G., and R. Seed (eds.). 1986. *The ecology of rocky shores.* New York: Columbia University Press. Comprehensive, worldwide review of the rocky intertidal zone.

Newell, R. C. 1970. *Biology of intertidal animals.* New York: Elsevier. Adaptations of animals to the intertidal environment.

Nybakken, J. W. 1997. *Marine biology: An ecological approach,* 4th ed. Menlo Park, CA: Benjamin/Cummings. Informative chapters on rocky, sandy, and muddy shores and coral reefs.

Paine, R. T. 1969. The *Pisaster-Tegula* interaction: Prey patches, predator food preference, and intertidal community structure. *Ecology* 59:150–961. An outstanding paper on the role of predation in the rocky shore community.

Pomeroy, L. R., and E. J. Kuenzler. 1969. Phosphorus turnover by coral reef animals. Pages 478–483 in D. J. Nelson and F. E. Evans, eds. *Symposium on radioecology conference.* 670503. Springfield, VA: National Technical Information Services.

Reaka, M. J. (ed.). 1985. *Ecology of coral reefs.* Symposia Series for Undersea Research, NOAA Undersea Research Program 3. Washington DC: US Department of Commerce.

Sale, P. F. 1980. The ecology of fishes on coral reefs. *Annual Review of Oceanography and Marine Biology* 18:367–421. Sweeping review.

Stephenson, T. A., and A. Stephenson. 1973. *Life between the tidemarks on rocky shores.* San Francisco: Freeman. A detailed description of the structure of intertidal life around the world.

Underwood, A. J., E. J. Denley, and M. J. Moran. 1983. Experimental analyses of the structure and dynamics of midshore rocky intertidal communities in New South Wales. *Oecologica* 56:202–219.

Wellington, G. W. 1982. Depth zonation of corals in the Gulf of Panama: Control and facilitation by resident reef fishes. *Ecological Monographs* 52:223–241.

Wilson, R., and J. Q. Wilson. 1985. *Watching fishes: Life and behavior on coral reefs.* New York: Harper & Row. Written for a general audience, this well-illustrated book is accessible and informative.

Yonge, C. M. 1949. *The seashore.* London: Collins. A well-illustrated classic.

CHAPTER 29 Wetlands

Bertness, M. D. 1984. Ribbed mussels and *Spartina alterniflora* production on a New England marsh. *Ecology* 65:1794–1807. Relationship between consumer and producer in the salt marsh.

———. 1999. *The ecology of Atlantic shorelines.* Sunderland, MA: Sinauer Associates.

Bildstein, K. L., G. T. Bancroft, P. J. Dugan, et al. 1991. Approaches to the conservation of coastal wetlands in the Western Hemisphere. *Wilson Bulletin* 103:218–254. Excellent overview of problems.

Chapman, V. J. 1976. *Mangrove vegetation.* Leutershausen, Germany: J. Cramer. An authoritative reference on mangroves and mangrove forests.

———. (ed.). 1977. *Wet coastal ecosystems.* Amsterdam: Elsevier. Covers salt marshes and mangals of the world. A basic reference.

Clark, J. 1974. *Coastal ecosystems: Ecological considerations for the management of the coastal zone.* Washington, DC: Conservation Foundation.

Cowardin, L. M., V. Carter, and E. C. Golet. 1979. *Classification of wetlands and deepwater habitats of the United States.* U.S. Department of Interior, Fish and Wildlife Service FWS/OBS-79/31. A revised classification of wetlands.

Dahl, T. E. 1990. *Wetland losses in the United States, 1780's to 1980's.* U.S. Department of Interior, Fish and Wildlife Service.

Dugan, P. (ed.). 1993. *Wetlands in danger: A world conservation atlas.* New York: Oxford University Press. Survey of world wetland ecosystems and human impacts on them.

Ewel, K. C. 1990. Multiple demands on wetlands. *Bioscience* 40:660–666. Societal benefits of wetlands, with cypress swamps serving as a case study.

Ewel, K. C., and H. T. Odum (eds.). 1986. *Cypress swamps.* Gainesville: University Presses of Florida. In-depth studies of the structure, function, and management of cypress swamps in southern United States.

Good, R. E., D. F. Whigham, and R. L. Simpson (eds.). 1978. *Freshwater wetlands: Ecological processes and management potential.* New York: Academic. A review of functional aspects of wetlands and their management implications.

Gore, A. P. J. (ed.). 1983. *Mire, swamp, bog, fen, and moor.* Ecosystems of the world 4A and 4B. Amsterdam: Elsevier. A review of the world of freshwater wetlands.

Greesen, P. S., J. R. Clark, and J. E. Clark (eds.). 1979. *Wetland functions and values: The state of our understanding.* Minneapolis: American Water Resources Association. An earlier work still of value.

Haines, B. L., and E. L. Dunn. 1985. Coastal marshes. Pages 323–347 in B. F. Chabot and H. A. Mooney, eds. *Physiological ecology of North American plant communities.* New York: Chapman and Hall.

Hopkinson, C. S., and J. P. Schubauer. 1984. Static and dynamic aspects of nitrogen cycling in the salt marsh graminoid *Spartina alterniflora. Ecology* 65:961–969.

Howarth, R. W., and J. Teal. 1979. Sulfate reduction in a New England salt marsh. *Limnology and Oceanography* 24:999–1013.

Jefferies, R. L., and A. J. Davy (eds.). 1979. *Ecological processes in coastal environments.* Oxford: Blackwell.

Josselyn, M. 1983. *The ecology of San Francisco tidal marshes: A community profile.* U.S. Fish and Wildlife Service Office of Biological Services. FWS/OBS–82/83.

Long, S. P., and C. F. Mason. 1983. *Salt marsh ecology.* New York: Chapman & Hall. A good, accessible introduction to the salt marsh ecosystem.

Lugo, A. E. 1990. *The forested wetlands.* Amsterdam: Elsevier. Excellent overview and discussion of structure and function.

Lugo, A. E., and S. C. Snedaker. 1974. The ecology of mangroves. *Annual Review of Ecology and Systematics* 5:39–64.

Mitsch, W. J., and J. C. Gosslink. 1993. *Wetlands,* 2d ed. New York: Van Nostrand Reinhold. A pioneering text and major reference.

Moore, P. D., and D. J. Bellemany. 1974. *Peatlands.* New York: Springer-Verlag. An outstanding introduction to the ecology and development of peatlands.

National Audubon Society. 1990. The last wetlands. *Audubon* 92(4). A highly informative issue devoted entirely to wetlands, their management, and preservation.

Niering, W. A., and B. Hales. 1991. *Wetlands of North America.* Charlottesville, VA: Thomasson-Grant. An exceptionally illustrated survey of North American wetlands, both freshwater and salt.

Nixon, S. W., and C. A. Oviatt. 1973. Ecology of a New England salt marsh. *Ecological Monographs* 43:463–498.

Nybakken, J. W. 1997. *Marine biology: An ecological approach,* 4th ed. Menlo Park, CA: Benjamin/Cummings. Chapters on estuaries, salt marshes, and mangrove forests.

Odum, W. E., C. C. McIvor, and T. J. Smith III. 1982. *The ecology of the mangroves of South Florida: A community profile.* U.S. Fish and Wildlife Service Office of Biological Services. FWS/OBS 81/24.

Payne, N. F. 1992. *Techniques for wildlife habitat management of wetlands.* New York: McGraw-Hill. Although emphasis is on management, this book contains a wealth of information on wetland types, structure, and function.

Perkins, E. J. 1974. *The biology of estuaries and coastal waters.* New York: Academic Press.

Pomeroy, L. R., and R. G. Wiegert (eds.). 1981. *The ecology of a salt marsh.* New York: Springer-Verlag. Synthesis of a 20-year study of all aspects of a southern U.S. coastal marsh.

Stout, J. P. 1984. The ecology of irregularly flooded salt marshes of northeastern Gulf of Mexico: A community profile. *U.S. Fish and Wildlife Service Biological Report* 85(7.1).

Teal, J. 1962. Energy flow in a salt marsh ecosystem of Georgia. *Ecology* 43:614–624.

Teal, J., and M. Teal. 1969. *Life and death of the salt marsh.* Boston: Little Brown. A classic.

Tiner, R. W. 1991. The concept of a hydrophyte for wetland identification. *Bioscience* 41:236–247. Reviews problems and means of identifying wetlands.

Valiela, I., and J. M. Teal. 1979. The nitrogen budget of a salt marsh ecosystem. *Nature* 47:337–371.

Valiela, I., J. M. Teal, and W. G. Denser. 1978. The nature of growth forms in salt marsh grass *Spartina alternifolia. American Naturalist* 112:461–370.

Van der Valk, A. (ed.). 1989. *Northern prairie wetlands.* Ames, IA: Iowa State University Press. Detailed studies of major wetlands rapidly disappearing.

Weller, M. W. 1981. *Freshwater wetlands: Ecology and wildlife management.* Minneapolis: University of Minnesota Press.

Zedler, J., T. Winfield, and D. Mauriello. 1992. *The ecology of Southern California coastal marshes: A community profile.* U.S. Fish and Wildlife Service Office of Biological Services FWS/OBS 81/54.

CHAPTER 30 Global Environmental Change

Adams, R. M., R. Alig, J. M. Callaway, B. A. McCarl, and S. M. Winnet. 1995. *The economic effects of climate change on U.S. agriculture.* Final report. Climate Change Impacts Program. Palo Alto, CA: EPRI.

Adams, R. M., R. A. Fleming, C. C. Chang, and B. A. McCarl. 1995. A reassessment of the economic effects of global climate change on U.S. agriculture. *Climate Change* 30:147–167.

Amthor, J. S. 1995. Terrestrial higher-plant response to increasing atmospheric CO_2 in relation to the global carbon cycle. *Global Change Biology* 1:243–274.

Bazzaz, F. A. 1996. *Plants in changing environments: Linking physiological, population, and community ecology.* New York: Cambridge University Press.

Butcher, S. S., R. J. Charlson, G. H. Orians, and G. V. Wolfe (eds.). 1992. *Global biogeochemical cycles*. New York: Academic Press.

Drake, B. G., and P. W. Leadley. 1991. Canopy photosynthesis of crops and native plant communities exposed to long-term elevated carbon dioxide. *Plant Cell Environment* 14:853–860.

Drake, B. G., G. Peresta, E. Beugeling, and R. Matamala. 1996. Long-term elevated CO_2 exposure in a Chesapeake Bay wetland: Ecosystem gas exchange, primary productivity, and tissue nitrogen. In G. Koch and H. A. Mooney, eds. *Carbon dioxide and terrestrial ecosystems*. San Diego: Academic Press.

Edmonds, J. 1992. Why understanding the natural sinks and sources of CO_2 is important: A policy analysis perspective. *Water, Air and Soil Pollution* 64:11–21.

Field, C. B., R. B. Jackson, and H. A. Mooney. 1995. Stomatal responses to CO_2: Implications from the plant to global scale. *Plant Cell Environment* 18:1214–1225.

Gates, D. 1993. *Climate change and its biological consequences*. Sunderland, MA: Sinauer Associates. A highly accessible introduction to climate change and its effects, past and present.

Harte, J., and R. Shaw. 1995. Shifting dominance with a montane vegetation community: Results of a climate-warming experiment. *Science* 267:876–880.

Heimann, M. (ed.). 1993. *The global carbon cycle*. Berlin: Springer-Verlag.

Holdridge, L. R. 1947. Determination of world formulations from simple climatic data. *Science* 105:367–368.

Houghton, J. T. 1997. *Global warming: The complete briefing*. Cambridge, England: Cambridge University Press.

Houghton, J. T., L. G. Meira Filho, J. Bruce, H. Lee, B. A. Callander, E. Haites, N. Harris, and K. Maskell (eds). 1995. *Climate change 1994*. Cambridge, England: Cambridge University Press.

Houghton, J. T., L. G. Meira Filho, B. A. Callander, N. Harris, A. Kattenberg, and K. Maskell (eds.). 1996. *Climate change 1995: The science of climate change*. Intergovernmental Panel on Climate Change. Cambridge, England: Cambridge University Press.

Houghton, R. A. 1995. Land-use change and the carbon cycle. *Global Change Biology* 1:275–287.

Iverson, L. R., and A. M. Prasad. 2001. Potential changes in tree species richness and forest community type following climate change. *Ecosystems* 4:186–199.

Kalkstein, L. S., and J. S. Green. 1997. An evaluation of climate/mortality relationships in large U.S. cities and possible impacts of a climate change. *Environmental Health Perspectives* 105:84–93.

Kalkstein, L. S., and G. Tan. 1995. Human health. In K. Strzepek and J. Smith, eds. *As climate changes: International impacts and implications*. New York: Cambridge University Press.

Kattenberg, A., F. Giorgi, H. Grassl, G. A. Meehl, J. F. B. Mitchell, R. J. Stouffer, T. Tokioka, A. J. Weaver, and T. M. L. Wigley. 1996. Climate models projections of future climate. Pages 285–357 in J. T. Houghton et al., eds. *Climate change 1995*. The Science of Climate Change. Intergovernmental Panel on Climate Change. Cambridge, UK: Cambridge University Press.

Keeling, C. D., T. P. Whorf, M. Wahlen, and J. van der Plicht. 1995. Interannual extremes in the rate of rise of atmospheric carbon dioxide since 1980. *Nature* 375:666–670.

Korner, C., M. Diemer, B. Schappi, P. Niklaus, and J. Arnone. 1997. The response of alpine grassland to four seasons of CO_2 enrichment: A synthesis. *Acta Oecologia* 18:165–175.

Marland, G., and T. Boden. 1993. The magnitude and distribution of fossil-fuel related carbon releases. Pages 117–138 in M. Heimann, ed. *The global carbon cycle*. New York: Springer-Verlag.

Miko, U. F. 1996. Climate change impacts on forests. In R. T. Watson, M. C. Zinyowera, and R. H. Moss, eds. *Climate change 1995: Impacts, adaptations and mitigation of climate change*. New York: Cambridge University Press.

Mitchell, J. F. B., R. A. Davis, W. J. Ingram, and C. A. Senior. 1995. On surface temperature, greenhouse gases and aerosols: Models and observations. *Journal of Climatology* 10:2364–2386.

Mitchell, J. F. B., T. J. Johns, J. M. Gregory, and S. B. F. Tett. 1995. Climate response to increasing levels of greenhouse gases and sulfate aerosols. *Nature* 376:501–504.

Nicholls, R. J., and S. P. Leatherman. 1995. Global sea-level rise. In K. Strzepek and J. B. Smith, eds. *As climate changes: International impacts and implications*. Cambridge, England: Cambridge University Press.

Oechel, W. C., and G. L. Vourlitis. 1996. Direct effects of elevated CO_2 on Arctic plant and ecosystem function. Pages 163–176 in G. Koch and H. A. Mooney, eds. *Carbon dioxide and terrestrial ecosystems*. San Diego: Academic Press.

Owensby, C. E., J. M. Ham, A. Knapp, C. W. Rice, P. I. Coyne, and L. M. Auen. 1996. Ecosystem-level responses of tallgrass prairie to elevated CO_2. Pages 147–162 in G. Koch and H. A. Mooney, eds. *Carbon dioxide and terrestrial ecosystems*. San Diego: Academic Press.

Peterjohn, W. T., J. M. Melillo, F. P. Bowles, and P. A. Steudler. 1993. Soil warming and trace gas fluxes: Experimental design and preliminary flux results. *Oecologia* 93:18–24.

Peters, R. L., and T. E. Lovejoy (eds.). 1992. *Global warming and biological diversity*. New Haven: Yale University Press. A sobering, readable examination of the potential effect of global warming on vegetation, soils, animals, parasites and diseases, and ecosystems.

Root, T. 1988. Energy constraints on avian distributions and abundances. *Ecology* 69:330–339.

Schneider, S. H. 1989. *Global warming*. San Francisco: Sierra Club Books. Highly readable examination of global warming and the debate engendered in politics and the media.

Smith, T. M., P. N. Halpin, H. H. Shugart, and C. Secrett. 1994. Global forests. In K. Strzpeck and J. Smith, eds. *As climate changes: International impacts and implications*. Cambridge, England: Cambridge University Press.

Smith, T. M., R. Leemans, and H. H. Shugart. 1992. Sensitivity of terrestrial carbon storage to CO_2-induced climate change: Comparison of four scenarios based on general circulation models. *Climatic Change* 21:367–384.

Strain, B. R., and J. D. Cure (eds.). 1985. *Direct effects of increasing carbon dioxide on vegetation*. Washington, DC: U.S. Department of Energy Publication DOE/ER-0238.

Trabalka, J. R. (ed.). 1985. *Atmospheric carbon dioxide and the global carbon cycle*. Washington, DC: U.S. Department of Energy Report DOE/ER-0239.

Trabalka, J. R., and D. E. Reichle (eds.). 1994. *The changing carbon cycle: A global analysis*. New York: Springer-Verlag.

VEMAP Participants. 1995. Vegetation/ecosystem modeling and analysis project: Comparing biogeography and biogeochemistry models in a continental-scale study of terrestrial ecosystem responses to climate change and CO_2 doubling. *Global Biogeochemical Cycles* 9:407–437.

Watson, R. T., M. C. Zinyowera, R. H. Moss, and D. J. Dokken (eds.). 1998. *The regional impacts of climate change* (a special report of IPCC Working Group II). Cambridge, UK: Cambridge University Press.

Woodward, F. I. (ed.). 1992. *Global climate change: The ecological consequences*. London: Academic Press. Emphasis on climate change research.

Yoshino, M., T. Horie, H. Seino, H. Tsujii, T. Uchijima, and Z. Uchijima. 1988. The effects of climate variations on agriculture in Japan. In M. Parry, T. R. Carter, and N. T. Konijn, eds. *The impacts of climate variation on agriculture*. Volume 1, *Assessments in cool temperate and cold regions*. Dordrecht, The Netherlands: Kluwer.

ECOLOGICAL APPLICATIONS: How a Lack of Mushrooms Helped Fuel the Industrial Revolution

Robinson, J. M. 1990. Lignin, land plants, and fungi: Biological evolution affecting Phanerozoic oxygen balance. *Geology* 15:607–610.

Swift, M. J., O. W. Heal, and J. M. Anderson. 1978. *Decomposition in terrestrial ecosystems*. Oxford: Blackwell Scientific Publications.

Glossary

A horizon Surface stratum of mineral soil, characterized by maximum accumulation of organic matter, maximum biological activity, and loss of such materials as iron, aluminum oxides, and clays.

abiotic Nonliving; the abiotic component of the environment includes soil, water, air, light, nutrients, and the like.

abrupt speciation Spontaneous rise of a new species, largely through polyploidy.

abundance The number of individuals of a species in a given area.

abyssal Relating to the bottom waters of oceans, usually below 1000 m.

acclimation Alteration of an individual's physiological rate or capacity to perform a function through long-term exposure to certain conditions.

acclimatization Changes in physiological state or tolerance that appear in a species after long exposure to different natural environments.

acid deposition Wet and dry atmospheric fallout with an extremely low pH, brought about when water vapor in the atmosphere combines with hydrogen sulfide and nitrous oxide vapors released by burning fossil fuels; the sulfuric and nitric acid in rain, fog, snow, gases, and particulate matter.

active transport Movement of ions and molecules across a cell membrane against a concentration gradient, involving an expenditure of energy, in a direction opposite to simple diffusion.

adaptation A genetically determined characteristic (behavioral, morphological, or physiological) that improves an organism's ability to survive and reproduce under prevailing environmental conditions.

adaptive radiation Evolution from a common ancestor of divergent forms adapted to distinct ways of life.

adiabatic cooling A decrease in air temperature when a rising parcel of warm air cools by expanding (which uses energy) rather than losing heat to the surrounding air; the rate of cooling is approximately 1°C/100 m for dry air and 0.6°C/100 m for moist air.

adiabatic lapse rate Rate at which a parcel of air loses temperature with elevation if no heat is gained from or lost to an external source.

adiabatic process A process in which heat is neither lost to nor gained from the outside.

aerenchyma Plant tissue with large air-filled intercellular spaces, usually found in roots and stems of aquatic and marsh plants.

aerobic Living or occurring only in the presence of free uncombined molecular oxygen either as a gas in the atmosphere or dissolved in water.

aestivation Dormancy in animals through a drought or dry season.

age distribution The proportion of individuals in various age classes for any one year. See also *age structure*.

age-specific mortality rate The proportion of deaths occurring per unit time in each age group within a population.

age-specific schedule of birth Average number of offspring produced per individual per unit time as a function of age class.

age structure The number or proportion of individuals in each age group within a population.

aggregate A group of soil particles adhering in a cluster; compare *ped*.

aggregative response Movement of predators into areas of high prey density.

aggressive mimicry Resemblance of a predator or parasite to a harmless species to deceive potential prey.

Alfisol Soil characterized by an accumulation of iron and aluminum in the B horizon.

alien species A species not native or endemic to an area.

allele One of two or more alternative forms of a gene that occupies the same relative position or locus on homologous chromosomes.

allelopathy Effect of metabolic products of plants (excluding microorganisms) on the growth and development of other nearby plants.

allogenic Term applied to successional change brought about by a change in the physical environment.

allogenic succession Ecological change or development of species structure and community composition brought about by some external force, such as fire or storms.

allopatric Having different areas of geographical distribution; possessing nonoverlapping ranges.

allopatric speciation The separation of a population into two or more evolutionary units by some geographical barrier that causes reproductive isolation.

alluvial soil Soil developing from recent alluvium (material deposited by running water), exhibiting no horizon development, and typical of floodplains.

alpha diversity The variety of organisms occurring in a given place or habitat; compare *beta diversity, gamma diversity*.

altricial Condition among birds and mammals of being hatched or born usually blind and too weak to support their own weight.

altruism A form of behavior in which an individual increases the welfare of another at the expense of its own welfare.

ambient Surrounding, external, or unconfined in condition.

amensalism Relationship between two species in which one is inhibited or harmed, while the other (the amensal) is unaffected.

ammonification Breakdown of proteins and amino acids, especially by fungi and bacteria, with ammonia as the excretory byproduct.

anaerobic Adapted to environmental conditions devoid of oxygen.

Andisol Soil derived from volcanic ejecta, not highly weathered, with a dark upper layer.

annual grassland Grassland in California dominated by exotic annual grasses that reseed every year, replacing native perennial grasses.

anticyclone An area of high atmospheric pressure characterized by subsiding air and horizontal divergence of air near the surface in its central region.

aphotic zone A deepwater area of marine ecosystems below the depth of effective light penetration.

apparent plants Large, easy to locate plants possessing quantitative defenses not easily mobilized at the point of attack, such as tannins.

area-insensitive species Species that are at home in any size habitat patch.

Aridisol Desert soil characterized by little organic matter and high base content.

arroyo A water-carved canyon in a desert.

asexual reproduction Any form of reproduction, such as budding, that does not involve the fusion of gametes.

assimilation Transformation or incorporation of a substance by organisms; absorption and conversion of energy and nutrients into constituents of an organism.

association A natural unit of vegetation characterized by a relatively uniform species composition and often dominated by a particular species.

atmospheric pressure The downward force exerted by the weight of the overlying atmosphere.

atoll A ring-shaped coral reef that encloses or almost encloses a lagoon and is surrounded by open sea.

ATP Adenosine triphosphate; major energy-transferring molecules in all biological systems.

aufwuchs Community of plants and animals attached to or moving about on submerged surfaces; compare *periphyton*.

autogenic Self-generated.

autogenic succession Succession driven by environmental changes brought about by the organisms themselves.

autotrophic community Community whose energy source is photosynthesis, thus based on primary producers.

autotrophic succession Succession in a predominantly inorganic environment with early and continued dominance of green plants (autotrophs).

autotrophs Organisms that produce organic material from inorganic chemicals and some source of energy.

available water capacity Supply of water available to plants in a well-drained soil.

B horizon Soil stratum beneath the A horizon, characterized by an accumulation of silica, clay, and iron and aluminum oxides and possessing blocky or prismatic structure.

bajadas A gentle sloping plain of unconsolidated rock debris resting against a mountain front in a semiarid environment.

basal metabolic rate The minimal amount of energy expenditure needed by an animal to maintain vital processes.

Batesian mimicry Resemblance of a palatable or harmless species, the mimic, to an unpalatable or dangerous species, the model.

bathyal Pertaining to anything, but especially organisms, in the deep sea, below the photic or lighted zone, and above 4000 m.

bathypelagic Lightless zone of the open ocean, lying above the abyssal or bottom water, usually above 4000 m.

behavioral defenses Aggressive and submissive postures or actions that threaten or deter enemies.

behavioral ecology The study of the behavior of an organism in its natural habitat.

benthic zone The area of the sea bottom.

benthos Animals and plants living on the bottom of a lake or sea, from the high water mark to the greatest depth.

beta diversity Variety of organisms occupying different habitats over a region; regional diversity; compare *alpha diversity*, *gamma diversity*.

biennial Plant that requires two years to complete a life cycle, with vegetative growth the first year and reproductive growth (flowers and seeds) the second.

biochemical oxygen demand (BOD) A measure of the oxygen needed in a specified volume of water to decompose organic materials; the greater the amount of organic matter in water, the higher the BOD.

biodiversity A measure of the different kinds of organisms within a given region.

biogeochemical cycle Movement of elements or compounds through living organisms and the nonliving environment.

biogeography The study of past and present geographical distribution of plants and animals at different taxonomic levels, the habitats at which they occur, and their ecological relationships.

biological clock The internal mechanism of an organism that controls circadian rhythms without external time cues.

biological magnification Process by which pesticides and other substances become more concentrated in each link of the food chain.

biological species A group of potentially interbreeding populations reproductively isolated from all other populations.

bioluminescence Production of light by living organisms.

biomass Weight of living material, usually expressed as dry weight per unit area.

biome Major regional ecological community of plants and animals; usually corresponds to plant ecologists' and European ecologists' classification of plant formations and life zones.

biosphere Thin layer about Earth in which all living organisms exist.

biotic Applied to the living component of an ecosystem.

biotic community Any assemblage of populations living in a prescribed area or physical habitat.

blanket mire Large area of upland dominated by sphagnum moss and dependent upon precipitation for a water supply; a moor.

bog Wetland ecosystem characterized by an accumulation of peat, acid conditions, and dominance of sphagnum moss.

border The line separating the edges of landscape elements such as forest and field.

bottleneck An evolutionary term for any stressful situation that greatly reduces a population.

bottomland Land bordering a river that floods when the river overflows its banks.

bottom-up control Influence of producers on the trophic levels above them in a food web.

boundary Zone composed of the edges of adjacent ecosystems.

boundary layer A layer of still air close to or at the surface of an object.

brood parasitism Laying eggs in the nest of another species or in the nest of another individual of the same species.

browse The part of current leaf and twig growth of shrubs, woody vines, and trees available for animal consumption.

bryophyte Member of the division in the plant kingdom of non-flowering plants comprising mosses (Musci), liverworts (Hepaticae), and hornworts (Anthocerotae).

buffer A chemical solution that resists or dampens change in pH upon addition of acids or bases.

bundle sheath Cells surrounding small vascular bundles in the the leaves of vascular plants.

C horizon Soil stratum beneath the solum (A and B horizons), little affected by biological activity or soil-forming processes.

C_3 plant Any plant that produces as its first step in photosynthesis the three-carbon compound phosphoglyceric acid.

C_4 plant Any plant that produces as its first step in photosynthesis a four-carbon compound, malic or aspartic acid.

calcicole Plant susceptible to aluminum toxicity, acidity, and other factors influenced by the absence of calcium.

calcification Process of soil formation characterized by accumulation of calcium in lower horizons.

calcifuge Plant with a low calcium requirement that can live in soils with a pH of 4.0 or less.

caliche An alkaline, often rocklike salt deposit on the surface of soil in arid regions; it forms at the level where leached Ca salts from the upper soil horizons are precipitated.

calorie Amount of heat needed to raise 1 g of water 1°C, usually from 15°C to 16°C.

CAM plant (Crassulacean Acid Metabolism) Plant (cactus or other succulent) that separates the processes of carbon dioxide uptake and fixation when growing under arid conditions; it takes up gaseous carbon dioxide at night, when stomata are open, and uses it during the day, when stomata are closed.

cannibalism Killing and consuming one's own kind; intraspecific predation.

canopy Uppermost layer of vegetation formed by trees; also the uppermost layer of vegetation in shrub communities or in any terrestrial plant community in which the upper layer forms a distinct habitat.

capillary water That portion of water in the soil held by capillary forces between soil particles.

carnivore Organism that feeds on animal tissue; taxonomically, a member of the order Carnivora (Mammalia).

carnivory The killing and eating of animals by another animal.

carrying capacity *(K)* Number of individual organisms the resources of a given area can support, usually through the most unfavorable period of the year.

catchment The area that draws surface runoff from precipitation into a stream; synonymous with watershed.

cation Part of a dissociated molecule carrying a positive electrical charge.

cation exchange capacity Ability of a soil particle to absorb positively charged ions.

chamaephyte Perennial shoot or bud from the surface of the ground to about 25 cm above the surface.

chaparral Vegetation consisting of broadleaved evergreen shrubs, found in regions of mediterranean-type climate.

character displacement The principle that two species are more different where they occur together than where they are separated geographically.

chemical defense The use by organisms of bitter, distasteful, or toxic secretions that deter potential enemies.

chemical ecology Study of the nature and use of chemical substances produced by plants and animals.

chemical weathering The action of a set of chemical processes such as oxidation, hydrolysis, and reduction operating at the atomic and molecular level that break down and re-form rocks and minerals.

chilling tolerance Ability of a plant to carry on photosynthesis within a range of +5°C to +10°C.

chromosome One of a group of threadlike structures of different lengths and sizes in the nuclei of cells of eukaryote organisms.

circadian rhythm Endogenous rhythm of physiological or behavioral activity of approximately 24 hours duration.

climate Long-term average pattern of local, regional, or global weather.

climax Stable end community of succession that is capable of self-perpetuation under prevailing environmental conditions.

climograph A diagram describing a locality based on the annual cycle of temperature and precipitation.

cline Gradual change in population characteristics over a geographical area, usually associated with changes in environmental conditions.

clone A population of genetically identical individuals resulting from asexual reproduction.

closed system A system that neither receives inputs from nor contributes output to the external environment.

coefficient of community Index of similarity between two stands or communities. The index ranges from 0 to indicate communities with no species in common to 100 to indicate communities with identical species composition.

coevolution Joint evolution of two or more noninterbreeding species that have a close ecological relationship; through reciprocal selective pressures, the evolution of one species in the relationship is partially dependent on the evolution of the other.

coexistence Two or more species living together in the same habitat, usually with some form of competitive interaction.

cohort A group of individuals of the same age.

cold resistance Ability of a plant to resist low temperature stress without injury.

collectors Feeding group of stream invertebrates that filter fine organic particles from flowing water or pick up particles from the stream bottom.

colluvium Mixed deposit of soil material and rock fragments accumulated near the base of a steep slope through soil creep, landslides, and local surface runoff.

commensalism Relationship between species that is beneficial to one, but neutral or of no benefit to the other.

community A group of interacting plants and animals inhabiting a given area.

community ecology Study of the living component of ecosystems; description and analysis of patterns and processes within the community.

compartment Major reservoir or component of an ecosystem.

compensation depth In aquatic ecosystems, the depth of the water column at which light intensity reaching plants is just sufficient for the rate of photosynthesis to balance the rate of respiration.

compensation level Light intensity at which photosynthesis and respiration balance each other, so that net production is 0; in aquatic systems, usually the depth of light penetration at which oxygen utilized in respiration equals oxygen produced by photosynthesis.

competition Any interaction that is mutually detrimental to both participants, occurring between species that share limited resources.

competitive exclusion principle Hypothesis that when two or more species coexist using the same resource, one must displace or exclude the other.

competitive release Niche expansion in response to reduced interspecific competition.

conduction Direct transfer of heat from one substance to another.

conductivity Ability to exchange heat with the surrounding environment.

conservation ecology A synthetic field that applies principles of ecology, biogeography, population genetics, economics, sociology, and other fields to the maintenance of biological diversity. Also called conservation biology.

consumer Any organism that lives on other organisms, dead or alive.

consumption efficiency Ratio of ingestion to production or energy available.

contest competition Competition in which a limited resource is shared only by dominant individuals; a relatively constant number of individuals survive, regardless of initial density.

continental shelf Gently seaward-sloping surface of a continent that extends to a depth of about 200 m.

continuum A gradient of environmental characteristics or changes in community composition.

convection Transfer of heat by the circulation of a liquid or gas.

convergent evolution Development of similar characteristics in different species living in different areas under similar environmental conditions.

coprophagy Feeding on feces.

Coriolis force Physical consequence of the law of conservation of angular momentum; as a result of Earth's rotation, a moving object veers to the right in the Northern Hemisphere and to the left in the Southern Hemisphere relative to Earth's surface.

corridor A strip of a particular type of vegetation that differs from land on both sides.

countercurrent circulation An anatomical and physiological arrangement by which heat exchange takes place between outgoing warm arterial blood and cool venous blood returning to the body core; important in maintaining temperature homeostasis in many vertebrates.

covalence Sharing of a pair of electrons between two atoms.

critical daylength The period of daylight, specific for any given species, that triggers a long-day or a short-day response in organisms.

critical thermal maximum Temperature at which an animal's capacity to move is so reduced that it cannot escape from thermal conditions that will lead to death.

crown fire Fire that sweeps through the canopy of a forest.

crown shyness Growth pattern of Asian tropical rain forest trees in which crowns are spaced about a meter apart, giving the impression of a jigsaw puzzle.

crude birthrate The number of young produced per unit of population.

crude density The number of individuals per unit area; compare *ecological density*.

cryptic coloration Coloration of organisms that makes them resemble or blend into their habitat or background.

cryptophyte Plant with overwintering buds buried in the ground on a bulb or rhizome.

cultural eutrophication Accelerated nutrient enrichment of aquatic ecosystems by a heavy influx of pollutants that causes major shifts in plant and animal life.

current Water movements that result in the horizontal transport of water masses.

cyclic replacement Succession in which the sequence of seral stages is repeated by imposition of some disturbance, so that the sere never arrives at a climax or stable sere.

day-neutral plant A plant that does not require any particular photoperiod to flower.

death rate Number of individuals in a population dying in a given time interval divided by the number alive at the midpoint of the time interval.

deciduous Of leaves, shed during a certain season (winter in temperate regions; dry season in the tropics); of trees, having deciduous parts.

decomposer Organism that obtains energy from the breakdown of dead organic matter to simpler substances; most precisely refers to bacteria and fungi.

decomposition Breakdown of complex organic substances into simpler ones.

deductive method In testing hypotheses, going from the specific to the general.

defensive mutualism Relationship in which one of the mutualists seems to protect the other from harm.

definitive host Host in which a parasite becomes an adult and reaches maturity.

degree-days A measure commonly used to integrate temperature over a single growing season to plant growth.

deme Local population or interbreeding group within a larger population.

demography The statistical study of the size and structure of populations and changes within them.

denitrification Reduction of nitrates and nitrites to nitrogen by microorganisms.

density Size of a population in relation to a definite unit of space; see *crude density, ecological density.*

density dependence Regulation of population growth by mechanisms controlled by the size of the population; effect increases as population size increases.

density independence Being unaffected by population density; regulation of growth is not tied to population density.

dependent variable Variable *y*, the second of two numbers in an ordered pair *(x, y)*; the set of all values taken on by the dependent variable is called the range of the function; compare *independent variable.*

desert grassland Grassland of hot, dry climates, with rainfall varying between 200 and 500 mm, dominated by bunchgrasses and widely interspersed with other desert vegetation.

desertification Process of desert expansion or formation as a consequence of climatic change, poor land management, or both.

deterministic model Mathematical model in which all relationships are fixed and a given input produces one exact prediction as an output.

detrital food chain Food chain in which detritivores consume detritus or litter, mostly from plants, with subsequent transfer of energy to various trophic levels; ties into the grazing food chain; compare *grazing food chain.*

detritivore Organism that feeds on dead organic matter; usually applies to detritus-feeding organisms other than bacteria and fungi.

detritus Fresh to partly decomposed plant and animal matter.

dew point Temperature at which condensation of water in the atmosphere begins.

diameter at breast height (dbh) Diameter of a tree measured at 1.4 m (4 feet, 6 inches) from ground level.

diapause A period of dormancy, usually seasonal, in the life cycle of an insect, in which growth and development cease and metabolism greatly decreases.

diffuse coevolution Coevolution involving the interactions of many organisms, in contrast to pair interactions.

diffuse competition Competition in which a species experiences interference from numerous other species that deplete the same resources.

diffusion Spontaneous movement of particles of gases or liquids from an area of high concentration to an area of low concentration.

dimorphism Existing in two structural forms, two color forms, two sexes, and the like.

dioecious Having male and female reproductive organs on separate plants; compare *monoecious.*

diploid Having chromosomes in homologous pairs, or twice the haploid number of chromosomes.

directional selection Selection favoring individuals at one extreme of the phenotype in a population.

disease Any deviation from a normal state of health.

dispersal Leaving an area of birth or activity for another area.

dispersion Distribution of organisms within a population over an area.

disruptive selection Selection in which two extreme phenotypes in the population leave more offspring than the intermediate phenotype, which has lower fitness.

distribution Arrangement of organisms within an area.

disturbance A discrete event in time that disrupts an ecosystem, community, or population, changing substrates and resource availability.

disturbance regime Pattern of disturbance that characterizes a landscape over a long period of time.

diversity Abundance of different species in a given location; species richness.

diversity index The mathematical expression of species richness of a given community or area.

division Subdivision of a domain that is characterized by definitive vegetation that falls within the same regional climate.

DNA A nucleic acid (deoxyribonucleic acid) mainly found in the chromosomes; the hereditary material of all organisms except certain viruses.

doldrums Oceanic equatorial zone that has low pressure and light variable winds; moves seasonally north and south of the equator.

domain A group of ecoregions with similar, related continental climates or oceanic water masses.

domestic grassland Grasslands that are seeded and maintained by human efforts; includes hayfields, pastures, golf courses, and lawns.

dominance In a community, control over environmental conditions influencing associated species by one or several species, plant or animal, enforced by number, density, or growth form; in a population, behavioral, hierarchical order that gives high-ranking individuals priority of access to essential resources; in genetics, ability of an allele to mask the expression of an alternative form of the same gene in a heterozygous condition.

dominant Population possessing ecological dominance in a given community and thereby governing type and abundance of other species in the community.

dominant gene An allele that is expressed in either the homozygous or heterozygous state.

dormancy State of cessation of growth and suspended biological activity, during which life is maintained.

drought avoidance Ability of a plant to escape dry periods by becoming dormant or surviving as a seed.

drought resistance Sum of drought tolerance and drought avoidance.

drought tolerance Ability of plants to maintain physiological activity in spite of the lack of water or to survive the drying of tissues.

dry deposition Pollutants—mainly sulfur dioxides and nitrogen oxides—that return to Earth in the form of particulate matter and airborne gases, thus introducing acidic material to the ground and surface waters.

dryfall Nutrients brought into an ecosystem by airborne particles and aerosols.

dynamic pool model Optimum yield model using growth, recruitment, mortality, and fishing intensity to predict yield.

dystrophic Term applied to a body of water with a high content of humic or organic matter, often with high littoral productivity and low plankton productivity.

E horizon Mineral horizon characterized by the loss of clay, iron, or aluminum and a concentration of quartz and other resistant minerals in sand and silt sizes; light in color.

early successional species Plant species characterized by high dispersal rates, ability to colonize disturbed sites, short life span, and shade intolerance.

easterlies A system of broad steady prevailing winds around Earth over the equatorial regions created by the westward deflection of air that follows the barometric pressure gradients from subtropical high to equatorial low; also called trade winds.

ecocline A geographical gradient of communities or ecosystems produced by responses of vegetation to environmental gradients of rainfall, temperature, nutrient concentrations, and other factors.

ecological density Density measured in terms of the number of individuals per area of available living space; compare *crude density.*

ecological efficiency Percentage of biomass produced by one trophic level that is incorporated into biomass of the next highest trophic level.

ecological pyramid A graphical representation of the trophic structure and function of an ecosystem.

ecological release Expansion of habitat or increase in food availability resulting from release of a species from interspecific competition.

ecology The study of relations between organisms and their natural environment, living and nonliving.

ecoregion A geographic group of landscape mosaics.

ecosystem The biotic community and its abiotic environment functioning as a system.

ecosystem ecology The study of natural systems with emphasis on energy flow and nutrient cycling.

ecotone Transition zone between two structurally different communities; see also *edge.*

ecotype Subspecies or race adapted to a particular set of environmental conditions.

ectomycorrhizae Mutualistic association between fungi and roots in which the fungi form sheaths around the outside of the roots.

ectoparasite Parasite, such as a flea, that lives in the fur, feathers, or skin of the host.

ectothermy Determination of body temperature primarily by external thermal conditions.

edaphic Relating to soil.

edge Place where two or more vegetation types meet.

edge effect Response of organisms, animals in particular, to environmental conditions created by the edge.

edge species Species that are restricted exclusively to the edge environment.

effective population size The size of an ideal population that would undergo the same amount of random genetic drift as the actual population; sometimes used to measure the amount of inbreeding in a finite, randomly mating population.

egestion Elimination of undigested food material.

elaiosome Shiny, oil-containing, ant-attracting tissue on the seed coat of many plants.

emigration Movement of part of a population permanently out of an area.

endemic Restricted to a given region.

endoparasite Parasite that lives within the body of the host.

endothermic reaction Chemical reaction that gains energy from the environment.

endothermy Regulation of body temperature by internal heat production; allows maintenance of appreciable difference between body temperature and external temperature.

energy Capacity to do work.

Entisols Embryonic mineral soils whose profile is just beginning to develop; common on recent floodplains and wind deposits, they lack distinct horizons.

entrainment Synchronization of an organism's activity cycle with environmental cycles.

entropy Transformation of matter and energy to a more random, more disorganized state.

environment Total surroundings of an organism, including other plants and animals and those of its own kind.

epidemic Rapid spread of a bacterial or viral disease in a human population; compare *epizootic.*

epifauna Benthic animals that live on or move over the surface of a substrate.

epiflora Benthic plants that live on the surface of a substrate.

epilimnion Warm, oxygen-rich upper layer of water in a lake or other body of water, usually seasonal.

epiphyte Plant that lives wholly on the surface of other plants, deriving support but not nutrients from them.

epizootic Rapid spread of a bacterial or viral disease in a dense population of animals.

equilibrium species Species whose population exists in equilibrium with resources and at a stable density.

equilibrium turnover rate Change in species composition per unit time when immigration equals extinction.

equitability Evenness of distribution of species abundance patterns; maximum equitability is the same number of individuals among all species in the community.

estivation Dormancy in animals during a period of drought or a dry season.

estuary A partially enclosed embayment where freshwater and seawater meet and mix.

euphotic zone Surface layer of water to the depth of light penetration where photosynthetic production equals respiration.

eutrophic Term applied to a body of water with high nutrient content and high productivity.

eutrophication Nutrient enrichment of a body of water; called cultural eutrophication when accelerated by introduction of massive amounts of nutrients from human activity.

eutrophy Condition of being nutrient-rich.

evaporation Loss of water vapor from soil or open water or another exposed surface.

evapotranspiration Sum of the loss of moisture by evaporation from land and water surfaces and by transpiration from plants.

evenness Degree of equitability in the distribution of individuals among a group of species; see *equitability.*

evolution Change in gene frequency through time resulting from natural selection and producing cumulative changes in characteristics of a population.

evolutionary ecology Integrated study of evolution, genetics, natural selection, and adaptations within an ecological context; evolutionary interpretation of population, community, and ecosystem ecology.

exothermic reaction Chemical reaction that releases heat to the environment.

exploitative competition Competition by a group or groups of organisms that reduces a resource to a point that adversely affects other organisms.

exponential growth (r) Instantaneous rate of population growth, expressed as proportional increase per unit of time.

extinction coefficient Point at which the intensity of light reaching a certain depth is insufficient for photosynthesis; ratio of intensity of light at a given depth to intensity at the surface.

F_1 generation The first generation of offspring from a cross between individuals homozygous for contrasting alleles; the F_1 is necessarily heterozygous.

F_2 generation Offspring produced by selfing or by allowing the F_1 generation to breed among themselves.

facilitation A situation where one species benefits from the presence or action of another.

facilitation model A model of succession in which a community prepares or "facilitates" the way for a succeeding community.

facultative Able to adjust optimally to different environmental conditions.

fecundity Potential ability of an organism to produce eggs or young; rate of production of young by a female.

fell-field Area within a tundra characterized by stony debris and sparse vegetation.

fen Slightly acidic wetland dominated by sedges, in which peat accumulates.

fermentation Breakdown of carbohydrates and other organic matter under anaerobic conditions.

field capacity Amount of water held by soil against the force of gravity.

field study A controlled experimental study carried out in a natural environment rather than in the laboratory.

filter effect Corridors that provide dispersal routes for some species but restrict the movement of others.

filtering collectors Stream insect larvae that feed on fine particulate matter by filtering it from water flowing past them.

finite multiplication rate Expressed as lambda, λ, the geometric rate of increase by discrete time intervals; given a stable age distribution, lambda can be used as a multiplier to project population size.

finite population growth rate Geometric rate of population increase by discrete time intervals.

first law of thermodynamics Energy is neither created nor destroyed; in any transfer or transformation no gain or loss of total energy occurs.

first-level carnivores Organisms that feed on first-level consumers or plant eaters.

first-level consumers Organisms that feed on plants.

first trophic level Producers; organisms that fix energy that becomes the basic source of energy for consumers.

fitness Genetic contribution by an individual's descendants to future generations.

fixation Process in soil by which certain chemical elements essential for plant growth are converted from a soluble or exchangeable form to a less soluble or nonexchangeable form.

fixed quota Harvest removal of a certain percentage of a population, based on maximum sustained yield estimates.

flashing coloration Hidden markings on animals that, when quickly exposed, startle or divert the attention of a potential predator.

floating reserve Individuals in a population of a territorial species that do not hold territories and remain unmated, but are available to refill territories vacated by death of an owner.

flux Flow of energy from a source to a sink or receiver.

foliage height diversity Measure of the degree of layering or vertical stratification of foliage in a forest.

food chain Movement of energy and nutrients from one feeding group of organisms to another in a series that begins with plants and ends with carnivores, detrital feeders, and decomposers.

food web Interlocking pattern formed by a series of interconnecting food chains.

foraging strategy Manner in which animals seek food and allocate their time and effort in obtaining it.

forb Herbaceous plant other than grass, sedge, or rush.

forest floor Term given to the ground layer of leaves, detritus.

formation Classification of vegetation based on dominant life forms.

founder effect Effect of starting a population with a small number of colonists, which contain only a small and often biased sample of genetic variation of the parent population; a markedly different new population may arise.

fragmentation Reduction of a large habitat area into small, scattered remnants; reduction of leaves and other organic matter into smaller particles.

free-running cycle Length of a circadian rhythm in the absence of an external time cue.

frequency In landscape ecology, the mean number of disturbances that occur within a time interval; in community ecology, the proportion of sample plots in which a species occurs relative to the total number of sample plots; the probability of finding the species in any one sample plot.

frost pocket Depression in the landscape into which cold air drains, lowering the temperature relative to the surrounding area; such pockets often support their own characteristic group of cold-tolerant plants.

frugivore Organism that feeds on fruit.

functional group A collection of species that exploit the same array of resources or perform similar functions within the community.

functional response Change in rate of exploitation of a prey species by a predator in relation to changing prey density.

fundamental niche Total range of environmental conditions under which a species can survive.

Gaia hypothesis The idea that the biosphere is a self-regulating entity controlling the physical and chemical environment.

gamma diversity Differences among similar habitats in widely separated regions.

gap Opening made in a forest canopy by some small disturbance such as windfall; death of an individual tree or group of trees that influences the development of vegetation beneath.

garrigue Shrub woodland characteristic of limestone areas with low rainfall and thin, poor dry soils; widespread in the Mediterranean countries of southern Europe.

gaseous cycle A biogeochemical cycle with the main reservoir or pool of nutrients in the atmosphere and ocean.

gathering collectors Stream insect larvae that pick up and feed on particles from stream bottom sediment.

Gelisol Soil that contains permafrost within 200 cm of the ground surface; it is characterized by perennial coldness rather than diagnostic horizons.

gene Unit material of inheritance; more specifically, a small unit of a DNA molecule, coded for a specific protein to produce one of the many attributes of a species.

gene flow Exchange of genetic material between populations.

gene frequency Relative abundance of different alleles carried by an individual or a population; allele frequency.

gene pool The sum of all the genes of all individuals in a population.

genet A genetic individual that arises from a single fertilized egg.

genetic drift Random fluctuation in allele frequency over time, due to chance alone without any influence by natural selection; important in small populations.

genotype Genetic constitution of an organism.

genotypic frequency The proportion of various genotypes in a population; compare *gene frequency*.

geographic isolates Groups of populations that are semi-isolated from one another by some extrinsic barrier; compare *subspecies*.

geometric rate of increase (λ) Factor by which the size of a population increases over a period of time.

gleization A process in waterlogged soils in which iron, because of an inadequate supply of oxygen, is reduced to a ferrous compound, giving dull gray or bluish mottles and color to the horizons.

gley soil Soil developed under conditions of poor drainage, resulting in reduction of iron and other elements and in gray or bluish colors and mottles.

global ecology The study of ecological systems on a global scale.

gougers Stream insect larvae that that burrow into waterlogged limbs and trunks of fallen trees.

grazers Stream invertebrates that feed on algal coating on rocks and other substrates.

grazing food chain Food chain in which primary producers (green plants) are eaten by grazing herbivores, with subsequent energy transfers to other trophic levels.

greenhouse effect Selective energy absorption by carbon dioxide in the atmosphere, which allows short wavelength energy to pass through but absorbs longer wavelengths and reflects heat back to Earth.

greenhouse gas A gas that absorbs long-wave radiation and thus contributes to the greenhouse effect when present in the atmosphere; includes water vapor, carbon dioxide, methane, nitrous oxides, and ozone.

gross primary production Energy fixed per unit area by photosynthetic activity of plants before respiration; total energy flow at the secondary level is not gross production, but rather assimilation, because consumers use material already produced with respiratory losses.

gross reproductive rate Sum of the mean number of females born to each female age group.

ground fire Fire that consumes organic matter down to the mineral substrate or bare rock.

groundwater Water that occurs below Earth's surface in pore spaces within bedrock and soil, free to move under the influence of gravity.

growth form Morphological category of plants, such as tree, shrub, or vine.

guild A group of populations that utilize a gradient of resources in a similar way.

gully erosion Form of surface erosion caused by torrents of water that bite deeply into topsoil and soft sediments.

gyre Circular motion of water in major ocean basins.

habitat Place where a plant or animal lives.

habitat selection Behavioral responses of individuals of a species involving certain environmental cues used to choose a potentially suitable environment.

hadal That part of the ocean below 6000 m.

halophyte Terrestrial plant adapted morphologically or physiologically to grow in salt-rich soil.

haploid Having a single set of unpaired chromosomes in each cell nucleus.

Hardy-Weinberg law The proposition that genotypic ratios resulting from random mating remain unchanged from one generation to another, provided natural selection, genetic drift, and mutation are absent.

harvest effort Approach to harvesting populations by manipulating or controlling hunting efforts, by means such as setting seasons and bag limits.

heat Energy in the process of being transferred from one object to another because of the temperature differences between the two.

heat dome Storage and reradiation of heat about and above urban areas, in which the temperature may be considerably higher than in the surrounding countryside.

hemicryptophyte Perennial shoots or buds close to the surface of the ground; often covered with litter.

hemiparasite Plant parasite that has chlorophyll, carries on photosynthesis, yet derives some nutrients from its host.

herb layer Lichens, moss, ferns, herbaceous plants, and small woody seedlings growing on the forest floor.

herbivore Organism that feeds on plant tissue.

herbivory Feeding on plants.

hermaphrodite Organism possessing the reproductive organs of both sexes.

heterogeneity State of being mixed in composition; can refer to genetic or environmental conditions.

heterotherm An organism that during part of its life history becomes either endothermic or ectothermic; hibernating endotherms become ectothermic, and foraging insects such as bees become endothermic during periods of activity; they are characterized by rapid, drastic, repeated changes in body temperature.

heterotrophic community Community that is dependent upon and supported by energy already fixed by the autotrophic community.

heterotrophic succession Succession that occurs on dead organic matter; detritivores feed in sequence, each group releasing nutrients used by the next group, until resources are exhausted.

heterotrophs Organisms that are unable to manufacture their own food from inorganic materials and thus rely on other organisms, living and dead, as a source of energy and nutrients.

heterozygous Containing two different alleles of a gene, one from each parent, at the corresponding loci of a pair of chromosomes.

hibernation Winter dormancy in animals, characterized by a great decrease in metabolism.

hierarchy A sequence of sets made up of smaller subsets.

Histosol Soil characterized by high organic matter content.

home range Area over which an animal ranges throughout the year.

homeostasis Maintenance of a nearly constant internal environment in the midst of a varying external environment; more generally, the tendency of a biological system to maintain itself in a state of stable equilibrium.

homeostatic plateau Limited range of maximum and minimum values of physiological tolerances within which an organism operates.

homeotherm Animal with a fairly constant body temperature; also spelled homoiotherm and homotherm.

homeothermy Regulation of body temperature by physiological means.

homologous chromosomes Corresponding chromosomes from male and female parents that pair during meiosis.

homozygous Containing two identical alleles of a gene at the corresponding loci of a pair of chromosomes.

horizon Major zone or layer of soil, with its own particular structure and characteristics.

horse latitudes Subtropical latitudes coinciding with a major anticyclonic belt, characterized by generally settled weather and a light or moderate wind.

host Organism that provides food or other benefit to another organism of a different species; usually refers to an organism exploited by a parasite.

humus Organic material derived from partial decay of plant and animal matter.

hybrid Plant or animal resulting from a cross between genetically different parents.

hydrogen bonding A type of bond occurring between an atom of oxygen or nitrogen and a hydrogen atom joined to oxygen or nitrogen on another molecule; responsible for the properties of water.

hydroperiod In wetlands, the duration, frequency, depth, and season of flooding.

hydrothermal vent Place on ocean floor where water, heated by molten rock, issues from fissures; vent water contains sulfides oxidized by chemosynthetic bacteria, providing support for carnivores and detritivores.

hygroscopic water Water held so tightly by soil particles that it is unavailable to plants in sufficient amounts for their survival.

hyperthermia Rise in body temperature to reduce thermal differences between an animal and a hot environment, thus reducing the rate of heat flow into the body.

hypertrophic Condition of lakes that have received excessive amounts of nutrients, making them highly and unnaturally eutrophic; compare *eutrophic*.

hypervolume The multidimensional space of a species niche; compare *niche*.

hypha Filament of a fungus thallus or vegetative body.

hypolimnion Cold, oxygen-poor zone of a lake, below the thermocline.

hypothesis Proposed explanation for a phenomenon; we should be able to test it, accepting or rejecting it on the basis of experimentation.

igneous rock Rock that has crystallized after Earth's crust melts.

immigration Arrival of new individuals into a habitat or population.

immobilization Conversion of an element from inorganic to organic form in microbial or plant tissue, rendering the nutrient unavailable to other organisms.

importance value Sum of relative density, relative dominance, and relative frequency of a species in a community.

inbreeding Mating among close relatives.

inbreeding depression Detrimental effects of inbreeding.

Inceptisol Mineral soil that has one or more horizons in which mineral materials have been weathered or removed and that is only beginning to develop a distinctive soil profile.

incipient lethal temperature Temperature at which a stated fraction of a population of animals (usually 50 percent) will die when brought rapidly to it from a different temperature.

inclusive fitness Sum of the fitness of an individual and the fitness of its relatives, weighted according to the degree of relationship.

independent variable Variable x, the first of two numbers of an ordered pair (x, y); the set of all values taken on by the independent variable is called the domain of the function; compare *dependent variable*.

index of abundance Estimates of animal populations derived from counts of animal sign, call, and number of animals observed along a prescribed route; useful in indicating trends of populations from year to year or habitat to habitat.

indirect mutualism Situation in which one species indirectly benefits another species by reducing the population size of its strong competitor.

individualistic concept The view, first proposed by H. A. Gleason, that vegetation is a continuous variable in a continuously changing environment; therefore no two vegetational communities are identical, and associations of species arise only from similarities in requirements.

induced edge Edge that results from some disturbance; adjoining vegetation types are successional, changing or disappearing with time, maintained only by periodic disturbances.

infauna Organisms living within a substrate.

infection Diseased condition arising when pathogenic microorganisms enter a body, become established, and multiply.

infiltration Downward movement of water into the soil.

infralittoral fringe Region below the littoral region of the sea.

inherent edge Stable, permanent edge determined by long-term natural features and conditions.

inhibition model Model of succession proposing that the dominant vegetation occupying a site prevents colonization of that site by other plants of the next successional community.

inputs Flow of energy and nutrients from the surrounding environment into an ecosystem.

instar Form of insect or other arthropod between successive molts.

integrated pest management Holistic approach to pest control that considers biological, ecological, economic, and social aspects; the object is to control pests before outbreaks can occur.

intensity A measure of the proportion of the total biomass or population of a species that a disturbance kills or eliminates.

interception The capture of rainwater by vegetation, from which the water evaporates and does not reach the ground.

interference competition Competition in which access to a resource is limited by the presence of a competitor.

interior species Organisms that require large areas of habitat, even though their home ranges may be small.

intermediate disturbance hypothesis The concept that species diversity is greatest in those habitats experiencing a moderate amount of disturbance, allowing the coexistence of early and late successional species.

intermediate host Host that harbors a developmental phase of a parasite; the infective stage or stages can develop only when the parasite is independent of its definitive host; compare *definitive host*.

internal cycling Movement or cycling of nutrients through components of ecosystems.

intersexual selection Choice of a mate, usually by the female.

interspecific Between individuals of different species.

intertropical convergence zone (ITCZ) The boundary zone separating the northeast trade winds of the Northern Hemisphere from the southeast trade winds of the Southern Hemisphere.

intraguild predation Predation among species occupying the same trophic level and using a similar food resource.

intrasexual selection Competition among members of the same sex for a mate, most common among males and characterized by fighting and display.

intraspecific Between individuals of the same species.

intrinsic rate of increase The per capita rate of growth of a population that has reached a stable age distribution and is free of competition and other growth restraints.

invasive species A nonnative species that successfully colonizes a disturbed area or empty niche, spreads, and outcompetes associated native species.

inversion In genetics, reversal of part of a chromosome so that genes lie in reverse order; in meteorology, increase rather than decrease in air temperature with height, caused by radiational cooling of Earth (radiational inversion) or by compression and heating of subsiding air masses from high-pressure areas (subsidence inversion).

ion An atom that is electrically charged as a result of a loss of one or more electrons or a gain of electrons.

island biogeography Study of distribution of organisms and community structure on islands.

isolating mechanism Any structural, behavioral, or physiological mechanism that blocks or inhibits gene exchange between two populations.

isotherm Line drawn on a map connecting points with the same temperature at a certain period of time.

iteroparity Having multiple broods over a lifetime.

K-selection Selection under carrying capacity conditions and a high level of competition.

keystone predation Predation that is central to the organization of a community; the predator enhances one or more inferior competitors by reducing the abundance of the superior competitor.

keystone species A species whose activities have a significant role in determining community structure.

kin selection Differential reproduction among groups of closely related individuals.

kinetic energy Energy associated with motion; performs work at the expense of potential energy.

Krummholz Stunted form of trees characteristic of transition zone between alpine tundra and subalpine coniferous forest.

landscape ecology Study of structure, function, and change in a heterogeneous landscape composed of interacting ecosystems.

lapse rate The rate at which the temperature decreases for each unit of increase of height in the atmosphere.

late successional species Long-lived, shade-tolerant plant species that supplant early successional species.

latent heat Amount of heat given up when a unit mass of a substance converts from a liquid to a solid state, or the amount of heat absorbed when a substance converts from the solid to liquid state.

latent heat of evaporation The amount of heat absorbed per gram of a liquid to convert it to a gas.

laterization Soil-forming process in hot, humid climates, characterized by intense oxidation; results in loss of bases and in a deeply weathered soil composed of silica, sesquioxides of iron and aluminum, clays, and residual quartz.

law of tolerance The idea that organisms live within a range between maximum and minimum amounts of substances or conditions that limit their presence or success; compare *Liebig's law of the minimum*.

leaching Dissolving and washing of nutrients out of soil, litter, and organic matter.

leaf area index The total leaf area of a plant exposed to incoming light energy relative to the ground surface area beneath the plant.

lek Communal courtship area males use to attract and mate with females.

lentic Pertaining to standing water, such as lakes and ponds.

lessivage The washing in suspension of fine clay and silt down into cracks and other voids in a soil body.

Liebig's law of the minimum The idea that the growth of an individual or a population is limited by the lowest amount needed of an essential nutrient.

life expectancy The average number of years to be lived in the future by members of a population.

life table Tabulation of mortality and survivorship of a population; static, time-specific, or vertical life tables are based on a cross section of a population at a given time; dynamic, cohort, or horizontal life tables are based on a cohort followed throughout life.

life zone Major area of plant and animal life, equivalent to a biome; transcontinental region or belt characterized by particular plants and animals and distinguished by temperature differences; applies best to mountainous regions where temperature changes accompany changes in altitude.

light compensation point Depth of water or level of light at which photosynthesis and respiration balance each other.

light saturation point Amount of light at which plants achieve the maximum rate of photosynthesis.

limit cycle Stable oscillation in the population levels of a species, usually reflecting predator-prey interactions.

limiting resource Resource or environmental condition that limits the abundance and distribution of an organism.

limnetic Pertaining to or living in the open water of a pond or lake.

limnetic zone Shallow-water zone of a lake or sea, in which light penetrates to the bottom.

Lithosol Soil showing little or no evidence of soil development and consisting mainly of partly weathered rock fragments or nearly barren rock.

littoral zone Shallow water of a lake, in which light penetrates to the bottom, permitting submerged, floating, and emergent vegetative growth; also shore zone of tidal water between high-water and low-water marks.

local diversity Number of species in a small area of homogeneous habitat. Also called alpha diversity.

locus Site on a chromosome occupied by a specific gene.

loess Soil developed from wind-deposited material.

logistic curve S-shaped curve of population growth that slows at first, steepens, and then flattens out at asymptote, determined by carrying capacity.

logistic equation Mathematical expression for the population growth curve in which rate of increase decreases linearly as population size increases.

long-day organism Plant or animal that requires long days—days with more than a certain minimum of daylight—to flower or come into reproductive condition.

lotic Pertaining to flowing water.

macromutation Mutation at the level of the chromosome.

macronutrients Essential nutrients plants and animals need in large amounts.

macroparasite Any of the parasitic worms, lice, fungi, and the like that have comparatively long generation time, spread by direct or indirect transmission, and may involve intermediate hosts or vectors.

mallee Sclerophyllous shrub community in Australia; most of the species are *Eucalyptus*.

mangal A mangrove swamp.

mangrove forest Tropical inshore communities dominated by several species of mangrove trees and shrubs capable of growth and reproduction in areas inundated daily by seawater.

maquis Sclerophyllous shrub vegetation in the Mediterranean region.

marine inversion Cool, moist air from the ocean moves inward over land beneath a layer of warmer, drier air.

marsh Wetland dominated by grassy vegetation such as cattails and sedges.

material cycling See nutrient cycling.

mating system Pattern of mating between individuals in a population.

matrix The background land-use type in a mosaic, characterized by extensive cover and high connectivity.

mattoral Sclerophyllous shrub vegetation in regions of Chile with mediterranean climate.

maximum sustainable yield The maximum rate at which individuals can be harvested from a population without reducing its size; recruitment balances harvesting.

mechanical weathering Breakdown of rocks and minerals in place by disintegration processes such as freezing, thawing, and pressure that do not involve chemical reactions.

mediterranean-type climate Semiarid climate characterized by a hot, dry summer and a wet, mild winter.

meiofauna Benthic organisms within the size range from 1 to 0.1 mm; interstitial fauna.

meiosis Two successive divisions by a gametic cell, with only one duplication of chromosomes, so the number of chromosomes in daughter cells is one-half the diploid number.

melanization A change in color value in a soil caused by the addition of organic matter.

melatonin Special hormone in animals that serves to measure time; associated with the biological clock.

meristem Region in a plant containing actively or potentially dividing cells.

mesic Moderately moist.

mesopelagic Uppermost lightless pelagic zone.

mesophyll Specialized tissue located between the epidermal layers of a leaf; *palisade mesophyll* consists of cylindrical cells at right angles to upper epidermis and contains many chloroplasts; *spongy mesophyll* lies next to the lower epidermis and has interconnecting, irregularly shaped cells with large intercellular spaces.

metabolism The chemical reactions in cells responsible for breaking down molecules to provide energy (catabolism) and building more complex molecules from simpler molecules (anabolism).

metalimnion Transition zone in a lake between hypolimnion and epilimnion; region of rapid temperature decline.

metamorphic rock An aggregate of minerals formed when heat and pressure recrystallize rocks.

metamorphosis Abrupt transition between life stages.

metapopulation A population broken into sets of subpopulations held together by dispersal or movements of individuals among them.

metatrophic Having a moderate amount of nutrients; stage in a nutrient-poor lake becoming eutrophic.

micella (micelle) Soil particle of clay and humus, carrying negative electrical charge at the surface.

microbial decomposition Decomposition processes performed by bacteria and fungi; involved in immobilization and mineralization of nutrients.

microbial loop Feeding loop in which bacteria take up dissolved organic matter produced by plankton and nanoplankton consume the bacteria; adds several trophic levels to the plankton food chain.

microbivore Organism that feed on microbes, especially in the soil and litter.

microclimate Climate on a very local scale, which differs from the general climate of the area; influences the presence and distribution of organisms.

microflora Bacteria and certain fungi inhabiting the soil.

microhabitat That part of the general habitat utilized by an organism.

micromutation A mutation at the level of the gene; point mutation.

micronutrient Essential nutrient needed in very small quantities by plants and animals.

microparasite Any of the viruses, bacteria, and protozoans, characterized by small size, short generation time, and rapid multiplication.

migration Intentional, directional, usually seasonal movement of animals between two regions or habitats; involves departure and return of the same individual; a round-trip movement.

mimicry Resemblance of one organism to another or to an object in the environment, evolved to deceive predators.

mineralization Microbial breakdown of humus and other organic matter in soil to inorganic substances.

minimum base temperature The temperature at which net photosynthesis is at or approaching zero; used to determine an index of degree days.

minimum viable population (MVP) Size of a population which, with a given probability, will ensure the existence of the population for a stated period of time.

mire Wetland characterized by an accumulation of peat.

mitosis Cell division involving chromosome duplication, resulting in two daughter cells with the full complement of chromosomes, genetically the same as the parent cell.

mixed-grass prairie Grassland in mid North America, characterized by great variation in precipitation and a mixture of largely cool-season shortgrass and tallgrass species.

model In theoretical and systems ecology, an abstraction or simplification of a natural phenomenon, developed to predict a new phenomenon or to provide insight into existing ones; in mimetic association, the organism mimicked by a different organism.

moder Humus in which plant fragments and mineral particles form loose netlike structures held together by a chain of small arthropod droppings.

modular organism Organism that grows by repeated iteration of parts, such as branches or shoots of a plant; some parts may separate and become physically and physiologically independent.

Mollisol Soil formed by calcification, characterized by accumulation of calcium carbonate in lower horizons and high organic content in upper horizons.

monoecious Having male and female reproductive organs separated in different floral structures on the same plant; compare *hermaphrodite, dioecious*.

monogamy In animals, mating and maintenance of a pair bond with only one member of the opposite sex at a time.

montane Related to mountains.

moor A blanket bog or peatland.

mor Humus in which unincorporated organic matter usually is matted or compacted or both and distinct from mineral soil; low in bases and acid in reaction.

morphological species Species described as monotypic, possessing little variation in color pattern, structure, proportion, and other features; the "field guide" species.

morphology Study of the form of organisms.

mortality rate The probability of dying; the ratio of number dying in a given time interval to the number alive at the beginning of the time interval.

mosaic A pattern of patches, corridors, and matrices in the landscape.

mull Humus that contains appreciable amounts of mineral bases and forms a rich layer of forested soil, consisting of mixed organic and mineral matter; blends into the upper mineral layer without abrupt changes in soil characteristics.

Mullerian mimicry Resemblance of two or more conspicuously marked distasteful species, which increases predator avoidance.

mutation Transmissible changes in the structure of a gene or chromosome.

mutualism Relationship between two species in which both benefit.

mycelium Mass of hyphae that make up the vegetative portion of a fungus.

mycorrhizae Association of fungus with roots of higher plants, which improves the plants' uptake of nutrients from the soil.

myrmecochory Dispersal by ants.

myrmecochores Plants that possess ant-attracting substances on their seed coats.

nanoplankton Plankton with a size range from 2 to 20 μm.

natality Production of new individuals in a population.

natural selection Differential reproduction and survival of individuals that results in elimination of maladaptive traits from a population.

neap tide Tide of small range that occurs at the first and last quarters of the moon when Earth, Moon, and Sun are at right angles.

negative feedback Homeostatic control in which an increase in some substance or activity ultimately inhibits or reverses the direction of the processes leading to the increase.

nekton Aquatic animals that are able to move at will through the water.

neritic Marine environment embracing the regions where land masses extend outward as a continental shelf.

net ecosystem production Difference between net primary production and carbon lost through consumer and decomposer respiration.

net mineralization rate Difference between the rates of mineralization and immobilization.

net photosynthesis The difference between the rate of carbon uptake in photosynthesis and carbon loss in respiration.

net primary production Energy accumulated in plant biomass.

net production Accumulation of total biomass over a given period of time after respiration is deducted from gross production in plants and from assimilated energy in consumer organisms.

net reproductive rate Average number of female offspring produced by an average female during her lifetime.

neutrophilic Preferring a habitat that is neither acid nor alkaline.

niche Functional role of a species in the community, including activities and relationships.

niche breadth Range of a single niche dimension occupied by a population.

niche compression Restriction of the use of a resource such as food or space because of intense competition.

niche overlap Sharing of niche space by two or more species.

niche preemption Procurement by a species of a portion of available resources, leaving less for the next.

nighttime surface inversion A temperature surface inversion that develops when nighttime radiational cooling of the ground and air above results in a layer of cool air overlaid by warm air.

nitrification Breakdown of nitrogen-containing organic compounds into nitrates and nitrites.

nitrogen fixation Conversion of atmospheric nitrogen to forms usable by organisms.

nonshivering thermogenesis Production of metabolic heat by burning highly vascular brown fatty tissue capable of a high rate of oxygen consumption.

northeast trade winds See trade winds.

nucleotide A compound formed by the condensation of a nitrogenous base with a sugar and phosphoric acid; structural unit of DNA.

null hypothesis A statement of no difference between sets of values formulated for statistical testing.

numerical response Change in size of a population of predators in response to change in density of its prey.

nutrient Substance an organism requires for normal growth and activity.

nutrient cycle Pathway of an element or nutrient through the ecosystem, from assimilation by organisms to release by decomposition.

nutrient spiraling In flowing-water ecosystems, the combined processes of nutrient cycling and downstream transport.

obligate Having no alternative in response to a particular condition or in way of life.

obligatory relationship A symbiotic relationship in which one symbiont cannot survive and reproduce without the other.

oceanic Referring to the regions of the sea with depths greater than 200 m that lie beyond the continental shelf.

old-growth forest Forest that has not been disturbed by humans for hundreds of years.

oligotrophic Term applied to a body of water low in nutrients and in productivity.

oligotrophy Nutrient-poor condition.

omnivore An animal (heterotroph) that feeds on both plant and animal matter.

open system System with exchanges of materials and energy to the surrounding environment.

operative temperature range Range of body temperatuires at which poikilotherms carry out daily activity.

opportunistic species Organisms able to exploit temporary habitats or conditions.

optimal foraging Tendency of animals to harvest food efficiently, selecting food sizes or food patches that supply maximum food intake for energy expended.

optimum yield Amount of material that can be removed from a population to produce maximum biomass on a sustained yield basis.

organismic concept of the community Idea that species, especially plant species, are integrated into an internally interdependent unit; upon maturity and death of the community, another identical plant community will replace it.

oscillation Regular fluctuation in a fixed cycle above or below some set point.

osmosis Movement of water molecules across a differentially permeable membrane in response to a concentration or pressure gradient.

osmotic potential The attraction of water across a membrane; the more concentrated a solution, the lower is its osmotic potential.

osmotic pressure Pressure needed to prevent passage of water or another solvent through a semipermeable membrane separating a solvent from a solution.

outbreeding Production of offspring by the fusion of distantly related gametes.

outbreeding depression Hybridization between two populations, each adapted to different environments, that destroys coadapted gene complexes, making the offspring maladapted to local conditions.

outputs Export of nutrients and energy from an ecosystem to the surrounding environment.

overdispersion Situation in which the distribution of organisms is random but clumped, with some areas empty and some heavily overpopulated; contagious distribution.

overturn Vertical mixing of layers in a body of water, brought about by seasonal changes in temperature.

Oxisol Soil developed under humid semitropical and tropical conditions, characterized by silicates and hydrous oxides, clays, residual quartz, deficiency in bases, and low plant nutrients; formed by laterization.

paleoecology Study of ecology of past communities by means of the fossil record.

pampas Temperate South American grassland, dominated by bunchgrasses; much of the moister pampas are under cultivation.

parasitism Relationship between two species in which one benefits while the other is harmed (although not usually killed directly).

parasitoid Insect larva that kills its host by consuming the host's soft tissues before pupation or metamorphosis into an adult.

parthenogenesis Development of an individual from an egg that did not undergo fertilization.

patch An area of habitat that differs from its surroundings with sufficient resources to allow a population to persist.

peat Unconsolidated material consisting of undecomposed and only slightly decomposed organic matter under conditions of excessive moisture.

peatland Any ecosystem dominated by peat; compare *bog*, *mire*, and *fen*.

ped Soil particles held together in a cluster of various sizes.

pedon a three-dimensionsal unit of soil to the depth of parent material; its lateral cross-section is large enough to allow the study of all horizon variability.

pelagic Referring to the open sea.

PEP The enzyme phosphoenolpyruvate carboxylase that catalyzes the fixation of CO_2 into four-carbon acids, malate and aspartate.

percent base saturation The extent to which the exchange sites of soil particles are occupied by exchangeable base cations or by cations other than hydrogen and aluminum, expressed as percentage of total cation exchange capacity; compare *cation exchange capacity*.

percent similarity An index of proportional similarity that considers the number of species in each community, the species common to both communities, and the abundance of species.

percolation The movement of water downward and outward through subsurface soil, often continuing down to groundwater.

perennating tissue Underground organs or buds that store food for new shoots of the next growing season.

periphyton In freshwater ecosystems, organisms that are attached to submerged plant stems and leaves; see *aufwuchs*.

permafrost Permanently frozen soil in arctic regions.

permanent wilting point Point at which water potential in the soil and conductivity assume such low values that the plant is unable to extract sufficient water to survive and wilts permanently.

pest An animal that humans consider undesirable; what constitutes a pest varies with time, place, circumstance, and individual attitudes.

phanerophyte Tree, shrub, or vine that bears perennating buds on aerial shoots.

phenology Study of seasonal changes in plant and animal life and the relationship of these changes to weather and climate.

phenotype Physical expression of a characteristic of an organism, determined by both genetic constitution and environment.

phenotypic plasticity Ability to change form under different environmental conditions.

pheromone Chemical substance released by an animal that influences behavior of others of the same species.

photic zone Lighted water column of a lake or ocean, inhabited by plankton.

photoinhibition The slowing or stopping of a plant process by light.

photoperiodism Response of plants and animals to changes in relative duration of light and dark.

photosynthate Energy-rich organic molecules produced during photosynthesis.

photosynthesis Use of light energy by plants to convert carbon dioxide and water into simple sugars.

photosynthetically active radiation (PAR) That part of the light spectrum between wavelengths of 400 and 700 nm that is used by plants in photosynthesis.

phreatophyte Type of plant that habitually obtains its water supply from groundwater.

physiognomy Outward appearance of the landscape.

physiological ecology Study of the physiological functioning of organisms in relation to their environment.

phytochrome A protein pigment in plants involved in photoperiodic responses and other photoreactions.

phytoplankton Small, floating plant life in aquatic ecosystems; planktonic plants.

pioneer species Plants that invade disturbed sites or appear in early stages of succession.

plankton Small, floating or weakly swimming plants and animals in freshwater and marine ecosystems.

playas Low, generally flat basins in deserts that receive waters that rush down from higher elevations and water-cut canyons.

Pleistocene Geological epoch extending from about 2 million to 10,000 years ago, characterized by recurring glaciers; the Ice Age.

pneumatophore An erect respiratory root that protrudes above waterlogged soils; typical of baldcypress and mangroves.

podzolization Soil-forming process in which acid leaches the A horizon and iron, aluminum, silica, and clays accumulate in lower horizons.

poikilotherm An organism whose body temperature varies according to the temperature of its surroundings.

poikilothermy Variation of body temperature with external conditions.

polar easterlies Easterly wind located at high latitudes poleward of the subpolar low.

polyandry Mating of one female with several males.

polygamy Acquisition by an individual of two or more mates, none of which is mated to other individuals.

polygyny Mating of one male with several females.

polymorphism Occurrence of more than one distinct form of individuals in a population.

polyploid Having three or more times the haploid number of chromosomes.

polyploidy The condition of a cell or an organism that has more than its normal set of chromosomes.

population A group of individuals of the same species living in a given area at a given time.

population bottleneck See *bottleneck*.

population cycles Recurrent oscillations between high and low points in a population in a manner more regular than we would expect to occur by chance. The two most common intervals between peaks are three to four in arctic voles and nine to ten in snowshoe hares.

population density The number of individuals in a population per unit area.

population dynamics Study of the factors that influence the number and density of populations in time and space.

population ecology Study of how populations grow, fluctuate, spread, and interact intraspecifically and interspecifically.

population genetics The study of changes in gene frequency and genotypes in populations.

population projection table Chart of growth of a population developed by calculating the births and mortality of each age group over time.

population regulation Mechanisms or factors within a population that cause it to decrease when density is high and increase when density is low.

positive feedback Control in a system that reinforces a process in the same direction.

potential energy Energy available to do work.

potential evapotranspiration Amount of water that would be transpired under constantly optimal conditions of soil moisture and plant cover.

practical salinity unit The total amount of dissolved material in seawater as expressed as parts per thousand (0/00).

precipitation All the forms of water that fall to Earth; includes rain, snow, hail, sleet, fog, mist, drizzle, and the measured amounts of each.

precocial In birds, hatched with down, open eyes, and ability to move about; in mammals, born with open eyes and ability to follow the mother, as fawns and calves can.

predation Relationship in which one living organism serves as a food source for another.

preferred temperature Range of temperatures within which poikilotherms function most efficiently.

primary producer Green plant or chemosynthetic bacterium that converts light or chemical energy into organismal tissue.

primary production Production by green plants over time.

primary productivity Rate at which plants produce biomass per unit area per unit time.

primary succession Vegetational development starting on a new site never before colonized by life.

production Amount of energy formed by an individual, population, or community per unit time.

productivity Rate of energy fixation or storage per unit time; not to be confused with production.

profundal zone Deep zone in aquatic ecosystems, below the limnetic zone.

promiscuity Member of one sex mates with more than one member of the opposite sex, and the relationship terminates after mating.

protective armor Hard outer covering of an animal body such as shells of turtles and spines of porcupines that deter or make the owner somewhat invulnerable to most enemies.

province Within a division, a subdivision that corresponds to broad vegetation regions that correspond to a climatic subzone.

pycnocline Area in the water column where the highest rate of change in density occurs for a given change in depth.

pyramid of biomass Diagrammatic representation of biomass at different trophic levels in an ecosystem.

pyramid of energy Diagrammatic representation of the flow of energy through different trophic levels.

pyramid of numbers Diagrammatic representation of the number of individual organisms present at each trophic level in an ecosystem; the least useful pyramid.

quaking bog Bog characterized by a floating mat of peat and vegetation over water.

qualitative inhibitors In plants, chemical defense by toxic substances that interfere with a consumer's metabolism; quickly synthesized.

quantitative inhibitors In plants, chemical defense by substances that reduce digestibility or potential energy from food.

r-selection Selection under low population densities; favors high reproductive rates under conditions of low competition.

rain forest Permanently wet forest of the tropics; also the wet coniferous forest of the Pacific northwest of the United States.

rain shadow A dry region on the leeward side of a mountain range resulting from a reduction in rainfall.

raised bog A bog in which the accumulation of peat has raised the surface above both the surrounding landscape and the water table; it develops its own perched water table.

ramet An individual member of a plant clone.

random distribution Distribution lacking pattern or order; placement of each individual is independent of all other individuals.

rate of increase Factor by which a population changes over a given period of time; compare *exponential growth, geometric rate of increase, intrinsic rate of increase.*

realized niche Portion of fundamental niche space occupied by a population facing competition from populations of other species; environmental conditions under which a population survives and reproduces in nature.

recessive gene Applies to an allele whose phenotypic effect is expressed in the homozygous state and masked in the presence of an allele in organisms heterozygous for that gene.

recombination Exchange of genetic material by independent assortment of chromosomes and their genes during gamete production, allowing a random mix of different sets of genes at fertilization.

recruitment Addition of new individuals to a population by reproduction.

redundancy model Relates to effect of loss of species on ecosystem stability. The loss of some species may have little effect because other species can expand their role and take up functions vacated by the lost species.

regional diversity Total number of species observed in all habitats within a geographic area. Also called gamma diversity.

regolith Mantle of unconsolidated material below the soil, from which soil develops.

regular distribution A pattern in which individuals are more widely separated from each other than would be expected by chance; underdispersion.

relative abundance Proportional representation of a species in a community or sample of a community.

relative humidity Water vapor content of air at a given temperature, expressed as a percentage of the water vapor needed for saturation at that temperature.

reproductive allocation Proportion of its resources that an organism expends on reproduction over a given period of time.

reproductive cost Decrease in survivorship or rate of growth when an individual increases its current allocation to reproduction; reflected in decreased potential for future reproduction.

reproductive effort Proportion of its resources an organism expends on reproduction.

reproductive isolation Separation of one population from another by inability to produce viable offspring when the two populations mate.

reproductive value Potential reproductive output of an individual at a particular age relative to that of a newborn individual at the same time.

reservoir Compartment of an ecosystem.

resource Environmental component used by a living organism.

resource allocation Apportioning the supply of a resource to specific uses.

respiration Metabolic assimilation of oxygen, accompanied by production of carbon dioxide and water, release of energy, and breakdown of organic compounds.

restoration ecology The application of principles of ecosystem development and function to the restoration and management of disturbed lands.

rete A large network or discrete vascular bundle of intermingling small blood vessels carrying arterial and venous blood that acts as a heat exchanger in mammals and certain fish and sharks.

rhizobia Bacteria capable of living mutualistically with higher plants.

rhizome A horizontal underground stem that branches and gives rise to vegetative structures.

richness A component of species diversity; the number of species present in an area.

riffle Stretch of shallow, fast, rough water flowing between pools in a stream.

rill erosion Surface erosion that forms numerous small channels several centimeters deep; can develop into gully erosion.

riparian woodland Woodland along the bank of a river or stream; riverbank forests are often called gallery forests.

rivet model Idea that the loss of a species in an ecosystem is analogous to the loss of a rivet from an airplane. When the loss reaches a certain threshold, there are major catastrophic effects.

root-to-shoot ratio Ratio of the weight of roots to the weight of shoots of a plant.

rotation time Interval between the recurrence of a disturbance event; or interval between harvests of a crop, such as trees.

rubisco Enzyme in photosynthesis that catalyzes the initial transformation of CO_2 into sugar.

ruminant Ungulate with a three-chamber or four-chamber stomach; in the large first chamber or rumen bacteria ferment plant matter.

salinity A measure of the total quantity of dissolved substances in water in parts per thousand (0/00) by weight.

salinization The process of accumulation of soluble salts in soil, usually by upward capillary movement from a salty groundwater source. Also called salination.

salt marsh Communities of emergent vegetation rooted in a soil alternately flooded and drained by tidal action.

saprophage Organism that feeds on dead plant and animal matter; mainly bacteria, fungi, and invertebrates such as insect larvae.

saprophyte Plant that draws its nourishment from dead plant and animal matter, mostly the former.

saturated Refers to air that contains the maximum amount of water vapor it can hold at a given temperature and pressure.

saturation vapor pressure Maximum amount of water vapor a volume of air can hold at a given temperature.

savanna Tropical grassland, usually with scattered trees or shrubs.

scale Level of resolution within the dimensions of time and space; spatial proportion as ratio of length on a map to actual length.

scavenger Animal that feeds on dead animals or on animal products, such as dung.

sclerophyll Woody plant with hard, leathery, evergreen leaves that prevent moisture loss.

scramble competition Intraspecific competition in which limited resources are shared to the point that no individual survives.

scrapers Aquatic invertebrates that feed on algal coating on stones and rubble in streams; also called grazers.

search image Mental image formed in predators, enabling them to find more quickly and to concentrate on a common type of prey.

seasonality Recurrence of biological events with the seasons.

second law of thermodynamics In any energy transfer or transformation, part of the energy assumes a form that cannot be passed on any further.

second-level carnivores Organisms that feed on first-level carnivores or second-level consumers.

second-level consumers Organisms that feed on first-level consumers or herbivores; carnivores.

secondary producers Organisms that derive energy from consuming plant or animal tissue and breaking down assimilated carbon compounds.

secondary production Production by consumer organisms over time.

secondary substances Organic compounds that plants produce for chemical defense.

secondary succession Development of vegetation after a disturbance.

sedimentary cycle Weathering of rock and leaching of its minerals, transport, deposition, and burial.

sedimentary rock Rock formed by the deposition and compression of mineral and rock particles.

selection Differential survival or reproduction of individuals in a population because of phenotypic differences among them.

selective breeding Selection by humans of individuals of animals or plants with a desired trait and mating them with other individuals exhibiting the same traits, resulting in populations of organisms with specific characteristics; analogous with natural selection.

selective pressure Any force acting on individuals in a population that determines which individuals leave more descendants than others; gives direction to the evolutionary process.

self-thinning Progressive decline in density of plants associated with the increasing size of individuals.

semelparity Having only a single reproductive effort in a lifetime over one short period of time.

semiarid Fairly dry in climate, with precipitation between 25 and 60 cm a year and with an evapotranspiration rate high enough that the potential loss of water to the environment exceeds inputs.

semispecies A group of organisms taxonomically intermediate between a race and a species with incomplete isolating mechanisms.

senescence Process of aging.

sequential hermaphrodite An individual organism that changes sex from female to male or male to female at some time in its life cycle.

seral Following a series of stages.

sere The series of successional stages on a given site that lead to a terminal community.

serotiny Release by heat of seeds tightly held in the cones of some species of coniferous trees.

serpentine soil Soil derived from ultrabasic rocks that are high in iron, magnesium, nickel, chromium, and cobalt and low in calcium, potassium, sodium, and aluminum; supports distinctive communities.

sessile Not free to move about; permanently attached to a substrate.

sex ratio The relative number of males to females in a population.

sex reversal A change in functioning so that a member of one sex behaves as the other.

sexual dimorphism The occurrence of morphological differences other than primary sexual characteristics that distinguish males from females in a species.

sexual selection Selection by one sex for an individual of the other sex based on some specific characteristic or characteristics; usually takes place through courtship behavior.

shade-intolerant Growing and reproducing best under high light conditions; growing poorly and failing to reproduce under low light conditions.

shade-tolerant Able to grow and reproduce under low light conditions.

sheet erosion Transport of soil material from slopes by a thin, mobile sheet of water.

shifting mosaic Constantly changing pattern of patches as each passes through successive stages of development.

short-day organisms Plants and animals that come into reproductive condition under conditions of short days—days with less than a certain maximum length.

shortgrass plains Westernmost grasslands of the Great Plains, characterized by infrequent rainfall, low humidity, and high winds; dominated by shallow-rooted, sod-forming grasses.

shredders Aquatic invertebrates that that feed on coarse particulate matter in streams.

siblicide Killing of an offspring by another offspring of the same parents.

sibling species Species with similar appearance but unable to interbreed.

sigmoid curve S-shaped curve of logistic growth.

simultaneous hermaphrodite An individual organism that possesses both male and female sex organs at the same time in its life cycle.

sink An unfilled, submarginal, or marginal habitat where the population can persist only by immigration from other habitats, because it experiences low reproduction or high mortality.

sink habitat Area of habitat that receives immigrants from a source habitat, but in which the subpopulation would continually decrease in size because of mortality and poor reproductive success without continual immigration from excess individuals in a source habitat.

sink patch Area where population of a species can be maintained only by immigration. See *sink habitat*.

site Combination of biotic, climatic, and soil conditions that determine an area's capacity to produce vegetation.

snag Dead or partially dead tree at least 10.2 cm dbh and 1.8 m tall; important habitat for cavity-nesting birds and mammals.

social dominance Physical dominance of one individual over another, usually maintained by some manifestation of aggressive behavior.

social parasite Animal that uses other individuals or species to rear its young, such as the cowbird.

soil association A group of defined and named soil taxonomic units occurring together in an individual and characteristic pattern over a geographic region.

soil horizon Developmental layer in the soil with characteristic thickness, color, texture, structure, acidity, nutrient concentration, and the like.

soil profile Distinctive layering of horizons in the soil.

soil series Basic unit of soil classification, consisting of soils that are alike in all major profile characteristics except texture of the A horizon; soil series are usually named for the locality where the typical soil was first recorded.

soil structure Arrangement of soil particles and aggregates.

soil texture Relative proportions of the three particle sizes (sand, silt, and clay) in the soil.

soil type Lowest unit in the system of soil classification, consisting of soils that are alike in all characteristics, including texture of the A horizon.

solar constant Rate at which solar energy is received on a surface just outside of Earth's atmosphere; current value is 0.140 watt/cm^2.

solifluction The downhill movement of soil that has been saturated with water.

source habitat Area of habitat in which a subpopulation of a species produces more individuals than needed for self-maintenance, thus contributing to emigration.

source patch Area where a population of a species reproductively produces more individuals than needed for replacement; these individuals emigrate to other areas. See *source habitat.*

speciation Separation of a population into two or more reproductively isolated populations.

species diversity Measurement that relates density of organisms of each type present in a habitat to the number of species in a habitat.

species evenness A component of species diversity index; it is a measure of the distribution of individuals among total species occupying a given area.

species richness Number of species in a given area.

specific heat Amount of energy that must be added or removed to raise or lower the temperature of a substance by a specific amount.

Sphagnum A genus of mosses that are most abundant in wet, acidic habitats; the dead cells rapidly fill with water, allowing the plant to hold many times its own weight in water.

spiraling Mechanism of retention of nutrients in flowing-water ecosystems, involving the interdependent processes of nutrient recycling and downstream transport.

Spodosol Soil characterized by the presence of a horizon in which organic matter and amorphous oxides of aluminum and iron have precipitated; includes podzolic soils.

spring tide A tide of greater than mean range that occurs every two weeks, when the Moon is full or new; maximum spring tides occur when Sun and Moon are in the same plane as Earth; compare *neap tide.*

stabilizing selection Selection favoring the middle in the distribution of phenotypes.

stable age distribution Constant proportion of individuals of various age classes in a population though population changes.

stable equilibrium Ability of a system to return to a particular point if displaced by an outside force.

stable limit cycle A regular fluctuation in the abundance of predator and prey populations, when stabilizing and destabilizing interactions balance.

stand Unit of vegetation that is essentially homogeneous in all layers and differs from adjacent types qualitatively and quantitatively.

standard deviation Statistical measure defining the dispersion of values about the mean in a normal distribution.

standing crop biomass Total amount of biomass per unit area at a given time.

static life table See *life table.*

stationary age distribution Special form of stable age distribution, in which the birthrate equals the death rate and age distribution remains fixed.

steppe Name given to Eurasian grasslands that extend from eastern Europe to western Siberia and China.

stoichiometry Branch of chemistry dealing with the quantitative relationships of elements in combination.

stomata Pores in the leaf or stem of a plant that allow gaseous exchange between the internal tissues and the atmosphere.

stratification Division of an aquatic or terrestrial community into distinguishable layers on the basis of temperature, moisture, light, vegetative structure, and other such factors, creating zones for different plant and animal types.

sublittoral zone Lower division of the sea, from about 40 m to below 200 m.

subsidence inversion Atmospheric inversion produced by sinking air.

subspecies Geographical unit of a species population, distinguishable by morphological, behavioral, or physiological characteristics.

succession Replacement of one community by another; often progresses to a stable terminal community called the climax.

successional sequence Pattern of colonization and extinction of plants on a given area over time; compare *sere.*

supercooling In ectotherms, lowering body temperature below freezing without freezing body tissue, by means of solutes, particularly glycerol.

supralittoral fringe The highest zone on the intertidal shore, bounded below by the upper limit of barnacles and above by the upper limit of *Littorina* snails.

surface fire Fire that feeds on the litter layer of forests and grasslands.

surface runoff The excess water flowing across the surface of the ground when the soil becomes saturated during heavy rains.

surface tension Elastic film across the surface of a liquid, caused by the attractive forces between molecules at the surface of the liquid.

survivorship The probability that a representative newborn individual in a cohort will survive to various ages.

survivorship curve A graph describing the survival of a cohort of individuals in a population from birth to the maximum age reached by any one member of the cohort.

sustained yield Yield per unit time equal to production per unit time in an exploited population.

swamp Wooded wetland in which water is near or above ground level.

switching Changing the diet from a less abundant to a more abundant prey species; see *threshold of security.*

symbiosis Situation in which two dissimilar organisms live together in close association.

sympatric Living in the same area; usually refers to overlapping populations.

sympatric speciation Production of a new species within a population or the dispersal range of a population.

system Set or collection of interdependent parts or subsystems enclosed within a defined boundary; the outside environment may provide inputs and receive outputs.

systems ecology Application of general systems theory and methods to ecology, with emphasis on sets of compartments linked by fluxes of energy and nutrients.

taiga The northern circumpolar boreal forest.

tallgrass prairie A narrow belt of tall grasses dominated by big bluestem that ran north and south adjacent to the deciduous forest of eastern North America; presence maintained by fire; largely destroyed by cultivation.

temperate rain forest Forest in regions characterized by mild climate and heavy rainfall that produces lush vegetative growth; one example is the coniferous forest of the Pacific Northwest of North America.

temperature A measure of the average speed or kinetic energy of atoms and molecules in a substance.

territory Area defended by an animal; varies among animal species according to social behavior, social organization, and resource requirements.

thermal conductance Rate at which heat flows through a substance.

thermal neutral zone Among homeotherms, the range of temperatures at which metabolic rate does not vary with temperature.

thermal radiation Heat transfer by long-wave or infrared radiation.

thermal tolerance Range of temperatures in which an aquatic poikilotherm is most at home.

thermocline Layer in a thermally stratified body of water in which temperature changes rapidly relative to the remainder of the body.

thermogenesis Increase in production of metabolic heat to counteract the loss of heat to a colder environment.

therophyte Plant that survives unfavorable conditions in the form of a seed; annual or ephemeral plant.

thinning law, 3/2 Self-thinning plant populations, sown at sufficiently high densities, approach and follow a thinning line with a slope of roughly –3/2; therefore in a growing population, plant weight increases faster than density decreases to a point where the slope changes to –1.

threshold of security Point in local population density at which the predator turns its attention to other prey because of harvesting efficiency; the segment of prey population below the threshold is relatively secure from predation; see *switching*.

throughfall That part of precipitation that falls through vegetation to the ground.

tidal overmixing Mixing of freshwater and seawater when a tidal wedge of seawater moves upstream in a tidal river faster than freshwater moves seaward; seawater on the surface tends to sink as lighter freshwater rises to the surface.

till Sediments laid down by the direct action of glacial ice without the intervention of water.

tiller In grasses, a lateral shoot arising at ground level.

time lag Delay in a response to change.

time-specific life table See *life table*.

tolerance model Model that proposes that succession leads to a community composed of species most efficient in exploiting resources; colonists neither increase nor decrease the rate of recruitment or growth of later colonists.

top-down control Influence of predators on size of lower trophic levels in a food web.

topography Physical structure of the landscape.

torpor Temporary great reduction in an animal's respiration, with loss of motion and feeling; reduces energy expenditure in response to some unfavorable environmental condition, such as heat or cold.

trace element Element occurring and needed in small quantities; see *micronutrient*.

trade winds Tropical easterly winds that blow in a steady direction from the subtropical high-pressure areas to the equatorial low-pressure areas between the latitudes 30° and 40° north and south; these winds are generally northeasterly in the Northern Hemisphere and southeasterly in the Southern Hemisphere.

translocation Transport of materials within a plant; absorption of minerals from soil into roots and their movement throughout the plant.

transpiration Loss of water vapor from a plant to the outside atmosphere.

trophic Related to feeding.

trophic level Functional classification of organisms in an ecosystem according to feeding relationships, from first-level autotrophs through succeeding levels of herbivores and carnivores.

trophic structure Organization of a community based on the number of feeding or energy transfer levels.

trophogenic zone Upper layer of the water column in ponds, lakes, and oceans, in which light is sufficient for photosynthesis.

tropholytic zone Area in lakes and oceans below the compensation point.

tundra Area in an arctic or alpine (high mountain) region, characterized by bare ground, absence of trees, and growth of mosses, lichens, sedges, forbs, and low shrubs.

turgor The state in a plant cell in which the protoplast is exerting pressure on the cell wall because of intake of water by osmosis.

turnover rate Rate of species lost and others gained.

Ultisol Low base soil associated with warm, humid climate and old terrain, taking on a reddish color from secondary iron oxides.

understory Growth of medium-height and small trees growing beneath the canopy of a forest; sometimes includes a shrub layer as well.

ungulate Any hoofed grazing mammal; usually refers to ruminants, such as cattle and deer.

unitary organism An organism, such as an arthropod or vertebrate, whose growth to adult form follows a determinate pathway, unlike modular organisms whose growth involves indeterminate repetition of units of structure.

upwelling In oceans and large lakes, a water current or movement of surface waters produced by wind that brings nutrient-loaded colder water to the surface; in open oceans, regions where surface currents diverge deep waters, which rise to the surface to replace departing waters.

vapor pressure The amount of pressure water vapor exerts independent of dry air.

vapor pressure deficit The difference between saturation vapor pressure and the actual vapor pressure at any given temperature.

vector Organism that transmits a pathogen from one organism to another.

vegetative reproduction Asexual reproduction in plants by means of specialized multicellular organs, such as bulbs, corms, rhizomes, stems, and the like.

veld Extensive grasslands in the east of the interior of South Africa, largely confined to high terrain.

vertical stratification Layering of physical conditions and life in a community.

vertical structure The arrangement of layers of vegetation in terrestrial and aquatic communities.

Vertisol Mineral soil that contains more than 30 percent of swelling clays that expand when wet and contract when dry, associated with seasonal wet and dry environments.

vesicular arbuscular mycorrhizae (VAM) A form of endomycorrhizae in which the fungus enters and grows within the host's cells and extends widely into the surrounding soil.

viscosity Property of a fluid that resists the force that causes it to flow.

Wallace's line Biogeographic line between the islands of Borneo and the Celebes that marks the eastern boundary of many landlocked Eurasian organisms and the western boundary of the Oriental region.

warning coloration Conspicuous color or markings on an animal that serve to discourage potential predators.

water potential Measure of energy needed to move water molecules across a semipermeable membrane; water tends to move from areas of high or less negative potential to areas of low or more negative potential.

water-use efficiency Ratio of net primary production to transpiration of water by a plant.

watershed Entire region drained by a waterway into a lake or reservoir; total area above a given point on a stream that contributes water to the flow at that point; the topographic dividing line from which surface streams flow in two different directions.

weather The combination of temperature, humidity, precipitation, wind, cloudiness, and other atmospheric conditions at a specific place and time.

weathering Physical and chemical breakdown of rock and its components at and below Earth's surface.

weed Plant possessing a high rate of dispersal, occurring opportunistically on land or water disturbed by human activity, and competing for resources with cultivated plants; a plant growing in the wrong place.

westerlies The dominant east to west motion of the winds centered over the middle latitudes of both hemispheres.

wet deposition Sulfur dioxide and nitrogen oxides that return as dilute solutions of nitric acid and sulfuric acid dissolved in rain and snow.

wetfall Component of acid deposition that reaches Earth by some form of precipitation; wet deposition.

wetland A general term applied to open-water habitats and seasonally or permanently waterlogged land areas; defining the extent of a wetland is controversial because of conflicting land-use demands.

wilting point Moisture content of soil at which plants wilt and fail to recover their turgidity when placed in a dark, humid atmosphere; measured by oven drying.

xeric Dry, especially in soil.

yield Individuals or biomass removed or harvested from a population per unit time.

zero net growth isocline An isocline along which the population growth rate is zero.

zonation Characteristic distribution of vegetation along an environmental gradient; this gradient may form latitudinal, altitudinal, or horizontal belts within an ecosystem.

zoogeography Study of the distribution of animals.

zooplankton Floating or weakly swimming animals in freshwater and marine ecosystems; planktonic animals.

zooxanthellae One-celled dinoflagellates that live symbiotically in the tissues of various marine invertebrates; closely associated with corals.

Credits

TEXT/ILLUSTRATION CREDITS

Chapter 2 **2.3:** Grant, P.R., *Ecology and Evolution of Darwin Finches.* © 1996 by Princeton University Press. Reprinted by permission of Princeton University Press.

Chapter 3 **3.4:** From R.G. Barry and R.J. Chorley, *Atmosphere, Weather and Climate*, 6th ed. © 1992 Routledge. Reprinted by permission of Taylor & Francis Books, Ltd. **3.9:** From T. Graedel and P. Crutzen, *Atmosphere, Climate and Change.* Scientific American Library. **3.13:** From Robert E. Coker, *This Great and Wide Sea.* © 1949 by the University of North Carolina Press, renewed 1977 by Robert M. Coker. Used by permission of the publisher. **3.20:** From Schroeder and Buck, "Fire Weather" *USDA Agriculture Handbook* 360, USDA Forest Service, Washington, DC. **3.21:** From Schroeder and Buck, "Fire Weather" *USDA Agriculture Handbook* 360, USDA Forest Service, Washington, DC.

Chapter 4 **4.1:** H.G. Haverson and J.L. Smith. "Solar Radiation as a Forcast Management Tool" *USDA Forest Service Gen. Tech Report* PSW 33. **4.2:** W.E. Reifsnyder and H.W. Lull, "Radiant Energy in Relation to Forests" *USDA Technical Bulletin* 1344 (1965): 21. **4.7:** B.A. Hutchinson and D.R. Matt, 1972. "The Distribution of Solar Radiation within a Deciduous Forest." *Ecological Monographs* 47: 205. © by the Ecological Society of America. Reprinted by permission. **4.15:** From *Biological Survey of the Raquette Watershed.* New York State Conservation Department, 1934.

Chapter 5 **5.4:** From *Nature and Properties of Soil*, 8th ed. by Nyle C. Brady, © 1974. Reprinted by permission of Prentice-Hall, Inc. Upper Saddle River, NJ. **5.6:** From Soil Conservation Service, USDA, Washington, DC.

Chapter 6 **6.8:** C.K. Augspurger, 1982. "Light Requirements of Tropical Tree Seedlings: A Comparative Study of Growth and Survival." *Journal of Ecology* 72: 777–795. Reprinted by permission. **6.14:** Mooney, Bjorkman, Ehrlinger, and Berry. "Photosynthetic Capacity of Death Valley Plants," *Carnegie Institute Yearbook* 75: 410–413. Copyright 1976. Reprinted by permission of Carnegie Institute. **6.16:** From *On the Economy of Plant Form and Function: Proceedings of the Sixth Maria Moors Cabot.* Edited by Thomas J. Givnish. Copyright © 1986. Reprinted by permission of Cambridge University Press. **6.19:** H.A. Mooney and S.L. Gulman, "Environmental and Evolutionary Constraints on the Photosynthetic Characteristics of Higher Plants," *Bioscience* 32 (1982). American Institute of Biological Science. **FOE 6.1:** From T.R. Parsons, M. Takahashi, and B. Hargrave. *Biological Oceanographic Processes*, 3e. Permagon Press, 1934.

Chapter 7 **7.2:** Swift et al., *Decomposition in Terrestrial Ecosystems.* © 1979. Reprinted by permission of Blackwell Science, Ltd. **7.6:** From Ivan Valiela, *Marine Biology Processes*, p. 310. © 1984. Reprinted by permission of the author. **7.15:** From *Freshwater Wetlands: Ecological Processes and Management Potential*, edited by R.E. Good, et al., © 1978, Elsevier Science (USA), reproduced by permission of the publisher.

Chapter 8 **8.3:** Reprinted by permission of The Wildlife Society. **8.10:** From *Animal Physiology*, 2nd ed. by Hill and Wyse. © 1989. Reprinted by permission of Pearson Education, Inc. **8.11:** Peterson et al., "Snake Thermal Ecology" in R.A. Siegel and J.T. Collins (eds.) *Snakes, Ecology and Behavior.* New York: The McGraw-Hill Cos., 1993. **8.14:** Peterson et al., "Snake Thermal Ecology" in R.A. Siegel and J.T. Collins (eds.) *Snakes, Ecology and Behavior.* New York: The McGraw-Hill Cos., 1993. **8.16:** K. Schmidt-Nielsen, *Animal Physiology*, 3e. Upper Saddle River, N.J.: Prentice-Hall, 1970. **8.21:** J.D. Palmer, "The Rhythmic Lives of the Crabs," *Bioscience*, 40 (1990). American Institute of Biological Sciences. **FOE 8.2a:** Edward Bunning, "Circadian Rhythms and the Time Measurement in Photoperiodism," *Cold Spring Harbor Symposium on Quantitative Biology*, vol. 25. © 1960. Reprinted by permission of Cold Spring Harbor Laboratory Press. **FOE 8.2b:** C.H. Johnson and J.W. Hastings, "The Elusive Mechanism of the Circadian Clock," *American Scientist* 74 (1986) 29–36. © 1986 Sigma Xi. Reprinted by permission.

Chapter 9 **9.6:** J. Wiens, "Patterns and Process in Grassland Bird Communities." *Ecological Monographs* 43 (1973):240, fig. 2. © 1973 The Ecological Society of America. Reprinted by permission. **9.8:** Reprinted by permission of The Botanical Society of America. **9.13:** H. Hett and O.L. Loucks, "Age Structure: Models of Balsam Fir and Eastern Hemlock," *Journal of Ecology* 64:1035. © 1976. Reprinted by permission of Blackwell Science, Ltd.

Chapter 10 **10.1:** R.R. Sharitz and J.R. McCormick, "Population Dynamics of Two Competing Plant Species." *Ecology* 54, 1973. © 1973 The Ecological Society of America. Reprinted by permission. **10.9:** U.S. Bureau of the Census. **10.11:** F.A. Pitelka, "Proceedings 18th Biology Colloquium," pp. 79–80 (Corvallis, OR: Oregon State University, 1957). **Table 10.4:** R.R. Sharitz and J.R. McCormick, "Population Dynamics of Two Competing Plant Species," *Ecology* 54. © 1973 The Ecological Society of America. Reprinted by permission.

Chapter 11 **11.2:** M.C. Dash and A.R. Hota, "Density Effects on Survival Growth Rate and Metamorphosis on Rana Tigrina Tadpoles." *Ecology* 61 (1980):1027, fig. 2. © 1980 The Ecological Society of America. Reprinted by permission. **11.5:** C.W. Fowler, "Density Dependence as Related to Life History Strategy." *Ecology* 62 (1981):607, fig. 4. © 1981 The Ecological Society of America. Reprinted by permission. **11.6:** From C.W. Fowler and T.D. Smith, *Dynamics of Large Mammal Populations.* © 1981. Reprinted by permission of John Wiley & Sons, Inc. **11.9:** T. Jones, "Dispersal Distance and Range of Nightly Movements in Merriam's Kangaroo Rats." *Journal of Mammalogy* 20:31. © 1989 American Society of Mammalogists. Used with permission. **11.11:** R.L. Smith, "Some Ecological Notes on the Grasshopper Sparrow," *Wilson Bulletin* 75 (1963). **11.12:** Mech et al., "Relationships of Deer and Moose Populations to Previous Winter's Snow." *Journal of Animal Ecology* 56:615–627. © 1987. Reprinted by permission of Blackwell Science, Ltd.

Chapter 12 **12.8:** Werner, P.A. and W.J. Platt, "Ecological Relationships of Co-occurring Golden Rods (*solidago: compositae*)" *American Naturalist* 110:959–971. Copyright 1976. Reprinted by permission of The University of Chicago Press. **12.13:** F. James, "Ordination of Habitat Relationships Among Birds," *Wilson Bulletin* 1971.

Chapter 13 **13.7:** R.H. Whittaker, "Vegetation of the Great Smokey Mountain," *Ecological Monographs* 26 (1956) 1–80. © 1956 The Ecological Society of America. Reprinted by permission. **13.8:** From *Metapopulation Biology: Ecology, Genetics, and Evolution*, edited by Hanski and Gilpin, © 1996, Elsevier Science (USA), reproduced by permission of the publishers.

13.11: From F. Bormann and G.E. Likens, *Pattern and Process in a Forested Ecosystem*, p. 116, fig. 4.4b. New York: Springer-Verlag, 1979. © 1979 Springer-Verlag. Used by permission. **13.13c:** Eric Alan Hobbie, "Nitrogen Cycling During Succession," Masters Thesis, University of Virginia, 1994. **13.16:** From R.H. Whittaker, "Quaternary History and the Stability of Forest Communities," *Forest Succession: Concepts and Applications*, D.C. West et al. (eds.) © 1981 Springer-Verlag. Reprinted by permission. **13.17:** P.A. Delcourt and H.R. Delcourt, "Vegetation Maps for Eastern North America, 40,000 yr BP to Present." In R. Romans (ed.) *Geobotany*. Copyright 1981. Reprinted by permission of Kluwer Academic/Plenum Publishers.

Chapter 14 14.3: Reprinted with permission from ASLO (*Limology Oceanography* 1025, 27, Tillman et al. "Competition Between Two Species of Diatoms for Silica.") © 1981 Society of Limology and Oceanography. **14.7:** M.P. Austin, R. H. Groves, L.M. Fresco, and P.E. Laye, "Relative Growth of Six Thistle Species along a Nutrient Gradient with Multispecies Competition," *Journal of Ecology* 73:667-684 (1965). Blackwell Science, Ltd. **14.9:** N.C. Emery, P.J. Ewanchuk, and M.D. Bertness, "Competition and Salt Marsh Plant Zonation: Stress Tolerators may be Dominant Competitors," *Ecology* 82(9) (2001). Ecological Society of America. **14.10:** H.C. Heller and D. Gates, "Altitudinal Zonation of Chipmunks (Eutamias): Energy Budgets," *Ecology* 52 (1971) 424, fig. 1. © 1971 by The Ecological Society of America. Reprinted by permission. **14.11:** N.K. Wieland and F.A. Bazzaz, "Physiological Ecology of Three Codominant Successional Annuals," *Ecology* 56 (1975) 686, fig. 6. © 1975 by The Ecological Society of America. Reprinted by permission. **14.12:** "Feline Canines," by Dayan from *American Naturalist* 136: 39–60. © 1990. Reprinted by permission of The University of Chicago Press. **14.14:** Courtesy of Krebs *Ecology* 5th ed., p. 192, © 2001 Benjamin Cummings; San Francisco, CA. **FOE 14.2b:** Grant, P.R., *Ecology and Evolution of Darwin Finches*. © 1996 by Princeton University Press. Reprinted by permission of Princeton University Press.

Chapter 15 15.3a: From *Ecology* 76. © 1995. Reprinted by permission of The Ecological Society of America. **15.6:** L.J. Mook, "Birds and Spruce Budworm," in R. Morris (ed.) *Entomological Society of Canada Memoirs* 31 (1963). **15.7:** Jedrzejewski et al., *Ecology* 76 (1995):192, fig. 11. © 1995 The Ecological Society of America. Reprinted by permission. **15.15:** N.B. Davies, "Prey Selection and Social Behavior in Wagtails," *Journal of Animal Ecology* 46:48. © 1977. Reprinted by permission of Blackwell Science, Ltd.

Chapter 17 17.3b: Courtesy of Krebs *Ecology* 5th ed., p. 448, © 2001 Benjamin Cummings; San Francisco, CA. **17.2, 17.6:** After Krebs et al. 2001. **17.7:** From T.M. Smith and M. Huston, "A Theory of Spatial and Temporal Dynamics of Plant Communities," *Vegetatio* 83:49–69 © 1989. Reprinted by permission of Kluwer Academic Publishers. **17.8:** Data from Whittaker 1960, courtesy of Krebs *Ecology* 5th ed., p. 394, © 2001 Benjamin Cummings; San Francisco, CA. **17.10:** N.C. Emery, P.J. Ewanchuk, and M.D. Bertness, "Competition and Salt Marsh Plant Zonation: Stress Tolerators may be Dominant Competitors." *Ecology* 82(9) (2001). Ecological Society of America. **17.12:** H.J. Oosting, *American Midland Naturalist* (Notre Dame, IN: University of Notre Dame, 1942). Reprinted with permission. **17.14:** R.H. McArthur and J. McArthur, *Ecology* 42:594–598.

Chapter 18 18.5: G.I. Murphy, "Vital Statistics of the Pacific Sardine and the Population Consequences," *Ecology*. © 1967 The Ecological Society of North America. Reprinted by permission. **18.6:** Reprinted with permission from H.A. Regier and W.L. Hartman, "Lake Erie's Fish Community: 150 Years of Cultural Stress," *Science* 180 (1973): 1248–1255. © 1973 American Association for the Advancement of Science. **18.11:** P. DeBach, *Biological Control by Natural Enemies*, p. 4, fig. 2. © 1974 Cambridge University Press. Reprinted by permission.

Chapter 19 19.8: Reprinted by permission of The Wildlife Society.

Chapter 20 20.3: H. Leith, "Primary Production Terrestrial Ecosystems," *Human Ecology* 1 (1973):303. Copyright 1973. Reprinted by permission of Kluwer Academic/Plenum Publishers. **20.4:** H. Leith, "Modeling Primary Productivity of the World," in H. Leith and R. Whittaker *Primary Production of the Biosphere*. © 1975. Reprinted by permission of Springer-Verlag. **20.5:** R.H. MacArthur and J.H. Connell, *The Biology of Populations*. © 1966. Reprinted by permission of John Wiley & Sons, Inc. **20.7a:** J. Pastor and J.D. Aber, C.A. McClaugherty and J.M. Melillo, "Above Ground Production and N&P Cycling along a Nitrogen Mineralization Gradient on Blackhawk Island, Wisconsin," *Ecology* (1984). Ecological Society of America. **20.16:** Begon, *Ecology: Individuals, Populations, and Communities*. 1996. Blackwell Publishers.

Chapter 21 21.3: J. Pastor and J.D. Aber, C.A. McClaugherty and J.M. Melillo, "Above Ground Production and N&P Cycling along a Nitrogen Mineralization Gradient on Blackhawk Island, Wisconsin," *Ecology* (1984). Ecological Society of America.

Chapter 22 22.7: Schlesinger, *Biochemistry: An Analysis of Global Change*, 2nd ed. © 1997. Reprinted by permission of Academic Press. **22.9:** Schlesinger, *Biochemistry: An Analysis of Global Change*, 2nd ed. © 1997. Reprinted by permission of Academic Press. **22.11:** Schlesinger, *Biochemistry: An Analysis of Global Change*, 2nd ed. © 1997. Reprinted by permission of Academic Press. **22.12:** Schlesinger, *Biochemistry: An Analysis of Global Change*, 2nd ed. © 1997. Reprinted by permission of Academic Press.

Chapter 24 24.13: Currie and Paquin, "Large Scale Biogeographical Patterns of Species Richness in Trees," *Nature*. 329. 1987. London: Macmillan, Ltd. **24.14:** From Currie, "Energy and Large Scale Biographical Patterns of Animal and Plant Species Richness," *American Naturalist* 137:27–49. © 1991. Reprinted by permission of The University of Chicago Press.

PHOTO CREDITS

Chapter 1 CO: © Ryan C. Taylor/Tom Stack and Associates. **1.1:** Tom Smith.

Chapter 2 CO: © Tim Davis/Photo Researchers Inc. **2.1:** Courtesy of Mr. G.P. Darwin/By Permission of Darwin Museum Down House. **2.5:** © Stephen J. Krasemann/DRK Photo. **2.10a:** © Jeff LeClere. **2.10b:** Walt Knapp. **2.11:** John Eastcott/Yva Momatiuk/DRK Photo. **2.12a:** © Dan Sudia. **2.12b:** © W. Perry Conway/Tom Stack and Associates. **2.15a:** UPI/Corbis-Bettmann. **2.15b:** © Archive Photos.

Chapter 3 CO: © Thomas Kitchin/Tom Stack and Associates. **3.7:** © Kennan Ward. **3.22a:** R.L. Smith. **3.22b:** © Richard H. Stewart/National Geographic Society.

Chapter 4 CO: © Michael S. Yamashita/Corbis. **4.6a:** © Thomas Kitchin/Tom Stack and Associates. **4.12a:** © John Shaw/Tom Stack and Associates. **4.12b:** Bob and Ira Spring. **4.13:** © Scott Blackman/Tom Stack and Associates.

Chapter 5 CO: © Terry Donnelly/Tom Stack and Associates. **5.9a:** © Richard Hamilton Smith/Corbis. **5.9b:** Courtesy of NASA. **EA-Figure A:** © Corbis. **EA-Figure B:** The Granger Collection, Ltd., New York.

Chapter 6 CO: © Ron Watts/Corbis. **FOE6.3a:** R.L. Smith. **FOE6.4b:** © Matt Bradley/Tom Stack and Associates.

Chapter 7 CO: © Papilio/Corbis. **7.1a:** Michael Fogden/Animals Animals/Earth Scenes. **7.1b:** Kevin Byron/Bruce Coleman Inc. **7.1c:** Oxford Scientific Films/Animals Animals/Earth Scenes. **7.1d:** (L) Gary Braasch/CORBIS (R) Bill Curtsinger. **7.7:** © Joe McDonald/Corbis. **7.8:** William E. Ferguson Photography. **FOE7.1:** R.L. Smith.

Chapter 8 CO: © Joe McDonald/Tom Stack and Associates. **8.4:** Courtesy of R.L. Smith. **EA-Figure A:** Library of Congress. **EA-Figure B:** (T) Corbis/Ralph A Clevenger (B) Corbis/Paul A. Sounders.

Index

Broadleaf evergreens, 478, **479**, 494, 502

Brood parasitism, 346

Browse, 306

Buffalo, African, **206–207**

Bunchgrasses, 518, **520**

Bundle sheath cells, **113**, 115

Bünning, E., 158

Buoyancy, in animals, 162–**163**

Bypass, in fisheries, 353

C

C_3 plants, 102, **113**, **117**

C_4 plants, **113**, 115, **117**

Calcification of soil, 88

Caliche, 88

CAM plants, 113, **115**

Candolle, Alphonse de, 468, 469

Cannibalism, 290, 331

Canopy, 248–249

Capillarity, 64

Capillary water, 85

Capture-recapture sampling, 181

Caraco, Nina, 457

Carbohydrates
decomposition of, 124, 126–127, **626**
photosynthesis of, 102
respiration and, 102–103, 124, 149
See also Glucose

Carbon balance, 103

Carbon cycle, 434–437, **435**, **437**, 445
human activities and, 450, 601, 602, 621

Carbon dioxide
acidity and, 73–**74**
in carbon cycle, 418, 434–**436**, 437
diurnal fluctuation of, 434–435, **436**
fertilization effect, 604
global rise in, 450, 598–600, **598**, **599**, 621–622
agriculture and, 618, **620**
measurement of, 600–**602**
ocean uptake and, 601, 602–603
plants and, 603–605, **604**, 618
rain forests and, 615, 622
soil respiration and, 615
tree species and, 611, **613**, 615
in photosynthesis, 102–105, 113, 115
from respiration, 103, 423
seasonal fluctuation of, 435–**436**
solar radiation and, 39
in streams, 552
See also Greenhouse effect

Carbon gain, net, 106–**107**, 110, 111, 116, 120

Carbonates, 434, 437, 444, 602

Carbonic acid, 73–74

Carnivores, 141, 143, 290, 306
marine zooplankton, 565
trophic level of, 336, 337, 407, 411, 413

Carnivorous plants, **302–303**

Carpenter, Charles, 7

Carpenter, Steve, 457

Carrier, 310, 311, 316

Carrying capacity *(K)*, 196, 197, 204
competition and, 271, 273, 278
human, 236
sustained yield and, 351
territoriality and, 214

Carson, Rachel, 458

Catchments, 544

Cation exchange capacity (CEC), **86–87**

Cations in soils, 86–87
acid precipitation and, 454, 455

Cause-and-effect relationship, 8–9

Cellulose
decomposition of, 127, 260, **626**
in herbivore diet, 138–139, 304, 318

Channelization, 557

Chaparral, **527–528**

Character displacement, 283

Chemical defenses
of animals, 296, 297
of plants, 303–304

Chemical ecology, 7

Chemical interactions, competitive, 270, 271

Chemical weathering, 81

Chemosynthetic bacteria, 567

Chlorinated hydrocarbons, **458–459**

Chlorofluorocarbons (CFCs), 456, 605

Chlorophyll, 102, 109, 118–119, 422

Chromosomal mutations, 19, 24

Chromosomes, **17**

Chronoseres (chronosequences), 343

Circadian rhythms, 60–61, 157–159, 161–162

Class, taxonomic, 21

Classifying organisms, 20, 21

Clausen, J., 27

Clay, 81, 82, 83, 84
cation exchange capacity and, **86**
erosion and, 92, 93
field capacity and, 85
soil orders and, 87, 88, 89, 90

Clear-cutting, 354, 355, 385, **386**
carbon dioxide and, 450
nutrient loss caused by, 425

Clements, Frederick, 6, 262, 264, 265, 341, 470

Climate
classification of, **469**, 470
decomposition and, 129–130, 615
defined, 38
ecosystem classification and, 468, **469–471**, **474**, 475–479, **488–489**

forest ecosystems and, 494–**495**, **496**

fossil fuel formation and, 625

global models of, **606–608**, 611, 612, 615, 618, 622

grassland ecosystems and, 518–521, **519**

interspecific competition and, 276

microclimates, 51–53

nutrient cycling and, 422

pollen and, **490–491**, 615

primary production and, **400–402**

secondary production and, 405–406

shrublands and, 526–527, 528

soil formation and, 80

species richness and, 481–484, **482**, **483**, **484**, 486

See also Precipitation; Temperature

Climate change, 598, **610–611**
agriculture and, 618–619, **620**
ecosystem effects of, 611–617, 621–622
human health and, 619–621, **620**, **621**
See also Global warming; Greenhouse effect

Cline, 26

Clonal growth, in plants, **173–174**

Cloning, 31–32, 33

Closed systems, 396, 398

Cloud forest, 497

Clouds, 39, 47, 56

Clumped distribution, **177**, 179, **180**

Clutch size, 231, 346

Coarse particulate organic matter (CPOM), 554, 555, 556

Coastal ecosystems, 480, 573–574
global warming and, 616–**617**, **618**
nutrient cycling in, 429–431, **430**
pollution and, 576–577
See also Seashore; Wetlands

Cody, Martin, 231

Coefficient of community, 254

Coevolution
in community, 264
in mutualism, 325
of parasite and host, 310, 315, 318
of predator and prey, 296

Coexistence, 273, 275, 276, 280–282, 283

Cohort, 186–187

Collectors, invertebrate, 133, **554**, **555**, 556

Coloration
of predator, 299, **300**
of prey, **296–297**, **298**
temperature regulation and, 166

Commensalism, 242, 310

Communities
biological structure of, 243–245
boundaries between, 253–254
contrasting views of, 262–**265**

Denitrification, 439, 440, 450, 456, 534

Density. *See* Population density

Density dependence, **204**

Density-independent influences, **204**, 215–**216**

Dependent variable, 8

Desert grassland, 520, 525

Desert scrub, **532–533**

Desertification, 94

Deserts, 531–535
animals of, **156–157**, 215–216, **534–535**
classification of, **531–533**
defined, 475, 531
geographic distribution of, 48, **519**
nutrient cycling in, 534
physical structure, 533
plant adaptations in, 112, 115, 532–**533**, 534
polar, 535
primary production in, 534
in rain shadows, 53, 532
saline, 112

Detrital food chain, 407–410, **408**, **409**, 412, 413
in streams, 555
See also Decomposition

Detritivores, 124, **125, 126**, 128, 129, 133, 138

Detritus
defined, 138
of seashore, 572–573
of streams, 554, 555
See also Plant litter

Dew point temperature, **47**

Diapause, 156, 159

Diffuse mutualisms, 322, 326, 331

Diffuse species interactions, 331

Diffusion, 64, 103

Digestive tracts, 138–141, **140**

Dinoflagellates, **564**, 590

Dinosaurs
ectothermy and, 151, 168, 198
extinction of, 389

Dioecious plants, 221

Diploid zygote, 19, 220

Directional selection, 13

Disease
insect-borne, 311, 315, 620–621
parasitic, 310

Dispersal, 177–179
outbreeding and, 226
population density and, 210–211
See also Alien species; Emigration; Immigration

Dispersion patterns, **177**, 180

Disruptive selection, 13

Dissolved organic carbon (DOC), 565

Dissolved organic matter (DOM), 133–134, 556, 565

Distribution of organisms, 173–175, **177**, 179–**180**
See also Biogeography

Disturbance, 378, 380–385
by animals, 378, 381, 383–**385**
to coral reefs, 575, 577
defined, 378
by humans, **385, 386**
shrub communities and, 526
succession and, 258, 344–345, 380, 386, 387
See also Restoration

Disturbance regime, 378

Diversity indexes, 245

Division
in ecosystem classification, 488
in plant classification, 21

DNA (deoxyribonucleic acid), 17, 25
mitochondrial, 321
See also Genetic engineering

Dodson, Stanley, 332, 334

Doldrums, 44

Domain, 488

Domestic animals, 356

Domestic grasslands, 518

Dominance
measures of, 243
social, 211–212

Dominant allele, 17, 18

Dominant species, 243
in plant succession, 340, 341, 342, 343

Doubling time, 195

Drainage classes, 85

Drought
in grasslands, 518, 524
seasonal, 496, 497, 527, 529

Drought-deciduous shrubs, 526

Drought-deciduous trees, 113, **478**, 479, 510

Dry deposition, 453, 454

Dry tropical forests, 494, **497–498**

Dryfall, 418, 438, 443

Dung beetles, **129**

Dye, Peter, 276

Dystrophic lakes, 550

E

Earth's rotation, 40, 44–**45**
estuarine salinity and, 75
seasons and, 60
tides and, 67

Easterlies, 44, 45

Ecliptic, 40

Ecological biogeography, 469

Ecological density, 177

Ecological pyramids, **413**

Ecology
defined, 4

global perspective, 621–622
history of, 6–7
as interdisciplinary science, **5**, 7
scientific methods in, 8–10

Economics
biodiversity and, 391
ecological concepts and, 4
of pest control, 358–359
sustainable harvest and, 353, 356

Ecoregions, **488–489**

Ecosystem ecology, history of, 6–7

Ecosystems
classification of, 468, **469–471**, **474**, **475–479**, **488–489**
climate change and, 611–617, 621–622
components of, 4, 5, **396–397**
defined, 4, 396
energy flow and, 396, 397–398, 408–413, **410, 412, 413**
outputs of, 396, 423–425, **424**

Ecotone, 369–370, 507, 580

Ecotypes, 26–27

Ectomycorrhizae, 318–**319, 326**
See also Mycorrhizae

Ectoparasites, 310, 314

Ectotherms, 148, 149–151, 166–168
energy efficiency, 407, **411**
See also Poikilotherms

Edge effect, 370, 371, 486

Edge species, **370**, 371, 373, **374**

Edges, 369, 370, **373**

Effective population size, 379

Efficiencies of conversion, 406–407, 410–**411, 412**, 413

Egler, F., 341

Ehrlich, Anne, 335

Ehrlich, Paul, 335

Elevation. *See* Altitude

Elton, Charles, 7, 28

Eluviation, 82

Emerson, A. E., 7

Emery, Nancy, 338

Emigration, 177, 186
of predators, 294, 295
between subpopulations, 378

Encounter competition, 270

Endangered species, 32, 362, 363, 594

Endoparasites, 310–311

Endothermic reaction, 397

Endotherms, 147–148, 149–151, 167
energy efficiency, 407, **411**
See also Homeotherms

Energy
carbon cycle and, 434
ecosystem and, 396, 397–398, 408–413, **410, 412, 413**
efficiencies of conversion, 406–407, 410–**411, 412**, 413
organism and, 70–71
plants and, 106, 398–400, **399**, 403

Food webs, 245–246
  of boreal forest, 331, **332, 336**
  community structure and, 334–337
  indirect interactions in, 331–332,
    333, 334
  of marine plankton, 564–565
  See also Food chains
Foraging strategies, 299–**301**
Forest floor, 248, 249, 250
  See also Plant litter
Forestry. *See* Timber harvesting
Forests
  air pollution and, **455,** 456–457
  carbon dioxide elevation and,
    599–600, **605, 622**
  classification of, 471, 477–478, 479
    climate and, 494–**495, 496**
  global carbon cycle and, 602
  primary productivity, 494–495,
    496
  species diversity in, 498, 500, 502,
    505–506
  See also Coniferous forests; Rain
    forests; Temperate forests; Trees
Formations, 468, **469,** 471
Forster, Johann Reinhold, 375
Fossil fuels
  combustion of, 442, 451, 463
    aerosols from, **609**
    carbon dioxide and, 450,
      598–**599,** 600–**602,** 603, 621
    formation of, 434, 437, 624–625,
      626
Fowler, Norma, 331
FPOM (fine particulate organic
  matter), 554, 555, 556
Fragipan, 83
Fragmentation, of organic matter, 124,
  132
Fragmented landscapes. *See* Patches
Frass, 260, 525
Frequency of disturbance, 378, **380**
Freshwater ecosystems, 479–480,
  544
  See also Lakes; Streams and rivers;
    Wetlands
Fringe wetlands, **581,** 582, 586
Frost hardening, 117
Frugivores, 323–**325,** 326
Fugitive species, 382
Functional groups, 335–336
Functional response, **292**–293, 295
Fundamental niche, 282, **284,**
  330–331
Fungi
  as decomposers, 82, 124, 127, 133,
    260, 525, 626
  mutualism and, 318–**319,** 320, **321,**
    326
  as parasites, 310, 314, 315, 316
  See also Lichens; Mycorrhizae

G

Gaia hypothesis, 5, 419
Galápagos Island finches, **13, 14,** 16,
  283
Gall formation, **314**
Gallery forests, 497, 513
Gamma (regional) diversity, 485–486
Gap, 378, 380
Garigue, 527
Gaseous cycles, 418, 434
Gathering collectors, 554
Gause, G. F., 273
Gause's principle, 275
Gelisols, 90
Gene frequency, 24–25
Gene pool, 18
General circulation models (GCMs),
  606–608, 611, 612, 615, 618, 622
Genes, **17**–19
  mitochondrial, 321
  mutation of, 19, 25, 487
Genet, 173
Genetic drift, 378, 379
Genetic engineering, 32–33
  for pest control, 361–362
Genetic variation, 19–20, 31–33,
  220–221
  outbreeding and, 226
Genetics
  basic principles, 17–19
  Mendel's experiments, 6, 16–17, 18,
    20, 25
  natural selection and, 13–16, 19–20,
    25, 31
Genotype, 18
Genus, 21
Geographic (allopatric) speciation,
  23–**24**
Glacial lakes, 544, **545**
Glacial periods, 261–262
  sea level and, 616
  species richness and, 485, 486
Glacial sediments, plant colonization
  of, 342, **343**
Gleason, H. A., 264–265
Gleization, 90
Global ecology, 7
Global warming, 605, 607–**608, 609,**
  610–611
  agriculture and, 618, **620**
  ecosystem effects of, 611, 612, **614,**
    615, 622
  nitrogen fertilizer and, 461
  sea level and, 616–**617, 618**
  See also Greenhouse effect
Glucose
  oxidation of, 102–103, 124, 126,
    149
  photosynthesis of, 102–103, 113,
    115, 403

Gougers, **554**
Grace, J. B., 283
Grant, Peter, 13
Grant, Rosemary, 13
Grasses
  ecosystem classification and, 471,
    475, 479, 480
  growth forms of, 518, **520**
Grasslands, 475, 479, 518–525
  animal life, **523**–524
  conversion to agriculture, 450
  desert, 520, 525
  geography, 518–521, **519**
  grazing on, 303, 305, 518, 522,
    523, 524–525
  of Great Plains, 97, 520
  primary production, **524**–525
  of savannas, 511–514
  soils, 89
  vertical structure, **522**–523
Grazers, invertebrate, 133, **554,** 555,
  556
Grazing
  in deserts, 534, 535
  disturbances caused by, 381, 383
  on grasslands, 303, 305, 518, 522,
    523, 524–525
  on public rangeland, 305
Grazing food chain, 407–**408, 409,**
  410, 412, 413
Great Plains, 97, 520
Greenhouse effect, 39, 456, 608, 622
  See also Global warming
Greenhouse gases, 450, 605, 609, 611,
  621
Grime, J. Phillip, 228, 382
Grinnell, Joseph, 28
Gross primary production, 398
Ground fire, **384**
Groundwater, 65
  contamination of, 462
  sea level and, 617
  wetlands and, 594
Groves, R. H., 277, 337
Growing season
  degree-days in, **118**
  ecosystem classification and, 470
  leaf morphology and, 477, 478
  primary productivity and, **400,** 401
Guilds, 242, 287, 336
Gully erosion, 93
Gurevitch, Jessica, 284
Gyres, 46

H

Habitat
  association and, 262, 265
  defined, 175
  host organism as, 310–311

Habitat (continued)
  human creation of, 356
  human restoration of, 363
  plant life histories and, 228
  population density and, 176–177
  source vs. sink, 210
  territories in, 214
  See also Environment; Patches
Habitat destruction
  agriculture and, 356
  endangered species and, 362
  extinction and, 199–200, 335, 356, 362, 371, 389, 390
Habitat fragmentation. See Landscape ecology
Habitat selection, 231–233, **232**, 345
Hadley, George, 45
Hadley cells, **45**
Haeckel, Ernst, 4
Hairston, Nelson, 337
Halophytes, **112**, 532
Haploid gametes, 19, 220
Hardy-Weinberg principle, 25
Harrison, J. L., 500
Headwater streams, 551, 554–555
Heat
  defined, 67
  as degraded energy, 397–398
  latent, 46, 63
  in organism's energy balance, 70–71, 106, 405
  specific heat, 62–63
  See also Temperature
Heat dome, urban, 51
Heat exchangers, countercurrent, **154–155**
Heathlands, **529**–530
Heavy metals, 457–458, 592, 594
Hedgerows, 368, 376, **377**, 530
Hemiparasites, 311–312, **313**
Herb layer, 248, 249
Herbicides, 359–360, 362, 458, 459, 461, 462–463, 592
Herbivores, 138–143
  below-ground, 408
  of deserts, **534**–535
  digestive flora of, 124, 138, 139, 140, 318
  marine zooplankton, 564–**565**
  predation on plants, 228, 290, 301–306, **304, 306**
  productivity of, 404, 405–**406**
  of savannas, **513**–514
  trophic level of, 336, 337, 407–410, 411, 413
Hermaphroditic organisms, **221–222**
Hesse, R., 7
Heterotherms, 148, 151
Heterotrophs, 102, 124, 336, 396, 397
  See also Secondary production
Heterozygous individual, 18

Hibernation, 151–152, 153
Hiesey, W. M., 27
Historical biogeography, 468–469
Histosols, 90
Holdridge, L. R., 471
Holdridge classification, 471, **474**, 615
Holling, C. S., 292
Holoparasites, 311
Home range, **212–213**
  body size and, 371
Homeostasis, 144–**146**
Homeostatic plateaus, 146
Homeotherms, 148, 149, 150
  insulation in, 153
  torpor in, 151–152
  See also Endotherms
Homologous chromosomes, 17
Homozygous individual, 18
Horse latitudes, 45, 48
Hosts, 310–316
Human disturbance, **385, 386**
Human health, climate change and, 619–621, **620, 621**
Human population, 236–238
  environmental problems and, 5, 7–8
Human-introduced species. See Alien species
Humans and biogeochemical cycles, 449–459
Humans and natural populations, 349–363
  economics and, 353, 356, 358–359, 391
  fisheries management, 350–354
  landscape patterns and, 356, 368–369, 370–371
  restoration ecology, 7, 363, 456
  sustained yield, 350–356
  weeds and pests, 358–362
  See also Agriculture; Timber harvesting
Humboldt, Friedrich von, 6
Humidity, 46–**47**
  heat transfer and, 71
  microclimates and, 52, 53
  in temperate forests, 502
  transpiration and, 104, 106
  in urban areas, 51
Humus, 82, 83
  carbon storage in, 434
  cation exchange and, 86
  erosion and, 92
  lignins in, 626
  soil orders and, 89
Hunter-gatherer societies, 236, 317, 598
Hunting tactics, 298–299
Huston, Michael, 343, 345
Hutchinson, G. E., 28, 382
Hybrid, **22**
Hydric soils, 85

Hydrogen bonding, 62
Hydrogen sulfide, 442, 443, 548, 566, 567
Hydrologic cycle. See Water (hydrologic) cycle
Hydrology, of freshwater wetlands, 586–587, 594
Hydroperiod, 586
Hydrophytic plants, 580–582
Hydrothermal vents, 567, **568**
Hygroscopic water, 85
Hypertrophic lake, 550
Hypervolume, 28
Hypolimnion, 68, 250, 426, 548, 550
Hypotheses, 8, 9

I

Ice, 62
Ice ages, 261–262
Illuviation, 83
Immigration, 177, 186
  extinction and, 199–200
  into sinks, 210, 378
  species richness and, **375–376**
Immobilization, 124, **131–132, 133**, 134, 422
Immune response, 314–315, 316
Inbreeding, 226, 378, 379
Inbreeding depression, 226
Inceptisols, 87–88
Independent variable, 8
Indexes
  of abundance, 180
  of diversity, 245
  of dominance, 243, 254
  of relative population size, 181
Indirect mutualism, 332, **334**
Individual(s), 10
  defined, 172–173
  density of, 175–177
Individualistic continuum concept, 264–**265**
Induced edges, 369
Industrial Revolution, 236, 237, 450, 453, 598, 624
Infauna, 566, 572
Infection, 310
Infective stage, 311, 313, 316
Infiltration, of water, 65
Inflammatory response, 314
Infralittoral fringe, 569, 572
Infrared radiation, 56, 57, 70
  See also Thermal radiation
Inherent edges, 369
Inputs, to ecosystem, 396, 418–419
Insect-borne diseases, 311, 315, 620–621
Insecticides. See Pesticides
Insulation, in animals, 153

M

MacArthur, Robert, 228, 375
Macronutrients, 71, **72**
Macroparasites, 310, 311, 316, 317
Magnesium deficiency, 143
Mallee, **527**, 528, 530–531
Malthus, Thomas, 6, 7, 236, 238
Mangals. *See* Mangrove forests
Mangrove forests, 480, 584–585, **587**, 590
  animals of, 161–**162**, 592
  global warming and, 617
  loss of, 593
Maquis, 527
Margalef's index of diversity, 245
Margulis, Lynn, 310, 321, 419
Marine ecosystems, 479, 480, 562, 563–564
  pollution in, 576
  species richness in, 484–485, **486**
  temperatures in, 69–70
  See also Oceans; Seashore
Marine inversion, **50**–51
Mark-recapture methods, 181
Marshes, 582, **583**
  carbon cycle and, 434
  See also Salt marshes
Mating systems, 223–224
Matrix, of landscape, 368
Mattoral, 527, 530
Maximum sustainable yield (MSY), 351–352, 353
Mayr, Ernst, 20, 25, 487
McGinnes, W. G., 526
Mechanical weathering, 81
Mediterranean-type shrublands, 526–528, **527**, 530
Meiofauna, 572
Melanization, 89
Melatonin, 157, 159, 160
Mendel, Gregor, 6, 16–17, 18, 20, 25
Merriam, G. Hart, 469–470
Mesopelagic zone, 562, 566
Mesophyll cells, 103, **113, 115**
Metalimnion, 68, 250, 426
Metapopulations, 177, 178, 377–**378**, 379, 387
Methane
  greenhouse effect and, 450, 605
  in lake bottom, 548
  ozone and, 456
Micelles, 86–87
Microbial decomposers, 82, 124, 126, 127, 130–131, 133, 260, 525
Microbial loop, **565**
Microbivores, 126
Microclimates, 51–53
  poikilothermy and, **152–153**
Microflora, 124
  See also Microbial decomposers

Micronutrients, 71, **72**
Microorganisms
  Gaia hypothesis and, 419
  in herbivore digestion, 124, 138, 139, 140, 318
  nitrogen fixation by, 318, **319**, 342, 439, 525, 534, 539
  in soil, 82, 91–92, 94
  See also Bacteria; Microbial decomposers
Microparasites, 310, 311, 316, 317
Migration, 178–179, 486, 593
Mimicry, 297, **299**
Mineral licks, **143**
Mineralization, **131**, 132, **133**, 134
  in coastal sediments, 430
  defined, 124
  in lakes, 548
  nutrient cycling and, 419, 422, **423**
  nutrient loss and, 425
  primary productivity and, 401
  in soil formation, 81, 82
Minerals, in animal diet, 71, **72**, 143
Minimum viable population, 379
Mires, 582–584, 587
Mitochondria, 103, 320–**321**
Mixed-grass prairie, 520, 525
Mobius, Karl, 7
Models, scientific, **9**
Mollisols, 89
Monoecious plants, 222
Monogamy, 223, 225, 226
Montane forests, 503
Moors, 583
Morisita's index, 254
Morphological species, 20
Mortality curves, **189**–190
Mortality rate, 186
  age-specific, 187
  population regulation and, 204, 207, 208
Mosaic, **368, 369**, 385–387, **386**
  See also Patches
Mountain rain forest, 497
Mountains
  atmospheric inversions and, 49–50
  rain shadows of, **52, 53**, 532
  species diversity and, 483–484
Mouthparts, **139**
MSY (maximum sustainable yield), 351–352, 353
Mudflats
  estuarine, 429, 442–443, 557, 558
  freshwater, 587
  seashore, 571, 572, 573, 574, 590
Mulch, 522–523, 525
Muller, Paul, 360
Mushrooms, lignin decomposition and, 626
Mutation, 16, 19, 25, 487

Mutualism, 318–326
  community structure and, 264, **326**
  defined, 242, 310, 318
  diffuse, 322, 326, 331
  facilitation as, 325
  of humans and food species, 356
  indirect, 332, **334**
  nonsymbiotic, 318, 322–325, 326
  in seed dispersal, 323–325, **324, 325**, 326
  symbiotic, 318–322, 326, 567
Mycorrhizae, 318–**319, 326**, 451, 499, 539

N

Nanoplankton, 564, 565
Natality, 191–192
Natural history, 265
Natural selection, 12–16, 19–20
  biogeography and, 486–489
  defined, 12
  by disturbances, 382
  in Galápagos finches, **13, 14**, 16, **283**
  interspecific competition and, 270, 281–283
  predation and, 290, 296, 297
  speciation and, 23, 25
  types of, 13, **15**
  See also Evolution
Neap tides, 67
Nectivores, 322
Needle-leaf evergreens, 478, **479**, 503
Negative feedback, 145–146
Nekton
  of lakes, 545, 548
  of oceans, 566
Neritic province, 480, **562**, 564, 569
Net carbon gain, 106–**107**, 110, 111, 116, 120
Net ecosystem production, 434
Net mineralization rate, 131
Net photosynthesis, 103, 105
Net primary production, 398, 434
  See also Primary production
Newbold, J. D., 429
Niche, **28**–29
  fundamental, 282, 284, 330–331
  realized, 282–285, **284, 285**, 337
Niche overlap, 283–284
Nighttime surface inversion, **48, 49**
Nitrates, 438, 439, 440, 445
  human activity and, 450, 451, 462
Nitric acid, 439, 453
Nitrification, 439, 450
Nitrogen
  from decomposition, 130–132, **131, 132**
  in herbivore diet, 141–142

internal cycling of, 422, **423**
  in grasslands, 525
mycorrhizae and, 318–319, 451
plant adaptations and, 118–120,
  119, 120
primary productivity and, 401, **404**
Nitrogen cycle, 418, **438–440**, 445
  human alteration of, 450–**451**, 453,
    463
Nitrogen fixation, 439, 450, 454
  in desert, 534
  in grasslands, 525
  by *Rhizobium*, 318, **319**, 342
  on tundra, 539
Nitrogen oxides, 450–451, 453, 456,
  461
  greenhouse effect and, 605
Nitrogen saturation, 451
Numerical response, 292, 294–296,
  295
Nutrient cycling, 417–431
  climate and, 422
  in coastal ecosystems, 429–431, **430**
  in deserts, 534
  in forest ecosystems, 494–495
  general model of, 418–**420**
  global warming and, 615
  in grasslands, 525
  ocean currents and, **430**, 431
  in open-water ecosystems, **425–427**
  radioisotope studies of, **421**
  rate of, 420, 421, **422–423**
  in savanna ecosystems, 514
  species differences in, **422–423**
  in streams and rivers, 424,
    427–429, **428**, 439
  in terrestrial ecosystems, **425**
  in tundra ecosystems, 539
  See also Biogeochemical cycles
Nutrient spiraling, 427–429, **428**, 556
Nutrients, 71, **72**
  animal acquisition of, 138–141
  animal requirements, 141–143
  export from ecosystem, 423–425,
    424
  in lakes, 420, 426, 549–550
  plant adaptations and, 118–120
  primary productivity and, 401, **404**
  symbiotic mutualism and, 318–321,
    326

O

Obligate anaerobes, 124
Obligate mutualism, 318, 325, 326
Obligatory relationship, parasitic, 310
Ocean currents, **46**, 603
  ecoregions and, 489
  nutrient cycling and, **430**, 431
Ocean ecoregions, **489**

Ocean sediments, 566, 567
  phosphorus in, 441, 442
Oceanic province, **562**
Oceanic zone, 480
Oceans
  benthic zone, 480, 484, 562,
    566–567, **568**, 570
  carbon dioxide uptake by, 437, 601,
    602–603
  coral reefs in, **574–575, 576–577,**
    615, 617
  ecosystems of, 563–564
  global warming of, 610
  hydrothermal vents in, 567, **568**
  nutrient cycling in, 426, **430**, 431
  pelagic zone, 480, 484, **486**, 562,
    563–566
  pollution of, 576–577
  primary productivity in, 401, 404,
    405, 484–485, 562, 564, 565
  stratification, **426**, 562–563
  zonation, **562**
  See also Aquatic ecosystems;
    Seashore
Oil pollution, 576, 593
Oligotrophic lakes, **549, 550**
Oligotrophy, 550
Omnivores, 141, 336, 407, 535
Open systems, 396, 398
Open-water ecosystems. *See* Lakes;
  Oceans
Operative temperature range, 148,
  166
Order, taxonomic, 21
Organismal concept, 262–264, **265**
Organophosphates, 361
Osmosis, **64**, 155, 157
Outbreeding, 226
Outbreeding depression, 226
Outputs, of ecosystem, 396, 423–425,
  424
Overcompensation, by grasses, 305
Overgrowth competition, 270
Overturn, 68, 73, 548
  in estuary, 558
Oxbow lakes, 544, **545**
Oxisols, 90
Oxygen
  animal acquisition of, 144, **145**
  in aquatic ecosystems, **68**, 71, 73,
    144, 426
    coasts, 429
    estuaries, 558
    lakes, 544, 548, 550
    streams, 553
  from photosynthesis, 102, 419, 444,
    445
  See also Respiration
Oxygen cycle, 418, 444–**445**
Oyster bed, 7, 558
Ozone, 56, 455, 456–457, 605

P

Paine, Robert, 332
Paleoecology, 261–262, **263–264**, 598
Palmer, J. D., 162
Pampas, 520–521, 523, 524
Paquin, V., 481
PAR (photosynthetically active
  radiation), 56, **60**, 102, 103,
  108, 109
Parasitism, 242, 290, 310–318
  adaptations in, **314–315**
  basic features of, 310–311
  coevolution and, 310, 315, 318
  evolution to mutualism, 318
  by hemiparasitic plants, 311–312,
    313
  by holoparasitic plants, 311
  host population dynamics, 313, 315,
    316–318
  social, 346
  transmission between hosts,
    311–314, **313**, 316
Parasitoidism, 242, 290
Parent material, 80, 81, 82, 83
Parental investment, **227–228**, 229
Park, Orlando, 7
Park, Thomas, 7, 273, 275
Parthenogenesis, 220
Particulate organic carbon (POC), 565
Particulate organic matter (POM),
  133, 554, 555, 556
Pastor, John, 401
Patches, 177, **368–369**
  corridors between, 376–377, 378,
    387
  edges of, 369–**370**, 371, **373**, 374,
    486
  island biogeography and, 375, 376,
    387
  metapopulations in, 177, 178,
    377–**378**, 379, 387
  shifting mosaic of, 386–387
  in shrublands, 530
  species diversity and, 370–374
Peat, 81, 90
Peat forests, tropical, 497
Peatlands, 582–**584**, 587, 592
  carbon cycle and, 434
Pedon, 80, 82
Peds, 84
Pelagic zone, 480, 562, 563–566
  species diversity in, 484, **486**
PEP (phosphoenolpyruvate carboxy-
  lase), 113, 115
Percent base saturation, 86
Percent similarity (PS), 254
Perennating tissues, 248
Perennial plants, 398–399
Periphyton, 548–549, 552
Permafrost, 90, 507–508, 535–536

Pesticides, **359**, 360–361, 458–459, 461, 462, 592
Pests, 358–362
PET. *See* Potential evapotranspiration
Petersen, C. D. J., 350
Petersen index, 181
Petrie, Marion, 224
pH, 73–**74**
　See also Acid deposition; Acidic soils
Phenotype, 18
Phenotypic plasticity, **19**
Pheromones, 210, 296, 304
Phosphorus
　acid precipitation and, 454
　in forests, 423
　in streams, 429
Phosphorus cycle, 418, **441–442**, 445
　human intrusion in, 457
Photic zone, 250, 426, 562, 566
Photoinhibition, 103
Photoperiod. *See* Daylength
Photosynthesis, 102–105
　in aquatic organisms, 73, 105, **109**, 250, 565
　bacterial, 442, 565
　biogeochemical cycles and, 445
　in C_3 plants, 102, **113**, **117**
　in C_4 plants, **113**, 115, **117**
　in CAM plants, 113, **115**
　carbon dioxide increase and, 603–605, **604**
　energy fixation in, 397, 398
　light availability and, **108–109**
　net carbon gain from, 106–**107**, 110, 111, 116, 120
　nutrient availability and, **119**, 120
　oxygen from, 102, 419, 444, 445
　temperature and, 105–**106**, **117–118**, 401
　See also Primary production
Photosynthetically active radiation (PAR), 56, **60**, 102, 103, **108**, 109
Phylum, 21
Physiological ecology, 7
Phytoplankton
　acid deposition and, 456
　carbon cycle and, 434, 437
　dissolved organic matter from, 133, 565
　of estuaries, 558, 590
　of lakes, 544, 546–548, **547**, 550
　light attenuation by, 58
　nutrient cycling and, 401, 404, 420, 422–423, 426
　of oceans, 562, **564**, 565, 566
　phosphorus cycle and, 441
　photosynthesis by, 73, **109**, 250
　production of, **406**, 411, 413
　of streams, 556
　See also Algae
Pianka, E., 228

Pickett, Stewart, 278
Pioneer species, 256
Plague, 317
Plankton. *See* Phytoplankton; Zooplankton
Plant adaptations, 107–108
　to climate, 475–479
　community structure and, 337–340
　to disturbance, 382
　life histories, 228–**229**
　to light availability, 108–111
　to nutrient availability, 118–120
　to salinity, **112**, 532, 588
　to temperature, 105–106, 117–118
　to water availability, 111–117
Plant litter
　decomposition of, 82, 126–128, 132, **133**
　detrital food web of, 408, **409**
　in forest ecosystems, 494–495, **497**
　　boreal, 508
　　coniferous, 82, 505
　nutrient cycling and, 420, **423**
　in shrublands, 530
　See also Mulch
Plants
　ages of, 181, **182**
　carbon dioxide elevation and, 603–605, **604**, 618
　carnivorous, **302–303**
　clonal growth in, **173–174**
　community structure and, 247–249, 337–340
　defenses of, 303–305
　ecosystem classification and, 468, **469**–470, 471, **474**, **475–479**, 488
　energy and, 106, 398–400, **399**, 403
　evolution of, 625, 626
　global warming and, 611, 612
　herbivore predation on, 228, 290, 301–306, **304**, **306**
　interspecific competition in, 214, 228, 275–276, **277**, 341, 342
　intraspecific competition in, 214, 215
　life form classification of, 248–249
　life tables, 188–**189**
　light attenuation by, 56, 57–**58**, 60
　polyploidy in, 24
　temperatures of, 105–**106**
　water transport in, 64, 104–**105**
　water-use efficiency, 105, 115, 604
　See also Trees
Playas, 533
Plinthite, 90
Pneumatophores, **114**, 585, 590
POC (particulate organic carbon), 565
Podzolization, 89, 90, 454
Poikilotherms, **148–149**, 150–151, **152–153**, 154, 155
　See also Ectotherms

Polar cell, 45
Polar desert, 535
Polar domain, 488, **489**
Polar easterlies, 44, 45
Polar front, 45
Pollen, climate and, **490–491**, 615
Pollination, mutualism and, 322–**323**, 326
Polyandry, 223, 224
Polygamy, 223, 224, 225, 379
Polygyny, 223
Polyploidy, 24
POM (particulate organic matter), 133, 554, 555, 556
Ponds, 544, **545**, 546, 548, 550
　See also Lakes
Pools, **552**
　behind dams, 556–557
　tidal, 570–**571**
Population(s)
　age structures, 180–183, **182**, **183**, 193–**194**
　defined, 5, 172
　effective size, 379
　geographic variations in, **26–27**
　individuals in, 172–173, 175–177
　metapopulations, 177, 178, 377–**378**, 379, 387
　sampling of, 179–**180**, 181
　spatial distribution in, 173–175, **177**, 179–**180**
Population cycles, **197–198**
Population density, 175–177, 179–180
　dispersal and, 210–211
　growth and, **205–208**
　reproduction and, **209**
　stress and, 209–210
Population ecology, 7
Population genetics, 6
Population growth, 186–197
　exponential model, **194–196**
　fecundity tables and, **191–192**
　human, **236–238**
　life histories and, 233
　life tables and, 186–191, **187**, **189**
　logistic model, **196–197**, 204, 270, 350
　of predator and prey, 290–292, **291**
　See also Extinction
Population projection table, 192–**193**, 195
Population regulation, 203–216
　density-dependent influences, **204**
　density-independent influences, 204, 215–**216**
　intraspecific competition and, 204–209
　mutualism and, 326
　parasitism and, 316–318
　social behavior and, 211–213

stress and, 209–210
territoriality and, **213–214**
Potential energy, 397
Potential evapotranspiration (PET)
 ecosystem classification and, 471
 species richness and, 482, **484**
Potholes, 586, 592, 594
Power, Mary, 336–337
Practical salinity units (psu), 62
Prairie, 479, 518–520
 restoration of, 363
 wetlands of, 586, 592, 594
Prasad, Anantha, 611
Precipitation, 64–65
 acid rain, 452, 453–454, 455
 ecosystem classification and, 470,
  471, **474**, 479, 488
 global changes in, 607, **608**, 611,
  615, 616, 622
 global pattern of, 47–49
 in grasslands, **524**
 nutrients carried by, 418, 419
 population dynamics and, 215–216
 primary productivity and, **400**, 401
 rain shadow and, **52**, **53**, 532
 in savanna ecosystems, 511–**512**
 secondary production and,
  405–406
 in tropical forests, 494, 496–498
 vapor pressure and, 47
 See also Climate
Precocial offspring, 228
Predation, 242, 289–307
 in benthic food chain, 567
 coevolution and, 296
 community structure and, 306–307,
  332, **333**, 336–337
 defenses against
  of animals, **296–298**, 299
  of plants, 303–305
 defined, 290
 diffuse interactions in, 331
 evolving to mutualism, 322
 food chain and, 305–**306**
 foraging strategies in, 299–**301**
 forms of, 290
 functional response in, **292–293**,
  295
 by human-introduced species,
  389–390
 hunting tactics in, 298–299
 mathematical models of, 290–**291**,
  292, 294
 numerical response in, 292, 294–296,
  295
 on plants, 228, 290, 301–306, **304**,
  306
 prey switching in, 293, **294**
 search image in, 293
 on seashore, 572
 in streams, 555, 556

Predator satiation, 298
Preemptive competition, 270
Prentice, Colin, 491
Pressure
 atmospheric, 41–42, 44
 osmotic, **64**
 of water vapor, 46–47
Prey switching, 293, **294**
Primary producers, 102, 396
 trophic levels and, 336, 337
Primary production
 defined, 398
 in deserts, 534
 energy fixed in, 398
 in grasslands, **524–525**
 grazing food chain and, 407–408,
  409
 in salt marsh, 429
 in streams, 552, 555–556
 on tundra, 538–539
 See also Photosynthesis
Primary productivity
 carbon cycle and, 434
 carbon dioxide increase and,
  604–605, 622
 of coral reefs, 575
 of cornfield, 403
 defined, 398
 environmental controls on, **400–401**,
  404
 evapotranspiration and, **401**, 481,
  482
 of forest ecosystems, 494–495, **496**
 global patterns, 401, **402**
 in lakes, 550
 nutrient cycling and, 419, 420, 422,
  423, **425**, 426
 in oceans, 401, 404, **405**, 484–485,
  562, 564, 565
 secondary production and,
  404–407
 species richness and, 481, 482
 temporal variations in, 404, **405**
Primary succession, 258–260, **259**
 See also Succession
Profundal zone, 545, 548
Promiscuity, 223
Protective armor, 297
Provinces, 488
Pycnocline, 431
Pyramids
 age, **182–183**
 biomass, **413**
 energy, **413**

Q

Quaking bogs, 584, **585**
Qualitative inhibitors, 304, 305
Quantitative inhibitors, 304, 305

R

r strategies, 228–229, 286
Radiation ecology, **421**
Rain forests
 mountain, 497
 subtropical, 479
 temperate, 504
 tropical, 479, 494, 496–**497**,
  498–500
  climate change and, 615, **616**, 622
  extinctions in, 390–391
Rain shadow, **52**, **53**, 532
Rainfall. *See* Precipitation
Raised bogs, 583
Ramets, 173
Random distribution, **177**, 179
Rate of increase, 195
Raunkiaer, Christen, 248
Realized niche, 282–285, **284**, **285**,
 337
Recessive allele, 17, 18
Recombination, genetic, 19, 220, 487
Red Queen hypothesis, 296
Redundancy model, 335
Regional (gamma) diversity, 485–486
Regolith, 81
Relative abundance, 243–244, 254
Relative humidity, **47**
 in temperate forests, 502
 transpiration and, 104, 106
 in urban areas, 51
 in valleys and depressions, 53
 See also Humidity
Reproduction, 220–221
 asexual, 19, 32, 173, **220–221**
 life histories and, 220, 228–**229**, 233
 See also Sexual reproduction
Reproductive cycles, 159–**161**
Reproductive effort, 226–228
 latitude and, 230–**231**
Reproductive isolation, 20, 22–24, 25
Reproductive rate, 191–192
Reproductive success, habitat selection
 and, 231–233
Reservoirs, 556–557
Resource management, ecoregions
 and, 489
Resource partitioning, 280–283, **281**
Respiration, 102–103
 carbon cycle and, 434, 437
 carbon loss in, 106, 423, 434
 by consumers, 405, 406, 434
 by decomposers, 124, 429, 434, 437
 ecosystem energy flow and, 408,
  409, 410, 411
 light availability and, 109
 mitochondria and, 103, 320–321
 oxygen cycle and, 444, 445
 in plant energy allocation, 398, 399,
  403

Z

Zero growth isoclines, 273, 291
Zonation, 250–**251**, 254, 258, 337, **338**
  in coral reef, 575
  in freshwater wetland, 586
  in ocean, 562

  in salt marsh, 251–253, **252**, 278, **279**, 338–340, **339**, **588–589**
  in tidal zone, **253**, 568–571, **569**, **570**, 572, **573**
Zoogeography, 476–**477**
Zooplankton
  dissolved organic matter from, 133–134

  in lakes, 546, **547**, 548
  nutrient cycling and, 420, 422–423, 430
  in oceans, 563, 564–**565**, 566, 575
  phosphorus cycle and, 441
  production of, **406**, 411, 413
  in streams, 556
Zooxanthellae, 320, 575